Contemporary Classics in the Life Sciences

Volume 1: Cell Biology

Contemporary Classics in Science

EUGENE GARFIELD, *Editor-in-Chief*

This volume is one of a series published by ISI Press®. The series is designed to bring together analyses of papers that have been designated Citation Classics because they are influential and widely quoted.

Books published in this series:

Contemporary Classics in the Life Sciences
    Volume 1: Cell Biology
    Volume 2: The Molecules of Life
    edited by JAMES T. BARRETT

Books to be published in this series:

Contemporary Classics in Clinical Medicine

Contemporary Classics in Plant, Animal, and Environmental Sciences

Contemporary Classics in Physical, Chemical, and Earth Sciences

Contemporary Classics in Engineering and Applied Science

Contemporary Classics in the Social and Behavioral Sciences

Contemporary Classics
in Science

Contemporary Classics in the Life Sciences

Volume 1: Cell Biology

Edited by
James T. Barrett

With a Foreword by
Joshua Lederberg

Preface by
Eugene Garfield

ISI PRESS®

Philadelphia

Published by

iSi PRESS®   A Subsidiary of the Institute
             for Scientific Information®
3501 Market Street, Philadelphia, Pennsylvania 19104 U.S.A.

Library of Congress Cataloging in Publication Data

Contemporary classics in the life sciences.

  (Contemporary classics in science)
  Selected articles originally appearing in *Current
  Contents.*
      Includes bibliographical references and indexes.
      Contents: v. 1. Cell biology—v. 2. The molecules of life.
      1. Life sciences—Addresses, essays, lectures.
  2. Biology—Addresses, essays, lectures.   I. Barrett,
  James T., 1927–   . II. Series.
  QH311.C644   1986   574   85-24840

ISBN 0-89495-053-3 (v. 1)
ISBN 0-89495-075-4 (v. 2)

Printed in the United States of America
92 91 90 89 88 87 86   7 6 5 4 3 2 1

Contents

Foreword

The interest shown by authors and readers in these vignettes is an evident response to a gap in contemporary scientific communication: that of personal historical reflection. However debatable the criterion of citations as a measure of intellectual value, they do speak to the intertextual continuity of scientific and methodological effort. Doubtless some uncited, even unpublished, productions are equally worthy of such reflection; their collection awaits new invention of how to unearth such cryptic gems. Meanwhile citation statistics, coupled with the good judgment of authors and editors, have helped in the selection of a panorama of discovery reminiscences unparalleled in compilations of the contemporary history of science.

ISI's instructions to authors account for the emphasis placed on certain themes: the inspiration of the work, the obstacles to its publication and acceptance, the most recent work by the author or a disciple elaborating on the same work. The multifarious glimpses of how so many scientists have perceived acts of discovery in day-to-day science, or more precisely put, how they tell the story today, are raw material for abundant reflection on the actual mechanism of science. It is especially rich on such sociological and psychological issues as academic organization, risk taking, the gatekeeping functions of journal publication, resistance to innovation, priority and credit. Between the lines, and sometimes in them, are many of the ambivalences and stresses adumbrated for the scientific career.[1,2] Some exhibit the tension between imagination and criticism, between the creation and destruction of worlds, that characterizes the most trenchant of intellectual and artistic advances.

The pieces are too brief to give more than hints of the internal technical history of the science itself, but these are also abundant. Especially interesting are the authors' own reflections on the fruition and elaboration of their earlier work, and their citations to current sources on the same subjects.

As useful as these vignettes are, especially as starting points for further inquiry, the collection has certain intrinsic limitations as serious history of science. The constraints include those that must be associated with any form of biography, especially autobiography.[3,4,5] The most meticulous of self-chronicles are tainted with the conflicting interests of authors likely to be over-involved with their biographical subject. The commentaries are casual productions, limited in space and documentation, and subjected to a bare minimum of editorial scrutiny. They are fascinating documents in themselves; they are an opening, not a final word. With the best of intentions, personal recollections may be fraught with conflation of timing and of motive. These are notoriously unreliable with respect to attributions of internal mental processes.[6,7] The commentaries should not bear an unwonted burden of being regarded as finished critical studies; they are valuable enough as vernacular, intimate statements of transparent face validity.

Still lacking in existing organs of scientific communication is any relaxed channel for the extended discussion of historical controversy. Anguish about misplaced priority of discovery[8] is fiercely felt, but mutedly published. Many authors of scientific articles may cite relevant prior work in the most patchy way—they do not always regard themselves as intellectual historians—and there is no place to repair the deficit without turning it into a federal case. There is no usable medium today to continue discussions of "what may really have happened" beyond "Contemporary Classics." Lacking such a critical forum, our authors may feel liberated to express their feelings more authentically than they have under the gauntlet of customary peer criticism; and we should be grateful that they have had that opportunity. Conversely, each article stands outside the discipline, characteristic of science, of public critical discourse. Whether an extended forum can be built without stifling easy self-expression is a challenge to architects of our communication system in science.

Meanwhile, there is much to savour and more to ponder in the menu before us.

Joshua Lederberg
Rockefeller University

REFERENCES

1. **Merton R K.** *Sociological ambivalence.* New York: Free Press, 1976.
2. **Eiduson B T & Beckman L, eds.** *Science as a career choice.* New York: Russell Sage Foundation, 1973.
3. **Runyan W M.** *Life histories and psychobiography.* Oxford: Oxford University Press, 1982.
4. **Zuckerman H A & Lederberg J.** From schizomycetes to bacterial sexuality: a case study of discontinuity in science. (Unpublished ms.), 1978.
5. **Woolgar S W.** Writing an intellectual history of scientific development. The use of discovery accounts. *Soc. Stud. Sci.* 6:395-422, 1976.

6. **Zimmerman D R.** *Rh: the intimate history of a disease and its conquest.* New York: Macmillan, 1973.

7. **Nisbett R E & Wilson T D.** Telling more than we know: verbal reports on mental processes. *Psychol. Rev.* 84:231-59, 1977.

8. **Merton R K.** Priorities in scientific discovery. *Am. Soc. Rev.* 22:635-59, 1957 (reprinted in: Merton R K. *The sociology of science.* Chicago: University of Chicago Press, 1973. p. 286-324).

Preface

For almost 20 years I've been writing about Citation Classics, my term for highly cited papers and books that are classics in their fields. In 1977 we began publishing in *Current Contents®* (*CC®*) the feature "This Week's Citation Classic"—an invited 500-word commentary by the author of a Citation Classic. Over 2,100 autobiographical commentaries have appeared so far. In requesting these commentaries, we asked authors of Citation Classics to describe their research, its genesis, and circumstances that affected its progress and publication. We encouraged them to include the type of personal details that are rarely found in formal scientific publication, such as obstacles encountered and byways taken. We also asked that they mention the contributions of co-authors, any awards or honors they received for their research, and any new terminology arising from their work. Finally, we asked them to speculate on the reasons for their paper or book having been cited so often.

With Volume One and Two we inaugurate the series "Contemporary Classics in Science." These two volumes contain the commentaries published in *CC/Life Sciences* from the beginning of 1977 to the end of 1984. The first volume, *Cell Biology,* includes Citation Classics from research on the cellular level, while the second volume, *The Molecules of Life,* includes those from research on the molecular level. Volume Two also contains commentaries on physical analysis and instrumentation, chemical analysis and preparative methods, and statistics—an artifact of publication from the years 1977 and 1978, when the same commentary, whatever the subject, appeared in each edition of *CC.* Volumes containing the commentaries published in other editions of *CC* from 1979 on are planned. The next volume will collect those published in *CC/Clinical Practice.* Subsequent volumes will cover the physical sciences and applied sciences, as well as plant, animal, and environmental sciences.

Although I have previously described how we choose a paper or book as a Citation Classic,[1,2,3] it is useful to review these procedures in order to make the purpose of this new monographic series clear. Not every scientific publication that deserves the designation "classic" is included in this series, which is limited to *Citation* Classics. There are other types of classics, including those that are rarely or only occasionally cited. More about these later. Our primary criterion for inviting a researcher to contribute a commentary is the number of citations that a particular work has accumulated.

We begin the selection process in gross terms by singling out the 300,000 papers and books most cited in our *Science Citation Index®* (*SCI®*) and *Social Sciences Citation Index®* (*SSCI®*) files (which presently span the years 1955 to 1985 and 1966 to 1985, respectively). We do not, however, rely solely on the highest number of citations in the entire population of publications, but make our selections by fields, in some of which, such as ecology, engineering, or mathematics, 100 citations or even fewer may qualify a work as a Citation Classic.

Searching for Citation Classics, therefore, is like fishing with nets. We seine the waters of the scientific literature in search of the biggest fish in a school. The big fish are the relatively few papers and books with the highest number of citations *in their field*. Now it's plain that the biggest fish in one school will be dwarfed by even the smallest fish in another school. So, too, the number of citations necessary to make a work a Citation Classic in radio astronomy, with its much smaller population of researchers and papers, is smaller than the number giving a work status as a Citation Classic in biochemistry. Realizing that one discipline is more populated than another, we have used different nets when searching different waters. In the Sea of Biochemistry we expect to encounter many giant fish, since in these waters the population of published papers is so great. The net we select has a wide mesh that captures only the big fish and allows the little ones to swim through. But the net we drop into the Bay of Radio Astronomy has a much smaller mesh and is designed to catch the largest of the relatively small fish found there. The mesh of the nets corresponds to the different thresholds of citation frequency we set depending on the fields in which we are searching for Citation Classics.

Along with the initial search for Citation Classics in terms of absolute and relative frequency of citation, we extend the search by creating a separate file for each journal to identify the most-cited papers published in that journal. If one assumes that a journal uniquely defines a field or specialty, then the list of most-cited papers for that journal will include many of the classics for that field. We have found that many classics were published in the first volumes of a specialty journal associated with the emergence of the then new field. But we have also found that the classic paper for a new field was sometimes published in a multidisciplinary journal such as *Nature* or *Science*. Thus, the most-cited paper published in a specialty journal may have been cited only 50 times, whereas the primordial paper for the same field may have appeared in *Nature* and received 100 citations. In cases such as these, we ask both the author of the paper published in the multidisciplinary journal and the author of the paper published in the specialty journal for commentaries. Not surprisingly, we often find that both papers were written by the same author(s).

We are further refining our selection process by relying increasingly on research-front data derived from co-citation analysis.[4] An analysis of most-cited, or core, papers in a research front provides a more sensitive classification of subjects than does citation analysis by journal.

Furthermore, and as a supplement to analytical methods, we ask for nominations from CC readers of works which they believe may qualify as Citation Classics.

* * * * *

We emphasize that this collection represents only a sample from the larger group we have identified as Citation Classics. About half of the authors invited to write commentaries actually do so. It follows, then, that the omission of a paper in this volume in no way signifies that it is not a Citation Classic. In the recent series of essays in CC devoted to the 1,000 most-cited papers in the SCI from 1961 to 1982, I identified papers for which we have received and published a *Citation Classics®* commentary.[5] Parts nine and ten of this ten-part series will appear in CC within the next few months. In Appendix A of Volume One I have provided a sampling of 100 most-cited works for which we have not yet received a commentary. I hope that some of the authors of these Citation Classics will put pen to paper to reminisce and interpret the citation impact of their research. Still, a remarkable number of Nobel laureates and other scientists I have described as *of Nobel class* have contributed commentaries appearing in both volumes. These are balanced by many from hundreds of other scientists who have received rather little formal recognition for their contributions.

* * * * *

Finally, a word about those classics I have reported as rarely or only occasionally cited. I have taken to describing these papers and books as uncited classics, and there are many reasons for a classic work having few or no citations. I have already described how a relatively low citation count can qualify a work as a classic when weighed against its companions in other fields, and how we use a flexible, composite, and, we hope, intelligent algorithm to compensate for this problem. It is most difficult, however, to compensate for the lifetime citation counts of papers and books published as long as 50 years ago or even more recently than that. We know that most older works receive fewer citations since typically most citations are received in the first decade after publication. Moreover, the exponential growth of the literature has changed the significance of any fixed threshold. In 1955 we processed 80,000 papers in about 600 journals in the SCI, whereas in 1985 we indexed about 600,000 from 3,000 journals in the natural and physical sciences. These numbers reflect the rapid expansion of modern science, which in turn affects the size of bibliographies and the potential for citation impact.

Furthermore, a recognized classic work may fail to be a Citation Classic because it has suffered what Robert K. Merton terms "obliteration by incorporation": "the obliteration of the source of ideas, methods, or findings, by their incorporation in currently accepted knowledge."[6,7] Thus, some works are no longer cited because their substance has been absorbed in the literature.[8] Just as neologisms and eponyms become part of scientific language, obliterated works become the common knowledge within a field, and explicit citation to them is viewed as unnecessary or pedantic. Quite often, a classic work is cited for a few years, excites rapid advances in its subject, and is then superseded by reviews or other papers containing new information. For a variety of reasons including little understood citation behavior, the primordial work is not mentioned; it nevertheless remains classic. On such matters I urge the reader to examine Joshua Lederberg's pithy remarks on obliteration in connection with the discovery that DNA is involved in the genetic transformation in bacteria.[9] The commentary published by Maclyn McCarty[10] on that milestone paper recently appeared and will be included in a future volume in this series for the life sciences.

*   *   *   *   *

I have always believed that these commentaries contribute to future historiography by preserving important biographical and behind-the-scenes information, otherwise generally unavailable. Working scientists reading this book will learn about unfamiliar aspects of otherwise familiar and classic research. These commentaries provide grist for the mill of historians and sociologists of science. They also help sensitize students and the public to the diverse nature and methods of science. The publication of these commentaries in this collected form adds to their value. Perhaps, too, the appearance of this book will stimulate other scientists to write *Citation Classics* commentaries. Since this is a continuing monographic series, their contributions will always be welcome.

Eugene Garfield
Institute for Scientific Information

REFERENCES

1. **Garfield E.** *Citation Classics*—four years of the human side of science. *Essays of an information scientist.* Philadelphia: ISI Press, 1982. Vol. 5. p. 123–34.
2. ———. The 100 most-cited papers ever and how we select *Citation Classics. Ibid.,* 1985. Vol. 7. p. 175–81.
3. ———. *Contemporary Classics in the Life Sciences:* an autobiographical feast. *Current Contents* (44):3–8, 4 November 1985.
4. ———. ABCs of cluster mapping. Parts 1 & 2. Most active fields in the life and physical sciences in 1978. *Essays of an information scientist.* Philadelphia: ISI Press, 1981. Vol. 4. p. 634–49.
5. ———. The articles most-cited in the *SCI,* 1961–1982. Pts. 1–5. *Ibid.,* 1985. Vol. 7. p. 175–81; 218–27, 270–6; 306–12; 325–35. Pts. 6–8. *Current Contents* (14): 3–10; (20):3–12; (33):3–11. Pts. 9–10 (in press).

6. **Merton R K.** *Social theory and social structure.* New York: Free Press, 1968. pp. 27–9, 35–8.

7. ———. Foreword. (Garfield E) *Citation indexing—its theory and application in science, technology, and humanities.* New York: John Wiley & Sons, 1979. p. vii–xi.

8. **Garfield E.** The "obliteration phenomenon" in science—and the advantage of being obliterated. *Essays of an information scientist.* Philadelphia: ISI Press, 1977. Vol. 2. p. 396–8.

9. **Lederberg J.** Foreword. (Garfield E) *Ibid.,* 1977. Vol. 1. p. xiv.

10. **McCarty M.** *Citation Classic.* Commentary on Avery O T, MacLeod C M & McCarty M. Studies on the chemical nature of the substance inducing transformation of pneumococcal types. Induction of transformation by a desoxyribonucleic acid fraction isolated from pneumococcus Type III. *J. Exp. Med.* 79:139–58, 1944. *Current Contents/Life Sciences* 28(50):26, 16 December 1985.

Introduction

Biographies and autobiographies of scientists have proliferated in recent years. Few of these, however, supply detailed information on a specific experiment or publication. Exceptional, therefore, is the paper published by the Nobel Prize-winning British physiologist Sir Alan L. Hodgkin entitled "Chance and Design in Electrophysiology: An Informal Account of Certain Experiments on Nerve Carried Out between 1934 and 1952." Hodgkin recalls the books, papers, people, and environments that determined his choice of research. He attempts to estimate the portion of his results obtained by planning and the portion obtained by accident. At the outset he explains why such an account is needed:

> I believe that the record of published papers conveys an impression of directness and planning which does not at all coincide with the actual sequence of events. The stated object of a piece of research often agrees more closely with the reason for continuing or finishing the work than it does with the idea which led to the original experiments. In writing papers, authors are encouraged to be logical, and, even if they wished to admit that some experiment was done for a perfectly dotty reason, they would not be encouraged to "clutter-up" the literature with irrelevant personal reminiscences. But over a long period I have developed a feeling of guilt about suppressing the part which chance and good fortune played in what now seems to be a rather logical development.[1]

The biologist Sir Peter B. Medawar, another British Nobel Prize winner, puts it rather more directly by asking if the scientific paper is not a fraud. His question is not meant to cast suspicion on the facts that are published in a scientific paper; rather, he asks if the universally accepted format of the scientific article, generally presented in terms of a fictional inductive method, does not systematically misrepresent the thought processes that led to scientific discoveries.[2] Medawar believes it does.

The sociologist Robert K. Merton notes that

Typically, the scientific paper or monograph presents an immaculate appearance which reproduces little or nothing of the intuitive leaps, false starts, mistakes, loose ends, and happy accidents that actually cluttered up the inquiry. The public record of science therefore fails to provide many of the source materials needed to reconstruct the actual course of scientific developments. . . . This practice of glossing over the actual course of inquiry results largely from the mores of scientific publication which call for a passive idiom and format of reporting which imply that ideas develop without benefit of human brain and that investigations are conducted without benefit of human hand.[3]

He then goes on to observe that Bacon, Leibniz, and Mach all took note of the difference between logical casuistry, based on the Euclidean and Cartesian ideals, and the often nonrational, nearly always circuitous course of discovery. Since the scientific article was invented 300 years ago there has evolved a format embracing these ideals and leading to typically immaculate, linear, and flawed accounts of discovery. There is room, therefore, for accounts, however brief, of the actual paths of inquiry.

The feature "This Week's Citation Classic" published in each weekly issue of *Current Contents®* (*CC®*) since 1977 has provided scientists with a forum in which they are permitted to describe their discoveries as they recall their having happened. In this way, *Citation Classics* commentaries continue to fill a lacuna in the history of science.

REFERENCES

1. **Hodgkin A L.** Chance and design in electrophysiology: an informal account of certain experiments on nerve carried out between 1934 and 1952. *The pursuit of nature: informal essays on the history of physiology.* Cambridge: Cambridge University Press, 1977. p. 1.
2. **Medawar P B.** Is the scientific paper a fraud? (Edge D, ed.) *Experiment.* London: BBC, 1964. p. 7–12.
3. **Merton R K.** On the history and systematics of sociological theory. *Social theory and social structure.* New York: Free Press, 1968. p. 4–6.

Chapter

1

Electron Microscopy

The first nine Citation Classics described in this chapter are founded on methodology. Though one of these papers was cited but 185 times, the total number of citations for the nine articles is 28,600, and the average is 3,178. These numbers attest to the burgeoning interest in electron microscopy in the 1960s.

The first Classic by Luft has been cited 6,953 times. His perfection of an epoxy embedding recipe required 5 years and was hampered by the different formulation of epoxy resins in Europe and the United States. His complaint that this labor placed the albatross label "methodologist" around his neck may be true, and if so, those who inhaled pyridine in his open laboratory for 5 years may think the label justified. Spurr's low-viscosity epoxy was designed for plant tissues, but is equally useful for dense tissues of animal origin. Trump's lifetime love affair with the microscope prepared him for the Classic which he describes. Fortunately, the all-night blind search for a suitable stain was not wasted when the unlabeled Coplin jar was identified by a colleague to contain toluidine blue. Toluidine blue staining of epoxy-preserved sections produces images not unlike those seen by electron microscopy. Lead citrate and uranyl acetate were stains designed to enhance the electron microscope image. Reynolds' lead stain has been cited 13,395 times and was popular because it minimized precipitation of lead carbonate, which spoiled many preparations. Venable admits that his modification was not the momentous achievement of the original method. His acceptance of two journal rejections, before the eventual publication of the article, is evenly handled in the last eight lines of his "This Week's Citation Classic." Uranyl acetate, as Stempak remarks, is most used in combination with lead staining although there is little benefit from the second stain in this circumstance.

Tissue autoradiography, freeze etching, and critical point drying are the subjects of the next three Classics. Each has made a significant contribution to electron microscopy. Adequate tissue autoradiography was required before firm structure–function relationships based on isotope incorporation could be established. The latter two procedures permitted a three-dimensional image

1

of cells or cell structures to be demonstrated on the customary photographic plate.

The next fourteen Classics deal with the ultrastructure of cells and cell organelles. As with the first nine Classics in this chapter, with one exception, all of the articles referred to were published prior to 1970. This accounts for the absence of any Classics on scanning electron microscopy. Consequently, these Classics are based on traditional transmission electron microscopy on such subjects as the lipid bilayer nature of the cell membrane, the filamentous nature of the cytoskeleton, the carbohydrate-rich outer cell coat and how it differs between normal and neoplastic cells, the rough endoplasmic reticulum, and the Golgi apparatus. Even the bacterial nucleus, bacteriophages, yeasts, and frog oocytes became of interest to electron microscopists. Kellenberger's contribution to the definition of the bacterial nucleus and of intracellular bacteriophages was based on a tedious adjustment of staining procedures first used for eukaryotes, and these publications have been acknowledged by microbiologists as outstanding cyto-morphologic studies of two of our smallest life forms.

Luft J H. Improvements in epoxy resin embedding methods.
Journal of Biophysical and Biochemical Cytology 9:409-14, 1961.

Epoxy resin embedding methods are presented consisting of variations of the original Araldite procedure of Glauert et al. The advantages of the methods are (1) rapid embedding, (2) easy sectioning of the embedded tissue, (3) good contrast in the electron microscope, and (4) a wide range of hardness, achieved by using two different anhydride curing agents. [The *SCI®* indicates that this paper was cited 6,953 times in the period 1961-1975.]

Professor John H. Luft
Department of Biological Structure
School of Medicine
University of Washington
Seattle, Washington 98195

February 22, 1977

"It is pleasant to be the author of a paper on a list of 'Citation Classics' and to be invited to comment on it. Part of the enjoyment comes from seeing that the product of one's laboratory works reliably around the world. Perhaps one reason that 'methods' are so frequently in this category is that many more scientists have need for a reproducible procedure rather than a novel idea or concept--it's hard to paraphrase a method. The project arose from the need for a good reliable embedding method so that I could get on with the main work--electric eels. The consequence of success was to find an albatross hung around my neck as a 'methodologist.'

"Back in the bad old days, electron microscopy of thin sections of biological tissue depended a lot on luck. Among other troubles, methacrylate embedding was erratic, requiring that 'good' embeddings be selected and the 'bad' thrown out. Selection was largely intuitive, which added variety to the final interpretations. In 1956, Audrey Glauert in Britain published promising results with the Ciba epoxy resin Araldite.[1] Our experiments with American Araldite gave brittle, useless blocks. A letter to Audrey (in January, 1958) revealed that we were not alone, and that she had struggled with Ciba/Basle to allow some English material to be sent to their American division. Samples arrived a month later, along with other supplies sent by Hugh Huxley, J.T. Randall and M.S.C. Birbeck in England to H.S. Bennett, then Chairman of the Anatomy Department in Seattle. The English material behaved properly, and by December, 1958, we found that a 3-stage cure (35-45-60°) gave equally good results with American Araldite. The clue came from the excellent book on epoxy resins by Lee and Neville.[2]

"Araldite was viscous, sticky stuff with an affinity for skin and clothes, and we looked for an alternative easier to live with. Shell produced an epoxy resin known as Epon 562 (later Epon 812) but their recipes, designed for maximum mechanical properties, gave blocks impossible to cut. Again we turned to epoxy chemistry in Lee and Neville. Keeping the 35-45-60° cure, we systematically varied the anhydride:epoxy ratio and the amount of accelerator. Each had an independent optimum for cutting. Hardness of the block could be varied over a wide range by any desired mixture of one soft and one hard liquid anhydride, still retaining the optimum anhydride:epoxy ratio. On July 20, 1959, the first embedding with the final recipe was done. Five days earlier, I had completed the epoxy titrations of the Epon 562 and 812 which we used, so that the anhydride:epoxy ratio could be set exactly. (The titration was done in pyridine in an open lab--very popular with the neighbors.) In August, I left for a year in London where I wrote up the paper, while the large and active electron microscopy group in Seattle was testing the method under a variety of conditions. R.L. Wood in Seattle adapted the optimum anhydride:epoxy ratio back to the original Araldite formula, and this was inserted into the manuscript on one of its many round trips between H.S. Bennett in Seattle and myself in London. The approved manuscript and I left London together ...."

---

1. **Glauert A M, Rogers G E & Glauert R H.** A new embedding medium for electron microscopy. *Nature* 178:803, 1956.
2. **Lee H & Neville K.** *Epoxy resins: their application and technology.* New York. McGraw-Hill, 1957.

Spurr A R. A low-viscosity epoxy resin embedding medium for electron microscopy.
J. Ultrastruct. Res. **26**:31-43, 1969.
[Dept. Vegetable Crops, Univ. California, Davis, CA]

The author reports a new epoxy-resin system for embedding tissue to be sectioned for observation by electron microscopy. Based on the low-viscosity epoxy resin, vinyl cyclohexene dioxide, and three other components selected mainly for low viscosity and long pot life, the medium is easily prepared and rapidly infiltrated into specimens. It is compatible with a wide range of dehydrating agents, and has been successfully used with diverse materials. [The *SCI®* indicates that this paper has been cited over 2,725 times since 1969.]

Arthur R. Spurr
Department of Vegetable Crops
University of California
Davis, CA 95616

September 26, 1979

"Advances in electron microscopy have been made through the innovations of numerous investigators. My contribution came at a time when many of my colleagues in ultrastructural research were receptive to a new epoxy-resin system of embedding materials. The primary advance made by the method is the outstandingly low viscosity achieved in the epoxy-resin mixture. Many have inquired about how the method was devised, and this 'Citation Classic' provides an opportunity to give a personal account about its development.

"My entry into ultrastructural research in the late 50s came when most procedures were geared to generally more pliable animal tissues, rather than to the heavily cutinized and hard, thick-walled, lignified elements so common in plants. Methods based on methacrylate or on the epoxy resins Araldite, Epon, and Maraglas, sometimes did not work well, or at all, with plant materials. So I started probing into the technical and industrial literature on epoxy resins. Had I been presented at the outset with the four components, the medium would have been comparatively easy to formulate, but the materials were found piecemeal. As the work progressed, trying to evaluate systems lacking in one or more of the final components was admittedly frustrating at times.

"Expectations reached a high point when it became apparent that the epoxy resin, vinyl cyclohexene dioxide, could be the basis for a system. Thereafter, the selection of additional components, all in keeping with a low-viscosity method, was based on it.

"Work was continued with numerous possible components, but primary attention was centered on the selection of an anhydride. Both solid and liquid forms were tested. Eventually, a liquid type from the Humphry Chemical Company, nonenyl succinic anhydride, yielded excellent castings. However, they were deeply colored, almost black. Through fractional distillations, a series of cuts was finally obtained that provided clear castings.

"At this point a reactive flexibilizer was needed, and many were evaluated. During this period of development the castings varied in the extreme, but finally D.E.R. 736 was found to give the best results. Now only the accelerator remained to be determined. Among many tested dimethylaminoethanol proved to be significantly better than other compounds because it conveyed a long pot life in the laboratory and yet a rapid cure with elevated temperature.

"To provide some insight into the physical properties of the medium and castings, the Dow Chemical Company cooperated in tests on viscosities and deflection temperatures. I later undertook further testing at Davis. Final evaluation was largely based on the quality of the thin sections when viewed in the electron microscope.

"Apart from the utility of the low-viscosity system with plant tissues, it has been successfully used with such diverse materials as human skin and bone, and even fish scales. The long developmental effort in searching out and testing new materials evidently proved to be worthwhile to many researchers in electron microscopy. The low-viscosity system became especially advantageous to hospital EM laboratories where time is of importance in diagnosis.

"I should comment that although the background literature on epoxy resins was helpful, it provided no specifics on applications in electron microscopy. The project accordingly involved empirical approaches, yet an element of judgment was essential in the evaluation of hundreds of formulations. A high quotient of persistence kept the project on track and to its ultimate fruition. Through the auspices of the University of California, the invention became the basis for a United States patent issued in 1976."

Trump B F, Smuckler E A & Benditt E P. A method for staining epoxy sections for light microscopy. *J. Ultrastruct. Res.* 5:343-8, 1961.
[Department of Pathology, University of Washington, Seattle, WA]

A technique for staining sections of osmium-fixed epoxy-embedded tissues without prior removal of embedment for light microscopy using aqueous toluidine blue at alkaline pH was presented. When sections are stained in this manner, their images under the light microscope are striking due to their great definition and resemblance to electron micrographs. They are useful, therefore, not only for the identification and mapping of areas seen in electron microscopy, but also because they permit better utilization of the full resolving power of the light microscope. [The *SCI*® indicates that this paper has been cited in over 820 publications since 1961.]

Benjamin F. Trump
Department of Pathology
School of Medicine
University of Maryland
Baltimore, MD 21201

February 23, 1984

"The stage was set in Kansas City, Missouri, in 1944, when my father and I were home together with chicken pox for two weeks. I had had microscopes since about the fourth grade; my main purpose was to study cells and tissues or, indeed, anything that I could see. It didn't matter that I hadn't acquired a commercial microtome because my father was my personal purchasing agent, mechanic, technician, and general supervisor who constructed the microtome and helped me shoot the rabbits and squirrels and catch the fish (the favorite experimental animals in those days) and then admired the results that I obtained.

"I think I was born destined to stain things to look at in a microscope but, in any case, I have had a lifelong love affair with microscopes that is not only amusing and soul-fulfilling but also useful in supporting my family and my research group.

"I readily accepted the advice of Robert E. Stowell in 1957 when he said that the foreseeable future in histopathology required expertise in electron microscopy (EM). So I set out for the University of Washington in August of 1959 to the laboratories of H.S. Bennett. Those times were exciting for EM since J. Luft was developing new epoxy resin embedding techniques.[1,2] This resulted in many solutions but also caused problems as epoxy sections were difficult to stain for light microscopy (LM). The only available technique for LM visualization was phase microscopy. This problem remained unsolved, and the need to stain epoxy sections for LM became acute, especially in pathology.

"We had the usual labs available, among them a histology-histochemistry lab which was always bustling with activity. One night, I was determined to solve this problem of staining epoxy sections before going to sleep. I cut a large number of semi-thin sections and proceeded to stain such with every stain-filled Coplin jar in the lab. By 7 a.m., the results were clear, amazing, and gratifying. I observed the same contrast in organelles as one could see in the electron microscope. This immediately extended the range of magnifications possible and gave images in photomicrographs which correlated well with those in low power electron microscopes. I could scarcely wait until Bennett and Benditt arrived. However, in analyzing my notes, I realized that the one Coplin jar which had worked was unlabeled and, thus, I had absolutely no idea what the responsible dye might be. Much to my delight, Smuckler identified the Coplin jar as one containing an alkaline solution of toluidine blue. I demonstrated the results to Bennett and Benditt and proceeded to write the paper, which was my first, including, of course, Smuckler and Benditt as coauthors. The paper was accepted immediately and has since been cited often because the technique was so useful.

"Since that time, of course, many variances of this technique and many additional techniques have been developed."[3]

1. **Luft J H.** Improvements in epoxy resin embedding methods. *J. Biophys. Biochem. Cytol.* 9:409-14, 1961.
2. ──────. Citation Classic. Commentary on *J. Biophys. Biochem. Cytol.* 9:409-14, 1961.
   Current Contents (20):8, 16 May 1977.
3. **Burns W A.** Thick sections: techniques and applications. (Trump B F & Jones R T, eds.) *Diagnostic electron microscopy.* New York: Wiley, 1978. Vol. 1. p. 141-66.

Reynolds E S. The use of lead citrate at high pH as an electron-opaque stain in electron microscopy. *J. Cell Biol.* 17:208-12, 1963.
[Dept. Anatomy, Harvard Medical School, Boston, MA]

A method for staining ultrathin sections for electron microscopy with solutions of lead citrate at high pH is described. Lead citrate solutions are stable for long periods of time and do not significantly contaminate the sections with unwanted precipitate. The physical-chemical basis of this staining is discussed. [The *SCI®* indicates that this paper has been cited over 13,395 times since 1963.]

Edward S. Reynolds
Department of Pathology
University of Texas Medical Branch
Galveston, TX 77550

June 26, 1981

"This oft cited 'Brief Note' in the *Journal of Cell Biology* describes a method for the staining of ultrathin sections for electron microscopy. It was really a rather simple thing to do—stain sections with lead ions where the lead is maintained in a complex with a multivalent organic anion (citrate) at a pH where the hydrated lead ion is positively charged and possibly exists in a polymeric complex. Complex formation between the lead cation and citrate multi-anion prevents its precipitation as lead carbonate and stains tissue sites (presumably phosphate groups of structural lipids RNA and DNA) which have a greater affinity for lead ion than citrate. Sound good? Yes, but unfortunately—or fortunately—the hypothesis stated above was devised after the fact to explain the phenomenon. Much of the hypothesis remains unproven, but the stain works!

"At the time that I stumbled across this phenomenon in 1961, I was a research fellow in the department of anatomy at Harvard Medical School. Electron microscopy was an all consuming technical *tour de force*. One had to be a longtime apprentice. The chairman of the department cut his sec- tions at home at night! Long hours were spent in the lab discussing the best ways of making glass knives. Milk bottle shards, heat broken knives, plier broken knives all had their advocates. Days and weeks were spent cutting the 'perfect' section. Staining with lead reminded one of the procedures of the medieval alchemist. Super pure lead oxide was boiled in 'CO$_2$ free' aqueous NaOH in a CO$_2$ free atmosphere. The product was diluted with CO$_2$ free water and sections stained for exact times under the most scrupulously clean conditions in an atmosphere elaborately purged of CO$_2$. Fresh stain had to be made daily and then was good only for several hours—if it worked at all. Yet, in spite of the most stringent precautions many 'perfect' sections were ruined by coarse black precipitates—almost always diagnosed as 'lead carbonate.' Having come from a postdoctoral fellowship in biophysics where I was aware of metal organic ligand binding constants, one morning I said something like, 'Why don't we stabilize the lead in solution with an organic ligand which will release it to binding sites in tissues, but which will keep it from reacting with CO$_2$ in the atmosphere.' Wiser heads nodded skeptically but said, 'Go ahead, but I don't think it will work.' The first vial of stain was ready within an hour. The effect was amazing. After a few days Fawcett came to me and said that I really must publish the method—and fast. So I did. The first vial of stain lasted three years. Sheer serendipity!

"By now the article has been cited several thousand times. Indeed, it has become such a classic that many authors no longer cite it at all. But one of my 'pet peeves' is those authors who, if they must cite it, cite it incorrectly. These people have become legion. In 1979—the last year for which there is an annual *Science Citation Index®* in our library—the article was miscited 22 different ways and the most frequent form of miscitation was repeated 14 times. Thus, I am often miscited more for this article than I am correctly cited for some of the articles which I consider my more significant publications.* *Sic transit gloria!*"

*Variant forms of citations to E.S. Reynolds's article will be corrected in the new *Science Citation Index®* 1975-1979 cumulation.

Citation Classics

Venable J H & Coggeshall R. A simplified lead citrate stain for use in electron microscopy. *J. Cell Biology* 25:407-08, 1965.

One of the more stable and reliable stains, commonly called the lead citrate stain or Reynolds' stain, is made by mixing lead nitrate and sodium citrate in distilled water, allowing time for lead citrate to form, then adding sodium hydroxide to raise the pH of the solution to 12. This paper reports the use of a commercially available lead citrate to eliminate the preparatory steps of Reynolds' procedure, thus saving considerable time. [The *SCI*® indicates that this paper was cited 2,120 times in the period 1961-1975.]

Professor John H. Venable
Department of Anatomy
College of Veterinary Medicine &
Biomedical Sciences
Colorado State University
Fort Collins, Colorado 80523

January 7, 1977

"A bit of sloth and a desire for speed prompted development of the simplified lead citrate stain. Certainly there was no meditation over the possibilities of creating anything original or of wide usefulness. A postdoctoral fellowship in the early 1960's to learn electron microscopy placed me across the hall from Ed Reynolds in Don Fawcett's department at Harvard. Reynolds had just concocted his new lead citrate stain, which solved much of our problems with lead precipitates. My new aggravation, however, was the elapsed one hour or more from the decision to make a new batch of Reynolds' formulation to the examination of the stained sections. Although the new stain was relatively stable, I seemed forced to make a new batch, hoping to eliminate any staining problems, each time I looked at my most recent flounderings at microtomy.

"A possible escape from this frustration stemmed from a thought that the chemical industry should be able to supply lead citrate more economically than we microscopists could ever achieve by tinkering individually.

The ingredients of Reynolds' formulation other than lead citrate, namely high pH and secondary reaction products, could be supplied directly and quickly. The secondary reaction products, however, could interfere with the staining, and thus be superfluous. To my surprise lead citrate was listed in K & K Chemical's catalogue.

"I talked *Sus Ito* out of the money to order the minimum quantity, 50 grams, a staggering amount for the use in mind. By the time it arrived, I had forgotten about it, Reynolds' stain serving me nicely. I still gave the simplified mixture a first trial. It yielded mainly despair. Twenty minutes in a high pH solution of lead citrate--the then average staining time--left the sections globbed with coarse precipitates. But the structure was stained. Eventually the staining time was reduced to seconds, the lead citrate concentration cut considerably and--*eureka*--we had a stain that could be made in ten minutes. In three more minutes the stained sections could be in the microscope.

"The use of this simple lead citrate modification did not sweep the local laboratories by any measure. Richard Coggeshall, my co-author, working on the nervous system of the leech, decided early to try it and gave it its first real test on a variety of tissues and spread its use in Sanford Palay's laboratory. By the time I finished my postdoctoral so many investigators there were using it that Coggeshall encouraged publication. My response was that no self-respecting journal would publish such an empirical, mundane recipe in the face of Reynolds' well conceived paper. Coggeshall's rebuttal was that we all were mailing the formulation to friends; placing it in the literature was a more efficient approach. *Stain Technology* and the *Journal of Histochemistry and Cytochemistry* rejected it dutifully on perfectly logical grounds.

"After some months Sanford Palay's influence brought the formulation to print in the *Journal of Cell Biology* and a 'Citation Classic' resulted, an extravagant honor for such minor intellectual effort."

Stempak J G & Ward R T. An improved staining method for electron microscopy.
J. Cell Biol. **22**:697-701, 1964.
[Dept. Anatomy, State Univ. NY Downstate Med. Ctr., Brooklyn, NY]

Uranyl acetate dissolved in methanol provides a faster, cleaner, and more intense stain for transmission electron microscopy than does aqueous solutions. It is effective for use in high-voltage electron microscopy and as a mordant in double-stain procedures, and it may stain relatively unreactive tissues more effectively. [The _SCI®_ indicates that this paper has been cited in over 705 publications since 1964.]

Jerome G. Stempak
Department of Anatomy
State University of New York
Downstate Medical Center
Brooklyn, NY 11203

July 12, 1984

"When I arrived at the State University of New York in 1962, I began a careful scrutiny of tissue preparation for electron microscopy. During that period, I dehydrated tissues with graded methanol solutions because I had duplicated some experiments of Bahr and others that indicated that less tissue shrinkage took place during methanol dehydration than during ethanol dehydration.[1] It occurred to me that _en bloc_ staining by a solution of uranyl acetate in methanol might eliminate the need to stain sections, so various solutions of uranyl acetate in 100 percent methanol were made, and the experiments attempted.

"The first experiments in _en bloc_ staining of adult rat liver with solutions of uranyl acetate in methanol yielded sections in which all elements were heavily stained, and the resulting micrography was uniformly dark. Micrographs of sections from unstained tissues that were routinely prepared and stained with solutions used for _en bloc_ staining yielded well-stained material in very short time periods, and the sections were uncommonly clean. It seemed sensible to document what was obviously successful, so we stained other tissues and embedments with similar success and reported on the stain. Refinement of the _en bloc_ staining experiments was put off until 'later,' a time much like 'tomorrow' since it seems not to come.

"The only modification in technique we have made over time is to insert the lid of a histology tissue-preparation capsule (we use 'Tissue-Tek') in-

to the Stender dish of solution and allow it to submerge. We place the grids in a vertical position in the perforations, which makes retrieval much easier.

"The technique enjoyed modest success for conventional transmission electron microscopy, and in 1973, Carasso, Delaunay, Favard, and Lechaire judged it a superior method for high-voltage electron microscopy.[2] Thick sections, re-embedded and sectioned perpendicular to their faces by the aforementioned authors, have confirmed our hypothesis that the excellent staining properties of the methanolic stain were due to better penetration of the epoxy sections.

"Today, most investigators believe that adequate contrast can be achieved only by double staining: uranyl acetate followed by lead. Our method provides sufficient contrast if used alone,[3,4] provided that time is varied to suit the particular embedding material. Sections from some Epon and Spurr blocks require up to 30 minutes, while the more resistant Maraglas may require 90 minutes. We prefer to avoid lead staining altogether so that we do not have to be concerned about the beam-induced granular coalescence and growth of lead particles during the observation of what turns out to have been one's best sections.

"However, for those who require double staining, we suggest methanolic uranyl acetate followed by Paley's lead stain. For ordinary use, lead staining time can be reduced to seconds, thus minimizing exposure to atmospheric CO_2, the principal cause of contamination. With longer exposure to lead, very intense staining of membranes occurs, rendering better viewing at low magnification, but excessive granularity at high magnification. Further, tissues or cells that are inherently difficult to stain may be adequately stained by this sequence.

"The publication has been well cited because the method stains rapidly and well, and is cleaner. The stain penetrates thick sections more deeply, making it excellent for high-voltage electron microscopy, and achieves a less granular image in high-resolution micrographs.[5] For those who require double staining, its use as the mordant yields superior results. For recent work in this field, see reference 6."

1. **Bahr G F, Bloom G & Friberg U.** Volume changes of tissues in physiological fluids during fixation in osmium tetroxide or formaldehyde and during subsequent treatment. _Exp. Cell Res._ **12**:342-55, 1957. (Cited 90 times since 1957.)
2. **Carasso N, Delaunay M C, Favard P & Lechaire J P.** Obtention et coloration de coupes épaisses pour la microscopie électronique a haute tension. _J. Microscopie_ **16**:257-68, 1973.
3. **Ward R T.** The origin of protein and fatty yolk in _Rana pipiens._ III. Intramitochondrial and primary vesicular yolk formation in frog oocytes. _Tissue Cell_ **10**:515-24, 1978.
4. —————. The origin of protein and fatty yolk in _Rana pipiens._ IV. Secondary vesicular yolk formation in frog oocytes. _Tissue Cell_ **10**:525-34, 1978.
5. **Stempak J G & Laurencin M.** High resolution microscopy of intracellular membranes. _J. Microscopie_ **9**:465-76, 1970.
6. **Ward R T.** The origin of protein and fatty yolk in _Rana pipiens._ V. Unusual paracrystalline configurations within the yolk precursor complex. _J. Morphology_ **165**:255-60, 1980.

Caro L G & van Tubergen R P. High-resolution autoradiography. 1. Methods. *J. Cell Biol.* **15**: 173-88, 1962.

The authors describe methods used in obtaining high resolution in autoradiography, with special emphasis on the technique of electron microscopic autoradiography, together with control experiments designed to establish the optimum conditions or procedures. These methods give a good localization of the label, at the subcellular level, and good reproducibility in relative grain counts. [The *SCI*® indicates that this paper was cited 637 times in the period 1962-1976.]

Professor Lucien Caro
Département de Biologie Moléculaire
Université de Genève
CH-1211 Geneva 4, Switzerland

July 26, 1977

This paper, together with its companion paper,[1] described techniques for autoradiography of thin sections in the electron microscope and control experiments designed to establish the optimum conditions and procedures.

"While we were graduate students in biophysics at Yale University, Bob van Tubergen and I often tried to devise ways of improving resolution in autoradiography. We had discussed a number of times the possibility of using the electron microscope but had dismissed, on theoretical grounds, the notion that it might result in improved resolution. After our separate first attempts had clearly shown a very high level of resolution we, of course, revised the theory.

"In 1960 I was a post-doctoral fellow in the laboratory of Prof. George Palade, at Rockefeller University, and engaged, in collaboration with him, in a study on the role of the Golgi complex in protein secretion. Bob van Tubergen's first results convinced me that electron microscopic autoradiography would be the best method for this study and I set out to develop techniques suitable for the study of thin sections. Soon after that, Bob was brought back from Europe, where he was spending his post-doctoral period, by an illness that would eventually interrupt a most promising scientific career. He then came to Prof. Palade's laboratory and, together, we developed the techniques described in this paper....

"I believe that the reason for the frequent citation of our paper is that it demonstrated clearly that high resolution autoradiography was possible without a drastic deterioration of the cytological structures in ultra thin sections. It also gave a set of simple recipes for the method. In my opinion, the main technical contribution was the introduction of the Ilford emulsion L4 to electron microscope autoradiography. Sixteen years later, and in spite of many attempts at finding a better product, it remains the most frequently used emulsion for this kind of work. Finally, I think the paper came at a time when electron microscope cytology was coming to the end of its purely descriptive phase and when cytologists were searching for experimental methods capable of giving indications as to the functions of the various intracellular organelles. Autoradiography provided one such method; hence the success of the paper.

"I would like to conclude these rambling notes by paying tribute to Bob van Tubergen who, while most of his energy was consumed in fighting, with admirable courage, a crippling disease, contributed so much, by his physical insight and his rigorous analytical mind, to the success of this work and of later experiments that utilized these techniques."

1. **Caro L G.** High resolution autoradiography. 2. The problem of resolution. *J. Cell Biol.* **15**: 189-99, 1962.

Steere R.L. Electron microscopy of structural detail in frozen biological specimens. *J. Biophys. Biochem. Cytol.* 3:45-60, 1957.

This paper describes a procedure for the preparation of pre-shadowed replicas which reveal fine ultrastructural details of frozen fractured, etched (by vacuum sublimation) biological specimens. Illustrations reveal the ordered arrays of plant virus particles within crystals, including tobacco mosaic virus crystals within infected cells. [The *SCI®* indicates that this paper was cited 185 times in the period 1961-1977.]

Russell L. Steere
Plant Virology Laboratory
Plant Protection Institute
Beltsville Agricultural Research Center
US Department of Agriculture
Beltsville, MD 20705

January 13, 1978

"I believe my pleasure in finding this report in the 'Citation Classics' list results from its appearance as the first successful application of an alternative approach (freeze-etching) to the visualization of biological ultrastructure by electron microscopy. This approach came about as a logical consequence of learned techniques, equipment, and information available to me at the time, and complete freedom to explore the unknown mixed with unsatiated curiosity, eternal optimism, and Steere stubbornness!

"The work was done under the supervision of Robley C. Williams in the virus laboratory at the University of California, Berkeley, where we had just demonstrated that the beautiful crystals found in cells of tobacco mosaic virus (TMV)-infected plants were composed ⅓ of TMV particles and ⅔ of a volatile matrix (probably water). We knew that chemical fixation destroyed these extremely fragile crystals so that thin section electron microscopy could not reveal their internal particle orientation.

"When I discussed the approach described in this paper with Robley in 1952 as a means of visualizing the true crystal structure, he remarked that it would hardly be worth the 3 to 4 years it might take to develop such a procedure just to visualize the ultrastructure of TMV crystals but that, as an alternative procedure for studying cytological ultrastructure, it might be extremely worthwhile. He cautioned me that he couldn't guarantee a promotion and that I might spend 3 to 4 years on such a development, then have to abandon it because of insurmountable obstacles. However, he did promise me that he would guarantee my job for at least 4 years if I wanted to tackle this important work.

"Now, after 20 years and numerous modifications to equipment and procedures, freeze-etching and related techniques have finally come of age. This approach to the study of biological ultrastructure has not only revealed the finite structure of TMV crystals and confirmed many results obtained by thin sectioning and other techniques, but has provided much visual information on membrane and other ultrastructural details unobtainable as yet by other approaches. Recent improvements have pushed the resolution of freeze-etching to the range of 1 nm, and permit the comparison of complementary faces of frozen fractured specimens.

"While doing the research leading to the publication of this paper, I had the rare pleasure of working under exceptional research administrators, whose main concerns were obtaining adequate funding for their programs, selecting qualified scientists, giving them the freedom to explore the unknown, and providing them with adequate protection from the 'publish or perish' philosophy, unnecessary 'busy work,' report writing, meetings, and other distractions.''

Anderson T F. Techniques for the preservation of three-dimensional structure in preparing specimens for the electron microscope.
Trans. NY Acad. Sci. 13:130-4, 1951.
[Johnson Foundation, University of Pennsylvania, Philadelphia, PA]

Distortions of biological and other specimens by surface tension are eliminated by first heating the ambient fluid above its critical point (where its surface tension vanishes) and then allowing the now gaseous fluid to escape. Thus prepared, delicate specimens retain their three-dimensional structures. [The SCI® indicates that this paper has been cited in over 1,060 publications since 1961.]

Thomas F. Anderson
Institute for Cancer Research
Fox Chase Cancer Center
Philadelphia, PA 19111

September 15, 1982

"Originally, most aqueous specimens were simply dried in air for study in the high vacuum of the electron microscope. When this is done, the surface tension of the evaporating water flattens most objects onto the supporting membrane. Thus, our early studies designed to show whether tailed bacteriophages adsorb to their host bacteria by their heads or by their tails gave equivocal results.

"It seemed to me that one could eliminate surface tension forces by heating the ambient liquid to a temperature above its critical point where it changes imperceptibly into a gas and then letting the gas escape. Since both the critical temperature and critical pressure of water are inconveniently high (374°C and 218 atm., respectively), it seemed desirable to replace water by a liquid like CO_2 ($T_c = 31°C$; $P_c = 78$ atm.). This could be done by passing the specimen through a series of miscible liquids like ethyl alcohol, amyl acetate, and finally liquid CO_2 in a pressure chamber. Then with the chamber completely filled with liquid CO_2 at room temperature, one would heat the chamber to 45°C or so and allow the CO_2 to escape.

"The day I got this idea, I assembled some high-pressure equipment borrowed from the generous people in the chemistry department at the University of Pennsylvania, and tried the method on the skin of an onion bought at the corner grocery. The method worked the first time it was tried! Whereas the cellular structures of the air-dried specimens were unrecognizable, wet and critical point-dried onion skin looked equally beautiful in the light microscope. We next took stereoscopic electron micrographs of critical point-dried ghosts of human erythrocytes, gels of tobacco mosaic virus, and cilia and trichocysts of paramecia. Each of them retained its three-dimensional structure.

"Finally, when the method was applied to mixtures of phages and bacteria, the infected bacteria looked like pincushions covered with pins.[1] T2 phages behaved like tiny syringes that infected bacteria by adsorbing tail first to the host and injecting DNA from their heads into them.

"Shortly after the idea of critical point-drying was first published,[2] two representatives of a major oil company visited my laboratory. They told me my method was very similar to one S.S. Kistler patented in 1932 for drying inorganic gels[3] to make tons of catalyst for cracking oils in the petroleum industry.

"Electron microscopists were very slow to adopt the method, even though everyone knew about it from the beautiful stereoscopic pictures of critical point-dried specimens I showed at meetings in both the US and Europe. And later, in 1956-1957, I took the critical point apparatus to Paris to study bacterial conjugation.[4] It wasn't until late-1960, when scanning electron microscopes became practical and useful, that the method became popular. Now, references to this paper still average about 100 per year, the reason being that it is the only drying method that conserves three-dimensional structure in fragile specimens.

"A recent review of this field can be found in Methods in Cell Biology."[5]

1. Anderson T F. Stereoscopic studies of cells and viruses in the electron microscope.
    Amer. Naturalist 86:91-100, 1952.
2. ————————. The use of critical point phenomena in preparing specimens for the electron microscope.
    J. Appl. Phys. 27:724, 1950.
3. Kistler S S. Coherent expanded aerogels. J. Phys. Chem. 36:52-64, 1932.
4. Anderson T F, Wollman E L & Jacob F. Sur les processus de conjugaison et de recombinaison
    chez Escherichia coli. III. Aspects morphologiques en microscopie électronique.
    Ann. Inst. Louis Pasteur 93:450-5, 1957.
5. Turner J N, ed. Methods in cell biology. Vol. 22. Three-dimensional ultrastructure in biology.
    New York: Academic Press, 1981. 363 p.

Robertson J D. The ultrastructure of cell membranes and their derivatives.
Biochem. Soc. Symp. **16**:3-43, 1959.
[Department of Anatomy, University College, London, England]

Electron microscopic (EM) observations on many biological membranes as seen in thin sections were reviewed and interpreted in molecular terms based on an analysis of nerve myelin. A general model of membrane molecular architecture was presented emphasizing the ubiquity of the lipid bilayer and chemical asymmetry. [The *SCI®* indicates that this paper has been cited over 455 times since 1961.]

J. David Robertson
Department of Anatomy
Duke University Medical Center
Durham, NC 27710

October 1, 1981

"The paper reviewed a number of observations that I had made during the mid-1950s regarding the molecular architecture of biological membranes. That decade began with many doubts about the exact role of the lipid bilayer in the molecular architecture of biological membranes, but it ended with a note of certainty expressed in this paper. Earlier work had suggested a crucial role for lipid bilayers in membrane structure. There was, however, doubt about how many were present, the manner in which proteins were associated with the lipids, and indeed whether or not any particular model could be applied generally. There were also doubts about whether or not there might be mosaic patches of pure protein contiguous with lipid regions. The EM studies reviewed dispelled the doubts about membrane thickness and led to the certainty that biological membranes contained one and only one, lipid bilayer and further that a useful general model of membrane molecular architecture could be proposed.

"About 1956 I applied two new developments in EM technique, permanganate fixation and epoxy embedding, and observed a triple-layered pattern in all cell membranes and membrane organelles studied. The membrane measured ∿ 7.5 nm in thickness and appeared as a pair of dense strata each ∿ 2 nm thick bordering a light central zone. The work provided the first direct evidence that nerve myelin consisted only of Schwann cell membranes to the resolution of the sections (∿ 2 nm). Partly on the basis of these observations I stated that the basic pattern resulted from the presence of one lipid bilayer as the fundamental core structure. A general model was proposed consisting of a lipid bilayer with protein at or in its polar surfaces in the dense strata. The idea of asymmetry due to polysaccharides in the outer surface was also advanced. This paradigm, called the unit membrane model, was the first one to make explicit the principle that biological membranes in general have *one* lipid bilayer as their basic structure. It resembled the earlier Danielli-Davson one but differed significantly in that only one bilayer was stipulated along with chemical asymmetry.

"At the time, these ideas met with considerable resistance partly because permanganate was a poor general fixative, but it is good and selective for membranes and revealed them in the manner that is now regarded as standard. Somehow the idea arose that I claimed all cell membranes were molecularly identical, a complete misinterpretation. Perhaps in my enthusiastic pursuit of getting across the idea that there was a *common molecular architecture* in membranes which could be defined, I gave the impression I was saying that all membranes were molecularly identical. Perhaps the problem here was that I was thinking in terms of molecules while others were not.

"The hydrophobic core of the bilayer is now known to be sometimes traversed by hydrophobic polypeptide chains. The model has been modified to take this into account but the main point was the establishment of the ubiquity of the bilayer, which I believe is now generally accepted. I have recently published work in this field."[1,2]

1. **Robertson J D.** The anatomy of biological interfaces. (Andreoli T E, Hoffman J F & Fanestil D D, eds.)
   Physiology of membrane disorders. New York: Plenum, 1978. p. 1-26.
2. ------------------. The nature and limitations of electron microscopic methods in biology. (Andreoli T E,
   Hoffman J F & Fanestil D D, eds.) *Physiology of membrane disorders.* New York: Plenum, 1978. p. 61-93.

Gray E G & Whittaker V P. The isolation of nerve endings from brain: an electron-microscopic study of cell fragments derived by homogenization and centrifugation.
J. Anat. Lond. **96**:79-87, 1962.
[Dept. Anat., Univ. Coll. London, and Biochem. Dept., Agr. Res. Council Inst. Animal Physiol., Babraham, Cambridge, England]

The authors present a method for the bulk isolation of synaptic complexes (synaptosomes) from brain which permits investigation of synaptic structure, biochemistry, and physiology in a way hitherto impossible, including the isolation of the various components of the synapse, e.g., synaptic vesicles, external presynaptic membranes, membrane proteins, and junctional complexes. Antibodies can now be prepared for study of the individual synaptic proteins. [The *SCI®* indicates that this paper has been cited over 1,100 times since 1962.]

E. G. Gray
Department of Anatomy and Embryology
University College London
London WC1E 6BT·
England
and
V. P. Whittaker
Max Planck Institute
D-3400 Göttingen
Federal Republic of Germany

October 30, 1980

"E. G. Gray: Traditionally at University College London, coauthors were placed in alphabetical order on publications. The concept of the senior author with its obvious advantages and disadvantages has, of course, changed all of this. Certainly, Victor Whittaker is the undisputed 'senior author' in the work described here.

"As I recall it, as soon as synaptic vesicles were discovered with the electron microscope (EM) and postulated to contain the transmitter substance (about 1954), Victor and Catherine Hebb seized on the feasibility of isolating the vesicles by subcellular fractionation. After some two years of effort they were finally able to produce a subfraction rich in postulated transmitters (acetylcholine, etc.). However, although particles resembling synaptic vesicles were present in the subfraction,[1] these were much smaller than sedimentation speed demanded.

"One afternoon in 1959, Victor visited my lab and explained his problem. In the first sections I looked at, under the EM, the anomaly immediately clarified. The subfraction, in fact, contained pinched-off nerve endings often with attached post-synaptic membranes—structures already so familiar to me from my cortex studies. Furthermore, their dimensions fitted exactly the sedimentation speeds. Victor, in fact, was seeing synaptic vesicles in his original studies, but these were derived from the nerve ending particles, which were breaking down during fixation because of his less sophisticated methods. At this stage I turned to other problems and Victor, having acquired his own EM unit, continued the work he and Hebb had pioneered.

"Victor P. Whittaker: I shall always remember the day, March 30, 1960, when I met George Gray's train at Cambridge Station and he said to me as we walked to my car: 'You have isolated vesicles, Victor, but inside pinched-off nerve endings.' As we examined George's micrographs later a vast vista opened before me and I knew what I should be doing for the next ten years. I was shortly leaving for America and the first announcement of the results was made in a lecture I gave in George Koelle's department of pharmacology in Philadelphia. During the US trip my technicians had instructions to repeat the experiments; this was done twice without success! When I got back I discovered that this was due to inadequate control of the temperature and osmotic pressure of the fixative. This confirmed the comparative lability of synaptosomes and explained why no one had identified them earlier. There was an unfortunate delay in publishing the full paper because it was rejected by the *Journal of Physiology*!

"Initially I used the acronym NEPs (nerve ending particles) for the detached nerve endings; then an American worker introduced PONEs (pinched-off nerve endings). While I was working on the isolation of synaptic vesicles,[2] I decided a more euphonious name was needed; one Saturday morning during a leisurely bath, the word synaptosome came to me. This was incorporated into the title of my paper, but the editors insisted on putting it between inverted commas. Now it has attained the distinction of being used unreferenced, like 'mitochondrion,' and has been incorporated into most European languages."

1. **Whittaker V P.** The isolation and characterization of acetylcholine-containing particles from brain. *Biochemical J.* **72**:694-706, 1959.
2. **Whittaker V P, Michaelson I A & Kirkland R J A.** The separation of synaptic vesicles from nerve-ending particles ('synaptosomes'). *Biochemical J.* **90**:293-303, 1964.

Huxley H E. Electron microscope studies on the structure of natural and synthetic protein filaments from striated muscle. *J. Mol. Biol.* 7:281-308, 1963.
[Medical Research Council Laboratory of Molecular Biology, Cambridge, England]

The author describes a technique for fragmenting striated muscle into its component thick and thin filaments by homogenization in a 'relaxing medium,' and discusses the functional implications of the polarity of the filaments thus revealed. [The *SCI®* indicates that this paper has been cited over 790 times since 1963.]

———————————————

Hugh E. Huxley
MRC Laboratory of Molecular Biology
University Postgraduate Medical School
Cambridge, CB2 2QH
England

January 18, 1978

"I am flattered to learn that one of my papers is a 'most cited article,' and I have been wondering why this should be. I suspect the reason is that, although the paper is primarily concerned with the structure of muscle, and therefore (at the time it was published anyway) of interest to a somewhat restricted audience, it does contain descriptions of a number of techniques which have had more widespread applications since then.

"One of these is a method for examining the filament-forming properties of the protein myosin. If the ionic strength of a solution of myosin in potassium chloride is lowered by dilution or dialysis from 0.6 to 0.1 M, or less, a precipitate or an opalescence forms, as was well-known at the time. If the precipitate is examined (at high dilution) in the electron-microscope by the negative staining technique, it can be seen very clearly and easily to consist of filaments somewhat similar to those in muscle and often showing a very characteristic bipolar appearance in the arrangement of the projections on them. This appearance arises because the myosin molecules have a 'head and tail' structure and because they

associate so that the tails form the backbone of the filament and the heads project sideways; the molecules are inserted with opposite polarity in either end of the filament. Thus there are projections on the lateral regions and a bare zone in the centre.

"The method is very simple and the appearance very characteristic, so that in fact it constitutes a good simple test for myosin or myosin-like molecules. With the great expansion of interest in contractile proteins from a wide variety of cells other than muscle, the method has been used quite often to confirm that a particular component was indeed myosin-like in its structural behaviour. The negative staining technique was itself relatively new at that time, being one that I had first used on tobacco mosaic virus in 1956,[1] and which was subsequently greatly improved by Brenner and Horne.[2] The description of the technique probably added to the popularity of the paper.

"Another technique first described in the paper, which has become very popular as a diagnostic tool, was the observation, again by the negative staining technique in the electron-miscroscope, of actin filaments which had been 'decorated' with myosin, heavy meromyosin, or HMM-subfragment 1, and which show a very characteristic 'arrowhead' structure. This arises in a very specific way from the precise angle of attachment of the myosin heads to actin, and it also shows up on the pitch and subunit repeat of the actin helix. Thus it can be used as a good simple test for either actin or myosin, and it also enables the polarity of the actin filaments to be established. More recently the technique has been used very extensively to identify actin in non-muscle systems.

"There were some other things in the paper of which I am quite fond, including, I believe, the first specific suggestion that cytoplasmic streaming might be due to a relative sliding or shearing interaction between polarized actin and myosin filaments in solution. But, I think, it must be the two methods to which I have referred which account for its popularity."

1. **Huxley, H E.** Some observations on the structure of Tobacco Mosaic Virus. *Proc. 1st European Regional Conf. Electron Microscopy.* Stockholm: Almqvist & Wiksell, 1956. p. 260.
2. **Brenner S & Horne R W.** A negative staining method for high resolution electron microscopy of viruses. *Biochem. Biophys. Acta* **34**:103-10, 1959.

Buckley I K & Porter K R. Cytoplasmic fibrils in living cultured cells: a light and
electron microscope study. *Protoplasma* 64:349-80, 1967.
[Biological Laboratories, Harvard University, Cambridge, MA]

Electron microscopy of cultured rat embryo
fibroblasts showed that cytoplasmic
filaments and microtubules are intimately
associated with components known to be
highly motile in the living cells. Seemingly
immobile cortical filament bundles appear
to be involved in the stabilization of cellular
attachment sites. [The *SCI*® indicates that
this paper has been cited over 185 times
since 1967.]

Ian K. Buckley
John Curtin School of
Medical Research
Australian National University
Canberra City, ACT 2601
Australia

March 28, 1979

"In 1963, as an experimental pathologist interested in the biology and
pathology of living cells, I went from
Australia to California on an Eleanor
Roosevelt Fellowship to work in
Charles M. Pomerat's laboratory. Keen
to learn the techniques of tissue
culture and cine microscopy for which
Pomerat was so well known, I wanted
to study the dynamic behavior of the
endoplasmic reticulum in living cells.
In the process, I became fascinated by
the variety and complexity of cellular
movements, a fascination which has remained ever since. The work on movements of the endoplasmic reticulum
led to correspondence with Keith
Porter, then professor of biology at
Harvard University.

"A year later when my fellowship
had come to an end I had the good fortune to spend ten months working in
Porter's laboratory where, surrounded
by an extraordinarily stimulating group
of cell biologists, I found myself in a
uniquely favorable environment for
studying cells and their movements.
With the excellent facilities and technical support staff in Porter's laboratory
it was possible, even as a complete
novice, to learn the basic elements of

electron microscopy and still continue
the light microscopic studies of living
cells. Our aim was to assess which fine
structural elements of the cytoplasmic
ground substance are most intimately
associated with the cell's moving parts.
Here we were fortunate in our cell
model, the cultured fibroblast, because
being very thinly spread on the cover
glass, the glutaraldehyde promptly
fixed all its cytoplasmic elements and
these were then clearly displayed
within sections made parallel to the
cell's plane of attachment. Consequently, we could show that everywhere throughout the cytoplasmic
matrix there were significant concentrations of microtubules and/or filaments. Since these structures were
always in the closest contact with the
moving parts, they appeared to be the
most likely elements involved in producing the movements. Although at the
time there were very few reports indicating the occurrence of muscle-like
proteins in nonmuscle cells, the concentrations and dispositions of filaments in our fibroblasts suggested the
likelihood of a useful analogy with the
muscle cell motility system.

"Our work has been cited most often
in relation to investigations which have
demonstrated the 'muscle' protein content of a wide variety of nonmuscle
cells and which, by the use of fluorescent antibodies, have shown the broad
distribution of these proteins in a variety of cultured cells. It may seem paradoxical that in such cells the motility
proteins are concentrated in the stress
fiber filament bundles, structures that
appear almost immobile in the living
cell, but this may make sense if these
fibers are concerned with slowly extending and stabilizing the cells' basal
attachments, a process probably requiring considerable force. In the long
run, however, solutions to the understanding of this and other questions
concerning the motility of nonmuscle
cells are likely to come only after a
very detailed knowledge of the intracellular distribution of the entire range
of motility proteins has evolved."

This Week's Citation Classic

Rambourg A & Leblond C P. Electron microscope observations on the carbohydrate-rich cell coat present at the surface of cells in the rat.
J. Cell Biol. **32**:27-53, 1967.
[Department of Anatomy, McGill University, Montreal, Canada]

Periodic acid-silver methenamine, a technique for glycoprotein detection on the electron microscope, stains the surface of a wide variety of cells. The absence of staining in the regions where adjacent plasma membranes fuse to form tight junctions indicates that the stained material is located at the outer surface of the plasma membrane. [The *SCI®* indicates that this paper has been cited in over 615 publications since 1967.]

Alain Rambourg
Département de Biologie du C.E.A.
Centre d'Études Nucléaires de Saclay
91191 Gif-sur-Yvette
France

July 5, 1983

"In the early 1960s, the literature on the features of the cell surface was in a state of confusion. Some claimed that a carbohydrate-rich cell coat was typical of cancer cells.[1] Others described a carbohydrate-containing fuzz on intestinal cells, but only at their apical surface,[2] although the lateral membranes had been shown to stain with the periodic acid (PA)-Schiff technique for carbohydrates.[3] Moreover, at that time, the long held view that a protein-rich 'cement' kept cells together was being abandoned by electron microscopists in favor of desmosomes and junctional complexes. The spark that made it possible to reach a unified view on the presence of carbohydrates at the cell surface arose from a fortuitous observation.

"In 1964, when, as a young MD, I joined C.P. Leblond's group at McGill University, Marian Neutra was investigating the biogenesis of glycoproteins by radioautography. Her histologic slides were stained by the PA-Schiff technique, but counterstained by hematoxylin. On a PA-Schiff stained section which I counterstained with toluidine blue instead of hematoxylin, the delicate plasma membrane appeared as a fine purplish line against a pale blue background. Thus, the PA-Schiff staining of the plasmalemma previously masked by hematoxylin became visible. This was confirmed in some 50 different cell types of the rat. We concluded that the plasmalemma of all cells included glycoproteins.[4] Moreover, since the staining disappeared along tight junctions, where the outer surfaces of adjacent plasma membranes were known to fuse, we further postulated that the layer staining for glycoprotein was exclusively located on the outside of the plasmalemma.[4]

"Since the work had been performed with the light microscope, the resolving power of which was thought by some authors not to allow the visualization of the plasma membrane, these conclusions were met with widespread criticism. It was thus decided to use the electron microscope and to adapt a modification of the PA-Schiff technique, i.e., the PA-silver methenamine method which had been used in the department.[5] The first attempts at staining the plasmalemma were plagued with silver precipitates and unspecific reactions. After many months of work, reproducible results were obtained, and the observations made on the light microscope were confirmed, as described in the present article.

"The success of this article is attributed to its presenting a broad but simple concept which unifies the views of microscopists postulating the existence of fuzz-like or cement-like carbohydrates at the cell surface, as well as the conclusions of biochemists extracting carbohydrates from the plasmalemma of cancer or blood cells. Our results, which were published in the widely read *Journal of Cell Biology*, indicated that carbohydrates were a constant feature of the cell surface and, indeed, it is now accepted[6] that the plasma membrane contains glycoproteins and glycolipids, the carbohydrate chains of which are exposed on its outer surface."

1. Gasic G & Gasic T. Removal of PAS positive surface sugars in tumor cells by glycosidases.
   Proc. Soc. Exp. Biol. Med. 114:660-3, 1963.
2. Ito S. The enteric surface coat on cat intestinal microvilli. *J. Cell Biol.* 27:475-91, 1965. (Cited 365 times since 1965.)
3. Puchtler H & Leblond C P. Histochemical analysis of cell membranes and associated structures as seen in the intestinal epithelium. *Amer. J. Anat.* 102:1-31, 1958.
4. Rambourg A, Neutra M & Leblond C P. Presence of a "cell coat" rich in carbohydrate at the surface of cells in the rat. *Anat. Rec.* 154:41-72, 1966. (Cited 175 times since 1966.)
5. Van Heyningen H E. Correlated light and electron microscope observations on glycoprotein-containing globules in the follicular cells of the thyroid gland of the rat. *J. Histochem. Cytochem.* 13:286-95, 1965.
6. Sturgess J, Moscarello M & Schachter H. The structure and biosynthesis of membrane glycoproteins.
   (Juliano R L & Rothstein A, eds.) *Current topics on membranes and transport. Volume 11. Cell surface glycoproteins: structure, biosynthesis and biological functions.* New York: Academic Press, 1978. p. 15-105.

Martínez-Palomo A. The surface coats of animal cells.
Int. Rev. Cytol. **29**:29-75, 1970.
[Laboratorio de Microscopía Electrónica, Instituto Nacional de Cardiología,
Mexico City, Mexico]

The paper describes ultrastructural and cytochemical investigations on the peripheral components of animal cells which allow the recognition of two types of surface layers: (1) cell coats located on the outer surface of cells, and (2) basal and external laminas, which border the surface of epithelial and mesenchymal cells, respectively. [The *SCI®* indicates that this paper has been cited in over 255 publications since 1970.]

A. Martínez-Palomo
Sección de Patología Experimental
Centro de Investigación y de Estudios
Avanzados
07000 Mexico City
Mexico

June 5, 1984

"This article was written during 1968, while I was establishing a new electron microscopy laboratory at the National Institute of Cardiology of Mexico. During my postgraduate training at the Cancer Research Institute at Villejuif, near Paris, France, I started, in 1965, working at the laboratory for electron microscopy headed by the late Wilhelm Bernhard. At that time, Bernhard's laboratory was steaming with new developments in the field of biological electron microscopy, including advanced techniques for autoradiography, immunocytochemistry, and cryomicrotomy.

"After completing an ultrastructural study of the replication of the oncogenic adenovirus 12, I started a project to analyze the structural modification of plasma membranes in cancer cells, both in solid tumors and in cell cultures. The finding of a striking deficiency in cell junctions in malignant cells[1,2] required only standard electron microscopic techniques. However, the study of the surface coats of tumor cells involved specialized cytochemical techniques, such as the ruthenium red method devised by Luft[3] and the phosphotungstic acid technique described by Rambourg.[4] With these techniques, we were able to demonstrate differences between normal and cancer cells in cultures, reported in this and other papers.[1,5]

"During the late 1960s, the information concerning the surface coat components of animal cells was sparse and dispersed among morphological, biochemical, and immunological reports. I felt the need for a critical review of the subject. The review was prepared back in my Mexico City laboratory. In retrospect, I think that some of the advantages that I had at that time, in order to complete what was going to become a relatively well cited article, were a peaceful setting, time to carefully review the literature and conduct my observations, and the lack of a deadline. These conditions are hardly found at present when experimentation, teaching, and the writing of grant proposals leave little spare time for thinking.

"The large number of reprint requests received—more than two thousand—indicated that timeliness was one of the possible assets of the article. As stated in the paper, the purpose was to critically review, in the light of my own experience, knowledge on the nature of cell surface layers of animal cells. After its publication in 1970, a bibliographic explosion in the field of surface components occurred, which continues even now. This subject, which 15 years ago could be covered in a single monograph, now requires several multiauthored series exclusively devoted to this important field of cell biology."

1. **Martínez-Palomo A, Braßlovsky C & Bernhard W.** Ultrastructural modifications of the cell surface and intercellular contacts in some transformed cell strains. *Cancer Res.* **29**:925-37, 1969. (Cited 135 times.)
2. **Martínez-Palomo A.** Ultrastructural modifications of intercellular junctions in some epithelial tumors. *Lab. Invest.* **22**:605-14, 1970. (Cited 65 times.)
3. **Luft J H.** Fine structure of capillary and endocapillary layer as revealed by ruthenium red. *Fed. Proc.* **25**:1773-83, 1966. (Cited 310 times.)
4. **Rambourg A.** An improved silver methenamine technique for the detection of periodic-acid reactive complex carbohydrates with the electron microscope. *J. Histochem. Cytochem.* **15**:409-12, 1967. (Cited 200 times.)
5. **Martínez-Palomo A & Braßlovsky C.** Surface layer in tumor cells transformed by Adeno-12 and SV40 viruses. *Virology* **34**:379-82, 1968. (Cited 50 times.)

Bruni C & Porter K R. The fine structure of the parenchymal cell of the normal
rat liver. I. General observations. *Amer. J. Pathol.* 46:691-755, 1965.
[Biological Laboratories, Harvard University, Cambridge, MA]

A group of observations, described for the first time in this report, is taken to indicate that liver secretory proteins are segregated by the membranes of the rough endoplasmic reticulum (RER). After transfer into the smooth endoplasmic reticulum (SER) and, subsequently, into the Golgi system, they are finally carried to the space of Disse by way of Golgi-released vacuoles. [The *SCI®* indicates that this paper has been cited in over 370 publications since 1965.]

————————————————

Carlo Bruni
Department of Neurology
University of Virginia Medical Center
Charlottesville, VA 22908

May 16, 1983

"This electron microscopic study of the normal mature rat hepatocyte, carried out in Porter's laboratory at Harvard University, was in part generated by our interest in searching for morphological expressions of the transport of liver secretory proteins into the bloodstream. For instance, according to biochemical studies,[1] liver albumin was synthesized by the ribosomes of the rough microsome fraction, but before final extrusion became associated with the smooth microsome fraction. Prior to 1965, very little work had, however, been published to clarify this transitional step at the level of cell fine structure.

"First, we saw that the lumen of the cisternae of the rough endoplasmic reticulum (RER) contained two kinds of substances, interpreted by us as proteins that had been synthesized by the membrane-associated ribosomes: one substance was in the form of highly electron dense spherical particles (30-100 nm in diameter), the other in the form of less dense fine fibrils. Second, the two types of substances detected within the RER were also detected within vesicles of the smooth endoplasmic reticulum (SER) and within the peripheral portions of the cisternae of the Golgi system. In this latter location, however, the granules often appeared to be remarkably more numerous than in the elements of the ER. This observation was in accordance with the concept that the Golgi apparatus plays a role in packaging proteins for export. Subsequently, granule-containing vacuoles were seen to lose continuity with the Golgi complex and to make contact with the hepatocyte plasma membrane at the space of Disse. Finally, this space was found to contain granules identical to those seen within membranous systems in the cytoplasm. Several blood proteins are known to be formed in the liver. It was not possible, however, to determine which proteins were segregated within the membranous cytoplasmic systems in our study, as we had used electron microscopic techniques alone. Nevertheless, it was speculated that the granules were representative of aggregates of albumin, but other possible interpretations were not ruled out.

"After publication of our report, investigations from other laboratories, using electron microscopy in conjunction with biochemical or immunological procedures to identify the products of liver secretory activity, were led to recognize that the granular material represented very low density lipoproteins[2] rather than aggregates of albumin. It was, however, similarly recognized and recently confirmed[3] that liver secretory proteins, both simple, like albumin,[3,4] and complexed either to lipids or carbohydrates,[5] all follow the pathway of transport formed by the RER, SER, and Golgi apparatus. We therefore think that a reason for the frequent citation of our report is that it clearly illustrated for the first time the basic fine structural aspects of the transport of blood liver proteins into the bloodstream. Other reasons were that the report enlarged on the role of the Golgi apparatus in separating proteins for export from proteins for intracellular use and that it presented both a description of additional previously unreported findings in the rat hepatocyte and an extensive review of earlier fine structural studies of this cell type."

1. Peters T, Jr. The biosynthesis of rat serum albumin. II. Intracellular phenomena in the secretion of newly formed albumin. *J. Biol. Chem.* 237:1186-9, 1962.
2. Jones A L, Ruderman B B & Herrera M G. Electron microscopic and biochemical study of lipoprotein synthesis in the isolated perfused rat liver. *J. Lipid Res.* 8:429-46, 1967.
3. Yokota S & Fahimi H D. Immunocytochemical localization of albumin in the secretory apparatus of rat liver parenchymal cells. *Proc. Nat. Acad. Sci. US—Biol. Sci.* 78:4970-4, 1981.
4. Peters T, Jr., Fleischer B & Fleischer S. The biosynthesis of rat serum albumin. IV. Apparent passage of albumin through the Golgi apparatus during secretion. *J. Biol. Chem.* 246:240-4, 1971.
5. Jamieson J C & Ashton F E. Studies on the acute phase of proteins of rat serum. IV. Pathway of secretion of albumin and α_1-acid glycoprotein from liver. *Can. J. Biochem.* 51:1281-91, 1973.

Friend D S & Farquhar M G. Functions of coated vesicles during protein absorption in the rat vas deferens. *J. Cell Biol.* **35**:357-76, 1967.
[Department of Pathology, University of California School of Medicine, San Francisco, CA]

The role of coated vesicles during horseradish peroxidase absorption was investigated by electron microscopy and cytochemistry. The results demonstrate that (a) this epithelium absorbs protein from the lumen, (b) large coated vesicles transport protein to lysosomes, and (c) some small coated vesicles move hydrolytic enzymes from the Golgi region to multivesicular bodies. [The *SCI®* indicates that this paper has been cited in over 710 publications since 1967.]

Daniel S. Friend
Department of Pathology
School of Medicine
University of California
San Francisco, CA 94143

May 23, 1984

"After completing two rich years as a postdoctoral fellow in Don Fawcett's department developing an ongoing love for microscopic precision and beauty, I sought Marilyn Farquhar in the department of pathology at the University of California, San Francisco, to have me for a third fellowship year. I wanted to learn enzyme cytochemistry at its best. I had been counseled on both coasts that descriptive microscopy had had its heyday: to be a contributor to the burgeoning field of cell biology, my dream, I must develop skills as an experimentalist. Cytochemistry, enjoyably tasted in OsO_4 impregnation studies,[1] was the type of tool I wished to master. The standards of excellence, ambience, and contemporary thinking in Marilyn's lab excited me.

"I brought with me some knowledge of using horseradish peroxidase (HRPase) as a tracer—then unpublished[2] know-how that Morris Karnovsky imparted to me. The stage for something exciting to happen was almost set—the facilities, laboratory knowledge, encouragement, the direction of a sound experimentalist, a new tracking procedure, and the freedom to work on a problem of my choice—a special delight granted me throughout my training. Two simple questions nagged me: was the epididymal epithelial cell absorptive? And if so, why did it have such a large Golgi apparatus?

"To begin, I infused HRPase into the lumen of the vas deferens trying to get retrograde flow to study its uptake by epididymal principal cells. Sometime between the sixth and thirteenth repetition of just-about-exactly this same experiment, I addressed myself to the equivalent question in the vas deferens. The thirteenth through the eighteenth repetitions all worked well: the cells took up the tracer in large, coated invaginations and vesicles. Comparing images from varying intervals indicated their direction and destination. Our improvements of the cytochemical and ancillary electron microscopic techniques for TPPase and ACPase, introduced by Alex Novikoff, Sidney Goldfischer, and Bob Smith, yielded comparable data on the flow of small coated vesicles. At that point, I began to write my findings, but had a hurdle to overcome: I did not fully grasp the quantitative relationship between two sets of vesicle-flow. The problem was resolved one sunny afternoon when George Palade patiently listened and intensely looked at all my data; then said one word—count. I counted the two vesicle populations at the different intervals and thereby cemented a relationship between them—a key to the success and popularity of the study. In addition to being among the first clearly illustrated and interpreted cytochemical studies of absorptive endocytosis, it was the first documentation of distinctive subsets of coated vesicles and an endosome compartment (the MVB), foreshadowing the now extensively explored areas of receptor-mediated endocytosis and membrane flow."[3-5]

1. Friend D S & Murray M J. Osmium impregnation of the Golgi apparatus. *Amer. J. Anat.* **117**:135-50, 1965. (Cited 125 times.)
2. Graham R C & Karnovsky M J. The early stages of absorption of injected horseradish peroxidase in the proximal tubules of mouse kidney: ultrastructural cytochemistry by a new technique. *J. Histochem. Cytochem.* **14**:291-302, 1966. (Cited 4,385 times.)
3. Farquhar M G & Palade G E. The Golgi apparatus (complex)—(1954-1981)—from artifact to center stage. *J. Cell Biol.* **91**:77s-103s, 1981.
4. Roth T F & Woods J W. Fundamental questions in receptor-mediated endocytosis. (Marchesi V T & Gallo R C, eds.) *Differentiation and function of hematopoietic cell surfaces.* New York: Liss, 1982. p. 163-81.
5. Bainton D F. The discovery of lysosomes. *J. Cell Biol.* **91**:665-765, 1981.

Majno G & Palade G E. Studies on inflammation. I. The effect of histamine and serotonin on vascular permeability: an electron microscopic study.
J. Biophys. Biochem. Cytol. **11**:571-605, 1961.
[Dept. Pathol., Harvard Med. Sch., Boston, MA and Rockefeller Inst., New York, NY]

It was well known that histamine increased the permeability of small blood vessels; we found the mechanism: the cells that line the smallest veins become partially disconnected, so that fluid can escape between them. This is one of the key events in acute inflammation. [The *SCI®* indicates that this paper has been cited over 625 times since 1961.]

Guido Majno
Department of Pathology
University of Massachusetts
Medical Center
Worcester, MA 01605

June 26, 1981

"*Citation Classic*? Would ISI® care to open a new category of *Refusal Classics*? Of all the scientific papers on which I was the senior author, only one was turned down: you know which one. This paper had been a labor of love. I had joined G.E. Palade's laboratory at the Rockefeller Institute in 1958, to be initiated in the new art of electron microscopy. We had carefully chosen a topic of common interest: Palade was an authority on blood vessels, and I was interested in any tissue as long as it was sick, so we settled for leaking blood vessels. Histamine was supposed to make capillaries leak, but nobody knew by what mechanism; so we injected histamine into rats and began to look for leaking capillaries. I remember spending days of frustration looking at capillaries that remained stubbornly normal—until we found out that the exciting events happened a little further downstream, in the venules, which had become riddled with tiny leaks. In the process we also stumbled upon an extremely cheap and beautiful biological trick: a drop of India ink injected into the bloodstream causes leaking vessels to become elegantly 'tattooed' in black; an experimental technique now known as vascular labeling (the phenomenon itself had been published at least 12 times before, but with the wrong explanation).

"When I was advised that the resulting paper would be turned down by the *Journal of Experimental Medicine*, I was half shocked, half amused, but completely sure that the reason could not be scientific. Sure enough, the startling reason given by that grand old gentleman, Peyton Rous, was that the paper contained *too much history*! Was I really supposed to chop out the historical roots of our work? 'Stick to your guns,' said Palade, lapsing into vernacular English (we used to speak French at that time). So we sent the paper to a new, competing journal, now called the *Journal of Cell Biology*. (It took me another eight years to go one step further, and find that the leaks were caused by endothelial contraction.)[1]

"Looking back upon that work, produced under the wing of a great master, I can say with a touch of nostalgia that we were bound to produce a paper of more than passing value because we found out—just in time—that we had merely confirmed, with more refined methods, a paper that had appeared in *Virchows Archiv* in 1875.[2] Since 1961, a mountain of literature has been appearing on the endothelium. A review of this field was recently published in *The Thrombotic Process in Atherogenesis*.[3]

"I believe that our work attracted attention because it solved a long-standing functional riddle by very simple (and very aesthetic) morphologic observations."

1. Majno G, Shea S M & Leventhal M. Endothelial contraction induced by histamine-type mediators.
   J. Cell Biol. **42**:647-72, 1969.
2. Arnold J. Ueber das Verhalten der Wandungen der Blutgefässe bei der Emigration weisser Blutkörper.
   Virchows Arch. **62**:487-503, 1875.
3. Majno G & Joris I. Endothelium 1977: a review. (Chandler A B, Eurenius K, McMillan G C, Nelson C B, Schwartz C J & Wessler S, eds.) *The thrombotic process in atherogenesis.* New York: Plenum, 1978. p. 169-225; 481-526.

Revel J P & Karnovsky M J. Hexagonal array of subunits in intercellular junctions of the mouse heart and liver. *J. Cell Biol.* **33**:C7-C12, 1967.
[Depts. Anat. and Pathol., Harvard Med. Sch., Boston, MA]

This paper reviews some of the circumstances which led us to conclude that structures similar to 'electric synapses' also exist between non-excitable cells. Now called 'gap junctions,' 'nexus,' and 'macula communicans,' they allow for exchange of low MW solutes between contacting cells. [The *SCI®* indicates that this paper has been cited in over 865 publications since 1967.]

Jean-Paul Revel
Division of Biology
California Institute of Technology
Pasadena, CA 91125

August 6, 1982

"Our paper was actually the result of a serendipitous discovery. For quite a while Morris Karnovsky and I had been interested in techniques for the ultrastructural demonstration of carbohydrate-containing moieties on cell surfaces. On the hypothesis that colloidal suspensions or polymers of heavy metals might be of use, we tried, among others, the lanthanides. The complexes formed could diffuse into and act as a tracer of intercellular spaces. Morris soon produced images in heart muscle where the whole intercellular space was filled with electron opaque material. He rushed over one evening with still-wet negatives and we saw, to our surprise, areas showing a beautiful quasi-crystalline pattern of ~70 Å particles when the cell membranes were viewed *en face*. I recalled the hexagonal arrays in electrical synapses after permanganate fixation,[1] and in negatively stained liver plasma membranes.[2] The subunit structure was not characteristic of the plasma membrane in general, and our 'chief,' D.W. Fawcett, agreed that we might be observing the so-called 'close' junctions in the heart. Shortly thereafter I found identical structures in a sample of liver, of course in a partially torn section which soon broke up altogether, but not before it had been triumphantly photographed.

"There followed the rapid accumulation of data, and we soon realized that depending on the angle at which one viewed the junctions, they could be seen as a series of striations, or as particles. We concluded that structures similar to electrical synapses could be seen in non-excitable tissues, and that they represented a unique type of cell junction. This, and the usefulness of the technique to demonstrate cell junctions, are two reasons our paper is still cited.

"I won the senior authorship by the toss of a coin. It should be pointed out that although we did not use the term 'gap junction,' somehow it has, over the years, been attributed to us!

"After our initial report was published we thought to put our observations into a detailed paper. Optical diffraction experiments confirmed the hexagonal packing of the particles. We made a number of plastic models with which we could reproduce all the appearances of the junctions as seen in the electron microscope. We never wrote up these results as we were soon pursuing instead the ultrastructure of the junctions as seen by the then newly introduced technique of freeze-cleaving, by which, it was hoped, further features of structure would be revealed. Besides many subtle details, it definitively showed that gap and tight junctions were clearly different structures. An excellent overview of the field is a series of short papers resulting from a symposium organized by Werner Loewenstein.[3] It is ironic to note, however, that recent rapid-freeze studies would indicate that the dramatic tight hexagonal packing we observed may represent, to some extent, an artifact of tissue preparation and fixation."

1. **Robertson J D.** The occurrence of a subunit pattern in the unit membranes of club endings in Mauthner cell synapses in goldfish brains. *J. Cell Biol.* **19**:201-21, 1963.
2. **Benedetti E L & Emmelot P.** Hexagonal array of subunits in tight junctions separated from isolated rat liver plasma membranes. *J. Cell Biol.* **38**:15-24, 1968.
3. **Loewenstein W R.** Introductory remarks to the symposium. *In Vitro* **16**:1007-68, 1980.

Ryter A & Kellenberger E. Etude au microscope électronique de plasmas contenant de l'acide désoxyribonucléique. I. Les nucléoides des bactéries en croissance active. (Electron microscope study of plasmas containing DNA. I. Bacterial nucleoids in active growth.)
Z. Naturforsch. Sect. B **13**:597-605, 1958.
[Laboratoire de Biophysique, Université de Genève, Switzerland]

The influence of fixation conditions on the bacterial nucleoid fine structure was studied by varying different parameters. It could be concluded that the finest fibrillar structure of the nucleoplasm was obtained when fixation was performed with one percent osmium tetroxide dissolved in Michaelis buffer pH 6.0 containing Ca^{++} and amino acids followed by post-fixation with uranyl acetate. [The *SCI®* indicates that this paper has been cited in over 890 publications since 1961.]

Antoinette Ryter
Unité de Microscopie Électronique
Département de Biologie Moléculaire
Institut Pasteur
75724 Paris
France

June 24, 1983

"This work was performed between 1953 and 1958 in the biophysics laboratory of the University of Geneva (Switzerland), which was directed by E. Kellenberger. At that time, electron microscopy and especially ultrathin sectioning were just beginning to be used. The osmium fixation conditions generally used for animal cells gave unsatisfactory results in bacteria, especially for nuclear material. The latter generally presented the appearance of rather coarse filaments. This pushed us to study carefully the influence of fixation conditions on the nucleotide fine structure and different DNA plasmas.

"This study was long and fastidious because several parameters intervened and interfered with each other. It finally appeared that phosphate buffer, currently used for eukaryotic cells, had to be absolutely avoided because calcium ions were indispensable for a good preservation. The slightly acidic pH, the presence of amino acids or small peptides, and at last the post-fixation with uranyl acetate all appeared to be necessary for a reproducible and good preservation of the nucleoid.

"These peculiar conditions are not required by eukaryotic nuclei because the histones present on DNA are rapidly cross-linked by the fixative, and maintain the three-dimensional structure of the chromosome. The fibrillar structure of the bacterial chromosome suggests that DNA association with protein partners, if it exists, is certainly much more discrete or labile than that of nucleohistones and is not sufficient to prevent DNA from collapsing during alcohol or acetone dehydration.

"One can add that this technical study was complicated by the fact that methacrylate, which was used at that time as embedding medium, frequently produced swelling and even 'explosion' of bacteria. To overcome this artifact we searched for a better embedding medium (a polyester[1]) which was improved at the same time as fixation. It is interesting to point out that this swelling artifact in methacrylate, which was especially obvious in bacteria, also led A.M. Glauert to find another embedding medium (Araldite).[2]

"The association of good fixation conditions and an improved embedding medium allowed Kellenberger and his collaborators to make major advances in the knowledge of the bacterial nucleoid[3] and to obtain, for the first time, information on intracellular bacteriophage multiplication.[4] The frequent citation of this paper is due to the fact that these fixation conditions were used by almost all electron microscopists working on prokaryotic cells. It is also sometimes only cited for post-fixation with uranyl acetate, which is often applied to eukaryotic cells because it improves membrane contrast and clearly shows its triple-layered structure."

1. Kellenberger E, Schwab W & Ryter A. L'utilisation d'un copolymère du groupe des polyesters comme matériel d'inclusion en ultramicrotomie. *Experientia* **12**:421-2, 1956.
2. Glauert A M & Glauert R H. Araldite as an embedding medium for electron microscopy. *J. Biophys. Biochem. Cytol.* **4**:191-4, 1958.
3. Kellenberger E, Ryter A & Séchaud J. Electron microscope study of DNA-containing plasms. II. Vegetative and mature phage DNA as compared with normal bacterial nucleoids in different physiological states. *J. Biophys. Biochem. Cytol.* **4**:671-8, 1958.
   [Citation Classic. *Current Contents/Life Sciences* **23**(7):12, 18 February 1980.]
4. Kellenberger E, Séchaud J & Ryter A. Electron microscopical studies of phage multiplication. IV. The establishment of the DNA pool of vegetative phage and the maturation of phage particles. *Virology* **8**:478-98, 1959.

This Week's Citation Classic

Kellenberger E, Ryter A & Séchaud J. Electron microscope study of DNA-containing plasms. II. Vegetative and mature phage DNA as compared with normal bacterial nucleoids in different physiological states. *J. Biophys. Biochem. Cytol.* 4:671-8, 1958.
[Laboratoire de Biophysique, University of Geneva, Geneva, Switzerland]

Under specific conditions—later called the 'R.K.-conditions'—bacterial nucleoids can be fixed with OsO_4 so as to contain fine fibrillar material when viewed on thin sections. Under other conditions, and—as was shown later—with aldehydes, aggregations occur. Those induced with EDTA (formerly under the trade name 'versene') allow a practical aggregation test. The pool of vegetative (=replicating and transcribing) T4 bacteriophage DNA is shown to be aggregation sensitive, as is the normal nucleoid. Artificial aggregation is different from physiological condensation as occurs, e.g., when T4-virions are packaged out of the DNA-pool. [The *SCI®* indicates that this paper has been cited over 915 times since 1961.]

Edward Kellenberger
Biozentrum der Universität Basel
Abteilung Mikrobiologie
CH-4056 Basel
Klingelbergstrasse 70
Switzerland

July 24, 1978

"The initial idea of a series of papers—of which this was the second—was to establish a test for distinguishing condensed DNA-plasms of eukaryotes, e.g., metaphase chromosomes from metabolically active, non condensed DNA, e.g., of DNA-plasms in the interphase of eukaryotes or during all times in prokaryotes. At that time the possibility was still open that the histones condense the DNA and render it thus metabolically inert. By removal of histones a decondensed DNA-plasm would arise which would be metabolically active. The decondensed DNA-plasms of bacterial nuclei and vegetative phage were known to be continuously replicating and transcribing. Some experiments then suggested that bacteria have no histones and in general fewer proteins bound to their DNA. My working hypothesis at that time was thus that during interphase the DNA-plasm of eukaryotes might be organised similar to that of prokaryotes; only the condensed metaphase chromosomes would have been specific for eukaryotes.

"The later applications of the methodology of this paper by others and ourselves showed that the DNA-plasms of mitochondria, chloroplasts, blue-green algae, dinoflagellates, and kinetoplasts of trypanosoma had a structure and behavior in fixation (coagulation-sensitive) analogous to that of bacteria. The DNA-plasm of interphase nuclei of eukaryotes, however, turned out *not* to be coagulation-sensitive. It also did not have the same fibrillar aspect. This result was suspected to be in relation to DNA binding proteins but only with our present knowledge can we understand it: in eukaryotes most of the histones always stay associated with the DNA of the chromatin. For bacteria, DNA binding proteins (different from histones) are demonstrated, but they are still found only in amounts which lead to much smaller protein-DNA ratios as in eukaryotes. Hence, it would be understandable that prokaryotic DNA-plasms behave similarly to naked DNA, while chromatin does not.

"Although we intended to draw the attention of the reader to the different organisational states of the nuclear material and how it can efficiently be investigated by electron microscopy of thin sections, I believe that the paper is mostly quoted as reference for a certain condition of fixation, which is mainly the result of an enormous effort of my collaborator Antoinette Ryter, then technician, now research director at the Pasteur Institute in Paris. It is only now, after the nucleosomes have been described, that some people are becoming aware again of possible differences between the prokaryotic and the eukaryotic type of organisation of the nuclear material. Others do not like differences and seize anything which could appear as evidence that bacterial DNA is organised with the same 'beads' as is the chromatin of eukaryotes. Has it not been demonstrated that SV40 virus is constructed by way of nucleosomes? As for most people viruses are lower organisms than bacteria, do we not have to expect nucleosomes also in bacteria? This wishful thinking reminds me of earlier times, when it became clear that microorganisms have genetics and when people thus tried very hard to demonstrate spindles and centrosomes in bacteria! Hopefully, further thorough investigation of bacterial chromosomes will not be inhibited by preconceptions and soon an explanation of the above discussed fundamental difference of prokaryotic and eukaryotic DNA-plasms will be provided."

Moor H & Mühlethaler K. Fine structure in frozen-etched yeast cells.
J. Cell Biol. 17:609-28, 1963.
[Laboratory of Electron Microscopy, Department of General Botany, Swiss Federal Institute of Technology, Zürich, Switzerland]

The freeze-etch technique involves cleaving a frozen specimen and 'etching' it by vacuum sublimation. The fine structure of the etched fracture face is preserved in a replica obtained by condensing evaporated platinum and carbon onto it. The application of this replica technique to yeast cells showed for the first time that the problems of freezing artifacts can be overcome and that freeze-fractured membranes exhibit many structural details which are not portrayed by other techniques. [The *SCI*® indicates that this paper has been cited over 945 times since 1963.]

Hans Moor
Institute for Cell Biology
Swiss Federal Institute of Technology
CH-8093 Zürich
Switzerland

April 10, 1981

"My interest in biology was first awakened by an uncle who is a plant ecologist, but hay fever blocked my way along this line; instead I decided to become a laboratory biologist. During my studies at the Swiss Federal Institute of Technology in the mid-1950s my enthusiasm for very small things was evoked by my teachers and later sponsors, A. Frey-Wyssling and K. Mühlethaler. In 1956 it was still an enormous challenge to enter the new and largely unexplored realm of cellular ultrastructure. In retrospect I can see that it was essential for my future work that from the beginning I became concerned with the problems of recognising and avoiding artifacts introduced by the preparation techniques used for electron microscopy. In those days freezing was suggested as being capable of solving the problems which arose from chemical fixation, embedding, and thin-sectioning. From the disappointing experiences of a colleague I learned that freeze-drying could not lead to success. But in 1957, a technique was published by Russell Steere which had all the features of a real alternative.[1]

"In the early 1950s several physicists had already used freezing in order to stabilize suspensions of diverse objects and, after cleaving the ice in order to disclose the inclusions, they produced a surface replica on the fracture face by condensing evaporated platinum and carbon onto it. The great merit of Steere was to have adapted this purely physical method for the preparation of biological objects. The pictures he published in 1957 showed well-preserved virus crystals in destroyed plant cells. These results and subsequent personal communications with him convinced me of the potentialities of the method and gave me insight into its technical limitations at that time. From this basis, I started the methodological development, a job which has kept me busy till now.[2]

"The work has proceeded along four different lines: development of an apparatus enabling fracturing, etching by vacuum sublimation of superficial ice, and coating of the frozen-fractured and etched specimen under high vacuum; precise physical control of every preparational step; freezing of cells and tissues without introducing artifacts due to ice crystal formation; and interpretation of the pictures which have given a completely new insight into cellular ultrastructure. A first breakthrough was the construction of a 'rather sophisticated' machine. Without the help of a bright technician, Heinz Waldner, and the commercial interest of a vacuum firm, this apparatus would never have been constructed and made generally available on the market.[3] The successful application of the technique to the investigation of yeast cell ultrastructure was published in 1963. It was evidently this second breakthrough which caught the attention of many scientists and—to judge from the large number of citations—was instrumental in the acceptance of freeze-etching as a standard preparation technique."

1. **Steere R L.** Electron microscopy of structural detail in frozen biological specimens. *J. Biophys. Biochem. Cytol.* 3:45-60, 1957.
[Citation Classic. *Current Contents* (47):17, 20 November 1978.]
2. **Moor H.** Recent progress in the freeze-etching technique. *Phil. Trans. Roy. Soc. London B* 261:121-41, 1971.
3. **Moor H, Mühlethaler K, Waldner H & Frey-Wyssling A.** A new freezing-ultramicrotome. *J. Biophys. Biochem. Cytol.* 10:1-13, 1961.

This Week's Citation Classic™

Balinsky B I & Devis R J. Origin and differentiation of cytoplasmic structures in the oocytes of *Xenopus laevis*. *Acta Embryol. Morphol. Exp.* **6**:55-108, 1963.
[Dept. Zoology, Univ. Witwatersrand, Johannesburg, South Africa]

The paper reports the results of a systematic electron microscopic investigation of the development of the frog oocyte, showing the origin of a variety of cytoplasmic structural elements in the oocyte: the yolk, the pigment granules, the cortical granules, and the vacuolated cortical cytoplasm. [The *SCI®* indicates that this paper has been cited in over 185 publications since 1963, making it the most-cited paper published in this journal.]

B.I. Balinsky
19 Oban Avenue
Blairgowrie
Johannesburg
2194 South Africa

September 15, 1983

"My work in electron microscopy (EM) is directly connected to a visit I paid to the US in 1956. After many years of work in embryology (mainly using microsurgery on embryos as a method of research) done in European countries, I found myself as head of the zoology department of the University of the Witwatersrand in Johannesburg, South Africa. Using my first sabbatical leave, I flew to the US. My aim was to see how new techniques could be applied to the study of embryonic development. One such new technique was EM. Application of EM to biological studies was at that time in its infancy. A senior colleague told me that EM is not of much use for an embryologist. 'One cannot cut serial sections; one picks up a section off the edge of a broken piece of glass.' Undaunted by this skeptical opinion, I went ahead. I had the good fortune of contacting the pioneers of

biological EM: George E. Palade[1] and Keith R. Porter[2] at the Rockefeller Institute in New York. Palade showed me how sections for EM are cut, and Porter encouraged me in my plans to apply EM to embryological research. At Yale University, which was my base in the US, I further studied the technique of embedding, cutting sections, and making EM photographs. On my return to Johannesburg, I was delighted to find that our university had acquired an electron microscope. This was done on the initiative of physicists, but I was the first to use the microscope for a biological project. Most of the time I was on my own, though during the work on oocytes I had the help of Rosemary Devis, my research assistant, who was paid by the Council for Scientific and Industrial Research of South Africa.

"I believe that the success of my paper was due to two circumstances. First, it was one of the early applications of a new technique to the study of animal development. Secondly, it was a matter of choosing as my object one of great and general interest: the egg in its formation. *Omne vivum ex ovo* — a statement obviously incorrect in its general meaning is true enough in application to the animal world. The more that is learned of animal development, the more attention is focused on the ovum that contains the mechanism on which depends all the subsequent developmental process. My paper embodied the result of the most comprehensive, at that time, systematic study, on the electron microscopic level, of the formation of an animal ovum. For a recent reference, see *The Vertebrate Ovary*."[3]

1. **Palade G E.** A study of fixation for electron microscopy. *J. Exp. Med.* **95**:285-98, 1952.
   (Cited 3,060 times since 1955.)
2. **Porter K R & Blum J.** A study in microtomy for electron microscopy. *Anat. Rec.* **117**:685-70, 1953.
   (Cited 220 times since 1955.)
3. **Jones R E,** ed. *The vertebrate ovary: comparative biology and evolution.* New York: Plenum Press, 1978. 853 p.

Chapter

2

Microbiology

Of the twenty-one Citation Classics represented in this chapter, twelve are directly based on studies of microbial genetics. And they make an impressive list, ranging from Beadle and Tatum's Nobel Prize-winning one gene–one enzyme concept derived from their study of *Neurospora* to bacterial transduction, transformation, and conjugation. Plasmids, transposons, or IS elements (insertion sequences) and the physical structure of the bacterial genome and how it is damaged by ultraviolet light and repaired are the subjects of the remaining Classics. These publications span three decades, from 1945 to 1976—hardly a burst of activity in what Zinder describes as a field where, in 1948, "discoveries [were] lying around waiting to be picked up." Still, these are Classics and we should not cloud the past with the current pace of articles on recombinant DNA, gene insertion, introns, exons, long terminal repeats (LTRs), etc. Nor should we ignore the contributions in eukaryotic genetics that are placed in a separate chapter.

The description by Beadle of his work with Tatum refers to a central publication in the path of their discoveries that led to their Nobel Prize in 1958. Beadle and Tatum's work, traditionally summarized in the phrase "one gene–one enzyme," was remarkable by several standards. Lederberg has pointed out how Beadle and Tatum aggressively searched for biochemical mutants in irradiated samples of their bread mold, in contrast to a passive acceptance of Nature's own mutation rate. Their first mutant, a vitamin B$_6$-requiring isolate, was derived from sample number 299, and their second, one requiring vitamin B, was derived from preparation number 1,085. Nevertheless, the ease with which their mutants could be prepared and analyzed soon yielded the information they had conceptualized.

Although we now accept a multi-gene association with the structure of many proteins, the progression of Beadle and Tatum's logic from an acceptance that biochemical processes are under genetic control and that biochemical pathways advance one step at a time to a demonstration that a mutation can effect a single event and to the one gene–one enzyme concept were important steps in the history of biochemical genetics. Their unique contribution was the development of an experimental method—nutritional mutation—for the study of gene control of metabolic pathways. It is interesting in this context that Beadle describes Tatum's recollection of the nutritional studies of fungi by Fries as an example of serendipity. Tatum's recall here was undoubtedly important, but since Fries and Tatum had worked together in Holland, it was not a serendipitous event.

The other Nobel Prize winner represented in this chapter is Lederberg, but the paper cited is only a small part of his Nobel award, which dealt with bacterial conjugation, the genomic fusion of unlike cells within a species which is a form of sex life among the bacteria that not all species enjoy. Zinder describes his discovery with Lederberg of the third method of bacterial gene transfer through bacteriophage transduction. This is the incidental transfer of a portion of the bacterial genome from one host cell to another by its packaging with the infectious bacteriophage particle. Bacterial transformation, discovered by Griffith (1928) and analyzed biochemically by Avery, MacLeod, and McCarty, and conjugation, discovered by Lederberg prior to his study with Zinder cited here, are the two earlier described forms by which bacterial genes can be moved from one cell to another. Spizizen's popular article was such because it took advantage of a special growth medium that altered the cell wall of a gram-positive bacterium to permit genetic transformation experiments in an otherwise resistant organism. Sexuality in bacteria, based on the transfer of genes from the F^+ donor male to the F' recipient female via conjugation, is dependent upon an extrachromosomal compartment of genes now known as the plasmid. Hirota found a method to delete this plasmid from the host cell by chemical treatment.

Another mobile genetic element, the IS reviewed by Starlinger and Saedler, is very similar to the larger transposon and is the source of "jumping genes" that can integrate with or disengage from the chromosome. The discovery of IS dates back to the late 1940s and the controlling elements first described by Barbara McClintock. Her description of these elements was largely unknown except to practicing geneticists until she received the Nobel Prize in 1983.

The next three Classics, those by Cairns, Clewell, and Bachmann, describe the morphology of the bacterial chromosome as a circular structure, one that could be viewed schematically as a clock where the time of any gene could be assigned. Replication of the chromosome appeared to represent a moving, dividing fork as the new DNA strand was formed.

The mutagenic role of UV light as a thymidine-dimerizing agent, how these dimers are excised in DNA repair, and the genetic basis of this mechanism of genome repair are the topics of the next three Classics. The very existence of DNA-repairing enzymes was a startling discovery, and variation in this potential among bacterial strains became a useful, though not the sole, explanation of the different mutagenic sensitivity of these strains.

The vast popularity of the Ames test to detect mutagens can be explained by a developing acceptance in the 1970s of a direct relationship between mutagenicity and carcinogenicity. The rapid evaluation of the mutagenic potential of a chemical by the Ames test, cited over 2,000 times, led to its application in thousands of laboratories. Ames' statement that undergraduate students assisted in the initial stages of this project is a good example of the type of contribution that aspiring, young scientists can make to contemporary research.

Fortunately, the tale described by Pastan—the rejection of three papers before an acceptance—is not common, but it is frequent enough to cast suspicion on the skill of editors, particularly when the paper becomes a Classic. The message apparently is that the third time is not the charm but the fourth time is.

Rickenberg's description of Monod's fixation on the word permease and how it resulted in their publication in a French language journal is an interesting sidelight on their Classic. The term permease, now indelibly printed in enzyme terminology, refers to an enzyme that facilitates the transport of a molecule across a cell membrane.

The remaining Classics, as is to be expected, refer to signal publications. The Hughes press is still in use upon certain gram-positive bacteria, fungi, and other cells that are refractory to disintegration by ultrasonifiers. The work of Stanier, Palleroni, and Doudoroff with the pseudomonads is indeed a Classic and is familiar to most microbiologists.

Beadle G W & Tatum E L. Neurospora. 2. Methods of producing and detecting mutations concerned with nutritional requirements.
Amer. J. Botany **32**:678-86, 1945.

The authors describe various methods of inducing, identifying and characterizing large number of gene mutations in the red bread mold Neurospora that are concerned with the synthesis of vitamins, amino acids and other essential metabolites. [The *SCI* ® indicates that this paper was cited 271 times in the period 1961-1977.]

George W. Beadle
5533 Dorchester Avenue
Chicago, IL 60637
December 27, 1977

"Our approach to a better understanding of genes, enzymes and chemical reactions goes back to the turn of the century when, shortly after the rediscovery of Mendel's work, the biochemist-physician Garrod demonstrated a relation between the genetic disease alcaptonuria and a postulated enzyme. Unfortunately his findings soon dropped out of the genetic literature.

"As a graduate student with the plant geneticist R.A. Emerson in the 1930's, I recall his futile attempts to interest plant physiologists and biochemists in making use of genetic traits in corn known to be concerned with the synthesis of chlorophyll and its role in photosynthesis.

"As a National Research Council Fellow at the California Institute of Technology, I was much influenced by researchers there, especially by Boris Ephrussi, a Rockefeller Foundation Fellow from Paris, who was interested in the role of genes in development and function. The organisms of choice for genetic studies, Drosophila for example, seemed less suited for studies of development than were sea urchins or frogs, favorites of embryologists.

"But the genetics of eye pigments in Drosophila did offer promise of bringing the two approaches together. By devising techniques of transplanting larval embryonic eye-buds we did identify two enzymatic steps in brown eye-pigment synthesis, with evidence that each was the immediate control of a single gene, a relation that prompted the slogan 'one gene-one enzyme.' We then attempted to understand the biosynthesis of brown eye pigment, Ephrussi with Kouvine in Paris and I with Tatum at Stanford. Tatum and Haagen-Smit isolated a crystalline precursor of brown pigment, later identified by Butenandt and Weidel as kynurenine.

"Serendipity then came to our aid. Tatum was familiar with the work of Mils Fries in Sweden who had worked out the nutritional requirements of a number of filamentous fungi. I knew the genetic advantages of the red bread mold Neurospora, also a filamentous fungus but one Fries had not investigated. Tatum soon determined that Neurospora would prosper on a simple chemically defined medium including biotin which had just become available commercially.

"We were thus in a position to induce an array of mutants as described in this paper. At the time it was written, we had identified the nutritional requirements of 43 mutant lines, these representing 23 vitamins, amino acids and other metabolites.

"The accomplishments recorded in this article represent the efforts of many investigators, most, but not all of whom are listed in the summary of mutant strains, Table 5 of the article, and in the literature cited. No longer could doubt continue as to the indispensable relations of genes, enzymes and biochemical reactions."

Zinder N D & Lederberg J. Genetic exchange in Salmonella.
J. Bacteriology 64:679-99, 1952.
[Department of Genetics, University of Wisconsin, Madison, WI]

A phenomenon by which genes in bacteria are exchanged and recombined has been discovered. The temperate bacteriophage, PLT-22, during growth has its genome replaced by bacterial genes. Each particle carries about one percent of the bacterial genome. These genes are carried to a new host cell in phage-like particles and can replace their homologous genes. The frequency is about 10⁻⁵/gene/particle. The phenomenon is called transduction. [The $SCI^®$ indicates that this paper has been cited in over 190 publications since 1961.]

Norton D. Zinder
Laboratory of Molecular Genetics
Rockefeller University
New York, NY 10021

January 10, 1983

"When I arrived in July 1948 in Lederberg's lab at the University of Wisconsin, it was one of those golden times in science. A new field, bacterial genetics, had just been born. It had no substantial past, no literature, just discoveries lying around waiting to be picked up.

"I had not been there three months when we developed the penicillin enrichment technique (negative selection) for bacterial auxotrophic mutants. This was important for me as my thesis work was to be an attempt to extend Lederberg's[1] finding of conjugation in E. coli K12. Many pairs of different mutants would be needed for the search. Auxotrophic mutants of S. typhimurium strains were to be mixed inter se and pairwise seeking prototrophic recombinants.

"For six months or so I diligently built up the set of mutants, doing an occasional cross and some coli genetics to relieve the monotony. Most of the attempts to cross the mutants failed. However, one particular mutant strain gave small numbers of prototrophs when crossed with several strains and a large number with one. Surprisingly, other markers, nine in all, did not segregate. There

appeared to be no linkage as existed in coli conjugation. The phenomenon was also asymmetric. Individual markers in one of the strains could be selected but not those of the other. Although resembling DNA-mediated transformation, the phenomenon was not DNase-sensitive.

"We turned to a procedure originally devised by Bernard Davis[2] — a U-tube with a bacterial sterile filter between the arms — to determine whether cell contact was necessary. It wasn't. When I told Lederberg, he sat down heavily and muttered that I should go work with Hotchkiss, who was working on transformation. Ultimately, I did. Quickly we determined that mixed cultures gave sterile filtrates containing large amounts of an activity that affected selective markers in many strains. The filtrates also contained a temperate phage. Because of how little we knew about phage or temperate phage at the time (1950-1951), this finding wasn't all that helpful. In a not quite straight path of research and analysis, we finally showed that the activity was in phage-like particles. Bacterial genes (soon to become DNA) were substituted for phage genes and transferred to the phage's next host. We called the phenomenon transduction — to lead across.

"Citation of this paper is not surprising as it documents the discovery and use of one of the three fundamental mechanisms of genetic exchange in bacteria — transduction; the others are conjugation and transformation. General transduction[3] provides the means for fine genetic analysis of genome structure. The discovery shortly thereafter of special transduction[4] complemented and added to the power of these procedures. The paper also describes phage P22 (then called PLT-22) which became, with lambda (special transduction), the two paradigms for the study of temperate bacteriophages.

"Today, we also recognize transduction in the oncogenic activities of some retroviruses and the effects of the constructs on recombinant DNA."

1. Lederberg J. Gene recombination and linked segregations in Escherichia coli. Genetics 32:505-25, 1947.
2. Davis B D. Nonfiltrability of the agents of recombination in Escherichia coli.
   J. Bacteriology 60:507-8, 1950.
3. Ikeda H & Tomizawa J. Transducing fragments in generalized transduction by phage P1.
   J. Mol. Biol. 14:85-109, 1965.
4. Morse M, Lederberg E & Lederberg J. Transduction in E. coli K-12. Genetics 41:121-33, 1956.

Spizizen J. Transformation of biochemically deficient strains of *Bacillus subtilis* by deoxyribonucleate. *Proc. Nat. Acad. Sci. US* 44:1072-8, 1958.
[Department of Microbiology, Western Reserve University School of Medicine, Cleveland, OH]

In this paper, published 26 years ago, I demonstrated DNA mediated genetic transformation in *Bacillus subtilis*. As this organism could grow in simple minimal media, it was possible to utilize a variety of auxotrophic markers. This made it possible to investigate the genetic controls of biosynthetic pathways as was being done in *Escherichia coli* using other gene transfer systems. [The *SCI®* indicates that this paper has been cited in over 1,010 publications since 1958.]

John Spizizen
Department of Microbiology
and Immunology
University of Arizona
Health Sciences Center
Tucson, AZ 85724

February 27, 1984

"We had previously attempted to transform *Escherichia coli* auxotrophs without success. However, DNA-protein complexes prepared from denatured T2 bacteriophage were found to 'transform' *E. coli* protoplasts,[1] that is, replication and maturation of the phage would occur. These experiments suggested to us that cell wall components were barriers to naked DNA entry. It was at this time that I speculated that *Bacillus subtilis* might be a promising organism, since during stages of outgrowth of germinating spores, cell wall components were minimal or entirely absent.

"Fortunately, Charles Yanofsky was able to provide me with several stable auxotrophic mutants of *B. subtilis* which had been isolated by Burkholder and Giles at Yale University.[2] Transformation of an indole-requiring strain (168) with DNA isolated from a wild strain (W23) was readily achieved. Although germinating spores were initially employed, vegetative cells grown in minimal medium suitably supplemented with a low concentration of casein hydrolysate were found to be highly transformable. Refinements to achieve optimal conditions for transformation were later introduced.[3] These studies demonstrated that specific growth conditions presumably affecting wall synthesis would allow cells to become 'competent' for transformation. The original concept that spore germination would provide the competent state was thus incorrect. Furthermore, other processes besides DNA permeability are now known to be involved in DNA transformation by competent cells. These include a series of coordinated enzymatic steps following DNA attachment, DNA cutting, separation of double strands, penetration of single strands, and pairing with homologous regions of the resident chromosome. The specific enzymes and genetic control of these reactions remain to be identified. *B. subtilis* strain 168 has the capability of carrying out this coordinated process. It was fortuitous that this strain was used as many if not most strains, including the donor strain W23, could not be made competent for chromosomal DNA transformation.

"Many papers describing modified procedures for transformation have appeared. However, continued reference to the original publication has been made mainly because of the minimal medium used. This simple medium has been employed for transformation in numerous laboratories, as well as for bacterial growth when supplemented with casein hydrolysate and required growth factors. It is ironic that this minor item in the publication is the one quoted most frequently.

"Nevertheless, this paper provided impetus for investigations of new areas of procaryotic genetics. These include the genetic control of sporulation, a primitive form of differentiation, and synthesis of extracellular proteins and wall polymers characteristic of gram-positive bacteria.[4]

"I was fortunate to have been stimulated by Yanofsky and Howard Gest to initiate and pursue these studies while at Western Reserve University."

1. **Spizizen J.** Infection of protoplasts by disrupted T2 virus. *Proc. Nat. Acad. Sci. US* 43:694-701, 1957.
2. **Burkholder P R & Giles N H, Jr.** Induced biochemical mutations in Bacillus subtilis. *Amer. J. Bot.* 34:345-8, 1947. (Cited 105 times since 1955.)
3. **Anagnostopoulos C & Spizizen J.** Requirements for transformation in *Bacillus subtilis*. *J. Bacteriology* 81:741-6, 1961. (Cited 615 times.)
4. **Dubnau D A,** ed. *The molecular biology of the bacilli. Volume I:* Bacillus subtilis. New York: Academic Press, 1982. 378 p.

Hirota Y. The effect of acridine dyes on mating type factors in *Escherichia coli.*
Proc. Nat. Acad. Sci. US **46**:57-64, 1960.
[Department of Genetics, Stanford University Medical Center, Palo Alto, CA]

Acridine dyes irreversibly convert F^+ (male) clones of *E. coli* into stable F^- (female) forms directly without selective growth. Hfr (male) clones are resistant to the action. These results are accounted for by the dual nature of F, plasmid F, and chromosomal F. [The *SCI®* indicates that this paper has been cited in over 475 publications since 1961.]

Yukinori Hirota
Department of Microbial Genetics
National Institute of Genetics
Yata 1,111 Mishima
Shizuoka-ken
411 Japan

September 21, 1983

"This paper is an extension of my two previous publications[1,2] which were done when I had just started graduate research in Hideo Kikkawa's laboratory at the department of biology, Faculty of Science, University of Osaka, Japan. In these papers, I reported my discovery of an efficient genetic alteration of *E. coli* from F^+ (male) to F^- (female). When F^+ was grown under the presence of either cobaltus salt or acridine dye in the medium, a large fraction of F^+ bacteria was converted to F^- (this phenomenon is known as 'F elimination,' 'disinfection of F,' or 'acridine curing'). I correctly interpreted this phenomenon as follows: F (sex factor) in F^+ bacteria is an extrachromosomal state replicating autonomously, and these reagents specifically inhibit autonomous replication of extrachromosomal F but do not inhibit chromosomal DNA replication of the host (*E. coli*). Upon growth of the host, F is diluted out from the F^+ cell and F^-

bacteria evolve. However, an alternate explanation was also possible: that is, spontaneously arised F^- by a rare mutational event is selected by these reagents.

"I needed to prove that my hypothesis was correct; however, I did not have enough experience to achieve my goal so I wanted to have the guidance of Joshua Lederberg at the University of Wisconsin. In 1959, Lederberg moved to Stanford University, and I followed him. He communicated my work to the January 1960 issue of the *Proceedings of the National Academy of Sciences of the USA.* From this work, I received the degree of Doctor of Science at the University of Osaka. A friend of mine said of my discovery and the previous great achievements in *E. coli* genetics by Lederberg and Jacob and Wollman,[3] 'An American discovered sexuality of bacteria, two Frenchmen enjoyed the bacterial sex, and a Japanese made bacteria enjoy two sexes.'

"My thesis work became a *Citation Classic* due to the nature of the work. It describes a simple method for strain construction of *E. coli*, and this organism has extensively been used in the fields of genetics, microbiology, molecular biology, and recombinant DNA technology. Also, it was the first clear demonstration of the existence of bacterial extrachromosomal factor, which was named plasmid.[4,5] Since publication of my work, the acridine curing technique has been applied to wide varieties of bacterial species to test for the presence or the absence of the other plasmid. It has been cited, in addition, because of its description of the methods to demonstrate direct conversion of F^+ to F^- without selective growth. It has often been cited in the textbooks of bacterial genetics."

1. **Hirota Y.** Artificial elimination of the F factor in *Bact. coli* K12. *Nature* **178**:92, 1956.
2. **Hirota Y & Iijima T.** Acriflavine as an effective agent for eliminating F factor in *Escherichia coli* K12. *Nature* **180**:655-6, 1957.
3. **Jacob F & Wollman E L.** *Sexuality and the genetics of bacteria.* New York: Academic Press, 1961. 374 p.
4. **Lederberg J.** Cell genetics and hereditary symbiosis. *Physiol. Rev.* **32**:403-30, 1952.
5. **Bukhari A I, Shapiro J A & Adhya S L,** eds. *DNA: insertion elements, plasmids, and episomes.* Cold Spring Harbor, NY: Cold Spring Harbor Laboratory, 1977. 782 p.

Starlinger P & Saedler H. IS-elements in microorganisms.
Curr. Topics Microbiol. Immunol. 75:111-52, 1976.
[Inst. Genetik, Univ. Köln, and Inst. Biologie III, Univ. Freiburg,
Federal Republic of Germany]

Mobile genetic elements have altered our view on the stability of the genome. They were discovered in maize by Barbara McClintock[1] in the 1940s. Biochemical studies began with the discovery of IS elements in *E. coli*. This paper describes the first decade of this research. [The *SCI®* indicates that this paper has been cited in over 190 publications since 1976.]

Peter Starlinger
Institut für Genetik
Universität Köln
5000 Köln 41
and
Heinz Saedler
Max-Planck-Institut für
Züchtungsforschung
5000 Köln 30
Federal Republic of Germany

April 3, 1984

"When we published our review, this research was ten years old. It started with the study of particular mutations in *E. coli* by Jim Shapiro, then in London, and by E. Jordan, H. Saedler, and P. Starlinger in Cologne. By the end of 1968, we knew that the mutations were caused by the insertions of DNA pieces into genes.[2,3] In 1972, we learned, together with the Szybalski laboratory in Madison, Wisconsin, that the insertions are not random pieces of DNA, but preformed mobile elements. They carry transcription stop signals and thus abolish the expression of genes located downstream in an operon ('polar' mutations). We thought that they deserved a name. Szybalski suggested PS ('It may read as polar sequence'). Caution prompted us to suggest IS (for insertion sequence).[4,5] Rightly so: not all insertions are polar, as was soon found out.[6]

"In 1972, we wrote our first review,[7] and pointed to the similarity of IS elements and McClintock's 'controlling elements.'[1] Not many people were interested then. Public recognition of the field came with the discovery of the larger transposons, first by British, and soon followed by US, workers. Transposons are very similar to IS elements, but they are larger and carry genes encoding the resistance to antibiotics. The famous Asilomar Conference on recombinant DNA research was the first major meeting where IS elements and transposons were talked about. It was during that meeting that Werner Arber asked us to write a review for *Current Topics in Microbiology and Immunology*.

"The review turned out to be a timely one, at least as judged by the disappearance rate of the reprints. The time for mobile genetic elements had finally come, and the first meeting on the topic, organized by the late Ahmad Bukhari, Shapiro, and Sankar Adhyia in Cold Spring Harbor, was already a very crowded one. Today mobile genetic elements have spread to virtually all classes of organisms and the research on them fills whole books."

1. McClintock B. Chromosome organization and genetic expression. *Cold Spring Harbor Symp.* 16:13-47, 1951.
   (Cited 305 times since 1955.)
2. Jordan E, Saedler H & Starlinger P. 0⁰ and strong polar mutations in the *gal* operon are insertions.
   Mol. Gen. Genet. 102:353-63, 1968. (Cited 130 times.)
3. Shapiro J A. Mutations caused by the insertion of genetic material into the galactose operon of *Escherichia coli*.
   J. Mol. Biol. 40:93-105, 1969. (Cited 100 times.)
4. Hirsch H J, Starlinger P & Brachet P. Two kinds of insertions in bacterial genes.
   Mol. Gen. Genet. 119:191-206, 1972. (Cited 130 times.)
5. Fiandt M, Szybalski W & Malamy M H. Polar mutations in *lac*, *gal* and phage λ consist of a few IS-DNA
   sequences inserted with either orientation. *Mol. Gen. Genet.* 119:223-31, 1972. (Cited 135 times.)
6. Saedler H, Reif H J, Hu S & Davidson N. *IS2*, a genetic element for turn-off and turn-on of gene activity in
   E. coli. Mol. Gen. Genet. 132:265-89, 1974. (Cited 150 times.)
7. Starlinger P & Saedler H. Insertion mutations in microorganisms. *Biochimie* 54:177-85, 1972. (Cited 100 times.)

Cairns J. The bacterial chromosome and its manner of replication as seen by
autoradiography. *J. Mol. Biol.* **6**:208-13, 1963.
[Dept. Microbiology, Australian Nat. Univ., Canberra, ACT, Australia]

Following the development of a method for visualizing DNA molecules by autoradiography after labeling with tritiated thymidine, the DNA of the bacterium *Escherichia coli* was shown in this paper to consist of a single molecule that is replicated at a moving locus (the replicating fork) at which both new DNA strands are being synthesized. [The *SCI®* indicates that this paper has been cited over 500 times since 1963.]

John Cairns
Imperial Cancer Research Fund
Mill Hill Laboratories
Burtonhole Lane
London, NW7 1AD
England

July 2, 1980

"This paper on the structure of the chromosome of *Escherichia coli* (plus a more complete picture, published in the *Cold Spring Harbor Symposium* in the same year[1]) was almost the last work I published on the autoradiography of DNA, before I moved from Australia to the US and became totally embroiled in fund-raising at Cold Spring Harbor. The project began in 1960 when I spent a year working in Hershey's laboratory. There was at that time great uncertainty about the real size of DNA molecules, not least of all because there were no precise ways of measuring their molecular weight. Hershey had just devised ways of extracting T2 DNA as an intact single molecule of known molecular weight, and one of the next things to do was, therefore, to measure its length in order to find out whether it had two strands or four. Having had experience with autoradiography, when I worked on the pattern multiplication of vaccinia virus,[2] it was a natural thought that I should try to observe these isolated T2 DNA molecules by autoradiography (although it took months of messing around with other projects before I realized what I should be doing). It was soon clear that molecules could be visualized this way, and I returned to Australia with the intention of studying the structure and replication of bacterial DNA—a subject that had attracted me since the fall of 1957 when I did the washing-up in a house in Pasadena where Meselson was doing the cooking (and, at the same time, was completing the original density-transfer experiment that established semiconservative replication[3]).

"The autoradiography of DNA was a rather slow business. Exposure times were over two months. Therefore it took a couple of years to work out a technique for minimizing DNA breakage during extraction, and then it was only after quite a long search that I found any molecules that were sufficiently untangled to be interpretable.

"The other part of the paper concerned the structure of the replicating fork—namely, the fact that short pulses of ^{3}H-thymidine were found to label two new strands, both apparently growing overall in the same direction. The autoradiography of pulse-labeled DNA is a much more straightforward procedure because it does not require that the whole chromosome is laid out in an untangled state, and so I also included a few experiments on the replication of HeLa cell DNA, the results of which, because of the exigencies of my circumstances in the US, were not published for 3 years.[4]

"Rather than being innovative, the work was really an act of tidying (a bit like washing-up, in fact) because it showed that the bacterial chromosome was simply a molecule of DNA—albeit one organized in a rather unexpected way. If my paper has been often quoted, that is probably because it describes something very simple."

1. **Cairns J.** The chromosome of *Escherichia coli. Cold Spring Harbor Symp.* **28**:43-6, 1963.
2. ----------. The initiation of vaccinia infection. *Virology* **11**:603-23, 1960.
3. **Meselson M & Stahl F W.** The replication of DNA in *Escherichia coli.*
   Proc. Nat. Acad. Sci. US **44**:671-82, 1958.
4. **Cairns J.** Autoradiography of HeLa cell DNA. *J. Mol. Biol.* **15**:372-3, 1966.

Clewell D B & Helinski D R. Supercoiled circular DNA-protein complex in
Escherichia coli: purification and induced conversion to an open circular DNA
form. *Proc. Nat. Acad. Sci. US* **62**:1159-66, 1969.
[Dept. Biology, Univ. California, San Diego, and Revelle Coll., La Jolla, CA]

The plasmid *ColE1* was isolated as a super-coiled DNA-protein 'relaxation' complex. Exposure to agents or conditions affecting protein structure triggered a nicking event which resulted in a conversion of the plasmid DNA to an open circular configuration. [The *SCI®* indicates that this paper has been cited in over 965 publications since 1969.]

Don B. Clewell
Department of Oral Biology
School of Dentistry
University of Michigan
Ann Arbor, MI 48109

December 27, 1982

"During 1967-1970, I was a National Institutes of Health (National Cancer Institute) postdoctoral fellow in Don Helinski's laboratory in the biology department at the University of California, San Diego, in La Jolla. Don was one of the few people in the world at that time involved in physical analyses of bacterial plasmids; he and his co-workers had recently characterized *ColE1* and a derivative of F by electron microscopy.

"As an approach to examining factors associated with *ColE1* maintenance, we embarked on an effort to isolate plasmid-protein complexes by sedimentation of lysates through sucrose density gradients. We made use of a lysis procedure based on recent work reported by Godson and Sinsheimer[1] on the use of various detergent mixtures to bring about the lysis of *E. coli* for different purposes.

" 'Gentle' lysis of *E. coli* (*ColE1*) cells labeled with H-thymine was accomplished by exposing lysozyme-EDTA treated cells (suspended in a sucrose solution) to a mixture of Brij 58 and deoxycholate. A 25 min.

centrifugation at 48,000 x g pelleted the majority of the cellular DNA; and when the supernatant ('cleared lysate') was fractionated on a 15 to 50 percent sucrose density gradient, we observed a prominent peak which contained *ColE1* DNA almost exclusively. Upon further analysis, this substance proved to sediment slightly faster (24S) than purified supercoiled 23S plasmid DNA; and when exposed to proteolytic enzymes the DNA shifted, surprisingly, to 17S. We were able to show that the 24S material consisted of a supercoiled DNA-protein complex; and agents or conditions known to affect protein structure (e.g., proteases, certain detergents, and heat) triggered a nicking event which gave rise to relaxed circular molecules.

"The initial observations were reported in the paper cited above; whereas a more thorough characterization of the 'relaxation complex' and its related proteins followed.[2,3] Interestingly, the amount of plasmid DNA that was present as a relaxation complex was highly influenced by the presence of glucose (and cAMP) in the medium and ranged from 30 to 80 percent.[4] Relaxation complexes were subsequently found in a number of plasmid systems, and later work in the laboratories of Joe Inselburg[5] and David Sherratt[6] showed that in the case of *ColE1*, the complex related to conjugative mobilization.

"The paper is cited frequently for its plasmid purification protocol. Significant volumes of the low viscosity, plasmid-enriched 'cleared-lysate' could be loaded directly into dye-buoyant gradients, affording a 'high-yield' purification step in the preparation of covalently closed circular (CCC) molecules. In the case of some plasmids (e.g., *ColE1*), however, this approach allows for the purification of only non-complexed CCC molecules; relaxation complex 'relaxes' under these conditions, and the DNA-protein complex ends up on the lower density side (shoulder) of the chromosomal band."

1. **Godson G N & Sinsheimer R L.** Lysis of *Escherichia coli* with a neutral detergent.
   Biochim. Biophys. Acta **149**:476-88, 1967.
2. **Clewell D B & Helinski D R.** Properties of a supercoiled deoxyribonucleic acid-protein relaxation complex
   and strand specificity of the relaxation event. *Biochemistry* **9**:4428-40, 1970.
3. **Helinski D R, Lovett M A, Williams P H, Katz L, Collins J, Kupersztoch-Portnoy Y, Sato S, Leavitt R W,
   Sparks R, Hershfield V, Guiney D G & Blair D G.** Modes of plasmid DNA replication in
   Escherichia coli. (Goulian M & Hanawalt P, eds.) *DNA synthesis and its regulation.*
   Menlo Park, CA: Benjamin, 1975. p. 514-36.
4. **Clewell D B & Helinski D R.** Effect of growth conditions on the formation of the relaxation complex of supercoiled
   ColE1 deoxyribonucleic acid and protein in *Escherichia coli*. *J. Bacteriology* **110**:1135-46, 1972.
5. **Inselburg J.** Studies of colicin E1 plasmid functions by analysis of deletions and TnA insertions of the plasmid.
   J. Bacteriology **132**:332-40, 1977.
6. **Warren G J, Twigg A J & Sherratt D J.** *ColE1* plasmid mobility and relaxation complex.
   Nature **274**:259-61, 1978.

Bachmann B J, Low K B & Taylor A L. Recalibrated linkage map of
Escherichia coli K-12. *Bacteriol. Rev.* **40**:116-67, 1976.
[Yale Univ. School of Medicine, New Haven, CT and Univ. Colorado
Medical Center, Denver, CO]

This paper presents a critical review of all
data on the linkage of genes on the *Escherichia coli* K-12 chromosome, a list of genes
and their functions, and a graphic representation of the circular chromosome. [The
SCI® indicates that this paper has been cited
over 1,185 times since 1976.]

Barbara J. Bachmann
Department of Human Genetics
Yale University School of Medicine
New Haven, CT 06510

October 27, 1981

"The series of linkage maps of
Escherichia coli K-12 was begun by A.L.
Taylor in 1964[1] and has served since as
an essential tool in experimental work
directed toward understanding the
genome of this organism. Early editions
of the 'map paper' presented extensive
mapping data from Taylor's laboratory,
in addition to providing syntheses of all
published data, a standardized system
of nomenclature, a list of gene functions, and a graphic representation of
the chromosome. By 1972,[2] the number
of gene loci on the map had increased
from 100 to 460. As explorations in
molecular biology came to be focused
to a large extent on this one organism,
the map paper became essential to
many who were not primarily microbial
geneticists.

"In 1975, another revision of the map
was badly needed and Taylor realized
that he could not spare the time required to continue the series. It seemed
reasonable for me to take over the review because, as curator of the *E. coli*
Genetic Stock Center, I need to keep
track of all of the loci on the map in
order to respond to requests for strains.
Taylor agreed to assist with this edition,
and came to Yale to help in the analysis
of the data.

"First, Brooks Low recalibrated the
map, which is still based on the time-of-
entry of markers by conjugation, using
the large set of Hfr strains which he had
assembled—and many of which he had
isolated. To our surprise and delight,
the result was a 100-minute map (which
is so convenient a figure that some
have assumed that this length was
chosen arbitrarily). This map paper contained around 680 gene loci and a bibliography of 763 citations. Obviously,
not every person using *E. coli* could afford to perform this synthesis for
himself. It is the drawing, the bibliography, and the table of gene functions
that are most often used. Apparently,
few people read the text, which contains many qualifications and exhorts
the reader to consult the original
literature.

"I later came to appreciate Taylor's
many reasons for wishing to relinquish
this task. Each time the map is drawn it
is necessary to go over all data ever
published and to reconcile the widely
varying, and sometimes conflicting,
results obtained in different laboratories. It is also necessary to arbitrate
disputes regarding gene symbols. Constructing the map is one way to offend
a great many people simultaneously.
When the entire map is fitted together,
almost no gene is put exactly where
any one person said it was and the position is seldom precisely known.

"I revised the map again in 1980,[3]
with the generous assistance of a great
many of the authors of the data, and
Low contributed a review of mapping
techniques. The number of gene loci
mapped has continued to increase lin-
early with time. The 1980 edition contains almost 1,000 loci, representing
possibly close to one quarter of the
genes of the best known of all
organisms."

1. **Taylor A L & Thoman M S.** The genetic map of *Escherichia coli* K-12. *Genetics* **50**:659-77, 1964.
   [The *SCI* indicates that this paper has been cited over 205 times since 1964.]
2. **Taylor A L & Trotter C D.** Linkage map of *Escherichia coli* strain K-12. *Bacteriol. Rev.* **36**:504-24, 1972.
   [The *SCI* indicates that this paper has been cited over 665 times since 1972.]
3. **Bachmann B J & Low K B.** Linkage map of *Escherichia coli* K-12, edition 6.
   Microbiol. Rev. **44**:1-56, 1980. [The *SCI* indicates that this paper has been cited over 200 times since 1980.]

Maaløe O & Hanawalt P C. Thymine deficiency and the normal DNA replication
cycle. I. *J. Mol. Biol.* 3:144-55, 1961.
[Institute of Microbiology, University of Copenhagen, Denmark]

The paper introduces the novel concept that, once initiated, a round of DNA replication will run to completion in the absence of protein synthesis. To start a new round, de novo protein synthesis is required. [The *SCI®* indicates that this paper has been cited in over 530 publications since 1961.]

Ole Maaløe
Institute of Microbiology
University of Copenhagen
1353 Copenhagen K
Denmark

May 25, 1984

"Our Institute of Microbiology at the University of Copenhagen was established in 1958, and since that time we have received numerous visitors from the US. Phil Hanawalt was one of them. He had just finished his PhD work at Yale University when he came over to work with me in November 1958. At first he continued doing radiation studies with strain 15T⁻ of *Escherichia coli*. Nothing striking happened until one day, looking at Phil's latest results, I experienced the only true flash of scientific intuition I can recall: somehow his results suggested to me that a round of replication might be autonomous in the sense that protein synthesis was needed to start the round but not to run it to completion. It is well known that '...chance only favours the prepared mind,' and I'm sure that in this case the element of preparedness was a keen awareness that a culture of bacteria must be viewed as a population of cells representing all stages in cycles such as the DNA replication cycle.

"The concept of autonomy in a round of replication was tested in various ways. In the paper discussed here the phenomenon of 'thymineless death' (a characteristic of the strain we worked with) was important. We could show that, in an exponentially growing population, a small fraction (two to three percent) of the cells were immune to thymineless death and that this fraction probably represented cells that had completed a round of replication but not initiated a new round. Incubation with thymine, but without arginine and uracil (strain 15T⁻ requires both for growth), for 40-60 minutes caused the entire population of cells to accumulate in the 'immune state.' Presumably, this long period was needed for cells that had just started a round of replication to complete it. We use the term 'run out' to describe this course of events.

"In the succeeding paper,[1] the analysis was pushed further by means of autoradiography of whole cells labeled in various ways by ³H-thymine. We could show that if cells more or less double their DNA content in the absence of protein synthesis, they do not initiate a new round of replication until protein (and mass) synthesis has caught up with replication. The studies presented in our two papers in the *Journal of Molecular Biology* have inspired much subsequent work on DNA replication *in vivo*.[2,3]

"In my own laboratory, the 'run out' type of experiment has also been used to study RNA and protein synthesis *in vivo*. The antibiotic rifampicin has been useful for studying run-out of RNA synthesis,[4,5] and the growth rate of polypeptide chains has been estimated by pulse-chase labeling with radioactive amino acids."[6]

1. **Hanawalt P C, Maaløe O, Cummings D J & Schaechter M.** The normal DNA replication cycle. II.
   J. Mol. Biol. 3:156-65, 1961. (Cited 150 times.)
2. **Lark K G, Repko T & Hoffman E J.** The effect of amino acid deprivation on subsequent deoxyribonucleic acid
   replication. *Biochim. Biophys. Acta* 76:9-24, 1963. (Cited 335 times.)
3. **Bird R E, Louarn J, Martuscelli J & Caro L.** Origin and sequence of chromosome replication in *Escherichia coli*.
   J. Mol. Biol. 70:549-66, 1972. (Cited 160 times.)
4. **Pato M L & von Meyenburg K.** Residual RNA synthesis in *Escherichia coli* after inhibition of transcription by
   rifampicin. *Cold Spring Harbor Symp.* 35:497-504, 1970. (Cited 170 times.)
5. **Molin S.** Ribosomal RNA chain elongation rates in *Escherichia coli*. (Kjeldgaard N C & Maaløe O. eds.)
   Control of ribosome synthesis: proceedings of the Alfred Benzon Symposium IX.
   Copenhagen: Munksgaard, 1976. p. 333-9.
6. **Ingraham J L, Maaløe O & Neidhardt F C.** *Growth of the bacterial cell.*
   Sunderland, MA: Sinauer Associates, 1983. p. 310.

Boyce R P & Howard-Flanders P. Release of ultraviolet light-induced thymine dimers from DNA in *E. coli* K12. *Proc. Nat. Acad. Sci. US* **51**:293-300, 1964.
[Dept. Radiology, Yale Univ. School of Med., New Haven, CT]

This paper presents evidence that the enzymatic excision of damaged bases in DNA, including ultraviolet photodimers, and the reconstruction of the DNA from information on the complementary DNA strand may be an important biological mechanism for the repair of DNA. [The *SCI®* indicates that this paper has been cited over 730 times since 1964.]

R.P. Boyce
Department of Biochemistry and
Molecular Biology
University of Florida
Gainesville, FL 32610

May 11, 1979

"Shortly before I joined Paul Howard-Flanders' radiobiology group at Yale in 1961, Ruth Hill at Columbia discovered an ultraviolet (UV) sensitive mutant of *E. coli* B which appeared unable to 'reactivate' UV damage in DNA.[1] Following this lead, Howard-Flanders, Lee Theriot, and Eva Bardwell (nee Simson) isolated and genetically mapped similar mutants in *E. coli* K12 which they named *uvr* mutants.[2] Reasoning that thymine dimers produced by UV irradiation might be repaired in normal *E. coli* K12, but not in the *uvr* mutants, we obtained data suggesting that dimers disappeared from the DNA during post-irradiation incubation of the bacteria. Initially we thought dimers in DNA were converted *in situ* to monomers. Then R.B. Setlow at Oak Ridge discovered that dimers were conserved in whole *E. coli* B after UV irradiation.[3] Our results were explicable if dimers were released from the DNA rather than repaired *in situ*. Testing this possibility we failed to detect dimers in the non-DNA fraction of irradiated *E. coli* K12 by paper chromatography. Informed of our results, Setlow suggested that we hydrolyze the non-DNA fraction in hot acid prior to chromatography. Upon doing this we immediately found the dimers and deduced that they had been released from the parental *E. coli* K12 DNA in a phosphorylated form which were retained by whole cells. However, dimers were not released in the *uvr* mutants. We surmised that (1) dimers are enzymatically excised from the DNA of normal *E. coli* causing a rupture in the DNA sugar-phosphate backbone; and (2) the UV sensitivity of the *uvr* mutants arises from their inability to excise dimers. Aided by E. Canellakis we arrived at the model of excision repair presented in this paper. In similar experiments, R.B. Setlow and W.C. Carrier independently arrived at the same conclusions and model.

"I think this work has been recognized for several reasons. First, it reinforced Hills' observation that radiation sensitivity is, in part, under genetic control and underscored the idea of the enzymatic repair of DNA. Our use of *E. coli* K12 strains, easily amenable to genetic analysis, laid the foundations for much subsequent work. Second, it proposed a general model for DNA repair that could be carried out by reasonable enzyme activities. It also proposed another *raison d'etre* for the DNA double helix not originally deduced by its clairvoyant discoverers, i.e., to permit the faithful reconstruction of an excised damaged region by copying the genetic information in an intact opposite strand. Third, it provided a rationale for the search of similar processes and enzymatic activities in higher cells.

"The paper by Setlow and Carrier[4] describing essentially the same results with Hills' strains of *E. coli* B deserves at least equal recognition. This work, published in the same issue of the *Proceedings*, predated our own by at least a month. Max Delbruck expeditiously communicated our paper in order that both would appear simultaneously."

1. **Hill R F.** A radiation-sensitive mutant of *Escherichia coli*. *Biochem. Biophys. Acta* **30**:636-7, 1958.
2. **Howard-Flanders P, Boyce R P, Simson E & Theriot L.** A genetic locus in *E. coli*: K12 that controls the reactivation of UV-photoproducts associated with thymine in DNA. *Proc. Nat. Acad. Sci. US* **48**:2109-15, 1962.
3. **Setlow R B, Swenson P A & Carrier W L.** Thymine dimers and inhibition of DNA synthesis by ultraviolet irradiation of cells. *Science* **142**:1464-6, 1963.
4. **Setlow R B & Carrier W L.** The disappearance of thymine dimers from DNA: an error-correcting mechanism. *Proc. Nat. Acad. Sci. US* **51**:226-31, 1964.

CC/NUMBER 36
SEPTEMBER 7, 1981

This Week's Citation Classic

Howard-Flanders P, Boyce R P & Theriot L. Three loci in *Escherichia coli*
K-12 that control the excision of pyrimidine dimers and certain other
mutagen products from DNA. *Genetics* 53:1119-36, 1966.
[Radiobiology Labs., Yale University School of Medicine, New Haven, CT]

Mutants of *E. coli* were isolated that were very sensitive to ultraviolet light or mitomycin C, and failed to reactivate UV-irradiated bacteriophages. They were unable to excise pyrimidine dimers from their DNA. The mutations mapped in three widely spaced loci, *uvrA,B,C.* [The *SCI®* indicates that this paper has been cited over 325 times since 1966.]

Paul Howard-Flanders
Department of Therapeutic Radiology
Yale University School of Medicine
New Haven, CT 06520 ·

August 18, 1981

"Prior to 1961, the concept that living cells might contain enzymes to repair DNA molecules was no more than a remote possibility that few scientists had even considered. The next five years brought the publication of several papers on the detection of genetic repair processes in microorganisms, and the application of the methods of genetics and biochemistry to their elucidation.[1-4]

"In 1961, there was no very obvious reason to study the effects of ultraviolet light on bacteria. However, multiplicity reactivation and photoreactivation had been discovered. Ruth Hill had isolated a mutant strain of the bacterium *E. coli* B which was of exceptional sensitivity to ultraviolet light and had lost the ability to reactivate ultraviolet irradiated bacteriophages.[5] The only plausible explanation was that bacteria normally contained enzymes for the repair of ultraviolet photoproducts in bacterial or phage DNA and that the sensitive mutant lacked one or more of these enzymes. This concept was novel in the early 1960s, since extensive experiments on radiation mutagenesis in fruit flies had shown X-ray damage in cells and tissues to be almost irreversible. Moreover, the scientific community had been influenced by Watson and Crick, who had discussed the biological implications of their famous model for the structure of DNA, but had not foreseen the possibility that enzymes might repair damaged DNA, or the significance of the complementary structure for the repair of damaged strands.

"Our goals were clear enough. We wished to investigate the fate of pyrimidine dimers in the DNA of bacteria, to explore the repair enzymes that act on the damaged DNA, and to study the genetic control of these processes. We also felt the need to bring the methods of biochemistry and genetics to bear directly on these problems, which formerly had been little more than radiobiological curiosities. Since this could not easily be done in strain B, we had to isolate new radiation sensitive mutants from the strain *E. coli* K12 in which fertile males and genetically marked females were available. Fortunately, the main ultraviolet photoproducts in DNA had already been identified as pyrimidine dimers, which could be detected by radio-chromatography, and the methods for genetic mapping in bacteria had been developed. R.P. Boyce and I (and R.B. Setlow working independently) found the repair to involve the excision of the pyrimidine dimers from the bacterial DNA (also a *Citation Classic*).[2,3]

"This work had shown bacterial DNA repair enzymes to act on damage induced by ultraviolet light and by mitomycin C, both known carcinogens. Subsequent work by many scientists has shown DNA repair enzymes in man to protect against the effects of environmental carcinogens and radiations, and to be necessary for keeping cancer incidence down to the present levels.[6]

"This paper provided proof that the products of three genes are necessary for excision repair and provided the first comprehensive genetic analysis of excision repair. For more recent work in the field see A. Sancar, et al."[7]

1. Howard-Flanders P, Boyce R P, Simson E & Theriot L. A genetic locus in *E. coli* K12 that controls the reactivation of UV-photoproducts associated with thymine in DNA.
Proc. Nat. Acad. Sci. US 48:2109-15, 1962.
2. Boyce R P & Howard-Flanders P. Release of ultraviolet light-induced thymine dimers from DNA in *E. coli* K-12.
Proc. Nat. Acad. Sci. US 51:293-300, 1964.
[Citation Classic. *Current Contents/Life Sciences* 23(35):12, 1 September 1980.]
3. Setlow R B & Carrier W L. The disappearance of thymine dimers from DNA: an error-correcting mechanism.
Proc. Nat. Acad. Sci. US 51:226-31, 1964.
4. Clark A J & Margulies A D. Isolation and characterization of recombination-deficient mutants of *Escherichia coli*
K12. *Proc. Nat. Acad. Sci. US* 53:451-9, 1965.
5. Hill R F. A radiation-sensitive mutant of *Escherichia coli.* Biochim. Biophys. Acta 30:636-7, 1958.
6. Ts'o P O P & Gelboin H V, eds. *Polycyclic hydrocarbons and cancer.* New York: Academic Press, 1978. 2 vols.
7. Sancar A, Warton R P, Seltzer S, Kacinski B M, Clarke N D & Rupp W D. Identification of the
uvrA gene product. *J. Mol. Biol.* 148:45-62, 1981.

This Week's Citation Classic

Kondo S, Ichikawa H, Iwo K & Kato T. Base-change mutagenesis and prophage induction in strains of *Escherichia coli* with different DNA repair capacities. *Genetics* **66**:187-217, 1970.
[Dept. Fundamental Radiol., Fac. Med., Osaka Univ., Kita-ku, Osaka, Japan]

Three types of repairless mutant strains of *E. coli* were tested for mutability by seven representative mutagens. One strain turned out more mutable than the wild type or nonmutable depending on the repair deficiency and the mutagen. This paper describes a simple method for obtaining information on the mechanism of mutation induction by various mutagens such as environmental ones. [The *SCI*® indicates that this paper has been cited over 230 times since 1970.]

Sohei Kondo
Department of Fundamental Radiology
Faculty of Medicine
Osaka University
Kita-ku, Osaka 530
Japan

July 24, 1981

" 'Incredibly small groups of atoms, much too small to display exact statistical laws, do play a dominating role in the very orderly events within a living organism.' This statement of E. Schrödinger in his famous book *What Is Life?*[1] had been a challenge to me. An idea occurred to me that the seemingly orderly events may be ascribed to self-controlled mechanisms which suppress the 'errors' in reactions of biomolecules which otherwise arise from statistical-mechanical fluctuations. Therefore, when I was offered the chair of the present professorship in the spring of 1963, I decided to switch from radiation physics to radiation biology to make a search for the biological mechanisms that correct genetic errors.

"In the fall of 1963, I was invited to work at the Oak Ridge National Laboratory, Tennessee, and started my first biological work on ultraviolet (UV) mutagenesis using *E. coli* strains isolated by E.M. Witkin. In the fall of 1964, on my way back to Japan, I visited Witkin in New York and learned a lot about the mutation work. After discussions with her, I thought that if we could isolate repairless mutants from her strains, they must be invaluable for elucidating the error-correction mechanisms. After some painstaking work, T. Kato was able to isolate various mutants. I decided to pick up three repairless (now known to be excisionless, DNA polymeraseless, recombinationless) mutants from them for detailed studies of mutagenesis by seven mutagens: UV and X rays, 4-nitroquinoline 1-oxide (4NQO), mitomycin C (MMC), ethylmethanesulfonate (EMS), methylmethanesulfonate (MMS), and nitrosoguanidine (NG). Haruko Ichikawa worked hard and produced a major part of the mutation data. Kazuhiro Iwo, then a postgraduate student, worked on prophage induction.

"In 1964, H. Endo, Kyushu University, generously gave me a water-soluble chemical carcinogen, 4NQO, discovered in Japan. To my surprise, 4NQO turned out remarkably UV mimetic in the sense that the excisionless strain is about 30 times more sensitive than the wild type to both killing and mutation by 4NQO as well as by UV. It turned out that the recombinationless strain is nonmutable by UV, 4NQO, X rays, and MMS but mutable by EMS and NG, even though, when compared with the wild type, it is similarly sensitive to killing by EMS, MMS, NG, and X rays. Those results demonstrate that induction of mutation depends on the biological factor in addition to physicochemical alteration of DNA and that lethal damage to DNA is not necessarily identical to mutagenic damage. We also obtained suggestive evidence that prophage induction shares some common steps with induction of mutation.

"Our paper was one of the earliest ones in which various chemicals in addition to UV were tested for their differential mutagenicities between repairless mutants and the wild type. Shortly after publication of our paper, environmental mutagens became the social concern, leading to the finding that carcinogens are mostly mutagenic in bacteria. This may be the reason for the frequent citation of our paper. More recent work in the field is reported by Witkin."[2]

1. **Schrödinger E.** *What is life? The physical aspect of the living cell.* Cambridge, England: Cambridge University Press, 1944. 91 p.
2. **Witkin E M.** Ultraviolet mutagenesis and inducible DNA repair in *Escherichia coli. Bacteriol. Rev.* **40**:869-907, 1976.

This Week's Citation Classic™

Ames B N, McCann J & Yamasaki E. Methods for detecting carcinogens and mutagens with the salmonella/mammalian-microsome mutagenicity test.
Mutat. Res. **31**:347-64, 1975.
[Department of Biochemistry, University of California, Berkeley, CA]

This paper describes the detailed methods and theory for the *Salmonella*/mammalian microsome mutagenicity test. It presents the techniques for handling and growing the histidine-requiring bacterial mutants, the function of the various mutations in the tester strains, and the preparation and storage of the rat liver homogenate and cofactor mix. [The *SCI®* indicates that this paper has been cited in over 2,705 publications since 1975.]

Bruce N. Ames
Department of Biochemistry
University of California
Berkeley, CA 94720

January 12, 1984

"One day in 1964 after reading a label on a potato chip package, I began to think about all of the new chemicals entering the environment. Being a geneticist as well as a biochemist, I thought it important to screen chemicals to which humans were exposed for mutagenicity. This led to our work on the development of a simple test for mutagens.

"My mutagenicity work started as an offshoot from my basic research on the molecular biology of *Salmonella* bacteria. I was studying how genes are switched on and off in response to the presence of the amino acid histidine and the effect of mutations which perturbed this control mechanism. I was using a large collection of histidine-requiring mutants, mostly made by Philip E. Hartman of Johns Hopkins University. The development of the test system for detecting mutagens using these mutants and our work on mutagenesis were done first at the National Institute for Arthritis and Metabolic Diseases, where I was until 1968, and then at Berkeley, where I have been since then.

"During the course of this work on mutagenesis, I became convinced that one essential aspect of carcinogens was their ability to damage DNA. We kept lists of popular carcinogens and kept working to make improvements in the test which could detect them as mutagens. One key development was the addition of a rat liver homogenate and cofactors to our petri plates. With this plus other improvements we had made in our tester strains we were able to detect over 80 percent of carcinogens as mutagens. The initial work on the test was mostly done with undergraduate students at Berkeley, particularly Frank Lee and Bill Durston. Joyce McCann, a postdoctoral fellow, then worked on the problem and made major contributions as did Edie Yamasaki, my unusually able technician.

"By 1975, over 500 laboratories were using our system (it is up to about 3,000 now) and we were inundated with letters and telephone calls. McCann, Yamasaki, and I decided to write a methods paper to help in dealing with these inquiries. We were reinforced in this decision because in 1975 McCann and I reviewed our own and other work on validating the test for detecting carcinogens as mutagens and we expected this review would lead to even more use of the test. We tried to make the methods paper, as we called it, definitive: it is the main paper now cited when people use our test. We recently wrote a new methods paper[1] incorporating modifications, by both ourselves and others, and discussing our new tester strains.[2,3] The new paper should make the 1975 paper obsolete. We have been surprised how few changes have been made in the procedure over eight years. We hope the additional tester strains[2,3] will make the test even more comprehensive. One of its major uses has been the detection of new classes of mutagens from complex mixtures, such as cooked food, plants, cigarette smoke, water, food, and urine, and man-made chemicals, such as hair dyes and flame retardants. Many of these mutagens have later been shown to be carcinogens. The test has been useful to industry as large numbers of chemicals are being screened in the development of new drugs and industrial chemicals. In addition, it has become widely used for investigating the metabolic pathways of carcinogens to their active forms. The most important theoretical contribution from our work has been support for the idea that damage to DNA is one essential aspect of carcinogenesis."

[This work led to the John Scott award in 1978.*]

1. **Maron D M & Ames B N.** Revised methods for the *Salmonella* mutagenicity test. *Mutat. Res.* **113**:173-215, 1983.
2. **Levin D E, Hollstein M, Christman M F, Schwiers E A & Ames B N.** A new *Salmonella* tester strain (TA102) with A:T base pairs at the site of mutation detects oxidative mutagens. *Proc. Nat. Acad. Sci. US—Biol. Sci.* **79**:7445-9, 1982.
3. **Levin D E, Yamasaki E & Ames B N.** A new *Salmonella* tester strain, TA97, for the detection of frameshift mutagens: a run of cytosines as a mutational hotspot. *Mutat. Res.* **94**:315-30, 1982.

*Garfield E.** Since 1816 the John Scott and other Philadelphia awards have recognized "useful" scientific discoveries—James Black and Benjamin Rubin head a list of recent distinguished recipients. *Essays of an information scientist.* Philadelphia: ISI Press, 1983. p. 686-94.

Hardy S J S, Kurland C G, Voynow P & Mora G. The ribosomal proteins of
Escherichia coli. I. Purification of the 30S ribosomal proteins.
Biochemistry 8:2897-905, 1969.
[Dept. Zool. and Lab. Molec. Biol., Univ. Wisconsin, Madison, WI]

This paper demonstrates that there are approximately 20 proteins in the 30S ribosomal subunits of *Escherichia coli* and describes their purification. It shows that the multiplicity of components is not due to contamination with nonribosomal proteins, nor to random disulfide bridge formation, nor to proteolytic degradation of fewer components. It also demonstrates that a similar multiplicity is found in intact cells. [The *SCI®* indicates that this paper has been cited over 590 times since 1969.]

Simon Hardy
Department of Biology
University of York
Heslington, York YO1 5DD
England

March.15, 1982

"The work described in this paper was the culmination of my career as a graduate student of Chuck Kurland. We were interested in the structure and function of the ribosomes of *E. coli*. Our earlier analysis of a ribosome-associated enzyme had forced on us the realization that mere association was, at best, only a weak indication of a protein's role in the structure or function of ribosomes. We therefore proposed several more rigorous criteria for identifying ribosomal proteins.[1] Then we wondered whether the pioneering work of Waller[2] showing a great multiplicity of ribosomal proteins was correct. Could it be instead a multiplicity of contaminants and other artifacts? Indeed, the idea that ribosomes were very simple structures similar to small viruses, containing only a few proteins, was both attractive and prevalent at that time.

"So we began the daunting task of purifying and characterizing all the ribosomal proteins, at each stage attempting to rule out all artifacts which would increase the number. Thus we began with rigorously washed particles, eliminated disulfide bridges, checked for proteolytic degradation, and showed that most of the proteins that we purified were present in intact cells. The purified proteins were then shown to be both physically and chemically distinct in the following paper.[3] Kurland's simple idea that phosphocellulose, being an analogue of ribosomal RNA, would be the ideal ion exchange resin for separating ribosomal proteins proved to be splendidly correct and was the greatest single contribution to the experimental work. Kurland, in fact, provided most of the ideas and direction. I worked out the purification schemes and Paul Voynow, in between measuring the molecular weights of the proteins,[3] provided large quantities of ribosomal subunits by developing our use of the then new zonal rotor. Guido Mora, a later arrival whose day was to come with the 50S subunit proteins, helped in the final purifications.

"This work and similar work from other laboratories destroyed the simple theory of ribosome structure. As a result, at the Nucleic Acids Gordon Conference of 1968, Jim Watson collected the two bottles of champagne he had bet Howard Dintzis at an earlier Gordon Conference, and I, along with many other contributors, was invited to share them. That thimbleful of champagne is still listed in the honors section of my C.V.

"Now comes the embarrassing part. Although the paper was an important and good contribution to the field at the time, it has been cited so many times only because I altered a magnesium concentration. Since we wanted to be sure that we were purifying *all* the ribosomal proteins, it was important to get the highest yields of protein in separating it from the ribosomal RNA. I discovered that by raising the magnesium concentration of the standard acetic acid extraction[4] from 0.01 M to 0.1 M, the recovery of protein increased from 80 percent to better than 95 percent. Since the acetic acid procedure is still regularly used and nobody has improved it, our paper has become a *Citation Classic*. A more recent review of this field may be found in 'Structure and function of the bacterial ribosome.'"[5]

1. **Hardy S J S & Kurland C G.** The relationship between poly A polymerase and the ribosomes. *Biochemistry* 5:3676-84, 1966.
2. **Waller J P.** Fractionation of the ribosomal protein from *Escherichia coli*. *J. Mol. Biol.* 10:319-36, 1964.
3. **Craven G R, Voynow P, Hardy S J S & Kurland C G.** The ribosomal proteins of *Escherichia coli*. II. Chemical and physical characterization of the 30S ribosomal proteins. *Biochemistry* 8:2906-15, 1969.
4. **Waller J P & Harris J.** Studies on the composition of the protein from *Escherichia coli* ribosomes. *Proc. Nat. Acad. Sci. US* 47:18-23, 1961.
5. **Kurland C G.** Structure and function of the bacterial ribosome. *Annu. Rev. Biochem.* 46:173-200, 1977.

Pastan I & Perlman R. Cyclic adenosine monophosphate in bacteria.
Science 169:339-44, 1970.
[Molec. Biol. Sect., Endocrinol. Branch, Natl. Cancer Inst., and Diabetes Sect., Clin. Endocrinol. Branch, Natl. Inst. Arthritis and Metabolic Dis., NIH, Bethesda, MD]

This paper reviewed the experiments that showed how cyclic AMP controls transcription in *Escherichia coli* and that elucidated the mechanism by which glucose represses the synthesis of a variety of inducible enzymes. [The *SCI®* indicates that this paper has been cited in over 420 publications since 1970.]

I. Pastan
Molecular Biology Laboratory
National Cancer Institute
National Institutes of Health
Bethesda, MD 20205

September 21, 1984

"In 1961, I arrived at the National Institutes of Health (NIH) and began to work in Ed Rall's department with Jim Field on the effects of thyroid-stimulating hormone (TSH) on the thyroid gland. Later, after a postdoctoral stay in Earl Stadtman's laboratory, I returned to my studies on TSH action. The discovery by Earl Sutherland[1] that many hormones activated adenylate cyclase and thus increased intracellular cyclic AMP levels prompted me to investigate cyclic AMP in the thyroid. I found that TSH activated adenylate cyclase in thyroid and that the addition of a cyclic AMP analog to the thyroid tissue reproduced many of the actions of TSH, suggesting cyclic AMP as the mediator of TSH action. A more fundamental problem was how cyclic AMP acted, and it occurred to me that a solution might come more readily from investigations on *E. coli*. Robert Perlman and I then joined forces to study this. Sutherland had shown that the addition of glucose to glucose-starved *E. coli* lowered cyclic AMP levels in the cells. One well-known effect of glucose in *E. coli* was to repress the synthesis of a variety of inducible enzymes. Therefore, Perlman and I began to examine the hypothesis that the role of cyclic AMP in *E. coli* was to stimulate the expression of a set of genes, and that glucose acted by regulating cyclic AMP levels. We were soon able to demonstrate that cyclic AMP not only stimulated the synthesis of some enzymes known to be under glucose control but also overcame glucose repression of these enzymes.

"We submitted a paper to a rapid-publication journal but it was rejected, as were papers to two other well-known journals. Finally, a paper submitted to *Biochemical and Biophysical Research Communications*[2] was accepted, as was a paper to the *Journal of Biological Chemistry*.[3] We breathed a sigh of relief, because by this time our unpublished results were well known, the early experiments were not hard to do, and many scientists were studying the same problem. Initially, it seemed possible that cyclic AMP might regulate gene expression in a very indirect manner, but a number of subsequent observations suggested that its action might be direct. These observations included our isolation of mutants that either could not produce cyclic AMP because of a defect in adenylate cyclase (*cya*) or could not respond to cyclic AMP because of a defective cyclic AMP receptor protein (CRP); our demonstration that *lac* promotor mutants were unresponsive to cyclic AMP;[4] and the demonstration by Zubay et al.[5] that cyclic AMP had an action in the cell-free synthesis of β-galactosidase. Finally, we were able to purify CRP and demonstrate in a test-tube reaction containing defined components (cyclic AMP, CRP, RNA polymerase, *lac* DNA) that transcription of the *lac* operon required cyclic AMP and CRP and was inhibited by purified *lac* repressor.[6,7] This was the first experiment that demonstrated the expression of a bacterial gene and its regulation using pure components.

"These studies provided direct confirmation of the regulatory models proposed by Jacob and Monod[8] and serve today as a model of gene regulation. I believe that our initial discoveries would have been more readily accepted and published if they had not contradicted the orthodox belief of how catabolites of glucose controlled gene expression. My fruitful collaboration with Perlman (now at the University of Illinois) and our joint excitement about the implications of our efforts in the newly emerging field of molecular biology kept us working with enthusiasm even though we had difficulty publishing our initial studies."

1. **Sutherland E W.** Studies on the mechanism of hormone action. *Les Prix Nobel.* Stockholm: Norstedt & Söner, 1972. p. 240-57.
2. **Perlman R & Pastan I.** Cyclic 3',5'-AMP: stimulation of galactosidase and tryptophanase induction in *E. coli*. *Biochem. Biophys. Res. Commun.* 30:656-64, 1968. (Cited 165 times.)
3. --------------------. Regulation of β-galactosidase synthesis in *Escherichia coli* by cyclic adenosine 3',5' monophosphate. *J. Biol. Chem.* 243:5420-7, 1968. (Cited 175 times.)
4. **Pastan I & Perlman R.** The role of the *lac* promotor locus in the regulation of β-galactosidase synthesis by cyclic 3',5'-AMP. *Proc. Nat. Acad. Sci. US* 61:1336-42, 1968. (Cited 70 times.)
5. **Chambers D A & Zubay G.** The stimulatory effect of cyclic adenosine 3'5'-monophosphate on DNA-directed synthesis of β-galactosidase in a cell-free system. *Proc. Nat. Acad. Sci. US* 63:118-22, 1969. (Cited 105 times.)
6. **Anderson W B, Schneider A B, Emmer M, Perlman R L & Pastan I.** Purification of and properties of the cyclic adenosine 3',5'-monophosphate receptor protein which mediates cyclic AMP dependent gene transcription in *Escherichia coli*. *J. Biol. Chem.* 246:5929-37, 1971. (Cited 150 times.)
7. **de Crombrugghe B, Chen B, Anderson W, Nissley P, Gottesman M, Pastan I & Perlman R.** *Lac* DNA, RNA polymerase and cyclic AMP receptor protein, *lac* repressor and inducer are the essential elements for controlled *lac* transcription. *Nature—New Biol.* 231:139-42, 1971. (Cited 140 times.)
8. **Jacob F & Monod J.** Genetic regulatory mechanisms in synthesis of proteins. *J. Mol. Biol.* 3:318-56, 1961. (Cited 2,740 times.)

Rickenberg H V, Cohen G N, Buttin G & Monod J. La galactoside-permease
d' *Escherichia coli. Ann. Inst. Pasteur* 91:829-57, 1956.
[Institut Pasteur, Service de Biochimie Cellulaire, Paris, France]

The experiments described in this paper demonstrated the existence in the bacterium *E. coli* of a 'system,' which mediated the stereospecific, concentrative uptake of β-galactosides. The use of mutants and of inhibitors of protein synthesis permitted the conclusion that the stereospecific component was a protein. [The *SCI®* indicates that this paper has been cited over 230 times since 1961.]

H.V. Rickenberg
Department of Molecular
and Cellular Biology
National Jewish Hospital
and Research Center
Denver, CO 80206

February 20, 1980

"In 1954 little was known about the mechanism by which the substrates of certain bacterial enzymes stimulated the synthesis of these enzymes. When I joined Monod's group at the Institut Pasteur as a postdoctoral fellow, he suggested to me and his colleagues, Georges Cohen and Gérard Buttin, that we attempt to trace the intracellular fate of ^{35}S-labeled thiomethylgalactoside, an analog of lactose and an effective inducer of β-galactosidase in *Escherichia coli*. β-Galactosidase was the most intensively studied inducible enzyme at the time. We found to our surprise that bacteria previously exposed to a β-galactoside concentrated the labeled galactoside several hundredfold, whereas bacteria which had not been pre-exposed to a β-galactoside did not concentrate the sugar. Bacteria which formed β-galactosidase constitutively also concentrated galactosides without prior exposure. Certain unusual mutants isolated earlier by Gabriel Lester

and myself[1] (when graduate students at Yale) on the basis of their inability to grow on lactose and yet endowed with β-galactosidase activity, we now found, did not concentrate galactosides. These and related observations led us to postulate an inducible, stereospecific transport system composed, at least in part, of protein. At about the same time, Georges Cohen and I found similar systems, specific for the transport of individual amino acids in *E. coli*.[2]

"Monod, in a fit of semantic exuberance, proposed the term 'permease.' Cohen, Buttin, and I were somewhat less enthusiastic about the term since, in addition to a certain lack of euphony, the suffix 'ase' carried the connotation of enzymic activity which we did not mean to imply in any strict sense. However, not feeling strongly about the matter, we let Monod have his way. When we submitted the paper to *Biochimica et Biophysica Acta*, the response was prompt and unequivocal: '...an exciting paper, but the term permease is inadmissible...' (I paraphrase). By this time Monod had become enamored of 'permease' and insisted on its use; and thus the paper was published (in French instead of the original English version) in the more compliant *Annales de L'Institut Pasteur*. Nine years later, Fox and Kennedy isolated a membranal protein with the ability to bind β-galactosides and other properties postulated by us earlier.[3]

"The relevance of our observation and the reason for the high citation of this paper are two-fold. The finding of the galactoside permease was at the origin of a trail of research that led, albeit often tortuously, to an ever more profound understanding of not only membranal transport in bacteria but, more importantly, of the role of the membrane in energy transduction in general. Secondly, the discovery of the permease and the finding that its synthesis was controlled coordinately with that of the β-galactosidase (and of the thiogalactoside transacetylase) gave rise to the concept of the operon."

1. **Rickenberg H V.** *β-galactosidase in Escherichia coli. Aspects of its formation and activity.*
    Unpublished thesis. New Haven, CT: Yale University, 1954. 114 p.
2. **Cohen G N & Rickenberg H V.** Concentration spécifique réversible des amino acides chez
    Escherichia coli. Ann. Inst. Pasteur 91:693-720, 1956.
3. **Fox C F & Kennedy E P.** Specific labeling and partial purification of the M protein, a component of
    the β-galactoside transport system of *Escherichia coli. Proc. Nat. Acad. Sci. US* 54:891-9, 1965.

This Week's Citation Classic

Torriani A. Influence of inorganic phosphate in the formation of phosphatases by *Escherichia coli. Biochim. Biophys. Acta* 38:460-9, 1960.
[Biological Laboratories, Harvard University, Cambridge, MA]

In this paper the existence of an alkaline phosphatase (AP) in *E. coli* was demonstrated and evidence was presented for a negative control of its synthesis by inorganic phosphate (Pi). An acid phosphatase was also studied; its synthesis was found to be independent from the level of phosphate in the growth medium. [The *SCI®* indicates that this paper has been cited in over 610 publications since 1961.]

Annamaria Torriani-Gorini
Department of Biology
Massachusetts Institute of Technology
Cambridge, MA 02139

June 8, 1982

"At the Pasteur Institute in Paris, with Jacques Monod, I was studying the rate of synthesis of enzymes to understand the kinetics of 'adaptation.' We chose some adaptive and some constitutive enzymes. One of these was the acid phosphatase known to be produced by *E. coli*. The cells were grown in a device (the Bactogene or chemostat[1]) which allowed continuous exponential growth in conditions limiting the growth rate. When we used phosphate as the limiting factor, I observed that a new phosphatase suddenly appeared. This suggested that inorganic phosphate (Pi) was exerting a negative control on the synthesis of the enzyme. The new phosphatase had an alkaline pH optimum and hydrolyzed all phosphomonoesters.

"Negative control was rather new at that time and was treated with distaste by Monod, who was geared toward positive control. Thus, my observation that Pi inhibited the synthesis of alkaline phosphatase (AP) remained untold for almost two years until I moved from the Pasteur Institute to Harvard University. I was slowly getting the facts organized into a paper which was very thoroughly criticized and corrected by A. Pappenheimer, when I received a letter from M. Pollock in London telling me that Horiuchi[2] in Japan had also found this enzyme and was publishing a note in *Nature*. I dashed the paper through! It has been cited often because it represents the initial observation and brings a complete proof of the existence of this enzyme. It also gives the method of limiting phosphate used to induce its synthesis. This finding has been at the basis of the study of the Pi regulon in *E. coli*.

"At that time (1959), Levinthal at the Massachusetts Institute of Technology was interested in demonstrating the colinearity of DNA and the amino acid sequence of protein. AP seemed an ideal protein to use since its activity is easy to measure and colonies, whether producing the enzyme or not, are also easy to screen. So, I moved to MIT. With Levinthal and Garen I started, by brute force, to analyze millions of colonies produced after mutagenesis and a collection of mutants with an altered enzyme was rapidly produced.[3] The protein was purified and its properties analyzed. The MW of each of the two subunits turned out to be small enough (43000) to make AP an interesting candidate for sequencing. However, the colinearity was proved by Yanofsky with tryptophan-synthetase and the sequence of AP was completed only recently.[4] But AP has continued to be an interesting protein. Its regulatory mechanism has been worked out in the last 20 years. It is now part of the phosphate regulon which involves porines, binding proteins, transport systems, proteins excretion (AP is a periplasmic protein), positive and negative control factors, and their corresponding genes. But...no one knows how Pi functions as a repressor yet!

"The enzyme is widely used in DNA analysis. Removal of the 5'-phosphate from DNA cleaved by endonucleases makes AP as very useful to scientists as it is to *E. coli*...under conditions of stress!"

1. **Monod J.** La technique de culture continue: théorie et applications. *Ann. Inst. Pasteur* 79:390-410, 1950.
2. **Horiuchi T, Horiuchi S & Mizuno D.** A possible negative feedback phenomenon controlling formation of alkaline phosphomonoesterase in *Escherichia coli*. *Nature* 183:1529-30, 1959.
3. **Levinthal C.** Genetic and chemical studies with alkaline phosphatase of *E. coli*. *Brookhaven Symp. Biol.* 12:76-85, 1959.
4. **Bradshaw R A, Cancedda F, Ericsson L H, Neumann P A, Piccoli S P, Schlesinger M J, Shriefer K & Walsh K A.** Amino acid sequence of *Escherichia coli* alkaline phosphatase. *Proc. Nat. Acad. Sci. US* 78:3473-7, 1981.

Hughes D E. A press for disrupting bacteria and other micro-organisms.
Brit. J. Exp. Pathol. 32:97-109, 1951.
[Med. Res. Council Unit for Res. in Cell Metab., Dept. Biochemistry, Univ. Sheffield, England]

A press is described consisting of a cylinder out of which a frozen cell paste is forced through a slit by pressure on a piston. It is used to disrupt bacteria and other microbes to obtain enzymes and other active intracellular components. [The *SCI®* indicates that this paper has been cited over 290 times since 1961.]

D.E. Hughes
Department of Microbiology
University College
Cardiff CF2 1TA
Wales

April 21, 1981

"That the Hughes Press is still finding such extensive use in studies in microbial physiology after the passage of some 30 years is probably because among physical methods for disintegrating cells it is the simplest to make and easiest to use. Doubts about its mode of action were often expressed by my colleagues as brute force and bloody ignorance. However, experiments have confirmed that its efficiency is, as I suggested, analogous to what was in 1950 known as regelation, a process by which glaciers move down a slope and entrained rock debris is fragmented. Subsequently, we have shown that both movement producing shear and pressure producing ice volume changes are necessary.

"The origin of the press arose from the frustration with the then available methods of microbial disintegration which older colleagues will remember, and particularly the variable results of grinding cell pastes with powdered glass by hand in a pestle and mortar. Cogitations on this during cycle rides every morning steeply uphill to my laboratory connected two apparently unrelated themes. One came from Buchner's classical use of a fruit press and sand to obtain a cell free fermenting juice from yeast.[1,2] The other connected this to the fate of living organisms and the process by which the 'delicate cell and the ponderous bone' were fossilised and described in a poem by Raine, just then published.

"Accordingly, the first press employed powdered glass, zircon, and even diamond dust as the abrasive which, to my mind, represented mineral grains which ruptured organisms under pressure: the soluble enzymes could then be extracted. Later thoughts on glaciers and freeze thaw disruption led to the substitution of abrasives by ice crystals. The temperature chosen (–25°) later appeared to be just that at which ice crystal phase changes occurred under the pressures achieved in a 'fly press.' I still have to explain that 'fly' means quick. This had another advantage in that particulate enzymes could also be obtained by differential centrifugation.

"In this simple, perhaps crude device, a wide range of cells have been crushed and new enzymes isolated over the years. 'Crude Hughes Crush,' is now accepted in some journals. The comma is sometimes misplaced, I suspect sometimes deliberately. A review of this field was published in 1971."[3]

1. **Buchner E.** Alkoholische Gährung ohne Hefezellen. [Vorläufige Mittheilung.] *Ber. Deut. Chem. Gies.* 30:117-24, 1897.
2. -------------. Alkoholische Gährung ohne Hefezellen. [Zweite Mittheilung.]
   Ber. Deut. Chem. Gies. 30:1110-13, 1897.
3. **Hughes D E, Wimpenny J W T & Lloyd D.** The disintegration of micro-organisms. (Norris J R & Ribbons D W, eds.)
   Methods in microbiology. London: Academic Press, 1971. Vol. 5B. p. 1-54.

Stanier R Y, Palleroni N J & Doudoroff M. The aerobic pseudomonads: a taxonomic study. *Journal of General Microbiology* **43**:159-71, 1966.

The authors describe a methodology for the taxonomic analysis of aerobic bacteria belonging to the genus *Pseudomonas*, based on the determination of biochemical, physiological and, in particular, nutritional characters. They show that it permits a satisfactory speciation of this bacterial group. [The *SCI*® indicates that this paper was cited 540 times in the period 1961-1975].

————◆━◆————

Professor Roger Y. Stanier
Institut Pasteur
25, Rue Du Docteur-Roux
75724 Paris, France

March 2, 1977

"Taxonomy is a subject which rarely arouses wide interest. Moreover, this paper is a particularly forbidding and arid example of taxonomic research: it is extremely long and detailed, with many lists of strains, and no less than 60 unreadable tables of data. I was therefore astonished to learn that it had been so widely cited. If the computer hasn't made a mistake, I can attribute this popularity only to its methodological content. The methods of characterization that we developed, described and applied here for the first time to define certain *Pseudomonas* species have subsequently been adopted by other workers in the field of bacterial taxonomy. The most useful of these techniques is undoubtedly the simple screening method that we described for determining the range of organic compounds utilizable as carbon and energy sources by heterotrophic bacteria. It revealed many taxonomically significant nutritional properties which had been completely overlooked by the traditional methods of characterization, which examined only the capacity of bacteria to utilize carbohydrates.

"Ironically enough, the authors of this paper did not then consider themselves to be professional bacterial taxonomists! Over the preceding 20 years, Michael Doudoroff, Norberto Palleroni and I had used many unidentified *Pseudomonas* strains for our biochemical and physiological investigations. These studies included the discovery of the Entner-Doudoroff pathway of carbohydrate dissimilation; the analysis of the pathways of oxidation of aromatic compounds; and investigations on enzyme induction. We had become exasperated by the impossibility of appending specific names to these strains through use of the manifestly inadequate existing taxonomic treatments. Since nobody else seemed able or willing to straighten out *Pseudomonas* taxonomy, we finally decided to take a fling at it ourselves; and this paper was the first outcome. Unfortunately, taxonomic research, if it is successful, becomes addictive. This proved to be the first of many papers, emanating from our laboratory at the University of California in Berkeley, which dealt with the phenotypic and genotypic characterization of *Pseudomonads* and other bacterial groups. In fact, these investigations came to an end only with the death of Professor **Doudoroff** in 1975. I have recently published a retrospective survey of their significance.[1] By 1975, the two other original authors had both left Berkeley; Dr. Palleroni joined the research staff of Hoffman-La Roche in Nutley, New Jersey, and I moved to the Pasteur Institute."

1. Stanier R Y. Reflexions sur la taxonomie des *Pseudomonas*. (Reflections on the taxonomy of *Pseudomonas*). *Bulletin de la Institut Pasteur* **74**:255-70, 1976.

Harold F M. Conservation and transformation of energy by bacterial membranes.
Bacteriol. Rev. **36**:172-230, 1972.
[National Jewish Hospital and Research Center, and Dept. Microbiology,
Univ. Colorado Medical Center, Denver, CO]

This article set out to describe, simply and clearly, what Peter Mitchell's chemiosmotic theory proposes and how it applies to bacterial physiology. It was perhaps the first major literature review based on the premise that the theory is correct in principle. [The *SCI®* indicates that this paper has been cited over 560 times since 1972.]

Franklin M. Harold
National Jewish Hospital
Division of Molecular and Cellular Biology
Denver, CO 80220

October 23, 1981

"Most scientists, even today, would rather praise Mitchell than read him. This was doubly true a decade ago, when few had made the effort to master his chemiosmotic theory [1] and barely a handful were prepared to base their own research upon it. The biochemical establishment, with some honorable exceptions, had dismissed the theory as incomprehensible and probably wrong: microbiologists were largely unaware of the ideas that would shortly transform our conception of how bacteria generate useful energy and perform work.

"My own induction into the chemiosmoticists' thin ranks had taken place just a few years before. In 1968, my laboratory at the National Jewish Hospital in Denver was engaged in research on antibiotics that affect bacterial membrane function. My attention was drawn to a paper by W.A. Hamilton,[2] who claimed that the commercial antibacterial agent tetrachlorosalicylanilide (TCS) dissociates active transport from metabolism. This seemed to me unlikely, as substituted salicylanilides were known to be potent uncouplers of oxidative phosphorylation. Jim Baarda and I therefore examined the effect of TCS on *Streptococcus faecalis*, an organism that lives by glycolysis alone, and found, to my surprise, that TCS dissociated metabolite accumulation from glycolysis.[3] At that time, the only mechanistic hypothesis of uncoupling was due to Mitchell and Moyle,[4] who had shown uncouplers to function as proton-conducting ionophores. We confirmed that TCS conducts protons and suggested that it may short-circuit a cellular proton circulation.[3] During the next three years we found that glycolyzing cells generate both the pH gradient and the electrical potential predicted by the chemiosmotic theory. Well before that point I was convinced that Mitchell was right and that his ideas would revolutionize bacterial physiology.

"Unfortunately, few microbiologists seemed to see it that way. I felt that this was chiefly a problem in communication (Mitchell's writing tends to be abstract and quite forbidding) and determined to bridge the gulf of incomprehension with a literature review of my own. The gratifying success of this effort may be due to two features: the article set out to explain, as clearly and simply as possible, what the chemiosmotic theory says and how it applies to bacterial physiology, and it was the first major review by someone besides Mitchell that was squarely based on the premise that the chemiosmotic theory is correct in principle. I may have succeeded too well, encouraging students to acquire at second hand a superficial acquaintance with a profound idea; one must hope that a little knowledge is still better than none.

"During the past decade the chemiosmotic theory has been universally accepted as the basis for research and reflection on the energetics of both bacterial and eukaryotic cells. In my own later writings[5] I have tried to explore some of these wider ramifications. The heroic era of bioenergetics is now past, and the academic one is under way. But the molecular mechanisms of energy coupling remain to be clarified, and provide ample fuel for the controversies that keep bioenergetics a frontier of research."

1. **Mitchell P.** Chemiosmotic coupling in oxidative and photosynthetic phosphorylation.
   Biol. Rev. Cambridge Phil. Soc. **41**:445-502, 1966.
   [Citation Classic. *Current Contents* (16):14, 17 April 1978.]
2. **Hamilton W A.** The mechanism of the bacteriostatic action of tetrachlorosalicylanilide: a
   membrane-active antibacterial compound. *J. Gen. Microbiol.* **50**:441-58, 1968.
3. **Harold F M & Baarda J R.** Inhibition of membrane transport in *Streptococcus faecalis* by uncouplers
   of oxidative phosphorylation and its relationship to proton conduction.
   J. Bacteriology **96**:2025-34, 1968.
4. **Mitchell P & Moyle J.** Acid-base titration across the membrane system of rat-liver mitochondria:
   catalysis by uncouplers. *Biochemical J.* **104**:588-600, 1967.
5. **Harold F M.** Ion currents and physiological functions in microorganisms.
   Annu. Rev. Microbiol. **31**:181-203, 1977.

Kratz W A & Myers J. Nutrition and growth of several blue-green algae.
Amer. J. Bot. **42**:282-7, 1955.
[Department of Zoology, University of Texas, Austin, TX]

Media and methods were developed that support-ed maximum growth rates of three blue-green algae (cyanobacteria). *Anacystis nidulans*, a new species brought into culture, had a temperature optimum of 41°C and the highest specific growth rate of any autotrophic organism. [The *SCI®* indi-cates that this paper has been cited in over 635 publications since 1955, making it one of the most-cited papers ever published in this journal.]

Jack Myers
Department of Zoology
University of Texas
Austin, TX 78712-1064

August 3, 1984

"This paper was part of the dissertation of my student, Bill Kratz, since deceased. He extended a line of work begun with the green alga, *Chlorella*. The rationale for the study was that the microalgae are microbes, rich in protein, with minimal skeletal crud; that such cell material is the end product of their photosynthetic metabolism; and that the specific growth rate, the first order rate constant for growth, measures performance of cellular machinery.

"Supporting methods had already been developed. Aeration with CO_2-enriched air is a requirement because of the high C con-tent (50 percent) of the cells produced. Light must be provided to thin layers of culture, which minimize self-shading. For these re-quirements, we had developed two culture systems. One was an elegant steady-state culture device. The other was the ultimate in simplicity: aerated test-tube cultures in which growth was estimated by periodic readings of absorbance.

"Extending such rationale and methods to the blue-green algae seemed a formidable task. These organisms were billed as sluggish growers, mostly limited to lower tempera-tures (<25°C). This reputation came from a microbiological study[1] and from a botanical study directed toward ecology.[2] Growth had been estimated in these studies after 7 to 14 days incubation. One important feature of these cells was a high pH (>7) requirement. This made the media problem more formida-ble because of the difficulty in avoiding precipitation of $HPO_4^=$, Mg^{++}, Ca^{++}, Fe^{+++}, and trace elements. Further, the classical 5-percent CO_2-in-air mixture was not compatible in practice with a high pH.

"With four species assembled from other laboratories, Bill began an attack on the problem of media composition. Meanwhile, he searched for other species by the simple procedure of selection culture: collect natu-ral waters, judiciously spike them with media components, and incubate in test-tube cultures. Because Texas waters are often warm, they were incubated at various temperatures up to 40°C.

"The technical media problems proved to be solvable. With the species already at hand, Bill could achieve cell concentrations and specific growth rates (5 to 8 percent per hour) comparable to those of many green algae. Meanwhile, there was an exciting development in one of the selection cul-tures. An alga sampled from a campus creek grew rapidly (30 percent per hour) at up to 40°C. Getting the unialgal culture from it was easy. Getting the bacteria-free culture was more difficult and resisted all Bill's at-tempts. Fortunately, he had sent a culture to M.B. Allen. When she graciously sent back an axenic culture, he quickly completed the paper.

"Bill went on to further study of his blue-greens, but his greatest impact had been made. He had brought the blue-green algae from the realm of exotic curiosities to exper-imentally usable microbes. More practical-ly, viewed in terms of its numerous citations, the paper presented a widely useful organ-ism and media for its culture. Although *Ana-cystis nidulans* probably was misnamed,[3,4] it came to be studied more than any other blue-green alga (or cyanobacterium) in cap-tivity."

1. Allen M B. The cultivation of Myxophyceae. *Arch. Mikrobiol.* **17**:34-53, 1952. (Cited 115 times since 1955.)
2. Gerloff G C, Fitzgerald G P & Skoog F. The mineral nutrition of *Coccochloris peniocystis*. *Amer. J. Bot.* **37**:835-40, 1950.
3. Stanier R Y, Kunisawa R, Mandel M & Cohen-Bazire G. Purification and properties of unicellular blue-green algae. *Bacteriol. Rev.* **35**:171-205, 1971. (Cited 395 times.)
4. Rippka R, Derulles J, Waterbury J B, Herdman M & Stanier R Y. Generic assignments, strain histories and properties of pure cultures of cyanobacteria. *J. Gen. Microbiol.* **111**:1-61, 1979. (Cited 215 times.)

Chapter

3

Immunology

3.1 Lymphocytes

To immunologists, the use of six Citation Classics on lectins to introduce a section on lymphocytes will not be surprising, although lectins could have been as logically included in volume 2 in the chapter on Proteins and Amino Acids or the chapter on Carbohydrates. The removal by phytohemagglutinin (PHA) of erythrocytes from whole blood in preparation for leukocyte culturing was the serendipitous cause, as Nowell states in his "This Week's Citation Classic" (TWCC), of his discovery that PHA is mitogenic to leukocytes. The reviewers of his manuscript indicated it was "an interesting observation but of no obvious significance." The same might have been said of the early descriptions of concanavalin A were it not for its ability to distinguish malignant from normal cells (p. 55). But the recognition of cancer cells was not the destined future of these plant glycoproteins, which were soon used to type lymphocytes or serve as cell surface probes, as indicated in the reviews by Sharon, Lis, and others, including Nicolson. Now the immunologists search not for lectins that distinguish B and T lymphocytes, but for those that will define T cell subsets and relatives of these lymphocytes such as the natural killer (NK) cells.

The three descriptions of lymphocyte purification contained in the TWCCs by Coulson, Böyum, and Shortman are truly delightful reading. Coulson's description of how he came to work in Chalmer's laboratory, his references to his favorite pubs, the "Rose" and the "Fountain," and his advice not to order too many reprints are woven into a one-page TWCC that must be read. Böyum, whose paper has been cited on 5,850 occasions, also exposes himself as a human being willing to describe his naïveté as a young scientist. His centrifuge-hoarding colleague is apparently also a person of good humor. And Shortman's last paragraph is a must for any scientific egocentric. It must be unusual to find three scientists working on a common topic to be so uncommonly good at writing.

The methods for purifying lymphocytes from other blood cells appeared in the mid-1960s, just when anatomical and functional studies of the thymus (TWCCs by Clark and Miller) demonstrated its lymphocytic nature and essential role in graft and tissue rejection. These thymus-derived cells could be identified by the specific antigens (Thy 1 was the first described) on their surface (p. 66) and were needed with other lymphocytes and macrophages (p. 67) for immunoglobulin formation, which was clearly shown by the *in vitro* experiments by Mitchell and Miller and by Mosier. Raff's rediscovery of immunoglobulin on one set of lymphocytes (see p. 71) and discovery of Thy 1 on the other set facilitated the enumeration and identification of T cells. His papers on this problem were initially rejected by *Science* and the *Journal of Experimental Medicine* as not sufficiently important. Bianco et al., in a much-cited paper, were also able to distinguish B cells and T cells since the former would bind complement and the latter would not.

Prior exposure of cells to antigens *in vivo* before their removal and *in vitro* cultivation stimulates them to produce antibody when incubated outside their host. The plaque-forming assay of Jerne and Nordin (1,805 citations) enabled the identification of the immunoglobulin-secreting cells and the development of a system for evaluating the influence of T cell subsets on this process. This initial publication of this phenomenon contributed somewhat to Jerne's 1984 Nobel Prize, but it was awarded primarily for his important contributions to immunological theory. Cunningham and Szenberg's report describing a modification of the plaque-forming assay became even more popular than the original article.

By the 1970s, an awareness that the T cells were not a homogeneous population had developed. The cytotoxic lymphocytes (CTL) were defined as those instrumental in graft and tumor rejection, and capable of destroying certain other cells (pp. 76 and 77). Blockade of these CTL by non-cytotoxic antibody attached to the target cell was subject to much disbelief, but the Hellströms were eventually proved correct. Antibodies of this kind actually increase the survival of neoplasms or transplanted tissues (pp. 78–80). The startling discovery that CTL attacked target cells through a dual recognition of the specific target antigen and a self-homing transplantation antigen has still not been resolved on biochemical grounds (p. 81), but is an important key to the specificity of these cells. Bloom and Bennett's description of T cell involvement in delayed hypersensitivity helped identify a second compartment of T cells and encouraged the study of delayed hypersensitivity *in vitro* where experimental manipulations could be controlled much easier than in an intact host. The suppressor T cell or T_s cell, first described by Gershon, was found by Waldmann and his co-workers to suppress immunoglobulin formation in one form of human hypogammaglobulinemia. The remarkable specificity of cyclosporin A as a cytotoxin for T cells (p. 86) makes it one of the most promising agents for use in transplantation immunology. Although this section contains many references to T cell lymphology, additional TWCCs on this subject will appear in a subsequent volume, *Contemporary Classics in Clinical Medicine*.

Citation Classics that emphasize the B cell include the oft-referenced papers by Sell and Gell describing the presence of immunoglobulin on the surface of these cells. Stewart Sell's TWCC crisply reviews the earlier work leading to his publication with Gell and indicates their paper's importance to other studies with B cells. Pernis, Forni, and Amante published one extension of the Sell and Gell paper by describing the orientation of this surface immunoglobulin of the B cell and identifying it as immunoglobulin M.

The TWCC by Basten describes one of the first studies of lymphocyte receptors. The receptor concept had not yet penetrated immunology, but immunologists were quick to follow the lead of biochemists searching for hormone and neurotransmitter receptors. At the present time, there is probably more information on receptors present on T cells than on B cells. The search for receptors for molecules of the complement system on lymphocytes, monocytes, and granulocytes was initiated by the report of Lay and Nussenzweig.

Nowell, Peter C. Phytohemagglutinin: an initiator of mitosis in cultures of normal human leukocytes. *Cancer Research* 20:462-6, 1960.

Mucoprotein plant extract, phytohemagglutinin (PHA), was found to be a specific initiator of mitotic activity. The paper suggested that the mitogenic actions of PHA did not involve mitosis *per se* but the alteration of circulating monocytes and large lymphocytes to a state in which they are capable of division. [The *SCI*® indicates that this paper was cited 917 times in the period 1961-1975].

Dr. Peter C. Nowell,
Department of Pathology G3
University of Pennsylvania
The School of Medicine
Philadelphia, Pennsylvania 19174
January 12, 1977

"The PHA story is another triumph for the Princes of Serendip. Upon joining the Penn faculty in 1957, I began working with short-term cultures of human leukemic cells, using a method developed by Edwin Osgood.[1] His technique employed PHA, a lectin extracted from Navy beans, to agglutinate and remove erythrocytes in preparing leukemic cells for culture. My technician and I traveled across town one day to obtain some leukemic blood, but found the patient was in remission. Rather than waste the trip, we cultured the leukocytes anyway, and to our surprise, found many mitoses. Culture of our own blood promptly followed, confirming the suspicion that normal leukocytes were proliferating in our cultures.

"My major interest then involved attempts to cause leukemic cells to differentiate in culture, but some time was spent on chromosome studies in leukemia, with David Hungerford, and on efforts to explain the growth of normal leukocytes in culture. The cytogenetic work produced the Philadelphia chromosome, but the studies of normal cells seemed initially less rewarding.

Variables in the culture system were eliminated one by one, and eventually PHA emerged as the initiator of mitosis. I agreed with one reviewer of the manuscript who indicated it was an interesting observation, but of no obvious significance. PHA- stimulated cultures did soon become widely used for human chromosome studies, although the chief supplier of PHA created a major crisis by marketing a non-mitogenic product just as many laboratories began to use it. Interestingly, the mitogenic component of PHA still has not been completely identified.

"I did not recognize that the responding cell was a small lymphocyte, believing large lymphocytes and monocytes were more likely candidates. Unfortunately, I chose blood from chronic lymphocytic leukemia to test the response of small lymphocytes, a population which we now know to consist largely of non-responsive B cells. Only when workers such as Gowans[2] demonstrated the proliferative capacity of the small lymphocyte *in vivo*, and its central role in immune responses, were others stimulated to use lymphocyte cultures triggered by PHA (and eventually many other agents) as *in vitro* models for studying various aspects of immunity and also the general problem of mitogenesis in resting mammalian cells. These investigations continue today in many laboratories, including my own.

"Now I write expensive, and I trust, adequately-focused grant proposals. The PHA story evolved in a laboratory which cost $5,000 annually to run, including technician, and in which the research goals were fuzzy, at best. One can only hope that the present research-academic establishment will continue to include enough money and time to allow young faculty the luxury of following up the *unexpected* result, which provides much of the excitement and many of the new directions in biomedical research."

1. **Osgood E E & Krippaehne M L.** The gradient tissue culture method. *Experimental Cell Research* 9:116-27, 1955.
2. **Gowans J L, Gesner B L & McGregor, D D.** The immunological activity of lymphocytes. *CIBA Foundation study group No. 10: Biological activity of the leucocyte.* (Wolstenholme G E W & O'Connor M. ed.) Boston: Little, Brown & Co., 1961, pp. 32-40.

This Week's Citation Classic

MacKinney A A, Jr., Stohlman F, Jr. & Brecher G. The kinetics of cell
proliferation in cultures of human peripheral blood. *Blood* 19:349-58, 1962.
[Natl. Inst. Arthritis and Metabolic Diseases, Natl. Insts. Health, Public Health
Serv., US Dept. Health, Education, and Welfare, Bethesda, MD]

Human peripheral blood cells in tissue culture increased their synthesis of DNA beginning at 24
hours and peaking at 72 hours. Morphologic and
kinetic analysis of the cells using tritiated
thymidine and colchicine indicated that small lymphocytes were induced to divide. [The *SCI®* indicates that this paper has been cited in over 325
publications since 1962.]

Archie A. MacKinney, Jr.
Hematology Section
William S. Middleton Memorial
Veterans Hospital
Madison, WI 53705

January 14, 1983

"The mature lymphocyte's capacity for division
had been the subject of controversy for 50 years.
Growth of thymus[1] or lymph nodes in some tissue
culture experiments had been described, but a
reproducible method for peripheral blood cells
did not emerge. Skeptics were convinced that the
peripheral blood lymphocyte was, like the erythrocyte and granulocyte, incapable of division.

"In the 1950s, plant lectins, e.g., phytohemagglutinin (PHA), had been used to sediment red
cells from peripheral blood. Purified white cells
from these preparations were capable of dividing
in vitro. Nowell[2] showed that the plant lectin PHA
was critical to cell growth. Hungerford[3] exploited
this technique to study human chromosome abnormalities.

"In 1960, I was working for Fred Stohlman and
George Brecher as a hematology research associate at the National Institutes of Health after
four years of internal medicine residency.
Stohlman returned from a conference in Edinburgh, Scotland, with news about this new peripheral blood tissue culture technique. In our discussion of this discovery, Brecher remarked that the
central question was which kind of white blood
cell was growing. It was a most attractive problem
and I was delighted with the chance to work on it.

"By morphologic analysis of the five different
white cells in the blood, we noted striking changes
in the two most abundant forms: granulocytes disappeared by 24 hours, while lymphocytes became
swollen and irregular, displayed nucleoli, and
were changed into clusters of rather malignant-
looking, large, mononuclear cells by 72 hours.

"Using tritiated thymidine and autoradiography
to identify cells in DNA synthesis, we found a
rapid increase in the number of lymphocytes
showing the nuclear thymidine label beginning at
24 hours, with a peak of about 50 percent of the
cells in DNA synthesis at 72 hours. Colchicine
(used to arrest cells in metaphase) showed that the
first divisions occurred at 40 hours. In other experiments, rare cells which were in division at the outset of the culture were tagged and their contribution to the large pool of replicating cells was excluded. We concluded that the replicating cells
were derived from a relatively large population of
previously nondividing lymphocytes. These data
were quickly confirmed.[4,5]

"Initially, lymphocyte culture was a biological
oddity since plant lectins could hardly be regarded as normal initiators of cell division. But antigens, bacterial and viral,[6,7] as well as foreign lymphocytes,[8] were soon added to the list of stimuli,
and lymphocyte culture became a major tool of
immunology. It was now possible to study lymphocyte competency and diversity with a high degree of sensitivity. It was found that cells of
thymic origin (T lymphocytes) responded to PHA,
while cells destined to make immunoglobulin (B
cells) responded to other plant lectins that could
induce immunoglobulin synthesis in culture.
There are now more than a dozen cells which fit
under the umbrella term 'lymphocyte' and the
variety of T lymphocytes continues to grow. Lymphocyte tissue culture is a standard technique of
the geneticist, transplant surgeon, immunologist,
virologist, oncologist, and cell biologist. It has
been estimated that over 10,000 papers have
emerged from our apparently innocent observation. Brecher predicted that 'we should get a little
mileage out of this paper.' None of us could have
imagined the scope of the present scientific
effort."

1. **Ball W D & Auerbach R.** In vitro formation of lymphocytes from embryonic thymus. *Exp. Cell Res.* 20:245-7, 1960.
2. **Nowell P C.** Phytohemagglutinin: an initiator of mitosis in cultures of normal human leukocytes.
   Cancer Res. 20:462-6, 1960. [Citation Classic. *Current Contents* (42):13, 17 October 1977.]
3. **Hungerford D A, Donnelly A J, Nowell P C & Beck S.** The chromosome constitution of a human phenotypic
   intersex. *Amer. J. Hum. Genet.* 11:215-36, 1959.
   [Citation Classic. *Current Contents/Clinical Practice* 8(28):12, 14 July 1980.]
4. **Carstairs K.** Transformation of the small lymphocyte in culture. *Lancet* 2:984, 1961.
5. **Cooper E H, Barkhan P & Hale A J.** Letter to editor. (Mitogenic activity of phytohemagglutinin.)
   Lancet 2:210, 1961.
6. **Pearmain G, Lycette R R & Fitzgerald P H.** Tuberculin-induced mitosis in peripheral blood leucocytes.
   Lancet 1:637-8, 1963.
7. **Elves M W, Roath S & Israëls M C G.** The response of lymphocytes to antigen challenge in vitro.
   Lancet 1:806-7, 1963.
8. **Bain B, Vas M R & Lowenstein L.** The development of large immature mononuclear cells in mixed leukocyte
   cultures. *Blood* 23:108-16, 1964. [Citation Classic. *Current Contents/Clinical Practice* 7(11):14, 12 March 1979.]

This Week's Citation Classic

Inbar M & Sachs L. Interaction of the carbohydrate-binding protein concanavalin A with normal and transformed cells.
Proc. Nat. Acad. Sci. US 63:1418-25, 1969.
[Department of Genetics, Weizmann Institute of Science, Rehovot, Israel]

This paper demonstrated that the interaction of the carbohydrate-binding protein concanavalin A (Con A) with membranes of intact normal and malignant transformed cells induces an agglutination of the transformed cells exclusively. Furthermore, it was also shown that this specific agglutination can be reversed by competition with the carbohydrate α-methyl-D-glucopyranoside that specifically binds to Con A. [The *SCI®* indicates that this paper has been cited over 650 times since 1969.]

Michael Inbar
Department of Cell Biology
Miles-Yeda Ltd. Kiryat Weizmann
Rehovot 76326
Israel

February 10, 1982

"After one year of studying the *in vitro* transformation of normal cells induced by X ray in Leo Sachs's laboratory at the Weizmann Institute of Science, it became necessary to change the subject of my PhD thesis and make other plans. It is evident that such a situation does not usually create a very friendly environment; however, after few, but long, discussions with my supervisor, he suggested that I find a new subject for my doctoral research.

"The work of M. Abercrombie, E.J. Ambrose, L. Weiss, A.B. Pardee, and others in the early-1960s suggested that the phenomena of malignant transformation, cell invasion, and metastasis may depend on the structure of the cell surface membrane. However, this area of research, known today as cellular membranology, was at that time in its early stages of development mainly due to the fact that not too many biochemical tools were available to study the complex structure of cell membranes (do we now have better tools?).

"In retrospect, the use of plant lectins as markers to study carbohydrate-containing membrane receptors on cell surfaces was initiated by J.C. Aub and co-workers[1] when they found that a lipase preparation from wheat germ agglutinates malignant cells exclusively. However, only a few years later, M.M. Burger and A.R. Goldberg[2] found that this differential agglutinability of normal and malignant cells is due to the presence of an agglutinin existing as an impurity in the lipase preparation. The purified wheat germ agglutinin (WGA) was found to be a glycoprotein that binds to the carbohydrate N-acetyl-glucosamine. These studies, together with a great deal of help from A.J. Kalb and J. Yariv, members of our institute who at that time were studying protein-carbohydrate interactions, led me to 'rediscover' the well-known protein concanavalin A (Con A), which was first isolated in a purified form from jack bean in 1936 by J.B. Sumner and S.F. Howell.[3] The next logical step was therefore to test the ability of Con A to interact with membrane receptors and agglutinate mammalian cells. This simple experiment has generated a tremendous amount of work since 1969, reaching a peak of interest with the publication of a book entitled *Concanavalin A as a Tool*.[4]

"I believe now that my initial work with Con A, together with the great deal of support I received from A. Dorfman of the University of Chicago, introduced the general concept that the interaction of lectins with membrane receptors in a reaction similar to the formation of an antigen-antibody complex is indeed a useful tool to study carbohydrate-containing membrane receptors in mammalian cells. Furthermore, I would also like to take the liberty of thinking that my first publication became a *Citation Classic* because of the many scientists who later used Con A as a tool in many different cellular systems and felt obligated to refer to our paper in which Con A was first used for such an application. It is also possible to assume that this paper was cited many times because the study on the interaction of Con A with mammalian cells was indeed a real 'push' for the concept of the dynamic structural organization of biological membranes."

1. **Aub J C, Tieslau C & Lankester A.** Reactions of normal and tumor cell surfaces to enzymes. I. Wheat-germ lipase and associated mucopolysaccharides. *Proc. Nat. Acad. Sci. US* 50:613-19, 1963.
2. **Burger M M & Goldberg A R.** Identification of a tumor-specific determinant on neoplastic cell surfaces. *Proc. Nat. Acad. Sci. US* 57:359-66, 1967.
3. **Sumner J B & Howell S F.** The identification of the hemagglutinin of the jack bean with concanavalin A. *J. Bacteriology* 32:227-37, 1936.
4. **Bittiger H & Schnebli H P,** eds. *Concanavalin A as a tool.* London: Wiley, 1976. 639 p.

Hadden J W, Hadden E M, Haddox M K & Goldberg N D. Guanosine 3':5'-cyclic monophosphate: a possible intracellular mediator of mitogenic influences in lymphocytes. *Proc. Nat. Acad. Sci. US* **69**:3024-7, 1972.
[Depts. Pathology and Pharmacology, Univ. Minnesota, Minneapolis, MN]

The observations implicate cyclic GMP as a positive effector of the proliferative process by virtue of the early increases found to be induced in lymphocytes by the lectin mitogens phytohemagglutinin (PHA) and concanavalin (Con A). [The *SCI*® indicates that this paper has been cited in over 520 publications since 1972.]

John W. Hadden
Program of Immunopharmacology
Department of Internal Medicine
College of Medicine
University of South Florida
Tampa, FL 33612

November 3, 1983

"A series of observations led us to query what the mechanism of lectin mitogen action in lymphocytes might be. The work of others had indicated that lectins might act like hormones. My wife and I were working with Robert A. Good, in the department of pediatrics at the University of Minnesota Medical School, investigating hormone action on lymphocytes. Having probed aspects of the antiproliferative action of cyclic AMP in lymphocytes, we were primed to learn of new mechanisms. I read with excitement of cyclic GMP in the work of Bill George and Nelson Goldberg of the department of pharmacology also at the University of Minnesota.[1] Since others had suggested that the lowering of cyclic AMP was involved with triggering nonlymphocytes to divide, it seemed logical to determine whether mitogen action might involve cyclic GMP in lymphocytes. I approached Nelson with the idea and he, too, thought it a good one.

"We set the experiments up and waited our turn for the cyclic GMP assay, for in those days Nelson and Mari Haddox were involved in measuring cyclic GMP by a time-consuming enzymic cycling assay. With excitement, Nelson presented me the initial results. We promptly completed a series of experiments. The results indicated that three different mitogen preparations induced early increases in cyclic GMP in lymphocytes. We suggested a working hypothesis that cyclic GMP represented at least one of the active signals to initiate cell proliferation and, based on our preliminary experiments with isolated nuclei, that its role might be as a 'membrane to nuclear signal,' a term coined by Good.

"I wrote the paper early one Sunday morning in the attic, interrupted only by Nelson's calls inquiring as to when I would finish. One of its reviewers for the *Proceedings of the National Academy of Sciences* was the late Nobel Laureate Earl Sutherland whose only comment was: 'These observations could be of great potential importance and besides Nelson Goldberg hasn't screwed up yet.' Once published, the paper yielded considerable excitement and controversy as it was the first to link cyclic GMP to cell proliferation. Our experiments with Nelson were extended with Carlos Lopez to show that cyclic GMP increased in serum and insulin-stimulated 3T3 cells in association with the induction of cell division.

"In the decade since, our work in this area has linked the action of lectins to calcium, as well as cyclic GMP, and a series of observations link both to a number of nuclear events including RNA synthesis, RNA polymerases I and II, and nuclear protein phosphorylation. Ron Coffey and I have probed the roles of calcium, phospholipid turnover, and lipoxygenase products in the activation of guanylate cyclase (for a review see reference 2). At least 21 other laboratories have confirmed and extended the original observations to other mitogens and related cyclic GMP increases to activation of cyclic GMP-dependent protein kinase. While many questions remain as to the mechanisms involved, the hypothesis to us still seems valid. Disturbing throughout this decade have been the several reports of failure to confirm. While a number of these may be explained by technical shortcomings in processing extracts and measuring the femtomole quantities of cyclic GMP involved, recent observations of Nelson indicate that the tightly linked (cyclase/phosphodiesterase) metabolic flux of cyclic GMP may be related to signal generation. This would help clarify how with mitogen stimulation of guanylate cyclase cyclic GMP steady state levels could undergo variable changes depending on the particular system."

1. George W J, Polson J B, O'Toole A G & Goldberg N D. Elevation of guanosine 3',5'-cyclic phosphate in rat heart after perfusion with acetylcholine. *Proc. Nat. Acad. Sci. US* **66**:398-403, 1970. (Cited 475 times.)
2. Hadden J W & Coffey R G. Cyclic nucleotides in mitogen-induced lymphocyte proliferation. *Immunol. Today* **3**:299-304, 1982.

Sharon N & Lis H. Lectins: cell-agglutinating and sugar-specific proteins.
Science 177:949-59, 1972.
[Department of Biophysics, Weizmann Institute of Science, Rehovot, Israel]

This article reviews the history of research on lectins since their discovery at the turn of the century, their specificity with respect to monosaccharides and cells, and the properties of purified lectins, in particular concanavalin A. Their enormous potential for studying the structure of complex carbohydrates, and especially of cell surfaces, is demonstrated. [The *SCI®* indicates that this paper has been cited over 930 times since 1972.]

Nathan Sharon and Halina Lis
Department of Biophysics
Weizmann Institute of Science
Rehovot 76 100
Israel

February 1, 1982

"Our involvement with lectins is another example of the unpredictability of scientific research. It started in the early-1960s in the course of studies on soybean proteins, carried out in collaboration with E. Katchalski (then head of our department) under a project supported by the US Department of Agriculture. The purpose of this project was to provide knowledge leading to improved utilization of these proteins in human nutrition. We chose to concentrate on the hemagglutinin known to be present in soybeans, not only because of its possible deleterious effect on the nutritional properties of raw soybeans and soybean oil meal, but more importantly because earlier work[1] indicated that it may be a glycoprotein. At the time, research on glycoproteins was in its infancy, and nothing was known about the occurrence of such compounds in plants.

"We soon proved that soybean agglutinin is indeed a glycoprotein,[2] and thus demonstrated for the first time that plants contain glycoproteins. Although this aspect of soybean agglutinin still occupies our attention,[3] by the late-1960s we became interested in its biological properties. This was prompted by reports in the literature that the lectins wheat germ agglutinin and concanavalin A agglutinated preferentially malignantly transformed cells. We found, in fact, that soybean agglutinin also possesses the same remarkable property.[4]

"This coincided with the surge of interest in cell membranes and in the key role of cell surface sugars in growth, differentiation, and malignancy. It became obvious to us that the sugar specificity of lectins makes them excellent cell surface probes, capable of giving new insights into the structure and function of the cell surface.

"Although there were a couple of books and several reviews on lectins, none of them dealt with their molecular properties, nor did they emphasize the enormous potential of lectins in biological research. The need for such a review was therefore apparent. In 1970-1971 one of us (Sharon) spent his sabbatical in the department of biochemistry at the University of California, Berkeley, where he discussed extensively with his host, C.E. Ballou, the possible role of carbohydrates as information and recognition molecules. He further suggested to D.E. Koshland, Jr., from the same department, who was also a member of the editorial board of *Science*, to write a review on lectins for that journal, a suggestion which was readily accepted. Writing was started in the fall of 1971 by Sharon in London, when he was Royal Society Visiting Professor in the laboratory of A. Neuberger at St. Mary's Hospital Medical School, and was completed by both authors early in 1972 in Rehovot.

"The review drew very favorable comments from our colleagues who found it both timely and of great interest to a broad spectrum of scientists. We believe that its continued high rate of citation is the result of the tremendous growth in studies on lectins and their applications in biology, immunology, and medicine. Indeed, within one decade these long neglected proteins with esoteric properties have become a household word in numerous laboratories."[5,6]

1. Wada S, Pallansch M J & Liener I E. Chemical composition and end groups of the soybean hemagglutinin. *J. Biol. Chem.* 233:395-400, 1958.
2. Lis H, Sharon N & Katchalski E. Soybean hemagglutinin, a plant glycoprotein. I. Isolation of a glycopeptide. *J. Biol. Chem.* 241:684-9, 1966.
3. Dorland L, van Halbeek H, Vliegenthart J F G, Lis H & Sharon N. Primary structure of the carbohydrate chain of soybean agglutinin; a reinvestigation by high-resolution ^{1}H-NMR spectroscopy. *J. Biol. Chem.* 256:7708-11, 1981.
4. Sela B A, Lis H, Sharon N & Sachs L. Different locations of carbohydrate-containing sites in the surface membrane of normal and transformed mammalian cells. *J. Membrane Biol.* 3:267-79, 1970.
5. Lis H & Sharon N. Lectins in higher plants. (Stumpf P K & Conn E E, eds.) *The biochemistry of plants: a comprehensive treatise.* New York: Academic Press, 1981. Vol. VI. p. 371-447.
6. Sharon N. Cell surface receptors for lectins: markers of murine and human lymphocyte subpopulations. (Fougereau M & Dausset J, eds.) *Immunology 80.* London: Academic Press, 1980. p. 254-78.

Lis H & Sharon N. The biochemistry of plant lectins (phytohemagglutinins).
Annu. Rev. Biochem. 42:541-74, 1973.
[Department of Biophysics, Weizmann Institute of Science, Rehovot, Israel]

This is an update and extension of our review entitled 'Lectins: cell-agglutinating and sugar-specific proteins,' published in *Science*.[1] [The *SCI*®
indicates that this paper has been cited in over 705
publications since 1973.]

Halina Lis and Nathan Sharon
Department of Biophysics
Weizmann Institute of Science
Rehovot 76 100
Israel

January 11, 1983

"Having just completed our sixth or seventh major review on lectins,[2] it is most gratifying to know that our first two reviews on the subject are still being cited very frequently.

"On a previous occasion, when our first review on lectins was identified as a *Citation Classic*,[3] we described briefly how we became involved in research in this area. Because of a lack of space, we could not pay tribute to Aaron Altschul, a distinguished plant protein chemist, then at the Southern Regional Research Laboratory, New Orleans, Louisiana, who as early as 1959 convinced us by his enthusiasm and zeal to embark on a study of plant proteins, a long neglected subject. For nearly ten years, our efforts were concentrated mainly on the purification and characterization of the lectin of soybean. Although we made several contributions which, in retrospect, appear to be significant, such as the first identification of a plant glycoprotein (soybean agglutinin) and the first demonstration that lectins occur as families of closely related isolectins,[4] our work attracted little attention. The situation changed dramatically with the realization, in the late-1960s, of the key role that cell surface sugars may play in cell growth and differentiation, in interactions of cells with their environment, as well as in a variety of pathological processes.

"Being aware of the enormous potential of lectins as tools for the study of glycoconjugates, both in solution and on cell surfaces, and the lack of a critical and informative review on the subject, we wrote our article for *Science*. We felt, however, that this did not do sufficient justice to the subject. We therefore proposed to E.E. Snell, the editor of *Annual Review of Biochemistry*, that we prepare an article for this series. Although the second review was completed less than a year after the first one, it contained nearly 100 references to papers published during that year. In addition to updating the literature, we discussed in this review the physicochemical properties of the half dozen lectins purified by then and dealt with some new topics, such as the structure of cell receptors for lectins. We also made the prediction that 'with the increased availability of purified lectins, extensive utilization of these proteins for preparative and analytical purposes may be envisaged.' Recent developments have more than justified this prediction. They include the application of lectins for the separation of protein variants that differ only slightly in their degree of glycosylation, for the fine resolution of complex mixtures of glycopeptides, and increasingly as reagents for histochemical and cytochemical studies. Most exciting is the demonstration, originally made in mice[5] and more recently in humans,[6] that lectins can be used in the fractionation of bone marrow cells for successful transplantation across histocompatibility barriers.

"We believe that the favourable response to the review is attributable not only to the tremendous growth of interest in lectins but also to the fact that we succeeded in conveying to the readers our fascination and enthusiasm for the subject."

1. **Sharon N & Lis H.** Lectins: cell-agglutinating and sugar-specific proteins. *Science* 177:949-59, 1972.
2. **Lis H & Sharon N.** Lectins—properties and applications to the study of complex carbohydrates in solution and on cell surfaces. (Ginsburg V & Robbins P, eds.) *Biology of carbohydrates.* New York: Wiley. Vol. II. In press, 1983.
3. **Sharon N & Lis H.** Citation Classic. Commentary on *Science* 177:949-59, 1972.
   Current Contents/Life Sciences 25(21):20, 24 May 1982.
4. **Lis H, Fridman C, Sharon N & Katchalski E.** Multiple hemagglutinins in soybean.
   Arch. Biochem. Biophys. 117:301-9, 1966.
   [The *SCI* indicates that this paper has been cited in over 55 publications since 1966.]
5. **Reisner Y, Itzicovitch L, Meshorer A & Sharon N.** Hemopoietic stem cell transplantation using mouse bone-marrow and spleen cells fractionated by lectins. *Proc. Nat. Acad. Sci. US* 75:2933-6, 1978.
6. **Reisner Y, Kapoor N, Kirkpatrick D, Pollack M S, Dupont B, Good R A & O'Reilly R J.** Transplantation for acute leukemia using HLA-A,B nonidentical parental marrow cells fractionated with soybean agglutinin and sheep red blood cells. *Lancet* 2:327-31, 1981.

Nicolson G L. The interactions of lectins with animal cell surfaces.
Int. Rev. Cytol. **39**:89-190, 1974.
[Cancer Council and Electron Microscopy Labs., Armand Hammer Ctr. for Cancer Biology,
Salk Inst. for Biological Studies, San Diego, CA]

This paper reviews the specificities and the many and varied uses in the biomedical sciences of lectins, proteins, or glycoproteins that bind to carbohydrate structures via bivalent or polyvalent interactions. [The *SCI®* indicates that this paper has been cited in over 700 publications since 1974.]

Garth L. Nicolson
Department of Tumor Biology
M.D. Anderson Hospital and
Tumor Institute
University of Texas
Houston, TX 77030

July 20, 1984

"I first became interested in lectins[1] as a graduate student at the University of California, San Diego, in the mid-1960s. These intriguing molecules had been used mainly to agglutinate red blood and other cells.[2,3] Their ability to distinguish untransformed from transformed cells[4] sparked an unusual interest in their interactions with a wide variety of different cell types. Various investigators were using lectins as cell mitogens, inducers, and probes for glycoconjugate structure, membrane dynamics, and asymmetry, as well as glycoprotein, cell, and virus purifications.

"My own graduate studies with S.J. Singer utilized lectins as ultrastructural probes for dynamic studies on cell membrane glycoconjugates.[5,6] After graduate school, I moved to the Salk Institute for Biological Studies where my interest in lectins grew to include their abilities to cause transmembrane perturbations,[7] their toxic actions on cells,[8] their modes of cell entry,[9] and, of course, their activities in binding and agglutinating transformed cells.[10] We also used lectins to probe the surfaces of specialized cells, such as the spermatozoan,[11] and to inhibit fertilization by blocking sites on mammalian eggs.[12]

"It is with amusement that I recall writing this paper, which was to become the first of several lengthy reviews I have written. At the time, I was quite inexperienced in organizing such a long paper, and the eleventh hour found my laboratory personnel working overtime on the references, pagination, tables, and figures. I owe special thanks to A. Brodginski, M. Lacorbiere, and G. Beattie for their efforts and for maintaining their humor when the boxes of reference cards were displaced to the floor.

"This review is probably cited so many times because of its extensive coverage of the field (it contained more than 100 pages and more than 700 references) and its comprehensive discussion of various aspects of the chemistry and biology of lectin interactions with animal cells. More recent reviews[13-15] have again dealt with these aspects of lectins and their interactions and uses, but I will always appreciate the fact that so many scientists have turned to my paper as one of the important reviews in this area."

1. **Boyd W C.** The lectins: their present status. *Vox Sang.* **8**:1-32, 1963. (Cited 155 times.)
2. **Lis H & Sharon N.** The biochemistry of plant lectins (phytohemagglutinins). *Annu. Rev. Biochem.* **42**:541-74, 1973.
3. ----------------, Citation Classic. Commentary on *Annu. Rev. Biochem.* **42**:541-74, 1973.
   Current Contents/Life Sciences **26**(11):19, 14 March 1983.
4. **Burger M M.** Surface changes in transformed cells detected by lectins. *Fed. Proc.* **32**:91-101, 1973.
   (Cited 340 times.)
5. **Singer S J & Nicolson G L.** The fluid mosaic model of the structure of cell membranes. *Science* **175**:720-31, 1972.
   (Cited 3,175 times.)
6. **Singer S J.** Citation Classic. Commentary on *Science* **175**:720-31, 1972. *Current Contents* (46):13, 14 December 1977.
7. **Ji T H & Nicolson G L.** Lectin binding and perturbation of the cell membrane outer surface induces a
   transmembrane organizational alteration at the inner surface. *Proc. Nat. Acad. Sci. US* **71**:2212-16, 1974.
   (Cited 95 times.)
8. **Nicolson G L, Lacorbiere M & Hunter T R.** Mechanism of cell entry and toxicity of an affinity purified lectin from
   Ricinus communis and its differential effects on normal and virus-transformed fibroblasts.
   Cancer Res. **35**:144-55, 1975. (Cited 380 times.)
9. **Nicolson G L.** Ultrastructural analysis of toxin binding and entry into mammalian cells. *Nature* **251**:628-30, 1974.
   (Cited 60 times.)
10. ----------------. Temperature-dependent mobility of concanavalin A sites on tumour cell surfaces.
    Nature New Biol. **243**:218-20, 1973. (Cited 225 times.)
11. **Nicolson G L & Yanagimachi R.** Mobility and the restriction of mobility of plasma membrane lectin-binding
    components. *Science* **184**:1294-6, 1974. (Cited 85 times.)
12. **Oikawa T, Yanagimachi R & Nicolson G L.** Wheat germ agglutinin blocks mammalian fertilization.
    Nature **241**:256-9, 1973. (Cited 75 times.)
13. **Goldstein I J & Hayes C E.** The lectins: carbohydrate-binding proteins of plants and animals.
    Advan. Carbohyd. Chem. Biochem. **35**:127-340, 1978. (Cited 380 times.)
14. **Roth J.** The lectins. Molecular probes in cell biology and membrane research. (Whole issue.)
    Exp. Pathol. **16**(Suppl. 3), 1978. 186 p.
15. **Barondes S H.** Soluble lectins: a new class of extracellular proteins. *Science* **223**:1259-64, 1984.

Coulson A S & Chalmers D G. Separation of viable lymphocytes from human blood. *Lancet* 1:468-9, 1964.
[Cambridge Univ., Dept. of Pathology, Cambridge, England]

The paper describes a technique for separating lymphocytes from human blood using defibrination and gelatine sedimentation. The lymphocytes so separated are viable and respond in tissue cultures. [The *SCI®* indicates that this paper has been cited over 290 times since 1964.]

Alan Coulson
3525 W. Benjamin Holt
#298
Stockton, CA 95209

March 25, 1978

"When as a second year student I went to see Professor Greaves in Cambridge in 1962 and told him I was extremely interested in doing research into lymphocytes in tissue culture he directed me to work with Dr. Chalmers, the University Haematologist. He said we should get along satisfactorily as we were both about the same size. Chalmers and I frequently discussed the optimal system for studying the small lymphocyte's repertoire. We received a lot of helpful contributions from our colleagues in the department during the happy hours spent at the 'Rose' and the 'Fountain' and other nearby hostelries during these discussions. We decided that the small lymphocytes should be isolated cells or at least a pure population in culture, certainly free of contamination by the macrophage/monocyte system which seems to serve as a glorified garbage collecting concern. In addition it would be a help if the medium was defined and the response quantitated in a meaningful way, i.e., not just by protein synthesis, or CO_2 production or DNA synthesis or some other indirect biochemical parameter, but by counting how many cells actually transformed into blast cells. In this way the interaction of lymphocyte and antigen would approximate a chemical reaction. Another entertaining thought was that in many respects small lymphocytes behaved as though they had evolved from parasitic protozoa, patrolling their host and keeping out all unrecognised material which might herald another parasitic intruder. An extension of this thinking subsequently led to the concept of an intracellular stimulation pathway.

"Initial attempts centered on getting pure small lymphocytes. In those early days the lab was cluttered with columns of glass beads and glass wool, and bottles of assorted dead and dying small lymphocytes. One day Professor Ceppelini stopped by to visit Dr. Coombs. While they were having tea with Chalmers, Ceppelini suggested gelatine might be helpful in lymphocyte separation as it had been used by French workers in the 1930s to this end. I went off in search of some gelatine and finally obtained some from the biochemistry department which was next door to us on Tennis Court Road. The gelatine came in a very grubby can but it had impeccable credentials, coming as it did from the British Glue and Gelatine Research Association (now sadly defunct). When combined with a preliminary defibrination process, gelatine sedimentation worked extremely well.

"Subsequently in the autumn of 1963 Chalmers and I presented this technique at a local Pathology Society meeting and it seemed to evoke a certain amount of interest. Chalmers decided it should be published and arranged this with one of his friends at the *Lancet*. At that time I was sharing a house on Panton Street with a variable number of physics and chemistry research students including my sister. I seem to remember my sister actually typed the paper, rewriting and editing it in the process. When it was published the other members of the house thought it quite amusing that anybody could burst into print with anything less than five years work. Professor Greaves told me that it would be bad form to order too many reprints and that the department would not pay for more than fifty.

"The most probable reasons why the paper has been cited frequently are that the paper and the title were succinct, the method was simple and usually worked, and it was published fortuitously at a time when lymphocytes immunology and tissue culture were beginning to become bandwagons."

Böyum A. Isolation of mononuclear cells and granulocytes from human blood. *Scand. J. Clin. Lab. Invest.* **21**(Suppl. 97):77-89, 1968.
[Norwegian Defence Research Establishment, Division for Toxicology, Kjeller, Norway]

A technique for isolation of lymphocytes is described. Blood is layered over a fluid with density of 1.077 g/ml. After centrifugation, red cells and granulocytes have formed a sediment at the bottom, and lymphocytes and monocytes are easily collected from the interface between plasma and the separation fluid. [The *SCI®* indicates that this paper has been cited in over 5,850 publications since 1968.]

Arne Böyum
Division for Toxicology
Norwegian Defence
Research Establishment
N-2007 Kjeller
Norway

October 6, 1982

"When the work started in 1961, the goal was to isolate bone marrow lymphocytes for studies of immune reactions following bone marrow transplantation. Fortunately, I was happily unaware of the obstacles ahead, and unfortunately, ignorant of Newton's law of motion. I figured that, with the appropriate gradient design, the slowly sedimenting lymphocytes could easily be picked up after centrifugation. Today, it is with understanding rather than self-irony that I reread in my proposed research protocol that the work would last one to two years. It took six years and I sure remember those hours at the microscope.

"The first two-year period was one of striking discrepancy between effort and progress. No matter how fancy the gradient design, the lymphocytes never behaved as predicted. I even constructed my own monstrous centrifuge. It never worked, and it is still there in the attic to remind me of my scientific infancy. With high polymer compounds as gradient material, the problem was that density and viscosity could not be varied independently. It occurred to me that this difficulty could be overcome using a mixture of two compounds. This turned out to be the first breakthrough. The choice of an X-ray contrast medium to adjust the density was a lucky one. Next, I ended up with different sugar polymers as partners for viscosity control. Moreover, for simplification, I switched from bone marrow to blood.

"A small episode in the lab radically changed the further work. On one occasion, when the gradient was already loaded and the centrifuge being used, while having to wait for a few minutes, I noted that the red cells started to aggregate at the interface, and fell rapidly to the bottom. So, I left the tube on the desk to see what finally happened. This is where a long study of sedimentation in a 1 g gravity field started. (My colleague who used the centrifuge still complains he never got the credit he deserved for his contribution.) I tested out every possible variable, and gradually learned something about the physicochemical mechanisms in a 1 g sedimentation process. This knowledge was then applied to centrifugal techniques. After a total of 3½ years I was able to obtain a pure suspension of mononuclear blood cells, but it took another year to perfect the technique. Altogether, it was a matter of finding the right density and composition of the separation fluid, and a suitable cell concentration. The technique is generally applicable to blood lymphocyte isolation.[1,2] This paper has been highly cited because it has the advantage of being a simple one-step procedure."

1. Böyum A. Separation of blood leucocytes, granulocytes and lymphocytes. *Tissue Antigen.* **4**:269-74, 1974.
2. ------------. Isolation of lymphocytes, granulocytes and macrophages. *Scand. J. Immunol.* **5**(Suppl. 5):9-15, 1976.

Shortman K. The separation of different cell classes from lymphoid organs. II. The purification and analysis of lymphocyte populations by equilibrium density gradient centrifugation. *Aust. J. Exp. Biol. Med. Sci.* **46**:375-96, 1968.
[Walter and Eliza Hall Institute of Medical Research, Melbourne, Australia]

Lymphocytes were separated according to their buoyant density, by centrifugation to equilibrium in continuous gradients of albumin. The procedure gave good resolution, high reproducibility, and good recovery of biologically active cells. Lymphocytes were separable into a series of discrete density subpopulations. [The *SCI®* indicates that this paper has been cited over 185 times since 1968.]

Ken Shortman
Biochemistry and Biophysics Unit
Walter and Eliza Hall Institute
of Medical Research
P.O. Royal Melbourne Hospital
Melbourne, Victoria 3050
Australia

January 30, 1980

"In 1964 when I began attempts to purify cells from the heterogeneous populations in lymphoid organs there was no technology available for cell separation: the best that could be done was a preliminary sorting of blood elements. After failing in attempts at affinity chromatography, I turned to methods that would separate on the basis of physical parameters, such as density. There was nothing new in the principle, the challenge being to reproducibly band and to recover in an active form labile entities whose density varied with environmental conditions such as osmolarity or pH, and which tended to aggregate, to 'stream' and to stick to the walls of the centrifuge tube. It was a bioengineering exercise.

"At the time most biologists would have considered cells as variable and imprecise entities. Encouraged by the work of Leif and Vinograd,[1] who used the analytical precision of physicists in their study of erythrocytes, we found that lymphocytes could also be studied this way, their physical parameters being specified with amazing precision.

"The main observation from our analytical approach was that lymphoid populations consisted of many physically separable subsets. Nowadays no immunologist working with the multiple subgroups of T and B cells would be surprised by this finding, but at the time everyone hoped life would be simpler. The possibility of experimental artifacts was then investigated by ourselves and others, but the published procedure had been well controlled and no artificial source of multiple peaks has been substantiated. Subsequent work has shown the main source of heterogeneity to be distinct metabolic or activation states of lymphocytes, probably occurring within each of the several lymphocyte subclasses.

"The frequency of citation of this article certainly does not reflect frequency of usage of the full procedure since the tedious analytical approach with 20-30 fractions is just too much hard work for most immunologists. However it has served as a basic reference for density separation of cells, especially since problems of general importance in cell separation, such as osmolarity control and cell aggregation, were considered in some detail. I must have accounted for a substantial proportion of the citation count myself, since in the past we have made extensive use of the technique to study lymphocyte differentiation. Nowadays even I think it is too much hard work."

---

1. Leif R C & Vinograd J. The distribution of buoyant density of human erythrocytes in bovine albumin solution. *Proc. Nat. Acad. Sci. US* **51**:520-8, 1964.

CC/NUMBER 15
APRIL 13, 1981

Clark S L, Jr. The thymus in mice of strain 129/J, studied with the electron
microscope. *Amer. J. Anat.* 112:1-9, 1963.
[Department of Anatomy, Washington University School of Medicine,
St. Louis, MO]

As viewed with the electron microscope, the thymus is a solid epithelial organ. Its numerous lymphocytes lie between the epithelial cells, separated by them from the external environment. Epithelial cells show abundant signs of secretion of putative hormones. [The *SCI®* indicates that this paper has been cited over 215 times since 1963.]

Sam L. Clark, Jr.
Center for Educational Resources
University of Massachusetts Medical Center
Worcester, MA 01605

March 25, 1981

"The first comprehensive description of the thymus as seen by electron microscopy, this paper appeared at a time when ignorance and confusion were suddenly being replaced by knowledge of the functions of the lymphoid system in immunity.

"My interest was aroused in the mid-1950s, while teaching histology to first-year medical students at Washington University. I was arrested by the assertion, attributed to Arnold Rich, that ignorance of the function of small lymphocytes, the most numerous cells in the body, was a major scandal in biology. Therefore, I set out to study the fragile tissues of lymph nodes by electron microscopy, a feat recently made possible by improvements in microtomes and the introduction of epoxy resins for imbedding. I found that the delicate collagenous fibers in lymph nodes were surrounded by the enveloping processes of reticular cells, and that the lymphoid cells lay in a sort of vascular space, continuous with the lymphatic sinuses, but separated from connective tissue by reticular cells. Furthermore, such isolation of lymphoid cells proved to be a widespread phenomenon in lymphoid tissues.

"Meanwhile, the late 1950s saw an explosive growth of knowledge concerning the roles of lymphoid cells in immunity, catalyzed by Sir Macfarlane Burnet's clonal selection theory,[1] which he unveiled during his Flexner Lectureship at Vanderbilt University in 1958. I met him there because he and Lady Burnet stayed in my father's house during those weeks in Nashville. I was struck with the idea that the sequestered environments I was seeing in lymphoid tissues might be the environments in which Burnet's quasi-evolutionary clonal selection takes place.

"At the same time, the function of the thymus was beginning to be understood through the work of J.F.A.P. Miller, who demonstrated the necessity for its presence during postnatal development of lymphoid tissues,[2] and Donald Metcalf, who produced evidence that a thymic factor stimulates lymphopoiesis. Marshall and White overcame the thymus's usual failure to produce antibodies by injecting antigen directly into the thymus, and postulated a blood-thymic barrier to penetration of antigens into the thymus.[3] My paper presented the first systematic study of the thymus with the electron microscope, although Hoshino[4] had described some of its epithelial cells. I confirmed its epithelial nature, and demonstrated a barrier of epithelial cells interposed between thymic lymphocytes and the rest of the body. There was abundant morphological evidence for secretion by thymic epithelial cells, the putative source of still incompletely defined thymic hormones.

"This paper, coming as it did at a pivotal point in the history of the field, brought together a widely scattered literature and emphasized to immunologists the potential value of a cellular point of view."

1. Burnet M. *The clonal selection theory of acquired immunity.*
   Nashville, TN: Vanderbilt University Press, 1959. 208 p.
2. Miller J F A P. Fate of subcutaneous thymus grafts in thymectomized mice inoculated with
   leukaemic filtrates. *Nature* 184 (Supp.23): 1809-10, 1959.
3. Marshall A H E & White R G. The immunological reactivating of the thymus.
   Brit. J. Exp. Pathol. 42:379-85, 1961.
4. Hoshino T. Occurrence of ciliated vesicle-containing reticular cells in the mouse thymus.
   Okajimas Fol. Anat. (Japan) 37:209-13, 1961.

Miller J F A P. Immunological function of the thymus.
Lancet 2:748-9, 1961.

Removal of the thymus of mice at birth was associated with atrophy of the lymphoid system, susceptibility to infections and inability to reject alien skin grafts. These findings established that the thymus is responsible for populating the lymphoid system with cells intimately involved in some immune reactions. [The *SCI*® indicates that this paper was cited 694 times in the period 1961-1976.]

―――――――――――◆――――――――――

J.F.A.P. Miller
The Walter and Eliza Hall Institute of
Medical Research
P.O. Royal Melbourne Hospital
Melbourne 3050, Australia

December 21, 1977

"In 1958-1960, I was a Ph.D. student at the Chester Beatty Research Institute in London. I was working on virus-induced leukemia, using mice and the Gross virus. As the thymus was known to be involved in spontaneous leukemia and leukemias induced by irradiation and chemicals, I wanted to determine if thymus removal (thymectomy) would prevent virus-induced leukemia. In those days the Gross virus had to be injected into newborn mice in order to induce leukemia. Yet thymectomy after weaning still prevented leukemia.

"I found that thymus implantation 6 months after thymectomy (which was performed at 1 month) restored the potential for leukemogenesis in mice inoculated with virus at birth. Clearly the virus must have remained latent and the next experiment was a logical follow up—that virus could be recovered from the non-leukemic tissues of thymectomized mice. I asked whether the virus could multiply *outside*

thymus tissue (it was subsequently found that it did so in marrow). Since, however, the virus had to be given at birth in order for leukemia to develop, the question could be studied only by thymectomizing mice *before* the virus was inoculated. At that time I had no idea that the thymus may have a role in establishing the immune system.

"The experiment met with some difficulties because many mice thymectomized at birth did not fare well but wasted away after 6 weeks and died. I found this intriguing, particularly because thymectomy after weaning had never been associated with untoward effects and had not curtailed life. It was clear that mice without a thymus from birth were susceptible to infection because, when they were kept in 'clean' conditions, the incidence of wasting was less.

"Postmortem examination showed 'atrophy' of the lymphoid system. Hence, I asked whether the lymphoid system depended for its development on an intact thymus in early life. It seemed an impertinent question in those days when the thymus was considered by most as being a vestigial structure filled with incompetent cells! I grafted neonatally thymectomized mice (those which had *not yet* wasted) with foreign skin. Unlike controls, they failed to reject this, even skin from rats. I concluded that the thymus must be responsible for producing the ancestors of the cells which would migrate out to function in various types of immune reactions.

"These data were accepted for publication in the *Lancet*. Subsequent work confirmed my findings and opened up a new chapter in immunology. We now have a better understanding of the cellular aspects of immunity and can envisage means of manipulating the immune system to our benefit in infections, vaccination procedures, immunological aberrations, transplantation of alien tissues and even in the fight against cancer."

Miller J F A P & Mitchell G F. Thymus and antigen-reactive cells.
Transplant. Rev. 1:3-42, 1969.
[Experimental Pathology Unit, Walter and Eliza Hall Institute of
Medical Research, Melbourne, Australia]

Immunodeficient neonatally thymectomized mice were restored to full responsiveness by thymus cells in contrast to heavily irradiated mice which responded only if bone marrow cells were also given. Genetically marked cells proved that antibody formers were derived not from thymus cells but from marrow precursors. [The *SCI®* indicates that this paper has been cited in over 700 publications since 1969.]

J.F.A.P. Miller
and
G.F. Mitchell
Walter and Eliza Hall Institute
of Medical Research
Royal Melbourne Hospital
Victoria 3050
Australia

February 23, 1983

"One of us[1] had previously shown that removal of the thymus from mice at birth was associated with lymphoid atrophy and immune defects. Yet thymus lymphocytes had hitherto been considered immunoincompetent in contrast to recirculating lymphocytes such as thoracic duct cells.[2] Perhaps some initial interaction with antigen was necessary to drive thymus cells to immunocompetence. We therefore studied the effects of injecting various cell types into either heavily irradiated or neonatally thymectomized mice. Marrow cells had no effect in these hosts. To our surprise, thymus cells were as effective as thoracic duct cells in restoring antibody formation when given simultaneously with antigen, but only in thymectomized, not in irradiated, mice. The latter required bone marrow to be given as well but responded better with thoracic duct than with thymus cells. If, however, the thymus cells had previously been exposed to antigen in another irradiated host, their ability to restore responsiveness in a second irradiated host given marrow cells was considerably enhanced for that antigen. This introduced the novel concept of 'thymus cell education' and indicated that some interaction took place between educated thymus cells and marrow cells.

"We took bets on which cell type was the precursor of the antibody-forming cell and one of us (JFAPM), who founded his career on the thymus, was certain that it would be the thymus cell. We used genetically marked cells, susceptible to destruction by specific antisera, to identify the precursors. The results were clear-cut: antibody-formers produced visible plaques but only in plates not incubated with antisera directed against the donor of the bone marrow cells. Antisera against thymus-derived cells had no effect. This was the first unequivocal demonstration that thymus-derived cells do not become antibody-formers but are required, at least in many responses, to 'help' antibody-forming precursors to produce antibody. Our data in thymus cell-injected, neonatally thymectomized mice were compelling and left no doubt that lymphocytes interacted in antibody production. Responses had been restored to normal levels. This contrasted with responses obtained by Claman et al.[3] in irradiated recipients of thymus and marrow cells which only reached one percent of those in intact mice.

"Our experiments were published in detail in four successive papers in the *Journal of Experimental Medicine*[4-7] and we were invited to review our findings by Göran Möller when he began his new and most successful series *Transplantation Reviews* (subsequently renamed *Immunological Reviews*). We believe our review has been frequently cited because it opened up the vast field of cell-to-cell interaction in immune responses and hence immunoregulation and immunomanipulation. Soon after, Roitt et al.[8] introduced a new terminology which shortened the terms thymus-derived and bone marrow-derived cells to T and B cells. There is hardly a single paper in immunology since then which does not mention T and B cells, letters which no longer need be explained in abbreviation lists."

1. **Miller J F A P.** Immunological function of the thymus. *Lancet* 2:748-9, 1961.
   [Citation Classic. *Current Contents* (24):11, 12 June 1978.]
2. **Gowans J L & McGregor D D.** The immunological activities of lymphocytes. *Progr. Allergy* 9:1-78, 1965.
3. **Claman H N, Chaperon E A & Triplett R F.** Thymus-marrow cell combinations—synergism in antibody production.
   Proc. Soc. Exp. Biol. Med. 122:1167-71, 1966.
4. **Miller J F A P & Mitchell G F.** Cell to cell interaction in the immune response. I. Hemolysin-forming cells
   in neonatally thymectomized mice reconstituted with thymus or thoracic duct lymphocytes.
   J. Exp. Med. 128:801-20, 1968.
5. **Mitchell G F & Miller J F A P.** Cell to cell interaction in the immune response. II. The source of hemolysin-forming
   cells in irradiated mice given bone marrow and thymus or thoracic duct lymphocytes.
   J. Exp. Med. 128:821-37, 1968.
6. **Nossal G J V, Cunningham A J, Mitchell G F & Miller J F A P.** Cell to cell interaction in the immune response. III.
   Chromosomal marker analysis of single antibody-forming cells in reconstituted, irradiated, or thymectomized
   mice. *J. Exp. Med.* 128:839-53, 1968.
7. **Martin W J & Miller J F A P.** Cell to cell interaction in the immune response. IV. Site of action of anti-
   lymphocyte globulin. *J. Exp. Med.* 128:855-74, 1968.
8. **Roitt I M, Greaves M F, Torrigiani G, Brostoff J & Playfair J H L.** The cellular basis of immunological responses.
   Lancet 2:367-71, 1969.

Reif A E & Allen J M V. The AKR thymic antigen and its distribution in leukemias and nervous tissues. *J. Exp. Med.* **120**:413-33, 1964.
[Biochemistry Sect., Dept. Surgery, Tufts Univ. Sch. Med., and Boston City Hosp., Boston, MA]

A clear-cut serological differentiation between AKR lymphocytes of thymic and nonthymic origin is reported: these two cell types are antigenically distinct. Thymocytes possess an antigen named θ-AKR in AKR and RF mice, a different antigen named θ-C3H in 16 other strains of mice. These antigens are present in high concentrations in thymus, nervous tissues, and some leukemias, and at low levels in other lymphoid organs and other leukemias. No exceptions were found. [The *SCI®* indicates that this paper has been cited in over 910 publications since 1964.]

Arnold E. Reif
Laboratory of Experimental
Cancer Immunotherapy
Mallory Institute of Pathology
Boston University School of Medicine
Boston City Hospital
Boston, MA 02118

January 11, 1983

"My coauthor was Joan Allen, who had just finished her training in a three-year junior college at the top of her class. Like many of my coauthors, she was a research assistant. I feel that coauthorship for able assistants increases their dedication and furthers their careers.

"Our purpose was to look for antibodies to detect leukemia-specific antigens. I had already developed a quantitative assay for antibody against surface antigens of tumor cells and determined the complement requirements.[1] As control cells for thymus-derived leukemia cells, normal thymocytes would have been ideal. However, the usual sources of complement lysed thymocytes, and only a single experiment on antibody lysis of thymocytes had been reported.[2] I de-

vised an assay by removing natural antibodies to thymocytes from sources of complement by absorption.[3]

"Now we could start our search for leukemia antigens. We cross-immunized mice from strains AKR and C3H with different types of lymphoid cells. Because these strains are compatible in H-2, at best a weak antibody response was expected. Instead, immunization of C3H mice with AKR lymphoid cells or leukemias produced powerful antibodies to AKR thymocytes, and inverse immunizations produced potent antibodies to C3H thymocytes. We had discovered two antigens, which I named theta-AKR and theta-C3H: my only use for two years of Greek in high school. Later, I was asked to rename theta (θ), and chose 'Thy.'

"Others were slow to use theta as a marker for thymus-derived (T) lymphocytes, even though we found that antisera reacted with and were absorbed by thymocytes to a much greater degree than by node or splenic lymphocytes, and concluded in 1964 that the Thy-1 antigen is specific for lymphocytes of thymic origin. As Snell *et al.* have remarked in this connection, 'This has been confirmed in numerous studies.'[4] Nor was attention paid to our paper of 1966, in part entitled '...The serologic detection of thymus-derived leukemia cells.'[5] It was not until 1969, that Schlesinger[6] and Raff[7] showed that the content of theta in non-thymic lymphoid organs resulted from its presence on peripheral T-cells rather than from its non-specific presence on various types of lymphocytes. Thereafter, the use of theta as a marker for T-cells mushroomed.

"In retrospect, theta could not have been discovered before an assay for the reaction of antibody against thymocytes had been developed. Once done, the eventual discovery of such a strong antigen was inevitable. Our paper is cited frequently because it gave the first description of the preparation and specificity of antibodies to Thy-1. While Thy-1 is not a tumor antigen, it has been useful for studying the immune response to tumors as well as to other diseases."

1. **Reif A E.** Immune cytolysis of three mouse ascites tumors. *J. Immunology* **89**:849-60, 1962.
2. **Gorer P A & Boyse E A.** Some reactions observed with transplanted reticulo-endothelial cells in mice.
(Albert F & Medawar P B, eds.) *Biological problems of grafting.*
Springfield, IL: Charles C. Thomas, 1959. p. 193-206.
3. **Reif A E.** Immune cytolysis of mouse thymic lymphocytes. *J. Immunology* **91**:557-67, 1963.
4. **Snell G D, Dausset J & Nathenson S.** *Histocompatibility.* New York: Academic Press, 1976. p. 70.
5. **Reif A E & Allen J M.** The antigenic stability of 3 AKR leukemias on isotransplantation and the serologic detection of thymus-derived leukemia cells. *Cancer Res.* **26**:123-30, 1966.
6. **Schlesinger M & Yron I.** Antigenic changes in lymph node cells following administration of antiserum to thymus cells. *Science* **164**:1412-13, 1969.
7. **Raff M C.** Theta isoantigen as a marker of thymus-derived lymphocytes in mice. *Nature* **224**:378-9, 1969.

Fishman M. Antibody formation in vitro. *J. Exp. Med.* 114:837-56, 1961.
[Division of Applied Immunology, Public Health Research Inst. of the
City of New York, Inc., NY]

Specific antibody against bacteriophage was initiated in cultures of lymph node fragments in response to their stimulation with a cell-free extract derived from macrophages which had been incubated with the antigen. Antibody production failed to occur if the antigen alone was added. [The *SCI®* indicates that this paper has been cited in over 545 publications since 1961.]

Marvin Fishman
Division of Immunology
St. Jude Children's Research Hospital
Memphis, TN 38101

October 6, 1983

"Producing antibodies *in vitro* was a major goal to be achieved in immunology during the late 1950s and early 1960s. Culturing immunocompetent organ tissues or cells with a variety of antigens had been generally unsuccessful. While working at the Public Health Research Institute of the City of New York in 1957, I entered this research arena after my interest was piqued by several investigators' observations on the anatomical intimacy between macrophages and lymphocytes.[1,2] The scavenger property of macrophages, which would result in degradation of antigens, provided the basis for the then current thinking that if macrophages had a role in antibody production, it was a passive one. The concept of a positive involvement of these cells in antibody formation was considered heresy. Yet it remained intriguing, despite the phagocytic characteristic of macrophages, that they might fill a helper role for lymphocytes in antibody production. This was put to the test and the results were published in the *Journal of Experimental Medicine*.

"From our experiments, we found that macrophage-antigen interaction was required to initiate a primary immune response in lymph node fragment cultures or in immunocompetent chick embryos. The product of this interaction was then reported to be sensitive in RNase digestion, thus introducing the phenomenon of immune RNAs.

"On a lighter note—in a separate experiment—some of the chick embryos were allowed to hatch with the eventual reward of a continuous supply of fresh eggs for everyone in the lab. The results of that experiment need not be discussed here.

"The results of the work with macrophages led to the positive personal recognition of my receiving the Waksman award for *in vitro* antibody production. It also began the stormy controversy over immune RNAs that today remains unresolved, even though mRNAs responsible for rabbit immunoglobulin synthesis are currently being used to obtain the cDNAs used for Southern and Northern blot analysis. The immune RNA story—thrust perhaps before its time on the immunological community in the early 1960s—may soon be buried under the more newsworthy and more rapidly arriving reports of progress in monoclonal antibody research and immunotherapy for a variety of diseases.[3-7]

"The work described in this paper was intended to test the role of the macrophage in the immune response, an unbelievable concept at that time. I did not intend to give birth to immune RNA, although I do not have any regrets about the appearance of the phenomenon.

"I feel that the most likely reason for the paper being cited so often is that it describes one of the first successful attempts to produce antibodies *in vitro*.

"Reports of recent studies of the macrophage-antigen relationship have come from the laboratories of Jakway and Shevach at the National Institutes of Health and Lu and Unanue at Harvard Medical School."[8,9]

1. Harris T N & Ehrich W E. The fate of injected particulate antigens in relation to the formation of antibodies. *J. Exp. Med.* 84:157-65, 1946.
2. Thiery J P. Microcinematographic contributions to the study of plasma cells. (Wolstenholme G E W & O'Connor M. eds.) *CIBA Foundation Symposium on Cellular Aspects of Immunity.* Boston: Little, Brown, 1959. p. 59-91.
3. Miller R A, Maloney D G, Warnke R & Levy R. Treatment of B-cell lymphoma with monoclonal anti-idiotype antibody. *N. Engl. J. Med.* 306:517-22, 1982.
4. Ritz J & Schlossman S F. Utilization of monoclonal antibodies in the treatment of leukemia and lymphoma. *Blood* 59:1-11, 1982.
5. Foon K A, Bernhard M I & Oldham R K. Monoclonal antibody therapy: assessment by animal tumor models. *J. Biol. Response Modifiers* 1:277-304, 1982.
6. Levy R, Stratte P T, Link M P, Oseroff A, Maloney D G & Miller R A. Monoclonal antibodies in leukemia therapy. (Murphy S B & Gilbert J R, eds.) *Leukemia research: advances in leukemia cell biology and therapy.* New York: Elsevier, 1983. p. 269-79.
7. Rosen S T, Winter J N & Epstein A L. Application of monoclonal antibodies to tumor diagnosis and therapy. *Ann. Clin. Lab. Sci.* 13:173-84, 1983.
8. Jakway J P & Shevach E M. Stimulation of T-cell activation by UV-treated, antigen-pulsed macrophages: evidence for a requirement for antigen processing and interleukin 1 secretion. *Cell Immunol.* 80:151-62, 1983.
9. Lu C Y & Unanue E R. Ontogeny of murine macrophages: functions related to antigen presentation. *Infec. Immunity* 36:169-75, 1982.

Mitchell G F & Miller J F A P. Cell to cell interaction in the immune response. II. The source of hemolysin-forming cells in irradiated mice given bone marrow and thymus or thoracic duct lymphocytes. *J. Exp. Med.* **128**:821-37, 1968.
[Walter and Eliza Hall Institute of Medical Research, Melbourne, Victoria, Australia]

Irradiated thymectomized mice injected with bone marrow cells responded well to antigen in terms of antibody production only when further injected with thymus-derived cells. Antibody-secreting cells were shown to be derived from precursors in marrow and not in thymus. [The *SCI®* indicates that this paper has been cited in over 550 publications since 1968.]

G.F. Mitchell
and
J.F.A.P. Miller
Walter and Eliza Hall Institute of
Medical Research
Melbourne, Victoria 3050
Australia

May 10, 1984

"Several questions concerning the immunological function of the thymus were prominent in the mid-1960s at the time when studies, subsequently referred to as the Miller and Mitchell experiments, were commenced. At that time, certain immune responses were known to be depressed profoundly in mice that had been thymectomized in the neonatal period.[1,2] The candidacy of the small lymphocyte as an antigen-reactive cell capable of initiating various immune responses was very strong.[3] Using the technique of thoracic duct cannulation, the recirculating pool of lymphocytes in neonatally thymectomized mice was calculated to be approximately one percent of the pool size in intact mice.[4,5] Was the thymus the major source of antigen-reactive lymphocytes found in the circulating pool? If so, why were cell suspensions from the thymus so inefficient at reconstituting immune responses in neonatally thymectomized mice, cells from spleen, lymph nodes, or thoracic duct being far better? Moreover, why were only some antibody responses defective in neonatally thymectomized mice?

"Results of several early experiments supported the notion that lymphocytes of relatively 'low immunocompetence' migrated from the thymus probably in small numbers, and that clonal expansion and 'education' of the migrants occurred through interaction with antigen in peripheral lymphoid organs.[4] Reconstitution experiments involving injections of cells and antigen into neonatally thymectomized mice were designed to examine lineage relationships between thymocytes, their presumed direct descendants in the recirculating pool, and antibody-secreting cells. Using F_1 hybrid-parental combinations and appropriate antisera, it was found that inoculated thymocytes were not the precursors of antibody-secreting cells.[6] However, these studies did not establish the immediate organ of origin of antibody formers. This was achieved through the use of adult thymectomized mice that had been irradiated and injected with bone marrow cells. An inoculum of thoracic duct cells from F_1 hybrid mice at the time of challenge with antigen resulted in high-level antibody production. It was already known that thoracic duct cells had a limited but definite capacity to form antibody in acutely irradiated recipients. In this regard they differed from the thymocytes or bone marrow cells. Anti-H-2 serum treatment of antibody-secreting cells from irradiated recipients of F_1 thoracic duct cells and parental bone marrow established the bone marrow as the origin of the bulk of antibody-secreting cells. We concluded that bone marrow contains precursors of antibody-secreting cells (now referred to as B cells), that thymus contains helper cells (now referred to as T_H cells) that promote antibody production by marrow-derived cells, and that the recirculating pool contains both cell types.

"We believe the paper has been cited frequently because it was a forerunner to the vast field of cell-to-cell interaction and immunoregulation in antibody production. The paper also provided a partial explanation for why some, but not all, antibody responses were defective in T cell-deprived mice."

1. **Miller J F A P.** Immunological function of the thymus. *Lancet* 2:748-9, 1961.
2. ―――――――, Citation Classic. Commentary on *Lancet* 2:748-9, 1961. *Current Contents* (24):11, 12 June 1978.
3. **Gowans J L & McGregor D D.** The immunological activities of lymphocytes. *Progr. Allergy* 9:1-78, 1965.
   (Cited 530 times.)
4. **Miller J F A P & Mitchell G F.** Thymus and antigen-reactive cells. *Transplant. Rev.* 1:3-42, 1969.
5. **Miller J F A P.** Citation Classic. Commentary on *Transplant. Rev.* 1:3-42, 1969.
   Current Contents/Life Sciences 26(17):21, 25 April 1983.
6. **Miller J F A P & Mitchell G F.** Cell to cell interaction in the immune response. I. Hemolysin-forming cells in neonatally thymectomized mice reconstituted with thymus or thoracic duct lymphocytes. *J. Exp. Med.* 128:801-20, 1968. (Cited 470 times.)

Mosier D E. A requirement for two cell types for antibody formation in vitro.
Science 158:1573-5, 1967.
[Department of Pathology, University of Chicago, IL]

Mouse spleen cells were separated into two populations by allowing one fraction to adhere to plastic culture dishes. Both the adherent and nonadherent fractions were required to support an in vitro primary antibody response to sheep erythrocytes. The adherent fraction consisted mainly of macrophages and the nonadherent fraction mainly of small lymphocytes. Only the adherent fraction had to be directly exposed to sheep erythrocytes to induce an antibody response. It was concluded that two cell types must interact to induce antibody formation and that the macrophage-rich population functioned to process or present antigen to the lymphocytic precursors of antibody-forming cells. [The *SCI®* indicates that this paper has been cited in over 565 publications since 1967.]

Donald F. Mosier
Research Pathology Section
Fox Chase Cancer Center
Philadelphia, PA 19111

April 26, 1983

"In the spring of 1966, I had just finished my first year of medical school and I was overcome with the ennui that only a year of gross anatomy taken among a herd of overachieving med students can produce. I took a leave of absence and sought refuge in the laboratory. I had become interested in immunology while working with Felix Haurowitz as an undergraduate at Indiana, and I was both intrigued and perplexed by the claims of Fishman and Adler[1] that 'immune RNA' from macrophages could stimulate antibody formation by naive lymphocytes. While considerable evidence suggested that macrophages bound antigen and that lymphocytes were the precursors of antibody-forming cells, that macrophage messenger RNA could encode specific immunoglobulin molecules or even that functional messenger RNA could be transferred from one cell to the next was viewed with considerable skepticism by molecular biologists.

"I began working on an *in vitro* model of macrophage-lymphocyte interaction using the (then) recently described method of Mishell and Dutton[2] for obtaining anti-sheep erythrocyte antibody responses from cultured spleen cells. This work was initiated with the help of Frank Fitch and Don Rowley in the department of pathology and I soon found myself a graduate student in that department. Rabinowitz[3] had just published a method for adhering macrophages to glass-bead columns, but I couldn't get enough cells that didn't stick to these columns to do my experiments. Not only that, but my unfractionated spleen cell cultures were giving meager antibody responses. I conferred with Fitch who made two key suggestions—first, that I go to San Diego to learn from Mishell and Dutton how to set up cultures correctly (buy the right fetal calf serum), and second, that I try just letting macrophages adhere to the standard culture dishes I was using. Thereafter, the experiments worked well and I was able to characterize both plastic adherent and nonadherent populations and show that they both were required for the immune response *in vitro*.

"We observed subsequently that both T and B lymphocytes were included in the nonadherent lymphocytes and that both were required for immunocompetence *in vitro*.[4,5] By this time, almost three years later, I was sufficiently rejuvenated to return to medical school, which I finished with as great dispatch as possible.

"This paper probably has been quoted frequently because it combines a direct demonstration of what was already widely suspected, that macrophages 'process' antigen for lymphocytes, and it introduced a simple method for cell separation. Subsequent improvements in lymphoid cell separation by adherence techniques have been made, most notably by Ly and Mishell."[6]

1. Fishman M & Adler F L. Antibody formation initiated in vitro. II. Antibody synthesis in X-irradiated recipients of diffusion chambers containing nucleic acid derived from macrophages incubated with antigen.
   J. Exp. Med. 117:595-602, 1963.
2. Mishell R I & Dutton R W. Immunization of normal mouse spleen cell suspensions in vitro. *Science* 153:1004-5, 1966.
3. Rabinowitz Y. Separation of lymphocytes, polymorphonuclear leukocytes, and monocytes on glass columns, including tissue culture observations. *Blood* 23:811-28, 1964.
4. Mosier D E & Coppleson L W. A three-cell interaction required for the induction of the primary immune response in vitro. *Proc. Nat. Acad. Sci. US* 61:542-7, 1968.
5. Mosier D E, Fitch F W, Rowley D A & Davies A J S. The cellular deficit in thymectomized mice.
   Nature 225:276-7, 1970.
6. Ly I & Mishell R I. Separation of mouse spleen cells by passage through columns of Sephadex G-10.
   J. Immunol. Method. 5:239-47, 1974.

Bianco C, Patrick R & Nussenzweig V. A population of lymphocytes bearing a
membrane receptor for antigen-antibody-complement complexes. I. Separation and
characterization. *J. Exp. Med.* **132**:702-20, 1970.
[Dept. Pathology, New York Univ. Sch. Med., New York, NY and Dept. Pathology,
Health Center, Univ. Connecticut, Farmington, CT]

Lymphocytes were divided into two discrete subpopulations: those bearing complement receptors (CRL) and those without the receptor (non-CRL). The two populations were physically separated using rosette formation and density gradients. CRL carried membrane immunoglobulin and adhered to nylon wool. They are known today as B-lymphocytes; non-CRL are known as T-lymphocytes. [The *SCI®* indicates that this paper has been cited over 1,170 times since 1970.]

Celso Bianco
Department of Pathology
Downstate Medical Center
State University of New York
Brooklyn, NY 11203

April 15, 1981

"This was my first scientific paper. I came to the laboratory of Victor Nussenzweig for a postdoctoral fellowship 22 months before its publication. The timing was right. Waltraut Lay had just finished her fellowship with Victor. They had shown that some white blood cells carried plasma membrane receptors for a complement component bound to immune complexes.[1] At that time, immunologists were starting to recognize a sharp functional division among lymphocytes. It seemed that lymphocytes which matured in the thymus participated in cell-mediated immune responses (delayed hypersensitivity reactions, transplantation immunity), while those derived from the bone marrow differentiated into antibody secreting plasma cells. (Presently these subpopulations are known as T and B lymphocytes.) My initial experiments were directed specifically to one question: Were the lymphocytes bearing complement receptors (CRL) a discrete cell population? I had to show that the observed percentage of CRL (for instance, 40 percent of spleen lymphocytes) was not the result of an artificial threshold created by limited sensitivity of the assays. I spent several

months developing procedures for the physical separation of the lymphocyte subpopulations. The successful method employed rosette formation between lymphocytes and complement coated red cells, and differential flotation in an albumin gradient. The procedures were reproducible and did not require unusual skills or reagents. We were then able to show that CRL expressed membrane associated immunoglobulins and adhered to nylon wool. Two subsequent papers[2,3] confirmed that the complement receptor was a marker for bone marrow derived, thymus independent lymphocytes, later called B cells.

"During our attempts to reproduce in humans some of the results obtained in mice we made another observation. One of our control reagents, sheep erythrocytes without added antibodies or complement, produced rosettes with a large population of human peripheral blood lymphocytes. I was deeply involved in the study of CRL and had no time to pursue this observation. We decided to call Waltraut, who had returned to Brazil. She worked intensively to standardize the assay and characterize the cell population. Another Brazilian immunologist, Nelson Mendes, searched for clinical materials, including thymuses of children undergoing cardiac surgery. After a few months we were all exultant. We had found a marker for human T cells (or, as we used to call them, non-CRL). We wrote a paper for *Nature*, hoping to have it published fast.[4] Meanwhile, Victor went to a conference in Finland, where he presented the findings on CRL. In a discussion period he presented our preliminary data on rosettes with human lymphocytes. Then everything went wrong. In the next year, two papers described the new T cell marker.[5,6] Our *Nature* paper was collecting dust in the printing office. It was published almost a year after submission. Probably, in our eagerness to divulge the findings, we lost our place in the pages of the *Science Citation Index®*.

"Waltraut is still in Brazil, now working in the immunology of *Trypanosoma cruzi*. Victor remained at New York University. I left the lymphocyte receptor area seven years ago to work on monocytes and macrophages. We are extremely happy with our *Citation Classic*. I recently published a review of this field."[7]

1. Lay W H & Nussenzweig V. Receptors for complement on leukocytes. *J. Exp. Med.* **128**:991-1009, 1968.
2. Bianco C & Nussenzweig V. Theta-bearing and complement-receptor lymphocytes are distinct populations of cells. *Science* **173**:154-6, 1971.
3. Dukor P, Bianco C & Nussenzweig V. Bone marrow origin of complement-receptor lymphocytes. *Eur. J. Immunol.* **1**:491-4, 1971.
4. Lay W H, Mendes N F, Bianco C & Nussenzweig V. Binding of sheep red blood cells to a large population of human lymphocytes. *Nature* **230**:531-2, 1971. [The *SCI®* indicates that this paper has been cited over 430 times since 1971.]
5. Brain P, Gordon J & Willetts W A. Rosette formation by peripheral lymphocytes. *Clin. Exp. Immunol.* **6**:681-8, 1970.
6. Coombs R R A, Gurner B W, Wilson A B, Holm G & Lindgren B. Rosette-formation between human lymphocytes and sheep red cells not involving immunoglobulin receptors. *Int. Arch. Allergy* **39**:658-63, 1970.
7. Bianco C. Plasma membrane receptors for complement. (Day N K & Good R A, eds.) *Biological amplification systems in immunology.* New York: Plenum, 1977.' p. 69-84.

Raff M C. Two distinct populations of peripheral lymphocytes in mice
distinguishable by immunofluorescence. *Immunology* 19:637-50, 1970.
[National Institute for Medical Research, Mill Hill, London, England]

In immunofluorescence studies using anti-θ (now
called Thy-1) and anti-immunoglobulin (Ig) antibodies on cell suspensions prepared from mouse
peripheral lymphoid tissues, thymus-dependent T
lymphocytes were shown to be Thy-1$^+$ and Ig$^-$,
while thymus-independent B lymphocytes were
shown to be Ig$^+$ and Thy-1$^-$. [The *SCI®* indicates
that this paper has been cited in over 570 publications since 1970.]

<a></a>

Martin C. Raff
MRC Neuroimmunology Project
Department of Zoology
University College London
London WC1E 6BT
England

May 7, 1984

"In 1968, I went to the National Institute for
Medical Research at Mill Hill, London, to work on
the immune system with Avrion Mitchison. I had
just completed my training in clinical neurology in
Boston, and this was my first real taste of science.
It was an exciting time in cellular immunology: it
was becoming clear that there were two classes of
lymphocytes—now called T and B cells—and several laboratories, including Mitchison's, were gathering evidence that T and B cells collaborated with
each other in making antibody responses. Since
the two types of lymphocytes looked the same and
were always found together in lymphoid tissues,
methods were badly needed for distinguishing and
separating them. Mitchison pointed me toward the
θ (Thy-1) antigen as a possible marker for T cells.

"Reif and Allen had discovered Thy-1 in 1964
and showed that it was on the surface of mouse
thymus lymphocytes by killing these cells with
anti-Thy-1 antibodies and complement.[1,2][3] (and,
independently, Schlesinger and Yron[4]) used a similar approach to show that T cells, but not B cells, in
peripheral tissues were also Thy-1$^+$. In order to

visualize Thy-1 on T cells, I turned to indirect immunofluorescence, using fluorescent anti-Ig antibodies to detect the binding of anti-Thy-1 antibodies. The method worked beautifully but turned up
an unexpected result: in control experiments
where the anti-Thy-1 antibodies were omitted, the
fluorescent anti-Ig labelled a substantial proportion of lymphocytes on its own. Roger Taylor and
Michel Sternberg, working across the hall from
me, independently found the same thing using radiolabelled anti-Ig antibodies, and we published
our observations together in *Nature* in 1970.[5]
These findings were exciting because they provided strong support for an important corollary of the
clonal selection hypothesis—that lymphocytes
have antibodies on their surfaces that function as
receptors for antigen. On the other hand, they
raised the question of why most lymphocytes were
Ig$^-$. Interestingly, in 1961, Möller had observed
that fluorescent anti-Ig antibodies labelled a small
number of lymphocytes but, since the concept of
antibody-like receptors on lymphocytes was not at
the forefront of immunological thinking, the implications were missed.[6,7]

"In the paper published in *Immunology* in 1970,
I showed that the Ig$^+$ lymphocytes were B cells
whereas the Ig$^-$ lymphocytes were T cells. The paper has been widely cited, I suspect, because it was
the first direct demonstration that B cells but not T
cells have detectable Ig on their surfaces. Since
the publication of this paper, the presence of surface Ig has been the defining characteristic of B
cells. Although not often cited in this regard, the
paper also raised the possibility for the first time
that T cell receptors for antigen may not be classical antibody molecules. This began a prolonged
and heated controversy concerning the nature of T
cell receptors, which has only been resolved
recently with the demonstration that these receptors are homologous to, but distinct from, Ig
molecules (reviewed in reference 8).

"It is ironic that both the *Nature* and *Immunology* papers on cell-surface Ig were originally rejected by *Science* and by the *Journal of Experimental Medicine*, respectively, because they were not
considered sufficiently important."

1. **Reif A E & Allen J M V.** The AKR thymic antigen and its distribution in leukemias and nervous tissues.
   J. Exp. Med. 120:413-33, 1964.
2. **Reif A E.** Citation Classic. Commentary on *J. Exp. Med.* 120:413-33, 1964.
   Current Contents/Life Sciences 26(5):17, 31 January 1983.
3. **Raff M C.** Theta isoantigen as a marker of thymus-derived lymphocytes in mice. *Nature* 224:378-9, 1969.
   (Cited 415 times.)
4. **Schlesinger M & Yron I.** Antigenic changes in lymph node cells after administration of antisera to thymus cells.
   Science 164:1412-14, 1969. (Cited 80 times.)
5. **Raff M C, Sternberg M & Taylor R B.** Immunoglobulin determinants on the surface of mouse lymphoid cells.
   Nature 225:553-4, 1970. (Cited 470 times.)
6. **Möller G.** Demonstration of mouse isoantigens at the cellular level by the fluorescent antibody technique.
   J. Exp. Med. 114:415-34, 1961.
7. -----------, Citation Classic. Commentary on *J. Exp. Med.* 114:415-34, 1961.
   Current Contents Life Sciences 27(27):20, 2 July 1984.
8. **Williams A F.** The T-lymphocyte antigen receptor—elusive no more. *Nature* 308:108-9, 1984.

Jerne N K & Nordin A A. Plaque formation in agar by single antibody-
producing cells. *Science* **140**:405, 1963.
[Dept. Microbiology, Univ. Pittsburgh Medical School, Pittsburgh, PA]

Distinct plaques, each of which is due to the release of hemolysin by a single antibody-forming cell, are revealed by complement after incubation, in an agar layer, of a mixture of sheep red cells and lymphoid cells from a rabbit immunized with sheep red cells. [The *SCI®* indicates that this paper has been cited over 1,805 times since 1963.]

Niels Kaj Jerne
Château de Bellevue
Castillon-du-Gard
F-30210 Gard
France

July 25, 1981

"Nineteen hundred and sixty-two was my sixth year in the World Health Organization, Geneva. Sitting there at my desk, I tried to devise a research plan for the fall of that year when I would join the department of microbiology in the School of Medicine of the University of Pittsburgh. At that time, I had become convinced that single lymphocytes responding to antigen will secrete about a thousand identical antibody molecules per second, and that they will continue to do so when suspended in a suitable medium. From my earlier research career, I had quite some experience with the assay of bacteriophage by plaque formation in semi-solid agar. In March 1962, I made the following entry in my diary: 'Antibody molecules diffusing from a single cell into agar containing red cells and complement should lyse a sufficient number of surrounding red cells to produce a visible plaque.'

"In Pittsburgh, in October that year, Albert Nordin (now at the National Institutes of Health), who had just obtained his PhD, became my research associate, working in the laboratory room next to my chairman's office. We adopted the classical phage plating technique in petri dishes. Nordin was extremely diligent and intelligent at the bench, and persisted in trying out minor modifications in the face of early failures. When I returned from my Christmas vacation in January 1963, he showed me a plate with hundreds of tiny plaques that, after staining the red cell background with benzidine, looked like stars in a cloudless sky. We spent a few weeks making spectacular photographs, and then, on February 2, sent our paper to *Science*.

"The importance of being able to count the number of cells secreting a specific antibody was immediately clear to everybody to whom we showed these pictures. These included Macfarlane Burnet, Max Delbrück, Peter Medawar, and Hilary Koprowski. The editors of *Science* proposed to place one of our pictures on the front page, but changed their mind at the last minute before publishing our paper in the April 26, 1963, issue.

"Claudia Henry and Hiroshi Fuji then joined our efforts. Henry improved the technique by adding DEAE-dextran to the agar,[1] and Fuji devised an 'indirect' plaquing technique for visualizing cells secreting immunoglobulins of the IgG-classes. Unfortunately, this improvement was first published by others.[2,3] The adoption of this experimental technique in all immunological laboratories has resulted in the high citation frequency of our original paper. The term 'plaque forming cell' (PFC) has become a standard term in immune response assays. An extensive review article[4] on the methodology and theory of plaque formation by antibody secreting lymphocytes, including certain technical modifications introduced by others, appeared in 1974.

"Our paper must be one of the shortest to become a *Citation Classic*. In fact, it was shorter than the present commentary!"

1. Jerne N K, Nordin A A & Henry C. The agar plaque technique for recognizing antibody-producing cells.
   (Amos B & Koprowski H, eds.) *Cell-bound antibodies.* Philadelphia: Wistar Institute Press, 1963. p. 109-25.
2. Sterzl J & Říha I. Detection of cells producing 7S antibodies by the plaque technique. *Nature* **208**:858, 1965.
3. Dresser D W & Wortis H H. Use of antiglobulin serum to detect cells producing antibody with low haemolytic
   efficiency. *Nature* **208**:859, 1965.
4. Jerne N K, Henry C, Nordin A A, Fuji H, Koros A M C & Lefkovits I. Plaque forming cells: methodology and
   theory. *Transplant. Rev.* **18**:130-91, 1974.

Cunningham A J & Szenberg A. **Further improvements in the plaque technique for detecting single antibody-forming cells.** *Immunology* 14:599-600, 1968.
[Walter and Eliza Hall Institute of Medical Research, Melbourne, Australia]

This paper described a simple modification of earlier techniques for detecting and counting single antibody-forming cells. [The *SCI®* indicates that this paper has been cited over 1,215 times since 1968.]

A.J. Cunningham
Division of Biological Research
Ontario Cancer Institute
Toronto, Ontario M4X 1K9
Canada

September 10, 1981

"Immunology in the early-1960s was moving from a preoccupation with serology to a focus on the cells responsible for immune functions. 'Cell mediated' reactions were still rather undefined, since T cells had not yet been identified as a separate group of lymphocytes, but there had been a great deal of interest for many years in the cells which produced antibody. This reached a climax with the independent development by Jerne and Nordin[1] and by Ingraham and Bussard[2] of simple techniques for detecting and counting single, living antibody-forming cells.

"In 1964, as a keen new graduate student, I became enchanted with the Jerne technique—areas of lysis looming up from the agar plate when complement was added, each around a single active cell which was responsible for lysis of thousands of erythrocytes. But when I tried to look at these cells more closely through a microscope, to get to know them better as it were, what disappointment! They could barely be seen through all the agar. It was obvious that the optical conditions could be improved by making the preparations very thin, perhaps one cell thick. Cedric Minns suggested leaving out the supporting medium and just letting cells fall to the bottom of a thin chamber. I made suitable chambers by cutting microtome sections of a block of paraffin ten microns thick, with a hole through the middle, and sticking these to a microscope slide. It worked beautifully, leading to my first paper, in *Nature* in 1965.[3] What excitement! (To a scientist that first paper ranks somewhere in importance between his first breath of air and his first love affair!)

"The technique was used for thesis studies quantitating cellular production of antibody in mice and sheep. Then in 1967, I went to the Walter and Eliza Hall Institute, Melbourne, for a postdoctoral year, and was enormously stimulated by many of the people there from its charismatic director, Gus Nossal, downward. They quickly persuaded Szenberg and me, working together, to make the plaque technique more robust and convenient to use. I think it was Gordon Ada who suggested the final form—two slides stuck together around double-sided sticky tape—making a simple chamber that many people have since used.

"Why is this paper cited so often? Only because it is a convenient technique for making a commonly needed measurement, and because there are a lot of immunologists. Its intellectual content is trivial, and the real technical advance was the original work—by Jerne and Ingraham. For me it was important because it led to ways of analysing the specificity of antibody released by single cells, and so to a discovery that these cells vary rapidly in phenotype.[4] Ironically, this has proved unpopular. Perhaps there are two correlations to be derived from citation data: small, appealing advances may be highly cited, while larger, threatening leaps of the imagination are ignored."

1. Jerne N K & Nordin A A. Plaque formation in agar by single antibody-forming cells. *Science* 140:405-7, 1963.
   [Citation Classic. *Current Contents/Life Sciences* 24(35):16, 31 August 1981.]
2. Ingraham J S & Bussard A. Application of a localised haemolysin reaction for specific detection of individual antibody-forming cells. *J. Exp. Med.* 119:667-84, 1964.
3. Cunningham A J. A method of increased sensitivity for detecting single antibody-forming cells.
   Nature 207:1106-7, 1965. [The *SCI* indicates that this paper has been cited over 205 times since 1965.]
4. ----------------. Evolution in microcosm: the rapid somatic diversification of lymphocytes.
   Cold Spring Harbor Symp. 41:761-70, 1977.
   [The *SCI* indicates that this paper has been cited over 5 times since 1977.]

Marbrook J. Primary immune response in cultures of spleen cells.
Lancet 2:1279-81, 1967.
[Walter and Eliza Hall Inst. Med. Res., Melbourne, Australia]

Spleen cells from normal unimmunized mice were cultured on dialysis membranes with heterologous erythrocytes above a reservoir of culture medium. Antibody-secreting cells appeared in significant numbers with a peak response after four to five days. [The *SCI®* indicates that this paper has been cited over 365 times since 1967.]

John Marbrook
Department of Cell Biology
University of Auckland
Auckland, New Zealand

January 29, 1981

"I went to the Walter and Eliza Hall Institute in Melbourne in 1966 with the idea that I should learn enough about immune cells to consider studying the molecular biology of antibody production. To contemplate using radioactive precursors of nucleic acids and proteins it was essential to move away from *in vivo* studies. When I joined Jacques Miller's group, he was keen to establish assays for small numbers of immunocompetent cells, particularly in relation to the effects of thymectomy. The idea was to culture lymphoid cells and use the haemolytic plaque assay to measure responsive cells.

"To obtain an *in vitro* immune response involved several miserable months in which immune spleen cells were cultured with sheep erythrocytes in every culture system I could find. We were able to modify the Bradley-Metcalf bone marrow culture system in soft agar to detect the secretion of haemolytic antibody secreting cells.[1] Although viable cells formed clusters in liquid medium there was a consistent loss of cell viability at high cell concentrations. The success with growing cells in agar suggested that cultures at high cell densities needed to be associated with a reservoir of nutrients. The first burst of haemolytic plaque-forming cells was generated when a single spleen cell suspension was placed in a dialysis bag with sheep erythrocytes, the bag being immersed in medium. It was from this experiment that the culture vessel was developed. It was obvious in 1966 that culture techniques were becoming mandatory for immunological studies so that the development of 'primary' responses *in vitro* was timely.

"In retrospect, the development of the work owed a lot, not only to encouragement from numerous colleagues at the Hall Institute, but also to the scientific climate that they generated. The work of Miller and Mitchell[2] on T and B cell collaboration and the cell separation techniques developed by Shortman[3] led naturally to detailed investigations of cell collaboration and B cell differentiation. It is clear that the site influences citations. If the 'collection of citations' can be compared to angling, it is a pursuit that flourishes in a busy stream, particularly if the stream is also a migratory route."

1. **Robinson W A, Marbrook J & Diener E.** Primary stimulation and measurement of antibody production to sheep red blood cells *in vitro. J. Exp. Med.* **126**:347-56, 1967.
2. **Miller J F A P & Mitchell G F.** Cell to cell interaction in the immune response. I. Hemolysin-forming cells in neonatally thymectomized mice reconstituted with thymus or thoracic duct lymphocytes. *J. Exp. Med.* **128**:801-20, 1968.
3. **Shortman K.** The separation of different cell classes from lymphoid organs. II. The purification and analysis of lymphocyte populations by equilibrium density gradient centrifugation. *Aust. J. Exp. Biol. Med. Sci.* **46**:375-96, 1968.
[Citation Classic. *Current Contents/Life Sciences* (14):12, 7 April 1980.]

Wybran J, Chantler S & Fudenberg H H. Isolation of normal T cells in chronic lymphatic leukaemia. *Lancet* 1:126-9, 1973.
[Sect. Haematology and Immunology, Dept. Medicine, Univ. California Sch. Medicine, San Francisco, CA]

This paper presented data showing that T cells isolated, by rosetting with sheep red blood cells, from the blood of patients with chronic lymphatic leukemia normally react to mitogens. This paper introduced two new techniques: a rosette technique to identify human T cells and a technique to isolate T cells from the peripheral blood. These T cell populations were rather pure and functional. [The *SCI®* indicates that this paper has been cited over 325 times since 1973.]

Joseph Wybran
Department of Immunology
Erasme Hospital
Free University of Brussels
808, route de Lennick
1070 Brussels
Belgium

April 22, 1981

"The work was done while I was a fellow in the laboratory of H. Hugh Fudenberg, at the University of California, San Francisco. At that time, Chantler was also spending one year in the department doing immunofluorescence on B cells. In fact, this paper was probably one of the first practical implications of the rosette system which we had previously described as a marker for human T lymphocytes.[1,2]

"The idea for this paper came rather obviously when I realized that rosettes (T lymphocytes surrounded by sheep red blood cells) should be heavier than nonrosettes. Thus, by first rosetting human lymphocytes with sheep red cells (a newly introduced terminology in human immunology) and layering them carefully over Ficoll-Hypaque, these rosettes should be recovered as T cells at the bottom whereas the nonrosetted cells (non T cells) would remain on top.

"The *Lancet* paper was the first application of this idea since we described the isolation of normally reactive T cells from the blood of patients with chronic lymphatic leukemia of B cell type. Other authors have since also approached this specific problem and using similar or different techniques have shown that T cells can be normal or slightly abnormal in B chronic lymphatic leukemia. This issue, although fundamental, will not be discussed here.

"The data presented in the *Lancet* article were very easily obtained. In fact, the experimental work lasted only two weeks. The writing was rapid and the article was accepted in two weeks by *Lancet*, even though it is probably, for its English style, my worst paper.

"I take the present opportunity to thank Fudenberg who always backed my efforts to prove that rosettes were a marker for human T cells.

"The concept was met with much scepticism by the immunological community and it took 18 months to have these fundamental observations published in the *Journal of Clinical Investigation.*[2]

"The reasons why this *Lancet* paper is so frequently quoted appear to be multiple. Indeed, this article contains two new techniques: the first, a fast (one hour) rosette technique to identify all human T cells, and the second, a simple procedure for isolating T cells. These techniques allowed the identification or isolation of T cells in disease and in health so that it was easily possible to study their functions. Furthermore, since the rosette techniques are simple, they very quickly became a much used tool for any investigator interested in human immunology. They do not require sophisticated equipment (a centrifuge, an incubator, and a microscope are sufficient). Therefore, I strongly suspect that this simple technique helped in the growing practical interest in clinical immunology by nonimmunologists. Not only have these techniques become powerful tools in understanding various immunological phenomena in medicine but appear to contribute to other new areas like immunopharmacology or more recently the identification of enkephalin receptors on human T lymphocytes.[3]

"The appearance of more sophisticated tools like the monoclonal antibodies against human T cells will probably take the place for identifying T cells, but I believe that the rosette technique will still remain much used in view of its simplicity, its low cost, and its requirement for only basic equipment."

1. Wybran J & Fudenberg H H. Rosette formation, a test for cellular immunity.
   Trans. Assn. Amer. Physicians 84:239-46, 1971.
2. Wybran J, Carr M C & Fudenberg H H. The human rosette forming cell as a marker of a population of thymus-derived cells. *J. Clin. Invest.* 51:2537-43, 1972.
3. Wybran J, Appelboom T, Famaey J P & Govaerts A. Suggestive evidence for receptors for morphine and methionine-enkephalin on normal human blood T lymphocytes. *J. Immunology* 123:1068-70, 1979.

Rosenau W & Moon H D. Lysis of homologous cells by sensitized lymphocytes in tissue culture. *J. Nat. Cancer Inst.* 27:471-83, 1961.

This investigation demonstrated that lymphocytes from one inbred-mouse strain, previously sensitized to cells from another strain with a background of different histocompatibility, would destroy the latter type of cells in tissue culture. The lymphocytes aggregated about the (homologous, allogeneic) target-cells, resulting in marked, progressive cytopathogenic changes, with extensive lysis of the targets. These events occurred without the demonstrable involvement of serum complement or antibody and permitted the direct study of cytolytic cellular immune reactions. [The *SCI®* indicates that this paper was cited 385 times in the period 1961-1977.]

Werner Rosenau
Department of Pathology
University of California
School of Medicine
San Francisco, CA 94143

January 11, 1978

"This work was originally conceived to develop an *in vitro* model for studying cellular immune reactions to examine possible cellular autoimmunity in certain progressive liver diseases. However, having successfully produced an *in vitro* model of cellular immunity, we subsequently abandoned the study of hepatic disorders and turned our attention to details of lympho-cyte/target-cell interaction, applying our findings to homograft immunity and additionally to tumor immunity in isogeneic animals.

At first this work met with great skepticism, since cellular immunity was not a popular field in the early '60's — its existence was even doubted by some scientists. Nevertheless, although I was at that time an unknown Assistant Professor, I was able to obtain grant support from the National Institutes of Health to extend this research.

"It was only natural that these earlier experiments with intact cells subsequently led to a search for cytotoxic chemical mediators of lymphocytes, an area pioneered in the late '60's by N. H. Ruddle and B. H. Waksman as well as G. A. Granger and associates. The isolation of highly purified cytotoxin from stimulated human lymphocyte or T-lymphocyte suspensions allowed us to study morphologic and biochemical alterations of target-cell membranes and also biochemical changes occurring within the cell. The role of lymphocyte toxins in cellular immune reactions in the intact organism remains to be conclusively proven, but evidence for their function(s) is increasingly accumulating. Now the field of cellular immunity has become deservedly respectable, and considerable progress has been made on many of its frontiers.

"Reflecting on the courage of the NIH in supporting an unknown Assistant Professor working in an area of then-dubious existence, I find it of interest to relate what recently happened to the same investigator when he tried to strike out into a new field on the basis of what he considered to be promising preliminary data. A majority critique by a granting agency concluded that 'the approach in this application is new and probably unreliable,' although a minority of the reviewers felt that the preliminary findings 'have already exceeded the results of many other studies.'

"How many new and potentially significant concepts are now lost to science because they are presented by an unknown researcher, an investigator unproven in a new field, or because they may be 'ahead-of-their-times'?"

This Week's Citation Classic™

Kay H D, Bonnard G D, West W H & Herberman R B. A functional comparison of human Fc-receptor-bearing lymphocytes active in natural cytotoxicity and antibody-dependent cellular cytotoxicity. *J. Immunology* 118:2058-66, 1977
[Lab. Immunodiagnosis, Natl. Cancer Inst., Natl. Insts. Health, Bethesda, MD]

When simultaneously tested in parallel assays, human peripheral blood lymphocytes active in 'spontaneous' or 'natural killer' (NK) cell lysis of cultured tumor cell lines were shown to be structurally and functionally similar to, if not identical to, Fc-receptor-bearing 'killer' (K) lymphocytes active in vitro in antibody-dependent cellular cytotoxicity (ADCC). [The *SCI®* indicates that this paper has been cited in over 320 publications since 1977.]

H. David Kay
Department of Internal Medicine
Section of Rheumatology and Immunology
University of Nebraska Medical Center
Omaha, NE 68105

April 25, 1984

"Often, major new developments in scientific understanding unfold when established 'dogma' is brought into question by unexpected but reproducible data. Such was the case in the early 1970s when it was found that lymphocytes from healthy donors frequently caused the lysis of tumor cells when cocultured in the same test vessel. This was unexpected, since at that time it was thought that cytolytic capacity was reserved for specifically 'sensitized' lymphocytes from patients with an active disease process. Strong in vitro reactivity by cells from normal donors was therefore viewed by many investigators as artifactual. However, during my postdoctoral training in 1972-1973 with Joseph G. Sinkovics in his tumor immunology laboratory at the M.D. Anderson Hospital and Tumor Institute in Houston, we found consistent evidence that the antitumor cytolytic activity of normal donor lymphocytes was a real phenomenon.[1,2] This independently corroborated earlier results from Ron Herberman's laboratory at the National Institutes of Health, reported by Rosenberg,[3] Oldham,[4] and McCoy,[5] who had also observed this in vitro cytotoxic phenomenon.

"In 1975, I had the very good fortune to continue my studies as a visiting scientist with Herberman's group at the National Cancer Institute. Because Herberman's lab was both active and crowded, I found my bench space was limited for some time to the work surface available inside an unused fume hood! This, however, had the advantage of keeping the next guy's books and beakers off my work area, and also added delightfully to my memories of a hardworking, productive environment, where ideas abounded and collaboration was easy. Working with Bill West (a physician/investigator already characterizing membrane receptors on 'natural killer' [NK] cells) and Guy Bonnard (a senior scientist in Herberman's group who was heading up the human NK studies unit), I began to explore the intriguing relationship between NK cells on the one hand, and 'killer' (K) cells on the other, which are identical to NK cells in every way, except that they require the Fc receptors on their surface membranes to bind IgG-sensitized targets. The results of our studies were submitted to the *Journal of Immunology* in November 1976.

"The popularity of this manuscript has been most surprising, and no doubt relates to: 1) the sound reputation established over the years by Herberman and his colleagues in NK cell studies; 2) the vigorous popularity which studies of NK cells have now come to enjoy (e.g., when I began my NK studies, there may have been two or three papers a year reporting tumor cell killing by normal-donor lymphocytes; today, there are at least ten times that many appearing every month!); and 3) the interesting results we obtained while addressing the NK/K cell relationship. For example, the paper has been frequently cited for our descriptions of the protease-sensitivity of NK activity compared to antibody-dependent cellular cytotoxicity (ADCC); our use of staphylococcal protein A to block binding of effector-cell Fc receptors to immobilized IgG immune complexes; our comparison of NK/ADCC function in T cell and non-T cell fractions; and our use of unique cell lines to simultaneously measure NK and ADCC functions in parallel but separate assays. A comprehensive update of the NK field was recently edited by Herberman."[6]

1. **Kay H D, Cabiness J R, Ervin F, Virgil W & Sinkovics J G.** Do lymphocytes presensitized to tumor antigens occur in normal individuals? *Abstracts of the annual meeting of the American Society for Microbiology.* Washington, DC: American Society for Microbiology, 1973. p. 111. Abstract no. M227.
2. **Kay H D & Sinkovics J G.** Cytotoxic lymphocytes from normal donors. *Lancet* 2:296-7, 1974.
3. **Rosenberg E B, Herberman R B, Levine P H, Halterman R H, McCoy J L & Wunderlich J R.** Lymphocyte cytotoxicity reactions to leukemia-associated antigens in identical twins. *Int. J. Cancer* 9:648-58, 1972. (Cited 80 times.)
4. **Oldham R K, Siwarski D, McCoy J L, Plata E J & Herberman R B.** Evaluation of a cell-mediated cytotoxicity assay utilizing ^{125}iododeoxyuridine-labeled tissue-culture target cells. *Natl. Cancer Inst. Monogr.* 37:49-58, 1973.
5. **McCoy J L, Herberman R B, Rosenberg E B, Donnelly F C, Levine P H & Alford C.** ^{51}Chromium-release assay for cell-mediated cytotoxicity of human leukemia and lymphoid tissue-culture cells. *Natl. Cancer Inst. Monogr.* 37:59-67, 1973.
6. **Herberman R B,** ed. *NK cells and other natural effector cells.* New York: Academic Press, 1982. 1,566 p.

Kaliss N. Immunological enhancement of tumor homografts in mice. A review.
Cancer Res. 18:992-1003, 1958.
[Roscoe B. Jackson Memorial Laboratory, Bar Harbor, ME]

The groundwork is described for the initial experiments establishing an immune basis for the enhancement of cancer allografts in actively or passively immunized mice. Immunological enhancement may be applicable to normal tissue allografts and may represent a broad range of other immunological manifestations. [The *SCI®* indicates that this paper has been cited over 390 times since 1961.]

Nathan Kaliss
The Jackson Laboratory
Bar Harbor, ME 04609

May 6, 1980

"I came to the Jackson Laboratory in July 1947 as a senior fellow of the American Cancer Society. My introduction to 'enhancement' was a seminar that summer by George Snell describing attempts to preimmunize mice with homologous freeze-dried tumor against subsequent tumor allografts. Paradoxically, some of the grafts, rather than experiencing accelerated rejection, grew progressively. This was intriguing since cancers were involved, and the immediate question was whether a 'cancer stimulating substance' was the effector. I addressed myself to this but my research was interrupted by the Laboratory's destruction in an October 1947 forest fire.

"With work resumed in the rebuilt laboratory, I found that pretreatment with freeze-dried normal tissues, specifically and necessarily from mice of strains indigenous to the test allografts, resulted in progressive graft growth. These results militated against the assumption of an 'enhancing substance.' My subsequent use of antiserum was suggested by literature reports of accelerated wound healing in rabbits receiving anti-rabbit spleen serum.[1] My treatment of prospective hosts with either rabbit anti-mouse or mouse anti-mouse tissue serum did ensure the tumor allografts' survival.

"The experiments which eventually established an immunological basis for enhancement, and related 'active' and 'passive' enhancement, were sparked by a report of intraperitoneal (ip) metastases appearing in mice given freeze-dried tumor and cortisone ip and a subcutaneous inoculum of live tumor.[2] (I could not confirm these results experimentally and concluded that live cells in inadequately freeze-dried tumor were the progenitors of the ip 'metastases.')

"Cortisone treatment involutes lymphoid tissues and inhibits antibody production, and this was the tool I used to demonstrate that antigraft antiserum in the host, actively or passively acquired, was the requisite for tumor allograft survival (hence the term 'immunological enhancement'). Indeed, subcutaneous allografts were rejected by mice pretreated simultaneously with a supernatant of freeze-dried homologous tumor and cortisone. The grafts survived in animals given supernatant alone or (and this was the decisive result) pretreated with supernatant and cortisone and in addition given antigraft alloantiserum at the time of tumor grafting. The grafts also grew progressively in mice given alloantiserum plus cortisone or antiserum alone.

"Immunological enhancement, whose expression involves fundamental immunological interactions still to be elucidated, is a definitive area of concern in immunobiology and tissue transplantation. The broad outlines of its experimental conditions for cancer allografts were detailed in my 1958 review, thus accounting for its frequent citation. Clinically, enhancement should promise the survival of normal grafts, seemingly the case with human kidney grafts,[3] but it poses a hazard in cancer immunotherapy. As the complexities of immunology are unraveled, however, the hope is to be able to manipulate reactions so as to avoid cancer enhancement and facilitate normal allograft survival."

1. Pomerat C M. A review of recent developments on reticulo-endothelial immune serum (R.E.I.S.). *Quart. Phi Beta Pi* 42:203, 1946.
2. Molomut N, Spain D M, Gault S D & Kreisler L. Preliminary report on the experimental induction of metastases from a heterologous cancer graft in mice. *Proc. Nat. Acad. Sci. US* 38:991-5, 1952.
3. Batchelor J R & Welsh K I. Mechanisms of enhancement of kidney allograft survival: a form of operational tolerance. *Brit. Med. Bull.* 32:113-17, 1976.

Hellström I, Hellström K E, Sjögren H O & Warner G A. Demonstration of
cell-mediated immunity to human neoplasms of various histological types.
Int. J. Cancer 7:1-16, 1971. [Depts. Microbiol. and Pathol., Univ. Washington Med.
Sch.; Dept. Immunol., Fred Hutchinson Cancer Ctr.; and Swedish Hosp. Tumor Inst.
and Med. Ctr., Seattle, WA]

Peripheral blood lymphocytes from 51 of 59
cancer patients were found to react *in vitro*
to autochthonous tumor cells, as were lymphocytes from 78 of 87 patients when tested
on allogeneic tumor cells of the same histological type as that of the lymphocyte
donor. Evidence for shared tumor antigens
was obtained for seven different tumor
groups, including melanomas and carcinomas of the colon and breast. [The *SCI*® indicates that this paper has been cited over 720
times since 1971.]

Ingegerd Hellström and Karl Erik Hellström
Division of Tumor Immunology
Fred Hutchinson Cancer Research Center,
Department of Microbiology and
Immunology, and Department of Pathology
University of Washington Medical School
Seattle, WA 98105

March 16, 1981

"A group headed by us had published, in
1968, that lymphocytes from some cancer
patients inhibit (or kill) plated tumor cells
from the respective patients or from other
patients with the same type of tumor.[1] We
tried to extend this work when Hans Olov
Sjögren (the brother of Ingegard Hellström
and now a professor at the University of
Lund, Sweden) spent two years in Seattle
around 1970. The 1971 paper presents our
findings.

"Our results were unexpected: not only
did we observe that tumors of the same type
share antigens but also that lymphocytes
from patients with growing tumors can be as
reactive as are lymphocytes from patients
whose tumors have been removed. We believed that the shared antigens were differentiation antigens,[2] and attributed the
growth of tumors in the face of an immune
reaction to serum factors suppressing
('blocking') lymphocyte reactivity.

"The major reason why our paper has
been much cited is that it provides some of
the first evidence that common human
tumors (e.g., carcinomas of the colon and
breast) can be recognized as foreign by the
host. The paper has also been controversial,
which contributes to the many citations,
too. Part of the controversy is due to the
fact that we had postulated that the shared
tumor antigens might be suitable targets for
diagnostic and therapeutic procedures—
'spontaneous' human tumors were thought
not to express any such antigens. The fact
that many colleagues had difficulties
reproducing our results did not help the matter. In retrospect we believe that the conflict
of findings had a technical explanation: we
had worked with tumors of short *in vitro*
passage and patients only treated with
surgery, and we had exposed the target cells
to lymphocytes for two to five days, while
the work of many colleagues was carried
out slightly differently. Destruction of
tumor cells by NK (natural killer) cells present in all lymphocyte populations was what
was primarily observed in their studies.

"Are the conclusions of our 1971 paper
valid today? We think so. Findings with the
leukocyte migration and adherence inhibition techniques have confirmed the presence of shared antigens in tumors of the
same type,[3] and monoclonal antibodies to
type specific differentiation antigens have
been recently raised.[4] Indeed, the diagnostic
and therapeutic procedures hinted at in our
1971 paper now attract much attention, using monoclonal antibodies as tools. Even
the postulated serum factors are well established—in the form of tumor antigens and
complexes activating suppressor T cells
(which in their turn form suppressor
factors).[3]

"Thus we feel happy about our 1971
paper. Of course, we would have felt even
happier, had we realized then that what was
attributed to 'cell-mediated tumor immunity' encompassed several mechanisms (T cell
killing, ADCC, NK effects) and that the 'tumor specificity' of most shared differentiation antigens is relative rather than absolute."

1. Hellström I, Hellström K E, Pierce G E & Yang J P S. Cellular and humoral immunity to different types of human
    neoplasms. *Nature* 220:1352-4, 1968.
2. Hellström I, Hellström K E & Shepard T H. Cell-mediated immunity against antigens common to human colonic
    carcinomas and fetal gut epithelium. *Int. J. Cancer* 6:346-51, 1970.
3. Hellström K E & Brown J P. Tumor antigens. (Sela M, ed.) *The antigens.*
    New York: Academic Press, 1979. Vol. 5, p. 1-82.
4. Herlyn M, Steplewski Y, Herlyn D & Koprowski H. Colorectal carcinoma-specific antigen: detection by means of
    monoclonal antibodies. *Proc. Nat. Acad. Sci. US* 76:1438-42, 1979.

This Week's Citation Classic

Hellström K E & Hellström I. Lymphocyte-mediated cytotoxicity and blocking
serum activity to tumor antigens. *Advan. Immunol.* **18**:209-77, 1974.
[Depts. Pathology, Microbiology, and Immunology, Univ. Washington
Medical School, Seattle, WA]

This paper reviews evidence that lymphocytes from animals and human patients with tumor are specifically reactive to cells from the same tumor *in vitro* and that their reactivity can be prevented by circulating 'blocking factors' such as tumor antigens and antigen-antibody complexes. [The *SCI®* indicates that this paper has been cited a total of 653 times of which 8 occurred in 1974, 74 in 1975, 119 in 1976, 113 in 1977, 126 in 1978, 87 in 1979, 66 in 1980, and 60 in 1981.]

Karl Erik Hellström and Ingegerd Hellström
Fred Hutchinson Cancer Research Center
University of Washington
1124 Columbia Street
Seattle, WA 98104

January 25, 1982

"In 1966, we both started working at the University of Washington Medical School in Seattle, having left George Klein's group at the Karolinska Institute in Stockholm, where we got our training. The project on which we embarked concerned lymphocyte reactivity to tumor-associated antigens as assayed *in vitro.*

"We found, rather to our surprise, that lymphocytes from mice with growing, chemically induced sarcomas were often as reactive to cells from the same sarcomas *in vitro* as were lymphocytes from mice whose tumors had been removed. Similar findings were made with other experimentally induced tumors and with human neoplasms.

"In an attempt to learn why tumors can grow progressively *in vivo*, in spite of the fact that the tumor-bearing individuals' lymphocytes can kill plated tumor cells *in vitro*, we tested serum from the respective tumor-bearers for any adverse effect on the ability of the lymphocytes to react. We observed that tumor-bearer serum could suppress ('block') lymphocyte reactivity, and we attributed this to circulating 'specific blocking factors.'[1] These factors were able to bind to

tumor cells from the donors of the respective sera and they disappeared shortly after tumor removal. In 1971, we obtained evidence that the circulating blocking factors were circulating antigen-antibody complexes and that free antigen could also serve as a blocking factor.[2]

"The findings that we had obtained were confirmed and extended in other laboratories.[3] They contradicted the prevailing view that lymphocyte clones that are reactive to a given tumor antigen are absent ('forbidden') from the tumor-bearing host. They indicated, instead, that lymphocyte reactivity must be regulated, and we proposed that the 'blocking factors' play an intricate part in this regulation. Further evidence supporting the view of regulation of lymphocyte activity rather than clonal loss came from studies which we performed on rats that had been made tolerant to skin allografts. These rats had lymphocytes that were reactive *in vitro* to the tolerated tissue, and they also had circulating blocking factors, inhibiting this reactivity.[4]

"Our 1974 paper in *Advances in Immunology* reviewed these findings. We believe that the reason our paper has been much cited reflects both the great amount of interest and the considerable controversy which our rather unexpected observations caused. Today, it is generally accepted that reactive lymphocytes occur in tumor-bearing animals, that their activity is subject to close regulation, that blocking factors in the form of tumor antigens and complexes turn on suppressor T cells, and that other blocking factors are the products of such cells.[5] The greatest advancement since 1973 is that it has become possible, using proper cell surface markers, to dissect subsets of lymphocytes with distinct functions, while in 1974 we did not know of NK cells and of various types of T killer, helper, and suppressor cells. Thus, the phenomenological framework in which we and others were then working is gradually being replaced by knowledge at the cellular and even at the molecular level."

1. Hellström I, Hellström K E, Evans C A, Heppner G, Pierce G E & Yang J P S. Serum mediated protection of neoplastic cells from inhibition by lymphocytes immune to their tumor specific antigens. *Proc. Nat. Acad. Sci. US* **62**:362-9, 1969.
2. Sjögren H O, Hellström I, Bansal S C & Hellström K E. Suggesting evidence that the 'blocking antibodies' of tumor-bearing individuals may be antigen-antibody complexes *Proc. Nat. Acad. Sci. US* **68**:1372-5, 1971.
3. Baldwin R W, Price M R & Robins R A. Inhibition of hepatoma-immune lymph node cell cytotoxicity by tumor-bearer serum, and solubilized hepatoma antigen. *Int. J. Cancer* **11**:527-35, 1973.
4. Bansal S C, Hellström K E, Hellström I & Sjögren H O. Cell-mediated immunity and blocking serum activity to tolerated allografts in rats. *J. Exp. Med.* **137**:590-602, 1973.
5. Hellström K E & Brown J P. Tumor antigens. (Sela M, ed.) *The antigens.* New York: Academic Press, 1979. Vol. 5. p. 1-82.

This Week's Citation Classic

Doherty P C, Blanden R V & Zinkernagel R M. Specificity of virus-immune effector T cells for H-2K or H-2D compatible interactions: implications for H-antigen diversity. *Transplant. Rev.* **29**:89-124, 1976. [Wistar Inst. Anat. Biol., Philadelphia, PA; Dept. Microbiol., John Curtin Sch. Med. Res., Canberra, Australia; Dept. Exp. Pathol., Scripps Clinic and Res. Found., La Jolla, CA]

We summarised here the concepts resulting from our discovery two years earlier that the so-called strong transplantation antigens function as recognition sites for self-monitoring cytotoxic T lymphocytes. This drastically changed thinking about both the nature of histocompatibility and immunological surveillance. [The *SCI®* indicates that this paper has been cited in over 650 publications since 1976.]

P.C. Doherty
Department of Experimental Pathology
John Curtin School of Medical Research
Canberra, ACT 2601
Australia

February 7, 1983

"The basic outline of this solicited review was written on a delayed British Airways flight to London, en route to spending six years at the Wistar Institute, Philadelphia. It presented the first comprehensive account of ideas and data generated during two years of intensive experimentation, starting with the discovery of major histocompatibility complex (MHC) restricted virus-immune cytotoxic T cell function by Rolf Zinkernagel and me in Canberra in October 1973.[1] Similar findings were made at about the same time by Gene Shearer at the National Institutes of Health, Bethesda, for the lymphocyte response to trinitrophenyl (TNP)-modified cells.[2] However, apart from the work in Shearer's laboratory, the isolation of Australia undoubtedly contributed to our having a 12-month-lead before the realization that a biological *raison d'être* had at last been found for the so-called strong transplantation antigens registered with the major groups in the Northern Hemisphere. The topic was dominating much of the debate in cellular immunology by the time that this article was published and it was immediately a key reference in the T cell field.

"The then current paradigm in the US was that the so-called immune response genes, which mapped to the I region of MHC, encoded part or all of the T cell receptor. We initially thought that we were studying something rather similar. However, we also proposed an alternative hypothesis that the virus-immune T cells might be recognising either some complex of virus and histocompatibility antigen, or a virus-induced alteration of MHC molecules themselves. This 'altered-self' model, which quite unknown to us reflected an earlier proposal made by Sherwood Lawrence[3] to explain the binding of transfer factor, came easily enough to mind when thinking about virus infections and focused attention onto the target/stimulator cells. A completely novel set of arguments could thus be made about the nature of alloreactivity, differential responsiveness, and MHC gene polymorphism. However, we gained the impression that our ideas were considered heretical by the established immunological community, who were then rolling on a different bandwagon.

"At the stage that this review was written, we found ourselves almost totally unable to generate any support at all for the idea that MHC genes were coding directly for the T cell receptor. Key experiments that were described here showed quite clearly that different sets of virus-immune T cells were associated with H-2K and H-2D, and that mere expression of a particular MHC product on the immune lymphocyte did not allow for recognition of virus-infected target cells. Also, the fact that mutations in relatively small pieces of DNA coding for the structural MHC gene product completely modified the spectrum of T cell recognition could not readily be accommodated with the earlier T cell receptor model for MHC. However, I still felt that we needed to be rather circumspect in the writing of this article, as there seemed no particular need to make powerful enemies. We had already argued a much more extreme case for the 'altered-self' model in an article written earlier for the 'Hypothesis' format of *Lancet.*[4] What we did not realise was that our ideas had made an impact: some of the leading proponents of the MHC-T cell receptor idea were already changing their ground and there would be evidence of a new emphasis at the Cold Spring Harbor meeting held early in 1976. Zinkernagel and I shared the Paul Ehrlich Prize for Medicine in 1983."

1. **Zinkernagel R M & Doherty P C.** Restriction of *in vitro* T cell-mediated cytotoxicity in lymphocytic choriomeningitis within a syngeneic or semiallogeneic system. *Nature* **248**:701-2, 1974.
2. **Shearer G M.** Cell-mediated cytotoxicity to trinitrophenyl-modified syngeneic lymphocytes. *Eur. J. Immunol.* **4**:527-33, 1974.
3. **Lawrence H S.** Homograft sensitivity: an expression of immunologic origins and consequences of individuality. *Physiol. Rev.* **39**:811-59, 1959.
4. **Doherty P C & Zinkernagel R M.** A biological role for the major histocompatibility antigens. *Lancet* **1**:1406-9, 1975.

Bloom B R & Bennett B. Mechanism of a reaction *in vitro* associated with delayed type hypersensitivity. *Science* **153**: 80-2, 1966.

The cell type responsible for inhibition by antigen migration *in vitro* of peritoneal exudate cells obtained from tuberculin-hypersensitive guinea pigs was studied. Exudate populations were separated into component cell types, the lymphocyte and the macrophage. Peritoneal lymphocytes from sensitive donors were the immunologically active cells in this system, the macrophages being merely indicator cells which migrate. Sensitized peritoneal lymphocyte populations upon interaction with specific antigen *in vitro*, elaborated into the medium a soluble material capable of inhibiting migration of normal exudate cells. [The *SCI®* indicates that this paper was cited 552 times in the period 1966-1977.]

Barry R. Bloom
Departments of Microbiology and
Immunology, and Cell Biology
Albert Einstein College of Medicine
Bronx, New York 10461

March 6, 1978

"It is particularly gratifying to learn that this paper has been frequently cited, because this represents the first independent research work done by my colleague, Boyce Bennett, and myself. We had each finished postdoctoral fellowships and arrived at about the same time at Albert Einstein College of Medicine. A good deal had already been learned about the mechanisms of delayed type hypersensitivity by the study of the transfer of immune cells into normal animals developed by my thesis advisor, Merrill Chase, but the problem was that once the immune cells were transferred into the recipient, it was impossible *in vivo* to analyze the biochemical mechanisms by which they mediated the passive transfer of reactivity.

"We entered into a collaboration to attempt to unravel the mechanism of delayed type hypersensitivity reaction solely by use of *in vitro* models. The principal one available at that time was the macrophage migration inhibition test originally discovered by Rich and Lewis in the 1930's, later refined by George and Vaughan and

shown to correlate with delayed type hypersensitivity reaction *in vivo* by John David.

"The question that we initially tried to approach, (and it seems inconceivable now that an answer was not known at the time) was whether both lymphocytes and macrophages possessed immunological specificity and the capability of recognizing and reacting to specific antigen. We found that purified macrophages from sensitized animals failed to react in the migration inhibition test, while the immune lymphocytes had the ability to recognize antigen. The extraordinary aspect of the results was that as few as 0.6% of lymphocytes from sensitized animals were capable of inhibiting the migration, in the presence of specific antigen, of 99.4% macrophages obtained from normal donors. This made it very unlikely that the *in vitro* reaction was mediated by a direct cell-to-cell contact, and suggested that the sensitized lymphocytes might be affecting the behavior of macrophages by secreting a soluble molecule.

"In the initial experiment in which we tested for such a soluble molecule, I recall the results were strikingly negative, although there was a peculiar wrinkling of the area of migrated macrophages. While it was tempting to dismiss this as trivial, we made one further effort to produce the *in vitro* reaction, using more concentrated supernatants of antigen-stimulated lymphocytes. The results were striking in that the supernatant of antigen-stimulated sensitive lymphocytes was able to transfer to perfectly normal peritoneal exudate cells the *in vitro* reaction hitherto seen only with sensitized cells. These results suggested for the first time that lymphocytes could communicate with other cell types by soluble non-antibody mediators. The first of these was described in this paper and termed migration inhibitory factor (MIF). Subsequently, a great number of other laboratories have both confirmed these results and added to the list over 52 such putative mediators or 'lymphokines.'

"It is amusing to reflect that in this work, we had between us grant funds of only $10,000, and did all of the technical work as well as all of the glassware ourselves. And yet we felt extraordinarily fortunate to have the luxury of being able to study *in vitro* a phenomenon which had almost exclusively been regarded as one which could be studied only in the intact animal, which would have been a much more expensive proposition."

This Week's Citation Classic

Bendixen G & Søborg M. A leucocyte migration technique for in vitro detection of cellular (delayed type) hypersensitivity in man. *Dan. Med. Bull.* **16**:1-6, 1969.
[Medical Depts. P and A, Rigshospitalet, Univ. Copenhagen, Denmark]

A capillary tube migration technique for investigation of specific antigen-induced migration inhibition of human peripheral blood cells is described. The method, which is a modification of similar preexisting techniques used in animal experiments, can detect lymphocyte-mediated immunity (delayed hypersensitivity) associated with, e.g., immunity of infection and autoimmunity. [The *SCI®* indicates that this paper has been cited over 270 times since 1969.]

Gunnar Bendixen
Medical Department TA
and
Laboratory of Medical Immunology
Rigshospitalet
University of Copenhagen
DK 2200 Copenhagen N
Denmark

October 20, 1981

"This paper describes the technique of antigen-induced peripheral blood leucocyte migration inhibition in man as it was performed when it was developed and fully established as scientific routine in our laboratory. The first publication on the subject appeared in 1967.[1]

"At the beginning of the 1960s, immunobiologists hypothesized that cell-mediated immune reactions as known from infectious disease and contact hypersensitivity were probably of pathogenetic importance also in tissue lesions in autoimmune, transplantation syndromes, and tumour diseases. Their assumptions were supported partly by logical deduction and partly by extrapolation from animal experimental evidence. So far, however, an appropriate method for such studies in clinical research has not been developed.

"The idea to develop this *in vitro* technique for detection of cell-mediated immunity in man was inspired by the studies by John David and his group[2] on tuberculin-induced, lymphocyte dependent migration inhibition of guinea pig peritoneal macrophages. Peripheral blood lymphocytes might be assumed representative of the cell-mediated immunity of any organism at any time, and antigen-lymphocyte interaction might cause inhibition of monocytes and/or granulocytes from peripheral blood in a way similar to the migration inhibition of guinea pig peritoneal macrophages. The capillary tube migration technique originally described by George and Vaughan[3] was therefore modified for use with peripheral human blood cells. At this time, my good friend Mogens Søborg obtained a position as research fellow and joined me in the laboratory.

"Results with tuberculin hypersensitivity were disappointing in these first experiments, but the *Brucella* model showed very clear results indeed. There was no doubt that this type of antimicrobial, cell-mediated immunity was associated with an antigen-specific reactivity of peripheral blood cells, demonstrable as antigen-induced migration inhibition *in vitro*. And the reactivity was closely related to the delayed type intracutaneous reaction to the same antigen. These results initially attracted only limited interest, since it was generally assumed that the system could only work with guinea pig macrophages. Søborg continued to explore the mechanism of the reaction, using *Brucella* hypersensitivity as a model, whereas, in my own studies, the technique was used to detect pathogenetic mechanisms in clinical immune disorders.

"The present paper was published at a time when significant results had shown that organ specific, cell-mediated immunity could be demonstrated with this technique in ulcerative colitis and glomerulonephritis. Later on, several other expectations were fulfilled: organ specific, cell-mediated immunity was demonstrated in other autoimmune conditions, and the method could detect cell-mediated immunity to tumours, transplants, and infection. Furthermore, this initially very crude technique could be refined and thereby contribute considerably to developments in the field of human lymphokine (lymphocyte hormone) research.[4,5] If the paper has been often cited, the reason may be both its clinical applicability within a field which at the time greatly needed an assay of this type and also its early initiation of improved techniques in basic lymphocyte research."

1. Søborg M & Bendixen G. Human lymphocyte migration as a parameter of hypersensitivity.
   Acta Med. Scand. **181**:247-56, 1967.
2. David J R, Al-Askari S, Lawrence H S & Thomas L. Delayed hypersensitivity in vitro. I. The specificity
   of inhibition of cell migration by antigens. *J. Immunology* **93**:264-73, 1964.
3. George M & Vaughan J H. In vitro cell migration as a model for delayed hypersensitivity.
   Proc. Soc. Exp. Biol. Med. **111**:514-21, 1962.
4. Bendixen G, Bendtzen K, Clausen J E, Kjaer M & Søborg M. Human leucocyte migration inhibition.
   Scand. J. Immunol. **5**(Suppl. 5):175-84, 1976.
5. Rocklin R E, Bendtzen K & Greineder D. Mediators of immunity: lymphokines and monokines.
   Advan. Immunol. **29**:56-136, 1980.

Strander H & Cantell K. Production of interferon by human leukocytes *in vitro*.
Ann. Med. Exp. Fenn. 44:265-73, 1966.
[State Serum Institute, Helsinki, Finland]

Leukocytes isolated from human blood were used for the production of large quantities of human interferon. Cells were incubated with inducing virus and the best inducers of interferon were established. The isolation procedure used for the leukocytes did not diminish their capacity to induce interferon and they could be stored prior to their use for the production of interferon. The procedure described allowed large-scale production of human interferon. [The *SCI*® indicates that this paper has been cited in over 165 publications, making it the most-cited paper ever published in this journal.]

Hans Strander
Department of Oncology
Karolinska Hospital
S-104 01 Stockholm 60
Sweden

April 4, 1984

"In 1963, I was working part-time at the Institute for Tumor Biology, headed by Georg Klein, at the Karolinska Institute. We were visited by Vainio, who was originally from Helsinki where he was a collaborator of Cantell. There was an interest in Helsinki in doing studies on both the production and action of interferon. Cantell had just arrived from the Henles' laboratory in Philadelphia. I had been following their research so I decided to join the Cantell/Vainio group in Helsinki.

"Interferon production was done in Cantell's laboratory. Cantell had already done a few pilot experiments where it was seen that human leukocytes could produce large amounts of interferon, as had previously been reported by Gresser.[1] I mostly studied the action of the interferon produced. Unfortunately, Vainio then died in a car accident and, after that, I devoted my time fully to the production of interferon in Cantell's laboratory during my remaining five-year stay in Helsinki.

"The work done there on the production of interferon by human leukocytes was also the basis for my thesis, which was later defended at the Karolinska Institute.[2] It was shown in the article discussed here that the large-scale production of human leukocyte interferon could be achieved by using buffy coats isolated from human blood bags. The cell products employed were waste products that did not diminish amounts of blood available for treating patients either with red cell concentrates or fresh plasma. It was also shown in the paper that the human leukocytes could be purified and still were able to produce optimal amounts of interferon. It was shown how the kinetics worked, and what type of viruses should be used as inducers. Also, the multiplicity of the viruses which had to be used was determined. The state of the virus preparations and the number of cells which had to be incubated to give optimal yields were also factors studied.

"It had been established that interferon had an effect on virus diseases and tumor diseases in experimental models.[3,4] Unfortunately, there was no human interferon available for clinical trials. The article in question showed that it is possible to obtain large quantities of human interferon for clinical work.

"The reason this article has been cited so much is probably that it started an era of interferon production in Helsinki, where the laboratory for ten years provided all the human leukocyte interferon which was available for clinical trials and also most of the human alpha interferon available for laboratory work in the world. It was also shown through collaboration between a virus laboratory and a blood center that it was possible to produce interferon on a large-scale basis. In many ways, the article encouraged the future production of large quantities of interferon for work on humans. At the time the article was published, however, the interest was not as strong as it became later, since it was not believed by many people that exogenous interferon therapy in humans would ever be a realistic possibility. Due to the findings in animals, we considered, however, that such work would be important. Also, such studies could provide some clues concerning the human defense mechanisms directed against viruses and tumors.

"For information concerning how far the interferon area has moved using such a product, the reader is referred to a special article."[5]

1. Gresser I. Production of interferon by suspensions of human leukocytes.
   Proc. Soc. Exp. Biol. Med. 108:799-803, 1961. (Cited 160 times.)
2. Strander H. *Production of interferon by suspended human leukocytes.* Thesis.
   Stockholm, Sweden: Karolinska Institute, 1971.
3. Finter N B. Interferon as an antiviral agent *in vivo*: qualitative and temporal aspects of the protection of mice against
   Semliki Forest virus. *Brit. J. Exp. Pathol.* 47:361-71, 1966. (Cited 50 times.)
4. Atanasiu P & Chany C. Action d'un interferon provenant de cellules malignes sur l'infection expérimentale du
   Hamster noveauné par le virus du polyome. *C.R. Acad. Sci.* 251:1687-9, 1960. (Cited 65 times.)
5. Strander H. Interferons and disease: a survey. (Burke D C & Morris A, eds.) *Interferons: from molecular biology to
   clinical application. Thirty-Fifth Symposium of the Society for General Microbiology held at the University of
   Cambridge. September 1983.* Cambridge, England: Cambridge University Press, 1983. p. 7-33.

This Week's Citation Classic

Waldmann T A, Broder S, Blaese R M, Durm M, Blackman M & Strober W.
Role of suppressor T cells in pathogenesis of common variable
hypogammaglobulinaemia. *Lancet* 2:609-13, 1974.
[Metabolism Branch, Natl. Cancer Inst., Natl. Insts. Health, Bethesda, MD]

In some cases, common variable hypogammaglobulinemia was shown to be caused by abnormalities of regulatory T cells which suppressed B cell maturation and antibody synthesis. This was the first demonstration of suppressor T cells in humans and the first description of a disease caused by suppressor T cell abnormalities. [The *SCI®* indicates that this paper has been cited in over 665 publications since 1974.]

Thomas A. Waldmann
Metabolism Branch
National Cancer Institute
National Institutes of Health
Bethesda, MD 20205

January 11, 1983

"The project on common variable immunodeficiency disease was part of a program of studies performed at the Metabolism Branch, National Cancer Institute (NCI), on immunoregulatory disorders in patients with immunodeficiency diseases. The intramural National Institutes of Health (NIH) environment was especially favourable for the study of these rare yet extremely instructive diseases. Common variable immunodeficiency is a heterogeneous group of diseases characterized by hypogammaglobulinemia due to different causes. Prior to the studies in this report it was assumed that these diseases were caused by intrinsic defects of the B lymphocytes and plasma cells. I was stimulated to consider an alternative pathogenetic mechanism by the seminal discovery made by Richard Gershon[1] of a new type of lymphocyte in mice, the suppressor T cell, that acts as a negative regulator of many immunological responses. We considered the possibility that certain patients with hypogammaglobulinemia might have excessive suppressor T cell activity as the cause of their hypogammaglobulinemia. To examine this hypothesis we developed an *in vitro* technique to study the terminal maturation of human B lymphocytes and described a new co-culture procedure to study suppressor T cell function. When we applied these techniques to the study of the pathogenesis of common variable immunodeficiency, we demonstrated that some patients had normal B cells but had an excessive number of activated suppressor T cells that inhibited B cell maturation and antibody synthesis. We suggested that in this subset of patients the hypogammaglobulinemia might be caused by these suppressor T cells. The basic observations were rapidly confirmed, but many questions were raised concerning their biological significance. These questions have largely been answered. However, the most critical issue, that is, whether the activation of suppressor cells is a primary pathogenetic mechanism causing the hypogammaglobulinemia or is a secondary event, will only be answered definitively when therapeutic techniques are developed that enable one to eliminate the suppressor T cells *in vivo* without affecting human B cell function.

"In the years since the publication of this study, we and others have demonstrated excessive suppressor T cell activity in association with an array of diseases including thymoma and hypogammaglobulinemia, selective IgA deficiency, and certain neoplasms including leukemias of suppressor T cells.[2]

"I believe that this paper is a *Citation Classic* for two reasons. It described the first method for the study of suppressor T cells in humans and was the first demonstration that a disorder of suppressor T cell activity can cause a human disease. Honors received as a result of these studies include the Stratton Medal from the American Society of Hematology; the Michael Heidelberger Lecture, Columbia University; and the G. Burroughs Mider Lectureship, NIH. Fredrickson, then director of NIH, introduced the Mider Lecture by stating, 'Dr. Waldmann's landmark discovery of active suppression of antibody synthesis by human suppressor T cells has revolutionized thinking about the pathogenesis of immunodeficiency disease and has generated a whole new field of research demonstrating the delicate balance of cell interactions in the homeostatic immune network.'"

1. **Gershon R K.** T cell control of antibody production. (Cooper M D & Warner N L, eds.) *Contemporary topics in immunobiology.* New York: Plenum Press, 1974. Vol. 3. p. 1-40.
2. **Waldmann T A & Broder S.** Suppressor cells in the regulation of the immune response. *Prog. Clin. Immunol.* 3:155-99, 1977.

Borel J F, Feurer C, Gubler H U & Stähelin H. Biological effects of cyclosporin A:
a new antilymphocytic agent. *Agents and Actions* 6:468-75, 1976.
[Biological and Medical Res. Div., Sandoz Ltd., Basel, Switzerland]

The fungus metabolite cyclosporin A is a small peptide which acts as a novel antilymphocytic agent. It strongly depressed antibody formation in mice. Skin graft rejection and graft-versus-host reaction in rodents were much delayed. It also prevented the development of experimental allergic encephalomyelitis and Freund's adjuvant arthritis. Cyclosporin A is not myelotoxic, and affects an early stage of mitogenic triggering of the immunocompetent lymphocytes. [The *SCI*® indicates that this paper has been cited in over 235 publications since 1976, making it the 3rd most-cited paper published in this journal.]

Jean F. Borel
Preclinical Research
Pharmaceutical Division
Sandoz Ltd.
CH-4002 Basel
Switzerland

September 27, 1983

"I had joined the Cellular Biology Division of Sandoz Ltd. in 1970 and was working mainly on immunosuppression and inflammation. Hartmann Stähelin, my then chief, and I used a complex mouse model for screening both cytostatic and immunosuppressive compounds. In collaboration with Zoltan L. Kis and his colleagues from the microbiology department, we had learned from experience that compounds of microbial origin often showed cytostatic or other pharmacological activities greater than the antimicrobial activity for which they had been selected. Of special interest for our pharmacological screen were purified or semipurified extracts with little antibiotic activity and low toxicity.

"It happened to be me who discovered, in January 1972, the marked immunosuppressive effect of the partially purified two-component mixture which contained cyclosporin A as an active principle. To our surprise, however, the compound had no effect on the survival time of leukaemic mice, indicating that immunosuppression was not linked to general cytostatic activity!

"In the following years, the compound was investigated in a battery of additional *in vitro* and *in vivo* experimental models. Meanwhile, my colleagues in microbiology were concentrating on purification and characterisation of the molecule, and on substantial improvement of the yield, an exceedingly difficult task. On the pharmacological side, Hans Ulrich Gubler was showing that cyclosporin A could both prevent and cure the chronic inflammatory process of Freund's adjuvant arthritis, but was ineffective in the models of acute inflammation. Camille Feurer was performing a number of crucial experiments demonstrating the almost complete lack of effect on haemopoiesis. Together with her and Dorothee Wiesinger, we were already starting to investigate the mechanism of action. All this work was published around the same time.[1-3]

"The reproducible immunosuppressive effects, and the remarkable lack of side effects as compared with reference drugs in clinical use, clearly supported the concept of a selective action on lymphoid cells and, together with the novel chemical structure, induced us to consider cyclosporin A as the prototype of a new generation of immunosuppressants.

"Because the paper was the first to report several of these results, it is not surprising that it is often cited, especially in view of the great impact that cyclosporin A is having today as a drug in the field of organ transplantation and as a probe in basic immunology. More extensive reviews have been published in references 4 and 5."

1. Borel J F. Comparative study of in vitro and in vivo drug effects on cell-mediated cytotoxicity.
   Immunology 31:631-41, 1976.
2. Borel J F, Feurer C, Magnée C & Stähelin H. Effects of the new antilymphocytic peptide cyclosporin A in animals.
   Immunology 32:1017-25, 1977.
3. Borel J F & Wiesinger D. Effect of cyclosporin A on murine lymphoid cells. (Lucas D O, ed.) *Regulatory mechanisms in lymphocyte activation.* New York: Academic Press, 1977. p. 716-18.
4. Borel J F. The history of cyclosporin A and its significance. (White D J G, ed.) *Cyclosporin A.* Amsterdam: Elsevier Biomedical Press, 1982. p. 5-17.
5. Morris P J. Cyclosporin A. (Overview.) *Transplantation* 32:349-54, 1981.

This Week's Citation Classic

Hirschhorn K, Bach F, Kolodny R L, Firschein I L & Hashem N. Immune response
and mitosis of human peripheral blood lymphocytes in vitro.
Science 142:1185-7, 1963.
[Department of Medicine, New York University School of Medicine, NY]

This paper described the initial experiments demonstrating the usefulness of peripheral blood lymphocyte cultures for immunologic studies. It demonstrated lymphocyte responses on exposure to antigens for which the donor was sensitized, as well as the mixed lymphocyte response found with co-cultivation of cells from unrelated individuals. [The *SCI®* indicates that this paper has been cited in over 555 publications since 1963.]

Kurt Hirschhorn
Department of Pediatrics
Mount Sinai School of Medicine
New York, NY 10029

May 20, 1983

"In 1959, we and others began using phytohemagglutinin (PHA) for the purpose of obtaining chromosome preparations from peripheral blood cultures. We noted, in addition to dividing cells, that many of the PHA stimulated cells resembled the graft versus host cells described by Gowans[1] in their morphology and staining properties, and we also showed that these cells were lymphocytes. It seemed possible to us that the then current dogma that peripheral blood lymphocytes are end cells might be wrong. We therefore began a series of experiments using cells from individuals who were tuberculin positive, had been vaccinated against diphtheria and pertussis, or had been shown to be sensitive to penicillin. Culture of such cells in the presence of the appropriate antigen resulted in the appearance of the same enlarged cells and mitoses seen with PHA, although in smaller numbers, while no such response was observed in cells from nonsensitized individuals. We interpreted these findings to indicate that peripheral blood lymphocytes were in fact not end cells but demonstrated immunologic memory.

"In a study of the effect of fibroblasts and their extracts on the lymphocytes of patients with eczema, we had noted that there was a small and variable amount of lymphocyte stimulation upon exposure to fibroblasts from unrelated individuals. We therefore postulated that peripheral blood lymphocytes may demonstrate a response to histocompatibility differences and proceeded to co-cultivate peripheral blood lymphocytes from unrelated individuals. We again found enlarged cells and mitoses and concluded that such mixed lymphocyte cultures represented an *in vitro* model of graft rejection.

"Thousands of papers have appeared since that time which have confirmed, utilized, and expanded upon these two fundamental principles. (See, for example, reference 2.) Such peripheral blood lymphocyte cultures have become standard techniques for the study of many aspects of normal immunology, as well as various immunologic abnormalities, in man as well as many other species. The mixed lymphocyte response has been shown to be the best correlate of donor-recipient compatibility in relation to organ transplantation, and has been found to be determined by an allelic series of surface antigens of the D and DR specificity.

"There was great excitement in our laboratory in those early days of the establishment of the field of cellular immunology. Of the many people working in my laboratory in the early-1960s, for short or long periods, several were coauthors on this paper. Fritz Bach, a postdoctoral fellow at the time, has gone on to carry the mixed lymphocyte response to a fine art and has become a leading immunobiologist, currently at the University of Minnesota. Roselyn Kolodny, a medical student on an elective is a pediatrician in Boston. Lester Firschein, also a postdoctoral fellow, is a geneticist and anthropologist at the City University of New York, and Nemat Hashem, a visiting scientist from Egypt, is now a leading human geneticist in that country and has continued to work on lymphocyte cultures.

"I am personally convinced that the work was done and that it succeeded because, as geneticists, we naively pursued an observation in another discipline, immunology, without the full realization that our results would question established dogma. In the years since then, I have consistently encouraged students and fellows not to fear a fresh viewpoint and to use their techniques in other fields. I believe that this paper has been frequently cited because the various preliminary findings reported attracted many proper immunologists to use a simple technique of cell culture for the study of numerous immunologic phenomena."

1. Gowans J L. The fate of parental strain small lymphocytes in F$_1$ hybrid rats. *Ann. NY Acad. Sci.* 99:432-55, 1962.
2. Hume D A & Weidemann M J. *Mitogenic lymphocyte transformation.*
Amsterdam: Elsevier/North-Holland Biochemical Press, 1980. 251 p.

Sell S & Gell P G H. **Studies on rabbit lymphocytes** *in vitro*. **I. Stimulation of blast transformation with an antiallotype serum.** *J. Exp. Med.* **122**:423-40, 1965.
[Dept. Experimental Pathology, Univ. Birmingham Med. Sch., England]

This paper describes the ability of antibodies to rabbit immunoglobulin allotypes to stimulate small resting peripheral blood lymphocytes to undergo 'blast' transformation, synthesize DNA, and divide. It is the first demonstration that some lymphocytes have surface immunoglobulin (sIg) and that this sIg can serve as a receptor for activation of the cell. [The *SCI®* indicates that this paper has been cited in over 435 publications since 1965.]

Stewart Sell
Department of Pathology and
Laboratory Medicine
University of Texas
Health Science Center
Houston, TX 77225

March 5, 1984

"In 1964, when I went to work with Philip Gell in Birmingham, England, studies in several laboratories that would lead to two decades of intensive exploration of cellular immunology were just beginning. Immunoglobulin antibody structure was being decoded, but the key question as to how lymphocytes recognized antigen was completely open. No one had been able to come up with a better idea than that proposed by Ehrlich's side chain theory at the turn of the century.[1] In 1961, Möller[2] described immunoglobulin on the surface of lymphocytes by immunofluorescence, but the technique used was not reproducible and it would only be later that vital staining procedures would give consistent results. Nowell[3,4] demonstrated that an extract of kidney beans, termed 'phytohemagglutinin,' stimulated peripheral blood lymphocytes to divide *in vitro*. George and Vaughan[5] described the phenomenon of inhibition of macrophage *in vitro* when antigen was added to cells from immunized animals and Pearmain et al.[6] reported that tuberculin could induce mitosis in peripheral

blood cells from humans with positive tuberculin skin tests. Dutton and Eady[7] were in the process of showing that cells from immunized rabbits could respond to antigen exposure *in vitro* by proliferation, and Ada and Nossal[8] were carrying out experiments to show that some lymphocytes from immunized animals would bind radiolabeled antigen.

"I had proposed to Gell that we try to develop better assays for studying delayed hypersensitivity *in vitro* based on lymphocyte activation. Before we were able to work out antigen specific stimulation, the phenomenon of anti-immunoglobulin activation of lymphocytes was observed and led our studies in a different direction. One day, I added an antiallotype serum to a culture of lymphocytes from a rabbit bearing that allotype. Two days later, a large number of cells had 'transformed' into blasts.

"During the next 20 years (for a review see reference 9), this system was used to describe a number of other original observations. A partial list includes: 1) The process of lymphocyte activation is reversible. Enlarged 'blast' cells change back to small cells when the stimulus for activation is removed. 2) Cross-linking is not absolutely required for activation, but cross-linking can enhance the activation process. 3) Single lymphocytes may bear more than one surface immunoglobulin (sIg) class at a time. 4) Single immature lymphocytes may express more than one light chain or VH region allelic marker (allelic inclusion). 5) Activation of lymphocytes by anti-immunoglobulin or lectin is associated with endocytosis of reactive surface markers. 6) Activation of proliferation of sIg positive cells is not necessarily followed by differentiation into immunoglobulin synthesizing cells. 7) There is a population of normal cells with an apparent overlap of T- and B-cell properties. Many of these observations have recently been verified with human and mouse lymphocytes but some remain controversial."

1. Ehrlich P. Croonian lecture. On immunity with special reference to cell life.
   Proc. Roy. Soc. London Ser. B **66**:424-48, 1900. (Cited 135 times since 1955.)
2. Möller G. Demonstration of mouse isoantigens at the cellular level by the fluorescent antibody technique.
   J. Exp. Med. **114**:415-34, 1961. (Cited 530 times.)
3. Nowell P C. Phytohemagglutinin: an initiator of mitosis in cultures of normal human leukocytes.
   Cancer Res. **20**:462-6, 1960.
4. ──────. Citation Classic. Commentary on *Cancer Res.* **20**:462-6, 1960. *Current Contents* (42):13, 17 October 1977.
5. George M & Vaughan J H. *In vitro* cell migration as a model for delayed hypersensitivity.
   Proc. Soc. Exp. Biol. Med. **111**:514-21, 1962. (Cited 645 times.)
6. Pearmain G, Lycette R R & Fitzgerald P H. Tuberculin-induced mitosis in peripheral blood leucocytes.
   Lancet **1**:637-8, 1963. (Cited 440 times.)
7. Dutton R W & Eady J D. An *in vitro* system for the study of the mechanism of antigenic stimulation in the secondary response. *Immunology* **7**:40-64, 1964. (Cited 205 times.)
8. Ada G L, Nossal G J V & Austin C M. Antigens in immunity. V. The ability of cells in lymphoid follicles to recognize foreignness. *Aust. J. Exp. Biol. Med. Sci.* **42**:331-46, 1964. (Cited 65 times.)
9. Sell S, Skaletsky E, Holdbrook R, Linthicum D S & Raffel C. Alternative hypotheses of lymphocyte surface immunoglobulin expression, B lymphocyte activation and B lymphocyte differentiation.
   Immunol. Rev. **52**:141-79, 1980.

Pernis B, Forni L & Amante L. Immunoglobulin spots on the surface of rabbit lymphocytes. *J. Exp. Med.* **132**:1001-18, 1970.
[Lab. Immunology, Clinica del Lavoro, Univ. Milan, and Clinica del Lavoro, Univ. Genoa, Italy]

Immunofluorescence shows immunoglobulins on the membrane of bone marrow and not thymus lymphocytes. About one half of lymphoid cells in blood and spleen have membrane immunoglobulins. The molecules show allelic exclusion, are oriented with the Fab toward the outside, and can bind antigens. [The *SCI®* indicates that this paper has been cited in over 645 publications since 1970.]

Benvenuto Pernis
Department of Microbiology
College of Physicians & Surgeons
Columbia University
New York, NY 10032

May 2, 1983

"The visual detection of immunoglobulins on the membrane of lymphocytes was an exciting finding since it provided direct support for the concept that specific receptors for antigens exist on the membranes of immunocompetent lymphoid cells.[1] The finding was published together with a number of facts that considerably clarified its biological relevance; these were: a) the allelic exclusion of the membrane immunoglobulins that gave the demonstration that these were actively synthesized by the cells that had them and could not be just passively absorbed; b) the absence of membrane immunoglobulins on thymus-derived cells and their presence in bone marrow lymphocytes; c) the orientation of the molecules with the Fab toward the outside and their capacity to bind antigen; and d) the prevalence of cells bearing membrane IgM over those with membrane IgG, an observation that strongly suggested the capacity of the cells to switch from IgM to IgG as subsequently demonstrated directly by immunofluorescence.[2]

"All this information, collected in one paper, provided the stimulus for much further research on the biological significance and medical relevance of the lymphoid cells that carry membrane immunoglobulins. This explains why this work has been cited so often. The work was performed in the laboratory of immunology of the Clinica del Lavoro of the University of Milan Medical School, Italy. This was a somewhat unusual setting since the Clinica del Lavoro is essentially a hospital dedicated to occupational diseases, and credit must be given to the then director of the Clinica del Lavoro, E.C. Vigliani, who supported basic research for many years, not only with his understanding, but also by devoting to it considerable financial resources. This created the appropriate environment for biological research, including immunology. In this environment, research was a joy and fun and fostered the enthusiasm of persons like L. Forni, who was actually responsible for most of the experiments described in our paper, and L. Amante, unfortunately prematurely deceased, who dealt masterfully with the necessary immunochemistry.

"We also had foreign guests like Martin Raff, who had seen immunoglobulin caps on mouse lymphocytes.[3] With Raff, we had a bet on whether the membrane immunoglobulins were distributed in multiple spots or in polar caps. So Raff came to Milan to see for himself and, of course, he saw spots. The reason was that in Milan, we did all our work in the presence of sodium azide in order to be sure that the immunoglobulin spots were not due to pinocytosis or secretion. Actually, azide inhibits capping, as was subsequently demonstrated.[4,5]

"In conclusion, the discovery of membrane immunoglobulins was a discovery for which the time was ripe, both for the existing technology and the conceptual background. It was one step in the continuously unfolding fascination of immunology."

1. **Burnet F M.** *The clonal selection theory of acquired immunity.* Cambridge: Cambridge University Press, 1959. 208 p.
2. **Pernis B, Forni L & Amante L.** Immunoglobulins as cell receptors. *Ann. NY Acad. Sci.* **190**:620-31, 1971.
3. **Raff M C, Sternberg M & Taylor R B.** Immunoglobulin determinants on the surface of mouse lymphoid cells. *Nature* **225**:553-4, 1970. [The *SCI* indicates that this paper has been cited in over 460 publications since 1970.]
4. **Taylor R B, Duffus P H, Raff M C & DePetris S.** Redistribution and pinocytosis of lymphocyte surface immunoglobulin molecules induced by anti-immunoglobulin antibody. *Nature New Biol.* **233**:225-9, 1971. [The *SCI* indicates that this paper has been cited in over 1,155 publications since 1971.]
5. **Loor F, Forni L & Pernis B.** The dynamic state of the lymphocyte membrane factors affecting the distribution and turnover of surface immunoglobulins. *Eur J. Immunol.* **2**:203-12, 1972. [The *SCI* indicates that this paper has been cited in over 455 publications since 1972.]

Basten A, Miller J F A P, Sprent J & Pye J. A receptor for antibody on
B lymphocytes. I. Method of detection and functional significance.
J. Exp. Med. **135**:610-26, 1972.
[Walter and Eliza Hall Inst. Med. Res., Melbourne, Victoria, Australia]

This paper contains the first description of the Fc receptor as a marker for B lymphocytes. The original method, or modifications of it, now forms the basis of standard assays for detection of the receptor and has been used to obtain enriched populations of B lymphocytes from mixed cell populations. [The *SCI®* indicates that this paper has been cited in over 550 publications since 1971.]

A. Basten
Clinical Immunology Research Center
University of Sydney
Sydney, NSW 2006
Australia

May 30, 1984

"The experiments described in this paper were carried out when I was a postdoctoral research fellow with Jacques Miller at the Walter and Eliza Hall Institute. At the time, I was sharing a laboratory with Jonathan Sprent, who was, and still is, the doyen of thoracic duct cannulation in the mouse. We were all interested in the phenomenon of immunological memory and, like others, wished to establish whether it was a manifestation of a qualitative or quantitative change in the recirculating lymphocyte pool. The obvious experiment was therefore to compare the proportion of thoracic duct lymphocytes from normal versus primed mice which bound radiolabelled antigen. Using [125]I FyG (fowl immunoglobulin G), autoradiographs were prepared and, to our great surprise, revealed that no less than 15 percent of the primed cells bound antigen, whereas less than one percent of normal cells did so. The immediate conclusion was that immunological memory represented a quantitative increase in the number of antigen binding cells. However, after a little reflection, and bearing in mind the rules of clonal selection, the result, we decided, was impossible. If true, then either clonal selection was wrong or the animals should have expired early in life from a surfeit of

memory cells of various specificities. Not surprisingly, the decision was made to repeat the experiment and on three separate occasions the original finding could not be reproduced.

"As a prelude to discarding the results as an unexplained artifact, each step in the experimental protocol was analysed with care. This revealed the only difference between the first and subsequent experiments to be the number of times the thoracic duct lymphocytes were washed before exposure to ^{125}I labelled antigen. When the cells were spun down and labelled immediately, as was done originally, 15 percent bound antigen. This figure dropped sharply following sequential washes. Since lymph plasma is known to contain antibodies, the possibility arose that a subpopulation of thoracic duct lymphocytes carried cytophilic antibody and therefore labelled spuriously with antigen. The subsequent studies which form the basis of this much cited paper confirmed the above hypothesis and provided conclusive proof for the existence of Fc receptors on B cells but not T cells.

"The major reason the paper is frequently cited is the fact that it was published at a time when the concept of Fc receptors was in its infancy and great interest was being shown in phenotypic differences between B cells and T cells. Subsequently, numerous groups have confirmed the original observations[1] although both we[1] and others[2] have found Fc receptors, albeit at lower density, on T cells as well. The receptor probably plays a key role in modulation of B cell responses since it serves as a trap for immune complexes irrespective of their specificity. In retrospect, it is amusing to note that the original work, as so often happens in research, was not directed toward discovering Fc receptors on B cells but had a completely different goal (i.e., memory). Interestingly, the original goal has yet to be achieved despite the fact that Fc receptors and perhaps even B cells have been relegated to second place behind the T cell and its antigen specific receptor."

1. Basten A, Miller J F A P, Warner N L, Abraham R, Chia E & Gamble J. A subpopulation of T cells bearing Fc receptors. *J. Immunology* **115**:1159-65, 1975. (Cited 90 times.)
2. Dickler H B. Lymphocyte receptors for immunoglobulin. *Advan. Immunol.* **24**:167-214, 1976. (Cited 290 times.)

CC/NUMBER 14
APRIL 2, 1984

Dickler H B & Kunkel H G. Interaction of aggregated γ-globulin with B lymphocytes. *J. Exp. Med.* **136**:191-6, 1972.
[Rockefeller University, New York, NY]

This paper describes a simple method for the detection of Fcγ receptors on human mononuclear cells. The binding of immunoglobulin complexes to Fcγ receptors on B lymphocytes was demonstrated and the requirements of this ligand-receptor interaction were partially characterized. [The *SCI®* indicates that this paper has been cited in over 800 publications since 1972.]

Howard B. Dickler
Immunology Branch
National Cancer Institute
National Institutes of Health
Bethesda, MD 20205

February 7, 1984

"This work was done in collaboration with, and in the laboratory of, Henry G. Kunkel. Kunkel's recent untimely and unexpected death has saddened us all. He will be sorely missed as a friend, mentor, and brilliant scientist.

"I arrived in Kunkel's laboratory at Rockefeller University following the completion of my training in internal medicine at New York Hospital. After the highly structured environment of medical training, and in view of my complete lack of research experience, I was shocked when Kunkel said, in essence, here is a desk and here is a bench and what would you like to do? I managed to reply that I would like to go to the library. During the first few months in the laboratory, the intoxication of freedom and responsibility kept me going when all my ideas and projects failed utterly.

"Kunkel, aware of the demonstration by Basten and colleagues of a receptor for immunoglobulin on murine B lymphocytes,[1] suggested that these cells might utilize Clq (a protein of the complement system) as the receptor. With the help of two other postdoctoral fellows, Vincent Agnello and Frederick Siegal, I set out to evaluate this possibility. The initial experiments showed binding of anti-Clq antibodies to a subpopulation of human peripheral blood mononuclear cells. However, subsequent experiments indicated that this binding was not due to the specificity of the antibodies but due to the fact that, through naiveté, I was using an acidic medium which caused aggregation of the antibodies. I then began using heat aggregated immunoglobulin as a model for studying the binding of complexes to mononuclear cells. These studies resulted in my first publication, which has now been selected as a *Citation Classic*.

"I think a number of factors result in this paper being cited frequently. 1) The method is simple. 2) Fcγ receptors are one of a number of markers used to characterize mononuclear cell subpopulations. 3) The functional role of Fcγ receptors on B lymphocytes and other cell populations is being actively investigated in a number of laboratories.[2,3]

"These initial studies piqued my curiosity as to the functional role of B lymphocyte Fcγ receptors and other surface membrane molecules in regulation of immune responses. I have sought answers to this question ever since."[4-6]

1. **Basten A, Miller J F A P, Sprent J & Pye J.** A receptor for antibody on B lymphocytes. I. Method of detection and functional significance. *J. Exp. Med.* 135:610-26, 1972. (Cited 550 times.)
2. **Kolsch E, Oberbarnscheidt J, Bruner K & Heuer J.** The Fc-receptor: its role in transmission of differentiation signals. *Immunol. Rev.* 49:61-78, 1980.
3. **Phillips N E & Parker D C.** Fc-dependent inhibition of mouse B cell activation by whole anti-μ antibodies. *J. Immunology* 130:602-6, 1983.
4. **Dickler H B.** Lymphocyte receptors for immunoglobulin. *Advan. Immunol.* 24:167-214, 1976. (Cited 270 times.)
5. ――――――――. Interactions between receptors for antigen and receptors for antibody—a review. *Mol. Immunol.* 19:1301-6, 1982.
6. **Shenk R R, Weissberger H Z & Dickler H B.** Anti-idiotype stimulation of antigen-specific antigen-independent antibody responses in vitro. II. Triggering of B lymphocytes by idiotype plus anti-idiotype in the absence of T lymphocytes. *J. Immunology.* In press, 1984.

Lay W H & Nussenzweig V. Receptors for complement on leukocytes.
J. Exp. Med. **128**:991-1009, 1968.
[Department of Pathology, New York University School of Medicine, NY]

Sheep red blood cells (SRBC), sensitized by 7S but not by 19S rabbit anti-SRBC antibodies, adhere to mouse macrophages. When, however, C factors are added to the antigen-19S antibody complex, the SRBC adhere to the membranes of most mouse peritoneal macrophages and blood polymorphonuclear cells ('rosette' formation). Similar rosettes are formed on 15 to 25 percent of lymph node lymphocytes but not on thymus lymphocytes. Rosette formation by AgAb C on macrophages and neutrophils depends on Mg^{++}. Adherence to lymphocytes is independent of divalent ions and occurs in the presence of EDTA. Mouse serum deficient in C_5 can be used as a source of C components. [The *SCI®* indicates that this paper has been cited in over 660 publications since 1968.]

Waltraut H. Lay
Department of Preventive Medicine
Institute of Tropical Medicine
São Paulo University
São Paulo
Brazil

June 13, 1984

"This was a paper that preceded another *Citation Classic* by Celso Bianco, R. Patrick, and Victor Nussenzweig on lymphocytes bearing complement receptors.[1,2] Both emerged under Victor's guidance in Baruj Benacerraf's pathology department at New York University and involved Brazilian fellowship holders: Bianco and me. It is perhaps worthwhile to note how often papers that become *Citation Classics* are the products of beginners under expert supervision.

"In 1966, I began working in P. Miescher's laboratory, where preliminary assays on the phenomenon of phagocytosis had led to the curious observation that human leukocytes fed with complement-coated erythrocytes stubbornly refused to engulf the meal, in spite of the tight adherence of the erythrocytes on their membranes. When I moved over to Victor's laboratory, this was the item chosen to be worked on in my scientific apprenticeship. We changed from human leukocytes to mouse macrophages mainly because mice do not charge $10 for volunteering. A secondary reason was that mouse complement is poorly lytic, and we could work in an entirely homologous system. Once again, the over-classic SRBC-anti-SRBC system was the tool used in our daily attack on the questions of how, when, and why complement-coated erythrocytes stuck to the membrane of phagocytic cells. Victor had been working on the immunochemical aspect of immunology, and this subject marked his 'debut' in cellular immunology. A previous paper by Berken and Benacerraf on cytophilic antibodies[3] and the contemporaneous work of Michel Rabinovitch on phagocytosis of aldehyde-fixed erythrocytes[4] guided our thoughts during our meticulous and slowly progressing work.

"No great discoveries led to the inclusion of our paper among the *Citation Classics*, just the priority of description and characterization of the complement receptors at the right moment, when people were concerned with the recognition functions of cell membranes. Bianco, who later worked out the subject of the complement receptors of lymphocytes, wrote a review of this field[5] and so did Victor.[6] I only enjoy the glory of being cited so often."

1. Bianco C, Patrick R & Nussenzweig V. A population of lymphocytes bearing a membrane receptor for antigen-antibody-complement complexes. I. Separation and characterization. *J. Exp. Med.* **132**:702-20, 1970.
2. Bianco C. Citation Classic. Commentary on *J. Exp. Med.* **132**:702-20, 1970.
   Current Contents/Life Sciences **24**(20):21, 12 May 1981.
3. Berken A & Benacerraf B. Properties of antibodies cytophilic for macrophages. *J. Exp. Med.* **123**:119-44, 1966.
   (Cited 540 times.)
4. Rabinovitch M. Phagocytosis: the engulfment stage. *Semin. Hematol.* **5**:134-55, 1968. (Cited 100 times.)
5. Bianco C. Plasma membrane receptors for complement. (Day N K & Good R A, eds.) *Biological amplification systems in immunology.* New York: Plenum Press, 1977. p. 69-84.
6. Nussenzweig V. Receptors for immune-complexes on lymphocytes. *Advan. Immunol.* **19**:217-58, 1974.
   (Cited 175 times.)

3.2 Immunological Methods and Serology

Clustered here are nine Citation Classics on radioimmunoassay (RIA), six on immunohistology, and nine others related to immunological methods. Since several of these Classics stem from the 1960s, an historical appreciation of the development of the RIA and fluorescent antibody (FAB) techniques, as well as early serological methods, can be acquired by reading the "This Week's Citation Classic" (TWCC) commentaries on them.

The denaturing effect of iodinating proteins by the chloramine T procedure (p. 96) was overcome by Marchalonis through the lactoperoxidase technique, a technique useful for any protein containing tyrosine. The Bolton and Hunter reagent, an iodinated succinimide, was even more versatile, since it could be conjugated to small peptides or proteins lacking tyrosine (p. 97). The commercially available Bolton–Hunter reagent is still in use.

The description by Yalow and Berson in 1960 of anti-insulin globulins in the sera of certain diabetics that opened the entire vista of RIA appeared prior to the availability of these gentle iodinating procedures. The next two Classics on insulin RIA are important for an additional reason other than their popularity. Morgan presents an almost historical review of the background to his publication, and Hales and Randle emphasize the role of commerce and the insulin measuring kit in contributing to their article's popularity. They are also quick to acknowledge Yalow's pioneering work and Berson's contributions to RIA.

When the great sensitivity of RIA for even picogram quantities of antigens became appreciated, hormones from many sources became quantifiable by this method. Classics on RIA determination of the luteinizing hormone (LH), gastrin, and prolactin follow those on insulin.

In 1968–1971, antigen purification methods were crude, and locating one rabbit in twenty-five that would respond exclusively to LH in a mixture of other antigens (p. 101) or of locating a serologically cross-reacting, more abundant antigen than the antigens of central interest, gastrin and prolactin (pp. 102 and 103), made these genuinely fortuitous events. Affinity chromatography of iodinated prolactin made use of another important immunological procedure, which was just being perfected in the late 1960s.

The fluorescent antibody (FAB) method developed by Albert Coons during World War II blossomed immediately thereafter as the first immunohistochemical procedure worthy of that name. Substitution of the isothiocyanate conjugate for isocyanate conferred a stability upon the linking reagent that was essential in its utility. The article by Riggs et al. is probably one of the few in which a thesis toward a master's degree became a Citation Classic.

Möller's description of his approach to Medawar's list of unsolved research problems eventually led to two important observations for which he has not received due credit. His determination that histocompatibility antigens were present on the cell exterior now seems all too obvious, but his was the first description of this fact. Likewise, the presence of immunoglobulin on the exterior of lymphocytes, and the ultimate capping of this immunoglobulin, is often erroneously credited to other investigators who made improvements on his procedures (see pp. 70 and 71).

The immunological sandwich devised by Sternberger as a histologic reagent relies on the capability of a first immunoglobulin (antibody) to react with two molecules of a second immunoglobulin. When one of the second immunoglobulin molecules is attached to a cell and the other second immunoglobulin is labeled with an enzyme, then an enzyme-linked antibody can be bound to a tissue or cell. This is the basis of Sternberger's immunoenzyme probe for histologic investigations.

Variants of immunodiffusion tests are legion and rely on the initial studies of Ouchterlony, Oudin, and Elek, and when an additional electrophoretic separation is included, rely upon the leading contributions of Grabar, Williams, and Courcon. That Laurell's rocket immunoelectrophoretic method did not reach the general utility predicted by Grabar does not detract from its generous use in research laboratories. The "Mancini" popularization of Feinberg's radial diffusion system is the clinician's current choice for quantitative immunodiffusion, although nephelometry is making fast inroads on immunodiffusion as a quantitative method.

The next Classics are less coordinated with one another. They record the important discoveries that IgM but not IgG is sensitive to disulfide dissociation and that $F(ab')_2$ fragments of immunoglobulins are likewise sulfhydryl sensitive. Mergenhagen et al.'s review on serum complement and bacterial endotoxin (lipopolysaccharide) was timely and united several seemingly unrelated observations into a single thesis. Both the complement system and bacterial endotoxins are complex and have multiple biological activities.

Citation Classics

Greenwood F C, Hunter W M & Glover J S. The preparation of
131I-labelled human growth hormone of high specific radioactivity.
Biochemical Journal 89:114-23, 1963.

The radio-immunoelectrophoretic technique for the assay of insulin in human sera developed by Yalow & Berson has been applied to glucagon and to growth hormone. This method uses low amounts of carrier-free [131I] iodide. No prior treatment of the iodide sample is required, and the iodide's high isotope content makes possible high specific radioactivity with a low degree of chemical substitution. [The *SCI®* indicates that this paper was cited 1,779 times in the period 1961-1975.]

Professor Frederick C. Greenwood
Pacific Biomedical Research Center
University of Hawaii at Manoa
1993 East-West Road
Honolulu, Hawaii 96822

December 15, 1976

"Some 15 years ago our aim was to develop a radioimmunoassay for human growth hormone. We regarded the development of a method for radio-iodinating growth hormone as a necessary prerequisite to achieve our goal. After watching with some amazement over the last 15 years the widespread application of our radio-iodination method, my memory of 'how it was' may be somewhat faulty. On many occasions I have been labelled as an expert on iodine chemistry, radio-iodine chemistry, and chemistry--labels of doubtful specificity. My expertise was in recognizing the importance of the classical radioimmunoassay paper of Berson and Yalow in 1960.[1]

"As a then full-time researcher with the Imperial Cancer Research Fund [London, England], my goal was to measure human growth hormone and human prolactin in plasma. With the assistance of Professor C.A. Gemzell in Stockholm, I had initiated isolation of growth hormone and its dreadful bioassay. Thence, having acquired some of the arts of protein chemistry from Professor H.F. Deutsch in Madison, Wisconsin, and particularly some feeling for immunochemistry, I

applied the tanned cell hemagglutination method developed for growth hormone by Professor C. Read of Iowa. In parallel studies Mr. Hunter had the back-breaking job of the bioassay for growth hormone. When it became apparent that neither the bioassay nor the tanned cell assay would work in human plasma we approached radioimmunoassay and radio-iodination with vim and vigor--it had to succeed.

"A critical point was our early visit to Amersham [The Radiochemical Centre, Bucks, England] to see how to handle 80 mCi of iodine 131 as then required by the Berson and Yalow technique, in the context of a normal lab in the center of London, only to be told shortly, but kindly, by Dr. Glover, 'don't.' Dr. Glover pointed out that Amersham was just making available carrier-free iodine 131. It may have been Dr. Glover, Mr. Hunter, or myself who picked out chloramine-T as a possible oxidizing agent. I am certain that the theoretical advantages of carrier-free iodine, together with an agent which would make available all the iodine for labelling, did not fully occur to me at the time. In this case, serendipity, motivation, and hard work substituted successfully for a *priori* reasoning.

"Certainly, I remember the initial experiments carried out by Mr. Hunter. We inadvertently radio-iodinated Sephadex, before realizing that we had to kill the reaction by metabisulfite, or risk losing, by absorption, microgram amounts of labelled material on columns. That was before we learned to pre-saturate with albumin. In one experiment we recovered 100% of the radioactivity by chasing with added potassium iodide. However, we repeatedly failed to show radio-iodination using *calculated* amounts of chloramine-T. The decision to start 'spooning in' chloramine-T was a desperate, but successful, last resort. Thereafter it was a case of smoothing out the methodology since we felt that without a routine method of radio-iodination, radioimmunoassay would become only a passing fad in a few research labs...."

1. **Yalow R S & Berson S A.** Immunoassay of endogenous plasma insulin in man.
Journal of Clinical Investigation 39:1157-75, 1960.

Marchalonis J J. An enzymic method for the trace iodination of immunoglobulins and other proteins. *Biochemical J.* **113**:299-305, 1969.
[Walter and Eliza Hall Inst. Med. Res., Parkville, Victoria, Australia]

A simple, gentle, and reproducible method for the trace iodination of immunoglobulins and other serum proteins by a system consisting of purified lactoperoxidase, hydrogen peroxide, and radioiodide is described. [The *SCI®* indicates that this paper has been cited in over 830 publications since 1969.]

John J. Marchalonis
Department of Biochemistry
Medical University of South Carolina
Charleston, SC 29425

May 24, 1983

"After obtaining my PhD in biochemistry with G.M. Edelman, I went to the laboratory of G.J.V. Nossal at the Walter and Eliza Hall Institute as a postdoctoral fellow to learn cellular immunology. At the time I joined the laboratory, Nossal and G.L. Ada were carrying out a series of studies using radioiodinated protein antigens designed to ascertain the distribution of antigen in lymphoid tissues during various stages of immunization.[1] I happened to subject labeled antigen preparations to polyacrylamide gel electrophoresis and found evidence that the proteins had been severely denatured during the process of radioactive labeling which employed the oxidizing agent chloramine-T. This observation gave impetus to the search for a method which would allow the radioiodination of proteins under conditions which did not cause noticeable denaturation. Initially, I planned to use a mince of thyroid because thyroid peroxidases catalyze the covalent binding of iodide to the phenol ring of tyrosine. However, it was my good fortune to discuss this problem with a group of graduate students at a pub close to the University of Melbourne whereupon I learned that G.R. Jago and D.McC. Hogg, a graduate student in his laboratory, had purified the enzyme lactoperoxidase from milk. It could be anticipated that lactoperoxidase would act in a fashion very similar to the original enzyme system I had chosen. The availability of the purified peroxidase was an obvious advantage over any system which would involve a tissue mince. I obtained the lactoperoxidase and carried out a straightforward series of experiments designed to show that, first, radioactive iodide could be covalently bound to proteins using this enzyme activated by peroxide, and, then, that the labeled proteins retained their normal electrophoretic characteristics and that the label was covalently bound to tyrosyl residues.

"Why has this paper become so highly cited? In the first place, the timing of its appearance was auspicious because many workers in various fields were becoming involved in the use and application of radioimmunoassay technology, and it was clear to them that the then currently used major labeling approaches were not adequate for a number of reasons. The use of lactoperoxidase-catalyzed iodination provided a simple and gentle means to label material for radioimmunoassay, tissue localization, and biochemical studies. This was the first paper in what was to become a burgeoning technology where various adaptations and modifications were made to the basic scheme allowing high-level uptake of iodide[2] and tailoring of the conditions for various proteins. The second reason for broad citation of this paper is that the lactoperoxidase-catalyzed radioiodination approach proved to be a general means of labeling exposed tyrosyl residues in proteins on the outer surface of plasma membranes of living cells.[3-5] This paper started initially as an exercise in 'problem solving' focusing upon the issue of externally labeling proteins in the absence of denaturation. The solution to the problem turned out to have general applicability to many biochemical systems as well as to be the first step in the identification of surface proteins of living cells."

1. Nossal G J V & Ada G L. *Antigens, lymphoid cells and the immune response.*
   New York: Academic Press, 1971. 324 p.
2. Thorell J I & Johansson B G. Enzymatic iodination of polypeptides with ^{125}I to high specific activities.
   Biochim. Biophys. Acta 251:363-9, 1971.
3. Phillips D R & Morrison M. Exposed protein on the intact human erythrocyte. *Biochemistry* 10:1766-71, 1971.
4. Baur S, Vitetta E S, Sherr C J, Schenkein I & Uhr J W. Isolation of heavy and light chains of immunoglobulin from the surfaces of lymphoid cells. *J. Immunology* 106:1133-5, 1971.
5. Marchalonis J J, Cone R E & Santer V. Enzymic iodination: a probe for accessible surface proteins of normal and neoplastic lymphocytes. *Biochemical J.* 124:921-7, 1971.

Bolton A E & Hunter W M. The labelling of proteins to high specific radioactivities by conjugation to a ^{125}I-containing acylating agent: application to the radioimmunoassay. *Biochemical J.* **133**:529-39, 1973.
[Med. Res. Council, Radioimmunoassay Team, Edinburgh, Scotland]

A method is described for labelling proteins to high specific activities with ^{125}I in which the protein is treated with ^{125}I-labelled 3-(4-hydroxyphenyl) propionic acid N-hydroxysuccinimide ester, resulting in the conjugation of the radioiodinated phenyl moiety to free amino groups in the protein by amide bonds. [The *SCI®* indicates that this paper has been cited in over 1,190 publications since 1973.]

Anthony E. Bolton
Department of Biochemistry
North East London Polytechnic
London E15 4LZ
England

May 14, 1984

"In 1970, I joined what was shortly to become the MRC Radioimmunoassay Team in Edinburgh, headed by Bill Hunter. I was introduced to the complexities of the radioimmunoassay (RIA) only to discover what an unreliable method it was at this time, principally because of problems of tracer preparation. For months before I arrived, some assays had been completely unusable for this reason, although the problem was sporadic. My task was to develop an alternative radioiodination system for proteins to supplement the chloramine-T oxidation method of Hunter and Greenwood,[1] then the only widely used method available.

"To avoid exposure of sensitive proteins to oxidising agents, Hunter suggested conjugating them to an ^{125}I-containing group. We met J. Rudinger, a peptide chemist from ETH Zurich, and a profitable collaboration ensued. The Zurich group synthesised a series of active ester derivatives which we tested as routes for incorporating ^{125}I into proteins. The N-hydroxysuccinimide ester of p-hydroxyphenyl propionic acid,[2] which re-acts with epsilon amino sidechains of lysine residues of proteins, showed promise. This had the additional advantage of altering different amino acid residues from tyrosine and histidine substituted in direct oxidative iodination.

"There followed a long period of development of the method at the end of which we approached a UK company about the commercial exploitation of the method; we were told there would be no market. We submitted the manuscript for publication. It was rejected on the grounds that new methods must be shown to be better in practise than those existing, not just new and theoretically better. At that point in time, this was not easy to prove — the sporadic problems of the Hunter-Greenwood method had spontaneously and unaccountably been resolved. Fortunately, one of our battery of antisera bound preferentially to tracer prepared by the new method — inclusion of these new data into the paper enabled its publication. Some time later, we heard that an American manufacturer wished to market the radioiodinated compound as 'Bolton-Hunter reagent.'

"I think the widespread use of this method, reflected in its high citation rate, results from the commercial availability of the Bolton-Hunter reagent, making this a simple method to use. I have prepared tracers of unstable proteins by this technique, e.g., in the RIA of the platelet-specific antigen PF4.[3] There are some proteins where the biological activity is retained after labelling by this but not direct oxidative iodination methods, and some peptides lack tyrosine and/or histidine. For these, conjugation labelling methods are necessary. Alternative active ester derivatives have been described since;[4] presumably these would have been developed independently under the pressure of a need for such methods.

"Rudinger sadly died in the mid-1970s. Hunter is now retired."

1. **Hunter W M & Greenwood F C.** Preparation of iodine-131 labelled human growth hormone of high specific activity. *Nature* **194**:495-6, 1962. (Cited 4,315 times.)
2. **Rudinger J & Ruegg U.** Preparation of N-succinimidyl 3-(4-hydroxyphenyl) propionate. *Biochemical J.* **133**:538-9, 1973. (Cited 35 times.)
3. **Bolton A E, Ludlam C A, Pepper D S, Moore S & Cash J D.** A radioimmunoassay for platelet factor 4. *Thromb. Res.* **8**:51-8, 1976. (Cited 115 times.)
4. **Langone J J.** Radioiodination by use of the Bolton-Hunter and related reagents. *Meth. Enzymology* **73**:112-27, 1981.

Citation Classics

Yalow R S & Berson S A. Immunoassay of endogenous plasma insulin in man.
Journal of Clinical Investigation 39:1157-75, 1960.

This paper describes in detail an immunoassay for plasma insulin in man, based on the reaction of human insulin added *in vitro* to plasma, which is after recovered and measured for endogenous insulin concentrations. Employing this method, the paper reports plasma insulin concentrations during glucose tolerance tests in nondiabetic and in early diabetic subjects, and plasma insulin concentrations in subjects with functioning islet cell tumors or leucine-sensitive hypoglycemia. [The *SCI®* indicates that this paper was cited 1,100 times in the period 1961-1975.]

Rosalyn S. Yalow, Ph.D.
Veterans Administration Hospital
130 West Kingsbridge Road
Bronx, New York 10468

January 7, 1977

"Radioimmunoassay methodology (RIA) has become the method of choice for the determination of the minute concentrations in plasma, other body fluids, or tissue extracts of hundreds of substances of biologic interest such as peptidal and non-peptidal hormones, drugs, enzymes, viruses, etc. It is in use in over 4,000 research and general clinical laboratories in the United States and in thousands of laboratories around the world. For instance, in 1973, a symposium on 'Radioimmunoassay and Related Procedures in Clinical Medicine and Research' was held in Turkey under the auspices of the International Atomic Energy Agency. Participants came from 42 countries, all concerned with the applicability of RIA to clinical problems and biomedical investigation in their own lands. Solution of public health problems such as those requiring increased sensitivity for detection of hepatitis B antigen in blood used for transfusion or early detection of neonatal hypothyroidism are dependent on RIA. RIA was first used in endocrinology to study the regulation of hormonal secretion and the diagnosis of states of hormonal excess or deficiency but its applications have since extended into virtually all medical specialities. At present RIA is considered so classic a method that relatively few of the scientific papers based on RIA refer to the original detailed description of the methodology presented in this 1960 paper in the *Journal of Clinical Investigation.*

"RIA developed out of our earlier work demonstrating the ubiquitous presence of insulin-binding antibodies in insulin-treated subjects.[1] In that paper we reported that the binding of labeled insulin is a quantitative function of the amount of insulin present when the antibody concentration is kept fixed. This provided the basis for the radioimmunoassay of insulin. However, several years were to pass before we were able to demonstrate the practical application of this principle to the measurement of plasma insulin in man. The 1960 *Journal of Clinical Investigation* paper gives detailed information concerning the problems, practices, and pitfalls in the development of a new RIA as well as presenting what was then new physiologic data on insulin levels in normal and pathologic states and in response to various stimuli.

"Perhaps one should be modest in a commentary on one's own work. However, of the more than 200 publications resulting from my collaborative work with Dr. Solomon A. Berson for the 22-year period before his untimely death in 1972, and the more than 50 papers from my laboratory since, the 1956 paper on the antigenicity of insulin in man and the 1960 paper on the assay of plasma insulin in man remain as the favorites. I am pleased, though not surprised, that the 1960 article has become a highly cited 'classic'."

1. Berson S A, Yalow R S, Bauman A, Rothschild M A & Newerly K. Insulin-I[131] metabolism in human subjects: demonstration of insulin binding globulin in the circulation of insulin-treated subjects.
Journal of Clinical Investigation 35:170-90, 1956.

Morgan C R & Lazarow A. Immunoassay of insulin: two antibody system, plasma insulin levels of normal, subdiabetic and diabetic rats. *Diabetes* **12**:115-26, 1963.

The authors present a detailed account of a two-step procedure for radio-immunoassay of insulin. In the first reaction, insulin forms a soluble complex with its specific antibody obtained from immunized guinea pigs. In the second reaction, this soluble complex is precipitated by an antibody to guinea pig serum obtained from immunized rabbits. Using a radioactive insulin tracer, the amount of radioactivity in the precipitate is dependent upon the concentration of insulin in the reaction mixture; i.e., with increasing concentrations of unlabeled insulin, the amount of radioactive insulin in the precipitate is decreased correspondingly. [The *SCI®* indicates that this paper was cited 1069 times in the period 1963-1976.]

Professor Carl R. Morgan
Department of Anatomy
Indiana University School of Medicine
Indianapolis, Indiana 46202
January 24, 1977

"It is gratifying to have developed a method that has found wide acceptance. This work was done as a part of my Ph.D. thesis. The late Arnold Lazarow, former Head of the Department of Anatomy at the University of Minnesota, was my mentor. I had completed my M.A. at the University of Nebraska. The following four years of experience as a technician in the laboratories of Dr. J.T. Syverton, then Head of Bacteriology at the University of Minnesota, served me well on this project. Thanks to the USPHS training grants program, in the reflected light of Sputnik, I resumed my graduate training.

"At the time these studies were initiated (1958), research on diabetes mellitus was inhibited by the lack of a specific, easily reproducible method of measuring large numbers of samples containing less than a nanogram of insulin.

"The ground work had been laid. Moloney & Coval (1955) reported that guinea pigs could be routinely and reliably immunized to other mammalian insulins.[1] Using a hemagglutination method, Arquilla & Stavitsky (1956) had shown the feasibility of using immunological procedures for assaying insulin.[2] (^{131}I)-insulin was available commercially. Skom & Talmage (1958) reported on the use of anti-human gamma globulin to precipitate the non-precipitating insulin antibodies in insulin-resistant human serum.[3] With these facts in hand, Lazarow and I proceeded to develop our two-antibody method for the immunoassay of insulin. There was a manual, single channel gamma radiation counter in the laboratory. With intense anticipation we watched the rows of blinking lights to see if our efforts were to be successful.

"Other immunoassay procedures were developed in other laboratories during this time. It is apparent that such a method was in great demand. The methodology has been adapted to the immunoassay of other hormones. I attribute the acceptance of our method and the interest in our paper to the relative ease, reproducibility, and precision of the two-antibody method.

"Arnold Lazarow was pleased with this work, and he considered it an important breakthrough. It was a satisfying experience for me to work with him, and I am pleased that our work has been useful in the continuation of research on diabetes."

1. **Moloney P J & Coval M.** Antigenicity of insulin: diabetes induced by specific antibodies. *Biochemical Journal* **59**:179-85, 1955.
2. **Arquilla E R & Stavitsky A B.** The production and identification of antibodies to insulin and their use in assaying insulin. *Journal of Clinical Investigation* **35**:458-66, 1956.
3. **Skom J H & Talmage D W.** Nonprecipitating insulin antibodies. *Journal of Clinical Investigation* **37**:783-6, 1958.

Hales C N & Randle P J. Immunoassay of insulin with insulin-antibody
precipitate. *Biochemical J.* **88**:137-46, 1963.
[Department of Biochemistry, University of Cambridge, Cambridge, England]

The authors describe a radioimmunoassay for insulin employing a solid phase reagent of guinea pig insulin antibody pre-precipitated with rabbit antiguinea pig γ-globulin serum. Sensitisation of the assay by pre-incubation with unknown or standard prior to addition of radiolabelled hormone is described. The assay proved suitable for the development of a commercial kit. [The *SCI*® indicates that this paper has been cited over 2,240 times since 1963.]

C. N. Hales
Department of Clinical Biochemistry
Addenbrooke's Hospital
Cambridge CB2 2QR
England
and
P.J. Randle
Nuffield Department of
Clinical Biochemistry
University of Oxford
Radcliffe Infirmary
Oxford OX2 6HE
England

October 20, 1980

"Whilst it is superficially pleasing to find this paper highly cited according to the *Science Citation Index*® *(SCI*®*)*, explaining the occurrence may serve to illustrate an important limitation of the *SCI* as an indicator of the real scientific importance of a paper. Our contribution describes one of the many modifications introduced to the original radioimmunoassay of Yalow & Berson.[1] Undoubtedly, if every paper dependent on results obtained by radioimmunoassay were to refer to the true originators of the concept, Yalow & Berson's paper would be at or near the top of the *SCI* and the latter would

then much more faithfully reflect the real value of their contribution. Fortunately, this achievement has been acknowledged by the award of the Nobel prize to Yalow, though it is a matter for regret that Berson did not live to share this recognition.

"Our own struggles to set up a radioimmunoassay for insulin led us at a very early stage to appreciate the need for a readily available technique for clinical research and to attempt to persuade commercial organisations to supply the reagents. The reluctance of the latter to countenance the technique as originally described, led us to explore the use of alternative approaches based on the anti-immunoglobulin precipitation technique of Sköm & Talmage.[2] A highly convenient solid-phase reagent was prepared along the lines pioneered by Moloney.[3] The work reported in this paper represents part of the PhD thesis by one of us (C.N.H.). The technique formed the basis of studies of insulin secretion in diabetes and of studies independently by both of us on the control and mechanism of insulin secretion *in vitro*.

"It is interesting, in the light of the current commercial interest in radioimmunoassay and the commercial success achieved by marketing our method, to recall the reluctance of the firms approached to enter this field. The fact that British firms were the first to commercialise an American discovery makes a pleasant change. However, it is disappointing that this early lead was not exploited more successfully. (It might perhaps be worth stating in these days of increasing commercial arrangements between academic scientists and industry, that our initiative which led to the first commercial radioimmunoassay kit was not motivated by and did not result in any financial benefit to us).

"Thus one can suggest that the position of the paper in the *SCI* is due to a number of factors: the proven value of radioimmunoassay, the large volume of work in diabetes and in particular on insulin secretion, the robustness of the technique we devised and its commercial availability."

1. **Yalow R S & Berson S A.** Assay of plasma insulin in human subjects by immunological methods. *Nature* **184**:1648-9, 1959.
2. **Sköm J H & Talmage P W.** Nonprecipitating insulin antibodies. *J. Clin. Invest.* **37**:783-6, 1958.
3. **Moloney P J.** Endogenous and pancreatic insulins. *Ciba Found. Coll. Endocrinol.* **14**:169-81, 1962.

Niswender G D, Midgley A R, Jr. , Monroe S E & Reichert L E, Jr. Radioimmunoassay
for rat luteinizing hormone with antiovine LH serum and ovine LH-131I.
Proc. Soc. Exp. Biol. Med. **128**:807-11, 1968.
[Dept. Pathol., Univ. Michigan, Ann Arbor, MI and Dept. Biochem., Div. Basic
Health Sci., Emory Univ., Atlanta, GA]

A radioimmunoassay procedure was developed for quantification of luteinizing hormone (LH) in the blood of rats utilizing an immunological cross-reaction between rat LH and an unique antiserum produced in a rabbit against ovine LH. The assay was characterized carefully to demonstrate that other pituitary hormones or components of serum did not interfere with the measurement of LH. The extreme sensitivity of the assay (5 pg NIH-LH-S1) allowed quantification of the hormone in as little as 0.02 to 0.2 ml serum. [The *SCI®* indicates that this paper has been cited over 970 times since 1968.]

———————————•———————————

Gordon D. Niswender
Department of Physiology and
Biophysics
College of Veterinary Medicine &
Biomedical Sciences
Colorado State University
Fort Collins, CO 80523

October 1, 1980

"This paper was the first of approximately 15 describing the unique properties of an antiserum (from rabbit number 15) which has been proven specific for quantification of LH in serum obtained from a large number of mammals including primates and numerous wildlife species. Literally hundreds of antisera against ovine LH have been tested by ourselves and others and to date none has demonstrated the great degree of cross-reaction with LH from other species demonstrated by antiserum from rabbit number 15. Since this antiserum appeared to be uniquely useful we have made it available for radioimmunoassay purposes to all scientists who have requested it. This is probably the reason the original paper is cited frequently. The assay system has been used by over 200 investigators in 33 countries to quantify LH in serum from over 30 species.

"Development of this antiserum is a classic example of serendipity in science. I had been funded by the National Institutes of Health for postdoctoral training with Rees Midgley at the University of Michigan. Therefore, while completing requirements for my PhD with Phillip Dzuik and A.V. Nalbandov at the University of Illinois, I immunized 25 rabbits with ovine LH. Based on present knowledge, the immunization techniques used were inappropriate and, even after repeated immunizations, only one rabbit (number 15) produced detectable antibodies to LH. I was allowed to move both the rabbit and the antisera I had collected to Michigan. Midgley was already recognized as one of the world's authorities on development of radioimmunoassay procedures and had been instrumental in establishing very rigid criteria for proving that radioimmunoassay procedures were reliable. Thus, the assay procedures which utilized this antiserum were all extensively characterized to demonstrate their reliability. Leo Reichert at Emory University supplied the highly purified LH for radioiodination. In subsequent research, he has supplied hormone preparations from a variety of species to test the hormonal specificity of several radioimmunoassay systems and has continued to be a valued colleague. Scott Monroe was a medical student working in the laboratory and had prepared the crude preparations of rat LH, follicle stimulating hormone (FSH), and thyroid stimulating hormone (TSH) used to test specificity of the radioimmunoassay procedure. He was also involved in development of the first assay for LH in Rhesus monkeys. Development of radioimmunoassay procedures gave reproductive biologists the capability to measure reproductive hormones in blood and was an important step in the advancement of our knowledge in reproductive biology."

Yalow R S & Berson S A. Radioimmunoassay of gastrin.
Gastroenterology **58**:1-14, 1970.
[Radioisotope Service, Veterans Administration Hosp., Bronx, and
Dept. Medicine, Mount Sinai School of Medicine, New York, NY]

Elevated plasma gastrin was noted in five patients with Zollinger-Ellison syndrome (ZE) and 17 pernicious anemia (PA) patients using a highly specific homologous porcine gastrin radioimmunoassay. Oral administration of HCl to PA patients produced an acute fall in plasma gastrin with a disappearance half-time of about seven minutes. [The *SCI®* indicates that this paper has been cited in over 620 publications since 1970.]

Rosalyn S. Yalow
Veterans Administration
Medical Center
130 W. Kingsbridge Road
Bronx, NY 10468

December 27, 1982

"In 1967, we had completed our studies on the immunochemical heterogeneity of plasma parathyroid hormone[1] and our detailed paper on the radioimmunoassay (RIA) of plasma ACTH.[2] RIA was already extensively employed by endocrinologists. It seemed an appropriate time for us to apply RIA to another field, gastroenterology. McGuigan had developed an RIA for gastrin[3] using antibodies to the carboxyl-terminal tetrapeptide of gastrin (G4) and ³H-labeled G4 as tracer, but this was too insensitive to measure plasma gastrin. Since purified pork gastrin or synthetic human gastrin was in short supply and/or expensive, we used for immunization two crude porcine gastrin preparations (one only ~0.5 per-

cent pure and the other only ~10 percent pure) kindly provided by Wilson Laboratories in May 1968. By September 1968, we had produced antisera suitable for assay of plasma gastrin using these preparations. Morton Grossman, who had stimulated our interest in this area, provided us with 50 μg each of purified porcine gastrin (G17) with sulfated and non-sulfated tyrosines that he had obtained from Rodney Gregory.

"During the next year we validated the assay, applied it to the diagnosis of gastrin secreting tumors, and noted the elevated gastrin levels in pernicious anemia and the rapid suppression by oral HCl in the latter condition. We also suggested that in older patients with an increased incidence of achlorhydria, there might be a modest elevation of basal gastrin. By the time we submitted this paper in September 1969, we had preliminary evidence that there was immunoreactive gastrin in plasma that differed from G17 and we were careful to note that 'the form in which endogenous human plasma gastrin circulates is not known.' We subsequently reported the discovery of 'big gastrin,'[4] now known to be a 34 amino acid peptide.

"I believe this paper has become a *Citation Classic* because of its unequivocal demonstration that immunization with a very low purity antigen is practical for RIA procedures and because it was the first to establish the low levels of gastrin in normal fasting serum. It has served as a model for gastroenterologists as to the need for careful validation of RIAs for gastrointestinal hormones."

1. **Berson S A & Yalow R S.** Immunochemical heterogeneity of parathyroid hormone.
   J. Clin. Endocrinol. Metab. 28:1037-47, 1968.
2. ———————————————. Radioimmunoassay of ACTH in plasma. *J. Clin. Invest.* 47:2725-51, 1968.
3. **McGuigan J E.** Antibodies to the carboxyl-terminal tetrapeptide of gastrin. *Gastroenterology* 53:697-705, 1967.
4. **Yalow R S & Berson S A.** Size and charge distinctions between endogenous human plasma gastrin in peripheral blood
   and heptadecapeptide gastrins. *Gastroenterology* 58:609-15, 1970.

This Week's Citation Classic

Hwang P, Guyda H & Friesen H. A radioimmunoassay for human prolactin.
Proc. Nat. Acad. Sci. US **68**:1902-6, 1971.
[McGill University Clinic, Royal Victoria Hosp., Montreal, Canada]

This paper reported the development of a radioimmunoassay for human prolactin, providing a simple, specific, and sensitive method of measuring prolactin concentrations in human circulation under various physiological and pathological conditions and facilitating the subsequent purification of human prolactin. [The *SCI®* indicates that this paper has been cited in over 645 publications since 1971.]

Peter Hwang
Department of Physiology
Faculty of Medicine
National University of Singapore
Republic of Singapore 0511

June 13, 1983

"Up to about 1970, there was still considerable doubt about the existence of a lactogenic hormone in man. Repeated attempts to purify prolactin from human pituitary glands had been unsuccessful. This failure, together with the observation that highly purified human growth hormone exhibited intrinsic lactogenic activity, led some investigators to propose that, perhaps, prolactin did not exist in man, its function being subserved by growth hormone.[1] This suggestion, however, was incompatible with other observations, such as the normal lactation seen in growth hormone deficient dwarfs.

"In 1970, the laboratory of Henry Friesen, among others, was attempting to demonstrate the separate existence of prolactin in primates. Harvey Guyda, working in the same laboratory, had by then shown that monkey pituitary glands secreted a lactogenic substance which was distinguishable immunologically from growth hormone. He had also succeeded in purifying a small amount of what was thought to be monkey prolactin from pituitary incubation media with an estimated purity of about ten percent by bioassay.[2]

"Soon after I joined Friesen's laboratory, I was asked to give some thought to the development of a radioimmunoassay for primate prolactin. This represented an act of considerable generosity and some courage on the part of Friesen, for I was then fresh out of the University of Singapore Medical School with no research experience whatever. The problem at that time was how one could develop a radioimmunoassay for a substance which was contaminated to the extent of 90 percent by other proteins, and which was available only in microgram quantities.

"Two rabbits were immunized with several hundred micrograms of the crude monkey prolactin preparation; what remained, however, was insufficient for us to attempt further purification in order to achieve a degree of purity suitable for radioiodination. Scaling up with a large number of monkey pituitaries would have been logical but prohibitively expensive.

"A way to bypass this difficulty was suggested by a paper published by Herbert and Hayashida[3] indicating that monkey prolactin and sheep prolactin might be immunologically related. Following up on this observation, we iodinated 5 μg of the crude monkey prolactin and applied the mixture, containing iodinated prolactin as well as other iodinated proteins, to an affinity column prepared by linking anti-sheep prolactin antibodies to Sepharose. Only the iodinated prolactin was retained by the column, yielding, on elution, an essentially homogeneous label suitable for use in a radioimmunoassay.

"The radioimmunoassay thus developed permitted easy measurement of circulating prolactin concentrations in humans for the first time. Perhaps more importantly, it also facilitated the subsequent chemical purification of human prolactin[4] which, together with a similar achievement by U.J. Lewis et al.,[5] not only firmly established the separate existence of human prolactin, but also made the radioimmunoassay available to many other investigators for extensive studies on the role of prolactin in human health and disease.[6] These factors have no doubt contributed to the number of citations."

1. Li C H. The chemistry of human pituitary growth hormone: 1956-1966. (Pecile A & Müller E E, eds.)
   Growth hormone. Amsterdam: Excerpta Medica, 1968. p. 3-28.
2. Guyda H & Friesen H. The separation of monkey prolactin from monkey growth hormone by affinity
   chromatography. *Biochem. Biophys. Res. Commun.* 42:1068-75, 1971.
3. Herbert D C & Hayashida T. Prolactin localisation in the primate pituitary by immunofluorescence.
   Science 169:378-9, 1970.
4. Hwang P, Guyda H & Friesen H. Purification of human prolactin. *J. Biol. Chem.* 247:1955-8, 1972.
5. Lewis U J, Singh R N & Seavey B K. Human prolactin: isolation and some properties.
   Biochem. Biophys. Res. Commun. 44:1169-76, 1971.
6. Flückiger E, del Pozo E & von Werder K. Prolactin: physiology, pharmacology and clinical findings.
   Monogr. Endocrinol. 23:1-224, 1982.

Coons A H & Kaplan M H. Localization of antigen in tissue cells. II.
Improvements in a method for the detection of antigen by means of fluorescent
antibody. *J. Exp. Med.* **91**:1-13, 1950.
[Dept. Bacteriology and Immunology, Harvard Med. Sch., Boston, MA]

A method employing the specificity of antibody labeled with fluorescein for the localization of antigen under the fluorescence microscope is presented. Included in the paper are a description of the synthesis of fluorescein isocyanate, the labeling material, and a method for removing over labeled proteins which bind indiscriminately to tissue elements and obscure specific reactions. [The *SCI®* indicates that this paper has been cited over 1,465 times since 1961.]

Albert H. Coons
Department of Pathology
Harvard Medical School
Boston, MA 02115

January 14, 1981
(revised)

"This paper described improvements in the method published earlier for the specific localization of foreign antigenic materials in tissue cells.[1] It was a general method for the histological localization of any antigen because it utilized specific antibody labeled with fluorescein as a histochemical reagent. Diluted specific antibody solutions so labeled were flooded over tissue sections. Any antigen present bound the antibody and fixed it in place. Excess reagent could be washed away leaving the bound antibody in place and it in turn could be localized by bombardment by light of appropriate wave length and visualized under the fluorescence microscope. Naturally, the critical step was the binding of the fluorochrome-labeled antibody by the antigen and the ability to wash away any excess fluorescent reagent. This principle joined the specificity of the antibody molecule to the resolving power of the light microscope; such a union

provided a general method now called immunohistochemistry for the investigation of native and foreign antigenic molecules in many locations and under many circumstances. Since then the same principle has been extended for use with the electron microscope by using antibody labeled with ferritin or with enzymes like horse radish peroxidase.

"Of course such reagents localizing and identifying antigens rapidly came to be used for the specific identification of various infectious agents: bacteria, rickettsiae, and viruses. It has also been applied to the study of autoimmune disease, e.g., nephritis, and in the detection of autoantibodies against tissue components. Immunofluorescence so-called therefore became a feature of the diagnostic, as well as the research, laboratory.

"Immunohistochemistry has gradually become useful in many areas of biology. Till now its weakness as a scientific method has been the difficulty of quantitating it.

"Recently however, computer activated light microscopes have made possible the rapid measurement of $1\mu^2$ areas of fluorescent cells. Such microscopes, attached to a computer, print out measurements of fluorescence intensity, allowing rapid comparison of the amount of antigen per unit area in various cells and parts of them. So far this ability is only beginning to be exploited.

"Surprisingly enough, and to the good fortune of anyone who wants to apply such a method, it has turned out that the antibody molecule is quite stable to many chemical manipulations and does not lose its specificity unless the label attaches itself close to the actual combining site.

"Addendum: Albert H. Coons died suddenly September 30, 1978. His commentary above is remarkably understated. The impact of this work on research in many biological disciplines, and, in particular, on studies of immunobiologic and pathologic processes is universally recognized. We remember him for the charm of his company, his penetrating wisdom, and his admonition: 'good research work stands on its own legs.' " —Melvin H. Kaplan

1. Coons A H, Creech H J, Jones R N & Berliner G. The demonstration of pneumococcal antigen in tissues by the use of fluorescent antibody. *J. Immunology* **45**:159-70, 1942.
[The *SCI®* indicates that this paper has been cited over 285 times since 1961.]

Riggs J L, Seiwald R J, Burckhalter J H, Downs C M & Metcalf T G. Isothiocyanate compounds as fluorescent labeling agents for immune serum.
Amer. J. Pathol. 34:1081-92, 1958.

The authors describe the synthesis of two fluorescent isothiocyanate dyes, fluorescein isothiocyanate (FITC), and rhodamine B isothiocyanate (RITC), substituting thiophosgene, a less toxic substance, for highly toxic phosgene gas during the synthesis of the compounds. These two fluorescent dyes were successfully coupled with antibodies which were subsequently used in the direct and indirect fluorescent antibody procedures for staining different bacterial, rickettsial and viral antigens. [The *SCI®* indicates that this paper was cited 546 times in the period 1961-1976.]

Dr. John L. Riggs
Viral and Rickettsial Disease Laboratory
California Department of Health
Berkeley, California 94704

February 10, 1977

"It is indeed flattering to be a 'most cited author' and to have one's paper referred to as a 'Citation Classic.' The paper was the result of a portion of the work done at the University of Kansas for my master's degree in bacteriology. I had been instructed by my major professors in the Department of Bacteriology, Dr. Theodore Metcalf and Dr. Cora Downs, to attempt to synthesize some fluorescein isocyanate following Dr. Albert Coon's procedure. The synthesis was to be carried out in the Laboratory of Pharmaceutical Chemistry under the direction of Dr. Joseph Burckhalter. Dr. Robert Seiwald, a postdoctoral student in Dr. Burckhalter's laboratory at that time, and I spent many hours in the laboratory attempting to produce isocyanates of many fluorescent compounds in addition to fluorescein. It was evident, however, that the isocyanates were so reactive that they were very unstable, and many precautions had to be taken in order to couple them to solutions of proteins.

One morning Dr. Burckhalter came into the laboratory and said, 'John, why don't you try to substitute a sulfur atom for the oxygen in the isocyanate portion of the molecule? Perhaps that will stabilize the compound.' Many more hours were spent in the laboratory by Seiwald and myself attempting to synthesize the isothiocyanate by different procedures suggested by Dr. Ray Brewster, then Chairman of the Department of Chemistry at the University of Kansas. We could not at that time find a commercial source of thiophosgene, and although many commercial laboratories would custom synthesize it for us, the cost was prohibitive. We were on the verge of attempting to synthesize it ourselves when an advertisement of Rapter Laboratories of Chicago was noted which stated that they had thiophosgene available. We immediately obtained some and carried out the experiments to produce the isothiocyanates. We were successful in producing the isothiocyanates (they were indeed more stable compounds) and, with Dr. Downs in the bacteriology laboratories, were also successful in coupling them to antibodies which were subsequently used in staining and specifically identifying different bacterial, rackettsial and viral antigens.

"Fluorescein and rhodamine isothiocyanate soon became available commercially and the fluorescent antibody procedure then became available as a tool in almost any laboratory that wished to use the technic. Quoting Dr. Coons,[1] 'Now it was no longer true, as Pressman had told me, that immunofluorescence was impossible because those who could do the chemistry couldn't do the histology, and *vice versa.*'

1. **Coons A H.** The development of immunohistochemistry.
Ann. N.Y. Acad. Sci. 177:5-9, 1971.

Marshall J D, Eveland W C & Smith C W. Superiority of fluorescein isothiocyanate (Riggs) for fluorescent-antibody technic with a modification of its application. *Proceedings of the Society for Experimental Biology and Medicine* 98:898-900, 1958.

The authors compare three methods for preparing fluorescein conjugated globulins by using two derivatives of fluorescein amine, and recommended one for fluorescent antibody staining. [The *SCI®* indicates that this paper was cited a total of 550 times in the period 1961-1976.]

Dr. John D. Marshall
Research Consultant
Letterman Army Institute of Research
San Francisco, California 94129
February 14, 1977

"I find great humor in being a 'most cited author' but the real credit should go to Coons and Kaplan[1] who developed the fluorescent antibody technic and to Riggs[2] who described the use of fluorescein isothiocyanate. Our work modified the technic for the conjugation of the dye to antibody and brought Riggs' dye to the attention of other workers.

"Perhaps you would be interested to know why and how our technic evolved. At the Armed Forces Institute of Pathology, our group was interested in the identification of microorganisms in tissues which had been processed for histological examination and were not suitable to standard microbiological processing. We were using the basic technics of Coons and Kaplan but because of potentially serious accident during the synthesis stage and the requirement for numerous high titered conjugated antisera, we were actively on the lookout for a safer, simpler, and less denaturing method.

"Coons and Kaplan published the first of a number of papers about their technic in 1950. Numerous investigators realized the value of being able to detect antigenic components at the cellular level; but the basic technic of Coons and Kaplan had three serious drawbacks: (1) the synthesis of fluorescein isocyanate required the use of phosgene gas (this alone was sufficient to deter many researchers); (2) the dye had a short shelf-life and required repeated synthesis; and (3) the method of conjugation resulted in denaturing the antibody of the finished product.

"In 1957, Riggs' masters degree thesis describing fluorescein isothiocyanate came to our attention. While the reagents used in his technic were toxic, thiophosgene was far less hazardous to handle than phosgene gas. The resultant product was quite stable, so a single lot of dye could be produced and used over several months, thereby eliminating the necessity for repeated synthesis required for fluorescein isocyanate. Earlier that year Goldman[3] had described a method for storing fluorescein isocyanate on filter paper which had the advantage of prolonging the shelf-life of the compound, alleviating one of the problems. Since fluorescein isothiocyanate was a stable dry powder, we chose to use it directly to eliminate the step of quantatively applying the dye to filter paper. Our procedure also eliminated the necessity of exposing the globulin solutions to organic solvents, reducing the possibility of denaturation of the proteins.

"We were delighted that the end result was a rather safe and simple method of preparing high titer reagents. Furthermore, our method did not compromise quality; the comparative study indicated that it had outstanding features—the fluorescein isothiocyanate was superior in stability and degree of fluorescence, and also the serum could be conjugated easily. Now our laboratory and other laboratories had a fluorescent antibody technic available which required minimum supplies, equipment, and personnel. With the increased interest, reagents soon became available commercially, which further stimulated the use of the method. We are certainly fortunate to have reported the work at a time when there was a need for the modification of this technic."

1. Coons A H & Kaplan M H. Localization of antigen in tissue cells. II. Improvements in a method for the detection of antigen by means of fluorescent antibody. *Journal of Experimental Medicine* 91:1-13, 1950.
2. Riggs J L. Synthesis of fluorescent compounds and their use for labelling antibody. Masters Thesis, University of Kansas, 1957.
3. Goldman M. Staining toxoplasma gondii with fluorescein-labelled antibody. I. The reaction in smears of peritoneal exudate. *Journal of Experimental Medicine* 105: 549-56, 1957.

This Week's Citation Classic ™

Möller G. Demonstration of mouse isoantigens at the cellular level by the fluorescent antibody technique. *J. Exp. Med.* **114**:415-34, 1961.
[Institute for Tumor Biology, Karolinska Institutet Medical School, Stockholm, Sweden]

The fluorescent antibody method applied to living cells in suspension demonstrated that H-2 and non-H-2 histocompatibility antigens were localized to the cell membrane. A proportion of lymphocytes, but not other cells, treated with only anti-immunoglobulin antisera exhibited staining of part of the membrane, giving rise to fluorescent crescents. This staining revealed surface bound immunoglobulin produced by the lymphocytes themselves [The *SCI*® indicates that this paper has been cited in over 535 publications since 1961.]

Year

'61	'62	'63	'64	'65	'66	'67	'68
0	7	11	8	17	12	24	34

Citations

Göran Möller
Department of Immunobiology
Karolinska Institute
Wallenberglaboratory
104 05 Stockholm
Sweden

May 16, 1984

"This was my second scientific publication and the result of nearly two years' frustrating work. In 1959, I started to do research in immunology and my first task was to find a suitable research project. I was reading the proceedings of a recent symposium,[1] which was introduced by Peter Medawar. The first paragraph of his paper listed a number of unsolved problems in the form of questions. This was excellent for me and I selected two questions from his list and made them into my research projects. The first was to determine the intracellular and histological localization of H-2 transplantation antigens and the second to study their phenotypic expression during ontogeny. I thought it would be easy to show the intracellular localization of the antigens by the fluorescent antibody method, which at that time was used only with tissue sections. After labelling antibodies with fluorescein — not so easy at a time when you had to start to prepare the fluoresceinisothiocyanate yourself — I applied them to tissue sections, but only obtained a diffuse nonspecific staining.

"My other Medawar-inspired project progressed better, and for various reasons I became interested in pinocytosis of cells during ontogeny. I used fluorescein labelled albumin to study pinocytosis and then observed that the great majority of cells did not take up the fluorescein and, after washing, they remained totally unstained in contrast to the tissue sections. It occurred to me that the use of living cells instead of tissue sections could solve the problem with the nonspecific staining. The first experiment with living cells worked well: all cells in the experimental group exhibited membrane staining and none in the control group. The use of living cells solved a technical problem and clearly demonstrated that H-2 antigens were membrane bound.

"However, a proportion of lymphocytes, in contrast to all other cells, stained when treated only with a fluorescent rabbit anti-mouse immunoglobulin antiserum. It was also membrane staining, but different from the ring staining seen with anti-H-2 sera. The antibodies were localized in one part of the lymphocytes only, making them look like the crescent of the moon. I showed that the stained structures were membrane bound immunoglobulin and that the immunoglobulin was produced by the lymphocytes and not passively picked up. I also found that they were absent in lymphocytes from embryos and appeared shortly after birth. This was the first demonstration of immunoglobulin receptors on the surface of B lymphocytes, although T and B cells had not yet been discovered.

"I am slightly surprised that the paper has been cited often for two reasons. First, it had little impact after its publication. It took seven years for the first confirmation of the membrane localization of H-2 antigens[2] and nine years before membrane immunoglobulin receptors were rediscovered and crescent formation renamed cap formation.[3]

"The second reason is that the fluorescent antibody method applied to living cells is now a routine method and I would not expect most immunologists to care about the original discovery. The same applies to the existence of membrane bound immunoglobulins. It is part of the common knowledge in immunology and reference to the original work done 23 years ago is not necessary."

1. **Albert F & Medawar P B,** eds. *Biological problems of grafting.* Oxford: Blackwell, 1959. 453 p.
2. **Cerottini J C & Brunner K T.** Localization of mouse isoantigens on the cell surface as revealed by immunofluorescence. *Immunology* **13**:395-405, 1967. (Cited 130 times.)
3. **Raff M C, Sternberg M & Taylor R B.** Immunoglobulin determinants on the surface of mouse lymphoid cells. *Nature* **225**:553-4, 1970. (Cited 470 times.)

CC/NUMBER 29
JULY 18, 1983

Lazarides E & Weber K. Actin antibody: the specific visualization of actin filaments in non-muscle cells. *Proc. Nat. Acad. Sci. US* 71:2268-72, 1974.
[Cold Spring Harbor Lab., Cold Spring Harbor, NY]

This paper describes a technique for the production of antibodies against cytoplasmic actin using as an antigen actin purified by sodium dodecyl sulfate (SDS) gel electrophoresis. The antibodies were then used in indirect immunofluorescence to reveal for the first time the cytoplasmic distribution of actin filaments in cells grown in tissue culture. [The *SCI®* indicates that this paper has been cited in over 605 publications since 1974.]

Elias Lazarides
Division of Biology
California Institute of Technology
Pasadena, CA 91125

May 3, 1983

"When Klaus Weber and I moved to Cold Spring Harbor from Harvard University, where I was a graduate student, we were fortunate to fall into the company of two scientists well versed in the problems of cell biology: Bob Pollack and Bob Goldman. Through numerous discussions with both of them, it became evident that what was needed in cell biology was a new way of studying cell structure and cell motility.

"From earlier work, it was evident that non-muscle cells contained actin and myosin, but the big question was where. Of course the choice would have been localization through antibodies, except that in the past a number of investigators had tried to raise antibodies against actin purified by conventional techniques but had failed, resulting only in the generation of antibodies to other antigens contaminating the actin preparation. This was presumably because actin was highly conserved and hence poorly immunogenic in its native state. I began to characterize what I thought was a new structural cytoplasmic protein from cells grown in tissue culture, trying to meticulously avoid actin. The protein was finally purified and characterized but I didn't know what it was or where it was in the cell.

"I got busy and started to purify enough of this protein to raise antibodies against it, but I realized that it was an uphill battle. I decided to purify enough of this protein from mouse 3T3 cells by sodium dodecyl sulfate (SDS) gel electrophoresis and in the process to also purify some actin from

these cells by the same technique to use as a negative control, since actin was so poorly immunogenic. Fred Miller, at the State University of New York in Stony Brook, was kind enough to do the immunizations for me. Several weeks later, we tested the sera using microcomplement fixation. The serum of the unknown protein I had purified unfortunately had no activity, but to our surprise the actin antiserum did. We then tested the actin antibodies by immunodiffusion and got a weak, but nonetheless real, precipitin line. Encouraged, we decided it would be worthwhile to do some immunofluorescence with these antibodies to see if we could detect any fluorescence.

"On December 24, 1973, Art Vogel, then a graduate student with Pollack, and I found ourselves late at night in the lab. Art was about to do some immunofluorescence for the SV40 T antigen and I asked him whether he could spare two coverslips to test my two antibodies. Two hours later, around midnight, we had a look in the fluorescent microscope. The antibody of the unknown protein gave essentially no fluorescence, but what the actin antibody revealed was indeed unbelievable. What a unique Christmas present. In the days to come, important suggestions on how to improve the immunofluorescence technique came from Pollack and Goldman and within two months Jim Watson communicated for us the first paper to the *Proceedings of the National Academy of Sciences.* This was soon followed by a set of papers showing that the filament bundles revealed by the actin antibodies coincided with the classically known microfilament bundles,[1] and contained also other known muscle proteins such as myosin,[2] tropomyosin,[3] and α-actinin,[4] further establishing the reality of the images revealed by the antibodies.

"The combination of SDS gel electrophoresis as a means of purifying structural proteins to be used subsequently as antigens, and the use of immunofluorescence as an assay for the cytoplasmic localization of these proteins provided a new methodological and conceptual approach to cell structure and motility. I think that this paper is quoted more often as being the first to demonstrate the distribution of actin using the immunofluorescence technique. For a recent review see reference 5."

1. **Goldman R D, Lazarides E, Pollack R & Weber K.** The distribution of actin in non-muscle cells. The use of actin antibody in the localization of actin within the microfilament bundles of mouse 3T3 cells. *Exp. Cell Res.* **90**:333-44, 1975.
   [The *SCI* indicates that this paper has been cited in over 205 publications since 1975.]
2. **Weber K & Groeschel-Stewart U.** Myosin antibody: the specific visualization of myosin containing filaments in non-muscle cells. *Proc. Nat. Acad. Sci. US* **71**:4561-5, 1974.
   [The *SCI* indicates that this paper has been cited in over 270 publications since 1974.]
3. **Lazarides E.** Tropomyosin antibody: the specific localization of tropomyosin in non-muscle cells. *J. Cell Biol.* **65**:549-61, 1975.
   [The *SCI* indicates that this paper has been cited in over 185 publications since 1975.]
4. **Lazarides E & Burridge U.** α-Actinin: immunofluorescent localization of a muscle structural protein in non-muscle cells. *Cell* **6**:289-98, 1975.
   [The *SCI* indicates that this paper has been cited in over 195 publications since 1975.]
5. **Weeds A.** Actin-binding proteins—regulators of cell architecture and motility. *Nature* **296**:811-16, 1982.

This Week's Citation Classic

Sternberger L A, Hardy P H, Jr., Cuculis J J & Meyer H G. The unlabeled antibody enzyme method of immunohistochemistry: preparation and properties of soluble antigen-antibody complex (horseradish peroxidase-antihorseradish peroxidase) and its use in identification of spirochetes.
J. Histochem. Cytochem. **18**:315-33, 1970. [Basic Sciences Dept., Medical Res. Lab., Edgewood Arsenal, and Dept. Microbiol., Johns Hopkins Univ. Sch. Med., Baltimore, MD]

Antigen can be localized immunocytochemically without use of covalently labeled antibodies by the sequence of primary antibody from species A, antibody to the immunoglobulin of species A produced in species B and applied in such a way that one antibody combining site reacts with the primary antibody and the other site remains free, soluble peroxidase-antiperoxidase complex (PAP) affinity purified from antiserum of species A, followed by histochemical reaction for peroxidase. The principle underlying this method yields high sensitivity because of inherently low background. [The *SCI*® indicates that this paper has been cited in over 1,280 publications since 1970.]

Ludwig A. Sternberger
Center for Brain Research
University of Rochester School of Medicine
Rochester, NY 14642

October 12, 1982

"In the fall of 1968, I was participating in the teaching of microbiology to medical students at Johns Hopkins University. Part of the laboratory requirement was an independent research project to be carried out in six two-hour periods. I felt that an assignment suitable for so short a time could be an attempt at immunocytochemistry without use of labeled antibodies. Such a technique would be simpler than conventional methods that used covalently labeled antibodies. I assumed that the experiment probably would fail, for the approach was so obvious that if successful, it should have been adopted long before. I prepared the reagents for the students, including affinity-purified antiperoxidase to be followed by peroxidase in a four-layer staining reaction. To my surprise, the students were able to visualize erythrocytes and spirochetes on their first attempt. This work was published[1] but has been quoted infrequently, perhaps for good reasons. Despite its high sensitivity, the method had the drawback of partial loss of peroxidase from low affinity antiperoxidase.'

"We' spent the ensuing months in futile attempts to resolve the low affinity problem that resulted from difficulties in dissociation of antiperoxidase from peroxidase during its purification, until one day it dawned on us, in the middle of another attempt, to change the protocol and aim to obtain from the immune precipitate soluble peroxidase-antiperoxidase (PAP) complex instead of antiperoxidase. We found that the addition of a small excess of peroxidase during the otherwise ineffective acid dissociation of peroxidase from antiperoxidase led to immediate solubilization of the precipitate, and thus intrinsically stable PAP has become readily available. The immunocytochemical use of PAP proved sensitive, because staining background was negligible. Much of the characterization of PAP was carried out in collaboration with Howard Meyer, who more than anyone else mastered not only the knowledge of structure and composition of PAP, but also the art of preparing it consistently at high concentration and activity, whether from antisera or from monoclonal antibodies.

"We had some difficulty in publishing these findings. One of the reviewers felt that the work was too insignificant to warrant a paper of its length. Influenced by such comments, I did not realize until 1974 that the method had the potential of wide applications (and frequent citations) because of its high sensitivity, low background, and consequent applicability to routinely fixed normal and pathologic tissue.

"Particularly interesting to me and my wife, Nancy, were the intriguing applications to neurobiology, so much so that we both decided to join the ranks of neuroscientists. Nancy soon discovered, during her work on myelin constituents, that with many (but not all) antigens, strong fixatives, such as osmium tetroxide, gave better structure and higher sensitivity, thus dispelling an old myth that in immunocytochemistry, fixation should be mild.

"My own interest in neurobiology was stimulated by the concept of the Scharrers[2,3] that peptidergic principles are fundamental to neuronal and vascular communication of cells. Our recent effort with monoclonal antibody immunocytochemistry confirms the universality of this concept and reveals a high diversity of such principles, sufficient to confer regional individuality to neurons. For a recent review, see reference 4."

1. **Sternberger L A.** Some new developments in immunocytochemistry. *Mikroskopie* **25**:346-61, 1969.
[The *SCI* indicates that this paper has been cited in over 30 publications since 1969.]
2. **Scharrer E & Scharrer B.** Secretory cells within the hypothalamus.
Res. Publ. Assn. Nerv. Ment. Dis. **20**:179-97, 1939.
3. **Scharrer B.** Neuroendocrinology and histochemistry. (Stoward P J & Polak J M, eds.)
Histochemistry: the widening horizons. New York: Wiley, 1981. p. 11-20.
4. **Sternberger L A.** *Immunocytochemistry.* second edition. New York: Wiley, 1979. 354 p.

Singer J M & Plotz C M. The latex fixation test. I. Application to the serologic diagnosis of rheumatoid arthritis. *American Journal of Medicine* 21:888-92, 1956.

The authors describe a latex fixation test for the serologic diagnosis of rheumatoid arthritis which improves on standard tests depending on erythrocyte agglutination by using biologically inert polyvinyl toluene and polysterene latex particles of uniform size. [The *SCI®* indicates that this paper was cited 566 times in the period 1961-1976.]

Dr. Jacques M. Singer, Director
Department of Microbiology & Immunology
Montefiore Hospital and Medical Center
Bronx, New York 10467

May 6, 1977

"Science is a puzzle-solving activity based upon one or more scientific achievements, while the perception of scientific novelties emerge as unpredictable events.

"Cecil, Nicholls and Stainsby (1931)[1] described the agglutination by rheumatoid arthritis sera of streptococci isolated from patients with RA. Later Waaler (1940)[1] and subsequently Rose and Ragan[2] showed that sheep cells coated with antibody against sheep blood would also agglutinate in the presence of RA sera. Thus two puzzles based on a similar agglutination system were presented.

"In 1946 Wallis[1] was able to demonstrate that a small percentage of patients with RA could agglutinate collodion particles, a substitute for sheep cells, previously suggested by Goodner (1941).[1] Although collodion particles were thought to react nonspecifically in sera of RA, Charles Plotz, in 1953, began a new study at Mount Sinai Hospital (New York) designed to utilize carrier particles. In 1954 Heller, Jaobson, Kolodny, and Kammerer[1] identified the antigen as being human IgG. When I started working as a fellow in 1955, I continued to work on this project, but for six months results were very discouraging. Prior to discontinuance of this work, Dr. Orenstein, from the electron microscopy section of Mount Sinai, was asked to look at collodion particles and determine their size if possible.

"In 1954 the first latex particles of uniform size were developed by Backus and Vanderhoff.[2] This followed the observations of Williams (1947)[2] that latex paints using particles of various sizes could be utilized in electron microscopy for secondary measurements of virus particles. Dr. Orenstein imbedded these particles in a micrograph grill. A print of the electron micrograph showed two or three perfect spherical, well dispersed latex particles, surrounded by a mass of particles which were clumped and unequal in size (our collodion particles). It was at this moment that the use of latex particles was visualized and their novel significance for many other applications fully perceived.

"Combining Heller's findings and our new latex, the latex fixation test was born. Clinical investigation of this test began, and in 1955 the second paper, 'Results in RA' by Plotz and Singer, appeared. Soon the last part of the puzzle was solved, the nature of the agglutinating factor of sheep cells or latex particles coated with HGG. Epstein, Johnson and Ragan (1956)[1] demonstrated that HGG reacts with sera of RA in a true immunological (precipitin) reaction. Franklin, Holman, Muller-Eberhard and Kunkel (1957)[1] found that the antibody in RA sera is a 19S and 22S component called rheumatoid factor (RF). This opened the vista for further development of the latex test and, in immunology, the concept of auto-antibodies in RA sera and further development of the latex test.

"The principle involved, the use of latex particles which can be coated with antigens or antibodies, found applications in a great variety of serological tests as well as many other applications (in medicoserological tests, in the study of phagocytosis *in vivo* and *in vitro*, as immunological markers, in electron microscope scanning, adjuvancy, etc.). A new industry has been created producing and using these particles for scientific purposes including the development of kits for use in doctors' offices and laboratories.

"With all the vast number of developments over the past 22 years it is of interest that the original latex fixation test utilizing the same 0.8 latex particles has remained unchallenged and the standard by which other tests are judged."

REFERENCES

1. **Singer J M.** Latex fixation test in rheumatic diseases — a review. *American Journal of Medicine* 31:766-79, 1961.
2. **Vanderhoff J W.** The use of monodisperse latex particles in medical research. American Chemical Society, Division of Organic Coatings and Plastic Chemistry. *Preprints* 24:223-32, 1964.

This Week's Citation Classic

Grabar P, Williams C A, Jr. & Courcon J. Méthode immuno-électrophorétique d'analyse de mélanges de substances antigéniques. (Method for immuno-electrophoretic analysis of mixtures of antigenic substances.) Biochim. Biophys. Acta 17:67-74, 1955.
[Service de Chimie Microbienne, Institut Pasteur, Paris, France]

The paper describes a simple method which in a single operation enables the definition of complex mixtures of antigens or haptens by three independent criteria: the chemical or biochemical properties of the antigens (using various dyes or enzyme substrates), their electrophoretic mobility, and their antigenic specificity. [The *SCI®* indicates that this paper has been cited over 680 times since 1961.]

P. Grabar
Institut Pasteur
25, Rue du Docteur Roux
Paris 75024
France

November 22, 1979

"It is a particular pleasure for me to be classified as one of the most-cited authors during the years 1961-75. I think that, generally, articles on methods are more often cited than papers on experimental works if the described method can be and is easily applied in various domains. The method called 'immuno-electrophoretic analysis' is now designated 'immuno electrophoresis' (which is grammatically incorrect), and in most recent publications in which this method is used the authors no longer cite my name, probably because they consider it a well-known method. Historically, I came to develop it when in 1951-52 I tried to establish a method for what is now known as electrofocusing, using paper-strips and a gradient of buffers. At that time no insoluble ampholites were available and my gradient was unstable. In order to stabilize this system I used agar-gels. In order to precipitate the electrophoretically separated proteins I tried to use specific antibodies, i.e., an immune-serum with which I covered the gel. Then, instead of such a covering, I disposed on the sides of the electrophoretic axis. gels containing the immune-serum in which the antigens were diffusing and thus forming precipitin arcs. In the last period of these experiments I was helped by Curtis A. Williams, at that time a young student. A short preliminary paper was published in the beginning of 1953[1] which was followed by this in the same journal in 1955. But the complete and detailed description of the method and of some applications was published in a monograph in collaboration with Pierre Burtin and several others. [2] The French edition appeared in 1960 and was followed a few years by English, German, Russian, Spanish, and Chinese translations. Very rapidly, the method came to be employed in many laboratories around the world, and we were then obliged to arrange many special short courses to teach it to the very large number of persons interested in its applications. Now it is applied as a diagnostic method in hospitals and in studies on immunology, microbiology, pathology, endocrinology, embryology, animal and plant physiology, etc.

"A dissertation for the degree of Doctor in Medical Sciences at the University of Rostock (East Germany) by M. Henker and G. Otto was entirely concerned with the bibliography on immuno-electrophoretic analysis from 1953 to 1970 and contained 5,275 classified citations. This shows that this method came to be used very rapidly in many studies.

"Several authors have proposed various modifications of the original method; the most interesting is the quantitative immuno-electrophoresis of Laurell." [3]

1. Grabar P & Williams C A. Méthode permettant l'étude conjuguée des propriétés électrophorétiques et immunochimiques d'un mélange de protéines. Application au sérum sanguin. (Method permitting the twin study of electrophoretic and immunochemical properties of a mixture of proteins. Application to blood serum.) Biochim. Biophys. Acta 10:193-4, 1953.
2. Grabar P & Burtin P, (eds.). Avec la collaboration de Chevance L G. Analyse immuno-électrophorétique: ses applications aux liquids biologiques humains. (Immunoelectrophoretic analysis: its applications to human biological fluids.) Paris: Masson, 1960. 294 p.
3. Laurell C B. Quantitative estimation of protein by electrophoresis in agarose gel containing antibodies. Analyt. Biochem. 15:45-52, 1966.

This Week's Citation Classic

Laurell C B. Quantitative estimation of proteins by electrophoresis in agarose gel containing antibodies. *Anal. Biochem.* 15:45-52, 1966.
[Department of Clinical Chemistry, Malmö General Hospital, Malmö, University of Lund, Sweden]

The paper describes a method for rapid quantitative analysis of proteins with a charge differing from that of the bulk of the immunoglobulins. The method utilizes the difference between the rate of electrophoretic migration of proteins and of their antibody complexes in agarose gel. [The *SCI** indicates that this paper has been cited over 1,640 times since 1966.]

Carl-Bertil Laurell
Department of Clinical Chemistry
University of Lund
Malmö General Hospital
S-214 01 Malmö
Sweden

October 24, 1980

"Quantitation of proteins by electrophoresis into agarose containing the corresponding antibodies has received the nickname 'rocket' (immuno) electrophoresis because the precipitates formed when the migrating protein antigens are precipitated by the antibodies resemble upright rockets. The method was later more thoroughly described under the name of electroimmunoassay,[1] whereas electroimmunodiffusion is more frequently used in the US. The original method was invented to minimize the effect of diffusion! The historical background is as follows.

"I had found a slight difference in mean electrophoretic mobility of α_1-antitrypsin in normal plasma and in plasma from patients with α_1-antitrypsin deficiency. I assumed that the deficiency cases had a slightly abnormal protein in low concentration or that their α_1-antitrypsin was in complexed form. I wanted to demonstrate the occurrence of molecules with normal and retarded mobility in plasma from heterozygotes with α_1-antitrypsin deficiency having only half the normal amount of α_1-antitrypsin in their plasma.

"On immunoelectrophoresis, the precipitation bows with rabbit anti-α_1-antitrypsin

were too extended to allow any conclusions about one or two precipitation maxima to be drawn. Our agar gel electrophoresis had high resolution, but development of immuno precipitates by diffusion—the second step in immunoelectrophoresis—caused blurring. Therefore, I invented (antigen-antibody) crossed immunoelectrophoresis. Here the diffusion step of immunoelectrophoresis was exchanged for an electrophoresis of the separate proteins into a gel with mainly stationary antibodies. The precipitation of proteins in an electric field by antibodies was rapid enough to give precipitates with peak heights related to the amount of antigen applied.

"Tristram Freeman from Mill Hill visited us and became enthusiastic about crossed immunoelectrophoresis. He announced the following spring that he had developed a quantitative method which he wanted to present in June.[2] Our old collaborator, J. F. Heremans, reported simultaneously that he had developed an immunochemical method for determination of plasma proteins together with Mancini.[3] Most plasma proteins had been isolated around 1960, creating an urgent need for fast methods of quantitative specific protein estimation. My experience from crossed immunoelectrophoresis suggested the logical development of a one-dimensional technique as a variant of crossed immunoelectrophoresis. Both the techniques could be used as alternatives to Mancini–Heremans single radial immunodiffusion.

"Of these methods, the 'Mancini' technique has been most used because of its simplicity and because the immunoplates are commercially available. Both will slowly lose in popularity and nephelometric methods will spread for a while, but 'radioimmunorockets' will probably have a good future in the protein concentration range of 5–0.1 mg/l. Variants of crossed immunoelectrophoresis will probably spread more, because of its capacity to reveal antigenic microheterogeneity, indicate complex formation, and unmask partial antigenic identity."

1. **Laurell C B,** ed. Electrophoretic and electro-immunochemical analysis of proteins.
   Scand. J. Clin. Invest. Suppl. 124 29:1-36, 1972.
2. **Clark H G M & Freeman T.** A quantitative immuno-electrophoresis method (Laurell electrophoresis) (Peeters H, ed.) *Protides of the biological fluids.* Amsterdam: Elsevier, 1967. p. 503-9.
3. **Mancini G, Carbonara A O & Heremans J F.** Immunochemical quantitation of antigens by single radial immunodiffusion. *Immunochemistry* 2:235-54, 1965.

Deutsch H F & Morton J I. Dissociation of human serum macroglobulins.
Science 125:600-1, 1957.
[Department of Physiological Chemistry, University of Wisconsin, Madison, WI]

When gamma globulins with molecular weights of about 1,000,000 are treated with mercaptans at neutral pH they are readily converted into subunits of about one-fifth the size of the parent molecules. Removal of the mercaptan leads to some reformation of the original protein and this reaggregation is blocked by alkylating agents. This demonstrated that the macromolecular type of antibody molecule contains subunits linked by disulfide bonds. [The *SCI®* indicates that this paper has been cited over 480 times since 1961.]

Harold F. Deutsch
Department of Physiological Chemistry
University of Wisconsin
Madison, WI 53706

April 4, 1978

"My interest in antibodies could be justly attributed to the 'fortunes of war' since my studies in this area were initiated in 1944 at the conclusion of my doctoral work in an unrelated field. I then participated in a project at the University of Wisconsin that was a subcontract of work on blood plasma fractionation at the Harvard Medical School sponsored by the Office of Naval Research. My initial task was to increase the yield of a human gamma globulin fraction that is now known as normal IgG. Concurrent studies at Wisconsin were being made to convert human IgG into smaller molecules by digestion with papain, bromelin, and pepsin. The latter enzyme had been previously employed commercially to diminish adverse effects of horse antitoxic sera. My attention was directed to reports in which cysteine activated papain had been employed to digest human IgG into what are today defined as Fc and Fab fragments. Since no cysteine control had been employed, I reduced IgG with it. Such treatment gave protein with an increased electrophoretic mobility but no significant changes in sedimentation behavior were noted. The heavy and light chain components most likely formed were not detected since dissociating conditions had not been employed.

"These studies were carried out concurrently with the isolation of a new human globulin fraction characterized by a different profile of antibody activities, greater electrophoretic mobility and containing considerable amounts of 19S protein that was relatively insoluble in water, i.e. euglobulin, as compared with the largely water soluble, i.e. pseudoglobulin, IgG fraction. The high molecular weight of the 19S component (approx. 1,000,000) eventually led to its designation as a gamma macroglobulin or IgM. Certain human pathologic sera had been noted to contain large amounts of euglobulin sedimenting at 19S but also containing from 15 to 25% of what appeared to be higher polymers of the 19S material. The electrophoretic homogeneity of one of these molecularly heterogeneous proteins that we had crystallized in 1956 had led us to suggest at this time that even the 19S material might be a polymeric form of a lower molecular weight unit. This was directly probed about one year later and it was found that reduction of both pathological and normal human IgM source proteins readily converted all of the IgM components into protein sedimenting near 7S. These IgM subunits of about 200,000 molecular weight reaggregated reversibly upon removal of the mercaptan and this could be blocked by alkylation of the reduced protein. At this time we also naturally tried the reduction and alkylation of 7S myeloma source Ig proteins. One of these was found to be converted into material sedimenting at 3.5S but this was not the usual result. The later demonstrated effects of detergents and of low pH on breaking non-covalent bonds between antibody subunits was not appreciated at this early date. Thus these experiments are cast in the category of 'ships that pass in the night.' However, our report on the chemical dissociation of IgM proteins essentially ushered in the era of the non-enzymatic production of antibody fragments designated as light and heavy chains and stimulated a great deal of the subsequent studies on the chemical nature of antibody subunits and their linkages. Numerous attempts to effect *in vivo* reductions of IgM components elevated in various pathologic states were also made following our report in *Science* but no significant improvement was noted in these patients."

This Week's Citation Classic

Nisonoff A, Wissler F C, Lipman L N & Woernley D L. Separation of univalent
fragments from the bivalent rabbit antibody molecule by reduction of disulfide
bonds. *Arch. Biochem. Biophys.* **89**:230-44, 1960.
[Depts. Biochem. Res. and Biophys., Roswell Park Mem. Inst., Buffalo, NY]

In this paper it was demonstrated that univalent fragments of rabbit antiovalbumin, having the same sedimentation coefficient and inhibitory effect on precipitation as the antigen-binding fragments produced by papain, can be prepared by treatment with pepsin in the presence of a reagent capable of breaking disulfide bonds. The reaction could also be carried out by successive treatments with pepsin and one of several such reagents. [The *SCI®* indicates that this paper has been cited over 685 times since 1961.]

Alfred Nisonoff
Department of Biology
Rosenstiel Research Center
Brandeis University
Waltham, MA 02154

September 28, 1981

"This work was initiated after Rodney Porter had demonstrated that papain cleaves rabbit IgG into two antigen-binding fragments (Fab) and one crystallizable fragment (Fc), roughly equal in size.[1] We wanted to study hapten-binding properties of active fragments derived from purified antibody to provide direct evidence for univalence and to ascertain whether there is cooperativity between the two binding sites. (We found later that there is none.) The work with proteolytic enzymes other than papain was initiated when it occurred to us that the cysteine present in preparations of papain to maintain its activity might be participating directly in the cleavage by reduction of interchain disulfide bonds. We therefore looked for a two-step cleavage, involving an enzyme and reducing agent, and found that this occurs when pepsin is used. Interestingly, our initial premise was wrong, i.e., that during cleavage by papain, cysteine was needed to reduce one or more interchain disulfide bonds. Papain and pepsin cleave on opposite 'sides' of the interheavy chain disulfide bond. Schematically, pepsin yields Fab-S-S-Fab + partially degraded Fc, whereas papain releases two Fab fragments from Fc, leaving the two sulfur atoms in the Fc fragment. After peptic digestion the two antigen-binding Fab fragments are linked only by an interheavy chain disulfide bond and not by additional noncovalent interactions; reduction of the disulfide bond yields univalent fragments, designated Fab'.

"Initially, the investigation was of interest because it provided information as to the topological arrangement of the three major fragments and the structural role of an interchain disulfide bond (subsequently shown to join two heavy chains). Peptic digestion has been used by others primarily because it provides a means of removing the Fc fragment without altering the bivalence of the antibody. The capacity to precipitate or agglutinate antigens is therefore retained. One can study various biological functions of the crystallizable (Fc) fragment by determining the effect of its removal, for example, on complement fixation. With the discovery of receptors for fragment Fc on various types of leukocytes, peptic digestion has been useful for preparing antibodies that will not bind to such receptors, or for determining whether binding of untreated molecules occurs through an Fc fragment.

"Peptic digestion followed by reduction and reoxidation also provides a means of artificially generating antibody molecules with two different combining sites of desired specificity (hybrid antibodies). So far this procedure has had limited application, principally in the staining of cell surfaces with ferritin by using a hybrid of anti-ferritin antibody and antibody to a cell surface antigen. The use of hybrid antibody has also been considered as a means of bringing a pharmacological agent specifically into contact with a desired tissue surface.

"This paper has been cited often because it provides a method for preparing bivalent antibody lacking the Fc fragment and an approach to the preparation of hybrid antibodies; also, because it helped to elucidate IgG structure.

"Collaborating in this research were Donald Woernley and Frank Wissler of the biophysics department at the Roswell Park Memorial Institute and Leah Lipman in the biochemistry department. For a recent review see *The Antibody Molecule*."[2]

1. **Porter R R.** The hydrolysis of rabbit γ-globulin and antibodies with crystalline papain. *Biochemical J.* **73**:119-26, 1959.
2. **Nisonoff A, Hopper J E & Spring S B.** *The antibody molecule.* New York: Academic Press, 1975. 542 p.

Levy H B & Sober H A. A simple chromatographic method for preparation of gamma globulin. *Proc. Soc. Exp. Biol. Med.* **103**:250-2, 1960.
[US Dept. Health, Education, and Welfare, NIH, Natl. Inst. Allergy and Infectious Diseases, Natl. Cancer Inst., Bethesda, MD]

A relatively simple method for separation of gamma globulin from serum has been presented. The procedure has been applied to a number of antisera, and it has been shown, with them at least, to give reasonably good localization of antibody in the gamma globulin fraction. [The *SCI*® indicates that this paper has been cited over 740 times since 1961.]

Hilton B. Levy
Section of Molecular Virology
Laboratory of Viral Diseases
National Institute of Allergy and
Infectious Diseases
National Institutes of Health
Bethesda, MD 20014

May 27, 1981

"This work is an example of the kind of serendipity that can develop when a group of people with a common interest meet. In the days when the National Institutes of Health was relatively small, the professional staff would meet informally in a small library in Building 7 and have lunch around an old mahogany table that had been with the US Public Health Service for many, many years. Topics of conversation were varied, including sports, politics, automobiles, and a heavy dose of science.

"In those days a major emphasis of the research of the National Institute of Allergy and Infectious Diseases was on the epidemiology of virus illness and the identification of new viruses of potential etiological importance. Tissue culture was coming into widespread use for the isolation of viruses.

The identification and relationship among these newly discovered agents involved the specific antisera against previously identified agents. If a given antiserum could neutralize the infectivity in tissue culture of a new agent, these two agents were either identical or closely related. However, frequently a crude whole antiserum would prove toxic to cells in tissue culture or, in other ways, would give effects which would confuse the results. I suggested that my colleagues purify the antiserum at least to the extent of using just gamma globulin. It was suggested, in no uncertain terms, that since there were no simple methods available for virologists to purify small amounts of antiserum, that I, a chemist, meaning a non-MD, devise such a method. Together with Herb Sober, this was done.

"I am pleased that the article was well received, but I do not feel that it was a major contribution. It does show: (1) how worthwhile ideas sometimes are born under unexpected circumstances, and (2) that 'methods' papers sometimes find widespread application. A corollary of this first point is the concept of a critical mass of people in scientific research. It is very frequently the subtle interplay of thoughts expressed by people in different disciplines that leads to the expression of an unfilled need by one element, and the realization by another element that he or they have the ability to meet this need. Isolated workers do not have the opportunity for this interaction.

"I regret to have to say that this was an isolated work of mine that apparently filled a need for such people, but bore no relationship to the molecular biology of viruses and resistance to virus infection that has been my major interest."[1]

1. **Levy H B, Levine A S, Engel K, Leventhal B, Krown S, Durie B & Stephen B.** Preliminary clinical results with a primate effective interferon inducer. (Hersch E M, Chirigos M A, and Mastrangelo M J, eds.) *Augmenting agents in cancer therapy.* New York: Raven Press, 1981.

Kessler S W. Rapid isolation of antigens from cells with a staphylococcal protein A-antibody adsorbent: parameters of the interaction of antibody-antigen complexes with protein A. *J. Immunology* 115:1617-24, 1975.
[Dept. Microbiology and Immunology, UCLA Sch. Med., Los Angeles, CA]

This paper introduced the concept of using fixed protein A-bearing staphylococci as an adsorbent for antibodies complexed with radiolabeled antigens from cell lysates. Antigen isolation procedures using this immunoprecipitation system were shown to be considerably faster and more quantitative, specific, versatile, and economical than conventional methods. [The *SCI®* indicates that this paper has been cited in over 1,390 publications since 1975.]

Steven W. Kessler
Departments of Medicine
and Biochemistry
Uniformed Services University of the
Health Sciences
Bethesda, MD 20814

January 18, 1983

' "In the early-1970s, relatively few workers had sufficient faith in the potential of immunoaffinity techniques for isolation of antigenic cell proteins to devote much time to the technology. I was convinced that it was just a matter of time until most details of membrane protein structure would be made accessible by this approach, and I had chosen a PhD project on the characterization of lymphocyte immunoglobulins that obliged me to justify my convictions. Most of my colleagues were more comfortable with their 'brute force' isolation methods, and felt dutybound to remind me of all the problems inherent in the double antibody immunoprecipitation system which reflected the current state of the art (see this and subsequent papers[1,2] for the grim details). Although my own system seemed to work as well as anyone's who published in my field, the results were typically and frustratingly variable.

" 'Salvation' came one day (my birthday) in the wake of a particularly grueling and largely inconclusive experiment. Making the best of a gloomy situation, I absentmindedly began to read an article on a subject that over the years I had studiously avoided, due to its apparent irrelevance to my interests. The topic was the affinity of staphylococcal protein A for IgG.[3] The realization that protein A-bearing staphylococci might substitute for my second antibody came immediately and seemed so logical that it simply had to work. Within a few weeks I had located a source for the appropriate strain (Cowan I) of *S. aureus* (at the time it wasn't listed in the ATCC catalog), obtained a seed culture, and prepared a test batch. My first experiment, designed purely empirically, gave the cleanest polyacrylamide gel patterns of lymphocyte immunoglobulins I had ever seen.

"As one of the first demonstrations of staphylococcal protein A being put to some practical use, this paper helped to hasten an awareness of the molecule's myriad other applications. Simplifying the immunoprecipitation approach to antigen isolation created opportunities for more workers to enter the field. Ultimately, it may have to accept some blame as well for having helped to spawn a generation of immunochemical 'experts' who have never mastered antigen-antibody equivalence point titrations."

1. **Kessler S W.** Cell membrane antigen isolation with the staphylococcal protein A-antibody adsorbent. *J. Immunology* **117**:1482-90, 1976.
2. ----------------. Use of protein A-bearing staphylococci for the immunoprecipitation and isolation of antigens from cells. *Meth. Enzymology* **73**:442-58, 1981.
3. **Kronvall G, Quie P G & Williams R C, Jr.** Quantitation of staphylococcal protein A: determination of equilibrium constant and number of protein A residues on bacteria. *J. Immunology* **104**:273-8, 1970.

Mergenhagen S E, Snyderman R, Gewurz H & Shin H S. Significance of
complement to the mechanism of action of endotoxin.
Curr. Topics Microbiol. Immunol. **50**:37-77, 1969.
[Immunol. Sect., Lab. Microbiol., Natl. Inst. Dental Res., NIH, Bethesda, MD; Depts. Pediat. and Surg.,
Univ. Minnesota, MN; and Dept. Microbiol., Johns Hopkins Univ. Sch. Med., Baltimore, MD]

This paper reviews data related to the role of complement in endotoxin action. Newer information on alternative pathways of complement activation as well as on the biologically active cleavage products of complement are discussed in terms of their relationship to the pathophysiology of endotoxin-induced inflammatory reactions. [The *SCI®* indicates that this paper has been cited in over 120 publications since 1969.]

Stephan E. Mergenhagen
Laboratory of Microbiology and Immunology
National Institute of Dental Research
National Institutes of Health
Bethesda, MD 20205

January 11, 1984

"In the late 1960s, two young physicians, Henry Gewurz and Ralph Snyderman, joined my laboratory at the National Institute of Dental Research as research associates in the US Public Health Service. Gewurz had significant training in immunology and considerable expertise in complement research because of his prior associations with Robert A. Good and Manfred Mayer. Snyderman, fresh out of an internship and residency program at Duke University, expressed an interest in investigating fundamental aspects of the inflammatory response and, more specifically, in studying the role of complement in leukocyte locomotion (chemotaxis). Because of a long-standing involvement in research on bacterial endotoxin, I was particularly interested in learning how inflammatory cells migrate into and become activated in inflammatory foci induced with gram-negative bacteria or their endotoxic lipopolysaccharide. In one of his first experiments, Snyderman showed that lipopolysaccharide, unlike other bacterial products, was not directly chemotactic for polymorphonuclear leukocytes when evaluated *in vitro* in a modified Boyden chamber. However, he found that a low molecular weight (15,000) chemotactic factor could be generated in serum by lipopolysaccharide. This factor could not be produced in heated (56°C for 30') serum or in serum deficient in the fifth component of complement. Earlier work by Peter Ward suggested that complement participated in the production of chemotactic activity; our data indicated that a cleavage product of C5 might be the chemoattractant. Further collaboration with Hyun Shin at Johns Hopkins University and Joerg Jensen at the University of Miami demonstrated conclusively that lipopolysaccharide, as well as antigen-antibody complexes, generated a low molecular weight chemotactically active peptide (C5a) from the fifth component of complement. From that point on, we studied various biological activities generated by the interaction of endotoxins with the complement system which included alterations in vascular permeability, smooth muscle reactivity, and mast cell and platelet degranulation. Many of these activities were mediated by the release of C5a or by other complement cleavage products, and their formation could explain certain of the inflammatory consequences which follow after an injection of endotoxin into a susceptible host. Indeed, C5a has proven to play a central role in mediating inflammation associated with complement activation in a number of human diseases.

"In searching the literature it became evident that others had suggested a role for complement in endotoxicity.[1] Lipopolysaccharides, like antigen-antibody complexes, were efficient activators of the complement system and many of the sequellae of immune complexes *in vivo* could be reproduced with endotoxin.[2] However, similar to the earlier observation of Louis Pillemer,[3] Gewurz showed that lipopolysaccharides, unlike immune complexes, consume each of the terminal components of complement (C3-C9) with minimal, if any, consumption of C1, C4, or C2. These findings, as well as the observation showing preferential consumption of late complement components by cobra venom factor, prompted us to suggest an alternative pathway for complement activation. Our notions became more of a reality when Ann Sandberg and Abraham Osler showed that guinea pig γ1 antibodies activated the late complement components with sparing of the earlier components.[4] The availability of guinea pigs lacking detectable C4 provided further evidence for an alternative pathway.[5] The purification and characterization of distinct serum proteins participating in the alternative pathway were provided by extensive studies from several laboratories.[6]

"The value of the review article published in 1969 and the reason that it has been frequently cited are twofold. First, it pointed to complement and its fragmentation products, such as C5a, as potential effectors of the endotoxin shock syndrome. Secondly, it stimulated others to provide precise experimental evidence for the existence of an alternative pathway of complement activation."

1. **Spink W W.** Endotoxin shock. *Ann. Intern. Med.* 57:538-52, 1962. (Cited 105 times.)
2. **Stetson C A.** Role of hypersensitivity in reactions to endotoxin. (Landy M & Braun W, eds.) *Bacterial endotoxins.* New Brunswick, NJ: Rutgers University Press, 1964. p. 658-62.
3. **Pillemer I, Schoenberg M D, Blum L & Wurz L.** Properdin system and immunity. II. Interaction of the properdin system with polysaccharides. *Science* 122:545-9, 1955. (Cited 255 times since 1955.)
4. **Sandberg A L, Osler A G, Shin H S & Oliveira B.** The biologic activity of guinea pig antibodies. II. Modes of complement interaction with γ-1 and γ-2 immunoglobulins. *J. Immunology* 104:329-34, 1970. (Cited 150 times.)
5. **Ellman L, Green I, Judge F & Frank M M.** *In vivo* studies in C4-deficient guinea pigs. *J. Exp. Med.* 134:162-75, 1971. (Cited 50 times.)
6. **Müller-Eberhard H J & Schreiber R D.** Molecular biology and chemistry of the alternative pathway of complement. *Advan. Immunol.* 29:1-53, 1980.

CC/NUMBER 3
JANUARY 21, 1980

Ada G L, Nossal G J V, Pye J & Abbot A. Antigens in immunity. 1. Preparation and properties of flagellar antigens from *Salmonella adelaide*.
Aust. J. Exper. Biol. Med. Sci. 42: 267-82, 1964.
[Walter and Eliza Hall Institute of Medical Research, Melbourne, Australia]

The paper describes the preparation and physical and chemical properties of the monomeric protein flagellin and of the formation *in vitro* of polymerised forms which appear similar in appearance to the parent flagella particles. [The *SCI*® indicates that this paper has been cited over 240 times since 1964.]

G.L. Ada
Microbiology Department
John Curtin School
of Medical Research
Australian National University
Canberra City, ACT 2601
Australia

August 7, 1979

"This work, together with the subsequent papers in the series, came about for two reasons. Spurred on by Burnet's concepts, my own interests were switching from virology to immunology. G.J.V. Nossal had recently returned to the Walter and Eliza Hall Institute after a two year period in Joshua Lederberg's new department at Stanford. After numerous discussions, Nossal and I were struck by the lack in the literature of experiments which attempted to determine how small, 'physiological' amounts of a bacterial or viral antigen might stimulate an antibody response. It seemed to us that the time had come to attempt to describe such events in molecular terms.

"The sort of questions we wished to answer included: Did the immunogen react directly with the precursor of the antibody forming cell? Did it become associated with the protein synthesizing units in the cell as a template, as many thought at that time? If so, how many molecules were involved? To answer such questions, we needed to use an antigen which had well-defined chemical and physical properties, was highly immunogenic, and could readily be labelled with a radioisotope to high specific activity.

"Fortunately for us, Hunter and Greenwood had recently described a direct oxidation method of iodination using a carrier-free preparation of iodide-131.[1] Two candidates for choice of immunogen were the influenza virus and the flagellar proteins isolated from *Salmonella*. The influenza virus, though highly immunogenic, was known to contain a number of different proteins so was less suitable. So the flagellar protein won the day. The paper describes the chemical and physical properties of the monomeric flagellin, and the original finding that it could readily be polymerised *in vitro* to form rod-shaped structures similar in appearance to the parent flagellin particles. Some of the immediate results from using this antigen were the demonstration of particular patterns of antigen localization in lymphoid tissues, factors which affected these patterns, and the failure to find antigen in antibody forming cells. The availability of the antigen led later to an appreciation of the role of multivalency in cell surface reactions.

"A major reason for the subsequent frequent citation of the paper was that the protein became widely used in immunological work, particularly by the many scientists who at some time or another worked at the Hall Institute."

1. **Hunter W M & Greenwood F C.** Preparation of iodine-131 labelled human growth hormone of high specific activity. *Nature* 194:495-6, 1962.

This Week's Citation Classic

CC/NUMBER 33
AUGUST 18, 1980

Vaitukaitis J L, Robbins J B, Nieschlag E & Ross G T. A method for producing
specific antisera with small doses of immunogen.
J. Clin. Endocrinol. Metab. 33:988-91, 1971.
[Reprod. Res. Br., Nat. Inst. Child Health Hum. Develop., NIH, Bethesda, MD]

A method for generating antisera with small doses of immunogen is described. With this technique, 100 ug or less of immunogen induced specific antibody production in rabbits injected intradermally. Moreover, animals injected with a single dose of the immunogen continued to produce antisera for several months in response to the single immunizing dose. [The *SCI®* indicates that this paper has been cited over 535 times since 1971.]

Judith L. Vaitukaitis
Departments of Medicine and
Physiology
Boston University School of Medicine
Boston City Hospital
Boston, MA 02118

June 26, 1980

"In the early 1970s, Griff Ross and I were studying structure-function relationships of human chorionic gonadotropin in terms of both its biologic and immunologic activities. This work was carried out in collaboration with Robert Canfield and Frank Morgan at the College of Physicians & Surgeons in New York City. The New York laboratory was purifying and biochemically characterizing hCG and had worked out a technique to separate hCG into its two subunits. A limited amount of highly purified subunit preparations was available to us for study. Moreover, the characterization of the subunits' amino acid composition and sequence was limited at that time. We thought the most rapid way of attaining some understanding of the structure-function relationship of the subunits was with an immunologic approach.

"Since relatively small quantities of the hCG subunits could be used for immunization, we decided to try our hand at generating antibody with small amounts of immunogen. Our laboratory and others had generally used mg quantities of antigens for immunization up to this time. Consequently, John Robbins, a microbiologist at NICHHD, was consulted, since he too used a variety of immunization techniques in his studies. Since no techniques incorporated microgram quantities of immunogen, we collectively decided to modify our immunization technique because of our restricted, precious supply of highly purified hCG subunits. In short, we rationalized that if an animal were injected intradermally over a wider area with increased concentrations of heat-killed tubercle bacillus in an emulsion containing microgram quantities of the subunit, antibody with high affinity and specificity might result. Since our initial studies were fruitful, we wished to know whether haptens conjugated to carrier proteins would also induce similar responses in animals immunized with 100 ug or less of conjugated haptens. Consequently, E. Nieschlag was invited to participate in our study with a steroid conjugated to bovine serum albumin or keyhole limpet hemocyanin.

"In short, the technique worked with generation of antibody of high titer and affinity being harvested between eight to 16 weeks after the primary immunization. Because the technique was so successful in our hands, we wished to share it with our scientific colleagues and, consequently, published it as a Rapid Communication. Just as highly purified hormonal preparations became available in the early 1970s, small quantities of highly purified preparations of non-hormonal substances were generated with the newer separation techniques. Consequently, investigators in other fields wished to generate antibody to their highly purified preparations and, consequently, used our approach.

"I continue to receive phone calls from investigators around the country who wish to use that technique to generate antibody to their precious, limited quantities of substances ranging from proteins extracted from viruses to subcellular enzymes. At the time we developed the technique, we had no idea how widely that approach would be utilized."

119

3.3 Immunity

The bactericidal immunoglobulins in blood can function alone or abet the action of phagocytic cells in the cytodestruction of pathogenic microbes. The "This Week's Citation Classic" (TWCC) commentaries by Miles, Clyde, and Biozzi touch on both these aspects of immunity. In commenting on one of the oldest Citation Classics covered in this volume, Miles relates his development with Misra of the surface count method for determining the bactericidal activity of defibrinated blood. It is amazing that in 1938 a staff biostatistician was available to assist in this project. Clyde relied on the growth inhibition of *Mycoplasma* species by antisera to establish a classification system for this genus of unusual microbes. The quantitative *in vivo* determination of phagocytosis reported by Biozzi, Benacerraf, and Halpern is still useful today, and their corrected phagocytic index α has been useful both in proving the critical role of spleen and liver cells in phagocytosis and in comparing phagocytosis in animals with different spleen–liver weights.

Phagocytic engulfment does not necessarily lead to intraphagocytic death. The latter relies on an oxidative burst by the phagocyte, as Sbarra describes, and the generation of reactive forms of oxygen. Myeloperoxidase is one agent that utilizes a toxic form of oxygen in phagocytes to achieve a lethal hit by utilizing H_2O_2 of phagocyte origin with a halide to kill the intracellular agent (p. 126).

The discovery of secretory immunoglobulins was an important key in explaining the resistance of the mucous membranes to invasion, the control of dental caries, and possibly also resistance to allergens. The discovery of secretory IgA was actually simultaneous in two separate laboratories. Since then a secretory form of IgM has also been described, but it is less abundant in secretions and probably has a lesser role in immunity.

The shared role of the thymus-derived and bone marrow-derived lymphocytes, the T and B cells, in immunity was uncovered gradually and represents a synthesis of ideas from many scientists. The group at Minnesota headed by R. A. Good made much of the "Experiments of Nature," which they embellished with controlled experiments to categorize immunodeficiency in terms of abnormalities in one or both of these lymphocyte populations.

The manipulations needed for successful skin grafting in laboratory animals and the perpetuation of lymphocytes in culture are the subjects of two of the most important of the remaining Classics. Medawar could not have been a Nobel Prize recipient if a method for the analysis of tolerance to skin grafts had not been perfected at the level of skin graft transplantation as a surgical procedure. The Marbrook system for perpetuating cells in tissue culture is based on the premise that larger culture volumes will dilute the toxic end products of cell metabolism and permit more extensive growth of the cells and extend their longevity (p. 74).

The remaining eight Classics described in this section stem from unrelated topics in immunity. Dausset's Nobel Prize-winning discovery of histocompatibility antigens in humans and how the designation HLA came to be are related in his TWCC. Tumor-specific transplantation antigens (TSTA) dependent upon host replication of virus antigens are the basis for the Classics by Klein and Purchase. Eva Klein acknowledges that the article published with her husband is oft cited because of the simplicity of the procedure, even though it is a second in the field of virus TSTA. Economics of Marek's disease in poultry, though of vast importance, is secondary to the discovery by Okazaki, Purchase, and Burmester that a viral vaccine could actually prevent cancer in chickens. This depends on the antigenic constancy of virus-induced, as opposed to chemically induced, tumors.

Schwartz's Classic neatly unifies a clinical observation with laboratory experiments, albeit in a separate species, for proof of a hypothesis based on the former, i.e., the concept that constant stimulation and replacement of cells permits that carcinogenic change to occur which is lacking in resting cells.

Immunity to transplanted cancer cells, not based on the response to TSTA but to differences in constitutive transplantation antigens, led to an eventual recognition that genetically matched donor and recipient are needed in cell-transfer studies (p. 135).

One of the most widely studied autoimmune diseases, systemic lupus erythematosus (SLE), was unrecognized as a systemic illness until 1954 (p. 136), despite Osler's report some 50 years earlier than the Classic of Harvey and his coauthors. Although the latter established a firm scientific basis for SLE as a systemic disease, the admonitions of Osler in his lectures in 1885 are an excellent reminder of the medical genius of Sir William Osler.

Miles A A & Misra S S. The estimation of the bactericidal power of the blood.
J. Hygiene **38**:732-49, 1938.
[Department of Pathology, British Postgraduate Medical School, London, England]

The survival rate, *p*, of a measured inoculum of *Staph. aureus* in a standard volume of defibrinated blood, is a reliable quantitative measure of the bactericidal power of blood. The number of viable organisms in the inoculum and in the blood-bacterium mixture may be estimated with the necessary accuracy by counts of colonies developing from measured volumes of the fluid let fall on to the surface of solid media. [The *SCI®* indicates that this paper has been cited over 740 times since 1961.]

Ashley A. Miles
Department of Medical Microbiology
London Hospital Medical College
London E 1, England

January 11, 1979

"I suspect, from the references I have come across, that the paper is most commonly cited for the description of the surface-viable count for bacteria. I devised this while at Cambridge in the early 1930s, in the first place for a class demonstration of the titration of phage preparations. The then recommended method was to seed an agar plate to produce a confluent bacterial lawn, and on each quarter of the plate to spread a volume of phage dilution with a glass spreader. This struck me as messy—a lot of spreaders were needed, the bacterial lawn that subsequently grew was smeared and uneven, and an unknown number of phage particles was removed on the spreader. Instead, I seeded 0.02 ml volumes of graded dilutions of phage suspensions on to the lawn from a dropping pipette, and counted the phages in the drop-areas containing discrete plaques.

"The method was obviously applicable to direct viable counts of bacteria, and had the advantage over both the pour-plate or roll-tube counts, not only because all time-consuming manipulations with molten agar at low temperatures were avoided, but because for organisms requiring them, opaque media like blood agar could be used. The disadvantage was statistical, because the coefficient of variation of mean counts tends to diminish with increase in the value of the mean. By the surface-viable method,

a maximum of 10-30 discrete colonies—depending on the colony size—can be accommodated in the small drop area. It is therefore less easy to obtain significant differences between the means of, say, 3 replicates of counts of 5-30 organisms than with the same number of replicates of counts in the region of the 300 or so countable in pour-plates or roll-tubes of less dilute suspensions. Accordingly, many more replicates of countable drop-areas are required with surface counts to attain the same level of significance. But even with these restrictions the method proved, for many research purposes, to be a useful substitute for the accepted counting methods.

"This was the state of affairs when at the British Postgraduate Medical School in London, S.S. Misra—who later became an eminent physician in Lucknow—began work with me on the bactericidal power of human blood for *Staph. aureus.* We used the surface count both for direct counts of inocula, and after incubation of blood-bacterium mixtures, for counts of the survivors in 0.02 ml drops of the mixture on blood agar plates.

"The biometrician, J.O. Irwin, then at the London School of Hygiene and Tropical Medicine, devised for us a formula for estimating the standard error of a ratio, which was needed for comparing differing proportions of the inoculum surviving the bactericidal action of blood. We spent a lot of time investigating the possible fallacies in the interpretation of bacterial survival rates in these conditions, and finally tested 26 normal adults and 6 sufferers from chronic staphylococcal infection; the bloods of the latter proved to be significantly far more bactericidal. But before we could apply the technique further, Misra returned to India, and I went to another job.

"From the scientific literature I have subsequently read, the bactericidal method and the statistical considerations in determining the significance of differences in killing rates appear, like the mule, to have nothing to show in the way of offspring; though as far as my own work is concerned they are coming home to roost in some current work on the bactericidal power of macrophage populations. The surface-viable count evidently—and gratifyingly—lives on."

Clyde W A, Jr. Mycoplasma species identification based upon growth inhibition by specific antisera. *J. Immunology* **92**:958-65, 1964.
[Dept. Pediatrics, Univ. North Carolina Sch. Med., Chapel Hill, NC]

Advantage was taken of the fact that mycoplasma growth is inhibited in the presence of species-specific antisera to perfect a new method of speciation. Variables involved were explored and the application to unknown isolates demonstrated. [The *SCI*® indicates that this paper has been cited in over 610 publications. It is among the ten most-cited papers for this journal.]

Wallace A. Clyde, Jr.
Department of Pediatrics
Infectious Disease/Pulmonary Division
School of Medicine
University of North Carolina
Chapel Hill, NC 27514

May 1, 1984

"The identification of Eaton's atypical pneumonia agents as a mycoplasma in 1962[1] led to extensive studies on the role of this organism as a cause of human respiratory disease. Cultures of patients for *Mycoplasma pneumoniae*, as the new organism was named,[2] yielded a variety of other mycoplasma species that had not been identified previously. Due to their small genome size (5 × 10[8] Daltons), mycoplasmas are much more limited in biochemical characteristics than are other conventional bacteria, and speciation consequently depends upon serological testing. The principal method that had been available was the complement fixation technique, a laborious process plagued by variable degrees of cross reaction.

"Edward and Fitzgerald in 1954[3] had observed that mycoplasma growth was inhibited if homotypic antibody was incorporated into the agar medium. A modification of these studies was presented by Huijman-Evers and Ruys,[4] who placed filter paper discs saturated with antisera on agar plates inoculated by the push-block technique. Further use of these ideas was made to simplify and better control the process, as well as to evaluate it for applicability to mycoplasma species identification.

"Antisera were prepared to eight different mycoplasma species. These organisms in broth cultures were used to inoculate agar plates, and antiserum-impregnated filter paper discs were applied. After incubation for several days to allow growth of the test organisms, zones of inhibition were observed around the homotypic, but not the heterotypic, antisera. The same mycoplasmas and antisera were studied also by the conventional complement fixation method for comparison. The growth inhibition technique proved to be much simpler and more specific as a means of speciation. However, the procedure was relatively insensitive compared to complement fixation and had no role as a serologic test.

"The growth inhibition technique was quickly adopted by other workers in the field of mycoplasmology and has enjoyed extensive application subsequently. This led rapidly to the identification of a wide variety of mycoplasma species which were previously known only by strain abbreviations. A critique of the procedure together with technical variations proposed by others is the subject of a paper from the World Health Organization.[5] A recent description of the growth inhibition technique appears in *Methods in Mycoplasmology*.[6]

"The report in *Journal of Immunology* played a major role in establishing me in a niche in the field. Currently the number of related publications is around 80, but no others have received the same citation popularity. It is of retrospective interest that the definitive experiments on which the work was based required less than one month to accomplish.

"The description of the growth inhibition test for mycoplasma speciation formed the basis of a ten-year Career Development Award from the National Institutes of Health to evaluate the role of mycoplasmas in human diseases. In turn, this work led to editorial board appointments for the *Journal of Bacteriology, Infection and Immunity, Journal of Clinical Microbiology*, and *Pediatric Infectious Diseases*. Currently, I am chairman-elect of the International Organization for Mycoplasmology, and will become the chief presiding officer following the forthcoming congress to be held June 24-29, 1984, in Jerusalem, Israel."

1. Chanock R M, Hayflick L & Barile M F. Growth on artificial medium of an agent associated with atypical pneumonia and its identification as a pleuropneumonia-like organism. *Proc. Nat. Acad. Sci. US* **48**:41-8, 1962. (Cited 710 times.)
2. Chanock R M, Dienes L, Eaton M D, Edward D G F, Freundt E A, Hayflick L, Hers J F P, Jensen K E, Liu C, Marmion B P, Morton H E, Mufson M A, Smith P F, Somerson N L & Taylor-Robinson D. *Mycoplasma pneumoniae*: proposed nomenclature for atypical pneumonia organism (Eaton agent). *Science* **140**:662, 1963. (Cited 70 times.)
3. Edward D G F & Fitzgerald W A. Inhibition of growth of pleuropneumonia-like organisms by antibody. *J. Pathol. Bacteriol.* **68**:23-30, 1954. (Cited 95 times since 1955.)
4. Huijman-Evers A G M & Ruys A C. Microorganisms of the pleuropneumonia group (family *Mycoplasmataceae*) in man. II. Serological identification and discussion of pathogenicity. *Anton. Leeuwenhoek J. Microbiol.* **22**:377-84, 1956. (Cited 60 times since 1956.)
5. World Health Organization. *The growth inhibition test.* VPH/MIC/76.7.
6. Clyde W A, Jr. Growth inhibition tests. (Razin S & Tully J G, eds.) *Methods in mycoplasmology.* New York: Academic Press, 1983. Vol. I. p. 405-10.

Biozzi G, Benacerraf B & Halpern B N. Quantitative study of the granulopectic activity of the reticulo-endothelial system. II: a study of the kinetics of the granulopectic activity of the R.E.S. in relation to the dose of carbon injected. Relationship between the weight of the organs and their activity. *Brit. J. Exp. Pathol.* 34:441-57, 1953.
[Lab. Exp. Med., Clin. Méd. Propedeutique, Hôp. Broussais & Ctr. Natl. Recherche Scientifique, Paris, France]

This article describes the first quantitative method for measuring the phagocytic function of the reticuloendothelial macrophages in contact with the circulating blood. This activity is measured from the rate of blood clearance of colloidal carbon particles injected intravenously. [The *SCI®* indicates that this paper has been cited in over 630 publications since 1961.]

Guido Biozzi
Service d'Immunogénétique
Section de Biologie
Institut Curie
75231 Paris
France

September 6, 1983

"After graduating from the medical faculty of the University of Rome (Italy) in 1947, I worked in collaboration with Zoltan Ovary (presently at the department of pathology of New York University) on the modification of endothelial cells of skin capillaries produced by locally injected histamine or by local anaphylactic reactions. This research led to the description of passive cutaneous anaphylaxis.[1,2]

"Among other colloidal dyes used in these investigations, we worked with suspension of colloidal carbon particles. These first experiments were actually carried out using the commercial India ink preparations.

"During these studies, I noticed that the carbon particles (diameter about 200 Å) injected intravenously did not filtrate through the capillary membrane and remained in the circulation until they were phagocytized by the reticuloendothelial system (RES) macrophages in contact with the circulating blood.

"In 1949, I obtained a fellowship to spend eight months in the laboratory directed by B.N. Halpern in Paris (France). There I began to speculate on the possibility of measuring the phagocytic activity of the RES macrophages from the rate of blood clearance of intravenously injected colloidal carbon particles. In the autumn of 1949, Baruj Benacerraf (1980 Nobel prizewinner, presently at Harvard Medical School in Boston) was working on histamine metabolism in Halpern's laboratory. He and Halpern were interested in the project of measurement of the phagocytic function of the RES macrophages.

"The first difficulty we encountered in this study was that commercial India ink preparations contained a shellac which produced intravascular fibrinogen clotting when large doses of carbon were injected. To overcome this inconvenience, which altered the rate of carbon particle phagocytosis, we contacted a commercial firm which specialized in India ink production (Pelikan Gunther Wagner, Hannover, Germany) in order to obtain a special colloidal carbon preparation with standardized carbon particle size stabilized with gelatin. This colloidal carbon preparation, designated C11/1431a, was perfectly suited for the study of the phagocytic activity of RES macrophages in mice, rats, guinea pigs, and rabbits.

"In order to express quantitatively this activity, we created a new terminology: the 'phagocytic index K' which measures the total activity of all the RES macrophages and the 'corrected phagocytic index α' which expresses this activity per unit of weight of liver and spleen that contains 96 percent of the RES macrophages in contact with the bloodstream.

"The reason why our publication has been so highly cited is that it described the first method of quantitative measurement of the phagocytic activity of the RES macrophages. This method has been very widely used, therefore, by all the authors studying this important function which plays a fundamental role in anti-infectious immunity.

"For a report of recent work in this field, see reference 3."

1. Biozzi G, Mené G & Ovary Z. Ricerche sui rapporti tra istamina e granulopessia dell'endotelio vascolare. Nota II. Sulla granulopessia dell'endotelio vascolare nella reazione anafilattica cutanea locale della cavia. *Sperimentale* 99:1-8, 1948.
2. ⸺. L'histamine et la granulopexie de l'endothélium vasculaire. *Rev. Immunol.* 12:320-34, 1948.
3. Stiffel C, Mouton D & Biozzi G. Kinetics of the phagocytic function of reticuloendothelial macrophages in vivo. (van Furth R, ed.) *Mononuclear phagocytes.* Oxford: Blackwell Scientific Publications, 1970. p. 335-81.

This Week's Citation Classic

Sbarra A J & Karnovsky M L. The biochemical basis of phagocytosis. I.
Metabolic changes during the ingestion of particles by polymorphonuclear
leukocytes. *J. Biol. Chem.* 234:1355-62, 1959.
[Depts. Bacteriology and Biological Chemistry, and Biophysical Lab.,
Harvard Med. Sch., Boston, MA]

In this study we report that the uptake of particulate material by leukocytes is accompanied by a number of stimulated metabolic events. It is concluded that phagocytosis is an energy requiring process. [The *SCI®* indicates that this paper has been cited in over 765 publications since 1961.]

Anthony J. Sbarra
Department of Medical Research
and Laboratories
St. Margaret's Hospital for Women
and
Department of Obstetrics and Gynecology
Tufts University School of Medicine
Boston, MA 02125

May 4, 1983

"From 1953 to 1956, I was an associate biologist in the Biology Division at the Oak Ridge National Laboratory. In the fall of 1955, my sister Lucia wrote to me from her home in New York; pointing out that I had been away from the East Coast long enough and that I should be thinking of moving back home!

"I mentioned my desire to return East to a colleague, who worked in the lab next to mine, and who had recently finished a doctorate at Harvard University. She immediately wrote her former mentor, asking him whether he might have a place in his laboratory for a person with my interest. He, in turn, referred her request to Suter and Karnovsky at Harvard Medical School, with the result that Suter offered me a post in his laboratory.

"At that time, this laboratory had an ongoing program concerned with a study of the interactions between *Mycobacteria* and guinea pig exudate cells! This was great; however, there was one complication. Suter had just accepted a new post, and would be leaving Harvard on July 1st. I headed East anyway.

"For the first month or two I read papers, but in the spring I began to do a few experiments with Karnovsky. Meanwhile, I wondered what I would do in July. In the midst of pondering my future, Karnovsky announced that he was planning to

stay in the phagocytosis field for two more years, and invited me to join him; I happily did.

"The study which led to our frequently cited paper was originally designed simply to look at the metabolic activities of guinea pig exudate cells, and specifically at the polymorphonuclear neutrophilic leukocyte (PMN), in both the absence and presence of particulate material. It was Karnovsky's contention that by using inert material, our system would be 'cleaner' and might generate more significant data. The first result of the addition of these particles to PMN was an immediate and dramatic increase in oxygen uptake, glucose-1-^{14}C oxidation, and lactic acid production. These findings were not entirely unexpected.[1,2] The use of inert particles and the omission of serum in the reaction, however, allowed us 'better' control of the ensuing reactions. And, most importantly, we noted a correlation between the stimulated metabolic events and actual particle entry.

"The actual significance of these metabolic stimulations was not obvious at the time. In order to learn more about them, the use of so-called 'respiratory' and 'glycolytic' inhibitors was introduced to the study for the first time. The finding that oxygen uptake, glucose-1-^{14}C oxidation, and particle uptake were not affected by the respiratory inhibitors, but were affected by the glycolytic inhibitors, was exciting. But again, the precise physiological significance of the findings was still not clear. A few years later, when Bob Good sent Art Page to Harvard to become familiar with the techniques used in our paper, we felt that something was 'breaking.' A few years later, Holmes, Page, and Good[2] reported, in a classic paper, that PMN collected from children with chronic granulomatous disease did not show stimulated oxidative activities when challenged with particles as control PMN did. In addition, they were not able to kill certain bacteria; and, finally, the children so afflicted experienced frequent and repeated infections with these organisms. These later observations shed considerable light on the physiological significance of our findings,[3-5] and were certainly contributory to the frequent citation of our paper.

"In light of all of the above, the opportunity to have worked in Karnovsky's laboratory was an experience I shall always remember and cherish."

1. Stähelin H, Suter E & Karnovsky M L. Studies on the interaction between phagocytes and tubercle bacilli.
   I. Observations on the metabolism of guinea pig leukocytes and the influence of phagocytosis.
   J. Exp. Med. 104:121-36, 1956.
2. Holmes B, Page A R & Good R A. Studies of the metabolic activity of leukocytes from patients with a genetic
   abnormality of phagocytic function. *J. Clin. Invest.* 46:1422-32, 1967.
   [The *SCI* indicates that this paper has been cited in over 485 publications since 1967.]
3. Sbarra A J & Strauss R R, eds. *The reticuloendothelial system. II. Biochemistry and metabolism.*
   New York: Plenum Press, 1980. 456 p.
4. Rossi F, Patriarca P L & Romeo D, eds. *Movement, metabolism, and bactericidal mechanisms of phagocytes.*
   Padua, Italy: Piccin Medical Books, 1977. 410 p.
5. Karnovsky M L & Bolis L, eds. *Phagocytosis—past and future.* New York: Academic Press, 1982. 592 p.

Klebanoff S J. Antimicrobial mechanisms in neutrophilic polymorphonuclear leukocytes. *Semin. Hematol.* **12**:117-42, 1975.
[Dept. Medicine, Univ. Washington Sch. Med., Seattle, WA]

The antimicrobial systems of neutrophils are divided into those dependent on oxygen and those which are not. The former include the myeloperoxidase-H_2O_2-halide system and highly reactive oxygen radicals, and the latter include granule cationic proteins, lysozyme, lactoferrin, and possibly a fall in intraphagosomal pH. [The *SCI®* indicates that this paper has been cited in over 355 publications since 1975.]

Seymour J. Klebanoff
Department of Medicine
School of Medicine
University of Washington
Seattle, WA 98195

August 23, 1982

"From 1957 to 1962, I was at Rockefeller University, an endocrinologist by trade and a thyroidologist by research interest. I had two projects under way with a graduate student, Cecil Yip, both involving peroxidases. One dealt with the mechanism of action of the thyroid hormones; thyroxine by virtue of its phenolic hydroxyl group was found to greatly stimulate reactions catalyzed by peroxidase.[1] The second project dealt with the biosynthesis of thyroxine, a reaction which required a thyroid peroxidase to iodinate the tyrosine residues of thyroglobulin.[2] This interest in peroxidases and their role in the mechanism of thyroxine action prompted a search for biologically important peroxidases which could be stimulated by thyroxine. Granulocytes are rich in peroxidase. We purified this enzyme (myeloperoxidase) and found that it, like horseradish peroxidase, was stimulated by thyroxine and, like thyroid peroxidase, iodinated proteins. Another peroxidase, lactoperoxidase, present in milk and saliva, was purified and found to react similarly.

"At the same time that this work was going on, Zanvil Cohn and James Hirsch at Rockefeller University had characterized the cytoplasmic granules of rabbit granulocytes[3] and demonstrated the release of their contents into the phagosome as a prelude to the death of the ingested organisms. I therefore approached Hirsch with a tube of green myeloperoxidase and a proposal that we determine if this granule enzyme could kill bacteria. If so, this biological action of a peroxidase might be stimulated by thyroxine. We found that myeloperoxidase was ineffective alone or when combined with H_2O_2. It was, however, known from thyroxine synthesis that peroxidase and H_2O_2 oxidize iodide to iodine, a well-known germicidal agent. So we added iodide; the solution turned light yellow and the bacteria were killed, all according to expectation. However, the key experiment, the stimulation of this reaction by thyroxine, was negative. We lost interest.

"The next several years were spent at the University of Washington on other things until I was made aware by Ray Luebke, an endodontics trainee, of an incompletely understood antimicrobial system in saliva, which required a heat-stable dialyzable component (thiocyanate ions) and an unknown heat-labile nondialyzable component. We demonstrated that the latter was salivary peroxidase and that H_2O_2 was an additional requirement.[4] This rekindled an interest in the antimicrobial properties of myeloperoxidase, which was found to have potent antimicrobial activity when combined with H_2O_2 and a halide (iodide, bromide, chloride). Evidence was found implicating this as one of the antimicrobial systems of phagocytes. Unfortunately, we were unable to come full circle and demonstrated a stimulation of this peroxidase-dependent reaction by thyroxine. Over the years these studies have been punctuated by reviews of the antimicrobial systems of phagocytes. The paper indicated here is one of these, and has been highly cited as it appeared at a time of exploding interest in the role of oxygen metabolites in the cytocidal mechanisms of phagocytes (see reference five for a more recent review of this area)."

1. **Klebanoff S J.** An effect of thyroxine and related compounds on the oxidation of certain hydrogen donors by the peroxidase system. *J. Biol. Chem.* **234**:2437-42, 1959.
2. **Klebanoff S J, Yip C & Kessler D.** The iodination of tyrosine by beef thyroid preparations. *Biochim. Biophys. Acta* **58**:563-74, 1962.
3. **Cohn Z A & Hirsch J G.** The isolation and properties of the specific cytoplasmic granules of rabbit polymorphonuclear leucocytes. *J. Exp. Med.* **112**:983-1004, 1960.
4. **Klebanoff S J & Luebke R G.** The antilactobacillus system of saliva. Role of salivary peroxidase. *Proc. Soc. Exp. Biol. Med.* **118**:483-6, 1965.
5. **Klebanoff S J.** Oxygen-dependent cytotoxic mechanisms of phagocytes. (Gallin J I & Fauci A S, eds.) *Advances in host defense mechanisms.* New York: Plenum Press, 1982. Vol. 1. p. 111-62.

Tomasi T B, Jr., Tan E M, Solomon A & Prendergast R A. Characteristics of an immune system common to certain external secretions.
J. Exp. Med. 121:101-24, 1965.
[Div. Exper. Med., Univ. Vermont, Coll. Med., Burlington, VT and Rockefeller Institute, New York, NY]

Abstract of techniques: Isolation of proteins from human saliva, milk, and serum by salt precipitation, ion exchange, and molecular sieve chromatography. Fluorescent antibody examination of tissues. *In vitro* culture synthesis of proteins using ¹⁴C leucine incorporation. Radiolabeling of IgA with iodine and injection into normals to quantitate distribution in serum and secretions. Measurement of specific antibody to blood group antibodies in different classes using hemagglutination and absorption with class specific antisera. [The *SCI®* indicates that this paper has been cited over 695 times since 1965.]

Thomas B. Tomasi
Department of Immunology
Mayo Clinic
Rochester, MN 55901

October 17, 1980

"This paper was the first to suggest the concept of an immune system common to mucosal fluids. It described in detail the chemical differences between serum and secretory antibodies. I am pleased that the work has been verified and extended in other laboratories and has stood the test of time. Other workers have showed its potential importance as a 'first line of defense' against potentially pathogenic microorganisms and the importance of stimulating this system in immunization against viruses and bacteria that enter the body via mucous membranes. More recently, evidence has been presented that the secretory system, as it is now called, may regulate the absorption of nonviable materials that are inhaled or ingested and that abnormalities of the mucosal system may occur in certain human diseases.

"You might be interested in the evolution of this work. Halsted Holman at Rockefeller University was examining the gastric secretions from patients with protein-losing enteropathies and found on immunological analysis that there was a good deal of gamma globulin in the fluids from these patients and also in normal controls. It occurred to me that perhaps some of these proteins were derived from swallowed saliva, and some weeks later I expectorated into a beaker and examined the whole saliva by gel diffusion using antialbumin and antigamma globulin antisera. Roughly quantitated by eye there appeared to be considerably more gamma globulin than albumin in saliva. This was unexpected if one assumed that the proteins in saliva were derived from serum by simple transudation. About that time I went to the University of Vermont, and during my first year a third year medical student, Sheldon Zigelbaum, now a psychiatrist in Boston, wished to do a summer research project in my lab. He was also a dentist, and the closest study I could think of to a 'dentally oriented' project was the observation on saliva. I suggested that he try to repeat my former observation and do more accurate quantitation of the proteins. With the assistance of a superb technician, Dolores Czerwinski, we showed by quantitative precipitation that there was four times more gamma globulin in normal parotid fluid than there was albumin. We subsequently continued this work using specific antisera and showed that the major gamma globulin in saliva and in most other external secretions was IgA.[1] These findings were first reported at the interim meetings of the Arthritis Association in 1963. I then went back to Rockefeller University, and it was there that I examined in a more sophisticated fashion, using methods involving *in vitro* synthesis, the fluorescent antibody technique, transport of labeled proteins from serum into secretions, and the measurement of specific antibodies and their classes in secretions, all with great assistance from my coauthors who were working primarily on other problems in Kunkel's laboratory. It is gratifying that this work has formed the basis for what is now an active area, and I think holds even more potential for the future."

1. Chodirker W B & Tomasi T B, Jr. Gamma-globulins: quantitative relationships in human serum and nonvascular fluids. *Science* 142:1080-1, 1963.

Peterson R D A, Cooper M D & Good R A. The pathogenesis of immunologic deficiency disease. *Amer. J. Med.* **38**:579-604, 1965.

On the basis of clinical observations and experimental laboratory studies, immunodeficiency diseases of man were categorized on the basis of their underlying cellular deficiency. This paper outlined the rationale for this classification. [The *SCI*® indicates that this paper was cited 271 times in the period 1966-1977].

Raymond D.A. Peterson
University of South Alabama
2451 Fillingim Street
Mobile, AL 36617
January 9, 1978

"It has been gratifying to see this paper cited so often during the past thirteen years, principally because the approach to research and discovery underlying the paper is one that evolved over many years and continues to be a powerful method of revealing nature's secrets. This brief commentary will focus on the background of the reported work, but the best take-home message is probably the philosophy of the approach.

"Irvine McQuarrie, a mentor of two of the authors of this paper, viewed diseases as 'experiments of nature' and admonished his students to take advantage of the opportunities they afforded as clues to understanding the normal state of affairs. This attitude toward diseases had been expressed in other ways by other investigators over the years, but now, coupled with modern technological advances, it becomes particularly formidable. Clinical observations taken to the laboratory provide perspectives on normal biological processes not otherwise available and often lead to the design of unique incisive experiments.

"Immunodeficiency disease, the subject of this commentary, is an example of this approach. Col. Ogden Bruton, an army pediatrician, made the initial critical observation of a boy with recurrent infections and agammaglobulinemia in 1952. My interest was kindled because I was a young medical officer stationed with Col. Bruton. Shortly after Bruton's observation, R.A. Good encountered a man with agammaglobulinemia and a thymic tumor. He and his colleagues took that observation to the laboratory, convinced that the thymus was somehow central to the sequence of events leading to gammaglobulin synthesis. In 1961 they finally hit pay-dirt when they discovered that thymectomy performed in newborn animals would render animals immunodeficient.

"Thymectomized animals were unable to make antibodies to certain antigens, but unlike the patient with agammaglobulinemia most of these animals had normal serum immunoglobulins. Cooper and I came on the scene at about this stage in the unraveling process. Taking important leads from the works of Bruce Glick and Noel Warner, we pushed on.

"We developed an experimental chicken model of human agammaglobulinemia and conclusively demonstrated that bursa-derived cells, now called B-lymphocytes, were responsible for immunoglobulin syntheses, while thymic lymphocytes, now called T-lymphocytes, were responsible for cellular immunity.

"Armed with this experimental data, we wrote the paper. We split human immunodeficiency diseases into three major categories: those due to abnormalities of T-cells, B-cells, or both.

"The story has come a long way from this rather simplistic classification but, for us at least, that was the start."

This Week's Citation Classic

Billingham R E & Medawar P B. The technique of free skin grafting in mammals.
J. Exp. Biol. **28**:385-402, 1951.
[Department of Zoology, University of Birmingham, Birmingham, England]

This paper was the first to provide a fully illustrated description of some consistently successful procedures for experimental skin grafting in laboratory mammals, together with an account of their rationale and an indication of their possible application for studying a variety of problems. [The *SCI®* indicates that this paper has been cited over 500 times since 1961.]

Rupert E. Billingham
Department of Cell Biology
University of Texas Health Science
Center at Dallas
Dallas, TX 75235

May 22, 1978

"This paper was written at a time when skin grafting was still almost the exclusive prerogative of plastic surgeons and rarely employed as a laboratory procedure. Our purpose was to make available to fellow biologists some simple procedures, with appropriate illustrations and an account of their underlying rationales for the preparation and consistently successful experimental transplantation of various types of skin grafts in some common laboratory animals. We presented information on the comparative anatomy and histology of the mammalian integument, on its regional variation within the individual, on the principles of wound healing as they apply to skin grafting, and on the phenomenon of transplantation immunity—a subject then in its infancy. We indicated the variety of problems for which skin grafting afforded a means of approach, often as an alternative to tissue culture. Some of these problems related to properties of skin itself, e.g., its pigmentation, hairgrowth, and sensitivity to contact allergens. Skin was the tissue of choice because of its ease of handling, accessibility for inspection, and removal of biopsy specimens without recourse to surgery, e.g., analysis of the resistance of living cells to procedures such as storage in the frozen state.

"I recall that we had misgivings about the acceptability of the manuscript for publication on the grounds that few of our contemporaries were interested in grafting or skin biology.

"It turned out that it was the application of the grafting techniques we described, or variants of them, that in our hands or those of others, led to the discovery of the principle of immunological tolerance and its experimental analysis, and to the widespread popularity of the skin allograft as the tissue of choice, or convenience for studies on transplantation immunity and the allograft reaction. The popularity of skin for such studies has only recently begun to wane, as increasing numbers of investigators master the techniques of microvascular surgery required to transplant kidneys, hearts, and other organs in rats and even in mice. This has helped to overthrow what Medawar aptly referred to as the 'doctrinal tyranny of skin grafts.'

"In reviewing my own research activities and those of various associates since this paper was published 27 years ago, I realize that any success I may have had as a scientist has derived largely from the application of a simple repertoire of methods and principles, worked out early in my career, to a reasonably broad range of problems, many of which were hinted at in the paper.

"To those who have cited this publication I am exceedingly grateful; it is always nice to have documentation that people read our papers. However, I must confess that my 'best seller' by the *Science Citation Index®* 's criteria certainly is not the publication I would have chosen to reminisce about."

Dausset J. Iso-leuco-anticorps. *Acta Haematol.* **20**:156-66, 1958.
[Centre National de Transfusion Sanguine, Paris, France]

This paper provides a description of the first human leucocyte (tissue) group. Polytransfused patients' sera agglutinated some but not all leucocytes. This group, initially called 'Mac,' is now known as HLA-A2. This work opened up the understanding of the human major histocompatibility complex HLA, a key to the immune system. [The *SCI®* indicates that this paper has been cited in over 165 publications since 1961, making it the 5th most-cited paper ever published in this journal.]

Jean Dausset
Unité INSERM U. 93
Hôpital Saint-Louis
75475 Paris
France

January 19, 1983

"In 1952, when only the red blood cell groups were known, I mixed on a glass plate the serum of an agranulocytic woman with the bone marrow of another individual. I observed a massive macroscopic agglutination. I soon understood that it was not due to autoantibodies but to alloantibodies (at that time called isoantibodies).

"This paper gathered together all the work that had been done since 1952, in collaboration with Gilbert Malinvaud, Hélène Brécy, and later, Jacques and Monique Colombani. (This work was also described in a book.[1])

"During this time, we organised the systematic detection of leucoantibodies in the sera of polytransfused patients tested against a panel of leucocytes from volunteers from the National Blood Transfusion Centre. The agglutinations were sometimes obvious and sometimes weak, of very dubious reproducibility. There were no computers at that time and our results were exposed on a large poster on the laboratory wall. We almost lost hope of ever making sense of it, so numerous were the different patterns. However, we noticed, after numerous repetitions, that the leucocytes from three panel donors were less often agglutinated than others. This gave us the idea that they lacked a certain antigen which is otherwise frequent in the population. Six sera did not agglutinate these three individuals, but agglutinated 11 others, thus dividing the population into two groups: one bearing the Mac antigen (not a Scottish name, but the initials of the surnames of the three non-agglutinated individuals) and the other without this antigen. An important fact was that these sera were unable to agglutinate their own leucocytes (they were not autoantibodies) nor the leucocytes of the other patients who produced similar antibodies.

"Formal proof was obtained when we observed that among the patients systematically transfused with Mac-positive blood, only the Mac-negative recipients developed an anti-Mac antibody.

"These leucocyte antigens are genetically determined since the pattern of agglutination with a battery of anti-leucocyte sera was strictly identical in monozygotic twins and different in dizygotic twins. Lastly, we demonstrated that leucocyte antibodies were responsible for the transfusion reactions.

"Thus, this work gathers together all the principles and first fruits of leucocyte immunohaematology. This was the starting point of an extraordinary biological adventure and is the reason why this paper has been highly cited. Two laboratories followed in our footsteps: J.J. Van Rood soon afterward described a supertypic antigen, 4a4b (Bw4,Bw6);[2] Rose Payne with W. Bodmer described the first two alleles, LA1 and LA2 (LA2 being identical to Mac).[3]

"Soon there were more than three musketeers: B. Amos, R. Ceppellini, F. Kissmeyer-Nielsen, P. Terasaki, and R. Walford joined the game, rapidly followed by numerous scientific communities. A unique collaborative research study was undertaken and has continued ever since. Thanks to ambitious international workshops which allow the exchange of reagents and information, the extraordinary skein of the human HLA complex was unraveled.[4-10] Its essential role in transplantation and, generally speaking, in immunology is well known, as well as its association with numerous diseases.

"Starting with the first serological data, it has been possible, in the space of 20 years, to decipher the biological composition and function of these molecules, protruding like antennae from the surface of our cells. We are now beginning to study the genes which govern these cells, and can expect still more marvels to come."

1. **Dausset J.** *Immuno-hématologie biologique et clinique.* Paris: Flammarion, 1956. 718 p.
2. **Van Rood J J & Van Leeuwen A.** Leukocyte grouping. A method and its applications.
    J. Clin. Invest. **42**:1382-90, 1963.
3. **Payne R, Trip M, Weigle J, Bodmer W & Bodmer J.** A new leukocyte iso-antigen system in man.
    Cold Spring Harbor Symp. **29**:285-95, 1964.
4. **Videbaek A,** ed. *Histocompatibility testing. 1965.* Copenhagen: Munksgaard, 1965. 4288 p.
5. **Curtoni E S, Mattluz P L & Tosi R M,** eds. *Histocompatibility testing. 1967.*
    Copenhagen: Munksgaard, 1967. 458 p.
6. **Terasaki P I,** ed. *Histocompatibility testing. 1970.* Copenhagen: Munksgaard, 1970. 658 p.
7. **Dausset J & Colombani J,** eds. *Histocompatibility testing. 1972.* Copenhagen: Munksgaard, 1973. 778 p.
8. **Kissmeyer-Nielsen F,** ed. *Histocompatibility testing. 1975.* Copenhagen: Munksgaard, 1975. 1035 p.
9. **Bodmer W F, Batchelor J R, Bodmer J G, Festenstein H & Morris P J,** eds. *Histocompatibility testing.*
    1977. Copenhagen: Munksgaard, 1977. 612 p.
10. **Terasaki P I,** ed. *Histocompatibility testing. 1980.* Los Angeles, CA: UCLA Tissue Typing Laboratory, 1980. 1227 p.

This Week's Citation Classic

CC/NUMBER 26
JUNE 25, 1979

Mittal K K, Mickey M R, Singal D P & Terasaki P I. Serotyping for
homotransplantation. 18. Refinement of microdroplet lymphocyte cytotoxicity test.
Transplantation 6:913-27, 1968. [Departments of Surgery and Biomathematics,
University of California, Los Angeles, CA]

The paper presents a microtest method for human leukocyte antigen (HLA) typing and tissue matching of human donors and recipients of organ or tissue transplants. The method was made simple and highly reproducible through critical evaluation of each step of the test procedure. [The *SCI®* indicates that this paper has been cited over 625 times since 1968.]

Kamal K. Mittal
Division of Blood and Blood Products
Bureau of Biologics
Food and Drug Administration
Bethesda, MD 20205

March 10, 1978

"In order to comment on the wide use of the human leukocyte typing method reported in this paper, it is important to mention some of the prior history of the HLA field. Soon after the discovery of the first alloantigen on leukocytes in 1958, a number of investigators attempted to resolve the degree of polymorphism and the mode of inheritance of these antigens. However, they encountered great technical difficulties with the reproducibility of the then current leukoagglutination methods.

"In 1964, Dr. Paul Terasaki made a major contribution to this newly emerging field by introducing a microcytotoxic assay for detection of these antigens.[1] Among several advantages, this method eliminated the risk of those false positive reactions which resulted from spontaneous agglutination of leukocytes in the leukoagglutination methods.

"About this time the investigators in this field initiated joint leukocyte-typing workshops. One such workshop was held in Torino, Italy, in June 1967.

An interesting feature of the workshop data was that in certain of the eleven families tested, it appeared as if an offspring had reacted positively with a typing serum, while both parents of that offspring had produced negative reactions with the same serum.

"I first met Terasaki when I was at the California Institute of Technology. Because of my interest in genetics, I was fascinated by the anomalies in the workshop data. Hoping naively that there might exist some 'hybrid' antigens (similar to those in rabbits) which are expressed only after appropriate alleles come together in an offspring, I came to Terasaki's laboratory to explore the phenomenon further.

"By performing rigidly controlled family studies, it did not take us long to discover that what we had suspected to be 'hybrid' antigens were in fact a consequence of variations in the methodology, and in the reactivity of cells from different donors. The need for a highly reproducible tissue typing method thus became quite apparent. A year's intensive effort at evaluating and refining every single step of the lymphocytotoxic test produced this paper.

"A number of modifications of this test have since been attempted; however, the great simplicity and high reproducibility of this method have resisted any significant alterations in the procedure. Its use has assisted in the identification of over 50 alloantigenic determinants of the HLA system. HLA typing is now being applied increasingly not only to organ or tissue transplantation, but also to platelet transfusions, and the prognosis and/or diagnosis of certain diseases. This recently increasing clinical application of this test has finally led to its standardization[2] by the Food and Drug Administration of the US Department of Health, Education and Welfare."

1. Terasaki P I & McClelland J D. Microdroplet-assay of human serum cytotoxins.
Nature (London) 204:998-1000, 1964.
2. Mittal K K. Standardization of the HLA typing method and reagents. *Transplantation* 25:275-79, 1978.

This Week's Citation Classic

Klein E & Klein G. Antigenic properties of lymphomas induced by the Moloney
agent. *J. Nat. Cancer Inst.* 32:547-68, 1964.
[Institute for Tumor Biology, Karolinska Institutet Medical School,
Stockholm, Sweden]

Mouse lymphomas induced by the Moloney agent
were strongly antigenic in isologous hosts, as indicated by transplantation and serologic tests.
Specific transplantation resistance was established
by pretreatment of the recipients with homografts
of other Moloney lymphomas, small isografts of
the same lymphoma, or virus-containing lymphoma homogenates. Extensive cross-resistance was
demonstrated between different Moloney lymphomas, while there were no certain cross-
reactions between Moloney lymphomas and lymphomas induced by the Gross virus, or between
Moloney lymphomas and a number of long-trans-
planted lymphomas. [The *SCI®* indicates that this
paper has been cited over 310 times since 1964.]

Eva Klein
Department of Tumor Biology
Karolinska Institutet
Box 60400
S-104 01 Stockholm
Sweden

September 18, 1981

"The important step that led to the work
presented in this paper was the discovery of the
tumor specific transplantation antigen (later
called TSTA) of polyoma virus induced mouse
tumors.[1] The discovery of the polyoma virus in the
1950s was unexpected and very surprising. In the
beginning, it was difficult to accept that a single
virus could induce so many different types of
tumors. Since polyoma virus was cytopathic for
mouse cells, we speculated that it may induce
tumors *not* by direct transformation but by
damaging cells that participate in central growth
controlling mechanisms. The possibility that
polyoma tumors were not autonomous was also
suggested by some of the original papers, claiming
that they were not transplantable. We know now
that this was an artifact, due to the use of
genetically heterogeneous mice.

"On the basis of the hypothesis, two young
medical students at the time (Sjögren and
Hellström) and one of us (GK) set up what we
thought a critical experiment. Since polyoma
was known to spread quickly by horizontal infection in mouse colonies, we set up a provisional
isolated quarantine facility, free of polyoma virus.
They all belonged to highly inbred, homozygous
strains. In another laboratory, polyoma tumors
were induced by the inoculation of newborn mice
of the same inbred strains. When tumors appeared, they were grafted to two groups of mice:
artificially polyoma-infected recipients and
critically polyoma-free controls, kept at the
isolated facility. The results were the opposite of
what we expected. According to our hypothesis,
the virus treated mice should have supported the
growth of the transplanted cells, in contrast to the
untreated mice. The opposite was found: all
tumors grew in untreated syngeneic hosts, but
were rejected by the polyoma-infected mice.
Subsequent analysis of the rejection mechanism
by Sjögren[2] and independently by Habel[3] showed
that the virus infected mice responded immunologically to a target antigen, present on the
surface of the polyoma transformed cells, later
designated as TSTA.

"We then asked the question whether similar
antigens also appeared on tumor cells transformed by oncogenic RNA viruses. Our first
choice was the Gross virus system where we did
find a similar transplantation resistance in virus
immunized mice, but a very weak one.[4] Around
the same time, Old, Boyse, and co-workers[5]
demonstrated a similar phenomenon with the
Friend and the Rauscher virus induced leukemias.
We then went on to the Moloney system which led
to the work that has become the *Citation Classic.*
The reason why this, rather than the polyoma or
the Gross papers, has become a *Citation Classic* is
due to the fact, we tend to believe, that more
workers use the Moloney system (or related
systems) for their work. Also, Moloney virus induced tumors are relatively strongly antigenic and
have therefore become more widely used.

"In summary, we believe that this history illustrates how a paper that does not represent the
first introduction of a new concept but rather a
follow-up on a related but different system can
become a *Citation Classic* because the system
described is more frequently used, for reasons of
convenience, than the actual experiment system
that served as the basis for the original discovery.
For a recent bibliography, consult the reviews in
Viral Oncology."[6]

1. Sjögren H O, Hellström I & Klein G. Resistance of polyoma virus immunized mice against transplantation of
   established polyoma tumors. *Exp. Cell Res.* 23:204-8, 1961.
2. Sjögren H O. Studies on specific transplantation resistance induced by polyoma virus infection.
   J. Nat. Cancer Inst. 32:361-74, 1964.
3. Habel K. Immunological determinants of polyoma virus oncogenesis. *J. Exp. Med.* 115:181-93, 1962.
   [The *SCI* indicates that this paper has been cited over 180 times since 1962.]
4. Klein G, Sjögren H O & Klein E. Demonstration of host resistance against isotransplantation of lymphomas
   induced by the Gross agent. *Cancer Res.* 22:955-61, 1962.
   [The *SCI* indicates that this paper has been cited over 165 times since 1962.]
5. Old L J & Boyse E A. Immunology of experimental tumors. *Annu. Rev. Med.* 15:167-86, 1964.
6. Klein G, ed. *Viral oncology.* New York: Raven Press, 1980. 842 p.

This Week's Citation Classic

Okazaki W, Purchase H G & Burmester B R. Protection against Marek's disease by vaccination with the herpesvirus of turkeys. *Avian Dis.* 14:413-29, 1970.
[US Dept. Agriculture, Regional Poultry Res. Lab., East Lansing, MI]

A virus isolated from turkeys, the herpesvirus of turkeys (HVT), which was nonpathogenic in chickens, was shown to protect chickens against Marek's disease (MD), a contagious, economically important cancer of chickens. Furthermore, the HVT infection persisted in the vaccinated chickens but did not spread to chickens in direct contact with them. [The *SCI®* indicates that this paper has been cited in over 205 publications since 1970.]

H. Graham Purchase
National Program Staff
Agricultural Research Service
Beltsville Agricultural Research Center
US Department of Agriculture
Beltsville, MD 20705

July 22, 1983

"The research that led to this paper began the most exciting and rewarding time of my life. The period started when Dick Witter came to me with the suggestion that, using the indirect fluorescent antibody test, I look at a new virus he had just isolated from turkeys[1] that was nonpathogenic for chickens. He asked others in the laboratory to examine the virus in the immunodiffusion test. The tests showed the turkey herpesvirus (HVT) to be related but not identical to Marek's disease (MD) virus, a virus causing devastating cancer losses in the poultry industry.

"A project to develop a vaccine against MD had just been started at our laboratory under the leadership of Bill Okazaki. He was attempting to attenuate the MD virus and to isolate naturally apathogenic viruses from the field. An attenuated vaccine against MD had just been reported in England. I recognized the potential of a vaccine for MD and with the encouragement of my close friend Dick, I approached the director, then Ben Burmester, to give me the leadership responsibility for the program to develop a vaccine. I became leader of a three-person team comprised of Okazaki, a visiting Israeli scientist, and myself. We designed and conducted a series of experiments that demonstrated that the HVT was a highly effective vaccine against MD. In the initial paper reported here, I was struck by the dramatic protection offered by HVT and it reminded me of the work of Pasteur on anthrax reported by Paul De Kruif in *The Microbe Hunters*.[2] I did most of the writing for this paper but made Okazaki first author in recognition of the fact that he was the initiator of the vaccine program at our laboratory.

"The paper received the Research Paper Award, now known as the P.P. Levine Award, for the best paper published in *Avian Diseases* in 1971. The laboratory and the team that conducted the research were recipients of several other awards. The government took out a general use patent and has licensed many vaccine producers to produce the HVT vaccine, which is now used worldwide. Expenditures by federal, state, and other agencies and industry on avian tumor virus research were almost $30 million since the turn of the century, yet the benefits in the US from the research amounted to over $168 million for 1974 alone, a benefit-cost ratio of 44.[3] In addition, MD is the first cancer of any animal for which a commercially applicable vaccine was developed. It is referred to as a model by workers with oncogenic herpesviruses of humans,[4] primates, and frogs. I feel sure that the originality and economic impact of the discovery were the reasons that it was cited so often."

1. **Witter R L, Nazerian K, Purchase H G & Burgoyne G H.** Isolation from turkeys of a cell-associated herpesvirus antigenically related to Marek's disease virus. *Amer. J. Vet. Res.* 31:525-38, 1970.
   [Citation Classic. *Current Contents/Agriculture, Biology & Environmental Sciences* 11(2):10, 14 January 1980.]
2. **De Kruif P.** *The microbe hunters.* New York: Harcourt, Brace, 1926. 363 p.
3. **Purchase H G.** The etiology and control of Marek's disease of chickens and the economic impact of a successful research program. *Virology in agriculture: invited papers presented at a symposium held May 10-12, 1976, at the Beltsville Agricultural Research Center, Beltsville, Maryland.* Montclair, NJ: Allanheld, Osmun, 1977. p. 63-81.
4. ─────────, Prevention of Marek's disease: a review. *Cancer Res.* 36:696-700, 1976.

Schwartz R S & Beldotti L. Malignant lymphomas following allogenic disease: transition from an immunological to a neoplastic disorder.
Science **149**:1511-14, 1965.
[New England Medical Ctr. Hospitals, and Tufts Univ. Sch. Med., Boston, MA]

The graft *versus* host reaction in mice was used to test the hypothesis that chronic stimulation of lymphocytes can lead to the development of malignant lymphoma. Long-term survivors of the reaction developed lymphomas that appeared to arise from lymphocytes of the recipients. [The *SCI®* indicates that this paper has been cited in over 255 publications since 1965.]

Robert S. Schwartz
Division of Hematology-Oncology
New England Medical Center Hospital
Tufts University School of Medicine
Boston, MA 02111

December 8, 1982

"My interest in immunological diseases began in 1953 when, as a New York University medical student working in Bellevue Hospital, I witnessed the dramatic response to cortisone of a woman with a flagrant psychosis caused by systemic lupus erythematosus. Four years later, while a medical resident at the Yale-New Haven Hospital, I heard a remarkable lecture on autoimmunity by William Dameshek. That experience made up my mind about future training, and in 1957 I went to Boston to study hematology under his direction at the New England Medical Center. The idea that autoimmunity even existed was at that time hotly debated, with Dameshek at the center of the controversy. He maintained that the immune system *could* produce autoreactive antibodies, a notion regarded with disdain by numerous 'authorities.' Dameshek's conviction was based principally on his own clinical observations, especially in cases of autoimmune hemolytic anemia and immunothrombocytopenia. There was, however, no experimental model that might convince the skeptics.

"During my fellowship with Dameshek, I began reading about graft *versus* host reactions because the possibility of bone marrow transplantation in humans was under intensive discussion. The immunological mechanisms of the reactions led me to consider the possibility of triggering an autoimmune disorder by the transplantation of a foreign immune system into a normal animal. This hypothesis was tested in mice by injecting parental strain spleen cells into F_1 hybrid animals. The recipients of the spleen cells developed not only immunohemolytic anemia, but also thrombocytopenia.[1] After this work was completed, we discovered (with mixed emotions) that Cock and Simonsen[2] had previously described the development of immunohemolytic anemia in chicks that were engrafted with foreign spleen cells.

"An intriguing clinical phenomenon, the development of autoimmune hemolytic anemia in patients with malignant lymphoma or chronic lymphocytic leukemia, prompted Dameshek and me[3] to postulate that lymphomas might arise following a sustained proliferation of autoantibody producing lymphocytes. The graft *versus* host model was also used to test this hypothesis because it provided a means to induce chronic *in vivo* stimulation of lymphocytes. The injection of C57B1/6 splenocytes into (C57B1/6 x DBA/2)F_1 mice caused malignant lymphomas in about 50 percent of the recipients, whereas none appeared in control mice. An interesting aspect of the results was that the lymphomas arose from lymphocytes of the *recipients*, and not from those of the donor. Thus, the mechanism really seemed to involve the proliferation of autologous lymphocytes.

"I think this work on the mouse has been highly cited because it explains some important aspects of the pathogenesis of human lymphomas. In Burkitt's lymphoma, for example, the Epstein-Barr virus seems to instigate a polyclonal proliferation of lymphocytes that terminates as a monoclonal neoplasm. Moreover, recent experiments suggest that retroviruses may act either by providing a sustained antigenic stimulus[4] or by causing the production of mitogenic factors by the T lymphocytes they infect.[5] Finally, a gratifying aspect of this work has been its innovative extension by *my* former postdoctoral fellow, Ernst Gleichmann."[6]

1. **Oliner H, Schwartz R & Dameshek W.** Studies in experimental autoimmune disorders. I. Clinical and laboratory features of autoimmunization (runt disease) in the mouse. *Blood* 17:20-44, 1961.
2. **Cock A G & Simonsen M.** Immunological attack on newborn chickens by injected adult cells. *Immunology* 1:103-10, 1958.
3. **Dameshek W & Schwartz R S.** Leukemia and autoimmunization—some possible relationships. *Blood* 14:1151-8, 1959.
4. **McGrath M S & Weissman I L.** AKR leukemogenesis: identification and biological significance of thymic lymphoma receptors for AKR retroviruses. *Cell* 17:65-75, 1979.
5. **Lee J C & Ihle J N.** Increased responses to lymphokines are correlated with preleukemia in mice inoculated with Moloney leukemia virus. *Proc. Nat. Acad. Sci. US* 78:7712-16, 1981.
6. **van Rappard-van der Veen F M, Rolink A G & Gleichmann E.** Diseases caused by reactions of T lymphocytes towards incompatible structures of the major histocompatibility complex. VI. Autoantibodies characteristic of systemic lupus erythematosus induced by abnormal T-B cell cooperation across I-E. *J. Exp. Med.* 155:1555-60, 1982.

Révész L. Detection of antigenic differences in isologous host-tumor systems by pretreatment with heavily irradiated tumor cells. *Cancer Res.* **20**:443-51, 1960.
[Institute for Tumor Biology, Karolinska Institute Medical School, Stockholm, Sweden]

Cells damaged lethally by irradiation show no change in either morphology or function, and maintain their specific antigenic character for a prolonged period after radiation exposure and before their final disintegration. Using such cells in transplanted grafts, a simple routine procedure was developed to test a given experimental tumor-host system for the presence or absence of antigenic differences. The paper describes the method and presents several examples for its practical use. [The *SCI®* indicates that this paper has been cited over 220 times since 1961.]

L. Révész
Department of Tumor Biology
Radiobiology Unit
Karolinska Institute
S-104 01 Stockholm 60
Sweden

January 9, 1978

"It is a paradox to find that of all my papers the most cited is that which reports experiments based upon some accidental observations of a phenomenon known for a long time, and of only indirect relevance to my current research. Clearly, this is another example of how futile it is to try to foresee the path of fundamental research, to say nothing of governing it.

"In the mid fifties, I was engaged in experimental work with the purpose of determining the dose-survival relationship in radiation-exposed tumor cell populations. Studies of the *in vivo* proliferation of an irradiated mouse tumor cell-line, and Ehrlich ascites tumor, met with some unforeseen obstacles. After transplantation of the irradiated cells to new hosts it appeared that, in addition to radiation, some reaction of the host organism interfered with cell multiplication. Thus, in many cases a host resistance developed which completely inhibited tumor growth. The resistance was found to be related to the size of the radiation-killed fraction in the transplanted cell population. The nature of this resistance was subsequently identified with the classical homograft reaction, well-known to operate in systems in which genetically determined iso-antigenic differences exist between the host and the transplanted tumor as in the case with Ehrlich cells. Indeed, in further experiments in which the transplant and the host were always matched genetically, no host resistance to the graft was induced by the presence of irradiated cells.

"The proliferation of antigenic tumors despite a concurrent homograft reaction in an immunogenetically foreign host can be viewed as a race between the development of the host response and the multiplication of tumor cells. Our observation could be interpreted as indicating that this race is influenced in favour of the host by lethally irradiated cells, which lose the ability to reproduce, but remain viable and undamaged in regard to their antigenic character for a prolonged time period after irradiation.

"As a model, our experiments demonstrated the fallacy of genetically non-matched tumor host-systems in therapeutic experiments. However, the findings have also an important positive aspect. Based upon the experience described here, a simple routine procedure can be developed using radiation-killed cells to test a given tumor-host system for the presence or absence of antigenic differences. It is most gratifying to learn from the high citation frequency of the paper which describes the method in detail that the technique has found a wide application."

Harvey A M, Shulman L E, Tumulty P A, Conley C L & Schoenrich E H. Systemic lupus erythematosus: review of the literature and clinical analysis of 138 cases. *Medicine* **33**: 291-437, 1954.

This analysis of the clinical and pathological findings in 138 cases of systemic lupus erythematosus (SLE) together with a review of the literature emphasized that with improved methods of diagnosis a broader concept of the character of this disease was necessary. Important features described were the episodic clinical course of the illness and the development of multiple serum protein abnormalities. [The *SCI®* indicates that this paper was cited 316 times in the period 1961-1976.]

A. McGehee Harvey, M.D.
The Johns Hopkins University
School of Medicine &
The Johns Hopkins Hospital
Baltimore, MD 21205

December 15, 1977

"In 1948 we observed a number of cases SLE in which the classical pathological changes were found at postmortem examination; however, the clinical manifestations and course of the illness suggested that the disease not only appeared as a fulminating, rather rapidly fatal illness but was often characterized by recurrent, seemingly unrelated mild illnesses, with prolonged asymptomatic intervals.

"We were stimulated to study this problem in more detail when asked to write an article for a memorial issue on the 100th anniversary of the birth of Sir William Osler for the *Bulletin of the Johns Hopkins Hospital* in 1949. In reviewing the series of articles which Osler published between 1895 and 1903 on 'The Visceral Complications of Erythema Exudativum Multiforme,' it was apparent that this brilliant clinician had described in detail what we were rediscovering some half a century later.

"In Osler's Gulstonian lectures on ulcerative endocarditis in 1885, he stated: 'It is of use from time to time to take stock, so to speak, of our knowledge of a particular disease, to see now exactly where we stand in regard to it, to inquire to what conclusions the accumulated facts seem to point and to ascertain in what direction we may look for fruitful investigations in the future.'

"By 1954, we had accumulated and analyzed our clinical and laboratory studies in 138 patients. These results fully confirmed the protean nature of the disease and supplied more details to the framework of the clinical picture contributed by Osler. It was shown that the course of the disease is frequently chronic, often characterized by recurrent, seemingly unrelated mild illnesses, with prolonged asymptomatic intervals. Emphasis was placed on the abnormalities of the serum proteins as illustrated by the LE cell factor.

"This review drew attention to the disease at a time when new advances in immunology provided novel approaches to the study of its pathogenesis. The article provided an important reference for the expanded concept of the clinical pattern of the disease and a complete description of its various manifestations. It served, as we had hoped, to point the direction for future fruitful investigations."

CC/NUMBER 28
JULY 9, 1984

This Week's Citation Classic™

Bourne H R, Lichtenstein L M, Melmon K L, Henney C S, Weinstein Y &
Shearer G M. Modulation of inflammation and immunity by cyclic AMP.
Science **184**:19-28, 1974.
[Depts. Med. and Pharmacol., Div. Clin. Pharmacol., Univ. Calif. Med. Ctr., San Francisco,
CA; Div. Immunol., Good Samaritan Hosp., Johns Hopkins Univ., Baltimore; and Immunol.
Branch, Natl. Cancer Inst., Bethesda, MD]

Drawing on experimental studies of basophils, mast cells, neutrophils, and lymphocytes, this review proposes the hypothesis that certain hormones and mediators of inflammation utilize cyclic AMP as an inhibitory second messenger to regulate the character and intensity of inflammatory and immune responses. [The *SCI*® indicates that this paper has been cited in over 570 publications since 1974.]

Henry R. Bourne
Departments of Pharmacology
and Medicine
and
Cardiovascular Research Institute
University of California
San Francisco, CA 94143

June 1, 1984

"By 1970, the discovery of cyclic AMP as an intracellular second messenger for hormones had been extended to embrace the actions of a multitude of first messengers acting on many types of mammalian cells. As a postdoc, I wanted to join the bandwagon by finding a hormone-responsive cell that had not yet been tested. I chose the human leukocyte after Ken Melmon, my adviser, showed me a paper[1] that suggested cyclic AMP might regulate at least one leukocyte function. β-adrenergic amines, prostaglandin-E$_1$, histamine, and cholera toxin turned out to stimulate leukocyte adenylate cyclase. A transcontinental collaboration with Larry Lichtenstein and Chris Henney at Johns Hopkins University then showed that the same agents, as well as a cyclic AMP analogue, inhibited allergic (IgE-mediated) release of histamine from basophilic leukocytes and killing of allogeneic cells by sensitized T lymphocytes. Melmon, in collaboration with Yacov Weinstein and Gene Shearer at the Weizmann Institute in Israel, found selective binding of functionally distinct lymphocyte subpopulations to histamine immobilized on Sepharose beads.

"The 1974 review summarized these experiments as the basis of a grand hypothesis regarding the inhibitory role of leukocyte cyclic AMP in regulation of inflammation and immunity. The hypothesis has held up remarkably well as a predictor of pharmacological effects *in vitro*, in that agents that elevate cyclic AMP in leukocyte subpopulations and mast cells do inhibit each cell's characteristic inflammatory or immune response.

"Our review emphasized a much broader prediction: that catecholamines, histamine, and prostaglandins—and probably other first messengers as well—would prove to play physiologically important roles in inhibiting inflammatory and immune responses *in vivo*. We proposed, for example, that histamine's ability to inhibit its own release might constitute the basis of an inhibitory feedback loop that would serve to inhibit the extent or intensity of allergic responses. Subsequent experiments (summarized in reference 2) provided suggestive evidence in favor of this broader implication, with respect to an inhibitory role of histamine in delayed hypersensitivity.

"My guess is that the review has been frequently cited because it highlighted the intersection of one extremely active area of investigation, cyclic AMP, with the even more rapidly expanding research fronts of inflammation and immunity, and did so in a journal that is widely read. In addition, investigators over the years may have continued to cite an attractive hypothesis that survived in part because it was so difficult to test *in vivo*.

"Now immunologists can clone and propagate functionally specific leukocytes and lymphocytes in tissue culture and study their responses to a large number of recently discovered chemical signals. Investigators of such responses must now consider a panoply of second messengers in addition to cyclic nucleotides. The results of their investigations will be much more complex and more interesting than we envisioned in 1974.

"Two recent reviews[2,3] cover more recent research in this area."

1. Lichtenstein L M & Margolis S. Histamine release in vitro: inhibition by catecholamines and methylxanthines. *Science* **161**:902-3, 1968. (Cited 415 times.)
2. Melmon K L, Rocklin R E & Rosenkranz R P. Autacoids as modulators of the inflammatory and immune response. *Amer. J. Med.* **71**:100-6, 1981.
3. Rocklin R E & Beer D J. Histamine and immune modulation. *Advan. Internal Med.* **28**:225-51, 1983.

Chapter

4

Virology and
Tissue Culture

Several exceptional discoveries are described in the few Citation Classics covered in this chapter. Two are especially dramatic: the first descriptions of Burkitt's lymphoma and of the hepatitis B virus. Other, less dramatic, but nevertheless extremely important discoveries reported in these Classics are the propagation of human diploid cells and the use of feeder cell cultures.

Burkitt's discovery of a lymphoma in African children as a central malignancy that could affect several parts of the body, thus confusing its true nature as a single disease, has been heralded as a masterly contribution to epidemiology. That it was also a study of cancer further embellished its importance, but, of course, it has been even more than a study of the epidemiology of cancer. Conservatives who demand a more stringent proof than Burkitt's lymphoma provides are now very few in number, and most virologists concede that Burkitt's lymphoma is the first human malignancy identified to have a viral etiology. (Some may insist that the human T cell leukemias are the first truly proven examples of human cancers with a viral etiology.) The C-shaped structures viewed by Epstein, Achong, and Barr in malignant Burkitt's lymphocytes compatible in morphology with herpesviruses and the later specific immunofluorescence identification of these particles as herpesviruses completed the initial scenario. From there, a second scene unfolded in which the Epstein-Barr virus (EBV or EBV particles) has become associated with the very common disease infectious mononucleosis and the much rarer disease nasopharyngeal carcinoma, found primarily in the Southeast Asian Chinese population. It has been suggested that infectious mononucleosis is the "normal" expression of infection by the EB virus and that Burkitt's lymphoma and nasopharyngeal carcinoma are "abnormal" expressions. The stimuli that convert the "normal" to the "abnormal" are conjectural.

The recognition of Baruch Blumberg by the Nobel Prize committee for his uncovering of hepatitis B virus as the Australia antigen is responsible for four Classics. The original search was for an antigen, eventually found in the serum of Australian aborigines, that would precipitate with the sera of certain individuals. This antigen was then recognized to exist in a high incidence in the blood of patients with hepatitis. Ultimately, this resulted in the identification of the Australia antigen as the hepatitis B virus surface antigen, designated HBsAg. HBsAg is a component of the larger Dane particle and exists separately and as a part, the outer cover, of the Dane particle in the blood of hepatitis patients.

But these two major studies aside, the first fourteen Classics, based on developments in tissue culture, trace from 1948 until 1984 advances such as the tedious, single-cell, capillary culture techniques, the formulation of Hanks' balanced salt solution and of Eagle's medium, cultivation of human diploid cell lines, open vessel culturing for free gas exchange, feeder cell methodology, and soft agar cloning. As the authors of these Classics describe, all of these save the capillary tube, single-cell culture method are in contemporary use. Hanks' "This Week's Citation Classic" (TWCC), beginning with his description of research on leprosy in the Philippines, makes excellent reading and clearly outlines how slowly (from 1939 to 1949) but how successfully scientific research can progress. Eagle's basal medium and minimal essential medium have been in use more than 25 years with very little change, although serum-free growth media adequate for certain mammalian cell lines are now available. His original publication was cited 2,255 times prior to 1977. Hayflick's description of the travails of publishing details of his senescent diploid cell line is a "Classic TWCC." His "reject editor" made the mistake of stating three bases for rejecting the paper, all of which were subsequently proved to be in error.

The concept of feeder cell nutrition for fastidious cell lines emerged from the earlier use of spent culture media as primers for cell growth (pp. 148–152). The selection of murine cell lines to condition media was fortuitous since these cells are quite readily passaged in artificial media. This was not extendable to short-lived cells of human origin, thus forcing the co-culture technique. The application of a soft agar layer over the feeder layer is essential to maintain the two cell lines physically separate.

The importance of a method to culture plant cells for progress in botany should not be overlooked. Murashige and Skoog's improved medium for this purpose has been cited more than 1,200 times.

Sanford K K, Earle W R & Likely G D. The growth in vitro of single
isolated tissue cells. *J. Nat. Cancer Inst.* 9:229-46, 1948.
[Natl. Cancer Inst., NIH, Public Health Service, Bethesda, MD]

This paper describes a procedure for the growth of single isolated tissue cells and the establishment of NCTC clone 929-L. By restricting the volume of culture medium and preconditioning the medium, conditions were devised that allowed the continued proliferation of single isolated cells of this line. [The *SCI®* indicates that this paper has been cited over 310 times since 1961.]

Katherine K. Sanford
Laboratory of Cellular
and Molecular Biology
National Cancer Institute
National Institutes of Health
Bethesda, MD 20014

June 16, 1981

"W.R. Earle would be pleased to know that our paper with G.D. Likely has been included on the list of most-cited papers. The frequent citation undoubtedly stems from the wide use of this cloned line in many fields of investigation.

"When I first went to the National Cancer Institute in 1947, the growth of single isolated cells was a challenging problem. A theory proposed to explain the many unsuccessful attempts was that the fixed tissue cell has a limited autonomy as a physiological unit and that exchange between cells in a tissue by bridges or contacts is required for initiation of proliferation. Our working hypothesis was that even the best culture media were so inadequate as to need extensive modification by the living cell before they would be suitable for single cell growth. We attempted to reduce the volume of culture medium bathing the single cell to that volume which the cell can adjust, as calculated from a mass culture. A capillary tube appeared to be the most practical culture vessel to allow handling of such a limited volume of medium without evaporation, pH change, or bacterial contamination. By immersing the capillary tube in a larger vessel of culture medium, a slow renewal of culture medium could be effected by diffusion, and ultimately the cells could migrate through the open ends of the capillary into the larger vessel. When emerging from the tube, a preconditioned medium was found necessary.

"Single cells were isolated by means of a capillary pipette. The terminal capillary tube was heat-sealed at each end, and scanned for well-separated cells. Segments approximately 5-8 mm long, each containing a single cell, were cut from the tube and inserted in a culture vessel. One technical problem, movement of fluid in the capillary tube during cutting, was solved later by using precut 8 mm-long capillaries.[1] With this capillary technique, we cloned cells from freshly isolated normal and malignant tissues of several species.

"Single cell isolation techniques have not been widely used in recent years, for many investigators have been satisfied with colony isolates as clones. However, the plating of single cell suspensions without scoring for single cells and without rigid segregation of such cells from other colonies or cells dislodged in the fluid medium provides no assurance that the colony is, in fact, a clone. The capillary tube is a useful microvessel that has certain advantages over other more recently introduced techniques. Besides protecting the cell from excessive evaporation and pH changes, the tube can be easily rolled over to examine all surfaces for contaminating cells."

1. Sanford K K. Capillary techniques. (Kruse P & Patterson M K, Jr., eds.) *Tissue culture, methods and applications.* New York: Academic Press, 1973. p. 237-41.

CC/NUMBER 5
JANUARY 30, 1984

Hanks J H & Wallace R E. Relation of oxygen and temperature in the preservation
of tissues by refrigeration. *Proc. Soc. Exp. Biol. Med.* 71:196-200, 1949.
[Leonard Wood Memorial, Dept. Bacteriology and Immunology, Harvard Medical Sch.,
Boston, MA]

During attempts to secure the growth of *Mycobacterium leprae* in cells cultivated from susceptible leprosy patients, an autoclavable balanced salt solution (BSS) was prepared. In long-term cultures, supersaturation with Ca xPi inhibited the growth of host cells and caused precipitates in the explants and embedding chicken plasma. These problems were circumvented by restricting Ca levels to 5 mg percent and Pi to 2.2 mg percent and by maintaining 10:1 ratios of mammalian serum to chick embryo extract. The BSS became known because it was adopted by John Enders et al.[1] in their propagation of the polio virus in cell cultures. [The *SCI*® indicates that this paper has been cited in over 1,045 publications since 1955.]

John H. Hanks
Department of Immunology and
Infectious Diseases
School of Hygiene and Public Health
Johns Hopkins University
Baltimore, MD 21205

October 17, 1983

"This story illustrates the art of producing a quoted paper by being an unremitting failure in one's primary purpose. The steps were: (a) stubborn investigation of a problem that has not yet been solved after 110 years of effort by scores of investigators, (b) making sense by making an autoclavable balanced salt solution (BSS), and (c) helping John Enders et al. to develop cell cultures for the propagation of polio virus.

"These accidents happened because in 1939 I had been induced by the Leonard Wood Memorial (American Leprosy Foundation) to investigate the microbiology of leprosy at the Philippine Culion Leprosarium (then the largest in the world), a remote 8 × 25 mile island outpost at the edge of the China Sea, just 12° north of the equator. I was given free rein to search for the central problems in bacteriology or immunology

and to investigate the problem I thought most important. It became evident that without cultivating the causative agent, one cannot gain the knowledge required to cope with a stubborn infectious disease.

"After about one year I had shot my bolt of brilliant ideas for inducing *M. leprae* to grow in bacteriologic media. The glowing report of Timofejewsky[2] suckered me into the belief that, if only I could learn to maintain human cells from susceptible patients in tissue culture, a useful (though more costly) tool could be made available. The trials and errors doubled sweat production in that sticky environment. The recommended filter sterilization of high-bicarbonate BSS opened a can of worms. Dissolved CO_2 (and changes in pH) were modified by pressures, temperature, and CO_2 concentrations. I terminated this foolishness[3] by adjusting the ratio of phosphage buffers to pH 6.8 to limit the hydrolysis of glucose, and by including phenol red color indicator so that pH could be seen across the room at any time and by autoclaving. Stimulation of cell respiration was ensured by one-tenth the usual bicarbonate or by the bicarbonate in serum. If interested in tissue cultivation, hundreds of biochemists, microbiologists, or country boys might have done the same.

"The third step arose from a long-standing friendship with Enders at Harvard University. He, Tom Weller, and/or Frank Robbins began coming to my lab to discuss uncertainties or problems in their efforts to obtain standard replication of the polio virus in an optimal cell culture system.[1] Their investigation resulted in replacing $30 monkeys with $.12 cell cultures (and a Nobel prize). One of their rare mistakes was to recommend that Hanks BSS was the way to go.

"The paper including BSS probably owes its popularity to the innate wisdom and laziness of humanity. It is useful to autoclave all possible reagents employed in cell cultivation. If contamination occurs, one has a specific list of components which need not interfere with sleeping at night."

1. Enders J F, Weller T H & Robbins F C. Cultivation of the Lansing strain of poliomyelitis virus in cultures of various human tissues. *Science* 109:85-7, 1949. (Cited 215 times since 1955.)
2. Timofejewsky A D. Explantationsversuche von leprösem Gewebe. *Arch. Exp. Zellforsch.* 9:182-202, 1930.
3. Hanks J H. Calcification of cell cultures in the presence of embryo juice and mammalian sera. *Proc. Soc. Exp. Biol. Med.* 71:328-34, 1949.

Citation Classics

Eagle H. Amino acid metabolism in mammalian cell cultures.
Science   130:432-37, 1959.

The present article "is a progress report rather than a review and in large part summarizes studies from a single laboratory" on the minimal essential medium for cultivation of mammalian cells in either monolayer or suspension. Every cell culture examined, whether human or animal in origin, required at least 13 amino acids for survival and growth. All the cultured human cells examined were found to contain large amounts of glutathione, taurine, glutamine, ammonia, and glutamic acid. [The *SCI*® indicates that this paper was cited 2,255 times in the period 1961-1975.]

Harry Eagle, M.D.
Cancer Research Center
Albert Einstein College of Medicine
New York, New York 10461

December 27, 1976

"The growth media characterized as BME (Basal Medium, Eagle) and MEM (Minimal Essential Medium) which have been used for 20 years, did not result from a planned attempt to develop a culture medium. The original objective was rather to define those components which were essential for the survival and growth of animal cells; and that objective was realized with the demonstration that 13 amino acids, 6 vitamins, 6 ionic species and glucose, plus an unknown complex of compounds present in serum, were both necessary and sufficient for the propagation of a wide variety of animal cells. The composite of these constituents constitutes the basal (1955)[1] or minimum essential (1959) medium, the latter differing primarily in the increased concentration of some of the factors described as being growth limiting. Obviously, whole serum contained all 26 of the essential defined components, but not in optimal or sufficient concentrations. A widespread practical application thus grew out of an interest in growth requirements, which was not initially directed at the practical objective of a growth medium.

"A number of nutritionally non-essential compounds which are essential for the cellular economy, and which are normally synthesized by the cells in amounts sufficient for growth, may be added to the minimal or basal medium; and some of these become essential at low population densities, or in cells with genetically determined defects in a specific biosynthetic pathway. Later studies from this laboratory dealt with such matters as the utilization of peptides and cofactors; the determination of the minimal intracellular concentrations of amino acids necessary for protein synthesis; amino acid transport; protein turnover; the synthesis of purines and pyrimidines; carbohydrate metabolism; *in vitro* screening of carcinolytic agents; population density effects on cellular metabolism, including contact inhibition of growth; pH effects; etc. Even now, however, 20 years after the description of the original basal medium, the role of serum, whether in terms of small molecules which may be bound to protein and slowly released to the medium, or in terms of the protein moieties themselves, has yet to be defined."

1. **Eagle H.** Nutrition needs of mammalian cells in tissue culture. *Science* 122:501-04, 1955.

Oxender D L & Christensen H N. Distinct mediating systems for the transport
of neutral amino acids by the Ehrlich cell. *J. Biol. Chem.* **238**:3686-99, 1963.
[Department of Biological Chemistry, University of Michigan, Ann Arbor, MI]

This paper developed a strategy for discriminating distinct transport systems. Based on the sensitivity of the transport activity to alkali metal ion, pH, and competitive interactions, two distinct but overlapping systems (A and L) for the neutral amino acids were described for the Ehrlich cell. [The *SCI®* indicates that this paper has been cited in over 580 publications since 1963.]

Dale L. Oxender
Department of Biological Chemistry
University of Michigan
Ann Arbor, MI 48109

November 3, 1983

"At a recent national meeting of biochemists, someone came up to me after listening to my presentation on the genetics and cloning of transport in bacteria and said, 'It is nice that you stayed in the same transport research field that your father was in. He was using more classical approaches with animal cells in the 1963 *Journal of Biological Chemistry* paper and you are now applying recombinant DNA techniques to study transport in bacteria.'[1] I had to tell him I was somewhat older than he thought and bore responsibility for both approaches.

"I was just finishing my postdoctoral training with Christensen when the research for the 1963 paper on the A and L transport systems was carried out. The main view in the transport field for animal cells prior to 1963 was that a broad specificity system served for essentially all of the neutral amino acids, even though looking back at results from our laboratory and others, heterogeneity in the systems was already evident. The main contribution of our 1963 effort was to develop a comprehensive characterization involving kinetics, competitive interactions, and

NA+-dependence, and to provide simple letter names for reference. The letters A and L were chosen at the time to denote alanine-preferring and leucine-preferring, respectively. We considered more descriptive names such as sodium dependent and sodium independent, and pH sensitive and pH insensitive, which describe one or more of the properties of the A and L transport systems. We abandoned these terms since we could not anticipate how many systems might later turn up and what their properties might be. As we have learned more about the systems, we stopped referring to A and L as alanine- and leucine-preferring since each system represents a broad specificity transport system and it is possible to find or prepare amino acid solutes for which each system shows more distinct preference.[2,3] The letters A and L should be considered as simple names for identification and not as indications of their specificity.

"Since that time, other systems for the neutral amino acids such as the ASC system have been distinguished.[4] We used the Ehrlich cell for the initial studies because it is convenient to propagate and has high levels of transport activity. By comparing transport activities in a variety of cells and tissues, we have found that the Ehrlich cell as well as other transformed cell types generally has increased System A activity while more normal cells have lower levels of System A and high levels of ASC (see references 2 and 3). Starvation of normal cells for amino acids results in increased System A activity.

"Some of the reasons the 1963 paper has been cited as much as it has may lie in the simple letter names for transport systems showing detailed characteristics, as well as describing in one place a comprehensive statement of the strategy of discriminating the several systems kinetically and by use of analogues, pH, and alkali-ion specificities."

1. **Landick R C & Oxender D L.** Bacterial periplasmic binding proteins in membranes and transport 1981: a critical review. (Martonosi A, ed.) *Membranes and transport.* New York: Plenum Press, 1982. p. 81-8.
2. **Christensen H N.** Exploiting amino acid structure to learn about membrane transport. *Advan. Enzymol. Relat. Areas Mol. Biol.* 49:41-101, 1979.
3. **Shotwell M A, Kilberg M S & Oxender D L.** The regulation of neutral amino acid transport in mammalian cells. *Biochim. Biophys. Acta* 737:267-84, 1983.
4. **Christensen H N, Liang M & Archer E G.** A distinct Na+-requiring transport system for alanine, serine, cysteine and similar amino acids. *J. Biol. Chem.* 242:5237-46, 1967. (Cited 140 times.)

Hayflick L & Moorhead P S. The serial cultivation of human diploid cell strains. *Exp. Cell Res.* **25**:585-621, 1961.

The authors describe the isolation and characterization of normal human diploid cell strains which have a limited capacity to replicate. The eventual degeneration of these strains leads to the hypothesis that the phenomenon is attributable to senescence at the cellular level. With these characteristics and their extremely broad virus spectrum, the use of diploid human cell strains for human virus vaccine production is suggested. [The *SCI®* indicates that this paper was cited a total of 949 times in the period 1961-1976.]

------------◆•◆•◆------------

Leonard Hayflick
Children's Hospital Medical Center
Bruce Lyon Memorial Research Laboratory
Oakland, California 94609

January 13, 1977

"The research described in this paper and the circumstances leading to publication produced several amusing incidents and instructive lessons. The studies began with the—now naive— intention of isolating cancer-causing viruses from cultured human cancer cells. The strategy was to look for changes in cultured normal human cells inoculated with fluids and extracts from cultured cancer cells. The prerequisite for normal human cells was met by utilizing tissue from aborted human fetuses. The normal cells multiplied vigorously for about ten months, and then, after undergoing about 50 population doublings, they died.

"We worried that the ultimate death of these cells might be caused by some trivial culture condition that we had overlooked. Several critical experiments ultimately convinced us that it was a programmed intracellular event. We also sent cultures to several skeptics who were told in advance when the luxuriating cultures given to them would die. Our predictions were met with disbelief, but when the telephone rang with the good news that the cultures had died when expected, we decided to publish.

"Our original suggestion that the phenomenon might be a manifestation of aging at the cellular-level seems even more tenable today. It has given rise to the new field of 'cytogerontology.'

"The cells were found to have other properties of immediate practical importance. We found them to be more sensitive to human viruses than any other cell and suggested that they had many advantages over monkey kidney cells for the production of human virus vaccines. Thousands of cultures were distributed to virologists, but our suggestion that the cells be used for human virus vaccines met with considerable resistance in this country. Ten years later, in 1973, the first poliomyelitis vaccine produced on our normal human cell strain WI-38 was distributed here.

"Efforts to publish the paper itself also met with resistance. *The Journal of Experimental Medicine* rejected the paper over the signature of a soon-to-be-named Nobel Laureate, giving as its reason the then prevalent dogma: 'The largest fact to have come out from tissue culture in the last fifty years is that cells inherently capable of multiplying will do so indefinitely if supplied with the right milieu *in vitro*.

"As to our data on virus sensitivity, the writer said, 'The observations on the effect of viruses on cultures of these (cells) seem extraneous...'; and as for our interpretation that the phenomenon represents aging, the letter commented, 'The inference that death of the cells...is due to "senescence at the cellular level" seems notably rash.' The manuscript was then sent to *Experimental Cell Research* where it was immediately accepted.

"It was also in 1961 that our laboratory first isolated and identified a new mycoplasma species, *Mycoplasma pneumoniae*, the etiological agent of primary atypical pneumonia in man. This paper is also on ISI® 's best-seller list of the 500 most cited papers.

"Both papers resulted from the use of resources 'bootlegged' from grants having entirely different purposes. If our work has had any value, it is a tribute to the then prevailing freedom to pursue interesting leads unfettered by preconceived expectations written into grant proposals. Regrettably, in recent years, such opportunities have become increasingly compromised by myopic administrative demands for strict accountability."

Leibovitz A. The growth and maintenance of tissue-cell cultures in free gas
exchange with the atmosphere. *Amer. J. Hyg.* 78:173-80, 1963.
[Sixth US Army Medical Laboratory, Fort Baker, CA]

Tissue cell cultures grew in free gas exchange with the atmosphere when the bicarbonate buffer was replaced by the free base amino acids, especially L-arginine. Glycolysis of the medium was significantly reduced when glucose was replaced by galactose, sodium pyruvate, and DL-alpha alanine. [The *SCI®* indicates that this paper has been cited over 270 times since 1963.]

Albert Leibovitz
Department of Internal Medicine
Section of Hematology and Oncology
Health Sciences Center
University of Arizona
Tucson, AZ 85724

November 16, 1981

"As an Army microbiologist, I was assigned to establish a diagnostic virology laboratory at the Fifth US Army Medical Laboratory in St. Louis. This was accomplished in 1958. Most commercially available human tissue cell lines were 'altered' cell lines, as HeLa or HEp II. In closed systems (screw-capped test tubes) they grew luxuriantly in such glucose-bicarbonate buffered media as Eagle's MEM, but several problems were obvious. These cell lines were highly glycolytic and unless the media were changed at least three times a week, the sharp drop in pH destroyed the cell monolayer. If the screw caps did not fit the test tubes properly, carbon dioxide escaped into the atmosphere and the residual sodium carbonate raised the pH to toxic levels. Maintenance of inoculated cultures was labor intensive; the technicians felt like 'apes' in feeding the cultures and in blind passing those that became toxic.

"The classical studies of Eagle and coworkers established the minimal nutritive requirements of tissue cells *in vitro*. Analysis of his data, especially his studies with the HeLa cell line,[1] revealed that the essential amino acids could be incorporated in media at much higher concentrations and that he utilized the basic amino acids as hydrochloride salts. My studies determined that the basic amino acids, in their free base form, could replace bicarbonate as a buffer. L-arginine, 3 μM/ml, yielded a pH 7.6 medium that permitted tissue cell growth in free gas exchange with the atmosphere.

"In 1960, I was transferred to the Sixth US Army Medical Laboratory in the presidio of San Francisco. Soon thereafter, I resumed my research to solve the glycolysis problem. Eagle et al.[2] and Chang and Geyer[3] noted that glycolysis was significantly reduced when galactose was substituted for glucose. Growth was haphazard, however, unless such carbohydrate saving agents as pyruvate and DL-alpha alanine were incorporated. Various combinations of galactose, sodium pyruvate, and DL-alpha alanine were tested and by my fifteenth formulation, the desired medium was attained.

"In collaboration with Charlotte John (unpublished observations), medium L-15 was used to establish cell cultures from normal human tissues. Such primary cultures are significantly less glycolytic than HeLa or HEp II and could be maintained at least 30 days without refeeding. Thus, the labor intensive side of diagnostic virology was eliminated, resulting in a significant savings in man-hours, media, and material. This proved most valuable in a collaborative study with the Walter Reed Army Institute of Research Virology Division in the development of adenovirus vaccines. In this nationwide study, not having to refeed inoculated cultures resulted in the savings of $50,000 the first year in just pipettes as well as thousands of man-hours of labor. I received an Army award and the Legion of Merit.

"I am naturally pleased that medium L-15 is being used internationally in a variety of fields including the establishment of cell lines from cold-blooded animals such as fish and amphibians. Since retiring from the Army, I have been engaged in the establishment of cell lines from human solid tumors using medium L-15 with certain additives."[4]

1. **Eagle H.** The specific amino acid requirements of a human carcinoma cell (strain HeLa) in tissue
   culture. *J. Exp. Med.* 102:37-48, 1955.
2. **Eagle H, Barban S, Levy M & Schulze H O.** The utilization of carbohydrates by human cell cultures.
   J. Biol. Chem. 233:551-8, 1958.
3. **Chang R S & Geyer R P.** Propagation of conjunctival and HeLa cells in various carbohydrate media.
   Proc. Soc. Exp. Biol. Med. 96:336-40, 1958.
4. **Leibovitz A, Stinson J C, McCombs W B, McCoy C E, Manzur K C & Mabry N D.**
   Classification of human colorectal adenocarcinoma cell lines. *Cancer Res.* 36:4562-9, 1976.

Elkind M M & Sutton H. Radiation response of mammalian cells grown in culture. 1. Repair of X-ray damage in surviving Chinese hamster cells. *Radiat. Res.* **13**:556-93, 1960.

The survival curve of cultured Chinese hamster cells was found to have a threshold. From biophysical principles, it may be inferred that damage must be accumulated to kill a cell and, therefore, a surviving cell is sublethally damaged. Dose fractionation was used to show that sublethal damage is rapidly repaired. [The *SCI®* indicates that this paper has been cited over 420 times since 1961.]

Mortimer M. Elkind
Division of Biological & Medical Research
Argonne National Laboratory
Argonne, IL 60439

September 8, 1978

"Having entered the field of radiobiology in 1953 from a background in physics, I had a natural interest in formal models of dose-effect curves such as those pertaining to cell killing. Because of the discrete and random nature of energy deposition events from ionizing radiation, inactivation models current in the 1950s derived primarily from hit-target theory. These usually presupposed a certain target content per cell and made assumptions connecting the inactivation of one or more targets and cell inactivation.

"In their simplicity, these formal inactivation schemes held for me an element of elegance. Still, they were essentially unsatisfying since no matter how extensive a story one might hope experimentally to develop, the significance of the resulting edifice would depend upon the validity of the particular formalism assumed.

"In 1956, T. T. Puck and P. I. Marcus, using HeLa cells, published the first quantitative X-ray survival curve of mammalian cells in culture.[1] Some of the features contained in their results proved, in the course of time, to be general for mammalian cells whether assayed *in vitro* or *in vivo*. The one that attracted me was the shoulder on the curve. I realized that, independent of any assumptions about target sensitivity or target organization, a survival curve of this shape indicated a requirement for damage accumulation to effect lethality. I also realized that if damage must be accumulated to kill, a cell surviving a moderate killing dose is likely to be sublethally affected. (In contrast, from formal target theory as applied in the 1950s one would conclude that the absence of a threshold on a survival curve means that a single hit suffices to kill and, consequently, that a surviving cell is an undamaged cell.)

"In the context of the foregoing, the question arose: Can surviving cells repair their sublethal damage or is it heritable? By 1958, inspired by the technical advances in mammalian cell cultivation due to Puck and his associates, I had converted my laboratory at the National Cancer Institute from one for yeast to one for mammalian-cell research. Ms. Sutton and I undertook to answer this question by using a dose fractionation technique.

"We reasoned that if cells surviving a dose large enough to surpass the shoulder repaired sublethal damage, this would be evident by a return of a shoulder on the survival curve of the survivors of the second dose. We had essentially no examples to draw upon for the time course of this putative repair since the study of such processes was in its infancy at that time. Hence, we applied the two-dose technique using intervals of minutes to hours. Our results showed not only that repair of sublethal damage was rapid, probably being completed before the first mitosis following an initial exposure, but also that cells were capable of repeated cycles of damage and repair. These results had important implications relative to radiation and public health and to the use of radiation in the treatment of cancer."

Reference

1. **Puck T T & Marcus P I.** Action of x-rays on mammalian cells. *J. Exp. Med.* **103**:653-66, 1956.

This Week's Citation Classic

Till J E & McCulloch E A. A direct measurement of the radiation sensitivity of normal
mouse bone marrow cells. *Radiat. Res.* 14:213-22, 1961.
[Dept. Med. Biophys., Univ. Toronto, and Divisions Biological Res. and Phys. of the
Ontario Cancer Inst., Toronto, Ontario, Canada]

The authors describe 'a technique for the measurement of the number of cells in a bone marrow suspension capable of continued proliferation.' It involves 'the formation of colonies of proliferating cells' which, in irradiated mice, 'appear as gross nodules in the spleen.' [The *SCI®* indicates that this paper has been cited over 1,125 times since 1961.]

J.E. Till
Division of Biological Research
Ontario Cancer Institute
500 Sherbourne Street
Toronto, Ontario M4X 1K9
Canada

August 16, 1979

"This paper was a report of work carried out in collaboration with Ernest McCulloch; my name appeared first on the paper simply because it was my turn. We thought the paper was a pretty good one, but certainly I for one did not expect it would be cited so often.

"The basis for the interest in this paper is that it described a method for detecting and counting pluripotent stem cells of the blood-forming system. The method is based on the ability of such stem cells, at least in mice, to multiply and differentiate to form localized colonies of descendants in the spleen of an irradiated mouse. These 'bumps on the spleen' had been described before, but in words such as 'local areas of regeneration.' A key contribution of our work was the realization that the 'bumps' could be regarded as 'colonies,' and that it was possible that the 'colonies' originated from single cells. Andrew Becker, in a subsequent publication, showed this view to be a reasonable one.[1] This established the validity of the method as an assay for the cell of origin of the colonies; this assay could then be used to study the properties of these cells. In the beginning, we were careful to avoid assigning an identity to these as yet uncharacterized cells, and only referred to them by the operational term 'colony-forming units' (CFU). This CFU terminology has persisted and proliferated, and I am sure it now confuses the uninitiated, since several kinds of different CFU have been described and are named in different ways using the term CFU with various prefixes or suffixes.

"When the colonies were examined, they were found often to contain more than one kind of differentiated cell. Subsequent work, especially by Alan Wu, provided evidence that these different cell types could be derived from the same CFU.[2] This provided reasonably strong evidence that the CFU were pluripotent in their potentiality for differentiation. The question was an important one to settle, because it had been debated by several generations of hematologists. Does the blood-forming system of the adult contain pluripotent (or only unipotent) stem cells? For the mouse, the answer was yes, pluripotent stem cells are still present in adult life.

"If we were indeed detecting pluripotent stem cells using the spleen colony assay, then the colonies should themselves contain new stem cells. In a rather important series of experiments initiated by Louis Siminovitch, this was shown to be the case.[3] CFU were capable of self-renewal, in that the colonies to which they gave rise contained significant numbers of new CFU. These properties, a capacity for extensive proliferation including self renewal, and a capacity to give rise to several types of differentiated descendants, are the essential properties of pluripotent stem cells."

1. Becker A J, McCulloch E A & Till J E. Cytological demonstration of the clonal nature of spleen colonies derived from transplanted mouse marrow cells. *Nature* 197:452-4, 1963.
2. Wu A M, Till J E, Siminovitch L & McCulloch E A. A cytological study of the capacity for differentiation of normal hemopoietic colony-forming cells. *J. Cell Physiol.* 69:177-84, 1967.
3. Siminovitch L, McCulloch E A & Till J E. The distribution of colony-forming cells among spleen colonies. *J. Cell Physiol.* 62:327-36, 1963.

Bradley T R & Metcalf D. The growth of mouse bone marrow cells *in vitro*.
Aust. J. Exp. Biol. Med. Sci. 44:287-300, 1966.
[Dept. Physiol., Univ. Melbourne; and Cancer Res. Unit, Walter and Eliza Hall Inst.,
Parkville, Victoria, Australia]

The article describes the quantitative clonal growth of mouse bone marrow cells *in vitro* using an agar culture system. Neonatal mouse kidney or whole embryo cells were used as feeder cells to stimulate colony development. The colonies were shown to be composed of large mononuclear cells probably phagocytosing agar and cells typical of the early stages of development of mouse granulocytes. [The *SCI®* indicates that this paper has been cited over 545 times since 1966.]

Ray Bradley
Cancer Institute
Peter MacCallum Hospital
Melbourne, 3000
Australia

December 12, 1978

"The work referred to was initiated by an observation made whilst attempting to start another investigation. I had returned to the Physiology Department, Melbourne University, from the National Cancer Institute where Lloyd Law had made me enthusiastic about thymus studies and Bob Roosa had taught me the elements of cell culture, particularly the importance of cloning techniques.

"I wanted to see whether one could detect clonogenic tumour cells in the thymus of AKR mice prior to the obvious tumour development. Don Metcalf and I were collaborating on another project involving thymus and spleen transplants and Don's AKR mouse colony was available, so I started trying to clone thymus, lymph node, and bone marrow cells, the latter two tissues as 'controls' for the thymus.

Two significant points in the technique were the use of agar to immobilize any progeny developed, which other workers had been starting to use, and the use of feeder cells which had also been recognized as often of use in cloning techniques.

"In the first experiments, the thymus cells failed miserably but small colonies developed from the marrow cells, the growth of which was greatly improved by a variety of feeder cells, particularly neonatal kidney and whole embryo cells. The morphology of the colony cells was an important issue. Don's experience in morphological studies was enlisted as we started trying to identify the cell types in the colonies, an exercise which is still necessary whenever new sources of proliferative factors are used in the system.

"The reason that the paper has been frequently cited is probably because it offered a simple technique for quantitation of both the progenitor cells in normal and pathological conditions and the proliferative factors elaborated by feeder cells or present in body fluids. As such it offered the possibility of resolving some of the problems concerning the control of granulopoiesis and, as emerged more clearly a little later, monocytopoiesis.

"In retrospect, two aspects of the work at this stage always seem important to me. The first point is that it was soon shown that identical colony formation could be obtained using mouse serum to replace the feeder cells which gave the system a more physiological basis. The second point has been that, had we started with any other species, it may well have not succeeded so easily. Perhaps this is a salutary lesson to those who want us to dispense with experimental animal model systems."

Pike B L & Robinson W A. Human bone marrow colony growth in agar-gel.
J. Cell. Physiol. **76**:77-84, 1970.
[Division of Hematology, Department of Medicine, University of Colorado
Medical Center, Denver, CO]

This article describes a method for the growth of colonies from human bone marrow aspirates *in vitro* in agar-gel medium. 'Feeder layers' containing normal human peripheral blood leukocytes were used to stimulate colony development. Colonies generated from normal human bone marrow cells were almost exclusively granulocytic. [The *SCI®* indicates that this paper has been cited in over 835 publications since 1970.]

Beverley L. Pike
Walter and Eliza Hall Institute
of Medical Research
Melbourne, Victoria 3050
Australia

April 27, 1983

"Both Bill Robinson and I learned the basics of *in vitro* culture of bone marrow cells at the clonal level from Donald Metcalf and Ray Bradley during time spent at the Walter and Eliza Hall Institute in Melbourne. Bill returned to Denver and I with him, and we took up the challenge of developing a simple culture system for colony development from single human bone marrow cells. Use of 'feeder layers' of a variety of cell types such as kidney tubules, as described by Bradley and Metcalf,[1] to support colony growth from murine bone marrow cells failed to stimulate colony growth from human cells. Sources of colony-stimulating activity (CSA), such as human urine, capable of stimulating colony development from murine bone marrow cells[2] were also totally inactive on human cells.

"At this time, we were documenting urinary CSA levels in leukemic patients to see if any correlation existed between CSA levels and the clinical course of the disease. It became readily apparent that a rough inverse correlation between the proportion of blast cells in the peripheral blood and CSA levels existed. Untreated patients with acute granulocytic leukemia exhibited low levels on presentation, which rose during remission following chemotherapy.[3] The knowledge that CSA activity rose concomitantly with the mature peripheral blood granulocyte levels led us to draw up a working model for the regulation of granulopoiesis by-products from mature granulocytes.

"To test our hypothesis, we titrated normal human peripheral blood buffy coat leukocytes into an agar-underlayer and added murine bone marrow cells in an overlayer in the presence and absence of a known source of CSA. As a human bone marrow aspirate sample was available at the time, it was set up simultaneously. We had tried so many variants and the best we hoped for was inhibition of colony growth of murine bone marrow with high numbers of peripheral white blood cells. I can well remember our excitement when a few days later we noted colony growth in the absence of added CSA. Furthermore, it appeared that the human marrow cells were also beginning to proliferate, and after a few more days, discrete colonies were visible.

"It was with great excitement, and some degree of courage, that we repeated the experiment. Our initial observations were confirmed. Human peripheral blood leukocytes provided adequate stimulation to allow single human bone marrow cells to proliferate to form granulocytic colonies. The solution to the problem was in fact quite simple.

"The reason this paper has been cited so frequently is that it provided the first simple, reproducible method for the generation of colonies from single human bone marrow cells *in vitro* (and subsequently also from human peripheral blood cells). This allowed studies of the regulation of granulopoiesis in both the normal and leukemic situations. The technique has been used clinically to classify leukemias by using the growth characteristics of cells as a criterion, to assess possible responses to chemotherapy, and also to assess remission status. Its overall contribution to the understanding of the growth characteristics of both normal and leukemic cells has been extensively covered in a publication by Metcalf."[4]

1. **Bradley T R & Metcalf D.** The growth of mouse bone marrow cells *in vitro.*
 Aust. J. Exp. Biol. Med. Sci. **44**:287-300, 1966.
 [Citation Classic. *Current Contents/Life Sciences* **22**(40):12, 1 October 1979.]
2. **Robinson W A, Stanley E R & Metcalf D.** Stimulation of bone marrow colony growth in vitro by human urine.
 Blood **33**:396-9, 1969.
3. **Robinson W A & Pike B L.** Leukopoietic activity in human urine: the granulocytic leukemias.
 N. Engl. J. Med. **282**:1291-7, 1970.
4. **Metcalf D.** Hemopoietic colonies: in vitro cloning of normal and leukemic cells.
 Rec. Res. Cancer Res. **61**:1-227, 1977.

This Week's Citation Classic

Puck T T, Marcus P I & Cieciura S J. Clonal growth of mammalian cells in vitro: growth characteristics of colonies from single HeLa cells with and without a "feeder" layer. *J. Exp. Med.* **103**:273-83, 1956.
[Dept. Biophysics, Florence R. Sabin Labs., Univ. Colorado Med. Ctr., Denver, CO]

A simple method is described for quantitative growth of single mammalian cells in tissue culture so that each cell forms a discrete colony. A feeder layer of X-irradiated cells can stimulate growth in suboptimal nutrient medium. The procedure permits mutant selection, and other genetic studies on somatic cells. [The *SCI®* indicates that this paper has been cited in over 660 publications since 1961.]

Theodore T. Puck
Eleanor Roosevelt Institute for Cancer Research
4200 East Ninth Avenue
Denver, CO 80262

November 22, 1982

"This paper describes experiments designed to furnish a new approach to the study of human genetics which would be free of the complications of human mating. Human genetics had been a weak science before 1956 because of the long human generation time and the impossibility of carrying out particular matings which would answer specific questions in human genetics. It occurred to me that a genetics based on somatic cells instead of germ cells, utilizing tissue culture techniques and applying the concepts of microbial genetics, would solve these problems. It became necessary to develop techniques for: taking biopsies of somatic cells from any person; producing reliable growth of such cells into large, genetically stable populations; and a rapid, quantitative, routine method for growing of single cells into discrete, macroscopic colonies. Such an approach would bypass the need for human mating and permit genetic experiments on somatic cells *in vitro* that should have all the simplicity and power of microbial genetics.

"I was influenced by the experiments of K.K. Sanford[1] and her colleagues, who had cloned occasional single mammalian cells though not in quantitative fashion, by Earle's concept of the need for cells to condition the primitive growth media available in those days, and by the simplicity of the culture techniques of Enders and his coworkers.

"I began experiments with Roshan Christensen, a postdoctoral fellow. We achieved reliable growth only with inocula of 500 cells or more. Philip Marcus then joined me as a graduate student. We considered ways of feeding single cells with nutrients from a mass culture. Leo Szilard, who was visiting the laboratory at this time, suggested suspending the single cells above a massive culture in the same dish. However, the problem was not solved until Marcus and I conceived of an X-irradiated feeder layer.[2] At first we separated the single cells from the feeder layer by a glass plate in the same dish, but then incorporated both cell types in a single layer. A simple quantitative operation suitable for genetic experiments resulted. Subsequently, we eliminated the feeder layer by improving the nutrient medium. However, the X-irradiated feeder layer has since found application in the induction of continuing growth of differentiated cells.

"Simple routine methods for initiating long-term cultures with stable karyotypes from any person were developed and single-cell survival curves were introduced as a means for study of the effects of physical, chemical, and biological agents on the reproduction of single cells.[3] Survival curve analysis has had many applications including initiation of quantitative mammalian cell radiobiology which contributed to a more effective radiotherapy and demonstrated the role of cell turnover in the mammalian radiation syndrome.[4] The human chromosomes were characterized and, with the aid of a study group which we organized in Denver, the classification system for the human chromosomes still used today was devised.[5] Subsequent developments in somatic cell genetics included preparation and characterization of single gene mutants,[4] cell hybridization,[6] development of recombinant DNA approaches, production of monoclonal antibodies, and mapping of the human genes.[7] Somatic cell genetics is now making important contributions to medicine and to the understanding of differentiation and development.[8]

"The work in this paper initiated the discipline of somatic cell genetics which has become the principal mode of study of human genetics and its applications to medicine and human molecular biology."

---

1. Sanford K K, Earle W R & Likely G E. The growth in vitro of single isolated tissue cells.
   J. Nat. Cancer Inst. 9:229-46, 1948.
   [Citation Classic. *Current Contents/Life Sciences* 24(29):19, 20 July 1981.]
2. Puck T T & Marcus P I. Rapid method for viable cell titration and clone production with HeLa cells on tissue culture: the use of X-irradiated cells to supply conditioning factors. *Proc. Nat. Acad. Sci. US* 4:432-7, 1955.
3. ----------------------. Action of X-rays on mammalian cells. *J. Exp. Med.* 103:653-66, 1956.
4. Puck T T. *The mammalian cell as a microorganism: genetic and biochemical studies in vitro.*
   San Francisco: Holden-Day, 1972. 219 p.
5. Book J A, Lejeune J, Levan A P, Chu E H Y, Ford C E, Fracarro M, Harnden D G, Esu T C, Hungerford D A, Jacobs P A, Makino S, Puck T T, Robinson A, Tjio J H, Catchside D G, Muller H J & Stern C. A proposed standard system of nomenclature of human mitotic chromosomes. *J. Amer. Med. Assn.* 174:159-62, 1960.
6. Harris H. *Cell fusion.* Cambridge, MA: Harvard University Press, 1970. 108 p.
7. Gusella J F, Jones C, Kao F T, Housman D & Puck T T. Genetic fine structure mapping in human chromosome 11 by use of repetitive DNA sequences. *Proc. Nat. Acad. Sci. US* 79:7804-8, 1982.
8. Puck T T & Kao F T. Somatic cell genetics and its application to medicine. *Annu. Rev. Genet.* 16:225-71, 1982.

Iscove N N, Senn J S, Till J E & McCulloch E A. Colony formation by normal and leukemic human marrow cells in culture: effect of conditioned medium from human leukocytes. *Blood* 37:1-5, 1971. [Inst. Med. Sci., Univ. Toronto; Ontario Cancer Inst.; and Sunnybrook Hosp., Toronto, Canada]

'Conditioned medium' obtained from cultures of human peripheral leukocytes promoted the growth of human marrow cells in cell culture. This material also permitted the growth of small colonies from the marrow of patients with acute myelogenous leukemia in relapse; in its absence, only occasional colonies were observed. [The *SCI*® indicates that this paper has been cited in over 390 publications since 1971.]

Norman N. Iscove
Basel Institute for Immunology
CH-4005 Basel
Switzerland

March 15, 1983

"Research into the precursors of blood cell formation was transformed in the 1960s by two developments. First, there was the introduction of the spleen colony assay for primitive hemopoietic precursors in the mouse by Till and McCulloch in 1961,[1] demonstrating the power of clonal analysis applied to precursor cells too rare to be examined directly. This development was soon followed, in the mid-1960s, by the successful adaptation of tissue culture methods to the clonal detection of granulocyte and macrophage precursors in the mouse by Pluznik and Sachs[2] and by Bradley and Metcalf.[3]

"By the time I began my graduate studies with McCulloch in 1968, many investigators were anxious to have culture techniques which would work with human material. In the previous year, Senn reported with McCulloch and Till the first success in culturing human granulocyte precursors from bone marrow.[4] The colonies were quite small, but within a short time Pike and Robinson had achieved significantly better growth conditions in agar over a 'feeder layer' of human blood leukocytes.[5] Tissue culture methodology was in rapid flux at the time. The requirement of mouse granulocyte and macrophage precursors for glycoprotein growth factors ('colony-stimulating activity,' CSA, or 'colony-stimulating factor,' CSF) was becoming apparent, and feeder layers producing CSA were in the course of being replaced by

medium 'conditioned' by appropriate cellular sources of these activities. It seemed likely that human cells would have the same requirement. In my host laboratory, methyl cellulose had been found to have certain technical advantages over agar, including a greater sensitivity of detection of CSA of mouse origin. I therefore set out, with Senn, McCulloch, and Till, to adapt the methyl cellulose method to the culture of human marrow cells and to replace the feeder layers with an appropriate conditioned medium.

"In our first attempts, colonies did not grow at all in the same conditions which worked so well with mouse cells. In particular, conditioned medium derived from mouse cells and active on mouse cells had no effect on human marrow cells. Finally, two elements, both inspired by the Pike and Robinson method, proved effective. The first was the inclusion of asparagine in our culture medium, a 'nonessential' amino acid thought at the time to be a requirement for leukemic but not normal cells. It proved essential for colony growth by normal human hemopoietic precursors. The second was the use of conditioned medium from human blood leukocytes as the source of CSA. This step was not straightforward, since medium from normal leukocytes incubated alone in liquid culture had only weak or undetectable activity. However, medium cultured over a feeder layer of leukocytes immobilized in agar did work. Such 'conditioned medium' was reported in our 1971 paper and became a standard constituent of our cultures. Later, phytohaemagglutinin was found to stimulate leukocytes to release activity in liquid culture, and more recently the original observations were explained by the finding by Hoang *et al.* that a stimulus to activity release can be extracted from the crude agar itself.[6]

"The high frequency of citation of this paper along with the one by Pike and Robinson[5] and similar studies in the same period by Chervenick and Boggs[7] and by Paran and Sachs *et al.*[8] reflects the fact that these studies initiated the era of application of clonal tissue culture to human hematopoietic precursors, demonstrated the existence of granulopoietic CSFs for both normal and leukemic human cells, and showed that leukemic cells could grow in culture and would therefore be amenable to direct study."

1. Till J E & McCulloch E A. A direct measurement of the radiation sensitivity of normal mouse bone marrow cells.
   Radiat. Res. 14:213-22, 1961. [Citation Classic. *Current Contents/Life Sciences* 22(43):12, 22 October 1979.]
2. Pluznik D H & Sachs L. The cloning of normal "mast" cells in tissue culture.
   J. Cell. Comp. Physiol. 66:319-24, 1965.
3. Bradley T R & Metcalf D. The growth of mouse bone marrow cells *in vitro*.
   Aust. J. Exp. Biol. Med. Sci. 44:287-300, 1966.
   [Citation Classic. *Current Contents/Life Sciences* 22(40):12, 1 October 1979.]
4. Senn J S, McCulloch E A & Till J E. Comparison of colony-forming ability of normal and leukaemic human marrow in cell culture. *Lancet* 2:597-8, 1967.
5. Pike B L & Robinson W A. Human bone marrow colony growth in agar-gel. *J. Cell. Physiol.* 76:77-84, 1970.
   [Citation Classic. *Current Contents/Life Sciences* 26(23):19, 6 June 1983.]
6. Hoang T, Iscove N N & Odartchenko N. Agar extract induces release of granulocyte colony-stimulating activity from human peripheral leukocytes. *Exp. Hematol.* 9:499-504, 1981.
7. Chervenick P A & Boggs D R. Bone marrow colonies: stimulation *in vitro* by supernatant from incubated human blood cells. *Science* 169:691-2, 1970.
   [The *SCI* indicates that this paper has been cited in over 120 publications since 1970.]
8. Paran M, Sachs L, Barak Y & Resnitzky P. *In vitro* induction of granulocyte differentiation in hemopoietic cells from leukemic and non-leukemic patients. *Proc. Nat. Acad. Sci. US* 67:1542-9, 1970.
   [The *SCI* indicates that this paper has been cited in over 255 publications since 1970.]

This Week's Citation Classic

Pluznik D H & Sachs L. The cloning of normal "mast" cells in tissue culture.
J. Cell. Comp. Physiol. **66**:319-24, 1965.
[Section of Genetics, Weizmann Institute of Science, Rehovot, Israel]

The findings published in this paper show that it is possible to obtain clones and clonal differentiation of normal hematopoietic cells in soft agar. The formation of these clones and their differentiation required specific inducing substance(s) produced by other cells that were seeded underneath the agar. It is concluded that a similar approach should be applicable to the *in vitro* cloning and clonal differentiation of all the different types of hematopoietic cells. [The *SCI*® indicates that this paper has been cited in over 590 publications since 1965.]

Leo Sachs
Department of Genetics
Weizmann Institute of Science
Rehovot 76100
Israel

July 11, 1983

"My research with hematopoietic cells began with the aim to develop cell culture systems for the cloning and clonal differentiation of specific cell types, as an approach to studying the controls that regulate growth and differentiation of normal and malignant cells. My first paper on hematopoietic cells *in vitro* in 1961, with a graduate student, H. Ginsburg, described the culture of leukemic cells using spleen feeder layers.[1] I also carried out experiments with normal cells and the next paper, in 1963, with the same student, described the clonal growth and differentiation of normal mast cells and granulocytes in liquid medium using fibroblast feeder layers to provide the necessary inducers for this clonal growth and differentiation.[2] I wrote as the concluding sentence in this paper: 'The described cultures thus seem to offer a useful system for a quantitative kinetic approach to hematopoietic cell formation and for experimental studies on the mechanism and regulation of hematopoietic cell differentiation.'[2]

"This 1963 paper led to such an approach. It was then shown in the 1965 paper with another graduate student, D.H. Pluznik (now at the National Institutes of Health), that normal hematopoietic cells can be cloned in soft agar and that the formation of these clones required an inducer(s) secreted by the feeder layer cells which had been seeded underneath the agar. Since the cells in agar showed some morphological resemblance to the mast cells which had previously been studied, they were provisionally labeled as 'mast cells' in this paper. However, our other results showed that they were macrophages.

"These two papers in 1963 and 1965 thus described the first *in vitro* system for cloning and clonal differentiation of specific types of normal hematopoietic cells, and showed that the induction of such clones was dependent upon inducers secreted by other cells. Our agar assay for cloning normal hematopoietic cells was confirmed by Bradley and Metcalf in 1966.[3] In the next step it was found that the inducers required for the production of macrophage and granulocyte clones were present in conditioned medium produced by the feeder cells, that there appeared to be a different inducer for macrophage and granulocyte clones, and that these cells could also be cloned in methylcellulose.[4] Cloning in agar and methylcellulose are the two main procedures which are now used. The use of appropriate human conditioned medium extended the cloning assay to human hematopoietic cells, showed that the appropriate normal inducer could induce differentiation of some human myeloid leukemic cells, and suggested that induction of normal differentiation of leukemic cells could be a useful approach to therapy.[5]

"This *in vitro* approach has led to the cloning and isolation of growth factors for all types of hematopoietic cells, including different types of lymphocytes. It also led to a further understanding of the controls that regulate growth and differentiation in hematopoiesis, how these controls are coupled in normal development, and the abnormalities in these controls in leukemia and other hematological abnormalities.[6] This can explain why this 1965 paper has been highly cited."

1. **Ginsburg H & Sachs L.** The long-term cultivation in tissue culture of leukemic cells from mouse leukemia induced by Moloney virus or by x-rays. *J. Nat. Cancer Inst.* 27:1153-71, 1961.
2. ⸻. Formation of pure suspensions of mast cells in tissue culture by differentiation of lymphoid cells from the mouse thymus. *J. Nat. Cancer Inst.* 31:1-40, 1963.
3. **Bradley T R & Metcalf D.** The growth of mouse bone marrow cells *in vitro.*
Aust. J. Exp. Biol. Med. Sci. 44:287-300, 1966.
[Citation Classic. Current Contents/Life Sciences 22(40):12, 1 October 1979.]
4. **Ichikawa Y, Pluznik D H & Sachs L.** *In vitro* control of the development of macrophage and granulocyte colonies. *Proc. Nat. Acad. Sci. US* 56:488-95, 1966.
[The SCI indicates that this paper has been cited in over 260 publications since 1966.]
5. **Paran M, Sachs L, Barak Y & Resnitzky P.** *In vitro* induction of granulocyte differentiation in hematopoietic cells from leukemic and non-leukemic patients. *Proc. Nat. Acad. Sci. US* 67:1542-9, 1970.
[The SCI indicates that this paper has been cited in over 250 publications since 1970.]
6. **Sachs L.** Normal developmental programmes in myeloid leukemia. Regulatory proteins in the control of growth and differentiation. *Cancer Surveys* 1:321-42, 1982.

Murashige T & Skoog F. A revised medium for rapid growth and bioassays with tobacco tissue cultures. *Physiol. Plant.* 15:473-97, 1962.

The paper describes an improvised nutrient medium which enabled substantially greater growth of tobacco tissue cultures. Our experiments differed from previous studies in examining several nutrient components simultaneously. The new formulation was conspicuously high in all macronutrient salts, also included all micronutrients, and provided iron in the slowly but more readily available chelated form. Among organic substances, sucrose was increased from 2 to 3% and *myo-inositol* was made a standard addendum. When suplemented with suitable additions of auxin and cytokinin, the medium enhanced the monthly yield of tobacco callus from 5 g to 125 g/culture. It further ensured minimal interference from inorganic and common organic nutrients when used in bioassays of growth-promoting plant or animal extracts. [The *SCI®* indicates that this paper was cited 1203 times in the period 1963-1977.]

Toshio Murashige
University of California, Riverside
College of Natural and
Agricultural Sciences
Riverside, CA 92521

January 16, 1978

"Tissue cultures are currently a major attraction among plant scientists, simply because of their special utility as a research tool and in commercial practice. Our paper contributes to basic tissue culture methodology, thus the frequent citation.

"The research described by our paper was an anticlimax of another project to which I had already devoted 3 years as a doctoral student. I began my dissertation work as Professor Skoog's student shortly after he and Professor C. O. Miller and their colleagues had discovered kinetin, the first of the cytokinin class of plant hormones. The auxin class was already common knowledge and the gibberellins were gaining familiarity in the western world. The discovery of kinetin inspired me greatly, and I was now determined to discover still other hormones that might remain hidden in plants. As an early reward of this determination, I discovered that an extract of tobacco leaves, when added in combination with auxin and kinetin to the culture medium of tobacco callus, increased tissue yield drastically, more than 500% over the controls.

"That autumn Professor Skoog hired several undergraduates to help me, and together we picked over 1,000 lbs of tobacco leaves from a field made available by Professor Ogden of Horticulture. We packed them in large plastic bags, normally used for turkeys, and hauled them to a large, rented food-storage locker where we froze them. Then, using facilities in Biochemistry and with the help of Professors Strong and Garver, we ground the leaves through a meat grinder that had seen better days, thawed them in a large stainless-steel vat, and squeezed out their juices with a hand-press. We pooled, boiled, chilled and filtered the juices. We obtained 150 gals. of clear extract, which we ultimately concentrated to 5 gals. of dark brown syrup in a Mojonnier evaporator.

"Innumerable fractionations of the extract followed. But after all the effort, I learned that much of the tobacco extract effect was located in the mineral fraction. Thus began the work to revise the composition of the nutrient formulation, which for many years had served as the basis of all plant tissue culture experiments.

"The revising took another 100 experiments and produced a new formula. But little enthusiasm remained for its publication. The blow of failing to discover a new hormone was still being felt. That is now past, and I am deeply appreciative of the recognition presently being given to the paper."

Singh K R P. Cell cultures derived from larvae of *Aedes albopictus* (Skuse) and
Aedes aegypti (L.). *Curr. Sci. India* **36**:506-8, 1967.
[Virus Research Centre, Poona, India]

This paper describes the establishment
of cell lines from the newly hatched
minced larvae of *Aedes albopictus* and
Aedes aegypti mosquitoes. Methods
employed for subculturing up to the fifteenth passage level, as well as preliminary characterization of the cell lines,
are given. [The *SCI*® indicates that this
paper has been cited in over 230 publications since 1967, making it the most-cited paper ever published in this journal.]

Vijai Dhanda
Division of Medical Zoology
and Entomology
National Institute of Virology
Indian Council of Medical Research
Pune— 411 001
India

November 29, 1983

"It is unfortunate that my predecessor
K.R.P. Singh, the author of the cited paper,
died within a decade of its publication. At
the request of the director of the National
Institute of Virology, Khorshed Pavri, I am
writing this commentary.

"In the early 1960s, soon after joining this
institute, Singh began studies on the vector
virus relationship of some of the arboviruses
which were important to public health. He
realized the importance of arthropod tissue
culture in studies on arboviruses and received full encouragement from T. Ramachandra Rao, then director of the institute,
and Charles R. Anderson, the chief scientific
representative of the Rockefeller Foundation. Singh was awarded a Rockefeller Foundation fellowship to visit the leading cell
culture laboratories all over the world to
gain experience in this discipline. On his return, he devoted himself wholeheartedly to
the problem and succeeded quickly in establishing cell lines from *Aedes albopictus* and
Aedes aegypti. The paper was sent to *Nature*
for publication, but was returned. It was
then sent to *Current Science* where it was
promptly accepted.

"Singh's success was mainly due to the
choice of his material, which consisted of
Aedes eggs capable of withstanding dry conditions and the rigors of surface sterilization. They can be accumulated in large
quantities to provide a sufficient amount of
tissue. Above all, Singh seemed to have possessed a 'green thumb.'

"Singh's cell lines were the first continuous mosquito cell lines developed anywhere
in the world. These were flown to Yale Arbovirus Research Unit in New Haven, Connecticut, where Sonja Buckley maintained a subline. Their distribution to scientists all over
the world was undertaken from this institute
and from Yale.

"Singh's *Aedes albopictus* cell line (ATC
15) found immense favor among arbovirologists because it was susceptible to several
mosquito-borne arboviruses. Distinctive
cytopathic effect with some flaviviruses facilitated their easy detection.[1,2] It is a rapidly growing cell line, requiring simple media
and easy maintenance. Thus, Singh provided
a tool which arbovirologists and cell biologists had been trying to obtain for a long
time.

"Successful cloning of ATC 15 cells by Igarashi[3] gave a further boost to Singh's work.
This clone is highly susceptible to both
dengue and chikungunya viruses and is now
popular all over the world.

"For his work, in 1968 Singh was awarded
the Shakuntala Amir Chand Prize of the Indian Council of Medical Research."

---

1. **Singh K R P.** Growth of arboviruses in *Aedes albopictus* and *A. aegypti* cell lines.
 Curr. Topics Microbiol. Immunol. **55**:127-33, 1971.
2. **Buckley S M, Hayes C G, Maloney J M, Lipman M, Aitken T H G & Casals J.** Arbovirus studies in
 invertebrate cell lines. (Kustak E & Maramorosch K, eds.) *Invertebrate tissue culture.*
 New York: Academic Press, 1976. p. 3-19.
3. **Igarashi A.** Isolation of a Singh's *Aedes albopictus* cell clone sensitive to dengue and chikungunya viruses.
 J. Gen. Virol. **40**:531-44, 1978.

This Week's Citation Classic

Burkitt D. A sarcoma involving the jaws in African children.
Brit. J. Surg. **46**:218-23, 1958.
[Department of Surgery, Makerere College Medical School,
and Mulago Hospital, Kampala, Uganda]

This was the first paper in which malignant tumours occurring in different parts of the body were recognised to be part of a single clinical syndrome. Emphasis was placed on the jaw lesions, the clinical, radiological, and histological features of which were described. [The *SCI®* indicates that this paper has been cited in over 310 publications since 1961.]

Denis P. Burkitt
Unit of Geographical Pathology
St. Thomas's Hospital
London SE1 7E7
England

March 8, 1983

"My interest was first roused by recognising that lesions occurring in the jaws of children, which had previously been considered to be primary tumours limited to the jaws, were almost if not invariably associated with tumours in other parts of the body. These tumours had not only a characteristic anatomical distribution, occurring commonly in the jaws, orbit, kidneys, adrenals, ovaries, testicles, and long bones, but also shared a common age distribution; a characteristic feature was the rarity of peripheral lymph node involvement.

"Previously, they had been histologically classified according to the site of the presenting lesion. When this was in the orbit, the tumour was a retinoblastoma; when in the kidneys or adrenals, a neuroblastoma; when in the long bones, a Wilm's tumour; and so on, all small-round-cell tumours. The tendency for these tumours to occur in the same patients and to share the same age distribution, and, as discovered subsequently, to have the same geographical distribution convinced me that they must all be part and parcel of the same tumour process. Not only was this confirmed initially by Davies and

O'Conor,[1] who identified these tumours as a form of lymphoma, but subsequently Wright[2] demonstrated how they could be distinguished, mainly on cytological criteria, from other forms of lymphoma. All the results of a common cause tend to occur in association with one another, and conversely, the recognition of an association between results, in this case diseases, suggested a common cause. It was precisely this line of argument that led to my other *Citation Classic*[3] on large-bowel cancer, which shares its pattern of geographical and socioeconomic distribution with that of other characteristically Western diseases.

"Soon after a recognisable clinical syndrome had been established, it became apparent that the tumour was common only in certain parts of Africa. A detailed study of the distribution of the disease in Africa initially revealed a belt across Africa lying for the most part between 10° N and 10° S of the equator. Later it was shown that the tumour only occurred in moist, warm areas.

"These observations led to the hypothesis that some insect vector might be implicated, and this in turn led to the discovery of the Epstein-Barr virus.[4]

"Some lessons to be learned were: 1. Dollars cannot be turned into ideas though the reverse can occur. My total research grants for my first 18 months that delineated the lymphoma belt amounted to £75 (under $150). 2. There is still a place for research in the absence of expensive equipment. 3. There is value in friendly cooperation with those of totally different abilities, background, and outlook.

"The reason why this publication has been cited so often is that it was the first description of the tumour that eventually became known as Burkitt's lymphoma. Moreover, it was in material from this tumour that the famous Epstein-Barr virus was isolated. My colleagues and I helped to construct the pad from which the Epstein-Barr virus rocket was launched. Epstein and his colleagues were dependent on my work to get off the ground. Our efforts would have been largely sterile without their intervention."

1. O'Conor G T & Davies J N P. Malignant tumors in African children, with special reference to malignant lymphoma.
   J. Pediatrics **56**:526-35, 1960.
2. Wright D H. Cytology and histochemistry of the Burkitt's lymphoma. *Brit. J. Cancer* **17**:50-5, 1963.
3. Burkitt D P. Epidemiology of cancer of the colon and rectum. *Cancer* **28**:3-13, 1971.
   [Citation Classic. *Current Contents/Clinical Practice* **9**(12):20, 23 March 1981.]
4. Epstein M A, Achong B G & Barr Y M. Virus particles in cultured lymphoblasts from Burkitt's lymphoma.
   Lancet **1**:702-3, 1964. [Citation Classic. *Current Contents/Life Sciences* **22**(14):10, 2 April 1979.]

Epstein M A, Achong B G & Barr Y M. Virus particles in cultured lympho-blasts from Burkitt's lymphoma. *Lancet* 1:702-3, 1964.

This paper reported the first finding of a virus in a continuous cell line cultured from Burkitt's lymphoma. Electron microscopy of thin sections showed that a small percentage of the cells contained immature and mature virus particles with the characteristic morphology of the herpes virus group. [The *SCI®* indicates that this paper has been cited over 490 times since 1964.]

M. A. Epstein
University of Bristol
Department of Pathology
University Walk
Bristol, BS8 1TD
England

February 16, 1978

"By the early 1960s many viruses causing tumours in animals were known, but there was no such agent with an aetiological link to human cancer. A fortunate chance took me in 1961 to a talk in London by Denis Burkitt, on the African childhood tumour which now bears his name—Burkitt's lymphoma (BL). For the first time outside Africa, Burkitt described climatic factors determining the tumour's distribution which suggested that it might be caused by a vector-borne virus.

"Having worked for several years with animal tumour viruses, this exciting possibility led me immediately to the decision to investigate BL for causative viruses. Despite practical difficulties, and obstruction stemming from scientific politics, the then British Empire Cancer Campaign generously financed a trip to East Africa which ensured a supply of BL material to my laboratory at the Middlesex Hospital Medical School in London. For two years, standard virological isolation techniques

and electron microscopy gave depressingly negative results. Two good things followed this period: 1) A US National Cancer Institute grant of $45,000 which allowed Yvonne Barr and Bert Achong to join me; 2) The realization that success might be achieved if BL cells could be grown *in vitro*, away from host defences so that an inapparent oncogenic virus might replicate. The prospects for establishing cells of a lymphoid tumour in culture were unpromising since this had not been accomplished with any human lymphocytic cell. Nevertheless, the first BL-derived cell line (EBI) was indeed established and when examined in the electron microscope a cell in the first grid square contained particles of obvious herpes virus morphology.

"Biological tests demonstrated that this was a new human herpes virus, and immunological and biochemical investigations confirmed this: The virus came to be known as Epstein-Barr (EB) virus after the cell line in which it was discovered.[1] At this stage, Yvonne Barr married an Australian and left to bring up a family outside Melbourne; Bert Achong still works with me.

"An immense body of information has now been accumulated on EB virus by many laboratories. In man the virus gives widespread inapparent infections, is also the cause of infectious mononucleosis, and has an astonishing association with two human cancers—nasopharyngeal carcinoma, as well as BL. Thus 14 years after its discovery EB virus, found by electron microscopy and at such small initial financial cost, has come to be the only convincing candidate for a human cancer virus, and work on it and the type of cell culture in which it was found has led in addition to great progress in many different disciplines. It is doubtless for these reasons that the paper describing the original finding of EB virus has been cited so frequently."

Reference

1. **Epstein M A, Henle G, Achong B G & Barr Y M.** Morphological and biological studies on a virus in cultured lymphoblasts from Burkitt's lymphoma.
   J. Exper. Med. **121**:761-70, 1965.

Henle G & Henle W. Immunofluorescence in cells derived from Burkitt's
lymphoma. *J. Bacteriology* 91:1248-56, 1966.
[Virus Labs., Children's Hosp. of Philadelphia, and Sch. Medicine, Univ. Pennsylvania,
Philadelphia, PA]

Immunofluorescence was used to detect the herpesvirus in cultured Burkitt's lymphoma (BL) and other lymphoblast cultures and to identify it as a previously unknown virus, now called Epstein-Barr virus. All sera from African BL patients elicited brilliant immunofluorescence, whereas sera from other donors gave weaker or no reactions. [The *SCI®* indicates that this paper has been cited in over 1,100 publications since 1966.]

Gertrude and Werner Henle
Joseph Stokes, Jr., Research Institute
Children's Hospital of Philadelphia
Philadelphia, PA 19104

January 15, 1984

"When C. Everett Koop, then surgeon in chief of the Children's Hospital of Philadelphia, now US Surgeon General, returned in 1963 from a conference in Africa, he urged us to work on Burkitt's lymphoma (BL) because of its presumed viral origin and relevance to our ten-year (*sic!*) research grant awarded by the National Cancer Institute under its Viral Oncology Program and entitled 'Interference phenomena in the detection of human cancer viruses.' Approaches to several physicians in Africa, among them D.P. Burkitt, revealed that all were already committed to other laboratories. A year later, M.A. Epstein and co-workers at Middlesex Hospital, London, reported growth of cell lines from BL biopsies and electron microscopic detection of herpesvirus particles in some of the cultured cells.[1,2] As no known herpesvirus had been shown to be oncogenic, most virologists considered the indigenous virus a passenger of no concern. When Epstein turned to us for help, our chance to work on BL had come. After implantation on monolayers of human embryonic kidney cells, cultured BL cells induced resistance to various viruses due to production of an interferon in line with the title of our grant.[3] The virus was not transmissible to routine host systems suggesting that it was heretofore unknown,[4] but immunology was needed for proof. Only immunofluorescence was feasible. As reported in this *Citation Classic* ™, the virus-producing cells in BL (as well as other) lymphoblast cultures were readily identified by sera from BL patients but also by commercial human gamma globulin and many sera collected anywhere in the world. Some reactive sera had no antibodies to herpes simplex, varicella zoster, or cytomegaloviruses proving this ubiquitous virus to be indeed new. It was henceforth called Epstein-Barr virus (EBV) after the EB-1 culture in which it was first observed.

"The indirect immunofluorescence test continues to be widely used today for titration of antibodies to EBV, more specifically its viral capsid antigen (VCA). By demonstration of elevated antibody titers as compared to controls, this test was instrumental in linking EBV not only with BL, but also nasopharyngeal carcinoma (NPC) and infectious mononucleosis (IM). Initial clues were obtained when NPC sera included among controls reacted as well as BL sera in double diffusion precipitation assays with BL cell extracts[5] and when an antibody-negative technician in our laboratory seroconverted in the course of IM.[6] These clues were substantiated subsequently by demonstration of EBV DNA in BL and NPC biopsies and of the EBV-associated nuclear antigen (EBNA) in the tumor cells and in lymphocytes of IM patients (cf. reference 7). With identification of the EBV-coded diffuse and restricted early antigens in addition to VCA and EBNA, and differentiation of the Ig class of the various antibodies, patients with IM, BL, and NPC were shown to develop characteristic antibody spectra which differ from each other and from the pattern of healthy seropositive donors.[7] EBV thus turned out to be the first human virus which is closely associated with human cancers."

1. **Epstein M A, Achong B G & Barr Y M.** Virus particles in cultured lymphoblasts from Burkitt's lymphoma. *Lancet* 1:702-3, 1964.
2. **Epstein M A.** Citation Classic. Commentary on *Lancet* 1:702-3, 1964. *Current Contents/Life Sciences* 22(14):10, 2 April 1979.
3. **Henle G & Henle W.** Evidence for a persistent viral infection in a cell line derived from Burkitt's lymphoma. *J. Bacteriology* 89:252-8, 1965. (Cited 65 times.)
4. **Epstein M A, Henle G, Achong B G & Barr Y M.** Morphological and biological studies on a virus in cultured lymphoblasts from Burkitt's lymphoma. *J. Exp. Med.* 121:761-70, 1965. (Cited 205 times.)
5. **Old L J, Boyse E A, Oettgen H F, de Harven E, Geering G, Williamson B & Clifford P.** Precipitating antibody in human serum to an antigen present in cultured Burkitt's lymphoma cells. *Proc. Nat. Acad. Sci. US* 56:1699-704, 1966. (Cited 270 times.)
6. **Henle G, Henle W & Diehl V.** Relation of Burkitt tumor associated herpestype virus to infectious mononucleosis. *Proc. Nat. Acad. Sci. US* 59:94-101, 1968. (Cited 730 times.)
7. **Henle W, Henle G & Lennette E T.** The Epstein-Barr virus. *Sci. Amer.* 241:48-59, 1979.

Blumberg B S, Alter H J & Visnich S. A "new" antigen in leukemia sera.
J. Amer. Med. Ass. **191**:541-6, 1965.
[Institute for Cancer Research, Philadelphia, PA and NIH, Bethesda, MD]

Australia antigen was discovered during the course of a systematic study of the sera of transfused patients to detect precipitating iso-antibodies against antigens found in serum. This approach had previously led to the discovery of inherited antigenic specificities on the low density lipoproteins (the Ag system). The Australia antigen was identified in the serum of an Australian aborigine and the precipitating antibody in the serum of a transfused hemophilia patient. Australia antigen was found to be rare in normal Americans but common in patients with leukemia. It was also quite common in normal people from Africa, Asia, and Oceania and there was family clustering. On the basis of these findings, several hypotheses were made related to leukemia, viral agents, and the genetic control of the presence of Australia antigen, and these were tested in an extended series of studies over the next ten years. [The *SCI®* indicates that this paper has been cited over 560 times since 1965.]

Baruch S. Blumberg
Institute for Cancer Research
7701 Burholme Avenue
Philadelphia, PA 19111

August 14, 1979

"This paper is the first in which the term 'Australia antigen' is used, although the discovery had been briefly reported the previous year.[1] Australia antigen was subsequently found to be the surface antigen of the hepatitis B virus (HBsAg). Hepatitis B virus also appears to be an important factor in the development of primary cancer of the liver, a disease which is a serious public health problem in many parts of Africa and Asia. If these inferences are correct, then this would be an example of a common human cancer in whose pathogenesis a virus is involved. Further, it may be possible to use measures to prevent this disease and we have developed a vaccine currently being tested. In subsequent papers in this series the methods for detecting carriers of hepatitis B virus were detailed. This led to the now common practice of testing the blood of potential donors for the presence of Australia antigen.[2] The presence of the antigen indicates that the donors have a high probability of transmitting hepatitis. This paper describes the original phase of the work, in which a large amount of data was collected on stored sera from patients and normal in-

dividuals. The hypotheses derived from these observations were tested in a subsequent series of papers.[3] This paper set the pattern for the research for the next ten years.

"It now appears that hepatitis B virus has many unusual characteristics, so much so that it has been suggested that for heuristic reasons, it may be useful to consider it as a member of a special group of infectious agents which we later termed 'Icrons' (after the Institute for Cancer Research). We described the geographic distribution of the carrier state which appears to be, at least in part, under genetic control. We also suggested that the antigen was associated with a virus; in this case, the virus of leukemia. There is still reason to believe that there may be a connection between hepatitis B virus and the etiologic agent of some forms of leukemia. The relation of the antigen to transfusion was also cited in the paper.

"The work began at the National Institutes of Health and continued at the Institute for Cancer Research in Philadelphia. Harvey J. Alter, who subsequently went on to complete his training as a hematologist, had come to work in our laboratory on this project. He is now at the Blood Bank of the National Institutes of Health and is an important investigator in the hepatitis field. Sam Visnich was a former Navy flyer and a commercial airline pilot who has now returned to flying.

"This work was done during a period when funding for non-directed research was more available than at present. At the beginning of this investigation it was impossible to see where it would lead. We had confidence that it would be important, since basic patterns of the distribution of Australia antigen began to emerge; in particular, the clustering of antigen positive individuals in families, the sex differences and the consistently high frequencies in tropical or underdeveloped areas as compared to northern Europe and the United States.

"In the 1960s research support was readily available to pursue apparently esoteric problems, and, in this case, it led to a practical application. Our work has been cited on several occasions as an example of the unpredictable but practical advantages of fundamental research, and I hope that it will have a continued effect on scientific directors and Congress on the importance of non-directed investigations."

1. **Blumberg B S.** Polymorphisms of serum proteins and the development of isoprecipitins in transfused patients. *Bull. NY Acad. Med.* **40**:377-86, 1964.
2. **Blumberg B S, Gerstley B J S, Hungerford D A, London W T & Sutnick A I.** A serum antigen (Australia antigen) in Down's syndrome, leukemia and hepatitis. *Ann. Intern. Med.* **66**:924-31, 1967.
3. **Blumberg B S.** Australia antigen and the biology of hepatitis B. *Science* **197**:17-25, 1977.

This Week's Citation Classic

Blumberg B S, Gerstley B J S, Hungerford D A, London W T & Sutnick A I.
A serum antigen (Australia antigen) in Down's syndrome, leukemia, and hepatitis. *Ann. Intern. Med.* **66**:924-31, 1967.
[Inst. for Cancer Research, Fox Chase, Philadelphia, PA]

This is the first paper in which it was stated that Australia antigen (Au) was the hepatitis virus. This subsequently led to the identification of the hepatitis B virus (HBV), the development of methods for the diagnosis of hepatitis B, the prevention of post-transfusion hepatitis, the control of hepatitis in high risk environments, a vaccine against hepatitis B, and the recognition of the role of HBV in the causation of primary hepatocellular carcinoma (PHC).[1] [The *SCI®* indicates that this paper has been cited in over 480 publications since 1967.]

Baruch S. Blumberg
Institute for Cancer Research
Fox Chase Cancer Center
7701 Burholme Avenue
Philadelphia, PA 19111

May 5, 1983

"Australia antigen (Au) was discovered in 1963, and described in 1964[2] and 1965.[3] It was present in many normal people particularly in the tropics. It was also found in leukemia patients. We hypothesized that some individuals had a trait, possibly inherited, which made them susceptible to leukemia and also to having persistent Au in their blood. We tested individuals who had a high likelihood of developing leukemia, including patients with Down's syndrome (DS). The hypothesis predicted a high frequency of Au in high risk groups. This prediction was fulfilled for DS. During the course of the study on DS, we found that one patient who had developed Au in the period he was under observation. During this period he had also developed hepatitis. This led to a systematic study of patients with hepatitis and in this paper, we reported an increased frequency of Au. We concluded, 'Most of the disease associations could be explained by the association of Au(1) with a virus, as suggested in our previous publications. The discovery of the frequent occurrence of Au(1) in patients with virus hepatitis raises the possibility that the agent present in some cases of this disease

may be Australia antigen or be responsible for its presence. The presence of Australia antigen in the thalassemia and hemophilia patients could be due to virus introduced by transfusions.'

"We continued to study ideas generated by earlier findings. Family studies showed that susceptibility to developing the carrier state for hepatitis B virus (HBV) was under genetic control. Recent studies indicate that HBV DNA is integrated into host DNA in carriers and in patients with chronic liver disease and primary cancer of the liver. Could a viral gene introduced in this manner be transmitted from generation to generation and explain the apparent Mendelian segregation we detected?[1]

"We found a striking difference in the age distribution of carriers and inferred that younger people are more likely to become carriers. We now believe that this is related to the differentiation of liver cells.[4] The role of maternal infection which is now known to make a major contribution to the carrier pool in many populations was first discussed in this paper. Vaccine programs are now based in large part on the vaccination of infants to prevent the development of carriers. We also questioned why DS patients are more susceptible to persistent infection than others. This led to studies on host immune differences which determine whether an infected individual will become a carrier with an increased risk of chronic liver disease or develop protective antibody. These studies also led to the finding that carriers have higher levels of serum iron and ferritin and lower levels of transferrin and a model of the pathogenesis of chronic liver disease and primary hepatocellular carcinoma (PHC).[5]

"The frequent citation of this paper may be a result of its having been the first published association of Au to what is now known to be HBV. We subsequently prepared a more extensive paper on the relation of Au to HBV, but it was rejected. The editors were reluctant to publish a paper asserting that the hepatitis virus had been found, and required additional studies and publications to have this concept accepted.

"London, Sutnick, and I continued to work on hepatitis. Gerstley continued in medical teaching and patient care. Sutnick became dean of the Medical College of Pennsylvania. Hungerford recently retired because of medical disability. This work eventually resulted in the award in 1976 of the Nobel prize and other recognitions."

1. **Blumberg B S.** Australia antigen and the biology of hepatitis B. *Science* **197**:17-25, 1977.
2. ------------------. Polymorphisms of serum proteins and the development of isoprecipitins in transfused patients.
   Bull. NY Acad. Med. **40**:377-86, 1964.
   [The *SCI* indicates that this paper has been cited in over 205 publications since 1964.]
3. **Blumberg B S, Alter H J & Visnich S.** A "new" antigen in leukemia sera. *J. Amer. Med. Assn.* **191**:541-6, 1965.
   [Citation Classic. *Current Contents/Life Sciences* **22**(51):14, 17 December 1979.]
4. **Blumberg B S & London W T.** Hepatitis B virus: pathogenesis and prevention of primary cancer of the liver.
   Cancer **50**:2657-65, 1982.
5. **Lustbader E D, Hann H W L & Blumberg B S.** Serum ferritin as a predictor of host response to hepatitis B virus
   infection. *Science*. In press, 1983.

Prince A M. An antigen detected in the blood during the incubation period of serum hepatitis. *Proc. Nat. Acad. Sci. US* **60**:814-21, 1968.
[Lab. Virology, New York Blood Ctr., and Dept. Pathology, New York Hospital - Cornell Med. Ctr., New York, NY]

This brief report describes the background for the finding of a hepatitis B virus specific antigen and the establishment of its identity with the 'Australia antigen' discovered by Blumberg. [The *SCI*® indicates that this paper has been cited over 685 times since 1968.]

Alfred M. Prince
New York Blood Center
310 East 67th Street
New York, NY 10021

October 3, 1980

"I became involved in the hepatitis story while serving as a virologist at the Medical General Laboratory 406 while in the US Army in Tokyo. In an attempt to demonstrate viral antigens characteristic of hepatitis by immunofluorescence, we surveyed 2,500 Korean troops for transaminase abnormalities and then hospitalized those with abnormal transaminase for further study. In 11 of 32 subjects with repeatedly abnormal transaminase, we detected by immunofluorescence nuclear and cytoplasmic antigens in their liver cells. We postulated that the antigens were hepatitis B specific, probably viral, antigens.[1] It thus became natural to postulate that Blumberg's 'Australia antigen,'[2] might represent the hepatitis B specific antigen which we had observed by immunofluorescence. In particular, the discrepancy between the frequency of detection of this antigen in West African and American blacks suggested to me that the genetic explanation then being proposed by Blumberg was probably incorrect.

"For these reasons, in 1966 I began a collaborative study with Blumberg to characterize the Australia antigen in order to determine whether it might have virus-like properties. Using a variety of density gradient centrifugation procedures, we were able to show that the Australia antigen was located on a small virus-like particle. I therefore wrote a paper in 1966, intending to publish it collaboratively to report these findings. Unfortunately, Blumberg was at the time unable to accept the proposed interpretation. We therefore decided to continue these investigations independently.

"Due to the importance of the findings, I insisted on being able to uncode coded reference serum collections. It required approximately one year's work to learn appropriate modifications in the agar composition, as well as the necessity for concentration of the immunoreactants. It then became relatively easy to define the system as outlined in this paper. Unfortunately, the relationship between our findings and the Australia antigen system was unclear since: (1) Blumberg had requested and received return of his reference reagents; (2) I had been informed that Blumberg's results on identical coded serum collections were different from my own; and (3) there were important differences in the agar gel diffusion techniques being used in the two laboratories. For this reason, the section initially included in the manuscript on the relationship between 'SH' and 'Australia antigen' was sufficiently weak to suffer editorial removal. However, with the aid of Lloyd Old, who provided a new set of reference Australia antigen reagents, we were able to investigate this relationship and reported that the two antigens were similar, if not identical, a few months after the appearance of this paper."

1. Prince A M, Fuji H & Gershon R K. Immunohistochemical studies on the etiology of anicteric hepatitis in Korea. *Amer. J. Hyg.* **79**:365-81, 1964.
2. Blumberg B S. Polymorphisms of the serum proteins and the development of iso-precipitins in transfused patients. *Bull. NY Acad. Med.* **40**:377-86, 1964.

Almeida J D, Rubenstein D & Stott E J. New antigen-antibody system in Australia-antigen-positive hepatitis. *Lancet* 2:1225-7, 1971.
[Dept. Virology, Royal Postgrad. Med. Sch., London, and Clinical Res. Ctr., Northwick Park Hosp., Harrow, Middlesex, England]

The paper reported that the double shelled Dane particle of what is now hepatitis B antigen, but which was then Australia antigen, could be degraded by detergents to reveal an internal component that was antigenically distinct from the surface of the particle. [The *SCI®* indicates that this paper has been cited in over 345 publications since 1971.]

June D. Almeida
Wellcome Research Laboratories
Beckenham, Kent BR3 3BS
England

December 22, 1982

"It has been said that Columbus, sailing in the wrong direction and using a false premise, nevertheless discovered America. The authors of this paper must admit that what happened to them had a certain similarity to his situation. The group included one electron microscopist and two tissue culture experts who had decided to have an attempt at the difficult task of growing the virus of hepatitis B in culture. Incidentally, over ten years later this still has not been achieved. Based on the fact that some viruses, for example, the reoviruses, have a higher infectivity for tissue culture if they are degraded with enzymes, it was felt that it might be possible to degrade the so-called Australia antigen with enzymes, and/or detergents, making available the essential nucleic acid. One year earlier, David Dane and his co-workers[1] had described a double shelled particle present in Australia antigen preparations that had much better viral characteristics than the smaller spheres and tubules of such preparations. This particle was later termed the Dane particle.

"We started with a preparation rich in Dane particles and subjected it to several degradative procedures. The end products of these experiments were always checked in the electron microscope by the negative staining technique as well as being inoculated onto a wide range of tissue cultures. Luckily, since one of my major interests is the visualisation of the interaction between virus and antibodies, we were able to recognise that Dane particles split with detergent released an inner component that became covered with antibody that must have been present in the original serum. Almost immediately we dropped the tissue culture aspect of the work that had yielded no positive results and turned our attention to this new phenomenon. It didn't take long to establish that sera positive for hepatitis B antigen always had antibody to this internal component or core, and that this system was quite distinct from the other antigenic components.

"The paper has been much quoted because it suggested that the Dane particle had the characteristics of other compound viruses such as the herpes group. It focused the attention of the biochemists on the inner component as the likely location for the viral nucleic acid, an assumption that turned out to be true. It also gave rise to the terminology that is still employed: that the outer covering of the Dane particle would be surface antigen, or HBsAg, and the inner component, the core antigen, or HBcAg.[2]

"One final comment about this paper is that the three authors took great pleasure in their work and whenever possible saw the funny side of things. For example, the micrograph that is figure 1 of the paper was carefully arranged so that there is a hepatitis B antigen cross marking a Western style grave at the foot."

1. Dane D S, Cameron C H & Briggs M. Virus-like particles in serum of patients with Australia antigen-associated hepatitis. *Lancet* 1:695-8, 1970.
2. Dienstag J L. Hepatitis viruses: characterization and diagnostic techniques. *Yale J. Biol. Med.* 53:61-9, 1980.

Rawls W E, Tompkins W A F & Melnick J L. The association of herpesvirus type 2 and carcinoma of the uterine cervix. *Amer. J. Epidemiol.* **89**:547-54, 1969.
[Dept. Virology and Epidemiology, Baylor Univ. Coll. Med., Houston, TX]

A method was developed to distinguish type 2 from type 1 herpesvirus antibodies in patients. Type 2 antibodies were detected in 78 percent of patients with cervical carcinoma, but in only 22 percent of matched control healthy women or women with other types of cancer. Type 1 antibodies begin to appear in the normal population soon after birth and by early adulthood almost all are infected. In contrast, type 2 antibodies do not appear until adolescence and the start of sexual relations, indicating venereal spread. [The *SCI®* indicates that this paper has been cited in over 270 publications since 1969.]

Joseph L. Melnick
Department of Virology and Epidemiology
Baylor College of Medicine
Houston, TX 77030
and
William E. Rawls
Department of Pathology
McMaster University
Hamilton, Ontario L8N 3Z5
Canada

July 7, 1983

"In the early 1960s, the department of virology and epidemiology at Baylor College of Medicine had begun a study of cervical cancer. Epidemiological studies had left little doubt that coitus was a prerequisite for the development of carcinoma of the cervix. The disease was virtually nonexistent among women of celibate religious orders. A higher incidence of the disease had been correlated with early intercourse, early marriage, multiple marriages, and multiple sexual partners. Pivotal among these variables was that of early intercourse. This suggested that the agent was maintained in the population primarily by venereal spread and that susceptibility to infection decreased with age or the susceptibility to the oncogenic effect of the agent was greatest in younger women. In either case, a venereally transmitted virus with oncogenic potential might then be a common viral agent.

"This led us to test smegma samples from young males for the presence of a transmissible venereal agent. Of the 220 samples tested, four yielded a previously unrecognized member of the herpesvirus group, now known as herpes simplex virus type 2 (HSV2), from work independently carried out by Dowdle, Nahmias, and their co-workers[1] in Atlanta and by us in Houston. We developed methods for determining antibodies to the type 2 virus and carried out the study reported in this paper, which showed that women with cervical cancer had been commonly infected with the newly recognized HSV2. Again, similar results were found independently at Emory University.[2]

"We believe this paper is cited frequently because it opened new avenues for investigating the role of viruses and cancer. The study bridged the gap often found between the clinic and the laboratory. New methods were developed and applied to an epidemiological problem; this resulted in focusing attention on a newly recognized herpesvirus as a potential cause of cervical cancer. It also pointed the way to a resolution of the problem: if the agent could be controlled, the disease should vanish. International efforts are now under way to develop effective chemotherapeutic drugs and viral vaccines; the results are promising in both directions.[3,4]

"In our paper we suggested that herpesvirus type 2 might also play a role in vulvar carcinoma. Recent evidence has supported this view. In the US, concomitant increases have occurred in prevalence of genital herpes infections and of vulvar carcinoma, especially in women under 40 years of age. In similar fashion to cervical cancer victims, these patients have type 2 antibodies in their blood and type 2 antigens that bind to DNA in their cancer cells.[5] With this new evidence that herpesvirus type 2 is associated with genital cancers, it is hoped that efforts will be intensified to control the virus infection and thus the cancer."

1. Dowdle W R, Nahmias A J, Harwell R W & Pauls F P. Association of antigenic type of herpesvirus hominis with site of viral recovery. *J. Immunology* **99**:974-80, 1967. (Cited 235 times.)
2. Nahmias A J, Josey W E, Naib Z M, Luce C F & Guest B. Antibodies to herpesvirus hominis types 1 and 2 in humans. II. Women with cervical cancer. *Amer. J. Epidemiol.* **91**:547-52, 1970. (Cited 220 times.)
3. Nahmias A J, Dowdle W R & Schinazi R F, eds. *The human herpesviruses, an interdisciplinary perspective.* New York: Elsevier, 1981. 721 p.
4. Myers M W, Glasgow L A & Glassco C J. Summary of a workshop on antiviral agents for genital herpesvirus infections. *J. Infec. Dis.* **145**:774-82, 1982.
5. Kaufman R H, Dreesman G R, Burek J, Korhonen M O, Matson D O, Melnick J L, Powell K L, Purifoy D J M, Courtney R J & Adam E. Herpesvirus-induced antigens in squamous-cell carcinoma in situ of the vulva. *N. Engl. J. Med.* **305**:483-8, 1981.

This Week's Citation Classic

CC/NUMBER 25
JUNE 22, 1981

Meléndez L V, Hunt R D, Daniel M D, García F G & Fraser C E O. *Herpesvirus saimiri.* II. Experimentally induced malignant lymphoma in primates.
Lab. Anim. Care **19**:378-86, 1969.
[New England Regional Primate Res. Ctr., Harvard Med. Sch., Southborough, MA]

Owl (*Aotus sp.*) **and marmoset** (*Saguinus sp.*) **developed a fatal malignant lymphoma of the reticulum cell type 13 to 28 days after inoculation with** *Herpesvirus saimiri,* **an indigenous viral agent from squirrel monkey** (*Saimiri sciureus*). [**The** *SCI*® **indicates that this paper has been cited over 175 times since 1969.**]

Luis V. Meléndez
Oficina Sanitaria Panamericana
Oficina Regional de la
Organizacion Mundial de la Salud
Centro Panamericano de Zoonosis
1000-Buenos Aires
Argentina

April 30, 1981

"The leukemogenic property of *Herpesvirus saimiri* was discovered when the team of workers at the Harvard Medical School Primate Research Center (R.D. Hunt, M.D. Daniel, F.G. García, and C.E.O. Fraser) led by me decided to inquire what would be the pathogenic action of an indigenous herpesvirus isolated from squirrel monkeys (*Saimiri sciureus*).[1]

"The remarkable result was to find out that 13 to 28 days post-inoculation the nonhuman primates developed a fatal malignant lymphoma of the reticulum cell type or reticulum cell sarcoma. It was in this casual way that it was proved for the first time that a DNA virus can induce leukemia in primate species. *Herpesvirus saimiri* is known to be today the DNA virus with the highest leukemogenic capacity for inducing malignancy in a variety of primates of the Old and New World[2] as well as in rabbits.[3] The malignancy was also found to be a natural disease in the natural environment of the nonhuman primates.[4]

"I think this paper has been highly cited because it deals with the first information of a herpesvirus with the capacity to induce the development of a neoplastic disease (malignant lymphoma) in nonhuman primates (New World monkeys). Besides, the malignant disease evolved in a very short period of time: less than one month. At the time that this finding was reported certainly it was a remarkable scientific discovery and today it still stands as perhaps one of the best experimental models to study induced leukemias in primates.

"I am presently director of the Pan American Zoonosis Center of the Pan American Health Organization in Buenos Aires, Argentina."

1. Meléndez L V, Daniel M D, Hunt R D & García F G. An apparently new herpesvirus from primary kidney cultures of the squirrel monkey (*Saimiri sciureus*). *Lab. Anim. Care* **18**:378-81, 1968.
2. Meléndez L V, Hunt R D, Daniel M D, Fraser C E O, Barahona H, King N W & García F G. Herpesviruses saimiri and ateles—their role in malignant lymphomas in monkeys. *Fed. Proc.* **31**:1643-50, 1972.
3. Daniel M D, Meléndez L V, Hunt R D, King N W & Williamson M E. Malignant lymphoma induced in rabbits by *Herpesvirus saimiri* strains. *Bacteriol. Proc.* **1970**:195, 1970.
4. Hunt R D, García F G, King N W, Fraser C E O & Meléndez L V. Spontaneous *Herpesvirus saimiri* lymphoma in owl monkeys. (Ito Y & Dutcher R M, eds.) *Comparative leukemia research 1973—leukemogenesis.* Tokyo: University of Tokyo Press, 1975. p. 351-5.

CC/NUMBER 30
JULY 27, 1981

Huebner R J & Todaro G J. Oncogenes of RNA tumor viruses as determinants of cancer. *Proc. Nat. Acad. Sci. US* 64:1087-94, 1969.
[Viral Carcinogenesis Branch, National Cancer Institute, National Institutes of Health, Bethesda, MD]

Sero-epidemiological and cell culture studies led us to the conclusion that vertebrate cells contained the genetic information for producing C-type RNA viruses. We also postulated that the viral information (the virogene) often included cellular oncogenes responsible for transformation and that the action of carcinogens, irradiation, and the normal aging process was also mediated by transforming proteins specified by the inherent oncogenes. [The *SCI®* indicates that this paper has been cited over 705 times since 1969.]

Robert J. Huebner
Laboratory of Cellular and
Molecular Biology
National Institutes of Health
Bethesda, MD 20205

June 22, 1981

"The viral oncogene theory was inescapable given the years of cumulative evidence culminating in the 1969 report in the *Proceedings of the National Academy of Sciences*. Actually, an earlier report[1] was given at an international conference held at the Abbaye de Royaumont in June 1969, where Andre Lwoff, as chairman, focused discussion on 'Les Oncogenes,' thus raising considerable controversy at the outset. This was appropriate since it was his germinal genetic theories, shared with Jacob and Monod, which helped formulate our concept that the proximal cause of cancer was mediated by endogenous transforming oncogenes residing within vertebrate cells. A number of earlier encounters with Sir McFarlane Burnet convinced me that he was right: the stochastic behavior of cancer in man could not be attributed to exogenous infectious viruses. This decision helped crystallize our view that cancer was the result of derepression of inherent species-specific endogenous oncogenes often linked in mice and chickens with endogenous oncorna-viruses. Genetic abnormalities or deficiencies, environmental carcinogens, and 'oncogenic' viruses were thus inducers of such oncogenes.

"Another key factor was our observation that the endogenous leukemia and sarcoma viruses in chickens, mice, rats, and cats actually carried 'movable' oncogenes, since 'sarc' genes could be rescued from primary hosts and transmitted as transforming oncogenes to cultured cells of a wide variety of animal hosts, including human. Subsequently, it was demonstrated by rescue experiments that the transforming and tumor-inducing leukemia and sarcoma oncogenes more often than not existed on tumor cells independently of virogene expression, and were rescuable by viruses. The fact that the oncogenes could be transmitted to cells of the same and other susceptible species clearly indicated that transforming proteins specified by endogenous sarc and leuk oncogenes provided species-specific determinants responsible for neoplasia and transformation, while exogenous factors (environmental chemicals, radiation, DNA tumor viruses) served as inducers of the postulated oncogene sequences. Definitive evidence establishing the role(s) of the transforming proteins specified by sarc and leuk oncogenes was provided by subsequent molecular, immunological, and immunoprevention studies in experimental animals, particularly in mice, rats, cats, and chickens.

"The viral oncogene theory expressed a new concept in which it was postulated that all carcinogenesis (radiation, carcinogens, aging, etc.) was mediated through activation of an endogenous host cell gene present in every vertebrate cell. This opened the way for subsequent discoveries of the sarc, onc, and leuk proteins by molecular biologists.

"Recently, it became apparent that the oncogene determinants of cancer represented also the required determinants for immunoprevention. Vaccines and IgG preparations incorporating immunity to sarc and leuk gene expressions have recently been used to prevent virtually all types of cancers in mice and rats, including cancers produced by chemicals and DNA tumor viruses.[2] These successful immuno-prevention studies we believe provide excellent models for development of comparable prevention of cancer in man."

1. Huebner R J, Todaro G J, Sarma P, Hartley J W, Freeman A E, Peters R L, Whitmire C E, Meier H & Gilden R V. "Switched-off" vertically transmitted C-type RNA tumor viruses as determinants of spontaneous and induced cancer: a new hypothesis of viral carcinogenesis. *Defectiveness, rescue and stimulation of oncogenic viruses. Second International Symposium on Tumor Viruses.* Paris: Editions du Centre National de la Recherche Scientifique, 1970. p. 33-57.
2. Huebner R J & Fish D C. Sarc and leuk oncogene-transforming proteins as antigenic determinants of cause and prevention of cancer in mice and rats. (Griffin A C & Shaw C R. eds.) *Carcinogens: identification and mechanisms of action.* New York: Raven Press, 1979. p. 277-98.

This Week's Citation Classic

Hsie A W & Puck T T. Morphological transformation of Chinese hamster cells by
dibutyryl adenosine cyclic 3':5'-monophosphate and testosterone.
Proc. Nat. Acad. Sci. US **68**:358-61, 1971.
[Sarah and Matthew Rosenhaus Lab., Eleanor Roosevelt Inst. for Cancer Research, and
Dept. Biophysics and Genetics, Univ. Colorado Med. Ctr., Denver, CO]

Chinese hamster ovary cells growing randomly in a compact, multilayered-colony were converted into a monolayer of elongated fibroblast-like cells when treated with dibutyryl cyclic AMP and testosterone suggesting a role of cyclic AMP in regulating cell shape and malignant conversion. [The *SCI*® indicates that this paper has been cited in over 560 publications since 1971.]

Abraham W. Hsie
Biology Division
Oak Ridge National Laboratory
Oak Ridge, TN 37830

July 20, 1983

"As a graduate student with Howard V. Rickenberg at the department of bacteriology (later changed to microbiology and now biology), Indiana University, Bloomington, I studied the role of phosphoenolpyruvate carboxykinase (PEPCK) in carbohydrate metabolism in *Escherichia coli*. I isolated a mutant (AB257pc-1) which formed high levels of PEPCK and other catabolic enzymes, such as beta-galactosidase and tryptophanase, when grown in a medium containing glucose as the sole source of carbon exhibiting the glucose effect or catabolite repression. The observation that the addition of glucose-6-phosphate severely repressed the formation of beta-galactosidase in this glucose-resistant mutant, AB257pc-1, led us to suggest the possible role of glucose-6-phosphate in catabolite repression.[1] A hypothesis proposing the regulatory role of adenosine 3':5'-monophosphate (cyclic AMP) as a 'derepressor' on catabolite repression was advanced by Ira Pastan and his colleagues.[2] This hypothesis was supported by subsequent analysis of AB257pc-1 by Rickenberg and his colleagues,[3] who showed that AB257pc-1 is defective in cyclic AMP phosphodiesterase, an enzyme which hydrolyzes cyclic AMP to 5'-AMP.

"I undertook my postdoctoral work with Theodore T. Puck, of the University of Colorado Medical Center, to learn somatic cell genetics. Initially, I studied the biochemistry and genetics of auxotrophic mutants of Chinese hamster ovary (CHO) cells. In the spring of 1970, my interest in the role of cyclic AMP in cellular function and the boredom of mutant characterization led me to investigate whether cyclic AMP or its derivatives would enable CHO cells to utilize lactose as a sole source of carbon, and to study the hypothetical mammalian equivalent of the lac operon. In my first experiment to determine the toxicity of N^6,O^2'-dibutyryl cyclic AMP to CHO cells, I found that after seven days incubation in the presence of 0.3-1.0 mM of dibutyryl cyclic AMP, CHO cells were converted from a randomly growing, compact, epithelial-like (transformed) morphology to a contact-inhibited, elongated, fibroblast-like (normal) form.[4] This was consistent with the reports that some tumors and transformed cells are defective in the cyclic AMP system; treatment with dibutyryl cyclic AMP apparently restores the normal behavior of the transformed (CHO) cells. Thus, cyclic AMP may play a role in regulating two fundamental biological processes: cellular morphogenesis and malignant conversion. These implications may have stimulated the later explosive research by various investigators in the continued search for the role of cyclic nucleotides in carcinogenesis. Similar findings concerning the effects of cyclic AMP have been made by others using other cell types.

"Meanwhile, Puck observed that steroid hormones, such as testosterone, caused effects similar to those produced by dibutyryl cyclic AMP, but they were less pronounced. Testosterone exerted a synergistic effect with dibutyryl cyclic AMP. I remember well the fine reception of these preliminary findings (with Charles A. Waldren)[4] at the annual meeting of the American Society of Cell Biology in 1970 and the full paper (with Puck) in 1971. In 1980, I received the Distinguished Alumni Service Award from Indiana University partly in recognition of this work.

"Looking back 14 years, I am indebted to my mentors Puck and Rickenberg, who provided the environment, foresight, and encouragement to make my work possible and interesting."

1. Hsie A W & Rickenberg H V. Catabolite repression in *Escherichia coli*: the role of glucose-6-phosphate.
   Biochem. Biophys. Res. Commun. 29:303-10, 1967.
2. Perlman R L & Pastan I. Cyclic 3':5'-AMP: stimulation of β-galactosidase and tryptophanase induction in *E. coli*.
   Biochem. Biophys. Res. Commun. 30:656-64, 1968.
3. Nielsen L D, Monard D & Rickenberg H V. Cyclic 3',5'-adenosine monophosphate phosphodiesterase of
   Escherichia coli. *J. Bacteriology* 116:857-66, 1973.
4. Hsie A W & Waldren C A. Conversion of Chinese hamster cells to fibroblastic form by dibutyryl adenosine
   3':5'-cyclic monophosphate. *J. Cell Biol.* 47:92a, 1970.

Research sponsored by the Office of Health and Environmental Research, US Department of Energy, under
contract W-7405-eng-26 with the Union Carbide Corporation.

NUMBER 10
MARCH 5, 1979

Bernhard W. The detection and study of tumor viruses with the electron microscope. *Cancer Res.* 20:712-27, 1960.

The ultrastructural aspects of all groups of RNA and DNA tumor viruses known at this date are presented on a series of electron micrographs and their evolutionary cycle is described in detail. The authors compare these findings with biological data on the same agents. They conclude that electron microscopy has now become an indispensable tool for further experimentation with known and still unknown tumor viruses. [The *SCI®* indicates that this paper has been cited 335 times since 1961.]

W. Bernhard
Institut de Recherches
sur le Cancer
B.P. 8, 94800 Villejuif
France

January 18, 1978

"Eighteen years after its publication, I believe that the success of this paper may be due to 4 different reasons: The first one was that it appeared *just in time*, where the *receptivity* of the reader for such a topic was suddenly exceptionally favorable. A previous similar paper, although less complete, had already been published in *Cancer Research* two years earlier but did not find a comparable resonance.[1] The decisive psychological turn in favor of the viral origin of cancer took place between 1958 and 1960, a period when a number of important papers was published, particularly on the *in vitro* transformation and improved *in vitro* titration methods.

"The second reason was due to the fact that our work, which we started with Charles Oberling as early as 1948, was published previously in *French* in periodicals such as the *Bulletin du Cancer* and the *Comptes Rendus de l'Academie des Sciences,* which were either not available or not read in the United States. When we published the first paper on C-type particles in 1958, we received from the whole world *two reprint requests!* There was an extraordinary skepticism until the late fifties against EM pictures showing 'so called' tumor viruses. At that time we were, of course, rather desperate that our studies did not find a better echo, but later we understood that this was a unique opportunity to accumulate still more convincing documents, *without too much competition.* Being *ignored* in science—at least for a short time—may sometimes turn out to be an *advantage!*

"Thirdly, during these years of darkness, we could steadily *improve* our techniques and select the very best pictures among many thousands. They had a more convincing effect than the earliest documents we published in French journals. In addition, we could classify the various particles according to merely morphological criteria and demonstrate their *ubiquity* in various strains of rodents or birds (A- and B-type particles in 1954/55, C-type particles in 1955/58). This simple nomenclature was later accepted by an international committee and has been used ever since.

"Finally, I was particularly fortunate to present this review at one of the 'breakthrough' meetings for the future of the viral theory, sponsored by the American Cancer Society and held at Rye, N.Y., November 19-21, 1959. Among the participants, there were most of the early great pioneers in this field: Peyton Rous, R. Shope, J.W. Beard, L. Gross, J. Furth, Sarah Stewart, J.T. Syverton, and outstanding virologists and phage specialists such as A. Lwoff, R. Dulbecco, F. Jacob, S. Luria, E.R. Stanley, A.B. Sabin, and other remarkable scientists.

"The contributions were published in a special Symposium volume of *Cancer Research* with the provocative title: *The Possible Role of Viruses in Cancer,* and for this reason were read by an unusually high number of investigators. Its publication in June 1960 coincided with the time when I was writing the obituary for Charles Oberling who died of bronchial carcinoma a few months earlier, after fighting with unusual courage throughout his scientific career for the recognition of the importance of oncogenic viruses in cancer research."

1. Bernhard W. Electron microscopy of tumor cells and tumor viruses. A review. *Cancer Res.* 18:491-509, 1958.

Rosen L. A hemagglutination-inhibition technique for typing adenoviruses.
Amer. J. Hyg. 71:120-8, 1960.
[Laboratory of Infectious Diseases, National Institutes of Health, Bethesda, MD]

The finding that 25 of 27 serotypes of adenoviruses agglutinated the erythrocytes of either rhesus monkeys or laboratory rats *in vitro* led to the development of a simple and rapid hemagglutination-inhibition technique for typing adenoviruses. [The *SCI*® indicates that this paper has been cited over 285 times since 1961.]

Leon Rosen
Pacific Research Unit
Research Corporation of the
University of Hawaii
Honolulu, HI 96806

March 31, 1980

"The hemagglutinating properties of adenoviruses were discovered as a by-product of investigations on hemagglutination by reoviruses and enteroviruses—the virus groups which were my major interest at the time. Agglutination of human erythrocytes had recently been reported for a few enterovirus serotypes.[1] In the course of attempting to demonstrate this property for other enteroviruses and reoviruses, I decided to test two adenovirus serotypes, one commonly isolated from humans and another rarer type—which I had noted gave a somewhat different type of cytopathic effect in cell cultures. Only the latter adenovirus agglutinated the human erythrocytes but this observation led to the systematic study of all the adenovirus serotypes and a great variety of erythrocytes. Others had attempted to demonstrate hemagglutination by adenoviruses previously but had failed to do so, probably because they used only the commonly isolated serotypes and only those erythrocytes (human, guinea pig, and chicken) usually employed in laboratories at that time.

"There probably are several reasons why this paper has been cited relatively frequently. First, the method described is still the easiest and fastest way to type a member of this large group of human viruses, which now includes 34 serotypes. Second, by employing a virus of known serotype the technique can be used to measure relatively type-specific antibody in man. Finally, and perhaps most important, hemagglutinating properties have been proven useful in subdividing the adenovirus family into subgroups which have other biologic properties in common. In this paper, the adenoviruses were subdivided for convenience of typing into several groups on the basis of the type of erythrocyte they agglutinated and the conditions of such agglutination. Later, it was found that these same subgroups also shared other characteristics such as antigenic relationships, epidemiologic behavior, and relative oncogenicity for laboratory animals.

"The research on which this paper is based is particularly memorable to me because a preliminary report was rejected by *Nature*. That in itself was not too bad, although I did, of course, question the judgment of the editor. What was infuriating, however, was that, after hearing nothing for several months, I received both the manuscript *and* the rejection note by sea mail! I confess that, perhaps irrationally, I avoided sending papers to that journal thereafter. The manuscript was next sent to *Virology* where it was accepted without change the day after it was received.[2] It is relevant to point out that the editor of that journal at the time was G.K. Hirst, a co-discoverer of hemagglutination by influenza viruses."

1. Goldfield M, Srihongse S & Fox J P. Hemagglutinins associated with certain human enteric viruses. *Proc. Soc. Exp. Biol. Med.* 96:788-91, 1957.
2. Rosen L. Hemagglutination by adenoviruses. *Virology* 5:574-7, 1958.

This Week's Citation Classic ™

Caspar D L D & Klug A. Physical principles in the construction of regular
viruses. *Cold Spring Harbor Symp.* 27:1-24, 1962.
[Children's Cancer Res. Foundation, Children's Hosp. Med. Ctr.; Dept. Biophys.,
Harvard Med. Sch., Boston, MA; and Med. Res. Council Lab. Molecular Biol.,
Univ. Postgrad. Med. Sch., Cambridge, England]

Starting with the postulate that regular virus capsids are constructed from identical subunits by a self-assembly process, the quasi-equivalence theory explained why icosahedral symmetry should be preferred for the design of isometric capsids, and the possible icosahedral surface lattices, defined by the set of triangulation numbers, were enumerated. [The *SCI*® indicates that this paper has been cited in over 640 publications since 1962.]

Donald L.D. Caspar
Rosenstiel Basic Medical Science
Research Center
Brandeis University
Waltham, MA 02254

December 27, 1983

"In 1956, Crick and Watson in Cambridge, England, predicted that isometric virus particles should be constructed from identical subunits arranged with cubic symmetry.[1] From the same laboratory, I reported X-ray crystallographic data providing the first evidence for icosahedral virus symmetry.[2] Aaron Klug and his colleagues in London soon showed that other isometric viruses were icosahedral, and suggested that some general principle might explain a preference for icosahedral symmetry.[3] I began a transatlantic collaboration with Klug in 1958, following Rosalind Franklin's untimely death. We first reviewed the work that she had started on the helical tobacco mosaic virus and then tried to understand the design of icosahedral viruses. Icosahedral symmetry requires 60 equivalently related parts, but for some icosahedral virus capsids, chemical data indicated *more* than 60 identical subunits, and the number of morphological units seen by electron microscopy was not a multiple of 60. The problem was to explain how to build the shells from a large number of identical units by repeating the same pattern of contact without the constraint of strict equivalence. An anticipatory key to our solution was Klug's recognition of an analogy with Buckminster Fuller's icosa-geodesic dome designs.[4]

"Early in 1962, Fuller came to lecture at Harvard University, which stimulated me to complete the models I had been building at the Children's Cancer Research Foundation in Boston, using his tensegrity principle, to demonstrate why icosahedral viruses *are* icosahedral and how the complete design could be built-in to the specific bonding properties of the parts. That year, Klug asked me to join him in writing a paper for *Cold Spring Harbor Symposium on Quantitative Biology* that was intended as an introduction to our major collaborative paper planned on design and construction of icosahedral viruses, based on my model building. In a hectic two weeks at the end of May 1962, we wrote our *Citation Classic* ™ paper, which ended up incorporating much of the theory it was to have introduced. We then changed the order of our names on this paper, and never completed the proleptically referenced theory paper. The models illustrating the theory finally appeared 18 years later, in a paper of mine subtitled, 'Quasi-equivalence revisited.'[5]

"We introduced three terms in our *Citation Classic* that have been useful for describing a variety of structures: 'triangulation number,' to define the possible icosahedral surface lattice designs; 'quasi-equivalence,' to describe nearly equivalent bonding of identical units; and 'self-assembly,' to identify assembly processes controlled by the specific bonding of the parts. Citations of these terms appear to account for most of the references to our paper.

"Starting with the simple presumption that specificity of bonding among identical, but adaptable, subunits should be conserved in the self-assembly of virus capsids, our quasi-equivalence theory explained icosahedral symmetry and enumerated the possible designs. The beauty of this theory and the success of the predictions made it appear that conservation of bonding specificity was a necessity in icosahedral virus architecture — until three years ago, when Ivan Rayment, working with me, established by X-ray structure analysis that the 60-hexavalent morphological units in the T=7 polyoma virus capsid are all pentamers,[6] instead of hexamers, as predicted on the expectation of quasi-equivalence. Understanding the significance of this result exemplifying the diversity of biological structures is now an experimental rather than a theoretical problem."

1. Crick F H C & Watson J D. The structure of small viruses. *Nature* 177:473-5, 1956. (Cited 170 times since 1956.)
2. Caspar D L D. Structure of tomato bushy stunt virus. *Nature* 177:476-7, 1956.
3. Finch J T & Klug A. Structure of poliomyelitis virus. *Nature* 183:1709-14, 1959. (Cited 90 times since 1959.)
4. Fuller R B & Marks R W. *The Dymaxion world of Buckminster Fuller.* New York: Doubleday, 1960. 246 p.
5. Caspar D L D. Movement and self-control in protein assemblies: quasi-equivalence revisited.
   Biophysical J. 32:103-35, 1980.
6. Rayment I, Baker T S, Caspar D L D & Murakami W T. Polyoma virus capsid structure at 22.5
   angstrom resolution. *Nature* 295:110-25, 1982.

This Week's Citation Classic

Summers D F, Maizel J V, Jr. & Darnell J E, Jr. Evidence for virus-specific noncapsid proteins in poliovirus-infected HeLa cells.
Proc. Nat. Acad. Sci. US 54:505-13, 1965.
[Depts. Biochemistry and Cell Biology, Albert Einstein Coll. Med., New York, NY]

This paper provides evidence that about 12 electrophoretically different virus-specific proteins are formed in poliovirus-infected cells and that many of these polypeptides are synthesized during most of the infectious cycle. [The *SCI®* indicates that this paper has been cited in over 455 publications since 1965.]

Donald F. Summers
Department of Cellular, Viral
and Molecular Biology
School of Medicine
University of Utah
Salt Lake City, UT 84132

July 6, 1983

"In 1964, I joined the laboratory of James E. Darnell in the department of biochemistry at Albert Einstein College of Medicine as a postdoctoral fellow. The lab was an exciting, active group working on the replication of poliovirus and on RNA metabolism in HeLa cells. I had just arrived from the National Institutes of Health (NIH) in Bethesda where I had spent a year working with Leon Levintow on the translation of poliovirus RNA *in vivo* in infected cells and *in vitro*, and when I arrived in Darnell's lab I continued to study the mechanism of translation of polio RNA and the products of translation of this large virus messenger RNA. Jake Maizel, a superb and innovative protein chemist who had worked on polio capsid proteins and virion structure for several years at NIH,[1] had also recently arrived at Einstein in cell biology as an assistant professor.

"We initially attempted to use anti-polio antisera to precipitate the cytoplasmic products of the translation of poliovirus mRNA, but these attempts were met with disappointments and little success. One day Jake and I were mulling over possible ways to isolate and/or separate the poliovirus-specific translational products in the infected cell when we decided to painstakingly review his card file on proteins (a mess of several hundred punch cards) item by item. We stumbled on a short paper by Criddle and Park[2] describing a method for solubilizing chloroplasts and separating the protein constituents in polyacrylamide gels containing urea and SDS. Since Jake had been working with acrylamide gels for several years, we took radiolabeled polio-infected cell cytoplasm and to our delight these cell extracts were completely solubilized by the SDS-urea. We then electrophoresed the extracts on tubular acrylamide gels containing SDS and urea and Jake sliced the gels by hand, and pulverized each slice in a mortar and pestle for assay in the scintillation counter. The first result we obtained showed quite clearly that the virus mRNA was directing the synthesis of the four capsid proteins plus the synthesis of at least ten noncapsid proteins. The combined molecular weights of all of these polio-specific proteins exceeded the coding capacity of the virus mRNA, and suggested an unusual translational mechanism for this very large eukaryotic message species which had already been shown to contain 40-60 ribosomes in polio-specific polyribosomes.[3] Our studies finally led to the observation that the polio mRNA is translated as a large 'polyprotein' which is subsequently cleaved into the individual capsid and noncapsid virus-specific proteins.[4]

"This paper is probably frequently cited because it was the first publication of the SDS-acrylamide gel system which was to develop into such a powerful tool for the resolution of cell and cell fraction proteins and nucleic acids, and also because this was the initial report showing that the picornavirus genome specified a large number of capsid and noncapsid proteins through a post-translational processing mechanism which is quite unique."

1. Maizel J V, Jr. Evidence for multiple components in the structural protein of type I poliovirus.
   Biochem. Biophys. Res. Commun. 13:483-9, 1963.
2. Criddle R S & Park L. Isolation and properties of a structural protein from chloroplasts.
   Biochem. Biophys. Res. Commun. 17:74-9, 1964.
3. Rich A, Penman S, Becker Y, Darnell J E & Hall C. Polyribosomes: size in normal and polio-infected HeLa cells.
   Science 142:1658-63, 1963.
4. Sangar D V. The replication of picornaviruses. *J. Gen. Virol.* 45:1-13, 1979.

Chapter

5

Genetics

The subjects of Citation Classics related to eukaryotic genetics lack the drama of contemporary molecular genetics and the genetic engineering that is occurring with the prokaryotes and to a lesser extent with the eukaryotes. Even so, three of the Classics described in this chapter were cited more than 1,000 times by other authors. These three included the highly practical technique for securing suitable chromosome preparations from human leukocytes, a method for band-staining these chromosomes, and a book on *Drosophila* genetics.

Since cytologic studies have been so instrumental in the diagnosis of congenital disease, methods for improving chromosome preparations for karyotyping have been of tremendous importance. That air drying, a technique used for decades to prepare prokaryotic material for staining, should be so vital to this is indeed startling. Rothfels admits that his paper with Siminovitch had no theoretical or philosophical merit but was the result of a simple observation that chance favored. The addition of phytohemagglutinin (PHA) to leukocyte preparations by Moorhead and his coauthor (3,455 citations) had the benefit of producing many cells in metaphase. This greatly improved human karyotyping simply by allowing an accurate enumeration of chromosomes. Banding of human chromosomes, a technique that facilitates recognition of each individual chromosome, was forgotten by Seabright for 4 years after failing to repeat the procedure and receiving the advice that his finding was artifactual. A later trial after learning that quinacrine could be used to band-stain chromosomes was successful. The trypsin method became the method of choice because of its high resolution and simplicity. Differential staining (banding) of chromosomes can be effected by numerous procedures: trypsin treatment, quinacrine staining, and even hybridization with DNA. These banding methods became useful in identifying abnormal chromosomes involved in congenital diseases even though they were not available when the first examples of human trisomy were discovered. Trisomy, the presence of a third chromosome with the normal pair, was associated with human congenital disease in 1959. This condition (trisomy) had been previously seen in lower organisms and its sudden association with the tragic Down's syndrome was totally unexpected. In 1960 trisomy 13 and trisomy 18 were discovered by the same group at the University of Wisconsin from a limited group of patients with multiple congenital abnormalities (pp. 179 and 180).

The need for efficient repair of DNA that had become damaged by radiation or faulty synthesis is the subject of Cleaver's Classic. Failure to repair ultraviolet-damaged DNA is the basis of the human disease xeroderma pigmentosa. Ataxia telangiectasia, initially thought to be an immunodeficiency disease, has since been demonstrated to be the result of DNA repair failure.

Chromatin, the DNA-histone complex of the chromosome, consists of discrete subunits that can be released by nuclease digestion of the exposed portion of the DNA. The beaded appearance of chromatin seen by electron microscopy represents the subunit now known as the nucleosome. These important discoveries described on pages 183 and 184 followed by six years a review on the biology of chromatin (p. 186) that may have been instrumental in stimulating a more chemical orientation to chromosome research.

Analysis of the X chromosome of *Drosophila* (p. 187) indicated that the gene size was 25-fold larger than is required to code for an average-sized peptide. Although operons were not then recognized, the unexpectedly large, non-gene DNA was thought to have some function in regulation. The non-histone proteins in the eukaryotic nucleoprotein complex are themselves regulatory since they influence the structure and thus the expression of the information encoded in the DNA (p. 188).

An important discovery in the early stages of genetic engineering was that *Xenopus laevis* oocytes make many copies of ribosomal RNA, in effect amplifying the DNA code. Artificial implantation of foreign DNA into these oocytes promotes a more extensive synthesis of the gene product than is possible in the intact, natural systems.

The rapid evolutionary pace of alterations in DNA, in excess of 10-fold that seen in proteins, means that most of these genetic changes are not expressed in a permanent gene product. The proteins that correspond to these altered DNA sequences have no necessary function in nature. This phenomenon of wasted changes in DNA and rejected proteins is the basis for the term non-Darwinian evolution used by King and Jukes.

The last two Classics described, one an enormous compilation of all genetic details of the fruit fly, *Drosophila,* and the other an important, though obviously less detailed report on the status of genetics in the human population, were published as books. The importance of *Drosophila* in the development of genetics as a science, and the faith that basic genetic principles can be applied to the human condition, account for the popularity of these two volumes. Indeed, the special interest in human genetics has been so intense that there is more knowledge about it than there is about *Drosophila* genetics.

Alfert M & Geschwind II. A selectve staining method for the basic proteins of cell nuclei. *Proc. Nat. Acad. Sci. US* **39**:991-9, 1953.

Treatment of formalin-fixed tissue sections with Fast Green at pH 8 after extraction of nucleic acids results in a specific chromatin stain. Model experiments indicate that histones or protamines are responsible for the staining picture. Some quantitative relationships between DNA and their associated histones are reported. [The *SCI®* indicates that this paper has been cited 494 times since 1961.]

Max Alfert
Zoology Department
University of California
Berkeley, CA 94720

March 19, 1978

"My doctoral thesis research in the late forties dealt with quantitative studies of DNA in individual cell nuclei, by staining and light absorption methods. The genetic role of DNA was then being established and there were speculations about the function of other nuclear components. Basic proteins were known to be associated with DNA and suspected to be involved in the regulation of gene expression; however, biochemical knowledge of those proteins was rudimentary and there existed no cytochemical test for their selective visualization, comparable to the Feulgen test for DNA.

"Against this background, I began to do a series of staining experiments on nuclear proteins and to quantitate them by light absorption measurements. I also engaged the advice and cooperation of my friend and colleague, Dr. I. I. Geschwind; as a professional biochemist, he was able to devise staining tests on natural and modified protein model systems which were necessary to elucidate the mechanism of the nuclear staining reaction. We were then able to publish the first cytochemical method for the direct visualization of histones (and related basic proteins), explain how it worked and provide some quantitative data, relating DNA to histone content of nuclei.

"Since then, histones have continued to attract the attention of cell biologists and they now figure prominently in the most recent theories of chromatin organization, involving nucleosomes. Much of the recent work is done at the fine structure level, supplemented by biochemical and biophysical studies. At the same time, many cytologists have remained interested in visualizing histones by staining and they have used and referred to our method in their numerous publications. (But despite the many citations of this method, it has managed to escape the attention of some authors of recent reviews on nuclear changes during the cell cycle, a subject to which this technique has been often applied.)

"Soon after the publication of this paper, Dr. Geschwind and I had mostly, but not exclusively favorable comments on it. A late, well-known biochemist from New York expressed his disdain for this and similar quantitative methods when they are used in conjunction with a microscope, rather than a test tube. However, subsequent applications of this method by us and many others have yielded interesting results on changes in nuclear composition, especially during spermiogenesis and in early embryonic development.

"This work was done at the Univrsity of California, with which we both remained associated until Dr. Geschwind's untimely death in July 1978. When we did that work, it was difficult to obtain funds for basic research; the same in true again today."

This Week's Citation Classic

Stent G S & Brenner S. A genetic locus for the regulation of ribonucleic acid synthesis. *Proc. Nat. Acad. Sci. US* **47**:2005-14, 1961.
[Medical Research Council Unit for Molecular Biology, Cavendish Laboratory, Cambridge, England]

Some amino acid auxotrophs of *E. coli* continue to synthesize ribosomal RNA during amino acid starvation, in contrast to others which stop such synthesis as soon as they are deprived of the required amino acid. Conjugation experiments showed that there exists a locus, RC, on the bacterial chromosome whose abnormal, or 'relaxed,' allele causes less stringent control by amino acids over ribosomal RNA synthesis than its normal, or 'stringent,' allele. [The *SCI®* indicates that this paper has been cited in over 535 publications since 1961.]

Gunther S. Stent
Department of Molecular Biology
University of California
Berkeley, CA 94720

June 24, 1983

"In 1961, Jacob and Monod had just revolutionized thinking about the mechanism of control of gene expression by publishing their operon model.[1] According to that model, the formation of specific classes of messenger RNA, and hence the quality of protein synthesis, is controlled by various metabolites which act at special regulatory genetic sites.

"So when I spent most of my 1961 sabbatical leave with Sydney Brenner at the Cambridge Medical Research Council (MRC) Unit for Molecular Biology, it seemed an obvious question to ask whether such special regulatory sites account also for the control of the overall rate of ribosomal RNA synthesis, and hence for the quantity of protein synthesis. This overall rate was known to depend on the presence of amino acids, because auxotrophic mutants requiring an amino acid for growth stop synthesizing not only protein but also ribosomal RNA as soon as they are deprived of that amino acid. Inasmuch as the amino acid could be shown to exert this regulatory role catalytically rather than stoichiometrically,[2,3] it was plausible to suppose that it acted as an inducer of RNA syn-

thesis. Accordingly, I set out to try to find operon theory-like regulatory sites involved in the amino acid-mediated control of ribosomal RNA synthesis, by isolating mutant strains of *E. coli* defective in normal ribosomal RNA control. Before I had got very far in my mutant hunt, Brenner pointed out to me that at least one such defective strain was already available, namely *E. coli* K12 strain 58-161, whose amino acid auxotrophs continue ribosomal RNA synthesis at a near-normal rate in the absence of the required amino acid.[4]

"Brenner and I then decided to cross Hfr donor variants of 58-161 to normal F⁻ recipients, in order to see whether the abnormal regulatory behavior of that strain might be attributable to some mutant locus on the *E. coli* chromosome. Since the facilities for bacterial genetics were quite poor at the MRC Unit (then still located in a hut in the courtyard of the Cavendish Physics Laboratory), I got nowhere in my crosses. Finally, in desperation I took the train down to London every morning to do the crosses in William Hayes's lab at the Hammersmith Hospital, with the help of Hayes's associate, Royston Clowes. And within a week or so, I had my first recombinants which showed that the phenotypes of normal (or 'stringent') and defective (or 'relaxed') control of ribosomal RNA synthesis are indeed attributable to different alleles of a genetic locus, which we designated as RC (now known as *relA*), situated somewhere between the *xyl* and *gal* genes.

"I imagine that the brief paper in which we published this simple finding came to be frequently cited only after 1969 when Gallant and Cashel[5] discovered that changes in intracellular concentrations of a hitherto unknown nucleotide tetraphosphate, which they designated 'magic spot,' are associated with the overall rate of RNA synthesis, and that the 'stringent' and 'relaxed' alleles of the RC locus have a differential effect on 'magic spot' levels. For this finding held out the promise of leading to an understanding of the control of the initiation of transcription by RNA polymerase. I have not followed these latter-day developments but it is my impression that, 'magic spots' notwithstanding, the problem of the overall control of bacterial RNA synthesis still remains to be fully solved."[6]

1. Jacob F & Monod J. Genetic regulatory mechanisms in the synthesis of proteins. *J. Mol. Biol.* 3:318-56, 1961.
2. Gros F & Gros F. Rôle des aminoacides dans la synthèse des acides nucleiques chez *Escherichia coli*. *Biochim. Biophys. Acta* 22:200-1, 1956.
3. Pardee A B & Prestidge L. The dependence of nucleic acid synthesis on the presence of amino acids in *Escherichia coli*. *J. Bacteriology* 71:677-83, 1956.
4. Borek E, Ryan A & Rockenbach J. Nucleic acid metabolism in relation to the lysogenic phenomenon. *J. Bacteriology* 69:460-7, 1955.
5. Cashel M & Gallant J. Two compounds implicated in the function of the RC gene of *Escherichia coli*. *Nature* 221:838-41, 1969.
6. Silverman R H & Ather A G. The search for guanosine tetraphosphate (ppGpp) and other unusual nucleotides in eucaryotes. *Microbiol. Rev.* 43:27-41, 1979.

Fraenkel-Conrat H, Singer B & Tsugita A. Purification of viral RNA by means of bentonite. *Virology* 14:54-8, 1961.
[Virus Laboratory, University of California, Berkeley, CA]

This paper represents the less important half of a piece of work published as two papers (p. 54 and 59). It demonstrates the efficacy of bentonite in diminishing residual traces of peptides in phenol-prepared viral RNA. The conclusion is that the upper limit for tobacco mosaic virus (TMV) protein contamination in such RNA preparations was one mole per 23 moles of the RNA. [The *SCI®* indicates that this paper has been cited over 500 times since 1961.]

Heinz Fraenkel-Conrat
Department of Molecular Biology
and Virus Laboratory
University of California
Berkeley, CA 94720

January 12, 1978

"The fact that this paper is cited with sufficient frequency to warrant its inclusion in this series is an illustration of the fallibility of citation frequency as a measure of the importance of a piece of work. All three authors have at least 10 other papers to their credit which they would list above this one in importance. And what this paper is quoted for is not its intrinsic point (which had some importance), but the fact that it contains a paragraph describing the method of washing and suspending the commercial bentonite clay. The purpose of preparing such bentonite suspensions is described in the companion paper: the adsorption and removal of ribonucleases leads to considerably more stable RNA preparations, and when tested in presence of bentonite TMV RNA shows much higher infectivity.[1] It is these findings by Singer and Fraenkel-Conrat which account for the popularity of this paper!

"What led to the research reported in these papers? As so frequently, and as it should be, it was a report in the literature seemingly unrelated to our work and yet of potential usefulness, namely that the clay, bentonite, binds and inhibits ribonucleases. Bea Singer and I had, since 1956, become convinced that the infectivity and thus the entire genetic information of TMV resided in its RNA.[2,3] (It is these papers one would expect to be most frequently cited, since TMV RNA was the first physico-chemically definable molecular species shown to represent a genome.) These studies were difficult because of the great sensitivity of RNA to the ubiquitous ribonucleases and the requirement of the intact molecule of 6390 nucleotides for TMV infectivity. Thus, not surprisingly, we were on the look-out for anti-nuclease agents and tested all that came to our attention. And of these, bentonite proved the most useful, as described above, notwithstanding the first reaction of a colleague also working with TMV RNA at Berkeley, 'I certainly would not put this mud into my RNA preparations.'

"On the basis of these findings it seemed only logical that removal of traces of ribonucleases by bentonite should cause a decrease in the 0.5-1% of protein seemingly present in such viral RNA preparations, and still representing a last hope of the die-hards who disbelieved RNA's potency. We were successful. Upon degrading ^{35}S labelled TMV in presence of bentonite, the amount of ^{35}S in the viral RNA was reduced by one order of magnitude. Also, Tsugita's analyses showed the amino acid composition of the residual 0.04% of protein(s) or peptide(s) to be quite unlike that of TMV protein. The finding that the residual trace of ^{35}S was displaced by reconstituting such RNA with TMV protein further proved that it did not represent residual virus or viral protein. Thus the much-quoted paper represented one more proof (than was needed) that viral RNA molecules represent fully competent genomes.

"B. Singer is currently at the Virus Laboratory, University of California, Berkeley, CA and A. Tsugita is at the EMBO Laboratory, Heidelburg, Federal Republic of Germany."

1. Singer B & Fraenkel-Conrat H. Effects of bentonite on infectivity and stability of TMV-RNA. *Virology* 14:59-65, 1961.
2. Fraenkel-Conrat H. The role of the nucleic acid in the reconstitution of active tobacco mosaic virus. *J. Amer. Chem. Soc.* 78:882-3, 1956.
3. Fraenkel-Conrat H, Singer B & Williams R C. Infectivity of viral nucleic acid. *Biochim. Biophys. Acta* 25:87-96, 1957.

CC/NUMBER 1
JANUARY 2, 1984

This Week's Citation Classic™

Taylor J H, Woods P S & Hughes W L. The organization and duplication of chromosomes as revealed by autoradiographic studies using tritium-labeled thymidine. *Proc. Nat. Acad. Sci. US* 43:122-8, 1957.
[Dept. Botany, Columbia Univ., New York, and Biology and Medical Depts., Brookhaven Natl. Lab., Upton, NY]

Autoradiographs of chromosomes prepared after cells of *Vicia* (broad bean) had incorporated ^{3}H-thymidine into DNA in the cell cycle before division showed that both daughter chromosomes (chromatids) were equally labeled. After a second cell cycle without ^{3}H-thymidine, the labeled DNA in each chromosome was segregated so that one chromatid was fully labeled and the other was unlabeled. Exceptions were produced by occasional exchanges of labeled and unlabeled segments between two daughter chromatids (sister chromatid exchanges). [The *SCI*® indicates that this paper has been cited in at least 680 publications since 1957, including 7, 29, 41, 42, 33, 36, 43, and 36 in 1957-1964, respectively, based on 1955-1964 *SCI* cumulation.]

J. Herbert Taylor
Department of Biological Science
and
Institute of Molecular Biophysics
Florida State University
Tallahassee, FL 32306

September 15, 1983

"Soon after Watson and Crick published a proposed structure for DNA and a scheme for its replication,[1,2] Friedkin *et al.* synthesized ^{14}C-thymidine and showed that it was incorporated exclusively into DNA.[3] When preliminary attempts to follow the distribution of new DNA in chromosomes labeled with ^{14}C-thymidine failed to give a decisive answer, I realized that autoradiographic resolution with tritium, a low energy beta emitter, would allow analysis of individual chromatids. With enthusiasm but little knowledge of how to label thymidine, I joined Philip Woods at the Brookhaven National Laboratory for a summer project in 1956.

"A few days after I arrived at Brookhaven, we learned that Walter (Pete) Hughes was preparing ^{3}H-thymidine to study its lethal effects on cancer cells. We sought his assistance and he agreed to share the ^{3}H-thymidine for the chromosome studies. We planned to grow *Vicia* roots in a solution of ^{3}H-thymidine for eight hours and then transfer them to a solution with colchicine and without ^{3}H-thymidine. Cells fixed within a few hours should show cells at the first division after their DNA was labeled during replication. Since colchicine blocks anaphase and cell division, but not chromosome reproduction in plants, the labeled cells which appeared after about 36 hours with 24 chromosomes would be at the second division after labeling of their DNA.

"The first half of our summer's work yielded no information, because the thymidine was not hot enough. Hughes soon had a second sample ready and we repeated the experiments, made the autoradiographs, and waited the two weeks necessary for the exposure. When the slides were developed, the dark grains were positioned over the red stained chromosome in many of the spread cells. In cells fixed a few hours after transfer to colchicine, both chromatids were fully labeled from end to end. However, some cells fixed after 36 hours had 24 chromosomes with one labeled chromatid lying beside an unlabeled chromatid in each metaphase chromosome. The segregation was occurring after the second replication and the only complication was an occasional exchange of segments between sister chromatids. Nevertheless, the result was convincing and many of our colleagues and visitors to the lab peered through our microscope during the next few days to see the first indication that the Watson-Crick scheme for DNA replication operated at the chromosome level. Reflection, however, revealed that we knew too little about chromosome structure to be sure that DNA chains were segregating.

"Our studies have been cited many times because they were the first significant studies with tritium-labeled molecules, which stimulated work on the structural organization of chromosomes, the time of replication of different parts of chromosomes, and the kinetics of the cell cycle. Sister chromatid exchanges which we discovered in these experiments were analyzed further to reveal the structure of chromosomes and these exchanges later proved to be good indicators for the effects of mutagens and carcinogens. A conference on this topic, the International Symposium on Sister Chromatid Exchange, to celebrate the twenty-fifth anniversary of the discovery and research on sister chromatid exchange as an indicator for mutagenic and carcinogenic agents, was to be held at Brookhaven National Laboratory, December 4-8, 1983. For a review of this field, see *Molecular Genetics*."[4]

1. Watson J D & Crick F H C. Molecular structure of nucleic acids. A structure for deoxyribose nucleic acid. *Nature* 171:737-8, 1953. (Cited explicitly over 1,055 times since 1955.)
2. ------------------------------. Genetic implications of the structure of deoxyribonucleic acid. *Nature* 171:964-7, 1953. (Cited 535 times since 1955.)
3. Friedkin M, Tilson D & Roberts D. Studies of deoxyribonucleic acid biosynthesis in embryonic tissues with thymidine-C[14]. *J. Biol. Chem.* 220:627-37, 1956. (Cited 165 times since 1956.)
4. Taylor J H, ed. *Molecular genetics.* New York: Academic Press, 1963-1979. Parts I-III.

Rothfels K H & Siminovitch L. An air-drying technique for flattening chromosomes in mammalian cells grown *in vitro*. *Stain Technol.* **33**:73-7, 1958.
[Department of Botany and Connaught Medical Research Laboratories, University of Toronto, Canada]

The technique facilitates karyotype analysis primarily of cells grown *in vitro*. Following standard preliminary treatments, the crucial step is the complete air-drying of cells on slides directly after fixation. This produces a more complete flattening of intact chromosome complements than is usually achieved manually. [The *SCI®* indicates that this paper has been cited in over 510 publications since 1958.]

Klaus Rothfels
Department of Botany
University of Toronto
Toronto, Ontario M5S 1A1
Canada

August 3, 1984

"Louis Siminovitch, coauthor of the paper, at the time was working in the laboratory of Raymond Parker (Connaught Laboratories) on cell-cycle aspects of monkey kidney cells. These were being isolated in quantity as substrate for polio vaccine production.

"In propagating these cells in tissue culture, Parker noted that such cultures tended to decline in proliferative capacity after serial transfers and many were lost. A few revived and formed distinct colonies, capable of indefinite *in vitro* proliferation and differing from the cells of origin in various morphological and biochemical traits.[1] Parker termed these cells 'altered' or 'transformed.'

"Siminovitch was interested in a karyotypic comparison of these cells and fresh monkey kidney cells. He invited me to collaborate, since I had a background in chromosome cytology—specifically, of grasshoppers, though sitting in a botany department. Initially, Siminovitch's responsibility was the cultivation of cells directly on slides while I did the chromosome work. Eventually, both of us became deeply involved in this aspect, as did A.A. Axelrad and E.A. McCulloch.

"Discovery of the crucial step in the method was quite accidental. Preliminary treatments such as hypotonic extension and mitotic inhibition (colchicine) were well established at the time. While making squash preparations, using acetoorcein, we noted that, in contrast to metaphases in the middle of slides, those at the edge frequently were absolutely flat with a completely two-dimensional display of chromosomes. We guessed that this was so because those cells had dried on removal of the slides from the fixative and before application of the stain solution. This interpretation was readily tested and proved to be correct. What resulted was a very simple technique, 'air drying,' which was described fully in the publication. The technique is simpler and more reproducible than later more elaborate ones and particularly valuable for people with an 'uneducated thumb.' The frequent citation of the paper is simply a consequence of the enormous amount of work being done that requires karyotypic analysis, particularly of human cells. Air drying presumably had been used prior to 1958, but we did not really search the literature for this, and the indication that the method might be useful came directly from our studies.

"In our own laboratory, the method proved immediately valuable, for it made possible the proof that most of the transformations noted by Parker in monkey kidney cell cultures were due to contamination with L cells.[2] It subsequently became apparent that transformations observed in other laboratories had a similar origin, HeLa cells being a frequent culprit.

"The contamination interpretation was supported by other studies, notably those of Axelrad, who demonstrated that the 'altered' cells behaved oncogenically like L-cells and not like monkey kidney cells.[2]

"Subsequent studies in our laboratory—primarily on mouse cells—showed that uncontaminated cultures can undergo karyotypic as well as other changes, but this is a gradual and progressive process.[3,4]

"In summary, the frequency of citation is clearly a result of the simplicity of the method and the very large number of studies, particularly of human cells, that require karyotypic analysis. The paper has no particular theoretical or philosophical merit nor an elaborate experimental basis.

"My own later interests centered largely on comparative banding-pattern studies in the polytene salivary-gland chromosomes of blackfly (simuliid) larvae. It is in this area that I have perhaps made some contributions (other than just letting slides dry out), and some of these are conceptual."[5]

1. Parker R C. Alterations in clonal populations of monkey kidney cells. *Poliomyelitis: papers and discussions presented at the Fourth International Poliomyelitis Conference.* Philadelphia: Lippincott, 1958. p. 257-67.
2. Rothfels K H, Axelrad A A, Siminovitch L, McCulloch E A & Parker R C. The origin of altered cell lines from mouse, monkey, and man, as indicated by chromosome and transplantation studies. *Can. Cancer Conf.* 3:189-214, 1959.
3. Rothfels K H & Parker R C. The karyotypes of cell lines recently established from normal mouse tissues. *J. Exp. Zool.* 142:507-20, 1959.
4. Rothfels K H, Kupelwieser E B & Parker R C. Effects of X-irradiated feeder layers on mitotic activity and development of aneuploidy in mouse-embryo cells *in vitro. Can. Cancer Conf.* 5:191-223, 1963.
5. Rothfels K H. Cytotaxonomy of blackflies (Simuliidae). *Annu. Rev. Entomol.* 24:507-39, 1979.

Harnden D G. A human skin culture technique used for cytological examinations. *Brit. J. Exp. Pathol.* **41**:31-7, 1960.

The culture of fibroblastic cells from tiny pieces of skin that could be taken from patients or volunteers is described in this paper, together with results of the use of these cells for studying the chromosomes of patients with a variety of developmental abnormalities. [The *SCI®* indicates that this paper was cited 201 times in the period 1961-1977.]

D.G. Harnden
University of Birmingham Medical School
Department of Cancer Studies
Birmingham B15 2TJ
England

January 12, 1978

"While working in Charles Ford's laboratory at Harwell, I found his work on human chromosomes more interesting than my own project, which involved culture of mouse tumours. The bone marrow technique used at that time for human chromosomes was hard to apply universally because obtaining the sample can be unpleasant. It seemed to me sensible to culture a more easily accessible tissue, and skin seemed the obvious choice. I asked a colleague to cut a tiny piece from my own forearm and I set it up in culture, using essentially the classical plasma clot technique for chick heart tissue that I had learned from Honor Fell in Cambridge. I also incorporated an idea from Henry Harris, who cultured rat cells for a short period before dissociation with the enzyme trypsin. It worked and I examined my own chromosomes (maybe I was the first to do so).

"The technique was reliable even in relatively unskilled hands, and since the sample required was small it could be obtained from anyone. It could be used for skin taken post mortem or sent long distances. (I examined the chromosomes of a pure blood aborigine from the Australian outback.) Human fibroblasts of embryonic origin or from surgical biopsies had been grown before and I was influenced by the studies, published while I was doing this work, of T.T. Puck in Denver, who cultured skin cells from biopsies using direct dissociation with trypsin. The cytogenetic results in the paper confirmed results already reported, but subsequently I used the technique for studying the first E trisomy, the first triple X female and other new chromosome complements.

"The paper was quoted, I am sure, because it describes a technique in detail. There was nothing tremendously original — as I have indicated it was an amalgam of previous work. It was, however, very exciting for me to have a powerful new technique, which was much in demand because of the upsurge of interest in human cytogenetics at that time. It has been largely superseded for this purpose by Paul Moorhead's blood culture technique. Possibly more important, cultured fibroblasts are now widely used in the study of human metabolic disease, and though there are now many different techniques, the one described in this paper gave a stimulus at an appropriate moment. The work was carried out when I was an inexperienced postdoctoral fellow. Knowing what I now know about the culture of cells, I realize that I was clearly very lucky to get the conditions right the first time. It may be important, however, that young workers should chance their arm a bit before they get inhibited by too much conventional wisdom."

Moorhead P S, Nowell P C, Mellman W J, Battips D M & Hungerford D A.
Chromosome preparations of leukocytes cultured from human peripheral blood.
Exp. Cell Res. **20**:613-16, 1960. [Wistar Inst. Anat. and Biol.; Dept. Pathol.,
Sch. Med., Univ. Pennsylvania; Dept. Pediat., Hosp. Univ. Pennsylvania and
Sch. Med., Univ. Pennsylvania; and Inst. for Cancer Res., Philadelphia, PA]

A combination of cytological and leukocyte culture techniques is described for a convenient, reliable approach to chromosome studies of humans. Advantages are: ease of obtaining a small volume of blood, adequate mitotic yield, and high proportion of quality metaphases for critical analysis of chromosome morphology. [The *SCI®* indicates that this paper has been cited in over 3,445 publications since 1961.]

Paul S. Moorhead
Children's Hospital of Philadelphia
34th Street and Civic Center Boulevard
Philadelphia, PA 19104

January 6, 1983

"This publication described the basic 'method' for cytological preparations of mitotic chromosomes from cultures of blood cells and contributed much to the explosion of published findings in human and animal cytogenetics in the 1960s. In essence, the method involved the combination of two prior contributions and a simple improvement of one aspect of each. Peter Nowell's discovery of phytohemagglutinin (PHA) as a mitogen[1] was combined with the technique of Rothfels and Siminovitch[2] for air drying to spread metaphase cells.

"The finding regarding PHA emerged from its original use in the separation of buffy coat from red blood cells. Osgood and Krippaehne,[3] pioneers in the cultivation of cellular elements of blood, had found mitoses in cells which attached to a slanted glass surface in their cultures. Mitotic activity seemed restricted to a certain level on the slide and this was attributed to an oxygen gradient in the culture. David Hungerford of the Institute for Cancer Research and Peter Nowell of the University of Pennsylvania were using such 'gradient cultures' for the study of human chromosomes. Bill Mellman, a pediatrician at the University of Pennsylvania, approached me regarding a similar collaboration and introduced me to Nowell and to gradient cultures. Disappointed with the small numbers of cells that attached to the glass slide, I examined the unattached cells on the floor of the culture vessel. Among them were high numbers of mitoses which had escaped attention, perhaps due to the wide acceptance of the importance of oxygen tension.

"Once it was established that the cells of interest did not attach to glass, a cytological method for suspended cells was sought. I had been impressed with the quality of chromosomes prepared by a cell suspension method published earlier by Rothfels and Siminovitch.[2] This method eliminated difficulties of the 'squash' technique and provided superior preparations, all chromosomes being in a single focal plane. However, air drying had repeatedly failed in my hands. Serendipity was involved in that cells of the mouse lymphoma, L-5178-Y, were available from Lionel Manson's laboratory near my own at the Wistar Institute. The ease of cultivation and high growth rate of these nonattaching cells permitted the testing of several variables each day. Surprisingly, minor departures from the usual air-drying procedure greatly affected cytological quality and it became apparent that rapidity of drying was a major factor. After weeks of trial and error involving hair dryers, dry ice, and other variations with little rationale, wetting the slide in cold distilled water prior to rapid evaporation of fixative dropped on the slide produced excellent results.

"These improvements were simple, but taken altogether they comprised a reliable and rapid method for karyological sampling of human subjects. The many citations to this paper were due to its practical value to a growing army of researchers, primarily pediatricians, who were drawn to cytogenetics by classic papers such as the determination of the true chromosome number in humans[4] and the demonstration that trisomy 21 was the cause of Down's syndrome.[5] A review in the field of cytogenetics has been published by H.J. Evans."[6]

1. **Nowell P C.** Phytohemagglutinin: an initiator of mitosis in cultures of normal human leukocytes. *Cancer Res.* **20**:462-6, 1960. [Citation Classic. *Current Contents* (42):13, 17 October 1977.]
2. **Rothfels K H & Siminovitch L.** An air-drying technique for flattening chromosomes in mammalian cells grown in vitro. *Stain Technol.* **33**:73-7, 1958.
3. **Osgood E E & Krippaehne M L.** The gradient tissue culture method. *Exp. Cell Res.* **9**:116-27, 1955.
4. **Tjio J H & Levan A.** The chromosome number of man. *Hereditas* **42**:1-6, 1956.
5. **Lejeune J, Gautier M & Turpin R.** Etudes des chromosomes somatiques de neuf enfants mongoliens. *C.R. Acad. Sci.* **248**:1721-2, 1959.
6. **Evans H J.** Some facts and fancies relating to chromosome structure in man. *Advan. Hum. Genet.* **8**:347-438, 1977.

Patau K, Smith D W, Therman E, Inhorn S L & Wagner H P. Multiple congenital anomaly caused by an extra autosome. *Lancet* **1**:790-3, 1960.

A syndrome of multiple developmental abnormalities is defined on the basis of a chromosome aberration, the presence of an extra D autosome in all cells. This paper relates the chromosomal abnormality to the process of nondisjunction and to the origin of congenital defects. [The *SCI*® indicates that this paper was cited 330 times in the period 1961-1977.]

Stanley L. Inhorn
State Laboratory of Hygiene
University of Wisconsin
Madison, WI 53706

January 17, 1978

"The field of human cytogenetics had experienced a rebirth in the late 1950's as a result of technical improvements that permitted clear delineation of somatic cell chromosomes. Recent reports from Europe had linked numerical chromosome abnormalities with known development syndromes — Turner's syndrome, Klinefelter's syndrome, and mongolism or Down's syndrome.

"A junior faculty man at the University of Wisconsin, David Smith, was just beginning to work in the field of developmental abnormalities. He was naturally excited by these discoveries in cytogenetics and was determined to apply chromosome studies to the cases of multiple congenital malformations he was examining. Thus, Smith corresponded with Charles Ford, in Great Britain, about the possibility of sending him bone marrow specimens from these cases. Dr. Ford graciously agreed to collaborate.

"At the time, I was a new faculty person in the Department of Pathology, working with quantitative cytologic methods. Smith asked me if I could prepare the marrows in tissue culture media for shipment to Britain. This I assured him would present no problem. As we were preparing our first material, we received a communique from Ford that his results with specimens sent from other investigators were unsatisfactory, and so he advised against sending any samples.

"I related this discouraging turn of events to D. Murray Angevine, chairman of Pathology. Dr. Angevine asked why we needed to send specimens to Europe when we had a prominent cytogeneticist, Klaus Patau, in the department. Although I shared a laboratory suite with Patau, I had regarded him as a quantitative cytologist and was unaware of his extensive experience in plant and animal cytogenetics.

"Within a short period of time, fruitful collaboration began between Patau, his associate Eva Therman (later to become Mrs. Patau), myself, Smith and his fellows. Smith selected cases of multiple congenital malformations for the chromosome study. Case 5 was the patient reported in this Citation Classic. The reason that this paper is so commonly cited is that it was the first delineation of a newly described syndrome of chromosomal etiology. Turner's, Klinefelter's, and Down's syndromes were well known conditions waiting for an etiologic explanation. The syndrome described in Case 5 was heretofore unrecognized. Also mentioned in this paper were two cases of another chromosomal syndrome, the trisomy 18 syndrome, the first being our Case 11.

"How does one name a new syndrome? Patau argued that we should call it the D_1 trisomy syndrome, since cytogenetic techniques at that time did not permit distinction between chromosomes 13, 14, or 15. Since we had discovered two trisomy syndromes in our first 11 cases, he reasoned that trisomies would be described for all chromosomes, so that trisomies D_2 and D_3 would soon be found. Time and better techniques have proven that D_1 trisomy is really trisomy 13, and that trisomies 14 and 15 are always lethal, resulting in spontaneous abortion. 13 trisomy is often referred to as Patau's syndrome, an appropriate tribute to one of the pioneers in human cytogenetics."

Smith D W, Patau K, Therman E & Inhorn S L. A new autosomal trisomy syndrome: multiple congenital anomalies caused by an extra chromosome.
J. Pediat. 57:338-45, 1960.

The recognition of the 18 trisomy syndrome as the second most common autosomal trisomy syndrome in man was not only new and important knowledge in its own right. It was also an early part of the rapid growth of knowledge about abnormal human morphogenesis which today is bringing us into the era of fetal preventive medicine. [The *SCI®* indicates that this paper was cited 138 times in the period 1961-1977.]

David W. Smith
Department of Pediatrics
University of Washington
School of Medicine
Seattle, WA 98195

January 17, 1978

"Back in 1960 there was little knowledge as to causes for congenital malformation problems. The average physician had only a vague concept as to what a chromosome was and had never heard the term trisomy used before. Though chromosomal abnormalities as a cause for altered morphogenesis had been recognized in plants, in the sea urchin by Bovari in the early 1900's, and in fruit flies by Morgan and his co-workers in the 1920's, little knowledge existed about chromosomal abnormalities in the human. In 1959 J. Lejeune and co-workers discovered that Down's syndrome (called mongolism at that time) was caused by an extra chromosome, designated trisomy 21. Thus a genetic imbalance, a triplicate dosage of genes, was responsible for one of the most common mental deficiency disorders in man. At the time I was at the University of Wisconsin Medical School. I reasoned that if a small extra autosome caused a moderate problem in morphogenesis, then a larger autosome in triplicate might cause a much more severe multiple defect problem and even be a cause of early death. As a clinician, I had experience with such patients and considered them to be prime candidates for other possible additional autosomal trisomy syndromes.

"Fortunately, I was able to find other investigators at the University of Wisconsin to assist in the study. Stanley Inhorn did the cell cultures and Klaus Patau and Eva Therman did the chromosomal studies. Klaus and Eva had between them a broad experience with chromosomes in insects and plants and became pioneers in the early development of cytogenetics in the human. The fifth patient we studied had an extra D chromosome, eventually shown to be 13 trisomy; and the eighth patient studied had an extra 18 chromosome. It was certainly exciting to look into the microscope and see that extra chromosome! A simple problem in chromosmal distribution resulting in 18 trisomy had resulted in this pattern of multiple anomalies in morphogenesis!

"As is usual, when you find the real answer, it is a simple one. Having evaluated one such patient, I realized that other 18 trisomy patients should have a similar pattern of malformation. I then searched for multiple-defect patients with a similar pattern. The pediatricians of Madison, Wisconsin, were an immense help to me in this search. Several months later I saw the second similar patient and within six months thereafter had evaluated a total of six patients with this same pattern of malformation, all of whom had the extra 18 chromosome, and hence the disorder was called the 18 trisomy syndrome. These infants tend to die in the early months after birth and, strangely, this syndrome had never been recognized before. This again emphasizes the point that you tend to recognize what you are looking for. It is now recognized everywhere as the second most common autosomal trisomy syndrome in man, with an incidence of about one in 3,000.

"The impact of the paper, however, was much greater than the recognition of a specific syndrome and its cause. It was a part of the beginning of an ever-increasing body of knowledge about the causes of problems of malformation in the human, which is now, thankfully, moving into a preventive phase. It helped in a small way to spawn or enhance the fields of medical genetics, dysmorphology and fetal preventive medicine."

Cleaver J E. Defective repair replication of DNA in xeroderma pigmentosum.
Nature 218:652-6, 1968.
[Lab. Radiobiology, Univ. California Medical Ctr., San Francisco, CA]

Normal skin fibroblasts can repair ultraviolet radiation damage to DNA by inserting new bases into DNA in the form of small patches. Cells from patients with the hereditary disease xeroderma pigmentosum carry a mutation such that repair replication of DNA is either absent or much reduced in comparison to normal fibroblasts. Patients with xeroderma pigmentosum develop fatal skin cancers when exposed to sunlight, and so the failure of DNA repair in the skin must be related to carcinogenesis. [The *SCI®* indicates that this paper has been cited over 610 times since 1968.]

James E. Cleaver
Laboratory of Radiobiology
School of Medicine
University of California
San Francisco, CA 94143

June 12, 1981

"In scientific research everyone travels in hope of that lucky break; rarely are we privileged to receive it. This paper, now a *Citation Classic*, describes one of those serendipitous discoveries that has been valuable in defining the role of damage to the genetic material (DNA) in human cancer. It sparked the growth of a large field of currently active, fruitful, and competitive research, and it is this which accounts for the article's frequent citation.

"In 1966 1968, we had the beginning of an interest in the way radiation damages human DNA. One direction was clear: to make significant progress we needed to select mutants that were altered in their capacity to recover from radiation damage. This direction was difficult, and has only recently been successful. The discovery of human mutants started from a most unlikely origin. In April 1967, the *San Francisco Chronicle* ran an article by David Perlman on a hereditary form of skin cancer in man induced by sunlight, xeroderma pigmentosum (XP).[1] From that article I guessed that perhaps XP could be a human mutant of the kind we wanted. The first experiment

on XP cells showed that the choice was right, and in just over a year this first report was published, and a second followed the next year.[2]

"The choice of the right disease was everything, because most of the experiments involved routine techniques. Anyone can repeat these first experiments in a matter of a few days, and the rapidity of the competition to harvest the cream from the subject testified to this. Ironically, an earlier report demonstrating hypersensitivity of XP cells to ultraviolet light by Gartler had been completely ignored.[3] That was fortunate for me, because if it had been pursued the pioneering work on XP would have been over before I left graduate school!

"There have been many ramifications from defining XP as a DNA repair deficiency in man. The etiology of a complex human disease involving both genetic and environmental components is clearer. Numerous researchers have been stimulated to search for other diseases that might have similar characteristics. XP studies have become a starting point and justification for many experiments investigating carcinogen action on DNA. The role of DNA repair mechanisms in carcinogenesis is a currently active field of research.

"A wide range of human diseases have now been identified that exhibit increases in sensitivity to environmental agents (chemicals, radiations) that damage DNA.[4] These hypersensitive diseases—XP, ataxia telangiectasia, Cockayne syndrome. Fanconi's anemia, etc.—form a diverse class within which XP is a subset involving defective DNA repair. The causes of hypersensitivity in the other diseases are more complex.

"Our work in XP has been recognized by two awards: the Radiation Research Society's Research Award in 1973, and the American Academy of Dermatology's Lila Gruber Award for Cancer Research in 1976. It has been generously supported by the Atomic Energy Commission and its successor, the Department of Energy."

1. Perlman D. A family pattern in cancer study. *San Francisco Chronicle* 12 April 1967, p. 4.
2. Cleaver J E. Xeroderma pigmentosum: a human disease in which an initial stage of DNA repair is defective. *Proc. Nat. Acad. Sci. US* 63:428-35, 1969.
3. Gartler S. Inborn errors of metabolism at the cell culture level. (Fishbein M, ed.) *The second international conference on congenital malformations.* New York: International Medical Congress, Ltd., 1964. p. 94.
4. Cleaver J E. DNA damage, repair systems and human hypersensitive diseases. *J. Environ. Pathol. Toxicol.* 3:53-68, 1980.

Seabright M. A rapid banding technique for human chromosomes.
Lancet 2:971-2, 1971.
[Cytogenetics Lab., Dept. Pathology, Salisbury General Hospital,
Wiltshire, England]

A method for banding human chromosomes using trypsin is described. The procedure is rapid and economical. Slides ready for observation can be obtained within ten minutes. [The *SCI®* indicates that this paper has been cited over 1,085 times since 1971.]

Marina Seabright
Wessex Regional Cytogenetics Unit
General Hospital
Salisbury, Wiltshire SP2 7SX
England

March 18, 1981

"The trypsin-bands story really began in 1967 while examining a chromosome preparation stained with Leishman. To my surprise I noticed the presence of strange stripes across the chromatids. Intrigued by this observation I sought the opinion of senior cytogeneticists regarding the possible significance of this phenomenon. Alas, they were not impressed. The prompt and unanimous verdict was: artifacts! Laboriously, I then tried to reproduce this picture by retracing every step of the staining schedule, even adding to the stain a few drops of Nescafé coffee because I remembered drinking it at the time. Nothing happened.... In retrospect, I must have used a pipette contaminated with trypsin which had been previously used for harvesting a culture of fibroblasts, hence the bands.

"Four long years went by before T Caspersson[1] discovered that quinacrine mustard (QM) had the property of producing fluorescent bands along the chromosomes. This method, which allowed the identification of each chromosome, was indeed a great step forward, but one felt that the resolution of the bands needed to be improved. Thus, among other things, I wondered if uncoiling the chromatid strands artificially before staining with QM would enhance the differentiation of the bands. I recalled that trypsin had been used in the past in order to 'relax' the coils of vicia faba chromosomes. This prompted me to try this method on human material. The first attempt resulted in a ghastly mess, but miraculously a few metaphases, which escaped the drastic effect of a very prolonged immersion in the trypsin solution, showed banding patterns similar to those observed on my ill-fated 1967 slide. And, moreover, they were comparable to the QM bands. This time I was aware of their significance! It took only a few days to establish the optimum conditions, obtain reproducible bands, standardize the method, and publish it. The immediate application was to determine the location of break points in naturally occurring chromosome rearrangements (translocations, etc.) in patients with congenital defects and to study the lesions and patterns of exchange induced by X-irradiation.[2]

"The main reasons why trypsin-banding is perhaps more widely used than other methods include its simplicity, low cost of reagents, speed of operation and, of course, the ability to produce good bands. These are important factors when choosing a technique for routine cytogenetics investigations. I have published several more recent articles in the field."[3-7]

1. **Caspersson T, Lomakka G & Zech L.** The 24 fluorescent patterns of the human metaphase chromosomes.
   Hereditas (Lund) 67:89-102, 1971.
2. **Seabright M.** High resolution studies on the patterns of induced exchanges in the human karyotype.
   Chromosome (Berlin) 40:333-46, 1973.
3. **Seabright M & Lewis G M.** Interstitial deletion of chromosome 7 detected in three unrelated patients.
   Hum. Genet. 42:223-6, 1978.
4. **Seabright M, Gregson N, Pacifico E, Mould S, Ryde J, Pearson J & Bradley A.** Rearrangements involving four chromosomes in a child with congenital abnormalities. *Cytogenet. Cell Genet.* 20:150-4, 1978.
5. **Pearson M D, MacLean N & Seabright M.** Silver staining of nucleolar organizer regions in the domestic cat, *Felis catus.*
   Cytogenet. Cell Genet. 24:247-54, 1979.
6. **Williamson E M, Miller J F & Seabright M.** Pericentric inversion (13) with two different recombinants in the same family.
   J. Med. Genet. 17:309-12, 1980.
7. **Seabright M, Gregson N M & Johnson M.** A familial polymorphic variant of chromosome 5.
   J. Hum. Genet. 17:444-6, 1980.

This Week's Citation Classic

Arrighi F E & Hsu T C. Localization of heterochromatin in human chromosomes. *Cytogenetics* 10:81-6, 1971.
[Dept. Biology, M.D. Anderson Hosp. and Tumor Inst., and Grad. Sch. Biomedical Sciences, Univ. Texas, Houston, TX]

Differentially stained chromosome areas can be obtained by subjecting cytological preparations to C banding procedures. The darkly stained areas are heterochromatin, which is composed of repeated DNA sequences. These sequences may be heterogeneous and of varying length. [The *SCI®* indicates that this paper has been cited in over 810 publications since 1971.]

Frances E. Arrighi
Departments of Molecular Biology
and Cell Biology
University of Texas
M.D. Anderson Hospital and
Tumor Institute
Houston, TX 77030

December 20, 1982

"The development of cytology into the cytogenetics that we know had a beginning, at least, in the last century. A recent highlight in this development was the successful *in situ* hybridization experiments of Pardue and Gall in 1969.[1] From their technique, the chromosome banding techniques developed that gave biologists insight into chromosome structure that was not previously possible. After learning the *in situ* hybridization technique in the laboratory of J.G. Gall, I returned to Houston to attempt to confirm their work with the mouse satellite DNA and to do other studies. Collaboration with Priscilla and Grady Saunders began. We spent considerable time isolating total and satellite DNA from mouse liver and then lost the satellite DNA. After discussing this discouraging event with Grady Saunders, we decided to try *in situ* hybridization on human DNA using repeated DNA sequences obtained by reassociation at various Cot values as the mobile component. We observed in these experiments that chromosomes had differentially stained areas and that these stained areas[2,3] also were composed of repeated sequences like those for the mouse reported by Pardue and Gall.[1] This subsequently led us to use portions of the *in situ* hybridization procedure on cells from a number of species, including rodents, bats, hoofed animals, carnivores, primates, and human beings. At the International Congress of Human Genetics in 1971, following the Paris Chromosome Standardization Conference, we reported that a satellite DNA from human cells had hybridized to the heterochromatin of human chromosome 9.[4]

"The development of these techniques has had a profound impact on diagnostic cytogenetics and the interpretation of chromosome evolution among species. Specific chromosome changes associated with congenital abnormalities and diseases can be assessed using Q, G, R, or C banding procedures. However, the function of heterochromatin and its cellular role is still unclear.

"We realized the potential usefulness of this procedure in biology and medicine, but the first journal to which the paper was submitted rejected it on the grounds that the work had no medical application. That this research work and technique have wide use and peer acceptance is a major satisfaction for us.

"These studies were completed when application for funding for basic research did not take so much time out of our day and therefore we had more time for scientific creativeness."

1. Pardue M L & Gall J G. Chromosomal localization of mouse satellite DNA. *Science* 168:1356-8, 1970.
2. Arrighi F E, Saunders P, Saunders G F & Hsu T C. Distribution of repetitious DNA in human chromosomes. *Experientia* 27:964-6, 1971.
3. Hsu T C, Arrighi F E & Saunders G F. Compositional heterogeneity of human heterochromatin. *Proc. Nat. Acad. Sci. US* 69:1464-6, 1972.
4. Arrighi F E, Getz M J, Saunders G F, Saunders P & Hsu T C. Location of various families of human DNA on human chromosomes. (de Grouchy J, Ebling F J G, Henderson I & François J, eds.) *4th International Congress of Human Genetics, Paris, 6-11 September 1971. Abstracts of papers presented.* Amsterdam: Excerpta Medica, 1971. p. 18.

This Week's Citation Classic

Noll M. Subunit structure of chromatin. *Nature* **251**:249-51, 20 September 1974.
[MRC Laboratory of Molecular Biology, Cambridge, England]

Evidence in favor of a subunit structure of chromatin is presented. The subunit is shown to contain about 200 base pairs of DNA and has been isolated by sedimentation through a sucrose gradient. As well as demonstrating the existence of a chromatin subunit, the methods described provide a means of isolating the subunit on a preparative scale and represent a test for chromatin structure. [The *SCI*® indicates that this paper has been cited in over 635 publications since 1974.]

Markus Noll
Department of Cell Biology
Biocenter of the University of Basel
CH-4056 Basel
Switzerland

December 23, 1982

"When I joined the small group working on chromatin at the Medical Research Council in Cambridge in 1973, I first discovered that the little I knew about chromatin was probably wrong. I arrived just in time for the annual lab lectures held during the first week of October. Although the lectures were captivating, covering a wide spectrum of biological structures, chromatin appeared to receive relatively little attention. However, this impression was deceiving as I soon learned from Francis Crick and Roger Kornberg at lunch. Roger proposed that chromatin is built on repetitive structural units of 200 base pairs of DNA and two each of the histones H2A, H2B, H3, and H4.[1]

"Because of its simplicity, the model was very appealing. It was consistent with experiments of Kornberg and Thomas on histone-histone interactions[2] and with results of Hewish and Burgoyne, who reported that DNA in rat liver nuclei was cleaved by an endogenous nuclease at sites multiples of a unit length apart.[3] On the other hand, the model seemed to be in conflict with the literature claiming that the four main types of histones did not occur in a simple stoichiometric relationship.

"After I had confirmed the result of Hewish and Burgoyne,[3] I started to purify the endogenous nuclease from rat liver. Soon, however, I realized that for any structural studies on a large scale, the amount of enzymatic activity that could be obtained was too small. In the hope that commercially available nucleases recognize the presumptive subunit structure, I tested DNase I and micrococcal nuclease which were in stock at the lab's stores. Whereas results obtained with DNase I were disappointing at first, micrococcal nuclease did indeed cleave chromatin at regularly spaced sites. Using this nuclease, I could demonstrate that at least 87 percent of rat liver chromatin consists of the proposed subunit structure and that the DNA length per subunit is 200 base pairs. Furthermore, I was able to isolate single subunits (later called nucleosomes) and define lengths of oligonucleosomes by sedimentation through sucrose gradients and thus provide the basis for detailed structural studies dependent on preparative amounts of nucleosomes.[4] Digestion with DNase I produced an equally startling result when the DNA fragments were analyzed in their single-stranded form in denaturing gels.[5] In contrast to micrococcal nuclease, DNase I cleaved the DNA within nucleosomes at sites spaced by the pitch of the DNA which was consistent with its regular arrangement on the outside of the histone core.[5] Digestion with either nuclease represented simple tests for chromatin structure which have since been used frequently.[6]

"The paper has probably been cited mainly for three reasons: it was the first demonstration of a well-defined subunit structure of chromatin as it had been proposed by Kornberg,[1] it described a method for the isolation of the subunits on a preparative scale, and it provided a test for chromatin structure. The work was honored by the Friedrich Miescher Award in 1978."

1. **Kornberg R D.** Chromatin structure: a repeating unit of histones and DNA. *Science* **184**:868-71, 1974.
2. **Kornberg R D & Thomas J O.** Chromatin structure: oligomers of the histones. *Science* **184**:865-8, 1974.
3. **Hewish D R & Burgoyne L A.** Chromatin sub-structure. The digestion of chromatin DNA at regularly spaced sites by a nuclear deoxyribonuclease. *Biochem. Biophys. Res. Commun.* **52**:504-10, 1973.
   [Citation Classic. *Current Contents/Life Sciences* 25(5):20, 1 February 1982.]
4. **Finch J T, Lutter L C, Rhodes D, Brown R S, Rushton B, Levitt M & Klug A.** Structure of nucleosome core particles of chromatin. *Nature* **269**:29-36, 1977.
5. **Noll M.** Internal structure of the chromatin subunit. *Nucl. Acid. Res.* **1**:1573-8, 1974.
6. **Igo-Kemenes T, Hörz W & Zachau H G.** Chromatin. *Annu. Rev. Biochem.* **51**:89-121, 1982.

This Week's Citation Classic

Olins A L & Olins D E. Spheroid chromatin units (v bodies).
Science 183:330-2, 25 January 1974.
[Univ. Tennessee, Graduate School of Biomedical Sciences, and Biology Div.,
Oak Ridge Natl. Lab., Oak Ridge, TN]

When isolated eukaryotic nuclei were spread at low ionic strength onto an electron microscope grid, they revealed a 'beads-on-a-string' appearance. This paper represented the first ultrastructural description of the repeating chromatin subunit. [The *SCI®* indicates that this paper has been cited in over 790 publications since 1974.]

Ada L. Olins and Donald E. Olins
Graduate School of Biomedical Sciences
University of Tennessee
and
Biology Division
Oak Ridge National Laboratory
Oak Ridge, TN 37830

January 6, 1983

"During the academic year 1970-1971, we enjoyed a sabbatical at the department of biophysics, King's College, London. The Pardon-Wilkins 'supercoil' model[1] of chromatin was widely accepted at that time, and we eagerly met these authors and other leading British scientists interested in chromatin. During that year, we became captivated with the beautiful electron micrographs of Howard Davies describing the ordered 30 nm chromatin 'unit threads' of avian erythrocyte nuclei.[2] Also during that year, Ruth Itzhaki[3] and Gary Felsenfeld[4] independently published evidence that a large amount of chromatin DNA was exposed. In addition, our own experiments demonstrated that isolated nuclei could be 're-versibly' decondensed by removing divalent metals and by reducing the buffer ionic strength.[5]

"Returning to Oak Ridge in September 1971, we set about to visualize the 'naked' stretches of DNA in chromatin, fully expecting to see very long regions between structures resembling supercoils We initially tried the method of critical point drying but the results were hopeless; chromatin became a tangled, twisted mess under these condi-

tions. Oscar L. Miller, Jr., was in Oak Ridge at that time, and we decided to try his method of centrifuging swollen nuclear contents onto a carbon film. For several months during the winter of 1972-1973, we accumulated micrographs of well-spread erythrocyte nuclei, never suspecting what could be revealed under a simple magnifying lens. One evening in February 1973, we happened to look closely at some negatively stained nuclei. To our excitement and surprise we saw that everywhere the chromatin looked like 'beads-on-a-string.' We called these particles *v* (nu) bodies because they were new in the nucleohistone field.

"By April 1973, we were confident of our discovery. We were teaching biophysics to graduate students at that time, and the model of hemoglobin led us to suggest the possibilities of a particle dyad axis, and of allosteric transitions. During May and June, we concentrated upon establishing that *v*-bodies could be seen in thymus and liver chromatin, and in July, we submitted our manuscript to *Science*. That summer we visited England and those scientists known to be interested in chromatin, describing the discovery and our speculations to our friends at King's College, to John Pardon and Brian Richards in High Wycombe, to Itzhaki in Manchester, and to Morton Bradbury in Portsmouth. We suggested to Davies that his unit threads might be a helical folding of *v*-bodies, rather than a folded 'supercoil.' Those were the last days before the field became charged with emotion and bristling with the claims of priority.

"In November 1973, at the annual meeting of the American Society of Cell Biology, we presented an abstract[6] on the discovery of *v*-bodies. Perusing the abstracts, we encountered a report by Chris Woodcock,[7] who had clearly been working on parallel lines. By 1974, the chromatin field had changed irreversibly. Ultrastructural and biochemical studies had converged, and the basic chromatin subunit was secure.[8] Our article in *Science* represented the publication of the first micrographs establishing the chromatin subunit (now called 'nucleosome'), coupled with speculations of its composition (two of each histone), a DNA packing ratio of 6:1, and its relationship to higher levels of chromatin. The high citation reflects its significance toward an understanding of chromosome structure and function."

1. Pardon J F & Wilkins M H F. A super-coil model for nucleohistone. *J. Mol. Biol.* 68:115-24, 1972.
2. Everid A C, Small J V & Davies H G. Electron-microscope observations on the structure of condensed chromatin: evidence for orderly arrays of unit threads on the surface of chicken erythrocyte nuclei. *J. Cell Sci.* 7:35-48, 1970.
3. Itzhaki R F. Studies on the accessibility of deoxyribonucleic acid in deoxyribonucleoprotein to cationic molecules. *Biochemical J.* 122:583-92, 1971.
4. Clark R J & Felsenfeld G. Structure of chromatin. *Nature-New Biol.* 229:101-6, 1971.
5. Olins D E & Olins A L. Physical studies of isolated eucaryotic nuclei. *J. Cell Biol.* 53:715-36, 1972.
6. Olins A L & Olins D E. Spheroid chromatin units (v bodies). *J. Cell Biol.* 59:252a, 1973.
7. Woodcock C L F. Ultrastructure of inactive chromatin. *J. Cell Biol.* 59:368a, 1973.
8. Igo-Kemenes T, Hörz W & Zachau H G. Chromatin. *Annu. Rev. Biochem.* 51:89-121, 1982.

This Week's Citation Classic™

Bonner J, Dahmus M E, Fambrough D, Huang R C, Marushige K & Tuan D Y H.
The biology of isolated chromatin. *Science* 159:47-56, 1968.
[Div. Biol., Calif. Inst. Technol., Pasadena, CA; Dept. Biol., Johns Hopkins Univ.,
Baltimore, MD; Dept. Biochem., Univ. British Columbia, Vancouver, Canada; and
Dept. Chem., Harvard Univ., Cambridge, MA]

This paper is a comprehensive review of chromatin biology and chemistry as seen in 1968, or eight years after the beginning of this science. Chromatin consists of DNA complexed with proteins known as histones. It is shown that histones are comprised of five species which are very much the same in all eukaryotic creatures and conserved in amino acid sequence to a remarkable degree, that all five histones are present in all organs of the same creature, and so forth. It is also shown that the complexing of histones to DNA in general makes the DNA a template for transcription of RNA. The paper also concerns the dissociation of chromatin into its constituent proteins and DNA and the reconstitution of the proteins and DNA into chromatin by gradient dialysis. [The *SCI*® indicates that this paper has been cited in over 445 publications since 1968.]

James Bonner
Phytogen
101 Waverly Drive
Pasadena, CA 91105

July 19, 1983

"I began work on the structure of chromatin in 1960, together with Ru-chih C. Huang.[1,2] In very short order we found that chromatin from eukaryotic organisms is readily prepared, we established criteria of purity, and we discovered RNA polymerase and partially purified it from chromatin of eukaryotic chromosomes of pea plants and rats. At that time, the field of study of isolated chromatin did not exist although some work had been done by Alfred E. Mirsky,[3] but little had been discovered about chromosomal chemistry or structure. This review paper summarizes some of our early findings, namely, that chromatin is DNA complexed with a typically chromosomal class of proteins, the histones; that chromatin serves as a very poor template for DNA synthesis by DNA-dependent RNA polymerase prepared from chromosomes, but that deproteinized DNA serves as an excellent template. Thus, one can imagine that the histones are interfering with the role of DNA as a template for RNA synthesis. We also showed in this paper that genes which are expressed in an organ are expressed in the chromatin of that organ, in the following sense. We prepared chromatin from pea plant apical buds and from pea plant cotyledons, and gave each RNA polymerase and a cell-free protein synthesizing system prepared from *E. coli* together with all of the essential amino acids, ATP, and the riboside triphosphates. This was then a chromosomally supported messenger RNA generating system, coupled to a protein translation system. We found that in the coupled system, chromatin of pea cotyledons makes messenger RNA that supports the formation of pea seed globulin, while chromatin from pea buds, which do not make pea seed globulin, does not form RNA which supports the synthesis of this protein. A very elegant experiment.

"Although the histones had been known for almost 100 years, their chemistry was quite obscure, and Huang and I decided to "grow" our own histone chemist, in the person of Douglas Fambrough. Fambrough, by early 1968, had made order out of chaos. He had shown that there are five species of histone molecules in both pea chromatin and calf thymus chromatin, and that the histones of the analogous classes from the two kinds of chromatin are very similar to one another. Later in 1968, we found by the sequencing of histone IV, the smallest and most readily purifiable of the histones, that the histones IV of pea and cow are essentially identical, each 102 amino acids long and differing only by two conservative substitutions, a valine for an isoleucine in one position and an arginine for a lysine in a second. Our review also showed that the histones of the chromatins of different organs of a given creature are all the same five classes of histones, that is, they are not organ-specific, and, as shown here, they are not species-specific.

"In 1968, with so little known of the chemistry or structure of histones, and with an emerging interest in this subject as a result of the great advances in our knowledge of DNA structure and chemistry, our review found a very great response.

"The members of my group associated with the study of isolated chromatin are all authors of this review, and their names appear in alphabetical order. Luckily, my name begins with 'B' so I'm first. Fambrough is now a staff member of the Carnegie Institution Laboratory of Embryology on the Johns Hopkins campus in Baltimore.

"There have been two reviews of the field of chromosome structure since our 1968 review."[4,5]

1. Huang R C & Bonner J. Histone, a suppressor of chromosomal RNA synthesis.
   Proc. Nat. Acad. Sci. US 48:1216-22, 1962.
2. Huang R C. Citation Classic. Commentary on *Proc. Nat. Acad. Sci. US* 48:1216-22, 1962.
   Current Contents (12):9, 20 March 1978.
3. Allfrey V G, Littau V C & Mirsky A E. On the role of histones in regulating ribonucleic acid synthesis
   in the cell nucleus. *Proc. Nat. Acad. Sci. US* 49:414-19, 1963. (Cited 545 times.)
4. Kornberg R D. Structure of chromatin. *Annu. Rev. Biochem.* 46:931-54, 1977. (Cited 705 times.)
5. Ts'o P O P, ed. *The molecular biology of the mammalian genetic apparatus.*
   Amsterdam: North-Holland, 1977. 2 vols.

Judd B H, Shen M W & Kaufman T C. The anatomy and function of a segment of
the X chromosome of *Drosophila melanogaster. Genetics* 71:139-56, 1972.
[Department of Zoology, University of Texas, Austin, TX]

A cytogenetic analysis of the organization of all the genes in a *Drosophila* chromosome segment shows essentially a one function-one polytene band relationship. This suggests that the average gene size is about 25 times larger than needed to encode an average polypeptide. [The *SCI*® indicates that this paper has been cited over 190 times since 1972.]

Burke H. Judd
Laboratory of Genetics
National Institute of Environmental
Health Sciences
Research Triangle Park, NC 27709

June 30, 1981

"The first series of experiments reported in this paper were done in 1962. I had been working on the white locus of *Drosophila melanogaster* and had given a talk to the genetics group at Texas about the interaction of white with the neighboring zeste locus. We were all stimulated by the concept of the bacterial operon that Jacob and Monod had recently formulated,[1] and I suggested that the zeste-white system might represent the eukaryotic equivalent of an operon. These loci are more than half a map unit apart on the chromosome and no mutants were known that mapped between them. Was it possible that the entire region was a huge operon-like cluster? It was from a conversation with Wilson Stone that the idea to saturate the region was generated. The rationale was to obtain a mutant representative of every gene in the segment and test each for an interaction with zeste or white.

"The first summer's work identified five complementation groups but none showed any zeste or white interactions. Margaret Shen joined me in 1964 and it was her endless patience and meticulous work that kept the project moving. By 1967, we had identified 12 groups and we knew that there was good correspondence between complementation groups and the bands in the polytene chromosomes. We were at first reluctant to accept this because there are only 5,000 to 6,000 bands and yet the amount of DNA in a *Drosophila* cell is the equivalent of about 100,000 genes the size of those in viruses and bacteria.

"Tom Kaufman became interested in the problem when he arrived as a graduate student in 1967. He set about trying to induce mutants with a series of chemical mutagens we had not previously used. Much hard work uncovered no new groups, and because of the large number of mutants we had produced, we became convinced that we were approaching saturation of the region.

"If our estimate of gene number proved correct, then gene size in *Drosophila* is very large indeed. If we were wrong, what is the nature of the genes for which we could get no mutants? Interesting questions about gene organization can be framed on either alternative, but we were confident that our efforts had indeed detected a majority of the genes in the region and we focused on why gene size was so unexpectedly large.

"To me the most satisfying aspect of the work is that it drew attention, particularly from molecular biologists, to the possibility that genes may consist of much more information than that which encodes a polypeptide. We had failed to uncover the sought after operon, only to discover that a single eukaryotic gene may, in some instances, be as large and complex as several operons or even an entire viral chromosome.

"I believe this paper is frequently cited because it reported one of the most direct measures of gene size and number in a eukaryote. It also raised questions about the organization and complexity of genes at a time when molecular analyses of DNA sequences were just beginning. Our suggestion that genes are much larger than expected because much of the information in them is used in regulation was somewhat controversial. These questions are not yet answered, but work with cloned genes makes it clear that many eukaryotic genes are large mosaic structures with sequences that contribute to a single mRNA interspersed with sequences of unknown functions."[2]

1. Jacob F & Monod J. On the regulation of gene activity. *Cold Spring Harbor Symp.* 26:193-211, 1961.
2. Axel R, Maniatis T & Fox C F, eds. *Eukaryotic gene regulation.*
New York: Academic Press, 1979. 661 p.

Stein G S, Spelsberg T C & Kleinsmith L J. Nonhistone chromosomal proteins
and gene regulation. *Science* **183**:817-24, 1974.
[Dept. Biochem., Univ. Florida, Gainesville, FL; Dept. Endocrine Res., Mayo Clinic,
Rochester, MN; and Dept. Zoology, Univ. Michigan, Ann Arbor, MI]

This review article summarizes the involvement of chromosomal proteins in the structural and functional properties of the eukaryotic genome. [The *SCI*® indicates that
this paper has been cited in over 485 publications since 1974.]

———◆———

Gary S. Stein
Department of Biochemistry and
Molecular Biology
J. Hillis Miller Health Center
College of Medicine
University of Florida
Gainesville, FL 32610

November 8, 1984

"In the early 1970s, considerable attention was focused on the nonhistone chromosomal proteins. Although it had for some
time been established that the eukaryotic
genome is a nucleoprotein complex consisting of DNA and two classes of chromosomal
proteins, histone and nonhistone proteins,
previous efforts had been largely directed
toward the structural and functional properties of DNA and histones. In July 1972, the
first Gordon Research Conference on Chromosomal Proteins was held at Beaver Dam,
Wisconsin. From the work presented at this
meeting and from the scientific literature, it
was becoming apparent that the nonhistone
proteins constituted a complex and heterogeneous class of molecules that could influence the packaging of DNA, enzymatic reactions that occur at the level of the genome,
and the expression of encoded genes.

"It was at this Gordon Research Conference that several colleagues encouraged
Lew Kleinsmith, Tom Spelsberg, and me to
write a review article for *Science* on the nonhistone proteins. The absence of a recent
review on this topic and the growing interest
in the nonhistone proteins led us to make
the decision that we would write the article.
During the course of that summer, several
lengthy telephone conversations took place
in conjunction with outlining the article.
Then, after exchanging several versions of
preliminary drafts, we got together that September in Ann Arbor, Michigan, to write the
final manuscript. To us, this was an extremely rewarding experience. It provided three
young investigators with an opportunity to
assess the development of an area, both
from a historical perspective and with respect to directions in which the field was going. Additionally, we were able to relate our
studies on the nonhistone proteins to the
studies of many other colleagues in this
field.

"Why has this paper been frequently cited? We believe that the explanation resides
largely in the general coverage of a topic
previously not reviewed in a comprehensive
manner. This has been coupled with an intense interest in the chromosomal proteins
by a broad spectrum of scientists.

"It has been rather gratifying to have followed research on chromosomal proteins
during the 12 years subsequent to writing
this review.[1-8] With the advent of recombinant DNA technology whereby the functional interactions of chromosomal proteins
with specific genes are now being examined,
our understanding of how the nonhistone
proteins mediate gene structure and gene
expression is rapidly increasing."

1. **Isenberg I.** Histones. *Annu. Rev. Biochem.* **48**:159-91, 1979. (Cited 230 times.)
2. **McGhee J D & Felsenfeld G.** Nucleosome structure. *Annu. Rev. Biochem.* **49**:1115-56, 1980. (Cited 365 times.)
3. **Felsenfeld G & McGhee J.** Methylation and gene control. *Nature* **296**:602-3, 1982. (Cited 80 times.)
4. **Weisbrod S.** Active chromatin. *Nature* **297**:289-95, 1982. (Cited 125 times.)
5. **Elgin S C R & Weintraub H.** Chromosomal proteins and chromatin structure.
   Annu. Rev. Biochem. **44**:725-74, 1975. (Cited 565 times.)
6. **Kornberg R D.** Structure of chromatin. *Annu. Rev. Biochem.* **46**:931-54, 1977. (Cited 760 times.)
7. **Stein G S & Kleinsmith L J,** eds. *Chromosomal proteins and their role in the regulation of gene expression:
   proceedings of the Florida Colloquium on Molecular Biology, March 13-14, 1975.*
   New York: Academic Press, 1975. 307 p.
8. **Stein G S, Stein J L & Kleinsmith L J.** Non-histone proteins and gene regulation. (MacClean N, Gregory F P &
   Flavell R A, eds.) *Eukaryotic genes: their structure, activity and regulation.*
   London: Butterworths, 1983. p. 31-52.

Brown D D & Dawid I B. Specific gene amplification in oocytes.
Science **160**:272-80, 1968.
[Department of Embryology, Carnegie Institution of Washington, Baltimore, MD]

The genes for 18S and 28S ribosomal RNA are amplified specifically in oocyte nuclei of amphibians forming more than a thousand nucleoli in each nucleus. These extra genes support enormous rates of ribosomal RNA synthesis during oogenesis. [The *SCI*® indicates that this paper has been cited in over 530 publications since 1968.]

Donald D. Brown
Department of Embryology
Carnegie Institution of Washington
Baltimore, MD 21210

July 7, 1983

"This paper and one published independently at the same time by Joseph Gall[1] were the first to demonstrate specific gene amplification — an event programmed into the development of a cell. The genes are those for ribosomal RNA in oocytes of the amphibian *Xenopus laevis*.

"My own involvement in this kind of research dates from 1960 when I first began to study gene expression in frog embryos at the Carnegie Institution's department of embryology in Baltimore. It had just become possible to measure RNAs as direct gene products, and the first RNAs to be purified were the stable ribosomal and transfer RNAs that comprise the bulk of cellular RNA. In 1964, John Gurdon and I found that a mutant of *Xenopus* that affects the number of nucleoli was defective in the synthesis of the two large ribosomal RNAs.[2] This confirmed that the nucleolus is the site of synthesis of ribosomal RNAs, a correlation that had already been made cytologically. Max Birnstiel, then in Edinburgh, demonstrated that this mutation was a deletion of the several hundred genes that encode these two ribosomal RNAs.[3] This crucial paper set the stage for the isolation of these genes, the very first instance of gene isolation from any living organism.

"An international meeting on the nucleolus was held in Montevideo, Uruguay, in 1965. Without a doubt, the highlight of that meeting was Birnstiel's demonstration of how he had used physicochemical techniques to isolate the ribosomal RNA genes.[4] At that conference I heard Oscar Miller, then a staff member at the Oak Ridge Laboratories, describe the presence of circular chromosomes in the many nucleoli of frog oocyte nuclei.[5] I knew instantly from the previous correlations of ribosomal RNA genes and the nucleolus that these must be extra copies of ribosomal RNA genes. Igor Dawid, a fellow staff member at Carnegie, and I set out to prove this idea.

"A key experiment described in our *Science* paper depended upon the isolation by hand of ten thousand nuclei from *Xenopus laevis* oocytes, a technique perfected by Mrs. Eddie Jordan, my colleague at Carnegie. We used buoyant density and hybridization methods to identify the amplified genes in extracts from these nuclei. Oocytes of amphibians, fish, and certain insects are known to amplify their ribosomal RNA genes, and there are now examples of amplification of genes for proteins. A phenomenon that I termed 'forced gene amplification' is a response by which cells become resistant to a drug by amplifying the gene whose product is interfered with by the drug. It is apparent that this is an important cause for resistance to chemotherapy. Our own search for other genes that might be amplified during development was negative causing us to focus on other kinds of gene control during development.

"The reference has been cited because it represents the first observation of one important mechanism for gene regulation in animal cells. Since that time, there have been numerous other examples of gene amplification in other animal cells. (For a review, see reference 6.)"

1. Gall J G. Differential synthesis of the genes for ribosomal RNA during amphibian oögenesis.
   Proc. Nat. Acad. Sci. US **60**:553-60, 1968.
   [The *SCI* indicates that this paper has been cited in over 280 publications since 1968.]
2. Brown D D & Gurdon J B. Absence of ribosomal RNA synthesis in the anucleolate mutant of *Xenopus laevis*.
   Proc. Nat. Acad. Sci. US **51**:139-46, 1964.
3. Wallace H & Birnstiel M L. Ribosomal cistrons and the nucleolar organizer.
   Biochim. Biophys. Acta **114**:296-310, 1966.
4. Birnstiel M L, Wallace H, Sirlin J L & Fischberg M. Localization of the ribosomal DNA complements in the nucleolar
   organizer region of *Xenopus laevis*. *Nat. Cancer Inst. Monogr.* **23**:431-47, 1966.
5. Miller O L, Jr. Structure and composition of peripheral nucleoli of salamander oocytes.
   Nat. Cancer Inst. Monogr. **23**:53-66, 1966.
6. Brown D D. Gene expression in eukaryotes. *Science* **211**:667-74, 1981.

Kimura M & Crow J F. The number of alleles that can be maintained in a finite population. *Genetics* 49:725-38, 1964.

If the number of possible neutral alleles is so large that each mutant is of a type not currently represented in the population, the equilibrium homozygosity is $(1 + 4N_e u)^{-1}$ where u is the mutation rate and N_e the effective population number. Formulae are also given for the heterozygosity and segregation load when all heterozygotes are selectively advantageous. [The *SCI®* indicates that this paper has been cited over 150 times since 1964.]

James F. Crow
Department of Genetics
University of Wisconsin
Madison, WI 53706

August 9, 1979

"This paper was written while Motoo Kimura was at the University of Wisconsin on leave from his regular position at the National Institute of Genetics in Japan. Earlier, he had been a graduate student with me at Wisconsin and we were continuing to work together.

"Kimura had been extremely successful, and still is, in applying diffusion methods to problems in population genetics. The Kolmogorov forward and backward equations offered solutions to a wide range of theoretical problems involving mutation, migration, and selection in the presence of random effects due to finite population size and fluctuating selection values.[1]

"By the early 1960s it was apparent that the gene had a nucleotide number of the order of 10^3, and that therefore the range of mutational possibility was enormous. This made possible the realistic, and greatly simplifying, assumption that each mutation was of a type not then existing in the population. The power of this assumption is

that it permits Sewall Wright's inbreeding coefficient F to be given an absolute meaning; 1 − F, instead of measuring heterozygosity only relative to an unknown standard, measures the absolute heterozygosity.[2] This conceptualization of the problem also permitted a simple, three line derivation that requires no advanced mathematics.

"The equilibrium formula, $F = (1 + 4N_e u)^{-1}$, where F is the fraction of homozygous loci, N_e is the effective population number, and u is the mutation rate, was not really new. The same equation had been given by Malécot, but not, I believe, with the absolute meaning.[3]

"Most of the paper consisted of Kimura's solution to the much more difficult problem when selection as well as random drift is involved. But the paper is much more often cited for the elementary neutral formula mentioned in the previous paragraph than for the much more difficult later parts.

"Why has this paper been widely cited? I can only guess, but I think there are four reasons.

"The first is the simplicity of the key formula, and the absolute meaning that it gives to Wright's inbreeding coefficient.

"The second is that the paper was written at a time of considerable discussion of the genetic load created by selectively maintained polymorphisms. This controversy has largely disappeared.

"The third is that, very soon after the appearance of this paper, the electrophoretic work of Harris and of Lewontin and Hubby demonstrated the high level of enzyme polymorphism in natural populations; this provided data to which our formula could be applied.[4,5]

"The fourth reason is the neutral theory of polymorphism, for which this is the starting point of further mathematical analysis. Of course the development of the theory and discussions of the possibility of testing and actual distribution of allele frequencies has now gone much further."

1. Kolmogorov A N. Über die analytischen Methoden in der Wahrscheinlichkeitsrechnung. *Math. Ann.* 104:415-58, 1931.
2. Wright S. The genetical structure of populations. *Ann. Eugen.* 15:323-54, 1951.
3. Malécot G. *Les mathématiques de l'hérédité.* Paris: Masson et Cie, 1948. 63 p.
4. Harris H. Enzyme polymorphisms in man. *Proc. Roy. Soc. London B* 164:298-310, 1966.
5. Lewontin R C & Hubby J L. A molecular approach to the study of genic heterozygosity in natural populations of *Drosophila pseudoobscura. Genetics* 54:595-609, 1966.

King J L & Jukes T H. Non-Darwinian evolution. *Science* 164:788-98, 1969.
[Donner Laboratory and Space Sciences Laboratory, Univ. California, Berkeley, CA]

Most evolutionary change at the DNA level is driven by mutation and drift, natural selection serving mainly to retard rates by screening out harmful changes. Total DNA and the most rapidly evolving proteins change at a rate ten times greater than average proteins; thus at least 90 percent of mutations resulting in amino acid changes are eliminated by natural selection. Genes and portions of genes most important to function are evolutionarily conserved. [The *SCI®* indicates that this paper has been cited in over 510 publications since 1969.]

Jack L. King
Department of Biology
University of California
Santa Barbara, CA 93106

April 27, 1983

"Tom Jukes came to me for help with the idea, which he had mentioned in his 1966 book, that most of the changes that have occurred in evolution at the level of protein sequences might be meaningless noise rather than adaptive change.[1] I found it easy to calculate that the rate of *neutral* evolutionary change per species per generation should be equal to the rate of selectively neutral mutations per gamete, regardless of population size.

"Before we had finished our manuscript, Motoo Kimura published the same idea in *Nature*.[2] His paper was much shorter than ours but mathematically much more sophisticated. For instance, he showed that selective neutrality did not have to be absolute; any gene with an advantage or disadvantage less than the reciprocal of the population size would be effectively neutral.

"We decided to go ahead. We buttressed the idea with observations, data, and testable predictions. For example, we predicted that in mammals, synonymous third-position changes in codons, and DNA changes in noncoding regions, would turn out to be about ten times more common than changes in other codon positions (and so

they are). We estimated that the human genome contained about 40,000 genes, with structural genes accounting for only about one percent of DNA. I held out for the provocative title, which indeed managed to provoke everyone. We were able to cite Kimura in our revision, and began what was to be a lengthy period of friendly and productive exchanges.

"*Science* rejected our manuscript. One referee said that we had merely set up and demolished a straw man and that the idea was obviously true and therefore trivial. The other said the idea was obviously false. We appealed; meanwhile our colleague, Jim Crow, was able to present most of our findings to the 1968 Genetics Congress.

"*Science's* next set of referees recommended publication. Even so, one referee objected to our suggesting that most allozyme polymorphisms were also probably neutral, so that passage had to be deleted.

" 'Non-Darwinian evolution' inflamed evolutionary biologists to action: most of the citations are reports of experiments that optimistically purported to have proved King and Jukes wrong at last. *Selectionists* felt that natural populations had enormous amounts of phenotypic variation of adaptive significance, due to polymorphism; *neutralists* felt that patterns of molecular change indicated selective neutrality. Both sides assumed that selectively neutral genes would not be expressed in the phenotype, and that adaptive change was due to a different class of genes—beneficial alleles—which were directly affected by natural selection. The controversy continued unresolved for a decade, apparently because both sides were essentially correct in the matters they were arguing about while both were mistaken in the ideas they held in common. Eventually, Kimura himself showed that the genes that determine polygenic variation may often, under stabilizing selection, have such small individual net advantages or disadvantages as to be effectively neutral.[3] Such genes determine phenotypic variation, and may bring about adaptation, by their massed effects, while their individual fates are subject only to mutation and drift. The neutral theory is now part of the accepted framework of evolutionary genetics."
[Jack L. King died on June 29, 1983.]

1. **Jukes T H.** *Molecules and evolution.* New York: Columbia University Press, 1966. p. 9.
2. **Kimura M.** Evolutionary rate at the molecular level. *Nature* 217:624-6, 1968.
   [The *SCI* indicates that this paper has been cited in over 400 publications since 1968.]
3. --------------, Possibility of extensive neutral evolution under stabilizing selection, with special reference to nonrandom usage of synonymous codons. *Proc. Nat. Acad. Sci. US* 78:5773-7, 1981.

Lindsley D L & Grell E H. *Genetic variations of* Drosophila melanogaster.
Washington, DC: Carnegie Institution, 1968. 472 p. Publication no. 627.
[Biology Division, Oak Ridge National Laboratory, Oak Ridge, TN]

This book is a compilation of all genetic variations reported for the fruit fly, *Drosophila melanogaster*, through 1967. It contains descriptive material on over 3,000 mutations, some 1,500 chromosome rearrangements, and sundry other special chromosomes and combinations of chromosomes. [The *SCI®* indicates that this book has been cited in over 1,935 publications since 1968.]

Dan L. Lindsley
Department of Biology
University of California
La Jolla, CA 92093

April 19, 1983

"In 1944, Calvin B. Bridges and Katherine S. Brehme[1] published the first compendium of known mutations and chromosome rearrangements in *Drosophila melanogaster* to appear since a review of the genetics of drosophila by Morgan, Bridges, and Sturtevant published in 1925.[2] By 1960 their volume, *The Mutants of* Drosophila melanogaster, was out of date and out of print, and there was a need for a revised edition. Sometime in 1963, when I was a member of the Biology Division of the Oak Ridge National Laboratory, I received a telephone call from Edward Novitski, then a member of the Genetics Society of America's Committee for the Maintenance of Genetic Stocks, asking me whether Ed Grell and I would undertake such a revision. We felt we could do the job within two years, so we accepted; it took us five years of full-time effort.

"This publication is not a report of a scientific experiment; rather it is a dictionary of genetic material used in drosophila research. We were plagued with the unforeseen necessity of making innumerable arbitrary decisions in matters of nomenclature and format. Although our predecessors had established conventions, we encountered many situations in which they were inadequate. Initially, we agonized over each decision, but in the end became quite cavalier in the matter. The conventions of nomenclature established in the book are in standard use today. We were fortunate in also having the editorial talents of Lucille Norton, Eileen Slaughter, and Elizabeth VonHalle, who were of enormous assistance in processing the several stages of manuscript production. The final manuscript was a gigantic deck of IBM type cards, each one containing a single line of the text, varityped in final format. These cards were then mechanically passed before a camera and photographed, producing a camera-ready negative for the printer. We chose this system because it allowed the manuscript to be easily updated by adding, subtracting, or rearranging cards; it was obsolete before we finished. Because its production was heavily subsidized by the Oak Ridge National Laboratory and it was published at cost by the Carnegie Institution of Washington, the book was initially available for $3.00 per copy.

"This volume appeared at a time when its predecessor had become unavailable; furthermore, its appearance coincided with a shift in emphasis of modern genetic research from prokaryotic to eukaryotic systems, among which *Drosophila melanogaster* was the genetically best-known example. The 1970s witnessed a tremendous resurgence of interest in drosophila as an experimental system, which was at least to some degree stimulated by the availability of our compilation, and which in turn accounts for its frequent citation. No worker can function effectively without the Red Book as we call it, and ready access to it is a *sine qua non* for aspiring drosophila geneticists. However, the explosion of information in the 14 years since its publication has dated the book; consequently, Georgianna Zimm and I have undertaken a revision. More recent work can be found in references 3 and 4. I suspect that this book was a crucial factor in my election to the National Academy of Sciences."

1. **Bridges C B & Brehme K S.** *The mutants of* Drosophila melanogaster.
    Washington, DC: Carnegie Institution, 1944. 253 p. Publication no. 552.
2. **Morgan T H, Bridges C B & Sturtevant A H.** *The genetics of* Drosophila.
    's-Gravenhage, the Netherlands: Martinus Nijhoff, 1925. 262 p.
3. **Ashburner M & Novitski E,** eds. *The genetics and biology of drosophila.* London: Academic Press, 1976.
    Vol. 1. Parts A-C.
4. **Ashburner M & Wright T R F,** eds. *The genetics and biology of drosophila.* London: Academic Press, 1980.
    Vol. 2. Parts A-D.

Cavalli-Sforza L L & Bodmer W F. *The genetics of human populations.*
San Francisco: W.H. Freeman, 1971. 965 p.
[Stanford University, Stanford, CA and Oxford University, Oxford, England]

Population genetics as applied to humans. Contents: Mendelian populations; mutation, and the elimination of deleterious mutants; coexistence of many forms of a gene (genetic polymorphisms); genetic variants detected by immunological techniques; demography and natural selection; inbreeding; consanguinity; effects of chance in evolution; continuous variation, environment versus genotype; sex chromosomes, sex ratio; human evolution, genetics, and society. [The *SCI*® indicates that this book has been cited over 565 times since 1971.]

L.L. Cavalli-Sforza
Department of Genetics
Stanford University School of Medicine
Stanford, CA 94305

March 17, 1981

"I was asked once by Josh Lederberg to give a course in genetics at Stanford in the summer of 1960 and chose a title similar to that of this book. This was the subject of the research in which he was engaged. Asked to repeat the course in 1962, my acceptance was made possible by sharing the task with Walter Bodmer. We had met in Cambridge, England, in the department of genetics, then directed by Sir Ronald Fisher, and although we had been there at different times this certainly generated a common background. The course was the foundation for the idea of writing the book that was actually published nine years after the cooperation in lecturing. During that period I was in Italy, at the University of Pavia,

and Bodmer was in Stanford. At the time the book was published, we had crossed over the Atlantic Ocean. I joined Stanford and Bodmer became a professor at Oxford. At the end of a second nine-year cycle of scientific activity there, he has become director of the Imperial Cancer Research Fund Laboratories. The book was written during times in which one of us came to Stanford, or the other to Italy, in Pavia, but more often, at a small village called San Michele di Pagana, not far from Portofino and Santa Margherita in the province of Genova.

"It is perhaps the fortunate choice of pleasant places during the time the book was written that has been responsible for its success in becoming a *Citation Classic*. Another reason is that the book was the first on a subject of importance to human geneticists, even more than to geneticists working on other organisms. The fact that practically no experimental genetics is possible on human individuals makes it necessary to resort to matings which occur in real populations. This requires a knowledge of population structure and of probabilistic treatments. Unfortunately, the mathematics necessary for a complete understanding of the subject is often above the limits of tolerance of most research workers in biology and medicine. With this problem in mind, at some later stage of preparation the book was rewritten, shifting all proofs and mathematical detail to the end of each chapter (under the prudent title of 'Worked Examples'). Did this contribute to the acceptance that the book received? The latest edition of this book appeared in 1978."[1]

1. Cavalli-Sforza L L & Bodmer W F. *The genetics of human populations.*
    San Francisco: W.H. Freeman, 1978. 965 p.

6

Cell Structure and
Function

Of the first twelve Citation Classics covered in this chapter, seven are review articles on different aspects of cancer. Review articles are often the least rewarding to write because the author must read many scientific reports, assign a priority of some type on the basis of worth, and attempt a coherent organization. Thus it was gratifying that seven reviews have become Classics, and moreover, two have been cited more than 1,000 times. The review described by Druckrey escaped some of the problems just mentioned because it was based so extensively on the research experiences of the four coauthors. Miller's report also shares this feature, although to a lesser extent. This Classic served as an avenue for the expostulation of the important discovery that enzymatic conversions *in vivo* may create a potent carcinogen from a compound that is non- or only weakly carcinogenic in its native state.

These scientific facts aside, the "This Week's Citation Classic" (TWCC) commentaries by Bullough and Shelton and at least the first paragraph and the final sentence of the TWCC by Williams should be read for their social commentary.

Although the cell receptor age is definitely upon us, only a few of the 564 TWCCs published in the Life Sciences series of *Current Contents* since 1977 can be classified in the cell receptor category. Seven of these are united here (pp. 209–215), and additional TWCCs on receptors will appear in other volumes of this series. Conception of the dual receptor system is credited to Ahlquist. His original manuscript was rejected and was later published in another journal as a result of a personal friendship with an influential editor. His idea was that the behavior of a drug, exemplified by norepinephrine, was dependent upon which of two receptors it acted upon. Likewise, a blocking agent for any agonist would be classified in the same way. The adrenergic receptors are defined as the α and β receptors, those for histamine as H1 and H2 (p. 210), and those for dopamine as D1 and D2 (p. 211). In some cases, subsets of these receptors exist (p. 212) or only one receptor is known. Borisy's tale of the colchicine receptor provides a worthy conclusion to this group of TWCCs.

Collagenase perfusion of tissues causes a dissolution of the intercellular cement and facilitates harvesting of individual biologically active cells. Rodbell applied this to the isolation of insulin-responding fat cells, Lacy and Kostianovsky applied it to the recovery of islets of Langerhans from the pancreas, and Berry and Friend applied it to the recovery of parenchymal cells from the brain. Lacy states in his TWCC that saline injection to disrupt pancreatic structure prior to collagenase infusion seemed a logical pretreatment, and Berry remarks that it was "one of the very few occasions in [his] research career when logic has been rewarded."

Interest in cell membranes received a stimulus from the review by Emmelot and co-investigators, who summarized its structure and function and identified the marker enzymes. The fluidity of the lipid bilayer described earlier by S. J. Singer was reemphasized by him and Nicolson in their review in *Science*. The capping phenomenon in lymphocytes and surface intermixing of antigens on cells served as examples of the fluid mosaic they proposed for cell membranes. Membranes that surrounded internal organelles of the cell were not being ignored. Hruban and Dallner summarize the Classics to which they contributed that delved into the behavior of intracellular membranes. The focal cytoplasmic degradation that Hruban notes following damage to cells would be equated with lysosomal degranulation today. The damage to organelles occurs because they become enclosed in a phagosome into which the degradative enzymes of lysosomes are discharged. Dallner, who served his post-doctoral with Siekevitz and Palade, published with them a dynamic study of the membrane of the rough endoplasmic reticulum. I wonder if he could walk up and down 66th Street safely in New York at night now, in order to check the birth labors of his pregnant rats at the Institute?

Among the several reviews in these Classics on the cell, that by Beams and Kessel was a key in establishing that the Golgi apparatus was a true cellular structure. As with the Golgi apparatus, it was the electron microscope that permitted a clear association between microfilaments or cytoskeletal filaments (pp. 227 and 228) of the cell and cell motility to be developed.

Pearse's great book *Histochemistry, Theoretical and Applied* is a heroic accomplishment cited over 11,500 times in its first twenty years. The missing epigram in the pirated Russian edition of his book is a feature attraction of his TWCC. It is a pleasure to note that his hope to be in the "CC Club" because of the value of one of his research publications has been met, and this Classic is the subject of the next TWCC.

The 80S maxi-ribosome better known as the polysome was discovered inadvertently when Warner, Knopf, and Rich tried unsuccessfully to force *E. coli* to synthesize poliovirus RNA. Warner's last paragraph is perhaps a truism of life, not just science.

Britton Chance's review on oxidative phosphorylation has been cited 1,705 times. Chance has been a brilliant scientist for nearly fifty years and at age seventy wrote an interesting and vivid description of his Classic.

Leblond's name appears on the next two Classics. The first describes the renewal rate of the intestinal epithelium from a histochemical approach and the second, from a radiographic view. The data in these papers are astonishing. Who could believe in 1948 or even 1960 that cells could appear, migrate along the microvilli, and be lost into the bowel lumen within 2 days or that one-fourth pound of such cells would be lost each day from the human being?

Dunn T B. Normal and pathologic anatomy of the reticular tissue in laboratory mice, with a classification and discussion of neoplasms.
J. Nat. Cancer Inst. **14**:1281-932, 1954.

The article presents a review of the hematopoietic and reticuloendothelial system, including an examination of the peripheral blood and blood-forming organs, normal anatomy, non-neoplastic and neoplastic changes, a proposed classification of tumors and a survey of the literature. [The *SCI®* indicates that this paper was cited 468 times in the period 1961-1977.]

Thelma B. Dunn
National Cancer Institute
1604 Jamestown Drive
Charlottesville, VA 22901

December 19, 1977

"When the Laboratory of Pathology at the National Cancer Institute was organized soon after World War II, members of the staff decided that each should concentrate on some laboratory species. I selected the mouse.

"I soon learned that the reticuloendothelial system was the most perplexing. I was dismayed to find how little had been published, and I realized that I would have to get the needed information for myself from the publications of others and from autopsies which I performed and the study of microscopic slides.

"After a few years I was bogged down in a chaotic mass of references, autopsy protocols, slides and photographs. I resolved to organize these into a comprehensive and reliable review. I had been trained in human pathology, and proposed to use the terminology and classification systems of medical pathology to make the work easier for others to follow.

"Finding a quiet place to work was a problem. Most laboratories are crowded with people and equipment. There is usually no chance to think or write. My assistant, Bill Stalters, found an unoccupied room in the Clinical Center which was completed but not yet fully occupied. He obtained a key and moved in a desk. We transported all of the material relating to the paper to this room. Every morning I went there and worked until noon. There was no telephone, and no one except Bill knew where I was. I think more and better papers would be written if every scientist were provided with a similar hide-away.

"The many excellent photographs made by Gebhard Gesell added greatly to the value of this paper, and the *Journal of the National Cancer Institute* placed no restriction on the number published.

"I think the frequent citation of this paper may be because it was the first comprehensive review in this field. It appeared when the virus-induced leukemias in mice were discovered and immunologic research was accelerating.

"I think there should be other reviews of this type, as I know many scientists hesitate to work with mice because there are no easily available publications on anatomy and pathology. I had hoped to write a comprehensive review on murine endocrinology and embryology which, I think, are more relevant to human cancer than molecular biology. I published a paper on the adrenal gland of the mouse, but because of failing eyesight and retirement, I got no further. I have published a book on experimental cancer research titled: *The Unseen Fight Against Cancer* which emphasizes the importance of basic research to human cancer. This did not require library work because it is based on my experience of nearly 30 years at the NCI."

Bennett J M, Catovsky D, Daniel M T, Flandrin G, Galton D A G, Gralnick H R & Sultan C. Proposals for the classification of the acute leukaemias. *Brit. J. Haematol.* **33**:451-8, 1976. [Univ. Rochester Cancer Ctr., Univ. Rochester Sch. Med. and Dentistry, NY; MRC Leukaemia Unit, Royal Postgrad. Med. Sch., London, England; Inst. Recherches sur les Leucémies et les Maladies du Sang. Hôp. Saint-Louis, Paris; Serv. Central Hématol.-Immunol., Hôp. Henry Mondor, Creteil, France; and Hematol. Serv., NIH, Bethesda, MD]

Based on an extensive study of blood and marrow films of over 200 cases of acute leukemia, seven French, American, and British hematologists (the FAB group) defined nine subgroups of leukemic morphologic variants. The originally proposed nomenclature (M1-M6 and L1-L3) has been adopted widely by numerous treatment groups. In addition, significant, nonrandom chromosome alterations (rearrangements, translocations) have been found to be associated with several of the defined FAB cell types. [Cited over 890 times, this is the most-cited paper ever published in this journal.]

John M. Bennett
University of Rochester Cancer Center
Medical Oncology Division
Rochester, NY 14642

August 14, 1984

"In 1974, G. Flandrin and I were discussing cytochemical staining of leukemia cells, and we realized that precise description of leukemic types of both acute myeloid and lymphoblastic leukemias were lacking. Further interactions with another hematologist, C. Sultan, prompted a decision to assemble a group of dedicated morphologists to discuss differences in concepts and practice in the diagnosis of the acute leukemias. The group's interest focused primarily on pathogenesis of the leukemias but also on therapeutic questions, particularly the availability to groups worldwide of a system to render data from different sources comparable. To achieve this goal, every effort was made to define each leukemia subtype as unambiguously as possible. Two extensive working meetings were held during 1974 and 1975, in which 150 cases of acute leukemia were reviewed and points of disagreement resolved. A uniformity of 85 percent agreement was achieved on a fresh set of slides, and the classification was published in 1976.

"The hematology community worldwide responded favorably. It became apparent, however, that there remained some areas of imprecision in both the myeloid and lymphoid classification schemas, which necessitated revisions. Subsequently, two additional papers appeared that described a variant type of promyelocytic leukemia and a scoring system for the two types of acute lymphoblastic leukemia (ALL)— L1, L2.[1,2] More recently, the FAB group has classified a very different group of hematologic disorders that include conditions variously described as preleukemia or subacute leukemia.[3] Five categories of 'myelodysplastic' syndromes were described, with significant differences in survival and leukemic evolution observed by us and others.

"Although it appears unlikely that substantial survival differences will be found between the major FAB myeloid subtypes, there is evidence that hypergranular promyelocytic leukemia (FAB M3) is associated with prolonged remission duration and acute monocytic leukemia (AML: FAB M5) with a short response duration and a high likelihood of central nervous system leukemia.[4] Of potentially greater importance is the association of certain myeloid subtypes with nonrandom chromosome abnormalities: FAB M2 (AML with maturation) with t(8:21) in 20 percent of cases; FAB M4Eo (acute myelomonocytic leukemia with eosinophilia) with C16 inversion; FAB M3 with t(15:17); and FAB M5 with chromosome 11 abnormalities.

"Concerning ALL in children, numerous groups have confirmed the prevalence of the L1 subtype over L2 and L3, whereas in adults there are more L2 cases, particularly in patients over age 30.[2] Moreover, patients with L2 morphology have a significantly shorter survival than those with L1, and those with L3 ('Burkitt cell' or B-cell ALL) have the worst survival (less than one year).[5] The only consistent chromosome abnormality associated with FAB lymphoid subtypes has been the translocation in FAB L3 of the c-myc gene locus translocated from chromosome 8 to the immunoglobulin gene locus on chromosome 14.[6]

"A decade ago, the FAB members could not predict the ultimate significance of our leukemia classification proposals. Intuitively, we believed that precise definitions of morphologic subtypes of the acute leukemias and myelodysplastic syndromes, incorporating understandable terms and excellent photomicrographs, would be of value to researchers. Overall, the acceptance of the 'proposals' indicates that the provision of an international classification provides a basis for useful discussion and comparative research. Certainly, we have not been without our critics, and their comments have led to appropriate responses and clarifications."

1. **Bennett J M, Catovsky D, Daniel M T, Flandrin G, Galton D A G, Gralnick H R & Sultan C.** A variant form of hypergranular promyelocytic leukemia (M3). *Ann. Intern. Med.* **92**:261, 1980.
2. ------------------, The morphological classification of acute lymphoblastic leukaemia. Concordance among observers and clinical correlations. *Brit. J. Haematol.* **47**:553-61, 1981.
3. ------------------, Proposals for the classification of the myelodysplastic syndromes. *Brit. J. Haematol.* **51**:189-99, 1982.
4. **Bennett J M & Begg C B.** Eastern Cooperative Oncology Group study of the cytochemistry of adult acute myeloid leukemia by correlation of subtypes with response and survival. *Cancer Res.* **41**:4833-7, 1981.
5. **Bloomfield C D.** The lymphocyte in clinical medicine: classification and prognosis of acute lymphoblastic leukemia. (Sell K W & Mill W V, eds.) *The lymphocyte.* New York: Liss, 1981. p. 167-83.
6. **Rowley J D.** Human oncogene locations and chromosome aberrations. *Nature* **301**:290-1, 1983.

Abercrombie M & Ambrose E J. The surface properties of cancer cells: a review.
Cancer Res. **22**:525-48, 1962.
[University College and Chester Beatty Res. Inst., Inst. Cancer Research, London, England]

The major deviations from normal of malignant cells—local invasion, metastasis, disorganisation and persistent growth—were considered as disorders of the cell surface; and the physical aspects of adhesiveness, locomotion and contacts of cells, and the biochemistry of their membranes, were discussed as background to these deviations. [The *SCI®* indicates that this paper has been cited over 470 times since 1962.]

Michael Abercrombie
Strangeways Research Laboratory
Worts' Causeway
Cambridge CB1 4RN
England

May 19, 1980

"The literature bearing directly on the surface properties of cancer cells is now so large that it is hard to think back to how little we had to go on when we wrote our review. There was the pioneering work of Dale Coman on the mutual adhesiveness of malignant cells.[1] There was the work of my own group suggesting a defect in contact inhibition of locomotion in malignant cells[2] (the virologists had not yet brought contact inhibition of multiplication on to the scene), studies of the surface charge of tumour cells by cell electrophoresis, begun by Ambrose and his colleagues,[3] were active at the time; and there had been a few investigations of the tumour cell surface by electron microscopy. Indeed, little was known about any aspect of the cell surface. The technical step that was to be so important, the isolation of the plasma membrane, had only just been taken by Neville[4] and the Herzenbergs.

"Nevertheless, sparse as were the data, an underlying mood of interest in the cell surface was abroad, doubtless owing a lot to the writings of such great figures in developmental biology as Johannes Holtfreter[5] and Paul Weiss.[6] The mood was felt within the world of cancer, and the editors of *Cancer Research* asked Ambrose for a review about the potentials of the field. We were both then in London, Ambrose at the Chester Beatty Research Institute and I at University College, and we had been meeting regularly for some years to talk about the cell surface. So Ambrose invited my collaboration in the review, and we decided to see what sort of case we could make for the supposition that the manifestations of malignancy were essentially the expression of a change in the cell surface, something not till then attempted. I led off with the biological evidence, putting most of the emphasis on invasion and metastasis, about which there was most to say, and Ambrose then reviewed the physical aspects that might be involved in explanation.

"The review appeared, and there quickly followed a surge of research on the cell surface in general, and its part in malignancy in particular, which has been mounting ever since. It was not, regrettably, cause and effect. Certainly we hope that our review encouraged the outburst a little, and some cancer research workers have told me that it did turn their thoughts towards the surface. It did not, as I had hoped, turn thoughts noticeably towards the problems of invasion and metastasis, which I still think, as I thought then, are relatively speaking badly neglected in cancer research. I believe in fact that the review was much less influential than its citation rate suggests. A realistic assessment would probably find that it served more as a kind of emblem of the newly popular idea than as a source of it. Whenever an author felt that he had to justify an experimental paper, with a ringing declaration that the cell surface is now considered to be important in the malignant transformation, there was our review conveniently ready to support him. It had been perfectly timed by the editors of *Cancer Research*.

"A more recent review of some of the biochemical aspects of the surface of transformed cells has been published by R.O. Hynes."[7]

1. Coman D R. *Cancer Res.* **4**:625-9, 1944.
2. Abercrombie M, Heaysman J E M & Karthauser H M. *Exp. Cell Res.* **13**:276-92, 1957.
3. Ambrose E J, James A M & Lowick J H B. *Nature* (London) **177**:576-7, 1956.
4. Neville D M, Jr. *J. Biophys. Biochem. Cytol.* **8**:413, 1960.
5. Holtfreter J A. *Ann. NY Acad. Sci.* **49**:709-60, 1948.
6. Weiss P. *Int. Rev. Cytol.* **7**:391-423, 1958.
7. Hynes R O. *Biochim. Biophys. Acta* **458**:73-107, 1976.

Citation Classics

Baserga R. Relationship of cell cycle to tumor growth and control of cell division; a review. *Cancer Res.* **25**:581-95, 1965.

The review article brings out the advantages of studying tumor growth in terms of cell cycle kinetics and opens the possibility of a new field of endeavor in biochemistry and cell proliferation. [The *SCI*® indicates that this paper was cited 236 times in the period 1965-1977.]

—————◆—————

Renato Baserga
Temple University
Health Sciences Center
Philadelphia, PA 19140

December 8, 1977

"Since this was a review article, there was no startling idea that suddenly struck me, and the only obstacles I had to overcome were the comments of the reviewers. The idea of writing a review relating our recently acquired knowledge of the cell cycle to tumor growth came to me as I was moving from Northwestern University in Chicago to Temple University in Philadelphia. It seemed to me the right moment to make the point on the direction that my research was taking.

"I had started out doing research 13 years before, in Chicago, and at that time I wanted to find out the mechanisms in the metastatic spread of tumors. In a few years I came to the conclusion that there was very little mystery about the metastatic spread of tumors, and that the problem with metastases was essentially a problem of cell proliferation. That is, that tumor cells grow indefinitely while normal cells do not. Thus I slowly drifted from a study of metastasis into a study of the mechanisms that control cell proliferation.

"This was about the time when the studies on the cell cycle were receiving strong impulses from the discovery of the wonderful uses that one could make of (^{3}H)-thymidine.

For a number of years, though, the cell cycle had been somehow the personal property of radiobiologists and I happened to stumble into it through my association with the Argonne National Laboratory. Cancer researchers were little aware of the cell cycle and its possible implications for tumor growth. In fact, a number of leading investigators in radiobiology simply refused to investigate the cell cycle of tumors on the ground that it was too difficult and complicated.

"Mort Mendelsohn and myself were probably the first to have the courage to tackle tumor growth in terms of cell cycle, and we found that tumors were amenable to a kinetic analysis. About this time, though, I felt that a kinetic analysis, while descriptive, really did not explain the basic mechanisms that control cell proliferation, and I therefore conceived the general idea that the cell cycle should be put in biochemical terms, rather than in purely kinetic terms. In writing the review, indicated above, I had exactly these two things in mind: 1) to show how the growth of tumors could be understood in terms of cell cycle kinetics, and 2) to point out that in the last analysis our understanding of life processes depends on our understanding of the underlying biochemistry.

"This review, therefore, was an attempt to open new fields of scientific endeavor. Since then our knowledge of the cell cycle has been extensively applied to basic and clinical studies of cancer drugs, and a whole field has developed on the biochemistry of cell proliferation that has been the object of a number of symposia, many reviews and even books. Personally I remember that when I was writing that review I felt elated about how much more we knew about cell division in 1965, than we did in 1950 when I started to be interested in biomedical research. Now in rereading that review I again feel how much more we know about cell division than we knew at that time, which is, I guess, another way of saying how little we know at any point in time, and how much more there is to know in the future."

Bullough W S. Mitotic and functional homeostasis: a speculative review.
Cancer Res. **25**: 1683-727, 1965.

This review put forward a new concept of the control of tissue growth by mitotic inhibitors, called chalones. Slowly this concept has become generally accepted, and it has also proved to have a significant bearing on the understanding of cancer. [The *SCI*® indicates that this paper was cited 249 times in the period 1965-1976.]

William S. Bullough
Birkbeck College
University of London
London, WC1E 7HX

December 13, 1977

"In retrospect the surprising popularity of this review may have had several causes. For one thing it was particularly comprehensive, which is always valuable, although someone, who much later confessed to being one of the referees, complained that great stamina was needed to read it. The comparable stamina needed to write it derived from the excitement of our previous discovery that cell division is controlled by tissue-specific mitotic inhibitors, christened 'chalones,' and not as was previously believed by mitotic stimulants. The main interest of the review thus arose from its complete re-interpretation of the available evidence on tissue growth in both normal and pathological conditions.

"This new angle of approach also made the review controversial, which may have added to its attraction. The old beliefs were so deeply entrenched that skepticism of the new ones was widespread; even *Cancer Research* was nervous and covered itself by adding the words 'a speculative review' to my original title! It has indeed been enlightening to experience for oneself how resistant scientists in general are to new ideas, and it has taken more than 10 years for the chalone concept to become generally accepted.

"Another obvious feature of the ideas expressed in the review was the light they might shed on the problem of cancer. For me part of the excitement of those days lay in the question: can tumour cells also be inhibited by the chalone of their tissue of origin? The unlikely answer soon proved to be 'yes,' which caused the following poetic outburst:

> We have found in murine urine
> And in hominal urinal
> A substance chalonal. Euphonal!
> A tissue controller, and droller,
> An answer to cancer!

However, it must now be added that urine is not to be recommended as a practical chalone source.

"The evidence linking chalones to cancer has gradually become overwhelming, and this too may have ensured a continuing interest in the original review. However, my attempt at a follow-up review of this evidence has proved too much for the nerves of *Cancer Research,* who this time have declined to publish it. Again one meets the 'new-idea syndrome,' which is particularly regrettable in a subject that is so short of original ideas. Fortunately, recent successful clinical experiments directed by my colleague Dr. T. Rytömaa, who has used a chalone against myeloid leukemia, may bring the ideas put forward in the 1965 review to a conclusion perhaps as significant in its own way as was the first use of hormones against hormone-deficiency diseases."

Wheeler G P. Studies related to the mechanisms of action of cytotoxic alkylating agents: *a review*. *Cancer Res.* 22:651-88, 1962.
[Kettering-Meyer Laboratory, Southern Research Institute, Birmingham, AL]

The purpose of this review was to present information from the published literature that related to the chemical and biochemical modes of action of five types of cytotoxic alkylating agents — mustards, ethylene-imines, sulfonic esters, epoxides, and N-alkyl-N-nitrosoureas. This information fell into the following categories: distribution and fate of the agents; effects upon glycolysis, respiration, and related processes; reaction with macromolecules; effects upon the properties and functioning of macromolecules; and antimitotic, cytologic, and mutagenic effects. [The *SCI®* indicates that this paper has been cited over 260 times since 1962.]

Glynn P. Wheeler
Kettering-Meyer Laboratory
Southern Research Institute
P.O. Box 3307-A
Birmingham, AL 35255

August 28, 1981

"The initial recognition of the potential use of alkylating agents as anticancer agents arose from studies of the biological, toxicological, pharmacological, and biochemical studies of the chemical warfare agents, the nitrogen mustards, during World War II. F.S. Philips reviewed these studies and others that were reported prior to 1950,[1] and A. Gilman reminisced about the first test of these agents as anticancer agents.[2] After the end of hostilities, C.P. Rhoads recruited some men who had been with him in the Chemical Corps and National Defense Research Council (Philips, D.A. Karnofsky, and C.C. Stock) to join him at Sloan-Kettering Institute for Cancer Research, where they continued their studies of cancer chemotherapy. Another member of the Chemical Corps team, H.E. Skipper, joined the staff of the newly founded Southern Research Institute, where he also initiated studies of cancer chemotherapy. The con-

tinuing interest of the Chemical Corps in the mustards and Skipper's experience in conducting field tests with them during the war resulted in the Corps contracting Southern Research Institute in the early 1950s to study further the chemistry and biochemistry of these agents. It was at this point that I, as a member of the biochemistry division under the direction of Skipper, became involved in studies of the biological alkylating agents.

"By the 1960s, many studies on the nitrogen mustards had been carried out in many laboratories, and several other types of alkylating agents had been found to have beneficial activity against experimental and clinical cancer. Many reports of the research on these agents had been published, but the reports were dispersed among many scientific journals. I was invited by the Cancer Chemotherapy National Service Center, with the concurrence of Skipper, to write a review article that would consolidate available information relating to the mechanisms of action of these agents. The above cited review was the product. The breadth of interest in the agents was reflected in the fact that we received 310 requests for reprints from persons in 35 states in this country and 228 requests from persons in 31 other countries. These figures and the number of citations to the article during subsequent years indicate that it did indeed fulfill the need for an updated, broad review. The fact that it was published in a widely circulated journal made it readily available to interested investigators.

"In the same year that the above review was published, a book[3] that covered much of the same material and covered some of it in much greater detail was published. During subsequent years a number of other authors have written updated reviews of work on these agents that supplement and complement the other reviews[4]

"It is an inherent property of reviews that they become obsolete. Although the facts they report endure, the interpretations of those facts must be altered or eliminated as new facts and understanding arise. Nevertheless, the older reviews evidently retain some lasting value, as this review is still occasionally cited in the current literature."

1. Philips F S. Recent contributions to the pharmacology of bis(2-haloethyl) amines and sulfides. *Pharmacol. Rev.* 2:281-323, 1950.
2. Gilman A. The initial clinical trial of nitrogen mustard. *Amer. J. Surg.* 105:574-8, 1963.
3. Ross W C J. *Biological alkylating agents.* London: Butterworths. 1962. 232 p.
4. Sartorelli A C & Johns D G, eds. *Antineoplastic and immunosuppressive agents.* Berlin: Springer-Verlag. 1975. Part II. Section C. p. 1-84.

Citation Classics

Earle W R, Schilling E L, Stark T H, Straus N P, Brown M F & Shelton E. Production of malignancy in vitro. 4. The mouse fibroblast cultures and changes seen in the living cells. *J. Nat. Cancer Inst.* 4:165-212, 1943.

This paper describes one of the earliest successful attempts to produce "cancer in a test-tube." Fibroblasts derived from the subcutaneous adipose connective tissue of a C₃H mouse were treated *in vitro* with 20-methylcholanthrene. Injection of the cultures into mice produced tumors. The first description of spontaneous malignant transformation of cells in culture was spawned when the control cultures also produced tumors. [The *SCI®* indicates that this paper was cited 601 times in the period 1961-1977.]

Emma Shelton
National Cancer Institute
Department of Health, Education and
Welfare
Bethesda, MD 20014

January 18, 1978

"This was Wilton Earle's magnum opus. The rest of us, whose names appear after his, lent but our willing hands to his project and he graciously placed us at the beginning of the article rather than the end. To have his article listed as a Citation Classic would be for him an encomium close to the Nobel Prize, for to him quantity was the ultimate essence of life.

"His project was conceived in the days when only a small coterie of scientists was engaged in cancer research, and it was a period when almost all of them believed that if normal cells could be converted to cancer cells by exposure to a pure chemical, the solution to the riddle of cancer was virtually assured. Wilton Earle set about to achieve that goal, constructing on the way an elaborate tissue culture suite rivaled, but not surpassed, only by the former laboratory of Alexis Carrel at the Rockefeller Institute.

These were the days when cells were grown in plasma clot and sustained by a soup consisting of an extract of macerated chick embryos and horse serum diluted in a specially formulated saline solution. It was an heroic effort to grow cells in culture. Chickens had to be bled for plasma (and were subsequently eaten by us, now it can be told, because it was World War II and meat was hard to get), horses bled for serum, 10-day-old chick embryos removed from the shell and macerated, and salts weighed out, dissolved and filtered. All this had to be done aseptically, for antibiotics were yet to come. We lost very few of the thousands of cultures to bacterial contamination.

"Wilton realized his goal when the carcinogen-treated cells altered their morphology and growth characteristics and ultimately produced tumors when injected into mice. But his triumph was short-lived for, in spite of his exquisite precautions, which stopped just short of bathing his technicians in acid, some of his control cultures transformed to the malignant state, and the riddle of cancer, instead of being solved, became more arcane than ever. Wilton was convinced until the day he died in 1964 that trace contamination with chemical carcinogen was the explanation for the transformation of the control cells. In spite of all our knowledge, the explanation still eludes us.

"Why has this paper been cited so frequently? Not because of the significance of its contents as outlined above, but because it describes the origin of the famous "L" cells that are now grown in laboratories around the world. Tissue culture techniques have changed so that even high school students can grow them now. At this writing, Strain L is 13,610 days old. Wilton would have loved that."

Miller E C, Miller J A & Hartmann H A. N-hydroxy-2-acetylaminofluorene: a metabolite of
2-acetylaminofluorene with increased carcinogenic activity in the rat.
Cancer Res. 21:815-24, 1961. [McArdle Memorial Laboratory for Cancer Research
& Department of Pathology, Medical School, University of Wisconsin, Madison, WI]

N-hydroxy -2 -acetylaminofluorene (N-hy-droxy-AAF), a major metabolite of 2-acetyl-aminofluorene (AAF) in the rat, was more carcinogenic than the parent amide in this species. The data provide strong evidence that N-hydroxy-AAF is a proximate carcinogen in the induction of tumors by AAF. [The _SCI®_ indicates that this paper has been cited over 240 times since 1961.]

Drs. Elizabeth C. Miller and
James A. Miller
McArdle Laboratory for
Cancer Research
University of Wisconsin Center
for Health Sciences
Madison, WI 53706

December 21, 1977

"This paper, in which we were fortunate to have the collaboration of Dr. Henrik A. Hartmann of the department of pathology of our medical school, was particularly exciting to us, since it reported the attainment of an intermediate goal of our research, i.e., the finding of a metabolite of a carcinogen that was more active than the parent compound. More importantly, it also provided a focus for future work on the elucidation of the mechanisms involved in the induction of tumors by AAF and other chemicals. On the other hand, this paper was just one of the series of papers from our laboratory and those of other investigators that, bit by bit, have led to a rather comprehensive knowledge of the metabolism, both in activation and deactivation, of chemical carcinogens and the means by which their metabolites can affect the functions of the target cells.

"Our interest in metabolic activation of chemical carcinogens stemmed from our studies in 1947 on the observation that hepatocarcinogenic dyes, though chemically unreactive, became bound covalently _in vivo_ to proteins in the liver of the rat. The amount of dye binding produced by different dyes appeared to correlate approximately with their hepatocarcinogenic activities. Subsequent studies in many labora-tories showed that covalent binding _in vivo_ of carcinogen residues to proteins and nucleic acids in target tissues was a common property of chemical carcinogens. These bound forms appear to be a necessary but not sufficient requirement for tumor formation. The facts that most chemical carcinogens are not reactive chemically and are carcinogenic only in certain tissues made it apparent that metabolism of the carcinogens to reactive intermediates must occur _in vivo_.

"Our efforts to find metabolites of the aminoazo dyes with increased carcinogenic activity failed,[1] and we turned to a study of the versatile carcinogenic amide AAF. With Dr. John W. Cramer (now professor of pharmacology, University of Nevada School of Medical Sciences), then a post-doctorate in our group, we observed a new urinary metabolite of AAF in the rat and found that the amounts excreted increased with the continued feeding of AAF for several weeks.[2] Isolation and characterization of this new metabolite showed that it was N-hydroxy-AAF, a new kind of metabolite of an aromatic amide. Once we had learned how to synthesize this compound, we administered it to rats. To our amazement it was more carcinogenic than AAF at the usual sites of tumor formation such as the liver, mammary glands, small intestine, and ear duct glands. In addition, it was car-cinogenic in the skin, subcutaneous tissue, forestomach, and peritoneum where AAF was inactive. _In vivo_ N-hydroxy-AAF gave rise to higher levels of macromolecule-bound AAF residues than did AAF.

"Subsequent work in many laboratories has shown that N-hydroxylation is an obligatory step in carcinogenesis by aromatic amides and amines. However, N-hydroxy-AAF was not reactive toward proteins and nucleic acids _in vitro_, and it finally became evident that a second step was required in the metabolic activation of AAF to an ultimate carcinogen. The ultimate car-cinogenic metabolites of most if not all chemical carcinogens now appear to be strongly reactive electrophilic and mutagenic compounds, but this is the story of another decade in research in chemical carcinogenesis."[3]

1. Miller J A & Miller E C. The carcinogenic aminoazo dyes. _Advan. Cancer. Res._ 1:339-96, 1953.
2. Cramer J W, Miller J A & Miller E C. N-hydroxylation—a new metabolic reaction observed in the rat with the carcinogen
2-acetylaminofluorene. _J. Biol. Chem._ 235:885-8, 1960.
3. Miller E C & Miller J A. The metabolism of chemical carcinogens to reactive electrophiles and their possible mechanisms
of action in carcinogenesis. _American Chemical Society Monograph No. 173: Chemical Carcinogens._ (Searle C E.
ed.) Washington. DC: American Chemical Society. 1976. p. 737-62.

Druckrey H, Preussmann R, Ivankovic S & Schmähl D. Organotropic carcinogenic effects of 65 different N-nitroso-compounds in BD-rats.
Z. Krebsforsch. **69**:103-201, 1967. [Forschergruppe Präventivmed., Max-Planck-Inst. Immunol., Freiburg, Federal Republic of Germany]

Almost all tested N-nitroso-dialkylamines and -acylalkylamides revealed striking organospecific carcinogenic effects, clearly dependent on chemical structure. The regular induction of cancer of the oesophagus, lungs, urinary bladder, or brain is described, providing reliable experimental models. [The *SCI®* indicates that this paper has been cited over 1,040 times since 1967.]

Hermann Druckrey
Reinhard-Booz-Str. 18
D 7802 Freiburg-Merzhausen
Federal Republic of Germany

April 6, 1981

"These systematic studies are based on earlier quantitative experiments with the liver-carcinogen 4-dimethylaminoazobenzene, known as 'butter yellow.' It was observed that at lower daily dosage the induction time increased up to the life span of rats. The total dose required did not increase, but became significantly smaller.[1] This surprising result indicated that the primary effects of all individual doses are not only irreversible but also transmitted to daughter cells and, thereby, 'genotoxic.'

"After the war, an otherwise precarious situation permitted a unique full-time cooperation with the electrophysicist K. Küpfmüller in elaborating the theoretical pharmacokinetic and ergokinetic bases for both the understanding of the former results and the systematic planning of new experiments. Their practical performance was then supported by the Deutsche Forschungsgemeinschaft.

"The great chance for comprehensive studies came in 1956, when the carcinogenicity of dimethylnitrosamine was discovered by P. N. Magee and J. M. Barnes.[2] The simple molecule, ON-N $(CH_3)_2$, is enzymatically demethylated in the liver, yielding an alkylating intermediate, which explains the production of liver cancer. However, preliminary experiments of D. Schmähl and R. Preussmann demonstrated that diethylnitrosamine has the same effect. This suggested that the first step in metabolic activation is probably an enzymic hydroxylation at one alpha C-atom, which then automatically leads to dealkylation.

Accordingly, all the various dialkylnitrosamines, susceptible to alpha C-hydroxylation, should be carcinogenic. Furthermore, since some of the hydroxylases might be organo- or substrate-specific, specific effects could also be expected. The idea was extremely fascinating, first, because organospecific effects, so fundamental in all pharmacology, are most easily recognizable in carcinogenesis and, second, because the group of N-nitroso-compounds is extraordinarily versatile, permitting an easy synthesis of innumerable derivatives and, thereby, systematic studies on relations between chemical structure and both efficacy and organospecificity. Expectations were surpassed by the experimental results. Surprisingly, striking organotropic effects were likewise observed with alkyl-acyl-nitrosamides, which don't need any enzymic activation to yield an alkylating intermediate. For example, with methylnitrosourea we succeeded for the first time in inducing cancer in the central organ, the brain, with an accuracy formerly considered unimaginable.

"The results of quantitative studies on the relations between dose (d) and induction time (t), extensively performed with diethylnitrosamine, confirmed that carcinogenesis obeys so simple a formula as $d \, t^n = const$. This is indeed perplexing, but natural regularities are necessarily simple. Since the numeric value of n was always greater than two, it corresponds to an accelerated process. Accordingly, the induction of cancer by one single dose or 'impulse' was demonstrated on several examples and, most convincingly, in transplacental experiments with ethylnitrosourea, mainly performed by S. Ivankovic. Although the dose was small and not toxic to the pregnant rat, practically all offspring later died with cancer of the brain and nervous system, which demonstrated the extreme vulnerability of the nervous organs during prenatal development.

"The worldwide interest the paper still enjoys is not surprising for four reasons: first, it provided reliable experimental models for almost all types of cancer; second, the biochemical mechanism of action is transparent; third, there are still numerous N-nitroso-compounds to be studied; and, last but not least, these carcinogens are easily formed by reaction of alkylamines or -amides with nitric acid, widely distributed in the human environment. It is hoped that the paper will contribute to the main goal: cancer prevention. More recent work is reported in 'Chemical carcinogenesis on N-nitroso derivatives.'"[3]

1. Druckrey H & Küpfmüller K. Quantitative Analyse der Krebsentstehung. *Z. Naturforsch.* **3**:254-66, 1948.
2. Magee P N & Barnes J M. The production of malignant primary hepatic tumours in the rat by feeding dimethylnitrosamine. *Brit. J. Cancer* **10**:114-22, 1956.
3. Druckrey H. Chemical carcinogenesis on N-nitroso derivatives. *GANN Monogr. Cancer Res.* **17**:107-32, 1975.

Miller J A. Carcinogenesis by chemicals: an overview—G.H.A. Clowes Memorial
Lecture. *Cancer Res.* 30:559-76, 1970.
[McArdle Lab. for Cancer Res., Univ. Wisconsin Med. Ctr., Madison, WI]

Chemical carcinogens comprise a wide variety of organic and inorganic compounds that share no common structural features. Recent work indicates that most, if not all, chemical carcinogens are, or are metabolized to, ultimate carcinogenic reactive and mutagenic electrophiles that bind covalently to informational macromolecules involved in the initiation of carcinogenesis by these agents. [The *SCI®* indicates that this paper has been cited over 1,100 times since 1970.]

James A. Miller
McArdle Laboratory for
Cancer Research
University of Wisconsin
Medical Center
Madison, WI 53706

September 9, 1981

"G.H.A. Clowes was director of research at Eli Lilly and Company for many years. He was a founder of the American Association for Cancer Research in 1907 and an early investigator of normal and malignant growth. The Clowes Memorial Lectureship Award, sponsored by Eli Lilly and Company, is awarded annually by the American Association for Cancer Research. The honor of being the ninth lecturer in this series is one that I share fully with my wife and co-worker, Elizabeth Cavert Miller. Our work together forms the greater part of this review. We owe very much to our many collaborators over the years, and especially to Harold P. Rusch, who founded and was director of the Mc-

Ardle Laboratory for Cancer Research for 33 years. His selfless work established a stable, stimulating, and productive research environment at the McArdle Laboratory.

"In this Clowes Lecture I reviewed research on the molecular mechanisms of chemical carcinogenesis, especially for the decade prior to 1969, the year of the award. This decade was particularly exciting to Elizabeth and me for it encompassed our finding of the first proximate carcinogenic metabolite, N-hydroxy-2-acetylaminofluorene,[1] and our realization, conceptually and experimentally, that this and many other carcinogens are converted metabolically to electrophilic ultimate mutagenic carcinogens.[2,3] Thus it appears that most, if not all, chemical carcinogens are, or are converted *in vivo* to, strong electrophiles as ultimate reactive and mutagenic carcinogens. This finding explains the great structural variety of chemical carcinogens and the nature of the covalent binding *in vivo* of residues of chemical carcinogens to DNA, RNA, and proteins, especially in target tissues. Much work in many laboratories in the past decade has extended these concepts to a wide range of synthetic and naturally occurring chemical carcinogens.[4]

"We were surprised that so many reprints of this review were requested by investigators in such a wide range of disciplines. This probably resulted in part from the increased concern and interest about chemical carcinogens in the environment and their control in modern life. Chemical carcinogens had, in a sense, 'arrived.' We well remember how lonely a field chemical carcinogenesis in the US was in our early research years."

1. Miller E C, Miller J A & Hartmann H A. N-hydroxy-2-acetylaminofluorene: a metabolite of 2-acetylaminofluorene with increased carcinogenic activity in the rat. *Cancer Res.* 21:815-24, 1961.
   [Citation Classic. *Current Contents/Life Sciences* 22(27):14, 2 July 1979.]
2. Miller J A & Miller E C. Metabolic activation of carcinogenic aromatic amines and amides via N-hydroxylation and N-hydroxyesterification and its relationship to ultimate carcinogens as electrophilic reactants. (Bergmann E D & Pullman B, eds.) *The Jerusalem symposia on quantum chemistry and biochemistry. Physiochemical mechanisms of carcinogenesis.* Jerusalem: Israel Academy of Sciences and Humanities, 1969. Vol. 1. p. 237-61.
3. Maher V M, Miller E C, Miller J A & Szybalski W. Mutations and decreases in density of transforming DNA produced by derivatives of the carcinogens 2-acetylaminofluorene and N-methyl-4-aminoazobenzene. *Mol. Pharmacol.* 4:411-26, 1968.
4. Miller E C. Some current perspectives on chemical carcinogenesis in humans and experimental animals: presidential address. *Cancer Res.* 38:1479-96, 1978.

This Week's Citation Classic

Heidelberger C. Chemical carcinogenesis. *Annu. Rev. Biochem.* **44**:79-121, 1975.
[McArdle Laboratory for Cancer Research, University of Wisconsin, Madison, WI]

Chemical carcinogenesis was reviewed from the perspective of a biochemist. Emphasis was on epidemiological evidence that chemicals in our environment are a major cause of human cancer; that most carcinogens require metabolic activation; and that new cell culture model systems were evolving. The most plausible mechanisms were emphasized. [The *SCI®* indicates that this paper has been cited over 610 times since 1975.]

Charles Heidelberger
Comprehensive Cancer Center
Kenneth Norris Jr. Cancer
Research Institute
and
School of Medicine
University of Southern California
Los Angeles, CA 90033

December 10, 1981

"It was mostly a matter of timing. *Annual Review of Biochemistry* has had its coterie of faithful readers over the decades, and the subject of chemical carcinogenesis had not been reviewed in that publication since 1959—and has not been since my 1975 review.

"The earlier review was written by my friends and colleagues Elizabeth and James Miller at the McArdle Laboratory for Cancer Research, University of Wisconsin.[1] I was fortunate to have spent many years in that laboratory, which, then and now, is considered to be one of the leading centers in the world for research in chemical carcinogenesis.

"In 1974, when I was invited to write the review, the field had undergone a number of exciting developments that clearly established the great importance of chemical carcinogenesis in the causation of human cancer and that unraveled some of the tangled complexities of mechanisms of action. These advances were recognized by many cancer researchers, but had not yet penetrated the consciousness of most biochemists.

"The first of these developments involved the increased recognition by epidemiologists that chemicals in our external and internal environments constitute a major cause of human cancer.[2] Since epidemiologists till the soil of an apparently nonbiochemical universe, I attempted to point out the importance and relevance of their observations and ever-increasing evidence.

"The second major development was the concept originated by the Millers[3] that most chemical carcinogens were not active *per se*, but required metabolic activation by cytochrome P450 monooxygenases to chemically reactive electrophiles. In my laboratory we produced the first experimental evidence that arene oxides are the activated form of polycyclic aromatic hydrocarbons.[4] My review in 1975 appeared to stimulate many 'pure' biochemists to conduct important research in this burgeoning field. It was Ames's recognition of the need for metabolic activation, which he supplied by suitably fortified liver homogenates to *Salmonella typhimurium*, that made his widely used test possible.[5] I reviewed the state of knowledge of the metabolic activation of several chemical classes of carcinogens.

"Berwald and Sachs[6] and my laboratory[7] had been the pioneers in developing model systems whereby normal rodent cells on treatment by chemical carcinogens underwent oncogenic transformation, and the implications and promises of this approach were called to the biochemists' attention. Finally, I reviewed the then status of mechanisms of action of chemical carcinogens with emphasis on somatic mutations vs. epigenetic mechanisms.

"I believe that the enduring popularity of the review stems from the breadth of perspectives and topics that I emphasized, which are still relevant today. Although I have received my share of honors and awards, these I believe were for my contributions to research and not for this review."

1. Miller E C & Miller J A. Biochemistry of carcinogenesis. *Annu. Rev. Biochem.* **28**:291-320, 1959.
2. Higginson J & Muir C S. Epidemiology. (Holland J F & Frei E, III, eds.) *Cancer medicine.* Philadelphia: Lea and Febiger, 1973. p. 241-306.
3. Miller E C & Miller J A. Biochemical mechanisms of chemical carcinogenesis. (Busch H, ed.) *Molecular biology of cancer.* New York: Academic Press, 1974. p. 377-402.
4. Grover P L, Sims P, Huberman E, Marquardt H, Kuroki T & Heidelberger C. In vitro transformation of rodent cells by K-region derivatives of polycyclic hydrocarbons. *Proc. Nat. Acad. Sci. US* **68**:1089-101, 1971.
5. Ames B N, Durston W E, Yamasaki E & Lee F D. Carcinogens are mutagens: a simple test system combining liver homogenates for activation and bacteria for detection. *Proc. Nat. Acad. Sci. US* **70**:2281-5, 1973.
6. Berwald Y & Sachs L. In vitro transformation of normal cells to tumor cells by carcinogenic hydrocarbons. *J. Nat. Cancer Inst.* **35**:641-61, 1965.
7. Chen T T & Heidelberger C. Quantitative studies on the malignant transformation of mouse prostate cells by carcinogenic hydrocarbons in vitro. *Int. J. Cancer* **4**:166-78, 1969.

This Week's Citation Classic

Williams E D, Karim S M M & Sandler M. Prostaglandin secretion by medullary carcinoma of the thyroid: a possible cause of the associated diarrhoea.
Lancet 1:22-3, 1968. [Dept. Pathol., Royal Postgrad. Med. Sch.; Inst. Obstet. and Gynaecol. and Bernhard Baron Mem. Res. Labs., Queen Charlotte's Maternity Hosp., London, England]

Significant levels of prostaglandins E_2 and $F_2\alpha$ were found in medullary carcinoma extracts, and peripheral and tumour venous blood, but not in appropriate controls. Tumour produced prostaglandins, possibly in concert with other agents, may play a role in production of the tumour associated diarrhoea. [The _SCI®_ indicates that this paper has been cited over 310 times since 1968.]

E.D. Williams
Department of Pathology
Welsh National School of Medicine
University of Wales
Heath Park, Cardiff CF4 4XN
Wales

July 22, 1981

"The request to write one's own history of some of one's own work seems too flattering to resist—I wonder how many refuse? In my case, the invitation came as a double surprise—that I should be asked at all, and that the article quoted should be the one it was.

"In 1963, I realised that medullary carcinoma of the thyroid was likely to be a tumour of C cells, and that C cells were the likely source of the recently discovered hormone, calcitonin. I therefore began a retrospective study of patients with medullary carcinoma to look for evidence of the effects of excess calcitonin production, starting with the obvious and rather naive idea that hypocalcaemia was the most likely. The cases were collected by reclassifying the pathology of several hundred cases of thyroid carcinoma from four different London hospitals. Disappointingly, hypocalcaemia only occurred following thyroparathyroidectomy, but, surprisingly, many patients had unexplained diarrhoea. A humoral link seemed possible, particularly as diarrhoea was more common in patients with wide-spread tumour, remitting in some after tumour removal only to recur with tumour spread. A paper describing these findings was, after earlier rejection, published,[1] and was quickly confirmed.

"The watery diarrhoea was accompanied by very rapid small intestinal transit, and I showed _in vitro_ that tumour extracts stimulated intestinal muscle contraction. As activity was destroyed by proteolysis, I became interested in the possibility that prostaglandins, then recently recognised as of potential clinical importance, were involved. Sultan Karim and Merton Sandler took over the assay work, and found prostaglandins E_2 and $F_2\alpha$ in tumour extracts and tumour venous effluent obtained at operation; we speculated that prostaglandin production might cause the diarrhoea.

"The work has been frequently quoted as the first report of prostaglandin production by tumours, the first to attribute a syndrome to this production, and the first reasonable explanation of the diarrhoea associated with medullary carcinoma of the thyroid. In the intervening 13 years, prostaglandin production by medullary carcinoma has been substantiated;[2] the cause of the diarrhoea remains uncertain.

"Medullary carcinoma is now known to produce many other substances, including somatostatin, serotonin, histaminase, nerve growth factor, ACTH, CRH, VIP, carcinoembryonic antigen, and substance P.[3] Substance P and VIP affect small gut motility and secretion; calcitonin itself increases jejunal water and salt secretion.[4] Prostaglandins, calcitonin, serotonin, and substance P and VIP may therefore all play a role in the diarrhoea associated with medullary carcinoma. The fact that one tumour produces such a variety of substances with a converging effect should stimulate further thought—after all, the differentiation of an endocrine cell can be regarded as directed towards a functional effect rather than hormone structure. It would be interesting, if improbable, if my next article in _Current Contents®_ could start, 'The idea came to me while I was writing my last _Citation Classic_...!'"

1. Williams E D. Diarrhoea and thyroid carcinoma. _Proc. Roy. Soc. Med._ 59:602-3, 1966.
2. Barrowman J A, Bennett A, Hillenbrand P, Rolles K, Pollock D J & Wright J T. Diarrhoea in thyroid medullary carcinoma: role of prostaglandins and therapeutic effect of nutmeg. _Brit. Med. J._ 3:11-12, 1975.
3. Williams E D. Medullary carcinoma of the thyroid. (DeGroot L J, ed.) _Endocrinology_. New York: Grune & Stratton, 1979. Vol. 2. p. 777-92.
4. Kennedy Gray T, Bieberdorf F A & Fordtran J S. Thyrocalcitonin and the jejunal absorption of calcium, water, and electrolytes in normal subjects. _J. Clin. Invest._ 52:3084-8, 1973.

Ahlquist R P. A study of the adrenotropic receptors.
Amer. J. Physiol. **153**:586-600, 1948.

This paper proposed that there were two different kinds of receptors for the neurotransmitter norepinephrine. This concept has resulted in new insights into physiological control mechanisms and a whole new class of drugs useful for cardiovascular disease. [The *SCI®* indicates that this paper was cited 1383 times in the period 1961-1977.]

Raymond P. Ahlquist
Charbonnier Professor of Pharmacology
Medical College of Georgia
Augusta, GA 30901

January 24, 1978

"The work on which my paper was based was carried out during the 'golden age' of biomedical research: no lab technicians, no statistical analysis, drugs of doubtful purity, smoked paper kymographs, inexpensive animals and a heavy teaching load. The only 'modern' instrument available was a home-made, high-frequency-response, optical manometer designed by W. F. Hamilton. The total budget for the project was about $3500; this included my annual salary.

"The concept of two receptors for the adrenergic neurotransmitter was the result of pure serendipity. The experimental work was done to find a drug that would relax the human uterus; the clinical objective was to find a cure for dysmenorrhea.

"I have described the concept of *alpha* and *beta* receptors in a short essay.[1] The following is a quotation from this. 'The original paper was rejected by the *Journal of Pharmacology and Experimental Therapeutics*, was a loser in the Abel Award competition, and was finally published in the *American Journal of Physiology* due to my personal friendship with the great physiologist W.F. Hamilton. It was ignored for five years. The reasons for this are obvious today; the concept did not fit with ideas developed since 1890 on the actions of epinephrine.'

"The utility of the concept was first recognized by pharmacology teachers. Adrenergic drugs could be classified according to which receptor they acted on. The actions, uses and side-effects of these drugs could be predicted. Physiologists started to use this concept to study control systems in the body. The *beta* receptor was the starting point for the discovery of cyclic AMP and adenyl cyclase by E.W. Sutherland and co-workers.

"At the time this paper was published, three classes of drugs that acted on these receptors were known: *alpha* agonists, *beta* agonists and *alpha* blocking agents. The fourth class, the *beta* blocking agents, was unknown, but their actions were completely predictable. Although I searched diligently for drugs of this type, someone else found them ten years later. The history of the *beta* receptor blockers is too long to cover here completely. Propranolol was introduced by James Black as a treatment for angina pectoris in 1963. Subsequently the *beta* blockers were found to be useful in arrhythmias, thyrotoxicosis, tremor, migraine, and essential hypertension. Although only one *beta* blocker, propranolol, is available in this country, at least 18 are available throughout the rest of the world.

"In my opinion, the most important contribution of my concept was to repopularize the idea of receptors. These had been described in the early part of this century but for some reason had been forgotten. Now there are receptors for hormones, peptides and drugs."

REFERENCE
1. **Ahlquist R P.** Adrenergic receptors: a personal and practical view.
Perspect. Biol. Med. **17**:119-22, 1973.

This Week's Citation Classic

Brimblecombe R W, Duncan W A M, Durant G J, Emmett J C, Ganellin C R & Parsons M E. Cimetidine—a non-thiourea H$_2$-receptor antagonist.
J. Int. Med. Res. 3:86-92, 1975.
[Research Inst., Smith Kline & French Labs. Ltd., Welwyn Garden City, Hertfordshire, England]

Data are presented on the chemistry, pharmacology, toxicology, and biochemistry of cimetidine, a histamine H$_2$-receptor antagonist. Pharmacologically, e.g., as an inhibitor of gastric acid secretion, cimetidine is similar to the previously described metiamide, but in toxicity studies with cimetidine, no haematological changes or renal damage were observed. [The *SCI®* indicates that this paper has been cited over 260 times since 1975.]

Roger W. Brimblecombe
Department of Pharmacology
Smith Kline & French Laboratories
Welwyn Garden City
Hertfordshire AL7 1EY
England

January 6, 1982

"The main pharmacological actions of histamine have been recognised since the first two decades of this century.[1] That these actions implied involvement of histamine in inflammatory and allergic reactions led to the search for drugs to antagonise the effects of histamine. Such drugs have been available and widely used clinically for about 40 years, but it soon became apparent that they were not capable of antagonising all the actions of histamine. This led to an idea, promulgated in the 1960s, that histamine exerted its effects through more than one set of pharmacological receptors.[2] This is now an accepted concept—the actions of histamine which can be antagonised by classical antihistamines are mediated through H$_1$ receptors, and other effects, including stimulation of gastric acid secretion, are mediated through H$_2$ receptors. The proof of the existence of H$_2$ receptors resulted from the discovery, in the UK laboratories of Smith Kline & French, of specific antagonists of histamine at these receptors.

"The first of these drugs, burimamide, was tested in human subjects, but was not sufficiently active by the oral route to be developed as a medicine. Its successor, metiamide, was also tested in healthy human subjects and, additionally, in extensive clinical trials in which it proved an effective agent in promoting the healing of duodenal ulcers. However, a reversible granulocytopenia, seen occasionally in dogs in toxicity studies, eventually manifested itself in a very low proportion of the treated subjects.

"This led to the search for a compound with essentially the same pharmacological profile as metiamide, but lacking the propensity for producing this undesirable effect. In cimetidine, the thiourea group of metiamide was replaced by a cyanoguanidine group with consequent elimination of the haematological side effect. Cimetidine has now been marketed in over 100 countries, has been used in over 11,000,000 patients, and is a highly effective agent in the treatment of peptic ulcer disease and other diseases associated with gastric hypersecretion.

"As well as proving to be a revolutionary new therapeutic agent in terms of clinical efficacy and marketing success, cimetidine has also proved to be a major new tool for studying the physiology, pharmacology, and pathology of histamine. These are the reasons for the high number of citations to this publication.

"The work to search for histamine H$_2$-receptor antagonists began at Smith Kline & French in 1964. The first publication, on burimamide, appeared in 1972,[3] and cimetidine was first marketed in 1976. Thus, 12 years elapsed during which hundreds of compounds were synthesized and tested, many scientists were involved in various aspects of research and development, and much money was spent.

"Individual authors of this paper and others connected with the research program have received a variety of awards for their work and in 1978 the Smith Kline & French research laboratories in the UK were granted the Queen's Award for Technological Achievement in recognition of this discovery. A book reporting recent work in this field has been published."[4]

1. **Douglas W W.** Histamine and antihistamines; 5-hydroxytryptamine and antagonists. (Goodman L S & Gilman A, eds.) *The pharmacological basis of therapeutics.* New York: Macmillan, 1975. p. 590-629.
2. **Ash A S F & Schild H O.** Receptors mediating some actions of histamine. *Brit. J. Pharmacol.* 27:427-39, 1966.
3. **Black J W, Duncan W A M, Durant G J, Ganellin C R & Parsons M E.** Definition and antagonism of histamine H$_2$-receptors. *Nature* 236:385-90, 1972.
4. **Torsoli A, Lucchelli P E & Brimblecombe R W,** eds. *Further experience with H$_2$-receptor antagonists in peptic ulcer disease and progress in histamine research.* Amsterdam: Excerpta Medica, 1980. 370 p.

Iversen L L. Dopamine receptors in the brain. *Science* **188**:1084-9, 1975.
[Medical Research Council Neurochemical Pharmacology Unit, Dept. Pharmacology, Univ. Cambridge, England]

This review summarised data on the dopamine-stimulated adenylate cyclase in brain as a model for central dopamine receptors. The potent inhibitory effects of a number of antischizophrenic drugs provided support for the hypothesis that such drugs act by blocking dopamine actions in the central nervous system. [The *SCI*® indicates that this paper has been cited in over 505 publications since 1975.]

Leslie L. Iversen
Neuroscience Research Centre
Merck Sharp & Dohme
Research Laboratories
Hoddesdon EN11 9BU
England

May 15, 1984

"By the early 1970s, the idea that the drugs used in treating schizophrenia (major tranquilisers or neuroleptics) acted by virtue of their ability to block dopamine receptors in brain had become increasingly attractive.[1] It had not been possible to test the hypothesis directly, however, as no simple means of testing drug actions on dopamine receptors in brain were available. The discovery, in 1972, by Brown and Makman[2] and Kebabian, Petzold, and Paul Greengard[3,4] that dopamine specifically stimulated cyclic AMP formation in homogenates of dopamine-rich areas of the central nervous system (retina and basal ganglia) was, therefore, immediately recognised as an important step forward, since it might offer a suitable model system.

"In the Neurochemical Pharmacology Unit in Cambridge, a PhD student, Richard Miller, Alan Horn, and I tested a number of antischizophrenic drugs as possible antagonists of dopamine responses. The results of

this and similar work carried out in Greengard's group[5] and by Brown and Makman were summarised in this short review article. The data at first sight seemed to support the hypothesis, as drugs in the phenothiazine and thioxanthene classes proved to be potent inhibitors of the dopamine-stimulated adenylate cyclase in brain; furthermore, pharmacologically inactive drugs in these classes were also inactive in the *in vitro* model.

"An important group of antischizophrenic drugs, however, the butyrophenones, which are pharmacologically very potent, proved to be only very weak antagonists of dopamine in the test-tube system. It is now recognised that the adenylate cyclase model represents only one class of dopamine receptors in brain, the so-called D1 type.[6] The D2 type of receptors which were subsequently characterised by their ability to bind radio-labeled butyrophenones[7] are not cyclase-linked, and are thought to represent the more likely target of antischizophrenic drugs, since compounds of all chemical classes interact potently at these sites.[8] The D1 receptors, nevertheless, may contribute to the actions of antischizophrenic drugs of the phenothiazine and thioxanthene classes. A highly specific antagonist of the D1 dopamine receptors was recently discovered,[9] SCH 23390, and this compound has a pharmacological profile very similar to that of classical neuroleptic drugs in animal tests.[10]

"The significance of being asked to write an invited review article for *Science* was not clear to me at the time, although the subsequent popularity of this review of an active and topical area of psychopharmacology research shows how widely read such articles are. I am indebted to my colleagues Horn and particularly Miller, whose experimental findings formed the basis of the review."

1. Snyder S H, Banerjee S P, Yamamura H I & Greenberg D. Drugs, neurotransmitters and schizophrenia. *Science* 184:1243-53, 1974. (Cited 550 times.)
2. Brown J H & Makman M H. Stimulation by dopamine of adenylate cyclase in retinal homogenates and of adenosine 3′5′-cyclic monophosphate formation in intact retina. *Proc. Nat. Acad. Sci. US* 69:538-43, 1972. (Cited 195 times.)
3. Kebabian J W, Petzold G L & Greengard P. Dopamine-sensitive adenylate cyclase in caudate nucleus of rat brain, and its similarity to the "dopamine receptor." *Proc. Nat. Acad. Sci. US* 69:2145-9, 1972.
4. Kebabian J W. Citation Classic. Commentary on *Proc. Nat. Acad. Sci. US* 69:2145-9, 1972. *Current Contents/Life Sciences* 26(11):18, 14 March 1983.
5. Clement-Cormier Y C. Citation Classic. Commentary on *Proc. Nat. Acad. Sci. US* 71:1113-17, 1974. *Current Contents/Life Sciences* 27(18):18, 30 April 1984.
6. Kebabian J W & Calne D B. Multiple receptors for dopamine. *Nature* 277:93-6, 1979. (Cited 775 times.)
7. Creese I, Burt D R & Snyder S H. Dopamine receptor binding predicts clinical and pharmacological potencies of antischizophrenic drugs. *Science* 192:481-3, 1976. (Cited 440 times.)
8. Seeman P. Brain dopamine receptors. *Pharmacol. Rev.* 32:299-313, 1980.
9. Iorio L C, Barnett A, Leitz F H, Houser V P & Korduba C A. SCH 23390, a potential benzazepine antipsychotic with unique interactions on dopaminergic systems. *J. Pharmacol. Exp. Ther.* 226:462-8, 1983.
10. Christensen A V, Arnt J, Hyttel J, Larsen J J & Svendson O. Pharmacological effects of a specific dopamine D1 antagonist SCH 23390 in comparison with neuroleptics. *Life Sci.* 34:1529-40, 1984.

Lands A M, Arnold A, McAuliff J P, Luduena F P & Brown T G, Jr.
Differentiation of receptor systems activated by sympathomimetic amines.
Nature 214:597-8, 1967.
[Sterling-Winthrop Research Institute, Rensselaer, NY]

On the basis of the relative agonist activities of norepinephrine and certain norepinephrine analogs, catechol ethanol amines in all instances, in selected autonomic systems, data have been presented to show that an α-, β_1-, β_2-adrenoceptor delineation is in better accord with experimental findings than can be encompassed within an α- and β-receptor one initially proposed. [The *SCI®* indicates that this paper has been cited over 875 times since 1967.]

Aaron Arnold
Terrace Avenue
Albany, NY 12203

October 6, 1981

"Minneman et al. generously noted recently, 'The pioneering work of Lands and his colleagues represented the first clear evidence that subtypes of β-adrenergic receptors exist.'[1] Accordingly, this seems an appropriate time to review the background for our studies.

"Autonomic responses to sympathomimetic amines were segregated into two groups by Ahlquist in 1948[2] with an inspired α- and β-receptor terminology suggestion. The α-receptor ones (vasoconstriction, nictitating membrane contraction, etc.) responded well to epinephrine and to norepinephrine. They responded only minimally, or not at all, to isoproterenol. The β-receptor ones (tracheal and bronchial smooth muscle relaxation, vasodilatation, etc.) responded best to isoproterenol, and well to epinephrine. They responded only weakly to norepinephrine. Ahlquist's suggestion circumvented the inconsistencies that resulted from attempts to classify the responses into excitatory and inhibitory groupings, for example.

"Ahlquist's proposal received important support from Powell and Slater's classical finding[3] that dichloroisoproterenol (DCI) could block the responses of agonists on the β-receptor. This finding complemented α-blocker data and was a reassuring tool in delineating α-receptor from β-receptor responses.

"However, Ahlquist's proposal left a troublesome finding open. Mammalian heart responds well to isoproterenol and as well to norepinephrine as to epinephrine.[4] In Ahlquist's scheme norepinephrine was only a minimally effective β-agonist. The discrepancy was minimized by listing the heart's β-receptors as 'exceptions.'

"In 1966 data from our group at Sterling-Winthrop Research Institute were presented[5] at the spring Pharmacology Society meetings which showed, with supporting correlation coefficient data, that the responses of the cardiac and lipolytic β-receptors to norepinephrine and certain norepinephrine analogs correlated. Similarly, the responses of the bronchial and vasodepressor β-receptors to the amines correlated. In no instances did the cardiac or lipolytic responses to the above amines correlate with those of the bronchial or vasodepressor ones. In more specific terms, relative to isoproterenol and epinephrine, norepinephrine and nordefrin were effective cardiac and lipolytic stimulants. They were only weak or minimally effective bronchodilators or vasodepressors. Contrariwise, certain vasodepressor amines (N-t-butylnorepinephrine, N-cyclopentyl-butanefrine, isoetharine) were effective bronchial and vasodepressor β-receptor agonists, but were relatively less so on the cardiac and lipolytic β-receptors. The data presented at the spring meetings were published in May 1967 in *Nature* (which is the subject paper of this *Citation Classic*) with the added suggestion that the heart and lipolytic receptor be termed β_1. The bronchial and vasodepressor one was termed β_2. The suggestion appears to have withstood the test of time and, as noted recently,[6] has specifically been termed by a number of investigators to have been 'widely accepted.' Additional instances of sympathomimetic amine responses subserved by the β_1-receptor (coronary and intestinal smooth muscle relaxation, kidney renin release, salivary gland excretion, etc.) and by the β_2-receptor (skeletal muscle contraction and glycogenolysis, pancreatic insulin release, rat diaphragm contraction, etc.) have been summarized.[6]

"The β_1-, β_2-receptor suggestion helps group autonomic responses into anticipated closely related and significantly unrelated effects. The suggestion yields an experimental basis for 'cardiac selective,' or anti-hypertensive selectivity. It provides a rationale for the possible elucidation of additional 'selective' agents while guarding against optimism wherein selectivity may less likely be achieved."

1. Minneman K P, Pittman R N & Molinoff P B. β-adrenergic receptor subtypes: properties, distribution, and regulation. *Annu. Rev. Neurosci.* 4:419-61, 1981.
2. Ahlquist R P. A study of the adrenotropic receptors. *Amer. J. Physiol.* 153:586-600, 1948.
   [Citation Classic. *Current Contents* (45):16, 6 November 1978.]
3. Powell C E & Slater I H. Blocking of inhibitory adrenergic receptors by a dichloro analog of isoproterenol. *J. Pharmacol. Exp. Ther.* 122:480-8, 1958.
4. Luduena F P, Ananenko E, Siegmund O H & Miller L C. Comparative pharmacology of the optical isomers of arterenol. *J. Pharmacol. Exp. Ther.* 95:155-70, 1949.
5. Arnold A, McAuliff J P, Luduena F P, Brown T G, Jr. & Lands A M. Lipolysis and sympathomimetic amines. *Fed. Proc.* 25:500, 1966.
6. Arnold A. Sympathomimetic amine-induced responses of effector organs subserved by α-, β_1- and β_2-adrenoceptors. *Handb. Exp. Pharmacol.* 54:63-88, 1980.

This Week's Citation Classic

Gavin J R, III, Roth J, Neville D M, Jr., De Meyts P & Buell D N. Insulin-dependent regulation of insulin receptor concentrations: a direct demonstration in cell culture.
Proc. Nat. Acad. Sci. US **71**:84-8, 1974. [Diabetes Sect., Clin. Endocrinol. Br., Natl. Inst. Arthritis, Metab., and Digestive Dis.; Sect. Biophys. Chem., Lab. Neurochem., Natl. Inst. Mental Health; and Immunol. Br., Natl. Cancer Inst., NIH, Bethesda, MD]

When cultured human lymphoid cells were incubated with native insulin, there was a time, temperature, and hormone concentration-dependent reduction in insulin receptor sites. This study represents a direct demonstration *in vitro* of the ability of a hormone to regulate its homologous receptor concentrations at the cellular level. [The *SCI*® indicates that this paper has been cited in over 770 publications since 1974.]

James R. Gavin, III
Metabolism Division
Department of Internal Medicine
Washington University School of Medicine
St. Louis, MO 63110

July 7, 1983

"The general field of peptide hormone-receptor interactions was still in its infancy when I joined Jesse Roth's lab in early 1971. Nevertheless, receptors were already viewed as rather passive participants in hormone action, while the dynamic regulatory aspects of hormone-mediated biologic responses were attributed entirely to fluctuations in the availability of potent circulating hormone. Our laboratory proposed the unconventional concept that receptors could play a much more direct role in the regulation of hormonal action in health and disease states. The initial data that strongly supported this view were the apparent inverse relationship between insulin binding to hepatic membranes and serum IRI in the insulin-resistant ob/ob mouse.[1] We therefore pursued the hypothesis that this reciprocal relationship between insulin and its receptors in the obese mouse reflected a major and perhaps a general mechanism for hormone-receptor regulatory interactions. This view was neither widely held nor readily embraced in the early 1970s.

"Jesse's laboratory was moving in the direction of developing methods for evaluation of receptors in man when I joined the group. My studies with various components of whole blood led to the description of insulin receptors in red and white blood cells in man. These findings led to the suggestion by David Neville that I screen some of the human lymphoid cell cultures for receptors in order to obtain a more plentiful (and homogeneous) tissue source for receptor characterization studies. Thus began our work with cell line #4265 and its more celebrated counterpart, the IM-9 cell.

"The availability of receptors in cultured cells provided the opportunity to test directly what effects of chronic exposure to insulin *per se* could be demonstrated on receptor binding activity in human cells. With help from Maxine Lesniak, it did not take long to design a series of experiments that could reasonably test the questions of interest to us.

"With our first few experiments we found that, indeed, cells chronically exposed to 10^{-7} M insulin (vs. acute exposure) had a markedly reduced ability to subsequently bind insulin. It was something of a shock when the data were not received well by our weekly data club. A strong sentiment was voiced that all our findings could be explained by contamination with residual insulin or some technical artifact. My only secure allies in the midst of vociferous skeptics were Jesse and Maxine. A series of rigorous control studies eventually quelled the earnest reservations of the data club and dispelled the contamination issue. The findings soon thereafter emerged as the first direct demonstration of regulation of a hormone receptor by its homologous peptide hormone. The concept of dynamic receptor regulation in hormone action was now placed on firm footing. The subsequent demonstration by Roth and Lesniak[2] of highly sensitive and specific regulation of GH receptors by hGH in these same cells was strong confirmation of the presence of this dynamic mechanism. Thus, it is all the more pleasing that from those stormy beginnings we have come to *Citation Classics.*

"I think this work is so frequently cited because it has contributed a great deal toward establishing equal status to the receptor as a key determinant in the dynamic regulation of hormone action. The term 'down-regulation' has been coined out of these observations, and mechanisms for this process have been described in recent literature.[3] While the concept of down-regulation does not account for the entire spectrum of known relationships between circulating hormones and tissue receptors, it has certainly been shown to be a fundamental mechanism with wide applicability, particularly in situations where receptors are chronically exposed to high levels of homologous hormones."

1. Kahn C R, Neville D M, Jr. & Roth J. Insulin-receptor interaction in the obese hyperglycemic mouse. *J. Biol. Chem.* 248:244-50, 1973. (Cited 390 times since 1973.)
2. Lesniak M A & Roth J. Regulation of receptor concentration by homologous hormone: effect of human growth hormone on its receptor in IM-9 lymphocytes. *J. Biol. Chem.* 251:3720-7, 1976. (Cited 245 times since 1976.)
3. Hizuka N, Gorden P, Lesniak M A, Van Obberghen E, Carpentier J L & Orci L. Polypeptide hormone degradation and receptor regulation are coupled to ligand internalization—a direct biochemical and morphologic demonstration. *J. Biol. Chem.* 256:4591-7, 1981.

Toyoda J, Nosaki H & Tomita T. Light-induced resistance changes in single
photoreceptors of *Necturus* and *Gekko*. *Vision Res.* 9:453-63, 1969.
[Department of Physiology, Keio University School of Medicine, Shinjuku-ku,
Tokyo, Japan]

Vertebrate photoreceptors respond to light with hyperpolarization, namely, with an inside negative potential change. Studies on their membrane properties indicate that the receptors are kept depolarized in the dark, as is common in nerve cells at their excited state, and become repolarized toward the resting membrane potential upon illumination. [The *SCI®* indicates that this paper has been cited over 180 times since 1969.]

Jun-ichi Toyoda
Department of Physiology
St. Marianna University
School of Medicine
Takatsu-ku, Kawasaki
Japan

July 31, 1981

"First of all, I would like to say how pleased we are to learn that our paper is listed as one of the most-cited articles in its field.

"The experiments described in this paper were performed using ordinary neurophysiological techniques and the results obtained were analysed on the basis of the neurophysiological principles widely accepted. In this respect, the paper is not especially epoch-making. The reason why it is cited so frequently is probably in its simple but unexpected conclusion that the vertebrate photoreceptors are excited in the dark and return to the resting state upon illumination. Generally, the membrane potential of the nerve cells is inside negative at the resting state, but shifts to positive upon excitation due to an increase in the permeability of the membrane to sodium ions which then tend to enter the cell. This positive potential change is called depolarization.

"The intracellular recording from vertebrate photoreceptors was first reported in 1965.[1] It was shown that they respond to light with hyperpolarization in contrast to most invertebrate photoreceptors which are known to respond to light with depolarization. This first attempt to record the intracellular response from photoreceptors met with many difficulties. The main problem was that even with microelectrodes of tip diameter less than 0.1 μm, the retinal cells easily escaped from the tip when the electrode was inserted gently into the retina. One of us, T. Tomita, tried to solve this problem by giving high acceleration to the electrode advancement. However, he happened to think of giving a high frequency vibration to the retina instead of giving an acceleration to the electrode. This method was found very effective. Even with the aid of such a jolting method, however, it took almost three months before we finally had the first sign of intracellular recording.

"The unusual response polarity of vertebrate photoreceptors immediately led us to the present experiments to solve the mechanisms underlying the response. The major problem was the size of the receptors. In order to study the permeability change of the membrane it was necessary to use double-barreled electrodes, with one barrel for recording the potential change and the other for passing electric current. The carp receptors we used at the beginning were too small to allow intracellular recording with such electrodes. The mud puppy and gecko, known to have large photoreceptors, were finally found to meet this purpose.

"The concept that the outer segment membrane of vertebrate photoreceptors becomes impermeable to sodium ions in response to illumination is now commonly cited in standard textbooks and monographs.[2] It served as one of the stepping-stones for later studies on the function of photoreceptors."

1. **Tomita T.** Electrophysiological study of the mechanisms subserving color coding in the fish retina.
   Cold Spring Harbor Symp. **30**:559-66, 1965.
2. **Rodieck R W.** *The vertebrate retina. Principles of structure and function.*
   San Francisco, CA: W. H. Freeman, 1973. 1044 p.

Borisy G G & Taylor E W. The mechanism of action of colchicine: binding of
colchicine-^{3}H to cellular protein. *J. Cell Biol.* **34**:525-33, 1967.
[Dept. Biophysics, Univ. Chicago, Chicago, IL]

^{3}H-labeled colchicine was used to identify
the colchicine-binding receptor in cells.
Colchicine-binding activity did not correlate
strictly with mitotic activity but did correlate with sources abundant in microtubules.
We suggest that the binding sites are the
subunit proteins of microtubules. [The *SCI*®
indicates that this paper has been cited over
595 times since 1967.]

Gary G. Borisy
Laboratory of Molecular Biology
University of Wisconsin
Madison, WI 53706

April 28, 1981

"Rosy said, 'It was all an artifact.' He
warned that I would never get my degree if I continued to study 'that
colchicine-binding (expletive deleted).'
'Rosy' was Joel L. Rosenbaum, now professor of biology at Yale University. At
that time (1964), we were both doing
research at the University of Chicago.
Rosy was a postdoctoral fellow working on ciliary regeneration with Frank
M. Child in the department of zoology
and I was a green, second-year graduate student working with Edwin W.
Taylor in biophysics.

"On lazy summer days we would sit
around Botany pond under a spreading
ginkgo tree, shooting the bull and going
over fresh results. It was on one of
those days that I announced that in a
survey of tissue types, I got my highest
level of colchicine-binding in extracts
of brain tissue.

"This was a most perplexing result.
Colchicine was a drug known to inhibit
mitosis in plant and animal cells by
disrupting the formation of the mitotic
spindle.[1] A few years earlier, Taylor[2]

had begun to study its mechanism of
action and inferred that it exerted its effects by binding to a specific target site
in dividing cells. Using ^{3}H-labeled colchicine, my project was to identify the
target site at the molecular level.

"We first developed an *in vitro*
colchicine-binding assay to quantitate
the amount of the target molecule. It
was quickly demonstrated that cultures
of dividing carcinoma cells contained a
lot of the colchicine-binding receptor.
But there was no reason to expect the
receptor to be present in the neuronal
cells of brain tissue.

"To support the hypothesis that the
colchicine-binding receptor was important in cell division, we attempted to
correlate colchicine-binding activity
with mitotic index. What a surprise,
then, to find in a survey of tissue types
that brain tissue, having the lowest
mitotic index, had the highest specific
activity of binding! How could anyone
believe that colchicine bound to a
component of the mitotic spindle if it
also bound so well to neurons?

"This was one of those happy instances where investigation of a paradox led to a deeper understanding of
the phenomenon in question. As further work showed, the reason brain tissue was so high in colchicine-binding
activity was because of its high content
of microtubules,[3] a major structural
component of neurons. Microtubules
also form the framework of the mitotic
spindle. The colchicine-binding receptor turned out to be tubulin, the subunit
of microtubules. Colchicine inhibited
mitosis by binding to tubulin, preventing its assembly into microtubules.
Thus, Rosy's 'colchicine artifact' led to
an understanding of the molecular
mechanism of action of the drug, to the
identification of the major component
of the mitotic spindle, and to the
discovery of the protein that microtubules are made of."

1. Dustin P. *Microtubules.* Berlin: Springer-Verlag, 1978. 452 p.
2. Taylor E W. The mechanism of colchicine inhibition of mitosis. *J. Cell Biol.* **25**:145-60, 1965.
3. Roberts K & Hyams J, eds. *Microtubules.* London: Academic Press, 1979. 595 p.

Grimelius L. A silver nitrate stain for α_2 cells in human pancreatic islets.
Acta Soc. Med. Upsal. 73:243-70, 1968.
[Department of Pathology, University of Uppsala, Sweden]

A simple silver-staining method was developed whereby most endocrine cells in the digestive tract, certain endocrine pancreatic cells, thyroid parafollicular cells, and so on, can be demonstrated. Only well-characterized and easily obtainable chemicals are required. The method is used mainly in routine histopathology for discriminating endocrine from nonendocrine tumors. [The *SCI®* indicates that this paper has been cited in over 475 publications since 1968.]

Lars Grimelius
Department of Pathology
University of Lund
Malmö General Hospital
S-214 01 Malmö
Sweden

May 16, 1984

"One of my first research projects as a young assistant at the Institute of Pathology in Uppsala was to investigate the endocrine pancreas in diabetic rats treated with antidiabetic agents. Good staining methods were available for demonstrating insulin-producing cells and this held for D(A)₁ cells, which were later found to contain somatostatin.[1,2] But how could the A(A₂) cells be visualized? I tested several silver techniques, but the results were not very satisfactory, except with one method developed by Bodian.[3] Here a silver proteinate was used for impregnation. Unfortunately, the quality of this substance varied considerably both between different manufacturers and between different batches from one producer. Out of about ten silver proteinates that I tested, only one gave a satisfactory result, and then only after modification of the Bodian technique. When I tried to order more silver proteinate from the manufacturer, I was informed that this batch was sold out. I was also told that one seldom succeeded in preparing silver proteinate with

the same staining properties from one occasion to another even when the manufacturing process was the same.

"Instead of dropping the research project, I decided to develop a silver technique of my own with the same staining properties as the Bodian method but where only well-characterized and easily obtainable chemicals were required. The different factors in the staining steps should also be well controlled, so that the results would be reproducible. After four years' work and more than 2,000 staining experiments, I managed to arrive at a simple silver (argyrophil) technique which comprised only two stages—silver impregnation and the reduction process. The staining was also developed such that counterstaining was unnecessary. Subsequent analyses of the technique showed that the silver-positive cells in fact represented A cells. Electron microscopic studies[4] revealed that the silver-positive reaction is due to precipitation of silver particles in the secretory granules (storage sites for hormones) of the endocrine cells. As with the Bodian technique, cells in the gastrointestinal mucosa were also stained. Later studies, by many other authors, showed that my technique stained almost all endocrine cell types in the digestive tract (cf. Grimelius and Wilander[5]). It was also found that the parafollicular cells in thyroid, and the endocrine cells in the respiratory tract, for example, could also be visualized. The demonstrable cells belonged to the APUD system.

"Originally, I developed the technique for the pancreatic A cells, but a few years later I modified it somewhat so that it would also give optimal staining of endocrine cells in the gastrointestinal tract.[5]

"The reasons this technique has become so widely established are probably that it is simple, that all chemicals are easily obtainable, and that the results are reproducible. The method has come into use in routine histopathologic diagnosis for discriminating endocrine tumors of APUD type from other endocrine and nonendocrine tumors.

"The research project with the diabetic rats was never completed!"

1. **Polak J M, Pearse A G E, Grimelius L, Bloom S R & Arimura A.** Growth-hormone release-inhibiting hormone in gastrointestinal and pancreatic D cells. *Lancet* 1:1220-2, 1975.
2. **Polak J M.** Citation Classic. Commentary on *Lancet* 1:1220-2, 1975.
   Current Contents/Life Sciences 25(34):18, 23 August 1982.
3. **Bodian D.** A new method for staining nerve fibres and nerve endings in mounted paraffin sections.
   Anat. Rec. 65:89-97, 1936. (Cited 400 times since 1955.)
4. **Grimelius L.** An electron microscopic study of silver stained adult human pancreatic islet cells, with reference to a new silver nitrate procedure. *Acta Soc. Med. Upsal.* 74:28-48, 1969.
5. **Grimelius L & Wilander E.** Silver stains in the study of endocrine cells of the gut and pancreas.
   Invest. Cell Pathol. 3:3-12, 1980.

Rodbell M. Metabolism of isolated fat cells. I. Effects of hormones on glucose
metabolism and lipolysis. *J. Biol. Chem.* **239**:375-80, 1964.
[Laboratory of Nutrition and Endocrinology, National Institute of Arthritis and
Metabolic Diseases, National Institutes of Health, Bethesda, MD]

A homogeneous preparation of isolated fat cells was prepared by treating adipose tissue with collagenase. The isolated fat cells were very sensitive to the actions of insulin on glucose metabolism and responded to a variety of lipolytic hormones with increased production of fatty acids. [The *SCI*® indicates that this paper has been cited over 1,485 times since 1964.]

Martin Rodbell
Laboratory of Nutrition
and Endocrinology
National Institute of Arthritis,
Metabolism, and Digestive Diseases
National Institutes of Health
Bethesda, MD 20014

October 14, 1980

"This paper was a turning point in the direction of my research career. Prior to this study, I had been concerned principally with the structure of lipoproteins and the mechanism by which chylomicrons were metabolized by liver and adipose tissue. It was during a sabbatical year with Brachet and Gaillard that I became interested in the metabolism and differentiation of animal cells in culture. From that experience I learned how important it was to have a single cell type and a chemically defined medium for such studies.

"From previous experience, I realized that adipose tissue was appropriate material for two reasons. Firstly, the metabolism of adipose tissue was known to be affected by a number of hormones. Secondly, adipose cells, being laden with fat, should be easily separated from other tissue cells since they should float to the surface once liberated from the tissue. On returning to NIH, my first task was to find some means of dispersing the tissue in a manner that would release the cells without affecting their normal behaviour toward hormones. I knew

from earlier cytochemical studies that adipose cells were embedded in a matrix of collagen fibers, which suggested that digestion of the collagen might liberate fat cells. Fortunately, a commercial preparation of crude collagenase was available for testing purposes (fortunately, also, that it was a crude enzyme since later studies showed that purified collagenase doesn't work).

"I recall vividly the first experiment, since Houssay was visiting the laboratory that day and observed with me the gradual digestion of the tissue and the release of the pearl-like objects of my desire. 'Great,' he shouted, 'but are they viable cells?' I replied by suggesting that the effects of insulin on glucose metabolism should be an excellent test of their viability. This was demonstrated a short time later. Immediately I recognized the importance of this finding since, until then, the actions of insulin were observed only with intact tissues. Moreover, the isolated cells were responsive to extraordinarily low concentrations of the hormone. From that time onward I have remained committed, for better or for worse, to investigate the molecular basis by which hormones interact with cell surface receptors and thereby alter the physiology and structure of their target cells.

"Peculiarly, the rat fat cell responds to numerous types of hormones. Thus, it is virtually a 'gold mine' for studies of hormone action. Moreover, unlike its tissue counterpart, the isolated fat cell is exposed directly and uniformly to ingredients in the incubation medium, thus providing the means of testing the effects of various agents on cell surface receptors. Undoubtedly, these are the primary reasons why the isolated fat cell preparation has been employed so often in studies of hormone action.

"Naturally, I am gratified by the response to this paper. With hindsight, it was a simple, straightforward exercise. Apparently, it often happens that a simple idea can engender consequences that are far beyond the intent."

Lacy P E & Kostianovsky M. Method for the isolation of intact islets of
Langerhans from the rat pancreas. *Diabetes* 16:35-9, 1967.
[Dept. Pathology, Washington Univ. Sch. Med., St. Louis, MO]

The paper describes a simple method for the isolation of intact, viable islets from the normal rat pancreas for *in vitro* studies on hormonal secretion. The method was based upon disruption of the exocrine pancreas prior to digestion of the tissue with collagenase. [The *SCI®* indicates that this paper has been cited over 685 times since 1967.]

Paul E. Lacy
Department of Pathology
Washington University
School of Medicine
St. Louis, MO 63110

January 27, 1981

"The islets of Langerhans are nests of endocrine cells which are scattered throughout the pancreas and comprise only a few percent of the entire organ. This anatomical fact placed a severe limitation on studies concerned with the mechanisms for stimulation and secretion of islet hormones. In our early studies on the ultrastructural events involved in insulin secretion, we attempted to circumvent the problem by serially sectioning blocks of pancreas until an islet was found. In order to correlate enzymatic changes in the islets with insulin secretion, we dissected islets from frozen-dried sections of pancreas, weighed the tissue on a quartz fiber balance and used it for quantitative enzymatic studies. Obviously, these approaches were extremely tedious and frustrating and provided new but yet limited information. Further progress required the development of a simple means for isolating viable islets for *in vitro* studies on hormonal secretion.

"In 1965, Moskalewski[1] reported that islets could be isolated by partial digestion of the guinea pig pancreas with collagenase. We tried the procedure and soon found that it did not provide a reproducible means of isolating viable islets in sufficient quantities for *in vitro* studies. In considering different approaches for improving the procedure, we knew that the exocrine portion of the pancreas was drained by ducts, whereas the islets were not associated with the ductal system. Therefore, it seemed logical to attempt to inject saline into the pancreatic duct in order to disrupt the exocrine pancreas prior to digesting it with collagenase. This is one of the rare instances where logic produced the expected results and a simple reproducible procedure was developed for the rapid isolation of viable islets for *in vitro* studies on hormonal secretion.

"This procedure for islet isolation quickly became a vital tool in our own investigations on the mechanism of glucose-induced insulin secretion. Utilizing isolated islets maintained *in vitro*, we were able to demonstrate that the microtubular-microfilament system of the beta cell was involved in insulin release following stimulation with glucose. Subcellular fractions of the islets were isolated and procedures were developed for obtaining a plasma membrane fraction for studies on the mechanism of recognition of glucose by beta cells. Isolated islets have also been used for studies on insulin synthesis, proinsulin synthesis, glucagon synthesis and secretion, morphology of the plasma membrane of islet cells, hexose and calcium transport, membrane potential of beta cells, and transplantation of islets into diabetic recipients.

"The diverse application of this procedure to many different problems would probably explain why this paper achieved the 'most-cited' list. From my standpoint, I am delighted that a logical conjecture in research was successful, and I am grateful to learn that others were also assisted in their studies on the endocrine pancreas."

1. Moskalewski S. Isolation and culture of the islets of Langerhans of the guinea pig.
Gen. Comp. Endocrinol. 5:342-53, 1965.

Berry M N & Friend D S. High-yield preparation of isolated rat liver parenchymal cells: a biochemical and fine structural study. *J. Cell Biol.* **43**:506-20, 1969.
[Div. Clinical Pathology and Dept. Pathology, Univ. California Sch. Med., San Francisco, CA]

A method is described for the high-yield preparation of suspensions of intact isolated rat hepatocytes, by perfusion of the liver with a Ca^{2+}-free medium containing collagenase. [The *SCI®* indicates that this paper has been cited in over 1,570 publications since 1969.]

Michael N. Berry
Department of Clinical Biochemistry
School of Medicine
Flinders University of South Australia
Bedford Park, South Australia 5042
Australia

September 2, 1983

"My attempts to prepare suspensions of intact isolated parenchymal liver cells (hepatocytes) began in New Zealand in 1958, with the encouragement of N.L. Edson, who recognized their potential value for metabolic studies. Employing Anderson's technique,[1] I obtained high yields of cells which, unfortunately, were virtually devoid of normal metabolic activity. This led to a three-year study in collaboration with a skilled electron microscopist, F.O. Simpson, which demonstrated that cells prepared by high-pressure perfusion of the liver suffered gross damage to cellular membranes and loss of cytoplasmic contents.[2,3] As a consequence of these studies, I developed a lasting interest in the relationship between cellular structure and function, as well as a determination to find a method for preparing intact cells.

"I continued my research under H.A. Krebs in Oxford. He discouraged further efforts directed toward isolated cell preparation, arguing very reasonably that if liver slices could not function as well as perfused liver, there was little likelihood that single cell suspensions would do so. At his direction, I spent a year establishing the perfused liver technique—experience that was subsequently to prove of great value for the preparation of isolated hepatocytes. I spent the next few years undertaking liver perfusion studies, now with an added incentive to find a satisfactory method for isolated liver cell preparation.

"In 1967, I took up an appointment at the University of California at San Francisco. R.B. Howard, then at Stanford University, had just published a key paper[4] showing for the first time that collagenase treatment of liver slices gave preparations of isolated intact hepatocytes, albeit in low yield. Recognizing the importance of Howard's observations, I set about trying to increase the yield. It seemed logical to attempt to perfuse livers with a medium containing collagenase, and it was highly gratifying to find that this approach produced large quantities of intact isolated cells. This was one of the very few occasions in my research career when logic has been rewarded! The outstanding electron micrographs of D.S. Friend made a major contribution to the paper and many of the citations refer to morphological aspects of the work.

"The bulk of the citations, however, reflect the usefulness of the preparation as a tool for the study of hepatic metabolism. It turned out, contrary to expectations, that isolated hepatocytes perform metabolically in most instances as well as the perfused liver, while being far easier to manipulate and permitting a single liver to be used for multiple studies. In fact, when Krebs came to appreciate this, he became a strong advocate for the method and this had much to do with its rapid rise in popularity. Since then, numerous modifications of the collagenase-perfusion technique have been published (not all of them, in my view, desirable), and in consequence the number of citations for the paper underestimates perhaps fivefold the usage of the method. Nevertheless, according to *Science Citation Index®*, the paper continues to be cited with increasing frequency, presumably reflecting the entry of new workers into the field. For a review, see the proceedings of a recent conference."[5]

1. **Anderson N G.** The mass isolation of whole cells from rat liver. *Science* 117:627-8, 1953.
   (Cited 165 times since 1955.)
2. **Berry M N.** Metabolic properties of cells isolated from adult mouse liver. *J. Cell Biol.* 15:1-8, 1962.
3. **Berry M N & Simpson F O.** Fine structure of cells isolated from adult mouse liver. *J. Cell Biol.* 15:9-17, 1962.
4. **Howard R B, Christensen A K, Gibbs F A & Pesch L A.** The enzymatic preparation of isolated intact parenchymal cells from rat liver. *J. Cell Biol.* 35:675-84, 1967. (Cited 125 times.)
5. **Harris R A & Cornell N W,** eds. *Isolation, characterization, and use of hepatocytes: proceedings of the International Symposium on Isolation, Characterization, and Use of Hepatocytes held at Indiana University School of Medicine, Indianapolis, October 22-24, 1982.* New York: Elsevier Biomedical, 1983. 660 p.

This Week's Citation Classic

Miller R G & Phillips R A. Separation of cells by velocity sedimentation.
J. Cell. Physiol. **73**:191-201, 1969.
[Dept. Medical Biophysics, Univ. Toronto, and Ontario Cancer Inst.,
Toronto, Ontario, Canada]

A simple and reproducible method for fractionating populations of living cells by velocity sedimentation in the Earth's gravitational field was described. Use of the procedure as an analytical tool rather than as a method for purifying cells was emphasized. [The *SCI®* indicates that this paper has been cited in over 915 publications since 1969.]

Richard G. Miller
Ontario Cancer Institute
Toronto, Ontario M4X 1K9
Canada

January 20, 1983

"After obtaining my PhD in nuclear physics in 1966, I decided to change fields and went to the lab of J.E. Till at the Ontario Cancer Institute to do a postdoc in cell biology. He and E.A. McCulloch were then deeply involved in studies of the hemopoietic stem cell, particularly the factors governing whether it underwent self-renewal or became committed to a particular differentiation pathway. It was felt that a population of 'pure' stem cells would greatly aid these studies and Till suggested I try to develop an appropriate separation procedure. The underlying concept was that cells differing in biological function might also differ sufficiently in various physical parameters that useful separations could be achieved.

"I chose to develop sedimentation separation. Cells are essentially spheres which will fall through a viscous fluid such as saline at a rate determined primarily by their size but also to some extent by their density. I first built an apparatus according to the design of Mel.[1] This evolved, in a number of discontinuous steps, into the much simpler and more powerful apparatus described in the highly cited article, an apparatus similar to but somewhat simpler than one independently developed by Peterson and Evans.[2]

"Why has this paper become so highly cited? There are, I think, two reasons. First, it describes a cell separation method which many people have found useful. Second, it embodies a new way of thinking about the analysis of cell populations, a way of thinking being developed at the same time (but using different separation procedures) by R.C. Leif in the US, K. Shortman in Australia, and K. Zeiller in the Federal Republic of Germany (reviewed in reference 3).

"By the time the 1969 article was written, R.A. Phillips, another postdoc in the lab, and I had begun to make extensive applications of the procedure (reviewed in reference 4). In the course of these experiments we realized that the primary use of the procedure was not so much in purifying a particular type of cell but in the analysis of differentiating cell populations. Thus, cells of a given type could be characterized with their sedimentation velocity in the same way that a biochemist can use the molecular weight estimated from a Sephadex column to characterize a molecule. To associate a function with a particular cell type, both function and morphology could be assessed independently for each fraction from the separation and a correlation sought between the two. Secondly, the procedure could be used to separate one type of cell from another, an often useful result even if neither cell type is obtained in very high purity. Thirdly, the procedure could be used to purify cells of a particular type. This last, although it was the initial objective, was seldom successfully achieved. Newer methods (e.g., fluorescence-activated cell sorting) can now often achieve this objective if the separation can be directly based on a biological marker for the function of interest (reviewed in reference 5). However, the first two objectives are often still most effectively achieved with the procedure of our 1969 paper (updated in reference 6)."

1. **Mel H C.** Sedimentation properties of nucleated and non-nucleated cells in normal rat bone marrow.
Nature 200:423-5, 1963.
2. **Peterson E A & Evans W H.** Separation of bone marrow cells by sedimentation at unit gravity.
Nature 214:824-5, 1967.
3. **Shortman K.** Physical procedures for the separation of animal cells. *Annu. Rev. Biophys. Bioeng.* 1:93-130, 1972.
4. **Miller R G, Gorczynski R M, Lafleur L, MacDonald H R & Phillips R A.** Cell separation analysis of B and T lymphocyte differentiation. *Transplant. Rev.* 25:59-97, 1975.
5. **Miller R G & Price G B.** Cell separation and surface markers. *Clin. Haematol.* 8:421-34, 1979.
6. **Miller R G.** Separation of cells by velocity sedimentation. (Bloom B R & David J R, eds.) *In vitro methods in cell-mediated and tumor immunity.* New York: Academic Press, 1976. Vol. 2. p. 283-307.

Wollenberger A, Ristau O & Schoffa G. Eine einfache technik der extrem schnellen Abkühlung grösserer Gewebestücke. (A simple technique for the extremely rapid cooling of larger tissue samples.) *Pflugers Arch. Physiol.* **270**:399-412, 1960.

Tissue and organs can be frozen in situ in a fraction of a second by being compressed to a thin layer between two aluminum blocks that are precooled in liquid nitrogen and for convenient handling form part of a clamp. [The *SCI®* indicates that this paper has been cited over 490 times since 1961.]

Albert Wollenberger
Central Institute of
Heart and Circulatory
Regulation Research
Academy of Sciences of the GDR
1115 Berlin-Buch, GDR

April 27, 1978

"This paper was previously classified in *Current Contents®* as belonging to the category of *Uncitedness III*,[1] thanks to a practice among authors of referring in their publications to its content without taking pains to cite it as a reference.

"The technique of rapid tissue fixation with the type of freezing clamp described in this 1960 paper was for the first time made public by myself and Bozkourt Wahler at the 20th International Physiology Congress in Brussels in August 1956. When I later asked Dr. Wahler to continue working with me in an attempt to verify our claims by actual measurements of tissue cooling rates, he declined because, as he said, he did not wish to waste his time on technical trivialities. I thereupon engaged the collaboration of two colleagues from a neighboring department—Georg Schoffa, who today is professor of biophysics at the Technical University of Karlsruhe, and his technician, Otto Ristau, now a research chemist at the Central Institute of Molecular Biology of our Academy here in Berlin-Buch. The thermoelectric measurements were done by Ristau and myself, both Schoffa and Ristau

were helpful in the theoretical analysis of the tissue cooling process, and I wrote the main part of the paper, which was submitted to *Pflugers Archiv*.

"Unfortunately, the paper appeared in print in a somewhat mutilated form, because a figure documenting the adequacy of our instrumental set-up was deleted at the insistence of one of the editors. This figure, which never was published, showed that as soon as the bare thermocouple used for monitoring tissue temperature was clamped between aluminum blocks precooled in liquid nitrogen to -196°C, its temperature fell without delay and at an initial rate of approximately 20,000°C/sec. There was thus no need to correct the thermo-oscillograms presented in our paper for instrumental inertia.

"The freezing clamp as a tool for the instant cryofixation of tissues owes its popularity partly to the simplicity of its design. Various modifications and more complicated freezing devices based on the clamping principle were introduced in the course of the years, the most sophisticated version being an apparatus that automatically freezes the heart of small open-chest animals at any predetermined point of the electrocardiogram. It was constructed in Moscow by Dr. A. N. Medelyanovski. Making use of this apparatus through collaboration with Medelyanovski's wife, Dr. Yenia Bogdanova, who was not exactly a newcomer to cryobiology after having been national women's champion of the USSR in figure ice-skating, my Berlin coworkers and I were privileged to have a part in the demonstration of systematic oscillations of myocardial cyclic nucleotide levels during the cardiac contraction cycle of the frog.[2] For less extravagant purposes, however, and in many laboratories throughout the world the primitive clamp described in 1960 continues to serve as a satisfactory tool."

1. **Garfield E.** Uncitedness III—the importance of *not* being cited.
   Current Contents (8):5-6, 21 February 1973.
2. **Wollenberger A, Babski E B, Krause E G, Genz S, Blohm D & Bogdanova E V.** Cyclic changes in levels of cyclic AMP and cyclic GMP in frog myocardium during the cardiac cycle.
   Biochem. Biophys. Res. Commun. **55**:446-52, 1973.

Emmelot P, Bos C J, Benedetti E L & Rümke Ph. Studies on plasma membranes. I. Chemical composition and enzyme content of plasma membranes isolated from rat liver. *Biochim. Biophys. Acta* **90**:126-45, 1964.
[Depts. Biochem., Electron Microscopy, and Immunol., Antoni Van Leeuwenhoek-Huis, The Netherlands Cancer Inst., Amsterdam, The Netherlands]

This paper reported for the first time a method for isolating *pure* liver plasma membranes, accompanied by a comprehensive account of chemical, enzymatic, structural, and immunological properties of the isolated membranes. Marker enzymes were thus established. [The *SCI*® indicates that this paper has been cited in over 695 publications since 1964.]

P. Emmelot
Divisions of Cell Biology and
Chemical Carcinogenesis
Antoni Van Leeuwenhoek-Huis
The Netherlands Cancer Institute
1066 CX Amsterdam
The Netherlands

April 23, 1982

"At the end of the 1950s, methods were available for the isolation of intracellular organelles but a method for plasma membrane isolation was still lacking. Consequently, the biochemistry of the cell surface was virtually a *terra incognita*. About that time, our interest in the plasma membrane was aroused by the finding that transplants of rapidly growing rat liver tumors readily yielded single, metabolically intact cells by suspending the finely divided tissue in a weakly acid phosphate buffer. This procedure did not work with rat liver.[1] The relative ease with which single tumor cells could be obtained was apparently due to a decrease in the mutual adhesiveness of the hepatoma as compared with the liver cells.

"Obviously for me as a biochemical oncologist, cell adhesiveness had a particular ring not only in view of the problem of metastasis but also because of the notions which were emerging at the time about functional relations between cell contact and the biosocial behavior of cells. These relations—pertaining to cell recognition and discrimination, tissue organization, and cell proliferation and differentiation—were mostly of a descriptive type being vague as to the molecular mechanisms concerned. Nevertheless, contact lesions conceptually fitted the asocial behavior of the cancer cell.

"We then set out to find a suitable method for the isolation of plasma membranes from liver and liver tumors of rats and mice. In a number of pilot experiments carried out with fractionated liver homogenates subjected to electron microscopy, we observed that large plasma membrane sheets, preserving intact bile space membranes and being interconnected by junctional complexes—poorly known at the time and qualified as desmosomes and terminal bars—could be observed, mainly as cosediments in the nuclear fraction. It was decided to use these morphological criteria as provisional markers for the isolation of plasma membranes. Early in the course of our work, a paper by Neville[2] describing the isolation of rat liver plasma membranes appeared. The data contained therein, plus our experience then, led to a procedure which was essentially a modification of Neville's method, specifically aiming at the removal of contaminating mitochondria. This method, which has sometimes been quoted as the Neville-Emmelot method, is the one featured in this *Citation Classic*, being our first full paper on the subject. An exhaustive description was provided later.[3]

"From the outset our interest and effort were not so much directed toward the normal plasma membrane per se, as a void area for study, but rather to the elucidation of surface properties which might distinguish tumor from normal cells. The aim of a comparative analysis, however, necessitated a thorough study of the chemical, enzymatic, structural, and immunological properties of the liver plasma membrane of which the first results were contained in our 1964 paper. This and following papers have contributed to the field of fundamental membranology, besides helping to elucidate our primary aim.

"The reason that our 1964 paper has been quoted so often is, I think, due to the fact that it presented a body of diversified data—provided by my coauthors in a multidisciplinary approach—which allowed a good appreciation of the method. I also believe that the elaboration of these data and the new findings reported in our subsequent papers have focused attention on the original 1964 paper. Furthermore, I have heard either by letter or personal contact that the method is easily reproducible, also with respect to the quantitative data. Most important, however, I have been informed more than once that the paper has been an impetus for others to study a particular problem at the level of the isolated plasma membrane."

1. Emmelot P. Comparative biochemistry of rat hepatomas. *Acta Union Int. Cancer* 20:902-8, 1964.
2. Neville D M. The isolation of a cell membrane fraction from rat liver. *J. Biophys. Biochem. Cytol.* 8:413-22, 1960.
3. Emmelot P, Bos C J, Van Hoeven R P & Van Blitterswijk W J. Isolation of plasma membranes from rat and mouse livers and hepatomas. *Methods Enzymol.* 31:75-90, 1974.

Singer S J & Nicolson G L. The fluid mosaic model of the structure of cell membranes. *Science* 175:720-31, 1972.

The paper presented a fluid mosaic model for the gross organization and structure of the proteins and lipids of biological membranes, and in light of this model suggested possible mechanisms for various membrane functions and membrane-mediated phenomena. [The *SCI®* indicates that this paper was cited 855 times in the period 1961-1975].

Dr. S.J. Singer
Department of Biology
University of California
La Jolla, California 92093

January 31, 1977

"This particular paper was the culmination of our ideas and experiments going back about 10 years. As a protein chemist, weaned on notions of the globularity of most soluble protein molecules and the importance of hydrophobic interactions, and because of our studies and those of others in the later 1950's of the conformational changes induced in proteins by non-aqueous solvents, the protein-lipid-protein sandwich model that was generally accepted as the membrane model in the early 1960's seemed wrong to me. My feelings were enhanced by the work from David Green's laboratory in the early 1960's, showing the importance of hydrophobic interactions in membrane protein systems. But it wasn't until 1965 that I attempted any experimental test of my ideas. At that point, I was able to obtain a commercially available circular dichroism-optical rotatory dispersion instrument, and John Lenard and I carried out some studies on the average conformations of pro-

teins in nearly intact membranes, showing that these were about 30-40% α-helical. Wallach and Zahler did similar work at the same time. We both published papers including such data and suggesting that membrane proteins were globular amphipathic molecules partially embedded in the membrane. We refined these ideas by additional experiments over the next few years, but I did not have the chance to put things together in my mind until I had a very pleasant and productive three-months' leave of absence at the Battelle Research Center in Seattle in the spring of 1970. There I wrote a long article on the molecular organization of membranes.[1] After a seminar at the Rockefeller Institute in early 1971, on my return to La Jolla, Dr. Simon Gordon wrote to tell me of a paper that had just appeared by Frye and Edidin,[2] which was a remarkable demonstration of the lateral mobility of protein components embedded in a lipid bilayer in the membrane. With my graduate student at the time, Garth Nicolson, I then wrote the *Science* paper which incorporated fluidity and mobility into the mosaic model. I guess that a major reason the paper has been cited so often is that it turned out to be very timely. At about the same time that it was published, immunologists were discovering the 'capping' phenomenon on lymphocyte surfaces,[3] in which antibodies specific for individual receptors on the surface collected their receptors into large clusters called caps. The fluid mosaic model provided a ready molecular explanation for this phenomenon, and for similar observations that were made soon after.

"The molecular biology of membranes has been an incredibly active and exponentially expanding field in the last decade, and it has been personally very rewarding to have been a part of that development."

REFERENCES

1. **Singer S J.** The molecular organization of biological membranes, in *Structure and function of biological membranes* (Rothfield L I, ed.). New York: Academic Press, 1971, p. 145-222.
2. **Frye L D & Edidin M.** The rapid intermixing of cell surface antigens after formation of mouse-human heterokaryons. *Journal of Cell Science* 7:319-35, 1970.
3. **Taylor R B, Duffus W P H, Raff M C & de Petris S.** Redistribution and pinocytosis of lymphocyte surface immunoglobulin molecules induced by anti-immunoglobulin antibody. *Nature New Biology* 233:225-9, 1971.

Hruban Z, Spargo B, Swift H, Wissler R W & Kleinfeld R G. Focal cytoplasmic degradation. *Amer. J. Pathol.* **42**: 657-83, 1963.

An intracellular process consisting of the sequestration of cytoplasmic components followed by the formation of complex dense bodies, which correspond to the lysosomes of biochemists, is established by ultrastructural studies. This natural process is enhanced by cellular injury. The noxious agent codetermines the structure of the bodies. [The *SCI*® indicates that this paper has been cited 274 times since 1963.]

Zdenek Hruban
The University of Chicago
Department of Pathology
Chicago, IL 60637

December 19, 1977

"While studying the ultrastructural changes induced in hepatocytes and pancreatic acinar cells by phenylalanine analogs, I observed a series of changes which I interpreted as the stages of a process. The results presented at the Federation meetings and the American Association of Pathologists and Biologists meetings in 1961 brought encouragement from Drs. Hans Popper and W. Bernhard. Full length articles appeared the next year. The principal finding was that damaged organelles are sequestered by membranes and degraded. The observed structures were interpreted as stages of a process which is beneficial in repairing cellular injury, and as stages in the formation of lysosomes. This interpretation changed the static concept of the lysosome as a particle into the dynamic concept of a degradative process. Rather than particles causing injury, lysosomes became byproducts of the cellular reaction to injury. We realized that focal degradation (FCD) is a general reaction of the cell to injury (starvation, deficiency, hypoxis) and reported it in the 'Classic' article.

"The relation of focal degradation to atrophy and phagocytosis was presented in *Federation Proceedings* in 1964, in an article equally frequently quoted.[1] Morphologists and pathologists responded positively to the new concept, but the dynamic concept of the lysosome as a process remains foreign to some biochemists even today. The term FCD has since been replaced by the simpler term autophagy.

"The results of our early morphological studies at the University of Chicago confirmed our belief that sequential ultrastructural studies at short time intervals and the arrangement of the images in logical sequences open new avenues for a dynamic interpretation of physiological and pathological processes which can not be fully understood by biochemistry alone. The application of these principles to studies of cellular alterations, as originally described in the 'Classic,' led to the concept of myeloid bodies, which are lysosomes sequestering amphophilic drugs bound to membranes. Although the full description of myeloid bodies appeared in 1965, the basic idea of drug-induced lysosomal changes was accepted slowly and has only recently been considered in drug safety evaluation. Other dynamic concepts arising from morphological investigations are the concept of nonlysosomal cytoplasmic degradation intranuclear layered inclusion (topolysis) and the concept of altered membrane flow, which accounts for the formation of intracytoplasmic vacuoles, the invaginations of cell membrane and the formation of cytoplasmic bullae.

"Looking back on morphological research in general, we find that in recent years many morphologists have been eliminated from active basic research by the denial of research grants, while a quasi-supernatural power is often attributed to pure biochemistry. The article on focal cytoplasmic degradation should continue to remind us that morphological studies yield interpretations as dynamic as those of other disciplines."

REFERENCE
1. Swift H & Hruban Z. Focal degredation as a biological process.
   Fed. Proc. **23**:1026-37, 1964.

Dallner G, Siekevitz P & Palade G E. Biogenesis of endoplasmic reticulum membranes. I. Structural and chemical differentiation in developing rat hepatocyte. *J. Cell Biol.* **30**:73-96, 1966.
[Dept. Cell Biology, Rockefeller Univ., New York, NY]

This paper and the accompanying paper[1] used the endoplasmic reticulum (ER) membranes in livers of newborn rats as models for membrane biogenesis. The investigation led us to the conclusion that new membranes are not synthesized *de novo* as one unit, but that the various enzymes are added at an individual rate in order to complete the membrane. Also, several components that are synthesized in the rough ER membranes are subsequently transferred to the smooth ones. [The *SCI®* indicates that this paper has been cited in over 565 publications since 1966.]

Gustav Dallner
Department of Pathology
Huddinge University Hospital
Karolinska Institutet Medical School
S-141 86 Huddinge
Sweden

September 21, 1984

"Considering the immense developments in our knowledge of lipid and protein structure, conformation, and synthesis in just the last 10 years, it is not easy to go back even farther and recount the problems involved in studies of membrane biogenesis in the mid-1960s. Most of the ideas about how membranes are synthesized were speculative then, and the experimental observations were limited. I arrived in New York on the *Queen Elizabeth I* to start postdoctoral studies at Rockefeller University (Institute, at that time). I met with Palade and Siekevitz on my first day at Rockefeller, and they told me that a week earlier they had done an experiment to test some ideas that would be a suitable starting point for me. They had bought a few pregnant rats, studied electron micrographs of the livers of the newborn rats, and measured glucose-6-phosphatase activity on isolated liver microsomes. The amount of membrane observed had increased directly with the enzyme activity. This was a system obviously suitable for analyzing membrane biogenesis.

"The experiments I performed were expensive and not so easy. I ordered and obtained 15 pregnant rats for every experiment. These rats had been mated for delivery on a defined day. There was, however, some discrepancy between my rhythm and that of the rats. Each delivery was spread out considerably but concentrated somehow in the nighttime hours. Since most of the important biological changes in this system occur just after birth, I had to collect a sufficient number of newborn rats to start an experiment. My re-education from a pathologist to a gynecologist took place during the endless nights when I walked up and down 66th Street between First and York Avenues (between my home and the laboratory) to look at my rats and estimate the approximate time of delivery. I would walk home for a short rest several times during the night. My wife would wake up sometimes and ask in a sleepy voice: 'Are you coming or going?'

"Our efforts yielded highly rewarding results, which is not always the case in research. I learned a great deal about the morphological, chemical, and enzymic pattern of the developing endoplasmic reticulum (ER). The ensuing paper contained a description of the isolation of rough and smooth microsomes that is widely used even today. The fact that many of the membrane components are synthesized in the rough ER before transport to the smooth ER was interesting *per se*. But this was a minor finding in comparison with the main conclusion that biological membranes are not synthesized *de novo* as a whole, like the ones obtained in reconstitution experiments.

"Our work indicated that individual proteins have individual turnover times and that they are placed in and removed from the membrane individually to construct the final product. We suggested the principle that membranes are synthesized in a 'multi-step process,' contrary to the hypothesis of a one-step assembly of a homogeneous membrane. Our hypothesis, which has been substantiated during the subsequent years of experimentation, is the reason for the frequent citation of this paper.

"Time has passed and I am no longer working in this particular field, but I have followed closely the immense development that has taken place in the last decade.[2-4] This work provided me, a postdoc from another continent, with the great opportunity to encounter the atmosphere and the people of Rockefeller Institute, which have contributed so richly to innovation and achievement in cell biology research."

1. Dallner G, Siekevitz P & Palade G E. Biogenesis of endoplasmic reticulum membranes. II. Synthesis of constitutive microsomal enzymes in developing rat hepatocyte. *J. Cell Biol.* **30**:97-117, 1966. (Cited 520 times.)
2. Blobel G & Dobberstein B. Transfer of proteins across membranes. I. Presence of proteolitically processed and unprocessed nascent immunoglobulin light chains on membrane-bound ribosomes of murine myeloma. *J. Cell Biol.* **67**:835-51, 1975. (Cited 1,435 times.)
3. Sabatini D D, Kreibich G, Morimoto T & Adesnik M. Mechanism for the incorporation of proteins in membranes and organelles. *J. Cell Biol.* **92**:1-22, 1982.
4. Schatz G. How mitochondria import proteins from the cytoplasm. *FEBS Lett.* **103**:203-11, 1979. (Cited 110 times.)

Beams H W & Kessel R G. The Golgi apparatus: structure and function.
Int. Rev. Cytol. 23:209-76, 1968.
[Department of Zoology, University of Iowa, Iowa City, IA]

In addition to summarizing the pros and cons of the Golgi apparatus controversy, evidence is given which led to its unequivocal establishment as an important cellular organelle and one that must be reckoned with in understanding many of the cellular functions. [The *SCI®* indicates that this paper has been cited in over 280 publications since 1968.]

H.W. Beams
Department of Zoology
University of Iowa
Iowa City, IA 52242

June 6, 1984

"In 1926, upon entering the Graduate College of the University of Wisconsin, I was confronted, like most beginning graduate students, with the problem of selecting an adviser and a research topic of interest to me that would also be suitable for partially fulfilling the requirements for the PhD. Among the courses in which I enrolled was a seminar in cytology taught by M.F. Guyer. The topic I was assigned to review for the class was the present status of the Golgi apparatus, a structure first described by Camillo Golgi in 1898 and termed by him the 'Apparato reticulare interno'.[1] In pursuing this subject, I found that it had been estimated that over 2,000 publications had appeared relative to its nature, varying from disbelief of its reality to a complete acceptance of its existence. In the heated and sometimes vehement debate concerning the Golgi apparatus, emotional statements concerning it were sometimes made as, for example, 'The first description was the first mistake.'[2] The controversy concerning the Golgi apparatus was largely due to the following: it could not be consistently demonstrated in most living cells, its demonstration was mainly limited to the somewhat

capricious metallic impregnation methods (osmium and silver), and its form and position vary considerably in different cell types and under different physiological conditions.

"My thesis dealt with the secretion process in the mammary glands of rats. From this study, I became convinced that the Golgi apparatus was a real organelle, probably involved in the secretion process. This work stimulated me to undertake further studies on the cytoplasmic organelles in the pancreatic and spinal ganglion cells of the rat where it was possible to demonstrate the 'vacuome[3] and the classical Golgi apparatus side by side in the same cell. In subsequent studies, various cells of both plants and animals were exposed to high centrifugal force (400,000 x g) in an air turbine ultracentrifuge developed by J.W. Beams and collaborators. The density of the Golgi apparatus differed from that of all the other cellular materials, adding further evidence in support of the view that it was a real cellular organelle. In 1934, I received a Rockefeller Foundation Fellowship to work in the laboratory of J.B. Gatenby at Trinity College, Dublin. There, in collaboration with Gatenby, a longtime advocate of the reality of the Golgi apparatus, cytological studies were continued on a number of different cell types.

"I think the paper received a large number of citations because it provided a readily available account of the Golgi apparatus controversy and pointed out how the development of the electron microscope paralleled the establishment of the Golgi apparatus as a bona fide cellular organelle.

"Much of my research cited in this review was done in collaboration with my colleagues, R.L. King and R.G. Kessel. The latter's name appears as coauthor of the cited paper. See references 4 and 5 for recent reports."

1. Golgi C. Sur la structure des cellules nerveuses. *Arch. Ital. Biol.* 30:60-71, 1898.
2. Hirsch G C. Introductory remarks. (Seno S & Cowdry E V. eds.) *Intracellular membraneous structure.* Proceedings of the First International Symposium for Cellular Chemistry. Okayama: Japan Society for Cell Biology. 1963. p. 193-5.
3. Parat M & Painlevé J. Appareil réticulaire interne de Golgi. trophosponge de Holmgren. et vacuome. *CR Acad. Sci.* 179:844-6. 1924.
4. Farquhar M G & Palade G E. The Golgi apparatus (complex)—(1954-1981)—from artifact to center stage. *J. Cell Biol.* 91:77s-103s. 1981.
5. Whaley W G. The Golgi apparatus. *Cell Biol. Monogr.* 2:1-190. 1975.

CC/NUMBER 52
DECEMBER 24-31, 1984

This Week's Citation Classic™

Wessells N K, Spooner B S, Ash J F, Bradley M O, Luduena M A, Taylor E L, Wrenn J T & Yamada K M. Microfilaments in cellular and developmental processes. *Science* 171:135-43, 1971.
[Department of Biological Sciences, Stanford University, CA]

This paper reported the correlations between cytoplasmic microfilaments and motile activities in an array of cellular and developmental systems using the mold metabolite cytochalasin B as the experimental probe. Cytochalasin B concomitantly disrupted microfilament arrays and inhibited motile activities This activity identified microfilaments as fundamental cytoplasmic structures in cellular and morphogenetic movements. [The *SCI®* indicates that this paper has been cited in over 1,325 publications since 1971.]

Brian S. Spooner
Division of Biology
Kansas State University
Manhattan, KS 66506

November 8, 1984

"By the late 1960s, electron microscopic analyses had led to the recognition that cytoplasmic arrays of microfilaments were present where they could provide the basis for the cell-shape changes involved in cytokinesis, cell motility, and morphogenetic movements of epithelia. In fact, the structural similarity between microfilaments and the thin filaments of muscle suggested that they could be components of a contractile apparatus powering various nonmuscle motility phenomena. However, experimental demonstration of the necessity of microfilaments in this process was lacking. Then, in 1967, Carter[1] published results showing that mold metabolites called cytochalasins caused binucleation of cultured cells. Next, a seminal experiment by Schroeder, which was first reported in 1969 and appeared later as a major publication in 1972,[2] demonstrated that cytochalasin caused disappearance of the 'contractile ring' of microfilaments normally present in the cleavage furrow, leading to failure of the furrow to form or progress, thus producing binucleation.

"In the late fall of 1969, I was a postdoctoral fellow analyzing tissue interactions in organ morphogenesis in N.K. Wessells's laboratory at Stanford University. Wessells had already published on microfilaments in developing epithelia. He had a graduate student, J.T. Wrenn, who discovered that the early event in estrogen-induced tubular gland formation in the chick oviduct was the appearance of arrays of microfilaments followed by cell-shape changes and tubular gland invaginations. The potential application of cytochalasin to this system was obvious from Schroeder's key discovery, and a supply was obtained. In addition, it was clear that microfilaments could act in the branching morphogenesis systems I was working on. Furthermore, another student, K.M. Yamada (presently chief, Membrane Biochemistry, Laboratory of Molecular Biology, National Cancer Institute), was investigating sensory ganglia neurite extension, a motile activity that could also involve microfilaments. Cytochalasin effects on these systems were dramatic, resulting in a series of papers in *Proceedings of the National Academy of Sciences of the USA* in 1970 showing disruption of microfilament systems and inhibition of salivary branching,[3] tubular gland formation,[4] and neurite extension.[5]

"Wessells's laboratory was a remarkable environment in which to work. He had an outstanding array of graduate students, including Wrenn and Yamada, M.O. Bradley, J.F. Ash, and M. Anderson (Luduena), and he ran his laboratory in a supportive and collaborative way. One worked with, rather than for, Wessells. His attitude was contagious and, in short order, these people had extended the motility-microfilament-cytochalasin correlation to a variety of systems. In addition to the organ systems and neurite extension, microfilament involvement was established for single-cell locomotion, ascidian metamorphosis, cytoplasmic streaming, contractions of egg cortices, platelet contraction, and other motile phenomena, all of which were reported in this *Citation Classic* manuscript.

"Response to the publication was rapid and sustained. There was controversy over the mode of action of cytochalasin, but, in general, microfilament involvement in motile phenomena was accepted and has been further documented. The succeeding decade saw an explosion of research[6] that continues today in the general area of microfilaments, cytoplasmic contractile proteins, and the cytoskeleton. The citation frequency of this paper stems from its timely appearance and the breadth of its coverage of motile phenomena involving microfilaments."

1. Carter S B. Effects of cytochalasins on mammalian cells. *Nature* 213:261-4, 1967. (Cited 665 times.)
2. Schroeder T E. The contractile ring. II. Determining its brief existence, volumetric changes, and vital role in cleaving *Arbacia* eggs. *J. Cell Biol.* 53:419-34, 1972. (Cited 160 times.)
3. Spooner B S & Wessells N K. Effects of cytochalasin B upon microfilaments involved in morphogenesis of salivary epithelium. *Proc. Nat. Acad. Sci. US* 66:360-4, 1970. (Cited 140 times.)
4. Wrenn J T & Wessells N K. Cytochalasin B: effects upon microfilaments involved in morphogenesis of estrogen-induced glands of the oviduct. *Proc. Nat. Acad. Sci. US* 66:904-8, 1970. (Cited 100 times.)
5. Yamada K M, Spooner B S & Wessells N K. Axon growth: roles of microfilaments and microtubules. *Proc. Nat. Acad. Sci. US* 66:1206-12, 1970. (Cited 245 times.)
6. Korn E D. Biochemistry of actomyosin-dependent cell motility (a review). *Proc. Nat. Acad. Sci. US* 75:588-99, 1978. (Cited 455 times.)

Pollard T D & Weihing R R. Actin and myosin and cell movement.
CRC Crit. Rev. Biochem. 2:1-65, 1974.
[Dept. Anatomy, Harvard Medical Sch., Boston, and Worcester Foundation for Experimental Biology, Shrewsbury, MA]

This is a review of the early biochemical and structural studies that established the existence of the contractile proteins, actin and myosin, in nonmuscle cells. This work provided the evidence for the generally accepted concept that contractile proteins generate the forces for many types of cellular movements. [The *SCI®* indicates that this paper has been cited in over 945 publications since 1974.]

Thomas D. Pollard
Department of Cell Biology and Anatomy
Johns Hopkins Medical School
Baltimore, MD 21205

May 15, 1984

"This article has been cited frequently because it was a timely, comprehensive review of a major new field in cell biology that was just on the threshold of its most rapid phase of growth. The subject is the molecular basis of cellular motility.

"Biologists have always been fascinated with cellular movements. By the late 1960s, microtubules and the ATPase protein dynein were identified as the motion generating system of eukaryotic flagella, but nothing concrete was known about the molecular basis of amoeboid movement, cytoplasmic streaming, cytokinesis, and so on. Then several pioneering papers[1-3] established that slime molds and other cells possess protein molecules quite similar to the actin and myosin that power muscle contraction. In the next few years, a handful of largely youthful investigators (including the authors of this review, who had been postdocs together with Ed Korn at the National Institutes of Health [NIH]) demonstrated the generality of these findings and established many of what remain today general principles in the field.

"Although the field was very young in 1973, our review was the size of a small book. It must have filled an information void. Since *CRC Critical Reviews in Biochemistry* was new and not generally available (for photocopying), we received many requests for reprints. This presented two problems: first, the publisher would not sell us reprints (I suppose that they were trying to promote the sales of their new journal); and, second, we did not have research funds to purchase reprints, since everyone's grants had been cut following President Nixon's impoundment of part of the NIH budget. Consequently, we decided to invest our honorarium and some personal funds in several hundred copies produced by the Harvard Printing Office. To make up for our losses, we requested a $1.00 donation for each reprint. A few kind souls replied.

"So, in the end, the review was distributed, not as a journal article, but largely by privately financed reprinting and mailing together with the widespread use of photocopying. Judging from the number of tattered photocopies that I have seen, the review appears to have been the primer for many new investigators who entered the field in the mid-1970s.

"Reviews on cytoplasmic contractile proteins are no longer a novelty (see, for example, reference 4) but short of writing a large monograph, no one could now attempt a comprehensive review like ours in 1974. I doubt that anyone has actually read the review for several years now, but it is still cited, perhaps as an historical landmark at the end of the dark ages."

1. Hatano S & Oosawa F. Isolation and characterization of plasmodium actin.
   Biochim. Biophys. Acta 127:488-98, 1966. (Cited 165 times.)
2. Adelman M R & Taylor E W. Further purification of slime mold myosin and slime mold actin.
   Biochemistry—USA 8:4976-88, 1969. (Cited 120 times.)
3. Ishikawa H, Bischoff R & Holtzer H. Formation of arrowhead complexes with heavy meromyosin in a variety of cell types. *J. Cell Biol.* 43:312-28, 1969. (Cited 750 times.)
4. Pollard T D. Cytoplasmic contractile proteins. *J. Cell Biol.* 91:156s-65s, 1981.

Pearse A G E. *Histochemistry, theoretical and applied.*
London: Churchill, 1960. 998 p.
[Royal Postgraduate Med. Sch., Univ. London, London, England]

This book covers histochemistry in its entirety including, as a separate chapter, its extension to the field of cytochemical ultrastructure. The text, with some 3,000 references and fully illustrated throughout, successfully achieves the delicate balance between theory and practice. [The *SCI®* indicates that this book has been cited over 11,580 times since 1961.]

Anthony G.E. Pearse
Royal Postgraduate Medical School
University of London
London W12 0HS
England

April 30, 1981

"The work cited belongs to a series which began in 1953[1] with the publication of a single volume treatise entitled *Histochemistry, Theoretical and Applied.* The cited second edition appeared in 1960, a two volume third edition in 1968/1972,[2] and the first volume of the three volume fourth edition in 1980.[3]

"Why did I ever embark on such a saga? Well, in 1947 I became assistant lecturer in pathology at the Postgraduate Medical School of London (old terminology) and one day, early in 1948, Henry Dible, the director of the department, suggested to me that I should construct a monograph on the histochemistry of the nucleic acids. He had (rightly) become excited by recent publications from the pen of the Belgian biochemist Jean Louis Brachet.[4] Histochemistry had been welded into some sort of a whole by another Belgian, Lucien Lison, with his book *Histochimie Animale.*[5] Readers of *Current Contents®* (*CC®*) will know that nobody (except me) read this because it was written in French.

"I replied that I would write a text embracing the whole subject or nothing and,

six days later, a publisher's contract to do just that was put before me. With its signing I embarked on a life of unceasing authorship, with all its entailed enquiry into the works of others. Hence my chosen epigram, the Horatian tag *Nam proprium est nihil,* in its true context not the modest 'nothing is original' but 'nothing is your own property.' From the pirated Russian first edition (1956), only this epigram was missing! So it is this, which in 1948 one of my colleagues called a future 'millstone around my neck,' which seems to have become a *Citation Classic.*

"This should be no matter for surprise, for it is evident that a goodly proportion of works which reach this elevated status are concerned with methodology. I can do no better by way of explanation for the apparent success of the 1960 edition than to quote from the 'Historical Introduction' to the fourth edition of *Histochemistry, Theoretical and Applied:*

> It may be that even in its modern form histochemistry has some way to go before it receives proper recognition from those workers in the basic sciences whom it serves so well. If verbal acknowledgement is still lacking, histochemists can take heart from the statistics provided by the records of the *Science Citation Index®* for the years 1961-72. During this period, in the total field of biochemistry and biomedicine, some 29 texts were cited more than 200 times. Five of the 29 were histochemical and these five amassed between them no less than 22.269 per cent of the total number of citations (13,543). The inference to be drawn from these statistics is obvious. It is that the techniques which comprise the discipline of histochemistry have been found useful, and that they are used, by workers in all the biological sciences.[3]

"While I am pleased to be a member of the prestigious CC Club, offered the choice I would certainly prefer election by virtue of one or other of my essays in applied histochemistry, dealing with the diffuse neuroendocrine (APUD) system and its tumors."[6,7]

1. **Pearse A G E.** *Histochemistry, theoretical and applied.* London: Churchill. 1953. 530 p.
2. --------------. *Histochemistry, theoretical and applied.* London: Churchill. 1968-1972. 2 vols.
3. --------------. *Histochemistry, theoretical and applied.* Vol. I. *Preparative and optical technology.*
   Edinburgh: Churchill Livingstone, 1980.
4. **Brachet J.** Nucleic acids in the cell and the embryo. *Symp. Soc. Biol. Nucleic Acid.* 1:207-24, 1947.
5. **Lison L.** *Histochimie animale.* Paris: Gauthier-Villars. 1936. 320 p.
6. **Pearse A G E.** The cytochemistry and ultrastructure of polypeptide hormone-producing cells of the APUD series and the
   embryologic, physiologic and pathologic implications of the concept. *J. Histochem. Cytochem.* 17:303-13, 1969.
   [The *SCI®* indicates that this paper has been cited over 735 times since 1969.]
7. --------------. The APUD concept and hormone production. *Clin. Endocrinol. Metab.* 9:211-2, 1980.

Pearse A G E. **The cytochemistry and ultrastructure of polypeptide hormone-producing cells of the APUD series and the embryologic, physiologic and pathologic implications of the concept.** *J. Histochem. Cytochem.* 17:303-13, 1969.
[Royal Postgraduate Medical School, London, England]

This paper is essentially the distillation of some four years of work on the common cytochemical and ultrastructural characteristics of a widespread and otherwise apparently unconnected collection of endocrine and presumed endocrine cells. Their most convincing (marker) characteristic gave rise to the acronym APUD, by which the whole series came ultimately to be recognized. [The *SCI®* indicates that this paper has been cited over 805 times since 1969.]

Anthony G.E. Pearse
Royal Postgraduate Medical School
University of London
London W12 0HS
England

January 11, 1982

"This paper was first presented at the Third International Congress of Histochemistry and Cytochemistry, held in New York in August 1968. It doubtlessly became the most cited of my contemporary papers because it was the first full and public expression of the APUD concept. Its content has remained valid, subject to minor modifications, up to the present time.

"Acquisition of the mass of facts leading up to the definitive formulation of the concept, and its feeding into my internal computer, had occupied 15 years of wide ranging cytochemical and ultrastructural studies. But one afternoon, late in 1964, I was examining under the microscope a series of dog thyroid preparations from which I hoped to derive information on the functional characteristics of the parafollicular cells. The latter had earlier in the year been identified as the source of Harold Copp's new hormone, calcitonin.[1,2] For the first and only time in my life I was able to cry εὐρηκα, for the parafollicular cells had all the cytochemical characteristics of those known endocrine cells, in the pituitary, pancreas, adrenals, and stomach, on which I had spent so many years of enquiry. Two years later, I had acquired sufficient courage to propound the concept[3,4] and a

further two elapsed before the acronym APUD (*not* Anthony Pearse's Ultimate Dogma, but Amine Precursor Uptake and Decarboxylation) appeared in print.[5]

"In these three papers a significant collective declaration was made that the APUD cells were derived from precursors of 'neural origin, perhaps coming from the neural crest.' For some of the cells, by then 40 in number, this proved to be the case but when studies by several groups of workers[6-8] showed that the 18 gastroenteropancreatic (GEP) APUD cells were not neural crest derivatives, the concept was modified to permit their origin from 'neuroendocrine programmed epiblast.'

"Marker studies[3-5,9,10] have now established with virtual certainty the neuroectodermal origin of all the APUD cells, GEP and non-GEP alike. They thus take their rightful place as constituents of the third, and oldest, division of the nervous system.[11]

"Recognition of the validity of the concept in some quarters has antedated its general acceptance. Perhaps because the diffuse neuroendocrine system is an acknowledged successor to the earlier 'diffuse endocrine epitheliale Organe' of the Austrian pathologist Friedrich von Feyrter (1895-1973),[12] election to membership of the Deutsche Akademie der Naturforscher Leopoldina (1973) and to the Deutsche Gesellschaft für Endokrinologie (1978) were followed by the award of the Ernst Jung Prize for Medicine (1979). The second Fred W. Stewart Prize, awarded to me in that same year by the Memorial Sloan-Kettering Cancer Center, New York, was clearly due to the vested interest of that establishment in oncology, and hence in my work on the neuroendocrine tumors (apudomas), while the clear acceptance of these as a real entity by the surgical fraternity led to the award of the 1976-1978 Triennial John Hunter Prize of the Royal College of Surgeons of England. But better than prizes is the knowledge that my intuitively derived views, expressed in the cited publication, have withstood the collective efforts of a dedicated band of falsifiers. To them, for their stimulating opposition, and to those who have supported me, I am equally grateful."

1. **Copp D H, Cameron E C, Cheney B A, Davidson A G F & Henze K G.** Evidence for calcitonin—a new hormone from the parathyroid that lowers blood calcium. *Endocrinology* 70:638-49, 1962.
2. **Foster G V, McIntyre I & Pearse A G E.** Calcitonin production and the mitochondrion-rich cells of the dog thyroid. *Nature* 203:1029-30, 1964.
3. **Pearse A G E.** 5-Hydroxytryptophan uptake by dog thyroid C cells and its possible significance in polypeptide hormone production. *Nature* 211:598-600, 1966.
4. ------------------, Common cytochemical properties of cells producing polypeptide hormones, with particular reference to calcitonin and the C cells. *Vet. Rec.* 79:587-90, 1966.
5. ------------------, Common cytochemical characteristics of cells producing polypeptide hormones (the APUD series) and their relevance to thyroid and ultimobranchial C cells and calcitonin. *Proc. Roy. Soc. B* 170:71-80, 1968.
6. **Le Douarin N & Teillet M A.** The migration of neural crest cells to the wall of the digestive tract in the avian embryo. *J. Embryol. Exp. Morphol.* 30:31-48, 1973.
7. **Andrew A.** Further evidence that enterochromaffin cells are not derived from the neural crest. *J. Embryol. Exp. Morphol.* 31:589-98, 1974.
8. **Pictet R L, Rall L B, Phelps P & Rutter W J.** The neural crest and the origins of the insulin-producing and other gastrointestinal hormone-producing cells. *Science* 191:191-2, 1976.
9. **Teitelman G, Joh T H & Reis D J.** Transformation of catecholaminergic precursors into glucagon (A) cells in mouse embryo pancreas. *Proc. Nat. Acad. Sci. US* 78:5225-9, 1981.
10. **Schmechel D, Marangos P J & Brightman M.** Neurone-specific enolase is a molecular marker for peripheral and central neuroendocrine cells. *Nature* 276:834-6, 1978.
11. **Pearse A G E.** Islet cell precursors are neurones. *Nature* 295:96-7, 1982.
12. **von Feyrter F.** *Über diffuse endokrine epitheliale Organe.* Leipzig: J.A. Barth, 1938. 62 p.

Warner J R, Knopf P M & Rich A. A Multiple Ribosomal Structure in Protein Synthesis. *Proceedings of the National Academy of Science* 49:122-9, 1963.

While it had been known for some years that the ribosome was the site of protein synthesis, this paper was the first demonstration that protein synthesis in rabbit reticulocytes takes place on a multiple-ribosomal structure, termed a polysome. The ribosomes are held together by a strand of RNA, presumably messenger RNA. The authors proposed a general model for the simultaneous translation of a single messenger RNA by several ribosomes. [The *SCI*® indicates that this paper was cited 547 times in the period 1961-1975].

Professor Jonathan R. Warner
Department of Biochemistry
Albert Einstein College of Medicine
Bronx, New York 10461

June 9, 1977

"This paper is a prime example of serendipity. I was a graduate student at MIT in Alex Rich's lab, interested in protein synthesis. Shortly after Nirenberg and Matthei's report on poly U stimulated protein synthesis, I had been working with Jim Darnell, trying to get *E. coli* ribosomes to translate poliovirus RNA. Reasoning that eukaryotic ribosomes might be more effective, I proposed a collaboration to Paul Knopf (now at Brown University). He had just finished his Ph.D. doing the Dintzis' experiment in a cell free extract of reticulocytes, and had two months to kill before setting out on the then obligatory post-doc at the MRC in Cambridge. We soon forgot the polio RNA, for in a control experiment analyzing ribosomes from reticulocytes which had been briefly labeled with ^{3}H leucine to identify active ribosomes, we observed peaks of optical density and radioactivity representing structures larger than 80S ribosomes. These structures were enriched in active ribosomes. Two possibilities occurred to us: Were these aggregates of ribosomes created by the high speed centrifugation used to prepare the ribosomes, or were they fragile structures partly destroyed by the homogenization needed to dissolve the ribosomal pellets?

"To resolve these two possibilities, we decided to sediment the reticulocyte ribosomes directly from a cell lysate on a sucrose gradient. Within a month we had done the experiments necessary to show that all protein synthesis in reticulocytes takes place on structures containing several ribosomes, usually five. The structures were insensitive to detergents, to ionic strength and to DNAse, but highly sensitive to RNAse and to shearing forces. While Paul and I were convinced of the importance of the finding, it remained for Alex Rich to develop the conceptual foundation of a stream of ribosomes flowing down a molecule of messenger RNA. A collaboration with Henry Slayter and Cecil Hall, also at MIT, provided the electron microscopic data to support this concept of a "polysome." One could see five ribosomes in a group, and occasionally a densely staining strand between them. It soon became apparent that protein synthesis in all organisms occurs on polysomes, which are a naturally efficient way to translate messenger RNA.

"Undoubtedly the time was ripe for finding polysomes. Messenger RNA had been clearly established, and the race was on to break the genetic code. Alfred Gierer in Tübingen, also using reticulocytes, and Hans Noll in Pittsburgh, using rat liver, independently came to the conclusion that protein synthesis takes place on multi-ribosome structures that are highly sensitive to RNAse. It is noteworthy that all the pioneering work in this area was carried out in mammalian cells. Polysomes are so sensitive to shear that they were, and still are, difficult to prepare from tough organisms like bacteria and fungi.

"For me, the experience was much more than a thesis. It was a "coming-of-age" as a scientist, giving me a sense of the excitement of science. Finally, it destroyed my illusion that science is a wholly rational pursuit. The fun is the unexpected observation which one follows to a discovery, and the challenge is distinguishing between the promising observations and the trivial artefacts."

Chance B & Williams G R. The respiratory chain and oxidative phosphorylation.
Advan. Enzymol. Relat. Areas Mol. Biol. **17**:65-134, 1956.
[University of Pennsylvania, Philadelphia, PA]

On the occasion of Dr. Chance's 70th birthday, we asked him to write a *Citation Classic* commentary. He graciously accepted with the following.

The identification of metabolic states of isolated mitochondria appears to have laid the foundation for quantitation of metabolic function in host cells and organelles. The mitochondria themselves provide consumer reports on oxygen and substrate metabolism. [The *SCI®* indicates that this paper has been cited in over 1,705 publications since 1961.]

Britton Chance
Department of Biochemistry
and Biophysics
School of Medicine
University of Pennsylvania
Philadelphia, PA 19104

September 9, 1983

"The postwar period was especially important to those biological scientists who had either willingly or perforce given up their chosen disciplines to contribute to their nations' efforts in World War II. This was especially true of myself, having volunteered before the US was engaged in the war to devote my maximal scientific effort over a period of six years to pursuing the war and to consolidating the scientific gains made in radar and circuitry developments. Thus, it was a breath of fresh air to resume my prewar research on enzyme-substrate compounds and visit in the laboratories of the 'giants' of the postwar years, Theorell and Keilin, who had continued their work during World War II. Shortly thereafter, I returned to Philadelphia to find a challenging position at the Johnson Research Foundation, University of Pennsylvania, available to me which I've held ever since.

"Given the security of tenure at a young age (36), a chair and an endowment, and satisfied that the work on enzyme-substrate compounds of peroxidases had established an experimental basis for the Michaelis-Menten theory, I decided to venture into the field of 'insoluble enzymes' and to develop rapid and sensitive spectrophotometric methods in order to understand the mechanism of their action.

"After some abortive attempts with Gene Kennedy, it became apparent that the quality of the mitochondrial preparations and the exhibiting respiratory control were critical to a successful experiment. The work of Lardy and Wellman showed the way and Lardy, G.R. Williams, and I prepared mitochondria which would respond to ADP, phosphate, or calcium addition with tenfold or even more increase of respiration (respiratory control). Williams and I worked together feverishly to obtain spectroscopic responses corresponding to various states of metabolic activity. Soon Joe Higgins joined us and was able to make a mathematical formulation explaining these responses, known as the crossover theorem. Verification of mitochondrial metabolic states in frog skeletal tissue was obtained with C.N. Connelly to extend the crossover theorem and gave me the necessary feeling of confidence to write up our work in detail in the *Journal of Biological Chemistry*[1,2] and to write the comprehensive review as well.

"The responses to the ideas were definitely 'not great'; the initial presentation before the American Society of Biological Chemists at San Francisco as a ten-minute presentation was disaster-ridden. The chairman, my good friend A. Lehninger, apparently misread the clock and told me that I was to sit down after four of the ten minutes allotted for the talk. However, confirmation of the work in other laboratories through use of the dual wavelength spectrophotometer and its extension to living tissues provided validation and extension of the basic principles. A significant influx of postdoctorals to my laboratory was especially helpful as was the publication of this comprehensive review. All these events gave the work a general acceptance and a large part of the work seems useful today, as applied not only to the respiratory chain but as applied to metabolic control in general.

"Perhaps the moral of this communication is that it really does pay to work at a new idea until it is fully developed, and, as in this case, not only published in the original literature, but also pulled together as a comprehensive review article. Obviously, I'm pleased that this article fared so well, apparently better than the full paper in the *Journal of Biological Chemistry!*"[1,2]

1. Chance B & Williams G R. Respiratory enzymes in oxidative phosphorylation. I-IV.
   J. Biol. Chem. **217**:383-93; 395-407; 409-27; 429-38, 1955.
   [These papers have been cited over 1,445 times in 1,317 publications since 1961.]
2. Chance B, Williams G R, Holmes W F & Higgins J. Respiratory enzymes in oxidative phosphorylation. V.
   J. Biol. Chem. **217**:439-51, 1955.

Leblond C P & Stevens C E. The constant renewal of the intestinal epithelium
in the albino rat. *Anat. Rec.* **100**:357-71, 1948.
[Dept. Anatomy, McGill Univ., Montreal, Canada]

At the time when this work was undertaken, it was known that cells arose from mitosis in the epithelial lining of small intestine, but the fate of these cells was obscure. Systematic comparison of the numbers of cells and mitoses led to the conclusion that epithelial cells turned over continually. A cell arising from mitosis in a crypt of the epithelium migrates out of this crypt, ascends the villus and, at the villus tip, drops into the lumen. The duration of the migration has been estimated at less than two days. [The *SCI®* indicates that this paper has been cited over 320 times since 1961.]

———————————●———————————

C.P. Leblond
Department of Anatomy
McGill University
Montreal, Quebec H3A 2B2
Canada

January 27, 1981

"The mitotic activity of the intestinal epithelium has often been considered as a regenerative process ensuring the replacement of cells damaged by the powerful enzymes and bacteria present in the gut lumen. Yet abundant mitoses seemed to be present even in the absence of detectable damage. Perhaps, instead of being the static tissue described by classical histologists, the intestinal epithelium might be the site of unsuspected dynamics. At the time when I considered this possibility, a student of British origin, Catherine Stevens, applied for graduate work. I proposed the study of the intestinal epithelium.

"Using animals in good health and impeccable methods of fixation, Stevens found mitotic activity in the crypts throughout the length of the small intestine in adult male and female rats, even after penicillin treatment or a five-day fast. Counts revealed that about three percent of the cells were undergoing mitosis at all times of day. Meanwhile, the duration of mitosis was measured by the colchicine method and found to approximate one hour. It was then readily calculated that the mitoses provide enough new cells to replace two-thirds of the population per day.

"The next step was to find what happened to the new cells. In one region of the epithelium—the villus tip—there were signs of death and extrusion of cells. It was, therefore, postulated that the cells arising in the crypts migrated out to become part of the villi and then ascended the villi to drop from the tips into the lumen. The loss should balance cell production to maintain steady state. The whole migration should take less than two days.

"This conclusion appeared surprising to fellow histologists. When Stevens presented these results at a meeting of the American Association of Anatomists, an old histologist from London, Ontario, commented, 'This is too silly for words,' to which she replied, 'It may be silly, but it is true.'

"This work is one of the few which I have published without the support of radioautography. Eventually, however, the rapid migration of epithelial cells from crypt to villus tip was fully confirmed by radioautography, using first ^{32}P-phosphate, later ^{14}C-adenine, and especially ^{3}H-thymidine.[1,2]

"Renewal of the intestinal epithelium was observed in man by Bertalanffy.[3] Recent unpublished calculations indicate that over a quarter of a pound of cells are daily shed into the gut. Recent work in the field has been done by both Cheng and myself.[4,5]

"Why is there continuous renewal of cells in the gut? With continuous exposure of the cells to enzymes, toxins, and bacteria, ordinary repair mechanisms may not be sufficient for protection, whereas constant supply of new healthy cells provides prompt replacement of weakened areas or gaps and thus anticipates damage."

1. Leblond C P & Messier B. Renewal of chief cells and goblet cells in the small intestine as shown by radioautography after injection of thymidine-H^3 into mice. *Anat. Rec.* **132**:247-59, 1958.
2. Leblond C P. The time dimension in histology. *Amer. J. Anat.* **116**:1-28, 1965.
3. Bertalanffy F D & Nagy K P. Mitotic rate and renewal rate of the epithelial cells of human duodenum. *Acta Anat.* **45**:362-70, 1961.
4. Cheng H & Leblond C P. Origin, differentiation, and renewal of the four main epithelial cell types in the mouse small intestine. *Amer. J. Anat.* **141**:461-80, 1974.
5. Leblond C P. The life history of cells in renewing systems. *Amer. J. Anat.* In press, February 1981.

CC/NUMBER 34
AUGUST 24, 1981

Messier B & Leblond C P. Cell proliferation and migration as revealed by
radioautography after injection of thymidine-H³ into male rats and mice.
Amer. J. Anat. **106**:247-85, 1960.
[Department of Anatomy, McGill University, Montreal, Canada]

Following a single injection of H³-thymidine into young rats and mice, the labeled nuclei were identified by radioautography at times varying from 20 minutes to three months thereafter. It was thus found that cell populations may be classified into three categories on the basis of H³-thymidine uptake: static, expanding, and renewing. [The *SCI®* indicates that this paper has been cited over 430 times since 1961.]

Bernard Messier
Département d'Anatomie
Faculté de Médecine
Université de Montréal
Montréal, Québec H3C 3J7
Canada

June 29, 1981

"This article was published at the dawn of the H³-thymidine era. Indeed, the recent commercial availability of tritiated thymidine in the late 1950s prompted investigators involved in DNA synthesis research to call on this DNA specific precursor for an answer to long-awaited questions. C¹⁴-thymidine had timidly pointed in that direction a few years previously, but the prohibitive cost of this precursor curbed its wide use. The competition between H³-thymidine and C¹⁴-thymidine quickly favored the former, for two main reasons: first, the possibility of obtaining samples with high specific activities and, second, the soft beta rays emitted by the decomposition of the tritium atoms. Although the latter property constituted, at the time, a difficulty for the biochemical assessment of the radioactivity levels in biological specimens, this same property proved to be a major asset for a good radioautographic localization.

"Because of the shortcomings of mitotic counts for the study of cell proliferation, the availability of a specific DNA precursor providing an excellent radioautographic localization allowed a new approach to the problem of cell kinetics. Indeed, the use of a pulse injection of labeled thymidine introduces a radioactive marker in the DNA of proliferating cells. The retention of the label in nuclei of various tissues for periods as long as six months offered the possibility of following the fate of newly-formed cells.

"With less than a dozen papers published on the subject in 1959, C.P. Leblond and I engaged in a comprehensive radioautographic survey of H³-thymidine uptake in several tissues of rats and mice to assess their rates of cell proliferation and migration. This work constituted the core of my PhD program. Coincident with this survey were the technical improvements put forward in collaboration with Beatrix Kopriwa to the then recently developed dipping technique of radioautography.¹ Counts of many thousands of radioactive and nonradioactive nuclei helped in classifying the cell populations into three categories: *static*, in which no labeled nuclei appear; *expanding*, in which a small number of indefinitely persisting labeled nuclei appear in proportion to the rate of growth; and *renewing*, in which the occurrence of large numbers of labeled nuclei indicates active cell production, while the subsequent rapid decrease and disappearance of the labeled nuclei indicates a corresponding cell loss.

"The atmosphere of scientific dedication in our laboratory played a major role in the success of this research project, but the timely coincidence of favorable circumstances was also helpful. This timely coincidence of the commercial availability of H³-thymidine and technical improvements of the radioautographic method probably account for the frequent citation of the paper. More recent work in the field has been published by Kopriwa."²

1. Kopriwa B M & Leblond C P. Improvements in the coating technique of radioautography.
   J. Histochem. Cytochem. **10**:269-84, 1962.
2. Kopriwa B M. A comparison of various procedures for fine grain development in electron microscopic radioautography. *Histochemistry* **44**:201-24, 1975.

Quastler H & Sherman F G. Cell population kinetics in the intestinal epithelium of the mouse. *Exp. Cell Res.* 17:420-38, 1959.
[Biology Dept., Brookhaven Natl. Lab., Upton, NY and Biology Dept., Brown Univ., Providence, RI]

Analysis of data obtained from autoradiographs of mouse intestinal epithelial cells labeled with tritiated thymidine led to a description of the kinetics of cell proliferation. The precise localization of the exposed emulsion allows cells synthesizing DNA to be identified. [The *SCI*® indicates that this paper has been cited over 1,025 times since 1961.]

Frederick G. Sherman
Department of Biology
Syracuse University
Syracuse, NY 13210

October 20, 1980

"The work of Leblond, Stevens, and Bogoroch[1] and other investigations published during the next decade stimulated the late Henry Quastler and me to explore methods for studying the dynamic equilibria of cell populations. The first attempts were made by analyzing autoradiographs of tissues which had been labeled with inorganic ^{32}P. Technical advances soon became available which markedly improved the quality of the autoradiographs and also enabled us to identify cells which were synthesizing DNA. One was the availability of liquid film emulsions; another, of great importance, was the successful preparation of tritiated thymidine by Walter L. Hughes who was then a staff scientist at the Brookhaven National Laboratory.[2] It was through his generosity and the cooperation of the Schwarz Laboratories that Quastler and I had access, almost from the first, to a portion of the meager supply of tritiated thymidine then available.

"The importance of this label for our investigation lay in the fact that administered thymidine is readily incorporated into those cells which are synthesizing DNA. The energy of the electrons emitted by tritium is so low that only those silver grains in the emulsion which lie directly over the labeled nuclei are exposed. The autoradiographs which result from this procedure enable the observer to identify precisely the cells which were synthesizing DNA at the time the labeled thymidine was administered.

"The data reported in this paper allow estimates to be made of the average time cells remain in the various compartments, the number of cells per compartment, and the number of cells per unit time in transit from one compartment to another. Additionally, it was found that in order for the transition from a proliferative to a 'functional' cell to occur, the cell must not only be in a certain neighborhood, but it must also be in a certain phase of its generative cycle.

"It is gratifying that our work has received the recognition that it has. Quastler's incisive analytical abilities and insights into the implications of the data expedited the development of a conceptual framework which not only enabled us to analyze the kinetics of intestinal epithelium but is also applicable to the analysis of other proliferating cell systems."

1. Leblond C P, Stevens C E & Bogoroch R. Histological localization of newly-formed desoxyribonucleic acid. *Science* 108:531-3, 1948.
2. Hughes W L, Bond V P, Brecher G, Cronkite E P, Painter R B, Quastler H & Sherman F G. Cellular proliferation in the mouse as revealed by autoradiography with tritiated thymidine. *Proc. Nat. Acad. Sci. US* 44:476-83, 1958.

Chapter

7

Physiology and Pharmacology

The large grouping of Citation Classics described in this chapter begins with a report from one of modern science's most famous figures: Hans A. Krebs, formulator of the Krebs cycle for the oxidation of carbohydrates and the Krebs urea cycle. With Henseleit in 1932, Krebs published the results of *in vitro* experiments with liver slices that established the basic organization of the urea cycle as we know it today. Their Classic, cited 2,180 times, was dually important: for the discovery of the urea cycle as such, and because, as Fruton stated, "the paper provided a clue to the organization of metabolic pathways in living cells."

Accompanying the Classic of Krebs are seven others pertaining to kidney physiology. One of the early uses of a cold carrier as a chaser to aid in the recovery of small quantities of a radioisotopically labeled molecule, in this case vitamin B_{12}, is a technique that has been applied to many similar problems. The remainder of these seven Classics are concerned with antidiuretic hormone and its regulations of sodium excretion, including the role of cAMP as a second messenger, the excretion of acids and bases, nephrotoxicity of an anesthetic, and a popular review on renin release.

The next few Classics focus on hormones: on insulin and its ability to increase cellular uptake of glucose, on the ying and yang relationship of glucose and fatty acid blood levels as influenced by insulin, on the inhibition of insulin secretion by somatostatin, and on the role of insulin in glucose homeostasis during fasting in human volunteers. The last experiment was undoubtedly performed (in 1965) well outside our current guidelines on human experimentation. Of the succeeding hormone-related "This Week's Citation Classic" (TWCC) commentaries, that written by Glick is especially readable. His glowing tribute to Yalow and Berson, who functioned almost as private tutors for him, and his fourth coauthor, Roth, describes an intimate teacher–student relationship at the peak of its perfection. The account by Pasteels is interesting both on the basis of his judgment in selecting an area of research with prolactin that was within his capability and outside that of his competitors and for the discovery itself of human prolactin. His effort to halt the traditional Belgian custom of publishing Ph.D. dissertations in their entirety as individual monographs in French may yet benefit young scientists of his country by allowing their accomplishments an earlier audience.

The bioassay for follicle-stimulating hormone (p. 254), cited 1,010 times, the review on calmodulin (p. 255), and the role of calcium ion in muscle contraction (p. 257) precede Classics on blood clotting and heart–lung research. Cheung's TWCC clearly states his good fortune in separating calmodulin from phosphodiesterase as the primer to his discovery of calmodulin and identification of its function with calcium-requiring enzymes.

Unfortunately, the complexities of the blood clotting cascade will not be clarified by reading the next four TWCCs (pp. 260–263), but at least it will be a relief to learn that coagulation scientists were once also in a cloud (in the 1950s), confused by the disordered naming and numbering of molecules in this system and uncertainties about the functional properties of some of the molecules.

The 90-page review on platelet function was written by Mustard and Packham at the invitation of the editor of *Pharmacological Reviews*. This article provided a significant impetus to research on platelets because Mustard was recognized as an authority in a field that was just emerging as a subject of research interest and because the review was so encompassing in scope.

Among the next eight Classics on heart–lung physiopharmacology, the most popular was the book by Hoffman and Cranefield on the *Electrophysiology of the Heart*, cited 1,380 times before 1981. The next most popular was the two-part review on the lesions of atherosclerosis (p. 267), cited 980 times in the 6-year period from the end of 1976 to January 1983. The venue of the *New England Journal of Medicine* for this review is undoubtedly important in this regard because it is one of the most widely read and respected of the medical journals.

The condemned engineer's shed in which $C^{15}O_2$ from the immediately adjacent cyclotron was analyzed must have presented an interesting architectural contrast to West's visitors, as he states in his TWCC. Recognition of the etiology of "respirator lung," the decrease in oxygen absorption due to inhalation of high concentrations of oxygen (also published in the *New England Journal of Medicine*), was an important discovery and instrumental in modifying oxygen therapy (p. 270). The rejection of the Classic Morris describes reflects, in an unfortunate way, how emotion can alter objectivity in reviewing manuscripts. Some authors would probably do as he did—seek a re-review by editors of the same journal—but most would probably submit the article to a second journal.

Some fascinating social comments are notable in several of the remaining TWCCs. Walshe's successful penicillamine treatment of the otherwise fatal Wilson's disease caused by the presence of excessive copper is an example. His work was aided significantly by the position of his father as Britain's leading neurologist and hence the availability of the needed patients. Mechoulam's TWCC contains a sad commentary on the status of research funding—namely, that research on a subject is not supported until that subject becomes a problem in some way, either of the public or, as is the case with marijuana, of a prominent politician.

Plant physiology and plant–insect relationships dominate the remaining Classics. This is one of the few chapters in this volume where advances in botany are recorded. Most of the botanical TWCCs were published in the Agriculture, Biology, and Environmental Sciences series of *Current Contents.* These early botanical Classics are based on the newly discovered photophosphorylation, CO_2 fixation by chloroplasts *in vitro,* and the identification that indole-3-acetic acid is a plant growth hormone and acts specifically on the plant cell wall to cause cell elongation. This last contrasts with the discovery of growth-retarding chemicals that encourage fruiting and dwarfing in plants, including the common house plants where this was of commercial significance.

The feeding habits of butterflies introduce us to the prospect of plant–insect co-evolution and the role of secondary plant substances. One study (p. 287) was generated from coffee break roundtables at the library. These plant molecules function as specific insect attractants or repellants and determine plant feeding specificity by insects, although most plants would suit their nutritional demands (p. 288). Some of these substances ingested by an insect are toxic to their predators and have thus influenced evolution of the predator–prey relationship (p. 289). These last three Classics compose an interesting grouping on a fascinating aspect of plant–insect relationships.

Krebs H A & Henseleit K. Studies on urea formation in the animal organism.
Hoppe-Seylers Z. Physiol. Chem. 210:33-66, 1932.
[Medical Clinic, University of Freiberg, Federal Republic of Germany]

The paper reports the outlines of the pathway of urea synthesis in the mammalian liver. It shows that ornithine promotes urea synthesis like a catalyst and that citrulline, arginine, and ornithine participate in a cyclic sequence, the net effect being the formation of urea from CO_2 and two molecules of ammonia. [The *SCI®* indicates that this paper has been cited over 2,180 times since 1961.]

Hans A. Krebs
Metabolic Research Laboratory
Radcliffe Infirmary
Oxford University
Oxford OX2 6HE
England

October 23, 1980

"Historians of science have rated this paper indeed as a major discovery. Joseph S. Fruton wrote: 'This work marked a new stage in the development of biochemical thought. Not only was an explanation of a biochemical synthesis offered for the first time in terms of chemical reactions identified in the appropriate biological system and not merely inferred by analogy to the known chemical behaviour of the presumed reactants, but also the paper provided a clue to the organisation of metabolic pathways in living cells. This became evident in 1937 with the appearance of the Krebs citric acid cycle, whose conceptual relation to the earlier ornithine cycle was obvious.'[1]

"This work was a first attempt to study a biosynthetic process in a tissue slice. When I started, it was uncertain whether a capacity for biosyntheses was retained on slicing, a technique developed by Warburg[2] for the study of degradative processes. To provide optimal conditions for the survival of tissue slices, I devised a medium which in respect to the inorganic constituents simulated blood plasma as closely as possible.

"Older saline solutions were grossly deficient in either bicarbonate and CO_2, or magnesium or phosphate. They had been based mainly on trial and error experiments which tested the ability of saline media to maintain the beating of the isolated frog heart. The new saline was based on the conviction that the concentrations of ions in blood plasma (almost identical in all mammalian species) have evolved to be optimally attuned to organ function.

"The new medium proved in fact superior to all earlier plasma saline substitutes in biochemical, physiological, and pharmacological work. This is the reason why it is now widely used and why the paper is frequently cited.

"People often enquire about my collaborator, Kurt Henseleit. He was an able and promising medical student. Having been associated with me, and not being a Nazi, he was told by the Hitler regime that there was no future for him in academic medicine. He became a successful internist in Friedrichshafen in South Germany, where he died in 1972."

1. **Fruton J S.** *Molecules and life: historical essays on the interplay of chemistry and biology.*
New York: Wiley Interscience, 1972. p. 436.
2. **Warburg O H.** *Über den Stoffwechsel der Tumoren.* Berlin: Springer, 1926. 263 p.

Schilling R F. Intrinsic factor studies. II. The effect of gastric juice on the urinary excretion of radioactivity after the oral administration of radioactive vitamin B₁₂. *J. Lab. Clin. Med.* 42:860-6, 1953.
[Dept. Medicine, Univ. Wisconsin Medical School, Madison, WI]

This paper describes a urinary excretion test for estimating a patient's ability to absorb vitamin B_{12}. The most common cause of failure to absorb vitamin B_{12} is pernicious anemia, but several other conditions also cause B_{12} malabsorption and can be diagnosed or suspected on the basis of studies of vitamin B_{12} absorption. [The *SCI®* indicates that this paper has been cited in over 740 publications since 1955.]

Robert F. Schilling
Department of Medicine
University of Wisconsin
Madison, WI 53792

May 22, 1984

"After spending two thoroughly stimulating years in Harvard's Thorndike Laboratory at the Boston City Hospital, I returned to the University of Wisconsin in 1951 and was given time and space to pursue my research interest in nutritional anemia. The use of radioisotopes as tracers in the study of biology, including human physiology, was relatively new but obviously capable of answering many questions. Rosenblum and Woodbury[1] had recently prepared radioactive vitamin B_{12} by providing ^{60}Co in the biosynthetic process. Heinle and his colleagues[2] had shown that the absorption of vitamin B_{12}, estimated by measuring fecal radioactivity for seven days, was greatly below normal in patients with pernicious anemia.

"I was attempting to measure absorption of the vitamin by determining radioactivity in plasma or urine in normal subjects, but no radioactivity was detected with the relatively insensitive apparatus available to me. I had been reading about cold carrier techniques in a radiochemistry text and decided to use what might be called an *in vivo* cold carrier technique: I injected a large amount of nonradioactive B_{12} after having drunk a physiologic quantity of radioactive B_{12}. The urine collected over the next 24 hours contained easily detectable and quantifiable radioactivity indicating that the orally ingested material had been absorbed. A trial of the technique on a patient known to have pernicious anemia produced no detectable urine radioactivity.

"The urine radioactivity test has been widely used as a test for vitamin B_{12} absorption because of its simplicity and utility. It has been carefully compared with fecal excretion and whole body counting methods, and the correlation of results is good. Vitamin B_{12} absorption tests have enabled studies which have significantly increased our understanding of gastrointestinal physiology.

"Physicians and patients are always hopeful that the patient's symptoms are due to a curable disorder. This urine radioactivity test is useful in sorting out the mechanisms of conditions leading to vitamin B_{12} deficiency, an ideally treatable or even preventable illness. For a recent review, see reference 3."

1. **Rosenblum C & Woodbury D T.** Cobalt 60 labeled vitamin B₁₂ of high specific activity. *Science* 113:215, 1951.
2. **Heinle R W, Welch A D, Scharf V, Meacham G C & Prusoff W H.** Studies of excretion (and absorption) of Co⁶⁰-labeled vitamin B₁₂ in pernicious anemia. *Trans. Assn. Am. Physician.* 65:214-22, 1952. (Cited 235 times since 1955.)
3. **Chanarin I.** *The megaloblastic anaemias.* Oxford: Blackwell Scientific, 1979. 800 p.

This Week's Citation Classic

Berliner R W & Davidson D G. Production of hypertonic urine in the absence of
pituitary antidiuretic hormone. *J. Clin. Invest.* **36**:1416-27, 1957.
[Lab. Kidney and Electrolyte Metabolism, Natl. Heart Inst., Natl. Insts. Health, Public Health
Service, Dept. Health, Education, and Welfare, Bethesda, MD]

Urine was collected separately from the two
kidneys of trained, unanesthetized dogs.
Constriction of one renal artery regularly led
to production of hypertonic urine by that
kidney while complete absence of antidiuretic hormone was established by maximally dilute urine obtained from the other. [The
SCI® indicates that this paper has been cited
in over 315 publications since 1961.]

Robert W. Berliner
School of Medicine
Yale University
New Haven, CT 06510

July 26, 1983

"At the time that this work was done,
the only demonstrated action of vasopressin had been that of increasing the
permeability of responsive membranes
to water. Furthermore, there was little
or no understanding of the mechanism
by which urine hypertonic to body fluids is produced; although the counter-
current hypothesis had been proposed
by Wirz and Kuhn,[1,2] it had not really
penetrated the thinking of most renal
physiologists.

"I had been interested in the possibility that the full range of physiologic action of vasopressin on the kidney might
be explained by its effect on membrane
permeability to water. If this were the
case, it might be possible to obtain hypertonic urine in the complete absence
of hormone provided the volume of water to be lost in achieving that hypertonic state were small enough. Indeed,
several years earlier, Jack Orloff and I
had studied a dog with surgically produced severe diabetes insipidus and

found that upon severe dehydration
from water deprivation, hypertonic
urine was produced. We realized, however, that it was impossible to establish
that there was not some residual capacity to secrete vasopressin. It seemed
that the best possibility of establishing
the absence of vasopressin would be to
have one kidney continue to produce
maximally dilute urine while manipulations were carried out on the other
kidney. The bladder-splitting operation
devised by Desautels made possible
chronic preparations in which urine
could be collected separately from the
two kidneys in trained, unanesthetized
dogs.[3]

"The rationale for the studies by
Davidson and me is quite precisely described in the introduction of the paper. The results were definitive in showing that hypertonic urine could regularly be produced by one kidney, when its
renal artery was constricted, while the
other kidney continued to put out maximally dilute urine. The effect was attributed to the reduction of glomerular
filtration in the artery-constricted kidney although the presumed key element was reduction in the volume and
salt content of the fluid delivered to
the diluting segment in the distal neph-
ron. It has since been found that other
means of limiting delivery will also produce the effect.

"I presume that the frequency with
which this paper has been cited relates
not only to its interest for renal physiology but because a similar phenomenon
may be involved in clinical states in
which salt retention is associated with
impaired ability to excrete dilute urine.

"Recent work in this field has been
reported in *Urinary Concentrating
Mechanism*."[4]

1. **Hargitay B & Kuhn W.** Das multiplicationsprinzip als grundlage der harnkonzentrierung in der niere.
   Z. Elektrochem. **55**:539-58, 1951. (Cited 160 times.)
2. **Wirz H, Hargitay B & Kuhn W.** Lokalisation des Konzentrierungsprozesses in der Niere durch direkte
   Kryoskopie. *Helv. Physiol. Pharmacol. Acta* **9**:196-207, 1951. (Cited 295 times.)
3. **Desautels R E.** Hemisection of the bladder for the collection of separate urine samples.
   Surg. Gynecol. Obstet. **105**:767-8, 1957.
4. **Jamison R L & Kriz W.** *Urinary concentrating mechanism: structure and function.*
   New York: Oxford University Press, 1982. 340 p.

de Wardener H E, Mills I H, Clapham W F & Hayter C J. Studies on the efferent mechanism of the sodium diuresis which follows the administration of intravenous saline in the dog. *Clin. Sci.* 21:249-58, 1961.
(Dept. of Medicine and the Isotope Lab., St. Thomas's Hospital and Medical School, London]

Dogs loaded with sodium-retaining and antidiuretic hormones were given a rapid saline infusion intravenously. Increased sodium excretion occurred with lowered filtration rate and renal denervation. Cross circulation between infused and non-infused dogs suggested the presence of a natriuretic hormone with a short half-life. [The *SCI*® indicates that this paper has been cited over 300 times since 1961.]

Ivor H. Mills
Department of Medicine
University of Cambridge Clinical School
Addenbrooke's Hospital
Cambridge CB2 2QQ, England

March 6, 1978

"In 1958 when I returned from working at the National Institutes of Health, Bethesda, on the afferent pathways in the regulation of aldosterone secretion, de Wardener was then working on the excretion of dilute urine in dogs loaded with antidiuretic hormone. I suggested that it might be related to the effect of the volume expansion on the rate of aldosterone secretion. He invited me to join him. We loaded dogs with sodium-retaining hormone (fluorocortisone), and studied sodium excretion.

"In our paper there were two sets of experiments which probably account for the frequency of citation of the paper. The first set dealt with the excretion of an infused saline load while the glomerular filtration rate was decreased. This we achieved by inflating a balloon in the aorta so that the pressure at the renal arteries was lowered to 90 mm Hg.

"The importance of this series of experiments was that it was the first time that a marked increase in sodium excretion had been produced while the glomerular filtration rate and, therefore, the rate at which sodium was filtered, was reduced to well below the rate of filtration during the pre-infusion period. Prior to that it had been argued—and even today some people still stick to the argument—that changes in filtration rate which were too small to be measured by available techniques could explain the increases in sodium excretion.

"The second set of experiments which were very important were those in which cross-circulation between two dogs was effected. One dog was on a beam balance and so could have its weight controlled during cross-circulation; the other dog at the table then received a rapid saline infusion. Although the blood of the dogs rapidly became the same for a variety of factors, only the infused dog had a pronounced diuresis.

"We concluded, especially from these two sets of experiments, that the regulation of sodium excretion depends upon a change in the concentration of a circulating substance, other than aldosterone, which had a short half life. It was assumed that we had for the first time demonstrated that a circulating natriuretic hormone existed, though we were careful not to use that term. After that, people all over the world started looking for the natriuretic hormone. Time has shown that there is a linked series of natriuretic substances and that some of these originate in the kidney itself."

Orloff J & Handler J. The role of adenosine 3',5'-phosphate in the action of
antidiuretic hormone. *Amer. J. Med.* 42:757-68, 1967.
[Lab. Kidney and Electrolyte Metabolism, Natl. Heart Inst., NIH, Bethesda, MD]

Antidiuretic hormone is responsible for the regulation of water balance in a variety of animals. In man it increases water reabsorption by the kidney, by stimulating production of adenosine 3',5'-phosphate (cyclic-AMP) in specific renal epithelial cells. The nucleotide increases the permeability of the apical membrane of these cells to water, thereby accelerating the reabsorptive process. [The *SCI®* indicates that this paper has been cited over 310 times since 1967.]

Jack Orloff
National Heart, Lung, and
Blood Institute
National Institutes of Health
Bethesda, MD 20205

March 12, 1981

"In 1940 Homer Smith, then professor of physiology at New York University Medical School, kindled my interest in the action of antidiuretic hormone (ADH) that has continued to this day. Ultimately my goal was to understand the biochemical and/or physical changes responsible for the structural alterations in the renal cell membranes that must account for the permeability effects of ADH. Although my initial studies performed at Yale University in the late 1940s and at the National Institutes of Health in the 1950s were in intact animals, Larry Early, who arrived in my lab in 1959, convinced me to switch to a simpler model, the isolated urinary bladder of the toad. The bladder responds to ADH by increasing its permeability to water in a manner indistinguishable from that of the isolated renal collecting tubule of the rabbit as shown by

Grantham and Burg in my lab.[1] At the time I was aware of Sutherland's second messenger hypothesis,[2] introduced to me by Martha Vaughan, my wife, who was studying the effect of hormones on the activity of glycogen phosphorylase. In addition, I knew that Jim Hilton had reported that high concentrations of ADH mimicked ACTH in the adrenal[3] and I wondered if cyclic-AMP, a second messenger in the action of ACTH, might be involved in the ADH effect in toad bladder and kidney. When I heard him present studies on the mimicry of glucagon by ADH in the liver at a Salt and Water Club meeting, I was convinced that my hypothesis was plausible.

"I returned to Bethesda and together with Joe Handler, who had joined my lab in 1960, began a fruitful collaborative effort that has lasted for more than a decade. In short order we proved that ADH elicited its effect on water permeability of toad bladder via the intermediacy of cyclic-AMP. The nucleotide alone or theophylline, which prevents its degradation, increased water permeability as did ADH. This was the first evidence that the second messenger could elicit a physiologic response in an intact tissue and lent credence to the Sutherland thesis which until then was supported only by biochemical evidence. Subsequently we showed that ADH increased the cyclic-AMP content of target cells.

"On rereading our review I realize how lucky and unduly optimistic we were at the time. It was written in the hope that our observations, our views on the problem, would stimulate us as well as our readers to engage in new studies that might accomplish the goal that we outlined, i.e., the elucidation of the biochemical and/or physical processes that alter the membrane and account for changes in water flow across cells. Why are we cited? Perhaps to emphasize that neither we nor others have yet succeeded."[4]

1. **Grantham J J & Burg M B.** Effect of vasopressin and cyclic AMP on permeability of isolated collecting tubules. *Amer. J. Physiol.* 211:255-9, 1966.
2. **Sutherland E W & Rall T W.** The relation of adenosine 3',5'-phosphate and phosphorylase to the actions of catecholamines and other hormones. *Pharmacol. Rev.* 12:265-99, 1960.
3. **Hilton J G, Scian L F, Westermann C D & Kruesi O R.** Direct stimulation of adrenocortical secretion by synthetic vasopressin in dogs. *Proc. Soc. Exp. Biol. Med.* 100:523-4, 1959.
4. **Strewler G J & Orloff J.** Role of cyclic nucleotides in the transport of water and electrolytes. *Advan. Cyclic Nucl. Res.* 8:311-61, 1977.

Milne M D, Scribner B H & Crawford M A. Non-ionic diffusion and the excretion of weak acids and bases. *Amer. J. Med.* 24:709-29, 1958.
[Dept. Medicine, Postgraduate Medical Sch. London, England]

Biological membranes are more permeable to the unionized fraction of lipid soluble weak bases and acids than to the ionized component. This accounts for more rapid excretion of many organic acids in alkaline urine and of weak bases in acid urine. Mathematical analysis allowing for a) slight permeability to the ionized component, b) slowness of full equilibration across the membrane, and c) limited capacity of available renal blood flow, gives a reasonable correspondence between observed and theoretical clearances of many such acids and bases. [The *SCI®* indicates that this paper has been cited over 315 times since 1961.]

M.D. Milne
12 York Avenue
East Sheen, London SW14 7LG
England

January 14, 1981

"My colleagues, B.H. Scribner and M.A. Crawford, and myself are flattered to have our paper and names included in the most-cited item list. This paper was an invited script in a group of papers devoted to advances in renal physiology, and stemmed from my visit to the US in 1957. My own interest in this field was stimulated by work reported from R.A. McCance's laboratories in Cambridge, England, where it was shown by use of the carbonic anhydrase inhibitor, acetazolamide, that increased excretion of ammonium in highly acidic urine was more related to urinary pH than to systemic changes in acid-base balance.[1] This prompted us to investigate clearances of many weak organic acids and bases at the extremes of urinary pH, both in man and experimental animals. It soon became clear that many lipid-soluble drugs showed a pH-dependent excretion, but that in the case of water-soluble acids of the Krebs cycle, e.g., citric and α-ketoglutaric acids, it was sys-

temic acidity and alkalinity which determined clearance rates.

"It seemed to us that the diffusion characteristics of the lipid-soluble compounds could easily be analysed mathematically even by individuals like ourselves who were relatively ignorant of advanced mathematical techniques. One of the main difficulties was that a somewhat naive and too literal application of diffusion theory gave the result that the clearance of a weak organic acid should increase ten-fold for every unit rise of urinary pH, whereas in most cases the observed rise was about tenfold for every three units of increase of urinary pH. This discrepancy was satisfactorily explained by limiting factors, detailed in the abstract to the paper. Defects in the argument which have been resolved in later papers were that it was not sufficiently emphasized that diffusion was mainly in the direction of tubular fluid to pericapillary blood,[2] and that there was a fundamental difference in the mechanisms of diffusion of lipid soluble compounds, e.g., quinine, and predominantly water-soluble bases, e.g., ammonia.[3] The latter compounds have later been shown to diffuse through membrane pores and not through the bimolecular lipid layer of the tubule cells.

"In retrospect, the appeal of this article was that it allowed prediction of the type of excretion of weak organic acids and bases, and thus greatly facilitated subsequent research. The principles involved have later been applied to the stomach, intestine, pancreas, brain and cerebro-spinal fluid, and to individual cells. Diffusion into cell cytoplasm has stimulated use of partition methods in the measurement of intra-cellular pH, particularly by use of the weak acid dimethyloxazolidine-dione and the weak base nicotine. The appeal of the article may well have been that the mathematical methods involved were relatively simple, and falsely gave both the writers and readers the pleasant, if temporary, delusion that they understood something of mathematical reasoning."

1. Ferguson E B, Jr. A study of the regulation of the rate of urinary ammonia excretion in the rat. *J. Physiology* 112:420-5, 1951.
2. Weiner I M & Mudge G H. Renal tubular mechanisms for excretion of organic acids and bases. *Amer. J. Med.* 36:743-62, 1964.
3. Bourke E, Asatoor A M & Milne M D. Mechanisms of excretion of some low-molecular weight bases in the rat. *Clin. Sci.* 42:635-42, 1972.

Crandell W B, Pappas S G & MacDonald A. Nephrotoxicity associated with methoxy-
flurane anesthesia. *Anesthesiology* 27:591-607, 1966.
[Veterans Administration Hospital, White River Junction, VT & Dartmouth
Medical School, Hanover, NH]

Inappropriate diuresis with consequent
metabolic changes occurred in 16 patients
of 94 exposed to methoxyflurane anesthesia.
Nothing similar was seen in 100 patients who
received other anesthetic agents for comparable operations during the same period.
Tests showed that the concentrating defect
originated in the kidney rather than in the
posterior pituitary. [The *SCI®* indicates that
this paper has been cited over 140 times
since 1961.]

———————◄█►———————

Walter B. Crandell
Veterans Administration Center
White River Junction, VT 05001

February 21, 1978

"It seems likely that our paper has been
cited frequently because it first documented
a problem of wide potential concern to surgeons and anesthesiologists and pointed
out that postoperative renal failure may
sometimes be characterized by diuresis
rather than the more usual oliguria. The
paper was not the product of a planned
research effort, but of a lucky meeting of inquiring minds and technical capabilities
The surgical residents were alert to
metabolic changes and fluid shifts; some
had spent several months of their training
working in the research laboratory That
laboratory, in turn, was geared to do the
tests needed to confirm their clinical
judgments.
"So when Chief Surgical Resident, Dr.
Stephen Pappas, on teaching rounds, asked
what had led up to a serum sodium of 162
mEq/L, the tools were at hand to find the
answer. The patient had had an abdominal
operation three days before. He was having
a large diuresis, with consequent metabolic
changes. On the next day diuresis and resultant dehydration appeared in another
patient; he too had had an abdominal
operation three days before. In order to distinguish between a disorder of the renal
concentrating mechanism and one of the
posterior pituitary, we measured the urine
osmolality in response to fluid deprivation,
rapid infusion, and Pitressin I-V. The defect,
in both patients, was shown to be primary
in the kidney.

"We saw and studied four more cases;
meanwhile we tried unsuccessfully to find
the cause of the disorder. None of the many
drugs used was common to all cases. Then
we heard that surgical residents at a large
city hospital had noticed that some of the
patients exposed to methoxyflurane required more intravenous fluids than usual
because of high urine output. A check of our
problem patients' charts showed that all
had received methoxyflurane.
"Dr. Pappas reviewed the charts of all 94
patients exposed to methoxyflurane, as
well as those of 100 patients who had had
comparable operations during the same
period, but with other anesthetic agents.
Ten more of the methoxyflurane recipients
had evidence of renal dysfunction—not a
single one of the other one hundred.
"Nevertheless, our anesthesiology
department rejected the implication that
methoxyflurane was responsible; although
this disturbed us, their skepticism was
useful in forcing us to strengthen our
evidence. We were concerned because the
use of methoxyflurane was expanding
rapidly—we were convinced that it could
damage the kidney and wanted to share our
information. We were encouraged that Dr.
Gilbert Mudge and Dr. Heinz Valtin, experts
in nephrology at the Dartmouth Medical
School, after reviewing our data,
acknowledged that our conclusions were
sound.
"A meeting was arranged with a representative of the distributor of methoxyflurane
and two prominent anesthesiologists. Reluc-
tantly accepting our data, they suggested
that it be shelved, since there had been favorable reports on methoxyflurane from
authoritative sources—moreover, the anesthesiologists wanted to avoid the kind of
predicament that they had been in with
halothane.
"Unwilling to suppress our information,
we submitted a paper to *Anesthesiology*.
The editor and his board were critical, but
eventually published the paper, accompanied by a qualifying editorial.
"Numerous reports corroborating our
findings have been published since then,
and the mechanism of the kidney damage
has been elucidated. Use of the drug is now
commonly limited to brief exposures. The
fair but skeptical editor of *Anesthesiology*
was a joint editor, six years later, of a
report of fatal renal failure ascribed to
methoxyflurane anesthesia."

Vander A J. Control of renin release. *Physiol. Rev.* 47:359-82, 1967.
[Department of Physiology, University of Michigan Medical School, Ann Arbor, MI]

The paper critically reviews the various theories proposed for the control of renin release. It then analyzes data from the many physiological and pathological situations associated with altered renin release, in the context of their consistency with these theories. [The *SCI*® indicates that this paper has been cited in over 605 publications since 1967.]

Arthur J. Vander
Department of Physiology
University of Michigan Medical School
Ann Arbor, MI 48109

October 1, 1984

"Until 1963, I had been a traditional renal physiologist at the University of Michigan, studying the handling of sodium and water by the kidneys. That year, I heard James O. Davis present Louis Tobian's intrarenal-baroreceptor hypothesis for the control of renin secretion, i.e., that the renin-secreting cells respond directly to the pressure within the renal arterioles.[1] It struck me, while listening to him, that changes in renal vascular pressure also caused important changes in the flow of fluid through the renal tubules and in sodium reabsorption, and I wondered whether one of these variables, acting via the macula densa, might be the actual controller of renin release. An experiment to distinguish between the baroreceptor and macula densa theories was easy to formulate, but I knew nothing about the methodology (then very crude) for measuring renin. At this time, a brilliant medical student, Richard Miller, came to work with me for the summer and began a totally unrelated project. However, we talked about the hypothesis and experiment I had been toying with, and without hesitation, Rick abandoned his original project, set up the necessary methods, and we were off.[2]

"My lab produced a large number of experiments dealing with the control of renin secretion and, in 1966, I was asked by James W. McCubbin and Irvine H. Page to contribute a chapter[3] on this subject to a book that they were editing. After reading my manuscript, Page stated that he thought it deserved a wider audience and he suggested I submit it to *Physiological Reviews*, explaining to the editors of this journal that a very similar version would be appearing in the forthcoming book. Happily, this was not deemed a problem and the review was accepted.

"I believe there are several reasons for the paper being cited so often. It was the first comprehensive review of a subject that is not only of basic physiological significance but has very important implications for a variety of common diseases, including hypertension. Also, during the previous five years, there had been remarkable advances in our understanding of the role of the renin-angiotensin-aldosterone system, and many of the system's functions and properties had been reviewed. The methodology for quantitative studies of the controls over renin secretion itself had also been rapidly improving, and the growing number of studies in this critical area had generated considerable controversy. Perhaps the major value of my review was its attempt to analyze the various theories that had been proposed and the body of literature underlying them. I hope that the theoretical analysis provided by the review helped to stimulate the deluge of experiments on this subject that began about that time and still continues. Of course, whether or not my review really played a stimulatory role, the authors of all these papers were obliged to cite it for the next nine years, until the next comprehensive review was published."[4]

1. **Tobian L.** Relationship of juxtaglomerular apparatus to renin and angiotensin. *Circulation* 25:189-92, 1962. (Cited 160 times.)
2. **Vander A J & Miller R.** Control of renin secretion in the anesthetized dog. *Amer. J. Physiol.* 207:537-46, 1964. (Cited 400 times.)
3. **Bunag R D, Vander A J & Kaneko Y.** Control of renin release. (Page I H & McCubbin J W, eds.) *Renal hypertension.* Chicago: Year Book Medical Publishers, 1968. p. 100-17. (Cited 40 times.)
4. **Davis J O & Freeman R H.** Mechanisms regulating renin release. *Physiol. Rev.* 56:1-56, 1976. (Cited 595 times.)

Morgan H E, Henderson M J, Regen D M & Park C R. Regulation of glucose
uptake in muscle. I. The effects of insulin and anoxia on glucose transport and
phosphorylation in the isolated, perfused heart of normal rats.
J. Biol. Chem. **236**:253-61, 1961.
[Dept. Physiology, Vanderbilt Univ. Sch. Medicine, Nashville, TN]

When rat hearts were perfused as Langendorff preparations, glucose transport limited
the rate of glucose uptake and was accelerated by insulin and anoxia. After transport
acceleration, glucose phosphorylation limited glucose uptake. Phosphorylation was increased by anoxia but not by insulin. [The
SCI® indicates that this paper has been cited
in over 515 publications since 1961.]

Howard E. Morgan
Department of Physiology
College of Medicine
Pennsylvania State University
Hershey, PA 17033

July 12, 1984

"This paper was the first in a series of six
papers published in the *Journal of Biological
Chemistry* concerned with regulation of glucose metabolism in hearts from normal and
diabetic rats.[1-5] This work was carried out in
the laboratory of Charles R. Park. At the
time, Margaret J. Henderson and I were
postdoctoral fellows. I was attracted to this
problem because of an interest in transport
and its regulation rather than cardiac metabolism, *per se*. This work represented a career
change for both David M. Regen and me and
resulted from Park's enthusiasm and encouragement toward a career in physiology. I

had been trained as an obstetrician/gynecol-
ogist, while Regen was a medical student
who had interrupted his training for a year
to gain experience in research.

"The major findings of this study were
that membrane transport was a major rate-
limiting step for glucose utilization in the
absence of insulin, and the rate of uptake
conformed to Michaelis-Menten kinetics. Insulin increased glucose uptake because of
an acceleration of the transport step. Under
these conditions, glucose phosphorylation
became the rate-limiting step. Anoxia accelerated glucose uptake by increasing rates of
both glucose transport and phosphorylation.
These studies confirmed earlier work in
Levine's[6] and Park's[7] laboratories that
showed an effect of insulin on glucose
transport and work by Randle and Smith[8] on
the effect of anoxia on glucose transport.

"The other contribution of this paper and
perhaps the reason for its frequent citation
was the introduction of an easy and effec
tive method for use of the isolated rat heart,
perfused as a Langendorff preparation, for
metabolic studies. Since this publication, a
method for perfusion of the rat heart as a
working preparation was developed in association with Neely, Liebermeister, and Battersby.[9] In addition, a model of cardiac
ischemia was introduced that employed the
working rat heart.[10] The isolated rat heart
perfused as Langendorff, working, or isch-
emic preparations is a frequently used
model for biochemical and physiological
studies of carbohydrate, fat, protein, and
RNA metabolism in cardiac muscle."[11]

1. **Morgan H E, Cadenas E, Regen D M & Park C R.** Regulation of glucose uptake in muscle. II. Rate-limiting steps
and effects of insulin and anoxia in heart muscle from diabetic rats. *J. Biol. Chem.* **236**:262-8, 1961.
(Cited 120 times.)
2. **Post R L, Morgan H E & Park C R.** Regulation of glucose uptake in muscle. III. The interaction of membrane
transport and phosphorylation in the control of glucose uptake. *J. Biol. Chem.* **236**:269-72, 1961.
(Cited 65 times.)
3. **Henderson M J, Morgan H E & Park C R.** Regulation of glucose uptake in muscle. IV. The effect of
hypophysectomy on glucose transport, phosphorylation, and insulin sensitivity in the isolated, perfused heart.
J. Biol. Chem. **236**:273-7, 1961. (Cited 25 times.)
4. ...., Regulation of glucose uptake in muscle. V. The effect of growth hormone
on glucose transport in the isolated, perfused rat heart. *J. Biol. Chem.* **236**:2157-61, 1961. (Cited 40 times.)
5. **Morgan H E, Regen D M, Henderson M J, Sawyer T K & Park C R.** Regulation of glucose uptake in muscle. VI.
The effects of hypophysectomy, adrenalectomy, growth hormone, hydrocortisone, and insulin on glucose
transport and phosphorylation in the perfused rat heart. *J. Biol. Chem.* **236**:2162-8, 1961. (Cited 55 times.)
6. **Levine R, Goldstein M S, Huddlestun B & Klein S P.** Action of insulin on the 'permeability' of cells to free hexoses,
as studied by its effect on the distribution of galactose. *Amer. J. Physiol.* **163**:70-6, 1950.
(Cited 225 times since 1955.)
7. **Park C R, Bornstein J & Post R L.** Effect of insulin on free glucose content of rat diaphragm *in vitro*.
Amer. J. Physiol. **182**:12-16, 1955. (Cited 110 times since 1955.)
8. **Randle P J & Smith G H.** Regulation of glucose uptake by muscle. I. The effects of insulin, anaerobiosis and cell
poisons on the uptake of glucose and release of potassium by isolated rat diaphragm.
Biochemical J. **70**:490-500, 1958. (Cited 235 times since 1958.)
9. **Neely J R, Liebermeister H, Battersby E J & Morgan H E.** Effect of pressure development on oxygen consumption
by isolated rat heart. *Amer. J. Physiol.* **212**:804-14, 1967. (Cited 415 times.)
10. **Neely J R, Rovetto M J, Whitmer J T & Morgan H E.** Effect of ischemia on function and metabolism of the isolated
rat heart. *Amer. J. Physiol.* **225**:651-8, 1973. (Cited 185 times.)
11. **Kira Y, Kochel P J, Gordon E E & Morgan H E.** Aortic perfusion pressure as a determinant of cardiac protein
synthesis. *Amer. J. Physiol.* **246**:C247-58, 1984.

Randle P J, Garland P B, Hales C N & Newsholme E A. The glucose fatty-acid cycle: its role in insulin sensitivity and the metabolic disturbances of diabetes mellitus. *Lancet* 1:785-9, 1963.
[Department of Biochemistry, University of Cambridge, Cambridge, England]

The 'glucose fatty-acid cycle' describes the reciprocal relationship between the catabolism of glucose and lipid fuels in animals. Evidence was given that release and oxidation of lipid fuels inhibits metabolic degradation of glucose in muscles, and conversely that metabolic effects of glucose inhibit release of lipid fuels and thereby facilitate uptake and oxidation of glucose. The physiologic and pathologic significance of this concept is discussed. [The *SCI®* indicates that this paper has been cited over 800 times since 1963.]

Philip J. Randle
Department of Clinical Biochemistry
John Radcliffe Hospital
Oxford OX2 6HE
England

June 16, 1981

"My collaborators were all graduate students in the department of biochemistry at Cambridge. They have each made distinguished careers. Peter Garland is head of the department of biochemistry in Dundee, Nick Hales is head of the department of clinical biochemistry at Cambridge, and Eric Newsholme is lecturer in biochemistry and fellow of Merton College in Oxford.

"As a graduate student at Cambridge, I had worked on the rat diaphragm assay for insulin and anti-insulin factors in blood plasma. This experience convinced me that the route to an understanding of the factors which modify the action of insulin in muscle was through identification of rate-limiting reactions in the tissue and of the biochemistry of their regulation. The pioneer of this general experimental approach was the late E.W. Sutherland.[1]

"In 1956, I had the good fortune to observe that glucose uptake and glycolysis in diaphragm muscle are increased by anoxia and (provided insulin was added) decreased by ketone bodies. Studies of the effects of anoxia between 1956 and 1961 identified membrane transport, hexokinase, and phosphofructokinase as rate-limiting reactions for glucose uptake and glycolysis in *in vitro* preparations of heart and diaphragm muscles. Work in the latter half of 1961 and early 1962 showed that metabolism of fatty acids (and ketone bodies) inhibited these reactions and additionally pyruvate dehydrogenase. The effects of fatty acids (and ketone bodies) were quantitatively very similar to the changes effected by starvation or induction of alloxan-diabetes in the rat.

"In consequence, much of 1962 was taken up with obtaining evidence for increased intracellular provision of fatty acids in heart and diaphragm muscles of diabetic or starved animals. By the end of 1962, we were convinced that we had sufficient evidence to justify writing a conceptual paper on regulatory interactions between glucose and fatty-acid metabolism and their physiologic and pathologic significance. This paper was written in January 1963.

"Perhaps this paper has been cited frequently because it summarised, at an opportune moment, a phase in the development of understanding of the process of fuel selection in mammalian muscle. I hope that it adequately represented and acknowledged the contribution of colleagues in other laboratories. The main controversy over the glucose fatty-acid cycle since then has been its applicability to skeletal muscle, and its quantitative importance in the whole animal. However, recent work by M.J. Rennie and J.O. Holloszy has shown its applicability to red skeletal muscle and re-emphasised its importance *in vivo*.[2] Much of my own work since 1963 has been concerned with detailed biochemical mechanisms relevant to the general concept."

1. Sutherland E W. The effect of the hyperglycaemic factor of the pancreas and of epinephrine on glycogenolysis. *Recent Progr. Hormone Res.* 5:441-59, 1950.
2. Rennie M J & Holloszy J O. Inhibition of glucose uptake and glycogenolysis by availability of oleate in well-oxygenated perfused skeletal muscle. *Biochemical J.* 168:161-70, 1977.

Cahill G F, Jr., Herrera M G, Morgan A P, Soeldner J S, Steinke J, Levy P L, Reichard G A, Jr. & Kipnis D M. Hormone-fuel interrelationships during fasting.
J. Clin. Invest. **45**:1751-69, 1966.
[Elliott P. Joslin Research Lab., Depts. Medicine & Surgery, Harvard Medical Sch. and Peter Bent Brigham Hosp.; and Diabetes Foundation, Inc., Boston, MA]

This paper reports that insulin plays as significant a role in glucose homeostasis in fasting as in feeding. It also emphasizes the progressive nitrogen and glucose sparing, with fatty acids and ketones becoming the dominant fuels. Finally, it shows that the brain diminishes glucose utilization during fasting. [The *SCI®* indicates that this paper has been cited in over 650 publications since 1966.]

George F. Cahill, Jr.
Howard Hughes Medical Institute
398 Brookline Avenue
Boston, MA 02215

August 6, 1984

"Since Berson and Yalow had introduced the immunoassay for insulin to the research community, and since one of us (J. Soeldner) had developed a very sensitive and accurate low-range insulin assay, a study on insulin levels in fasting man seemed warranted, particularly since it hadn't been done. Totally insulin-deprived animals and man develop fatal ketoacidosis, so we knew that even low levels of insulin must play some physiological role. Also, growth hormone had been suggested as a major lipid mobilizer and another of us (D. Kipnis) had a good immunoassay going in St. Louis thanks to colleagues W. Daughaday and C. Parker.

"Who could we get to volunteer to fast for a week while enduring C¹⁴-glucose turnover sampling, twice-daily Douglas bag breathing, blood sampling, precise water intake and urine sampling, and minimal bed-chair activity? It would be impossible to do it on ourselves like the previous one-day studies on insulin effects, exercise, utilization of various sugars as altered by insulin, and many other variables. This was a decade before the present regulations on self-experimentation or soliciting students for research. Nevertheless, we found the ideal, dependable, honest, motivated, and financially limited volunteers at the Harvard Divinity School who needed the money ($300 for the week) to go home for Christmas. There were one Baptist, four Congregationalists, and one Episcopalian (his glucose values were always highest!).

"I felt the paper to be a straightforward, not-too-imaginative piece of biomedical reporting, but it led to our subsequent studies showing that the brain uses ketoacids during starvation,[1] that alanine[2] and glutamine[3] were preferentially released from muscle, that insulin directly inhibited amino acid release *in situ* in forearm muscle,[4] and a number of others that were summarized.[5,6] Glucagon, growth hormone, and glucocorticoids were also studied, both by determinations and by infusion into obese subjects undergoing prolonged starvation for weight reduction, and found to have some interesting and yet unexplained effects, but their roles in controlling fuel flux and patterns were vastly subordinate to that of insulin. For this overall work, especially for the initial paper (the *Citation Classic*), I received recognition including the Goldberger Award in Nutrition, the Gairdner International Award of Canada, and the Banting Medal of the American Diabetes Association. My younger colleagues in these studies—Soeldner, Owen, Felig, Marliss, and Aoki, to name several—have all continued to provide significant contributions in the area.

"Perhaps most important, the quantitative schemes we generated for fuel flux in man in fasting, feeding, diabetes, and trauma have appeared as the standards in many physiological and biochemical texts. Also, the studies have served as a basis for the recently developed discipline of parenteral alimentation and hyperalimentation since they involved interorgan substrate and energy exchange. The most enjoyable part of the series, however, was the pleasure of doing straightforward, fundamental physiological biochemistry in man and having the results be directly applicable to clinical problems and human disease. They also led to a number of formal and informal collaborations and correspondence with scientists around the world interested in starvation in other species as well as man."

1. Owen O E, Morgan A P, Kemp H G, Sullivan J M, Herrera M G & Cahill G F, Jr. Brain metabolism during fasting. *J. Clin. Invest.* **46**:1589-95, 1967. (Cited 425 times.)
2. Felig P, Owen O E, Wahren J & Cahill G F, Jr. Amino acid metabolism during prolonged starvation. *J. Clin. Invest.* **48**:584-94, 1969. (Cited 435 times.)
3. Marliss E B, Aoki T T, Pozefsky T, Most A S & Cahill G F, Jr. Muscle and splanchnic glutamine and glutamate metabolism in postabsorptive and starved man. *J. Clin. Invest.* **50**:814-17, 1971. (Cited 145 times.)
4. Pozefsky T, Felig P, Tobin J D, Soeldner J S & Cahill G F, Jr. Amino acid balance across tissues of the forearm in postabsorptive man. Effects of insulin at two dose levels. *J. Clin. Invest.* **48**:2273-82, 1969. (Cited 285 times.)
5. Cahill G F, Jr. Starvation in man. *N. Engl. J. Med.* **282**:668-75, 1970. (Cited 365 times.)
6. ------------------. The Banting Memorial Lecture 1971. Physiology of insulin in man. *Diabetes* **20**:785-99, 1971. (Cited 215 times.)

Glick S M, Roth J, Yalow R S & Berson S A. The regulation of growth hormone secretion. *Recent Progress in Hormone Research* 21:241-83, 1965.
[Radioisotope Service, Veterans Administration Hospital, Bronx, NY]

Plasma radioimmunoassayable human growth hormone (HGH) levels fluctuate widely and rapidly in response to stimuli that have in common a shortage of carbohydrate energy substrate, and to stress. Glucose lowers plasma HGH in normals but not in acromegalics. The HGH response to insulin hypoglycemia distinguishes normals from hypopituitary subjects. [The *SCI*® indicates that this paper has been cited over 360 times since 1965.]

Seymour M. Glick
Soroka Medical Center
Department of Internal Medicine
P.O.B. 151
Beer-Sheba, Israel

March 14, 1978

"This publication represented the summary of two exciting years in 'partnership' with Jesse Roth in the now famous laboratory of Sol Berson and Ros Yalow. In 1961, just after their work on the radioimmunoassay (RIA) for insulin, we were given the 'easy' job of developing a RIA for HGH. But even under the guidance of the 'masters' and with a great deal of hard work, just the development of a reliable assay took us over a year.

"Being the 'only children' of the brilliant researchers and teachers, Berson and Yalow was an unforgettable experience. They gave of themselves to us with their characteristic intensity and excellence—well beyond three standard deviations from the norm. We staggered home twice weekly after a three hour lecture by Sol on differential equations with mimeographed notes prepared *just for the two of us*. Ros and Sol set an example of integrity, hard work and precision of thought that dazzled and depressed us. We appreciated most their willingness to let us struggle independently with our project, with little interference—but with the knowledge that the experts were at our beck and call.

"Struggle we did. For months we were in the 'paper business.' Whatman 3MM paper, fine for insulin, did not do for HGH. For a while we 'manufactured' our own composite brand —Whatman 3MM plus a site of application made up of DEAE paper. We spent hours on the floor of the lab gluing these components together.

"With the assay working, after improved hormone purification, we turned to physiology, in the search for stimuli and suppressors of HGH. The goal was the evaluation of clinical states of hyper- and hyposecretion. Fortunately, the first stimuli investigated included insulin hypoglycemia. The magnitude of the response surprised us, and we knew we had made a major discovery. The next few months were devoted to daily physiological experiments, many on ourselves and families. I learned that insulin hypoglycemia superimposed on a six day fast does not enhance one's feelings of well-being. But walking across the Bronx with Jesse, discussing our next experiments was an exhilirating way of testing the effect of exercise on HGH.

"The work altered drastically the then prevalent view of growth hormone as a 'slowly moving' hormone involved largely in growth. The work stimulated much new thought regarding the physiology of growth hormone and its neuroendocrine control. A direct means of measuring a pituitary hormone and its fluctuations was provided, and thereby a tool for assessing hypopituitarism and acromegaly. The breakthrough early in the course of the application of RIA to various new substances gave dramatic demonstration of the powerful tool developed by Berson and Yalow and its enormous potential for physiology and medicine."

This Week's Citation Classic

Alberti K G M M, Christensen N J, Christensen S E, Prange Hansen Aa, Iversen J, Lundbaek K, Seyer-Hansen K & Ørskov H. Inhibition of insulin secretion by somatostatin. *Lancet* 2:1299-301, 1973.
[Second University Clinic of Internal Medicine, Kommunehospitalet, Aarhus, Denmark]

Somatostatin was isolated from the hypothalamus and synthesized in 1973. Intravenous glucose tolerance tests in normal subjects demonstrated that somatostatin reduced insulin responses and decreased the glucose-disappearance rate. The effect was direct on the pancreatic B-cell as demonstrated in perfused pancreas. This was the first demonstration of an extrapituitary effect of somatostatin. [The *SCI®* indicates that this paper has been cited in over 395 publications since 1973.]

Hans Ørskov
Institute of Experimental Clinical Research
and Second University Clinic
of Internal Medicine
University of Aarhus
8000 Aarhus C
Denmark

September 18, 1982

"This study was a by-product of our hypothesis[1] that growth hormone is a causal factor in diabetic angiopathy. Prange Hansen had demonstrated that diabetics in ordinary clinical metabolic control had a two to three times elevated diurnal plasma growth hormone and this hyperproduction was metabolically dependent.[2] This fact, combined with the demonstration in the first randomized controlled clinical trial[3] that hypophysectomy delays the development of diabetic retinopathy, launched the hypothesis.

"A logical consequence was to find pharmaca able to suppress growth hormone secretion. After having administered about 100 different drugs during several more growth hormone stimulation (exercise) tests, the hypothalamic growth hormone inhibitor somatostatin was isolated by Guillemin and co-workers in 1973.[4]

"Norman Grant of Wyeth Laboratories, Philadelphia, heard of our efforts and favoured us with some synthetic somatostatin at a very early and very appropriate point in time. With this we succeeded by a hairbreadth to be the first to demonstrate in man the inhibitory effect on growth hormone secretion (active in diabetics as well as controls).[5] Somatostatin was destined not to be the ideal growth hormone suppressor taken daily by most diabetics and this is partly due to its other than growth hormone inhibitory effects — of which the world was totally unaware in autumn 1973.

"Although the experimental design had been less than ideal for the purpose, we noted that with overnight fast and exercise, suppressed plasma insulin seemed even lower in those receiving somatostatin infusion. We quickly examined our insulin responses to glucose with and without preceding injection of somatostatin and established with our last smidgen that this new insulin inhibition had direct action on the B-cell. This we published in December 1973; later followed a series of reports on somatostatin's inhibitory effects on glucagon and a lot of other hormones — and somatostatin became invaluable in studies of metabolic effects of hormones.

"I note with some regret that we concluded: 'These effects of somatostatin are definitely not physiological.... It is highly improbable that somatostatin ever reaches concentrations in the systemic circulation that would have any effect on insulin secretion.' But how were we to know that loads of readily releasable somatostatin was present in the pancreatic D-cell adjacent to the B-cell — and that the insulin inhibitory action would take on a new dimension and detonate a prolific series of studies of possible paracrine effects in the pancreas and elsewhere?

"So, our explanation that this paper reached the *Citation Classics'* hit list is that we happened to have some somatostatin very early; that we stumbled on its very first discovered extrapituitary action (when we urgently were looking for another effect); that the somatostatin field of research exploded thereafter; and that we ranked the order of coauthors alphabetically so it glared at observers from the top of reference lists."

1. Lundbaek K, Christensen N J, Jensen V A, Johansen K, Steen Olsen T, Prange Hansen Å, Ørskov H & Østerby R. Diabetes, diabetic angiopathy, and growth hormone. (Hypothesis.) *Lancet* 2:131-3, 1970.
2. Prange Hansen Aa. Normalization of growth hormone hyperresponse to exercise in juvenile diabetics after 'normalization' of blood sugar. *J. Clin. Invest.* 50:1806-11, 1971.
3. Lundbaek K, Malmros R, Andersen H C, Rasmussen J H, Bruntse E, Madsen P H & Jensen V A. Hypophysectomy for diabetic angiopathy. A controlled clinical trial. (Östman J, ed.) *Diabetes: proceedings of the Sixth Congress of the International Diabetes Federation, Stockholm, Sweden, 30 July-4 August 1967.* Amsterdam: Excerpta Medica, 1969. p. 127-39.
4. Brazeau P, Vale W, Burgus R, Ling N, Butcher M, Rivier J & Guillemin R. Hypothalamic peptide that inhibits the secretion of immunoreactive pituitary growth hormone. *Science* 179:77-9, 1973.
5. Prange Hansen Aa, Ørskov H, Seyer-Hansen K & Lundbaek K. Some actions of growth hormone release inhibiting factor. *Brit. Med. J.* 3:523-4, 1973.

Polak J M, Pearse A G E, Grimelius L, Bloom S R & Arimura A. Growth-hormone release-inhibiting hormone in gastrointestinal and pancreatic D cells. *Lancet* 1:1220-2, 1975.
[Depts. Histochem. and Med., Royal Postgrad. Med. Sch., London, England and Vet. Admin. Hosp., and Tulane Univ. Sch. Med., New Orleans, LA]

This paper provided morphological evidence for the production of an active peptide, somatostatin (or growth-hormone release-inhibiting hormone), by an endocrine cell type of the pancreatic islets which had been described 44 years before as separate and distinct from the A and B cells.[1] The peptide product of this 'third' type was then unknown. [The *SCI*® indicates that this paper has been cited in over 380 publications since 1975.]

Julia M. Polak
Hammersmith Hospital
Royal Postgraduate
Medical School
University of London
London W12 OHS
England

May 27, 1982

"In 1931 a 'third' type of endocrine cell (the D cell of the pancreas) was recognised in human pancreatic islets.[1] D cells were later found to be present in the islets of many other species.

"In spite of its distinctive morphology, the function and putative product of the third cell type remained mysterious. Many peptides were proposed as possible products of the D cell of the pancreatic islets. Among them, gastrin was the favourite candidate, in view of the repeated finding of pancreatic gastrinomas (responsible for the classical features of the Zollinger-Ellison syndrome) and the need to find their cell of origin, this type of tumour being one of the most common islet cell tumours. All efforts proved fruitless. In fact, many workers argued about the real identity of the D cell of the pancreas and suggested that this distinct cell type might merely be a modified A cell; some even proposed the term A_2 cell because of its alleged close morphological resemblance to the glucagon producing A_1 cell of the pancreas.

"In 1974, Lars Grimelius came to the department of histochemistry for a two-year sabbatical. He came from Uppsala with a Royal Society scholarship to work on the growing points of immunocytochemistry. Grimelius joined the department shortly after the revolutionary discovery of the potent hypothalamic inhibitor for the release of growth hormone (or somatotrophin) named growth-hormone release-inhibiting factor,[2] later renamed somatostatin. Among its inhibitory actions, its powerful effect as an inhibitor of the release of pancreatic hormones was soon realised. The potent inhibitory actions of extracts of pancreatic D (A_1) cells on the secretion of insulin had been reported as early as 1969.[3]

"In view of this, it seemed appropriate to investigate whether this newly discovered peptide originated from the D cells of the human pancreas. We were indeed in possession of the most appropriate technology. Grimelius, the inventor of the famous Grimelius silver impregnation technique for staining the A, glucagon, cells of the pancreas, was able to use not only his own technique but also that proposed by Hellman and Hellerstrom for the visualisation of D (A_2) cells of the pancreatic islets.[4] I myself was involved in the development of refined and specific immunocytochemical methods for the localisation of peptide hormones in endocrine cells, and Akira Arimura was generous enough to allow us to use his excellent antibodies to somatostatin. The winds of fortune were blowing in our direction, and in 1975 we were able to demonstrate confidently the presence of somatostatin-like immunoreactivity in the 'third' endocrine cell type of the pancreas, the D cells, and thus the long sought peptide product for the D cells was at last recognised. The finding of somatostatin-like immunoreactivity in the D cells of the pancreas has been repeatedly reported by many workers throughout the world.

"Since its discovery, somatostatin has generated an overwhelming interest from most scientific disciplines and it is clear that the knowledge of its precise cellular origin must have set up the ground basis for the understanding of its mode of action and putative role in human pathology; without doubt this is the reason why the paper is so popularly quoted."

1. **Bloom W.** A new type of granular cell in the islets of Langerhans of man. *Anat. Rec.* 49:365-84, 1931.
2. **Brazeau P, Vale W, Burgus R, Ling N, Butcher M, Rivier J & Guillemin R.** Hypothalamic polypeptide that inhibits the secretion of immunoreactive growth hormone. *Science* 179:77-9, 1973.
3. **Hellman B & Lernmark A.** Inhibition of the *in vitro* secretion of insulin by an extract of pancreatic a_1 cells. *Endocrinology* 84:1484-8, 1969.
4. **Hellman B & Hellerstrom C.** The specificity of the argyrophil reaction in the islets of Langerhans in man. *Acta Endocrinol.* 36:22-30, 1961.

Pasteels J L. Recherches morphologiques et expérimentales sur la sécrétion de
prolactine. *Arch. Biol.* 74:439-553, 1963.
[Labs. Histologie et Microscopie Électronique, Faculté de Médecine,
Université Libre de Bruxelles, Belgium]

The paper demonstrates autonomous production
of prolactin by rat pituitaries in tissue culture, and
the existence of a hypothalamic prolactin-inhibit-
ing factor. Similar experiments, performed on human hypophyses, were proof of the existence of a
separate human prolactin, distinct from growth
hormone, because addition of hypothalamic extracts to the cultures inhibited prolactin release
along with stimulation of growth hormone secretion. [The *SCI*® indicates that this paper has been
cited in over 170 publications since 1963, making
it the 2nd most-cited paper ever published in this
journal.]

J.L. Pasteels
Histology Laboratory
Faculty of Medicine
Free University of Brussels
B-1000 Brussels
Belgium

March 17, 1983

"I was already doing research as a student in
medicine in Herlant's laboratory, when I read the
elegant work of Guillemin and Rosenberg,[1] who
demonstrated that the addition of hypothalamic
fractions to tissue cultures of the anterior hypophysis restored their corticotropin production. For
more than one year, I dreamed of performing similar experiments on prolactin. From what was
known from experiments on pituitary grafts, it
could be expected that the cultures would secrete
prolactin autonomously, and that evidence of an
inhibitory neuroendocrine control could be obtained by addition of hypothalamic extracts.

"At the end of my studies, I had the opportunity
to work several months with Guillemin, when he
was in the laboratory of Courrier, at the Collège de
France. Guillemin was not doing tissue culture at
that time, and I got somewhat involved in his LRF
program. This further delayed my work on prolactin, but Guillemin, when I candidly told him my in-

tentions, gave me good advice on the preparation
of hypothalamic extracts.

"Back in Brussels, I benefited from the hospital-
ity of Mulnard in his tissue culture unit, and I got
the expected results on rat pituitaries in a few
weeks. I then found that J. Meites[2] and C.S. Nicoll,[3] in the US, were also studying prolactin production by rat pituitaries *in vitro.* Working alone
with small means, I was in no position to compete
with such brilliant people. Thus, I left them to perform the obvious experimentation that could be
done on prolactin physiology, and I made use of
the only asset that they did not share with me, i.e.,
proximity of a university hospital and therefore access to human material.

"At that time, it was generally believed that
there was no human prolactin, because prolactin
could not be extracted from human pituitaries,
and because human growth hormone (HGH) had
prolactin-like activities. Working with H. and
J. Brauman, who devised some precursor immunoassay of HGH, I could demonstrate that prolactin
and HGH were distinct molecules, because they
were submitted to reverse hypothalamic control.[4]

"Because it was a thesis, my paper had to be
published as a whole and in French. Much later, as
vice-dean and as dean of the University of Brussels School of Medicine, I took personal care in
changing that rule. It was a long time before my
work was confirmed. It was finally reported by
F. Greenwood, using his well-known radioimmunoassay system. Greenwood handsomely acknowledged my paper,[5,6] and this came in good time,
when human prolactin was finally purified (1971).[7]

"Another reason why my thesis is frequently
mentioned is the discovery, by E. Flückiger, of
bromocryptine, a dopamine agonist now widely
used in medicine to inhibit prolactin secretion.[8]
When starting his work in this field, Flückiger
gambled a lot of his career on the existence of
human prolactin. He once told me that he did so
because he read my thesis and was convinced.
That was my best reward."

1. **Guillemin R & Rosenberg B.** Humoral hypothalamic control of anterior pituitary: a study with combined
   tissue cultures. *Endocrinology* 57:599-607, 1955.
2. **Meites J.** Hypothalamic control of prolactin section. (Wolstenholme G E W & Knight J, eds.)
   Lactogenic hormones. London: Churchill Livingstone, 1972. p. 325-38.
3. **Nicoll C S.** Secretion of prolactin and growth hormone by adenohypophyses of rhesus monkeys *in vitro.*
   (Wolstenholme G E W & Knight J, eds.) *Lactogenic hormones.* London: Churchill Livingstone, 1972. p. 257-68.
4. **Brauman J, Brauman H & Pasteels J L.** Immunoassay of growth hormone in cultures of human hypophysis
   by the method of complement fixation: comparison of the growth hormone secretion and the prolactin activity.
   Nature 202:1116-18, 1964.
5. **Bryant G D & Greenwood F C.** The concentrations of human prolactin in plasma measured by radioimmunoassay:
   experimental and physiological modifications. (Wolstenholme G E W & Knight J, eds.) *Lactogenic hormones.*
   London: Churchill Livingstone, 1972. p. 197-206.
6. **Siler T M, Morgenstern L L & Greenwood F C.** The release of prolactin and other peptide hormones from
   human anterior pituitary tissue cultures. (Wolstenholme G E W & Knight J, eds.) *Lactogenic hormones.*
   London: Churchill Livingstone, 1972. p. 207-17.
7. **Lewis U J & Singh R N P.** Recovery of prolactin from human pituitary glands. (Pasteels J L & Robyn C, eds.)
   Human prolactin. Amsterdam: Excerpta Medica, 1973. p. 1-10.
8. **del Poso E & Flückiger E.** Prolactin inhibition: experimental and clinical studies. (Pasteels J L & Robyn C, eds.)
   Human prolactin. Amsterdam: Excerpta Medica, 1973. p. 291-301.

CC/NUMBER 21
MAY 21, 1984

This Week's Citation Classic™

Steelman S L & Pohley F M. Assay of the follicle stimulating hormone based on the augmentation with human chorionic gonadotropin.
Endocrinology 53:604-16, 1953.
[Fundamental Res. Dept., Armour Labs. and Res. Division, Armour and Co., Chicago, IL]

This paper describes a simple, specific method for the bioassay of the follicle stimulating hormone (FSH) based upon the augmentation of the ovarian weight response to FSH with human chorionic gonadotropin (HCG). [The *SCI®* indicates that this paper has been cited in over 1,010 publications since 1955.]

Sanford L. Steelman
Merck Sharp & Dohme
Research Laboratories
Rahway, NJ 07065

April 12, 1984

"As a result of the findings of Hench[1] that cortisone was useful in the treatment of rheumatoid arthritis, the Armour Laboratories began the production and sale of ACTH as an alternative therapy. Literally hundreds of pounds of porcine pituitaries were processed each week. As a result, there was potentially available a large quantity of pituitary by-products from which one could recover hormones. It was found that the residue, after extraction of the ACTH and posterior pituitary hormones, contained gonadotropins (predominantly follicle stimulating hormone [FSH]).

"After reviewing the literature, it was readily apparent that there was no simple, specific assay for FSH. In order to purify FSH, we needed a better assay. After examining many possible methods in a variety of animals, we decided that the interaction between FSH and LH on ovarian weight was the most promising. By administering a large excess of LH (as human chorionic gonadotropin [HCG]), it was shown that any LH contamination in the sample would not affect the ovarian weight response to FSH. Many variables were examined including normal vs. hypophysectomized rats, dose of HCG, frequency of administration, interfering hormones, etc. As a result of these exploratory studies, a simple, specific assay for FSH was developed using immature female rats. In the development of the method, Florence Pohley, a statistician, analyzed each experiment. When we arrived at a workable procedure, it was found that the response was not a function of the logarithm of the dose as is the case in almost all bioassays. The slope ratio method for calculation of potencies was utilized and was described in detail in the publication. Several other investigators have conducted extensive mathematical analyses of the dose response curve and confirmed our original data [personal communications]. However, most investigators now use the log dose calculation method employing a narrower portion of the dose response curve. More recent studies indicate that by using 40-50 IU of HCG, the frequency of administration can be reduced.

"With a good assay method in hand, the purification of porcine FSH progressed rapidly and the product was eventually tested in humans and animals and found to be active. It was marketed for animal use. We were able to prepare, in 1952, many grams of a purified FSH (264-151-X) which was well characterized with regard to contamination with other hormones. This preparation was widely distributed to investigators and was used as a reference standard until the National Institutes of Health began preparing and distributing purified hormone preparations.

"The paper has been frequently cited because the method was used to bioassay pituitary preparations as well as human and animal biological fluids. Until radioimmunoassays became commonplace, the method was the simplest, most sensitive, and specific available."

1. **Hench P S, Kendall E C, Slocumb C H & Polley H F.** The effect of a hormone of the adrenal cortex (17-hydroxy-11-dehydrocorticosterone: compound E) and of pituitary adrenocorticotropic hormone on rheumatoid arthritis. *Proc. Staff Meetings Mayo Clinic* 24:181-97, 1949. (Cited 260 times since 1955.)

This Week's Citation Classic

Cheung W Y. Calmodulin plays a pivotal role in cellular regulation.
Science 207:19-27, 1980.
[Dept. Biochemistry, St. Jude Children's Research Hosp., and Univ. Tennessee Ctr. for
Health Sciences, Memphis, TN]

The role of calcium ions (Ca^{2+}) in cell functions is beginning to be unraveled at the molecular level as a result of recent research on calcium-binding proteins and particularly on calmodulin. These proteins interact reversibly with Ca^{2+} to form a protein-Ca^{2+} complex, whose activity is regulated by a cellular flux of Ca^{2+}. Many of the effects of Ca^{2+} appear to be exerted through calmodulin-regulated enzymes. [The *SCI®* indicates that this paper has been cited in over 740 publications since 1980.]

Wai Yiu Cheung
Department of Biochemistry
St. Jude Children's Research Hospital
Memphis, TN 38101

August 26, 1983

"This article originated from my postdoctoral studies at the Johnson Research Foundation, University of Pennsylvania. In 1964, after completing a doctoral dissertation in some aspect of carbohydrate metabolism of an alga under Martin Gibbs at Cornell University, I joined Britton Chance as a postdoctoral fellow. He introduced me to a fascinating phenomenon—oscillation of NADH in a cell-free extract of yeast. Chance and his colleagues had shown that cyclic AMP produces striking effects on the oscillatory profile of NADH in yeast extract. Commercial preparations of cyclic AMP at that time were rather crude; it was my job to determine its purity and to find out if the cyclic nucleotide retains effectiveness after treatment with phosphodiesterase, which degrades it to 5'-AMP. While purifying the enzyme from bovine heart, I noticed that its activity was precipitously reduced. Subsequent experiments showed that the activity of the purified, but not the crude, enzyme varied widely depending on the assay procedure used.[1] Since phosphodiesterase regulates the extent and duration of cyclic AMP action, I decided to pursue this observation to learn why the enzyme lost activity upon purification and what substances might be used to restore it. The loss turned out to be the result of my inadvertent removal of an activator protein, now known as calmodulin. These early experiments with phosphodiesterase unexpectedly opened a fertile new area of research in cellular regulation.

"Despite several preliminary reports of an activator protein of phosphodiesterase,[2-6] and confirmation by other laboratories,[7,8] the finding stimulated only mild interest. At that time, much of the research on cyclic AMP was centered on adenylate cyclase, which catalyzes the synthesis of cyclic AMP.

"One aspect of calmodulin seemed especially puzzling. Although it regulates the activity of phosphodiesterase, its distribution is much wider than the enzyme's, implying additional functions. By the mid-1970s—I had moved to St. Jude Children's Research Hospital in 1967—calmodulin had been found in all eukaryotes, and workers in various laboratories began to realize that calmodulin regulates a broad spectrum of Ca^{2+}-dependent cellular processes, including the regulation of Ca^{2+} itself. Therefore, in 1978, I coined the name 'calmodulin' to denote that the protein is modulated by Ca^{2+} and that it also modulates Ca^{2+} concentration.

"The importance of calcium ion in cellular physiology has been appreciated for a century. The mechanism of its action remained unclear until the discovery and recent knowledge of calmodulin. Why is it that this ubiquitous protein, which mediates the action of Ca^{2+} in many cellular processes, remained hidden from investigators through several decades during which those processes were under intensive study? The answer lies primarily in calmodulin's ubiquity and abundance. Since it is always present, in cells and in cell extracts, it is never missed. Its role was disclosed only by its inadvertent removal from phosphodiesterase during routine purification of the enzyme

"My article reviews the salient features of calmodulin: its discovery, its molecular mechanism of action, its central role in cellular functions, criteria for calmodulin-regulated reactions, and some future directions. This wide-ranging coverage, coming at a time when interest in calmodulin was building rapidly, may be the reason for the paper's frequent citation."

1. **Cheung W Y.** Cyclic 3',5'-nucleotide phosphodiesterase. Pronounced stimulation by snake venom.
   Biochem. Biophys. Res. Commun. 29:478-82, 1967. (Cited 80 times.)
2. ----------------. Activation of a partially inactive cyclic 3',5'-nucleotide phosphodiesterase. (Abstract.)
   Fed. Proc. 27:783, 1968.
3. **Cheung W Y & Jenkins A.** Regulatory properties of cyclic 3',5'-nucleotide phosphodiesterase. (Abstract.)
   Fed. Proc. 28:473, 1969.
4. **Cheung W Y.** Cyclic 3',5'-nucleotide phosphodiesterase. Preparation of a partially inactive
   enzyme and its subsequent stimulation by snake venom. *Biochim. Biophys. Acta* 191:303-15, 1969.
   (Cited 105 times.)
5. **Cheung W Y & Patrick S.** A protein activator of cyclic 3',5'-nucleotide phosphodiesterase. (Abstract.)
   Fed. Proc. 29:602, 1970.
6. **Cheung W Y.** Cyclic 3',5'-nucleotide phosphodiesterase. Demonstration of an activator.
   Biochem. Biophys. Res. Commun. 38:533-8, 1970. (Cited 180 times, 1970-9; cited 280 times, 1980-3.)
7. **Kakiuchi S, Yamazaki R & Nakajima H.** Properties of a heat-stable phosphodiesterase activating factor
   isolated from brain extract: studies on cyclic 3',5'-nucleotide phosphodiesterase. II.
   Proc. Jpn. Acad. 46:587-92, 1970. (Cited 210 times.)
8. **Goren E N & Rosen O R.** The effect of nucleotides and a non-dialyzable factor on the hydrolysis of cyclic AMP
   by a cyclic nucleotide phosphodiesterase from beef heart. *Arch. Biochem. Biophys.* 142:720-3, 1971.
   (Cited 70 times.)

This Week's Citation Classic

CC/NUMBER 29
JULY 19, 1982

Hasselbach W & Makinose M. Die Calciumpumpe der "Erschlaffungsgrana" des Muskels und ihre Abhangigkeit von der ATP-Spaltung.
Biochemische Zeitschrift 333:518-28, 1961.
[Inst. Physiologie, Max-Planck-Inst. für Medizinische Forschung, Heidelberg, Federal Republic of Germany]

The findings published in this paper demonstrate that the calcium uptake of muscle microsomes is coupled to the activity of a calcium-dependent ATPase and that calcium accumulation results in the formation of steep calcium concentration gradients. [The *SCI*® indicates that this paper has been cited in over 590 publications since 1961.]

Wilhelm Hasselbach
Abteilung Physiologie
Max-Planck-Institut für
Medizinische Forschung
D-6900 Heidelberg
Federal Republic of Germany

March 17, 1982

"In the late-1950s, after *in vitro* experiments performed with isolated contractile proteins had furnished conclusive evidence[1] that ATP was the sole energy source of the contractile apparatus, the problem emerged as to how the ATP hydrolysing activity of the contractile proteins in the living muscle was switched on and off during a contraction relaxation cycle. The road was opened by the discovery of Marsh[2] showing that in aqueous muscle extracts a factor was present which could suppress the contraction of isolated contractile proteins and that this effect could be abolished by calcium ions.

"In 1957, we found that the microsomal particles were very similar to the vesicular fragments of the just rediscovered sarcoplasmic reticulum. After we had observed that the relaxing factor was only transiently inactivated by small quantities of calcium ions, we intended to find out what had occurred to the added calcium. Using radioactive calcium, we found that the calcium had been completely taken up by the vesicles. We then demonstrated that the accumulation of calcium by the vesicles was a process causally linked to ATP hydrolysis. It was shown that during calcium accumulation an extra ATPase becomes active. Later we verified that the translocation of two calcium ions requires the splitting of one molecule of ATP. This analysis became possible only due to our trick of using oxalate at correct concentrations for trapping calcium ions inside the vesicles. It allowed us to estimate the calcium concentration gradient established by the transport system and therefore to prove that the calcium ions were transported uphill and, furthermore, to measure simultaneously calcium uptake and ATP splitting.

"The accumulation of calcium oxalate excludes that calcium removal was brought about by ATP-dependent calcium binding. On the other hand, due to the use of oxalate we were attacked by physiologists for producing physiologically irrelevant artifacts with unphysiological reagents. The biochemists were reserved and connected our findings with the simultaneously observed energy-dependent mitochondrial calcium uptake which in the long run proved to be of minor physiological importance. The priority of our finding was temporarily shadowed by the fact that one year later results of Ebashi and Lipmann[3] were published with the claim of also having demonstrated an ATP-dependent concentration of calcium, although their results only supported the retention of minute amounts of calcium by the membranes.

"The paper is highly cited because in the presence of ATP the sarcoplasmic calcium transport system had become, in conjunction with the calcium sensitivity of the contractile system pioneered by A. Weber[4] and Ebashi,[5] a basic element in our concept of excitation contraction coupling and, furthermore, a system most suitable for studying the chemical and structural events connected with ion translocation.[6] I received the Feldberg Award in 1963 for this work."

1. Weber H H. *The motility of muscle and cells.* Cambridge, MA: Harvard University Press, 1958. 69 p.
2. Marsh B B. The effects of adenosine triphosphate on the fibre volume of muscle homogenate.
   Biochim. Biophys. Acta 9:247-60, 1952.
3. Ebashi S & Lipmann F. Adenosine triphosphate-linked concentration of calcium ions in a particulate fraction of rabbit muscle. *J. Cell Biol.* 14:389-400, 1962.
4. Weber A. On the role of calcium in the activity of adenosine 5'-triphosphate hydrolysis by actomyosin.
   J. Biol. Chem. 234:2764-9, 1959.
5. Ebashi S. Third component participating in the superprecipitation of 'natural actomyosin.' *Nature* 200:1010, 1963.
6. Hasselbach W. Calcium-activated ATPase of the sarcoplasmic reticulum membranes. (Bonting S L & de Pont J J H H M, eds.) *Membrane transport.* Amsterdam: Elsevier/North-Holland, 1981. p. 183-208.

CC/NUMBER 9
MARCH 1, 1982

Ebashi S & Endo M. Calcium ion and muscle contraction.
Progr. Biophys. Mol. Biol. 18:123-83, 1968.
[Dept. Pharmacology; Fac. Medicine, Univ. Tokyo, Tokyo, Japan]

The contemporary concept (1967) of the roles of Ca ion in muscle contraction is introduced. In the absence of Ca ion, troponin in collaboration with tropomyosin exerts an inhibitory effect on the actin filament, not to interact with myosin. Ca ion discharged from the sarcoplasmic reticulum under the influence of the action potential affects troponin and releases the actin filament from its depressed state, resulting in contraction. The sarcoplasmic reticulum then removes Ca ion from troponin at the expense of ATP and induces relaxation. The importance of Ca ion in other intracellular processes is also discussed. [The *SCI®* indicates that this paper has been cited over 965 times since 1968.]

Setsuro Ebashi
Department of Pharmacology
Faculty of Medicine
University of Tokyo
Bunkyo-ku, Tokyo 113
Japan

November 4, 1981

"What is described in the summary above is common knowledge and found in ordinary textbooks for college students. However, not so many years have elapsed since the establishment of the Ca concept in muscle contraction. In a symposium on muscle, held in 1962, the above proposal that the Ca uptake of the sarcoplasmic reticulum should be the key mechanism of relaxation was very unpopular; strong ardor for an imaginary 'soluble relaxing factor' was still dominating. A. Weber and I, perhaps the only persons who were convinced of the essential nature of Ca ion in muscle contraction at that time, were having a hard time.

"One of the criticisms offered to me at this meeting was that I should explain the reason why Ca ion was effective only on crude systems such as glycerinated muscle fibers or natural actomyosin (myosin B), but not on pure actomyosin. I was told, 'I cannot accept such a mysterious idea; if Ca ion is the real factor, it must act on the pure system.'

"If I had been allowed to speak in Japanese, I could have somehow refuted this argument. Unfortunately, or fortunately, my poor English could not afford it. As a consequence, I had to answer it by presenting experimental data. This eventually led me to the discovery of the third component[1] other than myosin and actin and eventually of troponin.[2]

"Now Ca ion is no longer the factor unique to muscle contraction, but 'a common mediator between function and metabolism,' as clearly predicted in this article (see p. 160). Perhaps this expression was too modest in view of its universal roles in fundamental biological processes. Now the enthusiasm of biochemists for Ca ion is somewhat similar to that created by cyclic nucleotides some years ago. I am a little worried that such enthusiasm could bring about transcendental belief in Ca ion, as we have often experienced in the history of science.

"This article has two distinct features. One is that it boldly emphasized the importance of Ca ion when most biochemists did not pay any attention to this ion. The other is that regulatory processes were conceptually separated from contractile processes; now the regulatory mechanism is one of the main fields not only in muscle research,[3] but also in biology in general.

"When Endo and I began writing this article at the beginning of 1967, the worldwide students' rebellion had already started in Japan, particularly in our medical school at the University of Tokyo. I was extremely pessimistic about the future of the university and Japanese science. Even such a rational, prudent person with keen insight as Endo did not oppose my desperate opinion. I said inwardly, 'This article could be my last scientific work.' Fortunately we were wrong, but this unusual mental state might have added something to this article.'"

1. Ebashi S. Third component participating in the superprecipitation of 'natural actomyosin.' *Nature* 200:1010, 1963.
2. Ebashi S & Kodama A. A new protein factor promoting aggregation of tropomyosin. *J. Biochemistry* 58:188-90, 1965.
3. Ebashi S, Maruyama K & Endo M, eds. *Muscle contraction, its regulatory mechanisms.*
    New York: Springer-Verlag. 1981. 549 p.

Schatzmann H J. Herzglykoside als Hemmstoffe für den aktiven Kalium- und Natriumtransport durch die Erythrocytenmembran. (Cardiac glycosides as inhibitors for the active potassium and sodium transport across the red cell membrane.) *Helv. Physiol. Pharmacol. Acta* 11:346-54, 1953.
[Pharmakologisches Institut, Universität Bern, Switzerland]

It was shown that k-strophanthoside (10^{-5} g/ml) and the aglycones strophanthidin and digitoxigenin completely block the uphill Na^+ and K^+ movements in human red cells and that this inhibition is likely to be due to a direct effect on the Na^+-K^+ pump. [The *SCI®* indicates that this paper has been cited in over 620 publications since 1955.]

H.J. Schatzmann
Department of Veterinary Pharmacology
University of Bern
3000 Bern, Switzerland

July 10, 1984

"After graduating from medical school, I worked in the Bernese Department of Pharmacology with W. Wilbrandt. With a primordial flame photometer (run in complete darkness and going awry every so often), I observed the red cell Na^+-K^+ pump. Its existence, at the time, was something more than a lingering feeling and less than a textbook truth.[1] Wilbrandt suggested that mineralocorticoids should be tested on it, the idea being that they supplied the ionophoric group. Intracellular receptors for steroids and their effects were unknown, and the idea seemed worth examining. I tried it—in vain.

"It was known that Na^+ reduced the force of the heart,[2] but the Na^+-Ca^{++} exchange system not having been discovered yet, one could think that internal Na^+ was responsible. My loose thinking went as follows: steroid hormones stimulate Na^+-K^+ transport (in the kidney); cardiac glycosides are steroids and might be cardiotonic by virtue of a similarity with steroid hormones. Unaware of the structural differences between the two, I exposed my red cells to strophanthoside, hoping that it would do what the corticoids failed to do. The inhibitory action was soon obvious. Thus, by looking for a steroid requirement, I had found an inhibitor; the three princes from Serendip smiled.

"Strophanthoside clearly did not ruin transport by causing a leak. To decide whether the inhibition acted on the pump or the energy supply, I measured glycolysis in a Warburg apparatus (another exasperating machine; Wilbrandt kept telling me that it was 'the very model Sir Hans Krebs was using'). There was no reduction of glycolysis, and I drew the correct conclusion, namely, that the inhibition acted directly on the pump. Whittam *et al.*[3] later showed that the experiment was not good enough. (A drop in glycolysis [by 15 percent] follows whenever the pump is stopped.)

"The discovery produced some interest probably for two reasons. First, owing to extreme specificity, cardiac glycosides identify whatever is connected to the Na^+-K^+ pump:[4] abnormal modes (reverse, exchange, uncoupled Na flux) as well as secondary transports. With their help, Skou's $Na^+$$K^+$-ATPase[5] was easily recognized as the Na^+-K^+ pump. Second, the finding fostered the hope that the mechanism of the positive inotropic action of cardiac glycosides, elusive as it was, was around the corner.[6]

"These scientific pursuits were pleasantly punctuated by little excitements. One day, an undesirable spatial arrangement between a lit Bunsen burner and the dustcover (a piece of pink cotton) set the flame photometer ablaze when nobody was present. Wilbrandt had the bad luck to step into the room when the spectacle was at its best. He fought the conflagration successfully without the convenience of a CO_2 fire extinguisher. Unsuspecting, I met him a minute later in the hall, dishevelled, agitated, and with a charred blanket in his hands. He did not conceal his disapproval of what he called an unduly literal interpretation of the word '*Flammenphotometer.*' "

1. **Maizels M.** Factors in the active transport of cations. *J. Physiol.—London* 112:59-83, 1951. (Cited 160 times since 1955.)
2. **Wilbrandt W & Koller H.** Die Calciumwirkung am Forschherzen als Funktion des Ionengleichgewichts zwischen Zellmembran und Umgebung. *Helv. Physiol. Pharmacol. Acta* 6:208-21. 1948. (Cited 145 times since 1955.)
3. **Whittam R, Ager M E & Wiley J S.** Control of lactate production by membrane adenosine triphosphatase activity in human erythrocytes. *Nature* 202:1111-12, 1964.
4. **Glynn I M.** The action of cardiac glycosides on ion movements. *Pharmacol. Rev.* 16:381-407. 1964. (Cited 545 times.)
5. **Skou J C.** The influence of some cations on adenosine-triphosphatase from peripheral nerve. *Biochim. Biophys. Acta* 23:394-401, 1957. (Cited 1,205 times since 1957.)
6. **Akera T & Brody T M.** The role of Na^+, K^+-ATPase in the inotropic action of digitalis. *Pharmacol. Rev.* 29:187-220, 1978. (Cited 150 times.)

Beutler E. The glutathione instability of drug-sensitive red cells. A new method for the in vitro detection of drug-sensitivity. *J. Lab. Clin. Med.* 49:84-94, 1957.

Normal, but not primaquine-sensitive red cells incubated for two hours with acetylphenylhydrazine in the presence of glucose, are able to maintain their level of reduced glutathione (GSH). Destruction of GSH in primaquine-sensitive cells occurs only in the presence of oxygen. [The *SCI®* indicates that this paper was cited 379 times in the period 1961-1977.]

Ernest Beutler
Chairman, Division of Medicine
City of Hope National Medical Center
Duarte, California 91010

November 29, 1977

"Military service appeared inevitable to me as a second year medical resident at the University of Chicago in 1953. Since I was interested in hematology, the opportunity to work at the Army Malaria Research Project at the Stateville Penitentiary on primaquine-induced hemolytic anemia seemed ideal. The new antimalarial drug primaquine produced a severe hemolytic anemia in some black subjects. Yet when their red cells were examined by methods then available, they seemed to be entirely normal. I observed Heinz bodies in the red cells during the course of hemolytic crises, and was able to show that chemicals such as acetylphenylhydrazine produced a different pattern of Heinz body formation in vitro in primaquine-sensitive red cells than in normal red cells. Using inhibitors, I observed that normal cells treated with iodoacetate or arsenite behaved like primaquine-sensitive cells with respect to Heinz body formation. This turned my attention to red cell glutathione (GSH); the levels were lowered in primaquine-sensitive red cells, and moreover fell abruptly when primaquine was administered to sensitive subjects.

"Army life is not always predictable, and after one year's service at Stateville, I was transferred to Camp Detrick to serve out the remainder of my two-year term of active duty. When I returned to a junior faculty position at the University of Chicago in 1955, I attempted to unravel the biochemical basis of primaquine-sensitivity in red cells. It occurred to me that incubating blood from primaquine-sensitive donors with acetylphenylhydrazine might result in an abrupt fall in their GSH content. I proposed this project to a postdoctoral Fellow in the department, but he was not interested, and I undertook these studies myself.

"It was much more difficult to obtain blood from primaquine-sensitive subjects in civilian life than it had been in the prison. However, one of our subjects (a con-man) had been released from prison and volunteered to come to our clinic and donate blood for $5. One day my donor happily told me that he had found a job but needed $25 for new clothes. I advanced him the money for 5 donations; I have never seen him since. In spite of such difficulties, I was able to pursue these studies.

"After incubating blood from a primaquine-sensitive patient and a normal subject with acetylphenylhydrazine, I prepared a filtrate and added nitroprusside and cyanide. I can still remember my exhilaration (and almost disbelief) when, on my first attempt, no color developed in the filtrate from the incubated primaquine-sensitive sample. The 'GSH stability test' reported in the 1957 paper was the first reliable means for in vitro detection of primaquine-sensitivity. It quickly led to the discovery that the defect was sex-linked and that its basis was a deficiency in the enzyme glucose-6-phosphate dehydrogenase. Its principal effect was perhaps to produce awareness that the metabolism of red blood cells might be important in the origin of hemolytic disease."

Koller F, Loeliger A & Duckert F. Experiments on a new clotting factor
(Factor VII). *Acta Haematol.* 6:1-18, 1951.
[Department of Medicine, University of Zurich, Switzerland]

Evidence for the existence of a factor in normal serum and plasma that accelerates thrombin formation is presented. It is not consumed during coagulation and is absorbed on barium sulfate. Its concentration is lowered very rapidly by oral anticoagulants. We designated it Factor VII. [The *SCI*® indicates that this paper has been cited in over 365 publications since 1955.]

Fritz Koller
Wenkenhaldenweg 16
CH-4125 Riehen (Basel)
Switzerland

June 17, 1984

"The starting point for the recognition of Factor VII as a new clotting factor was a contradiction: Owren declared in 1950 at the meeting of the European Society of Haematology in London that prothrombin was not consumed during normal blood coagulation but could be demonstrated in serum in high concentrations for weeks at a time.[1] This finding was in strict opposition to the classical coagulation theory and particularly to the work of Brinkhous (prothrombin utilisation test)[2] and Quick (prothrombin consumption test).[3] Obviously, Owren, on the one hand, and Brinkhous and Quick, on the other, were not referring to the same clotting factor. Loeliger, in our laboratory, therefore analysed both factors using the same assay methods as the authors mentioned, and Duckert isolated and purified Owren's factor using barium sulfate adsorption, elution with citrate, dialysis, and so on. With various concentrations of this purified factor (all other clotting factors being kept constant), they showed that the quantity of thrombin formed did not change, but that the velocity of its formation varied in pro-

portion to the concentration. Owren's factor, which under normal conditions is present in about the same concentration in plasma as in serum, was therefore recognized as an accelerator of prothrombin conversion — as a new clotting factor that we designated Factor VII. Moreover, it could be demonstrated that the factor that is consumed during coagulation and therefore does not exist in normal serum (or only in traces) is the real precursor of thrombin and therefore has to be designated as prothrombin. Variations in its concentration correspond exactly to variations in the quantity of thrombin formed. The velocity of thrombin formation is, however, not influenced by this factor.

"Concerning terminology, we adopted the designation of clotting factors by Roman numerals inaugurated by Owren (Factor V). In 1954, an International Congress on Thrombosis and Embolism was held in Basel, Switzerland, where mutual understanding was almost impossible because of the many different names proposed for the same clotting factors. Therefore, on the initiative of Irving Wright in New York, a committee for the standardisation of the nomenclature of clotting factors was founded at the congress. Four years later, an agreement was reached: Roman numerals were recommended by the committee for the designation of all clotting factors, a proposal that has been almost universally accepted. Our designation of Factor VII therefore had important consequences.

"The reason this paper has been cited often is perhaps the thoroughness with which the characteristics of the new factor had already been studied in 1951 (purification, assay method, its role in thrombin formation, comparison with prothrombin, behaviour during anticoagulant treatment, and so on). The fact that Factor VII is lowered by anticoagulants more rapidly than any other vitamin K-dependent factor has practical implications in the beginning of anticoagulant therapy (sensitivity of thromboplastin to Factor VII in the Quick-test).

"For recent review articles on the present status of Factor VII, see references 4 and 5."

1. Owren P A. Parahaemophilia: haemorrhagic diathesis due to absence of a previously unknown clotting factor.
   Lancet 1:446-50, 1947. (Cited 85 times since 1955.)
2. Brinkhous K M. A study of the clotting defect in hemophilia: the delayed formation of thrombin.
   Amer. J. Med. Sci. 198:509-16, 1939.
3. Quick A J. On the quantitative estimation of prothrombin. *Amer. J. Clin. Pathol.* 15:560-6, 1945.
   (Cited 125 times since 1955.)
4. Zur M & Nemerson Y. Tissue factor pathways of blood coagulation. (Bloom A L & Thomas D P, eds.)
   Haemostasis and thrombosis. Edinburgh: Churchill Livingstone, 1981. p. 124-39.
5. Rapaport S I. The activation of Factor IX by the tissue factor pathway. (Menaché D, Surgenor D, Mac N &
   Anderson H D, eds.) *Haemophilia and haemostasis.* New York: Liss, 1981. p. 57-76.

Mustard J F & Packham M A. Factors influencing platelet function: adhesion, release, and aggregation. *Pharmacol. Rev.* 22:97-187, 1970.
[Dept. Pathology, Fac. Medicine, McMaster Univ., Hamilton, and Dept. Biochemistry, Fac. Medicine, Univ. Toronto, Ontario, Canada]

This review article brought together most of the information about blood platelets that was available in 1970 concerning aggregation, the release of granule contents, inhibitors, morphology, and metabolism. Mention was also made of the role of platelets in hemostasis and thrombosis. [The *SCI®* indicates that this paper has been cited in over 665 publications since 1970.]

J. Fraser Mustard
Canadian Institute for Advanced Research
434 University Avenue
Toronto, Ontario M5G 1R6
Canada

July 6, 1984

"In 1938, Tocantins had reviewed all the literature on platelets up to that time and discussed such controversial issues as the origin of platelets and the role of platelets in hemostasis and thrombosis.[1] Studies of platelets, however, were relatively few until the late 1950s. In that decade, a few investigators recognized that platelets were probably involved in the development of vascular disease and its complications, such as thromboembolism, and were therefore worthy of intensive study. It was during this period that I did my doctoral thesis and became interested in the role of platelets in the response of blood to vessel injury. At about the same time, interest in hemostasis broadened as it was realized that a detailed knowledge of both blood coagulation and platelet function was required.

"Between 1965, when the book *The Physiology of Blood Platelets* by Marcus and Zucker appeared,[2] and 1970, when our review was published, the information about platelets had expanded exponentially. In addition, several investigators, including ourselves, had observed that the nonsteroidal anti-inflammatory drugs inhibited some aspects of platelet function. This observation led to considerable interest in them as agents that might modify the contribution of platelets to the complications of vascular disease. It was clear that the information that had been growing so rapidly should be brought together, and we accepted the invitation of the editor of *Pharmacological Reviews* to do so.

"The review probably would not have been written if Marian Packham had not joined my research group seven years earlier and become very knowledgeable about the field. Her ability to write and her passion for accurate detail ensured that the article was properly prepared. Indeed, the review was probably a good symbiosis of our talents. Writing this review was not an easy task. It consumed many evenings and parts of weekends for many months. As I recall, we worked on this review in a wide variety of settings, often under conditions that were not conducive to concentration. The editor rapped our knuckles for adding new information to the galley, but we were allowed to include it.

"Writing a review of a field in which I had been involved at the beginning of its growth, particularly as my approach covered a broad spectrum, gave me a strong advantage because I could bring together and synthesize the large amount of information that I had watched develop and to which I had contributed. This article obviously coincided with the needs of the many new investigators who were entering the field for a comprehensive coverage of the subject to serve as a base for their projects and provide references to the key findings. We have been told that the review was very useful to new investigators, and this may partly account for the number of times it has been cited. Other reasons for the frequency of citation may be that no one has attempted as comprehensive a review of platelet function in a single article since 1970 and that much of the material is still relevant."

1. **Tocantins L M.** The mammalian blood platelet in health and disease. *Medicine* 17:155-260, 1938.
   (Cited 135 times since 1955.)
2. **Marcus A J & Zucker M B.** *The physiology of blood platelets: recent biochemical, morphologic, and clinical research.* New York: Grune & Stratton, 1965. 162 p.

Langdell R D, Wagner R H & Brinkhous K M. Effect of antihemophilic factor
on one-stage clotting tests. *J. Lab. Clin. Med.* **41**:637-47, 1953.
[Department of Pathology, School of Medicine, University of North Carolina,
Chapel Hill, NC]

This article describes a laboratory procedure that has been applied as a screening test for a number of hemostatic disorders and is the basis for the assay of what at the time was called the antihemophilic factor (AHF) and is now known as factor VIII:C. The principle of the bioassay has been the basis for the measurement of several other plasma coagulation factors. [The *SCI®* indicates that this paper has been cited in over 655 publications since 1955.]

Robert D. Langdell
Department of Pathology
School of Medicine
University of North Carolina
Chapel Hill, NC 27514

May 30, 1984

"At the time these studies were done, much of the research activity in the Department of Pathology at the University of North Carolina centered on a bleeding disorder that arose apparently as a mutation in a group of inbred dogs. Genetic, clinical, and coagulation studies indicated that the disorder was very similar to human hemophilia.[1] As in humans, transfusion of normal plasma corrected the clotting defect and controlled the hemorrhagic phenomena. We attributed this effect to the antihemophilic factor (AHF) that could be fractionated along with fibrinogen from normal plasma. Efforts to characterize the trace protein were impeded by the methodology for measuring AHF based on the utilization of prothrombin in clotting hemophilic blood. To simplify the measurement of AHF, we utilized an observation that in one-stage clotting tests, some extracts of normal or hemophilic tissue could compensate for the factor missing in hemophilia (complete thromboplastin), while others, such as natural or synthetic cephalins, could not (partial thromboplastin).

"The partial thromboplastin time, a new procedure based on these observations, was used in our laboratory for over three years before the article was submitted for publication. This was in part due to observations indicating that the method was sensitive to coagulation factors other than the plasma factor deficient in hemophilia, reports that other hemorrhagic states could easily be confused with hemophilia, and studies that were in progress indicating that there were mild forms of human hemophilia.[2] It was ultimately determined that the corrective effect of a test sample on canine or human hemophilic plasma used as a substrate was dependent on what is now known as factor VIII:C.

"The partial thromboplastin time is now widely used in clinical laboratories as a screening test not only for hemophilia but also for coagulation factors that were unrecognized at the time the procedure was published.[3] Although the bioassay method was developed for measuring factor VIII:C, by using plasma deficient in any one of several other clotting factors as a substrate, it is used as an assay method for other coagulation factors. The original method has been modified by a number of investigators, and reagents are now available commercially. In view of the tendency of authors to cite their own work, it is surprising that the original publication has been cited frequently.

"It is of interest that our efforts to characterize the coagulation defect in hemophilia have general applicability and have contributed to the recognition of other clotting factors and to the recently announced production of factor VIII:C through genetic splicing techniques."

1. **Graham J B, Buckwalter J A, Hartley L J & Brinkhous K M.** Canine hemophilia: observations on the course, the clotting anomaly, and the effect of blood transfusions. *J. Exp. Med.* **90**:97-111, 1949.
(Cited 75 times since 1955.)
2. **Brinkhous K M, Langdell R D, Penick G D, Graham J B & Wagner R H.** Newer approaches to the study of hemophilia and hemophilioid states. *J. Amer. Med. Assn.* **154**:481-6, 1954. (Cited 90 times since 1955.)
3. **Brinkhous K M & Dombrose F A.** Partial thromboplastin time. (Schmidt R M, ed.) *CRC handbook series in clinical laboratory science. Section I: hematology.* Boca Raton, FL: CRC Press, 1980. Vol. 3. p. 221-46.

CC/NUMBER 13
MARCH 30, 1981

Bachmann F, Duckert F & Koller F. The Stuart-Prower factor assay and its clinical significance. *Thrombos. Diath. Haemorrh.* 2:24-38, 1958.
[Coagulation Lab., Dept. Medicine, Univ. Zürich, Switzerland]

A simple one-stage method is described for the assay of blood coagulation Factor X (Stuart-Prower-factor), using Seitz's filtered bovine plasma, Russell's viper venom, and cephalin as a source of lipid. Assay values of Factor X are not influenced by different concentrations of fibrinogen, prothrombin, or Factors V or VII. [The *SCI®* indicates that this paper has been cited over 390 times since 1961.]

Fedor Bachmann
Laboratoire Central d'Hématologie
Université de Lausanne
Ecole de Médecine
1011 Lausanne
Switzerland

November 22, 1980

"During the early 1950s several new blood coagulation factors were discovered in rapid sequence. Their simultaneous discovery by several groups of investigators resulted in a profusion of nomenclature. Malicious critics maintained that the field of coagulation was but one unholy mess and that the main purpose of coagulationists at international meetings was to disagree. From 1951 to 1954, no less than four new coagulation factors were discovered (Proconvertin, SPCA, VII; Christmas factor, PTC, IX; PTA, XI; Hageman, XII).

"In 1955, Duckert *et al*[1] postulated the existence of Factor X on the basis of *in vitro* experiments. Patients hitherto thought to suffer from Factor VII deficiency but whose coagulation defect resembled more closely that caused by the postulated Factor X were soon discovered independently in 1956 by Telfer *et al.*[2] (Prower-factor) in England, by Hougie[3] (Stuart-factor) in the US, and in 1957 by our research group[4] in Zürich,

Switzerland (Delia factor). During this exciting time I worked on my MD thesis in the lab of Koller, a physician who had early realized the importance of basic biochemistry in the field of hemostasis. His foresight led to the hiring of Duckert, a biochemist. The coagulation laboratory in Zürich initiated many physicians into the basic concepts of research work and the puzzling pedagogic tactics of Duckert. I remember how I was laboring to produce some fraction devoid of Factor X, trying various adsorption and salt precipitation methods, only to conclude, several weeks later, that all my attempts had been futile. At one point, Duckert smilingly observed: 'Don't continue along this line. It won't work. I too have tried all this without result.' To my astonished and dismayed question why he had not stopped me earlier, he calmly replied, 'You learn by your mistakes, not by your successes.'

"The observation of Hougie,[3] that Russell's viper venom directly activates Factor X, i.e., acted like the combination of tissue thromboplastin plus Factor VII, led to the development of our assay. I was systematically exploring the optimal conditions for the preparation of the reactifs used in the assay, their optimal concentration and stability on storage at different temperatures, and the effect of surface contact. Rereading the article I realise that I have apparently succeeded in delineating all the possible pitfalls encountered in the preparation of the reagents and in revealing all the little tricks of the trade upon which the successful duplication of a method often depends.

"The paper had quite a success shortly after its publication and our 300 reprints were soon exhausted. Factor X was to become a coagulation factor of crucial importance, because it takes a central position in the coagulation system, and its activated form, Factor Xa, is inhibited by small doses of heparin. This latter observation provided the basis for the successful and widely utilized low dose heparin prophylaxis in patients at risk to develop deep vein thrombosis."

1. Duckert F, Flückiger P, Matter M & Koller F. Clotting Factor X. Physiologic and physicochemical properties. *Proc. Soc. Exp. Biol. (NY)* 90:17-22, 1955.
2. Telfer T P, Denson K W & Wright D R. A 'new' coagulation defect. *Brit. J. Haematol.* 2:308-16, 1956.
3. Hougie C. The role of the Russell's viper venom on the Stuart clotting defect. *Proc. Soc. Exp. Biol.* 93:570-3, 1956.
4. Bachmann F, Duckert F, Geiger M, Baer P & Koller F. Differentiation of the Factor VII complex. Studies on the Stuart-Prower factor. *Thrombos. Diath. Haemorrh.* 1:169-94, 1957.

Grotte G. **Passage of dextran molecules across the blood-lymph barrier.**
Acta Chir. Scand. **211**(Suppl.):1-84, 1956.
[Institute of Physiology and Department of Clinical Chemistry,
University of Uppsala, Sweden]

This paper was the first study of the passage of a noncharged *polymer* (dextran) across capillary membranes from plasma to lymph. Here not only total concentration of a test substance was measured simultaneously in samples of plasma and lymph, but also the distribution of molecular sizes could be compared. This gave a quite clear view of the functional ultrastructure of capillary membranes in various regions of the body. [The *SCI®* indicates that this paper has been cited in over 295 publications since 1961.]

Gunnar Grotte
Department of Pediatric Surgery
University Hospital
751 85 Uppsala
Sweden

March 15, 1983

"During the years after World War II, surgical research in Sweden was much centered on surgical shock and blood volume restoration in surgery. During my surgical residency at the Serafimer Hospital in Stockholm, Sweden, where the surgeon, Gunnar Thorsén, was one of the leading scientists, I became interested in clinical trials of dextran as plasma 'substitute.' After a short first visit to the Mayo Clinic in Rochester, Minnesota, I went back there to work as a fellow in 1949. One evening, I was called to the home of Charles W. Mayo where one of the guests was one of the two inventors of dextran, B. Ingelman. Walking back to the hotel that night we intensely discussed if it was ever possible that molecules penetrate capillary walls somehow in relation to their molecular size. At the same time, a new valuable technique in experimental surgery had been developed: a technique of cannulation of various lymphatics in the rat. With this new lymph sampling technique it would be possible to inject fractions of dextran of different molecular sizes and to follow their appearance in lymph. I then started a series of experiments with J.L. Bollman and the master of lymphatic calculation at the division of experimental surgery of the Mayo Clinic, Emery van Hook. This pilot study indeed showed that the transport of dextrans from plasma into lymph and urine was related to the molecular sizes of the test substance.[1]

"After my return to Uppsala University in Sweden in 1951, these studies were continued at the Institute of Physiology with T. Teorell, who was well known in permeability research, and with the other inventor of dextran, A. Grönwall,[2] with whom G. Wallenius[3] at that time had developed a method for determination of molecular weight distribution of dextran in micro amounts. Wallenius then published the first permeability curves for the human glomerular membrane in his thesis of 1954.[3] The collaboration between these two institutions provided the necessary background and the work continued. The lymph cannulation technique was adapted for dogs and lymph could soon be collected from various regions of the dog, i.e., from the legs, heart, intestines, liver, and thoracic duct. Curves could soon be produced showing the different permeability characteristics of these various regions but the interpretation of these curves was still quite speculative.

"Much of the earlier work in this field had been done at Harvard's famous Institute of Physiology (by, among others, Drinker[4] and Landis[5]). Then in 1951, Pappenheimer and co-workers[6] at this same institution published their contribution to the pore theory of capillary permeability.[3] They suggested that capillary membranes have intercellular pores of 35-45 Å radius while my own investigation suggested a two-pore system with one set of pores of similar sizes as those suggested by Pappenheimer and co-workers, but also the presence of larger 'leaks,' probably situated at the venous ends of the capillaries. Then in 1953, Palade[7] suggested that macromolecules could be transported across capillary walls mainly by 'pinocytosis,' i.e., macromolecules are carried across the capillary membrane by small vesicles traversing endothelial cells.

"This experimental work on dextran was later confirmed and widely extended. A summary of this work was discussed at a symposium held in Uppsala, called Lymph Circulation,[8] in 1977 to commemorate both the 500 year jubilee of the University of Uppsala and also Olaus Rudbeck about 300 years after his discovery of the lymphatic circulation. A short summary of this complex problem is not possible."

1. **Grotte G, Knutson R C & Bollman J L.** The diffusion of dextrans of different molecular sizes to lymph and urine.
   J. Lab. Clin. Med. **38**:577-82, 1951.
2. **Grönwall A.** *Dextran and its use in colloidal infusion solutions.* Uppsala: Almqvist & Wiksell, 1957. 156 p.
3. **Wallenius G.** Renal clearance of dextran as measure of glomerular permeability.
   Acta Soc. Med. Uppsala **59**(Suppl. 4):1-9, 1954.
4. **Drinker C K & Field M E.** *Lymphatics, lymph and tissue.* Baltimore, MD: Williams & Wilkins, 1933. 254 p.
5. **Landis E M.** Capillary pressure and capillary permeability. *Physiol. Rev.* **14**:404-81, 1934.
6. **Pappenheimer J R, Renkin E M & Borrero L M.** Filtration, diffusion and molecular sieving through peripheral capillary membranes. A contribution to the pore theory of capillary permeability.
   Amer. J. Physiol. **167**:13-46, 1951.
7. **Palade G E.** The endoplasmic reticulum. *J. Biophys. Biochem. Cytol.* **2**(Suppl.):85-98, 1956.
8. **Lewis D H,** ed. Proceedings from the Symposium on Lymph Circulation. Basic morphological and physiological aspects of water and solute exchange in the microcirculation. (Whole issue.)
   Acta Physiol. Scand. **463**(Suppl.), 1979. 127 p.

Crone C. The permeability of capillaries in various organs as determined by use of the 'indicator diffusion' method. *Acta Physiol. Scand.* 58:292-305, 1963.
[Institute of Medical Physiology, University of Copenhagen, Denmark]

The paper showed how permeability of blood capillaries could be determined quantitatively after a single passage of test solutes lasting a few seconds. The method was applied to different organs, with special emphasis on exchange between blood and brain. [The *SCI®* indicates that this paper has been cited in over 355 publications since 1963.]

Christian Crone
Department of Medical Physiology
Panum Institute
University of Copenhagen
2200 Copenhagen
Denmark

October 22, 1983

"Pappenheimer and associates[1] introduced quantitative studies of capillary permeability. Their method, however, required isolation and artificial perfusion of organs. Chinard and his collaborators[2] devised a principle which allowed organs to be studied *in situ* but their method was not quantitative. I solved this problem by making certain simplifying assumptions which made it possible to arrive at a simple mathematical expression from which capillary permeability could be calculated. The 'inaccessible' capillary membrane thus became accessible for quantification under *in vivo* conditions. The paper obviously filled a gap because very little was known about capillary permeability at that time. Interestingly, another solution to the same problem appeared almost in the same year from the other part of the globe, from Martin de Julián and Yudilevich[3] in Chile.

"I was doing my thesis work in the late 1950s. The peculiar fact that the blood-brain barrier has an extremely low permeability aroused my interest. How could this be reconciled with knowledge that D-glucose virtually pours into the brain from the blood? The blood-brain barrier had, so far, largely been characterized with semiquantitative methods having a very poor time resolution. The 'indicator diffusion' method now made it possible to approach this structure. However, it was necessary to cut through a lot of mystifications about the blood-brain barrier to postulate that it was just another capillary with properties of its own. This immediately made it possible to apply the strict analysis used in capillary physiology in general — a reductionist view which paved the way for an impressive development in this area.

"With the method established, I went on to assess brain capillary permeability — a true gold mine — and I was lucky enough to find a real piece of gold: the nonlinear, facilitated, glucose transport across the blood-brain barrier[4] — perhaps the best spin-off of the method.

"For quite a few years, there was considerable resistance against using the indicator diffusion technique, but things changed, and in the 1970s a surge of papers appeared based on this approach. The first to give the work full credit were my American colleagues, who chose me as the first recipient of the International Zweifach Award in 1979[5] — given for work in microcirculation."

1. **Pappenheimer J R, Renkin E M & Borrero L M.** Filtration, diffusion and molecular sieving through peripheral capillary membranes. *Amer. J. Physiol.* 167:13-46, 1951. (Cited 495 times since 1955.)
2. **Chinard F P, Vosburgh G J & Enns T.** Transcapillary exchange of water and of other substances in certain organs in the dog. *Amer. J. Physiol.* 183:221-34, 1955. (Cited 165 times since 1955.)
3. **Martin de Julián P & Yudilevich D L.** A theory for the quantification of transcapillary exchange by tracer dilution curves. *Amer. J. Physiol.* 207:162-8, 1964. (Cited 80 times.)
4. **Crone C.** Facilitated transport of glucose from blood into brain tissue. *J. Physiol.—London* 181:103-13, 1965. (195 cites.)
5. ----------. Ariadne's threat—an autobiographical essay on capillary permeability. *Microvascular Res.* 20:133-49, 1980.

Ross R & Glomset J A. The pathogenesis of atherosclerosis.
N. Engl. J. Med. 295:369-77; 420-5, 1976.
[Depts. Pathology, Medicine, and Biochemistry, Sch. Med., and
Regional Primate Research Ctr., Univ. Washington, Seattle, WA]

The structure of the normal artery wall and data on our understanding of the cell biology of endothelium and smooth muscle *in vitro* and *in vivo* are covered. Three hypotheses of atherogenesis are discussed including the 'response to injury hypothesis,'[1] the 'monoclonal hypothesis,'[2] and the 'clonal senescence hypothesis.'[3] These are each evaluated, compared, and contrasted, and the potential role of lipids and connective tissues in atherogenesis is discussed. [The *SCI®* indicates that these papers have been cited over 980 times in 652 publications since 1976.]

Russell Ross
Department of Pathology
School of Medicine
University of Washington
Seattle, WA 98195

July 1, 1982

"The two-part paper on the pathogenesis of atherosclerosis written by John Glomset and myself began as a result of a request from the *New England Journal of Medicine* to write a review on studies that I had been pursuing on wound healing and inflammation. By the time that request had been received, John and I had, for a number of years, been very much involved in studying a number of aspects of the biology of arterial smooth muscle and endothelium and had become very much interested in the problems of atherogenesis and the state of the field. We spent many hours talking about various ideas and decided if we could convince the *New England Journal of Medicine* to change their invitation from one dealing with a review of wound healing to one dealing with atherosclerosis that we would tackle the problem of trying to put into perspective many of the ideas that we had tossed around over the preceding years, with a particular view to examining the question from the viewpoint of the cell biologist.

"One unique feature of the school of medicine at the University of Washington was the fact that at particular points in time at least three hypotheses of atherogenesis had been developed,[1-3] surprisingly, all emanating from the same department! Since all three of these hypotheses had generated a fair amount of interest, we decided that after discussing the cell biology of the problem, those notions and ideas should be related to the hypotheses at that particular state of their development, with, we must admit, some bias toward the 'response to injury hypothesis of atherosclerosis' that we had proposed to test.

"We have been fortunate to receive wide recognition for our work on the 'response to injury hypothesis.' More important, we hope that this paper served as a catalyst to help change directions in this field. Our ideas have changed quite a bit since this review was written in 1976 and although some of the notions have proved to be correct, a number of them have changed with the advent of new information concerning the biology of endothelium, smooth muscle, and, in particular, of the monocyte/macrophage and the platelet and their potential role in this entire process. Therefore, the 'response to injury hypothesis' today appears somewhat different from the one published in the cited paper and probably in another five years' time, the one that we would propose today would again appear different based on new information as it becomes available. I have recently published a paper in this field."[4]

1. **Ross R & Glomset J A.** Atherosclerosis and the arterial smooth muscle cell. *Science* 180:1332-9, 1973.
2. **Benditt E P & Benditt J M.** Evidence for a monoclonal origin of human atherosclerotic plaques. *Proc. Nat. Acad. Sci. US* 70:1753-6, 1973.
3. **Martin G, Ogburn C & Sprague C.** Senescence and vascular disease. (Cristafalo V J, Roberts J & Adelman R C, eds.) *Exploration in aging.* New York: Plenum Press, 1975. p. 163-93.
4. **Ross R.** George Lyman Duff Memorial Lecture. Atherosclerosis—a problem of the biology of arterial wall cells and their interaction with blood components. *Arteriosclerosis* 1:293-311, 1981.

Hoffman B F & Cranefield P F. *Electrophysiology of the heart.*
New York: McGraw Hill, 1960. 323 p.
[Dept. Physiology, Coll. Medicine, State Univ. New York Downstate Med. Ctr., Brooklyn, NY]

The first comprehensive monographic review of the then new field of the study of the electrophysiology of the heart by intracellular recording, this monograph has appeared in Russian and Japanese and remains in print in English 20 years after its initial publication. [The *SCI*² indicates that this book has been cited over 1,380 times since 1961.]

,Paul F. Cranefield
Rockefeller University
New York, NY 10021

October 1, 1980

"Two events in the period 1949-1952 profoundly affected the field of electrophysiology: the introduction of the intracellular microelectrode by Gilbert Ling and Ralph Gerard and the development of an ionic theory of the action potential by Bernard Katz, Alan Hodgkin, and Andrew Huxley. These events immediately affected cardiac electrophysiology; the 1950s produced many important studies of the shape of the action potential in different parts of the heart, of the origin and spread of electrical activity in the heart, of the excitability of the heart, and of the ionic basis of the cardiac action potential.

"Hoffman and I, probably in late 1956, decided to write a review article with the deliberate intention of expanding it into a monograph. The review appeared in January, 1958;¹ the book was completed in July, 1959. By the end of the 1950s a great deal had been learned, yet the field could still be encompassed in a monograph of reasonable length. It is not easy to determine to what extent the success of this monograph depended on its appearing at a moment when the field was about to expand rapidly and to what extent, if any, its appearance created interest in the subject. The field certainly expanded greatly; in addition it soon took on clinical implications. In 1963 we were asked to contribute to an important clinical journal an article reviewing abnormal rhythms of the heart in terms of the newer knowledge of the electrophysiology of the heart;² that article has been cited over 315 times. Another article we have written³ that has often been cited (over 135 times) also had clinical implications. By now a whole new field called clinical electrophysiology of the heart has emerged from the clinical applications of laboratory studies made in the period 1950-1970.

"The frequent citation of our monograph is reflected in its publishing history. McGraw-Hill published it in 1960 and reissued it in unaltered form in the mid 1960s. A pirated Russian translation appeared in 1962.⁴ An authorized Japanese translation appeared in 1977.⁵ A facsimile reissue of the 1960 monograph appeared in 1976.⁶ The fact that the monograph remains in print and continues to be cited probably reflects the fact that many basic areas in the field had been explored by the time we wrote the book so that it remains a useful introduction to the subject. It also reflects the fact that no compact and comprehensive monograph has displaced it; although one of us has since published a monograph in the area,⁷ it, like other recent books in the field, deals only with a particular aspect of the subject."

1. Cranefield P F & Hoffman B F. Electrophysiology of single cardiac cells. *Physiol. Rev.* 38:41-76, 1958.
2. Hoffman B F & Cranefield P F. The physiological basis of cardiac arrhythmias. *Amer. J. Med.* 37:670-84, 1964.
3. Hoffman B F, Moore E N, Stuckey J H & Cranefield P F. Functional properties of the atrioventricular conduction system. *Circ. Res.* 13:308-28, 1963.
4. Hoffman B F & Cranefield P F. *Elektrofisiologia cerdtsa.* Moscow: Publishing House of Foreign Literature, 1962. 390 p.
5. ----------------------------------, *Electrophysiology of the heart.* Nishinomiya: Nishinomiyahoseikan, 1977. 305 p.
6. ----------------------------------, *Electrophysiology of the heart.* Mount Kisco, NY: Futura, 1976. 323 p.
7. Cranefield P F. *The conduction of the cardiac impulse. The slow response and cardiac arrhythmias.* Mount Kisco, NY: Futura, 1975. 404 p.

Graham F K & Clifton R K. Heart-rate change as a component of the orienting response. *Psychol. Bull.* 65:305-20, 1966.

The argument is advanced, on theoretical grounds, that the direction of heart rate change in response to simple stimuli should distinguish an orienting-attentional process from defensive and startle reactions. The available research, reviewed in terms of defining criteria, supports the argument. [The *Science Citation Index®* (*SCI®*) and the *Social Sciences Citation Index™* (*SSCI™*) indicate that this paper was cited a total of 217 times in the period 1966-1976.]

Frances K. Graham, Ph.D.
University Hospitals
University of Wisconsin
Madison, Wisconsin 53706

December 5, 1977

"This paper was written when my collaborator and I decided that we might be working with an incorrect assumption. Rachel Clifton had joined me a year before as a postdoctoral fellow and we were trying to study orienting behavior in the newborn infant, using heart rate (HR) acceleration as the dependent measure. Soviet work on orienting, especially by E.N. Sokolov, had become available in translation only recently and promised to be a powerful tool for investigating the capabilities of young infants. The theory specified that an orientation reflex which enhanced stimulus processing occurred following any kind of discriminable change in stimulation. While orienting would be elicited by the first presentation of a stimulus, it would disappear after a few repetitions if the stimulus had no important consequences, but could be made to reappear by a minor change of stimulus characteristics. A particularly useful part of the theory was that orienting included automatic components which did not require a subject to cooperate in responding.

"Since some stimuli might also evoke defensive reflexes which would reduce rather than enhance the effects of stimulation, it was important to be able to distinguish orienting from such a defensive reaction. There were a number of functional differences between the two kinds of reactions but employing these was cumbersome and might not be valid for studying an immature organism. What was needed were reaction components which changed differentially depending on which reaction was evoked. Although Sokolov had stated that HR acceleration was part of orienting, this appeared to conflict with an hypothesis of 'directional fractionation' advanced by the Laceys. If the Laceys' hypothesis applied to the kind of situations used to study orienting, then HR should decelerate with orienting and accelerate with protective reactions.

"To solve the puzzle, we searched the literature for HR studies where the functional characteristics allowed classifying a response as orienting or protective. Once we had posed the problem, work on the paper went quickly and provided daily suspense. We had not known whether the existing literature was sufficiently consistent to provide any concusive answer. In the end, it was, and it supported our inference.

"The paper thus provided the rationale for a methodology that would prove useful in a number of areas. The paper's popularity was certainly due, in part, to the fact that the methodology was relatively simple and many laboratories were equipped to exploit it. More important, in the long run, was the linking of two theoretical approaches which had not previously been related to one another. This widened the range of problems that might fruitfully be explored with the methodology."

This Week's Citation Classic

West J B & Dollery C T. Distribution of blood flow and ventilation-perfusion
ratio in the lung, measured with radioactive CO_2.
J. Appl. Physiol. 15:405-10, 1960.
[Department of Medicine, Postgraduate Medical School, Hammersmith Hospital,
London, England]

Human volunteers inhaled radioactive CO_2 labeled with [15]O (half-life, two minutes) and counters over the chest measured the rate of removal of the radioactive gas from the lung during breath-holding. The results showed that blood flow increased markedly from apex to base in the upright lung. Regional ventilation was also measured and the topographical inequality of gas exchange in the lung was calculated. [The $SCI^®$ indicates that this paper has been cited in over 385 publications since 1961.]

John B. West
Section of Physiology
Department of Medicine
University of California
La Jolla, CA 92093

June 15, 1983

"During the late 1950s, my colleagues and I were presented with a remarkable opportunity at the Postgraduate Medical School, London. The British Medical Research Council had recently installed the first medical cyclotron in a hospital and it was a relatively simple matter to produce large quantities of [15]O by bombarding nitrogen with deuterons. Because [15]O has a half-life of only two minutes, large amounts (several millicuries) can be inhaled for a small radiation dose.

"When radioactive CO_2 was inhaled and its rate of removal from the lung was measured with counters over the chest, we were astonished to find the clearance rate from the apex of the lung was much less than from the base. This difference was abolished when the subjects lay supine, and both the apical and basal clearance rates increased during exercise in the upright position. It was clear that we were demonstrating for the first time the enormous topographical inequality of blood flow in the normal lung.

"In the same paper we were able to show that the ventilation of the lower zones of the lung exceeded the upper. This measurement was obtained from the initial increase in counting rate when the breath of radioactive CO_2 was inhaled. Then, armed with these new data on the inequality of ventilation and blood flow in the lung, we calculated the regional differences of gas exchange. These calculations were subsequently reported in a more sophisticated fashion[1] using better data on the distribution of ventilation obtained with radioactive xenon[2] but the message was essentially the same.

"The reason why this paper is quoted so often is that it marked the beginning of an extensive research program into the regional differences of function and structure in the lung. It is now known that there are marked regional differences of blood flow, ventilation, alveolar P_{O_2} and P_{CO_2}, intrapleural pressure, alveolar size, and mechanical stresses within the lung.[3] Elucidation of the causes of these regional differences has been a very fruitful area of respiratory physiology over the last 15 years. Moreover, these topographical differences have been shown to play a role in a number of lung diseases.

"A note might be added about the laboratory in which this work was done. Space at Hammersmith Hospital at that time was extremely tight. Since we had to have a laboratory near the cyclotron because of the very short life of the isotope, we chose to do the work in a small cottage alongside the cyclotron building, which had previously been used as a house for the hospital engineer. I believe it had been condemned because it was structurally so unsound. However, it suited us well and what used to be the engineer's living room was packed with counting equipment. I still remember the astonishment of American visitors who came to see this exotic research and could not believe that it was being done in such appalling conditions.

"A final note. Since the topographical inequality of blood flow in the lung is caused by gravity, we are anxious to see what happens in the weightless environment of space flight. At the present time, we are working very hard on an experiment to do just that on Spacelab 4 which will be launched in 1985."

1. West J B. Regional differences in gas exchange in the lung of erect man.
J. Appl. Physiol. 17:893-8, 1962. (Cited 200 times.)
2. Ball W C, Jr., Stewart P B, Newsham L G S & Bates D V. Regional pulmonary function studied with
xenon[133]. J. Clin. Invest. 41:519-31, 1962. (Cited 470 times.)
3. West J B, ed. Regional differences in the lung. New York: Academic Press, 1977. 488 p.

Nash G, Blennerhassett J B & Pontoppidan H. Pulmonary lesions associated with oxygen therapy and artificial ventilation. *N. Engl. J. Med.* **276**:368-74, 1967.
[Depts. Pathology, James Homer Wright Labs., Anesthesia Labs., and Respiratory Unit, Harvard Med. Sch., Massachusetts Gen. Hosp., Boston, MA]

Characteristic pathological changes were found in the lungs of a group of patients who died after prolonged mechanical ventilation. The alterations did not correlate with duration of mechanical ventilation but appeared to be associated with prolonged inhalation of high concentrations of oxygen. Pulmonary oxygen toxicity was implicated as a possible cause of morbidity and mortality in patients treated with mechanical ventilators. [The *SCI®* indicates that this paper has been cited in over 380 publications since 1967.]

Gerald Nash
Divisions of Anatomic Pathology
and Pulmonary Medicine
Cedars-Sinai Medical Center
Los Angeles, CA 90048

July 19, 1984

"In 1965, as a first-year resident in pathology at the Massachusetts General Hospital, I became intrigued with a problem that was troubling my clinical colleagues who were caring for patients requiring mechanical ventilation. They were puzzled by the occasional development of a progressive deterioration of pulmonary function that was apparently unrelated to the patient's underlying disease. The patients typically did well for a few days, then developed a progressive reduction in pulmonary compliance and vital capacity. They could not be weaned from the ventilator, and they eventually died of respiratory failure. Some physicians at the Massachusetts General Hospital believed that the mechanical ventilator was somehow the culprit, and they referred to the problem as the 'respirator lung syndrome.' H. Pontoppidan, of the Respiratory Unit, thought that the ventilator was being accused unjustly, and he was keenly interested in unraveling the mystery.

"J.B. Blennerhassett and I, in the Department of Pathology, were struck by unusual gross and microscopic appearances of the lungs of patients who died of this syndrome. With the enthusiastic support of Pontoppidan, we decided to compare the morphological findings of a group of patients who died in the Respiratory Unit with a control autopsy population. We found three major differences that characterized the study group: heavy lungs, hyaline membranes, and early interstitial fibrosis. These changes did not correlate with duration of mechanical ventilation, but they appeared to be related to prolonged inhalation of high concentrations of oxygen. Moreover, a review of the literature revealed that the lesions of pulmonary oxygen toxicity as described in animals were similar to those seen in our patients. We concluded that some of our patients with the so-called respirator lung syndrome probably had succumbed to oxygen toxicity.

"At the time this study was performed, oxygen was routinely administered in this country without concern for its possible toxic effects on the lung, and many patients were undoubtedly given toxic levels unnecessarily. This paper warned the medical community that pulmonary oxygen toxicity could develop during therapy for acute respiratory failure. We also recommended that the inspired oxygen concentration should be monitored, and if toxic concentrations must be given to sustain life, the dose should be lowered as soon as possible.

"After publication of this paper and others on the same topic, there was general acceptance of the notion that oxygen is potentially dangerous and its administration should be closely monitored. The paper helped pave the way to a more judicious use of the gas, and in the process it became highly cited. The subject has been recently reviewed by Deneke and Fanburg.[1] Another reason the paper is so highly cited is that it contains the first description of the evolution of a common, nonspecific morphological reaction of the lung to a variety of deleterious agents in addition to oxygen. The lesion is now well recognized and is known as 'diffuse alveolar damage.' "[2]

1. **Deneke S M & Fanburg B L.** Normobaric oxygen toxicity of the lung. *N. Engl. J. Med.* **303**:76-86, 1980.
2. **Katzenstein A A, Bloor C M & Lefbow A A.** Diffuse alveolar damage—the role of oxygen, shock, and related factors. *Amer. J. Pathol.* **85**:210-24, 1976. (Cited 105 times.)

Morris J F, Koski A & Johnson L C. Spirometric standards for healthy nonsmoking
adults. *Amer. Rev. Resp. Dis.* 103:57-67, 1971.
[Depts. Medicine, Univ. Oregon Med. Sch. and Veterans Admin. Hosp.,
Portland, and Dept. Health Education, Oregon State Univ., Corvallis, OR]

A sample of 988 healthy nonsmoking men
and women were tested for routine spirometric function. Ventilatory function correlated positively with height but, despite the
absence of known cardiopulmonary disease,
correlated negatively with age. Linear regression equations and nomograms were obtained from the data to provide predicted
normal standards. [The *SCI®* indicates that
this paper has been cited in over 435 publications since 1971.]

―――――――――――――――・―・――――

James F. Morris
Veterans Administration Medical Center
P.O. Box 1034
Portland, OR 97207

May 31, 1983

"During the 1840s, the Reverend John
Hutchinson in London was intrigued by
breathing function. He had a spirometer
devised, constructed, and titled the vital capacity procedure, and tested about 2,000 assorted Londoners. He studied their physical
measurements to determine which correlated best with the vital capacity. Very little
happened in this field until 100 years later
when normal values were obtained at Bellevue Hospital, and two decades later by the
VA-Armed Forces Cooperative Study. All
these and other studies suffered from including cigarette smokers or patients with
nonpulmonary diseases. It was obvious
when testing a healthy nonsmoker, especially those middle-aged or older, that the available predicted normal standards were too
low.

"In conjunction with Arthur Koski of
Oregon State University, we decided to test
approximately 1,000 healthy nonsmoking
adults living in a relatively pollution-free
region of western Oregon. We thought that
the greatest yield would come from studying religious groups whose tenets forbade

tobacco smoking. Lavon Johnson arranged
the testing sessions with officials of the Mor-
mon and Seventh-Day Adventist churches.
All testing was performed in conjunction
with regular church meetings. Unfortunate
aspects of these ecclesiastical locations
were limitation of the complexity of the
testing and skewing the age distribution. It
also eliminated the ability to perform physical examinations and chest roentgeno-
grams. Linear regression equations were
easily derived but construction of a nomogram proved to be an arcane, laborious process. The present level of electronics and
data processing were not available to us in
1969 which made our data gathering and
analysis and nomogram construction more
tedious than it would be today. We are currently retesting the original population after
14 to 15 years and will take advantage of the
electronic advances. This retesting will convert the original cross-sectional study to a
longitudinal one.

"Publication of the original article proved
to be difficult. It was initially reviewed by a
Nobel laureate, who took umbrage at an un-
intended slur and rejected the manuscript.
After my personal entreaty to the editor, the
manuscript was sent to two reviewers who
accepted it. The original data were supplied
to five medical centers where other investigators validated the regression equations
and derived additional information and
tests. We reexamined the forced vital capacity volume-time curves and derived a new
measurement, the forced end-expiratory
flow or FEF75-85 percent.[1]

"A more recent review was published in
the *Western Journal of Medicine.*[2] It added
two measurements to the original nomograms for men and women. The original article is frequently cited because it represents
the first study of ventilatory function in a
large number of normal American nonsmok-
ers. It was recently cited by the Section on
Respiratory Pathophysiology of the American College of Chest Physicians for providing predicted normal standards which are
the most widely used in general pulmonary
function laboratories."[3]

1. **Morris J F, Koski A & Breese J D.** Normal values and evaluation of forced end-expiratory flow.
   Amer. Rev. Resp. Dis. 111:755-62, 1975.
2. **Morris J F.** Spirometry in the evaluation of pulmonary function. *West. J. Med.* 125:110-18, 1976.
3. **Zamel N, Altose M D & Speir W A, Jr.** Statement on spirometry: a report of the Section on Respiratory
   Pathophysiology (ACCP). *Chest* 83:547-50, 1983.

CC/NUMBER 13
MARCH 31, 1980

Miller L L, Bly C G, Watson M L & Bale W F. The dominant role of the liver in plasma protein synthesis. A direct study of the isolated perfused rat liver with the aid of lysine-ε-C14. *J. Exp. Med.* 94:431-53, 1951. [Depts. Radiation Biol. and Pathol., Univ. Rochester Sch. Med. and Dentistry, Rochester, NY]

The need for a method allowing the protracted direct study of the liver uncomplicated by the contributions of other organs led to development of the isolated rat liver perfusion technique. The perfused rat liver closely simulated the physiological behavior of the liver *in vivo* with respect to metabolism of glucose and amino acids, synthesis of plasma proteins, and bile secretion. [The *SCI®* indicates that this paper has been cited over 600 times since 1961.]

Leon L. Miller
University of Rochester
School of Medicine and Dentistry
Rochester, NY 14642

February 22, 1980

"From 1938 to 1946, in the laboratory of George H. Whipple, I participated in studies of protein nutritional factors affecting the production of plasma proteins in intact dogs. Interpretation of results invoked the hypothesis that the liver was the site of albumin and fibrinogen synthesis, but there was no direct evidence to support this view.

"In 1948 I returned to Rochester to join William F. Bale in further studies of plasma protein biosynthesis with the new tool, ^{14}C-lysine, prepared by R.W. Helmkamp, an organic chemistry professor. The question of the liver's role in plasma protein synthesis became more challenging. A direct answer to that question seemed obtainable experimentally by isolating the liver and maintaining its circulation with an artificial pump oxygenator. The idea of isolated liver perfusion dated back to Claude Bernard[1] and to German physiologists.[2] Trowell[3] had perfused the rat liver with aqueous media in retrograde fashion and Lupton[4] briefly described an effect of vitamin K on prothrombin synthesis by the rat liver perfused for a few hours.

"Our first perfusion apparatus was a 'do it yourself' project. The 'heart-lung' was made from the multi-bulbed tube of an Allihn condenser, hand-ground glass valves, and fingers of surgical gloves. Our first rat liver perfusions were encouraging, but short-lived because of blood clotting; a pre-liver filter made of a piece of nylon stocking removed small clots and extended perfusion time to six or seven hours. Within a short time, with ^{14}C-lysine, we demonstrated that the liver was producing ^{14}C-labeled albumin, fibrinogen, and plasma globulins. Collaborating in this early work were graduate students Chauncey G. Bly and Michael L. Watson. Bly's PhD thesis was based on these early studies.[5]

"There are two reasons for the frequent citation of this paper. (a) It afforded the first unequivocal demonstration of the dominant role of the liver in the biosynthesis of serum albumin, fibrinogen, and approximately 80% of the remaining plasma globulins. Several years later, by utilizing preparative zone electrophoresis to fractionate perfusates from liver perfusions[6] and from rat hindquarters,[7] we were able to document the view that the normal liver is the site of synthesis of virtually all of the plasma proteins with the notable exception of the gamma globulins. (b) It afforded a relatively inexpensive reproducible system for studying the direct interaction of various agents, on and with the liver, under conditions closely approximating the physiological, and with the aid of isotopically labeled metabolites. Over the intervening years the originally described operative technique and the apparatus have been substantially improved so that perfusions 24 hours in duration are routine.[8]

"The technique of isolated rat liver perfusion has been widely and fruitfully applied to problems in biochemistry, physiology, and pharmacology; it will continue to be used by those seeking to explore further the unknowns of liver metabolism and function."

1. Bernard C. Sur le mecanisme de la formation du sucre dans le foie. *C.R. Acad. Sci.* 41:461-9, 1855.
2. Asp G. Zur Anatomie und Physiologie der Leber.
   Arbeiten aus der physiologischen Anstalt zur Leipzig 8:124-58, 1873.
3. Trowell O A. Urea formation in the isolated perfused liver of the normal rat. *J. Physiology* 100:432-58, 1942.
4. Lupton A M. The effect of perfusion through the isolated liver on the prothrombin activity of blood from normal and dicoumarol treated rats. *J. Pharmacol. Exp. Ther.* 89:306-12, 1947.
5. Bly C G. *Some studies in the biosynthesis of tissue and plasma proteins.*
   Unpublished PhD thesis. University of Rochester, 1952.
6. Miller L L & Bale W F. Synthesis of all plasma protein fractions except gamma globulins by the liver.
   J. Exp. Med. 99:125-32, 1954.
7. Miller L L, Bly C G & Bale W F. Plasma and tissue proteins produced by non hepatic rat organs as studied with lysine-ε-C14. *J. Exp. Med.* 99:133-53, 1954.
8. Miller L L. Technique of isolated rat liver perfusion. (Bartosek I, Guaitani A & Miller L L. eds.)
   Isolated liver perfusion and its applications. New York: Raven Press, 1973. p. 11-52.

Campbell R M, Cuthbertson D P, Matthews C M & McFarlane A S. Behaviour of 14C-
and 131I-labelled plasma proteins in the rat.
Int. J. Appl. Radiat. Isotop. 1:66-84, 1956.
[Rowett Inst., Bucksburn, Aberdeenshire, Scotland and Nat. Inst. Med. Res., Mill
Hill, London, England]

From the urinary excretion curves of 131I of separately labelled pure proteins by the rat and from total activities in the extravascular compartment it seems the metabolic breakdown of albumin and γ-globulin occurs in the intravascular compartment, but these proteins are metabolised independently. [The *SCI®* indicates that this paper has been cited over 190 times since 1961.]

David P. Cuthbertson
Department of Biochemistry
Royal Infirmary
Glasgow G4 OSF
Scotland

August 21, 1980

"During 1954-55 a longstanding friendship and common interest in the metabolism of the plasma proteins drew together myself, director of the Rowett Research Institute (RRI), Aberdeen, and A. S. McFarlane, head of biophysics at the National Institute for Medical Research (NIMR), London. We then had the assistance, respectively, of Rosa Campbell and Christine Matthews.

"I was primarily keen to study the metabolism of the plasma proteins following injury, for in collaboration with the late S. L. Tompsett I had earlier published a note.[1] Does injury 'produce both the disposition and means of cure'?[2] McFarlane, on the other hand, was more interested in the behaviour of the plasma proteins in disease. How would they differ? It was therefore decided to pool our respective techniques. Because of some 500 miles between the two centres it was arranged that the animal work would be done at RRI and the preparative and analytical work on the proteins at NIMR. It was first necessary to determine the metabolism of the plasma proteins in normal conditions using 131I-albumin and globulins of demonstrable purity. In the rat, which is more constant in behaviour than the rabbit, we found that the 14C and 131I-labelled proteins behaved in substantially the same way. But the specific activity curve of the total proteins in rat plasma could not at any time be expressed as a single exponential curve. Curves for individual plasma proteins were almost truly exponential for five to 15 days after injecting the labelled molecules. Applying the necessary corrections for growth and plasma sampling, equivalent apparent replacement rates were calculated. The excretion curves for 131I in the first few days after injecting 131I-albumin suggested that albumin and γ-globulin are broken down mainly, if not exclusively, in close proximity to the circulating plasma. Urinary and plasma data were interpreted to mean that rat plasma proteins are metabolised by essentially independent processes.

"In this research a new method was introduced for determining the mass distribution of proteins in both intra- and extravascular compartments and an illustration was provided of the use of total body γ-radiation measurement for studying the metabolism of 131I-labelled plasma proteins.

"Readers have obviously been impressed by the meticulous care exercised in this research and have used it as a pattern, thus accounting for its frequent citation."

1. Cuthbertson D P & Tompsett S L. Note on the effect of injury on the level of the plasma proteins.
   Brit. J. Exp. Pathol. 16:471-5, 1935.
2. Hunter J. *A treatise on the blood, inflammation and gunshot wounds.* London: Nicol, 1794. 575 p.

Matthews C M E. **The theory of tracer experiments with 131I-labelled plasma proteins.**
Phys. Med. Biol. **2**:36-53, 1957.
[National Institute for Medical Research, Mill Hill, London, England]

The mathematical theory of Rescigno for exchange of tracer in a system of compartments is applied to the distribution of 131I labelled plasma proteins injected intravenously. A method is given for finding catabolic rate, extravascular protein mass and exchange rates between intra and extravascular compartments. [The *SCI®* indicates that this paper has been cited over 350 times since 1961.]

C.M.E. Matthews
YMCA Tribal Development Project
Yellagiri Hills P.O.
N. Arcot District
Tamil Nadu 635 853
India

May 22, 1980

"This was the first paper I wrote in the field of biophysics. Cooperation between those trained in different disciplines is not easy, and physicists and biologists have very different points of view. To the physicist, the biologist seems to have an unscientific disregard for the mathematical exactitudes of the situation, whereas to the biologist the physicist seems unnecessarily fussy over trifles. However, in the end we managed to find a method which was sufficiently mathematically correct to satisfy me and also practical enough to satisfy my colleagues. I am grateful to these colleagues, especially A.S. McFarlane, S. Cohen, and A.H. Gordon, who helped me to express the mathematical equations in a form that was meaningful to biologists.

"Owing to the well worked out iodination method of McFarlane,[1] we were able to use labelled protein whose properties were the same as those of unlabelled protein, so that the mathematical analysis was not spoiled by initial elimination of a rapidly degraded component. I think this paper has been cited often because it establishes the basis of a method which has since been widely used.

"In research requiring knowledge of more than one discipline, it is not often that a single person can be found who has sufficient knowledge of all the different fields, and, as I have already said, cooperation between those from different disciplines is not easy. However, an attempt at cooperation seems well worthwhile as the bringing together of comparatively simple theory from different fields may lead to useful new ways of looking at the problem.

"I wrote this paper a few years after obtaining my PhD in physics, and at that time I had never done any systematic study of biology or physiology. I went into medical research as I wanted to make some useful positive contribution to human welfare. Since then my field of work has changed considerably. After working for about 13 years in the field of biophysics and publishing more than 50 papers, I then changed to the even more inexact social sciences and have been working on health education as a missionary in India for the past eight years. I feel now that this kind of work is much more important and needed, considering the actual health problems in most of the world today. But I still use mathematics, and I find my physics background and interest in research is useful. It helps me to use more scientific methods in my new field."

1. **McFarlane A S.** Labelling of plasma proteins with radioactive iodine. *Biochemical J.* **62**:135-43, 1956.

Borgström B, Dahlqvist A, Lundh G & Sjövall J. Studies of intestinal digestion and
absorption in the human. *J. Clin. Invest.* **36**:1521-36, 1957.
[Swedish Med. Res. Council, Unit for Metabolic Studies, and Dept. Physiological
Chemistry, Univ. Lund, Sweden]

Intestinal content, sampled over the length
of the human intestine after feeding a test
meal including a nonabsorbable reference
substance, was analysed for food products,
bile constituents, digestive enzymes, etc., to
reveal the conditions for digestion and the
sites of absorption. [The *SCI®* indicates that
this paper has been cited in over 620 publications since 1961.]

Bengt Borgström
Department of Physiological Chemistry
University of Lund
S-220 07 Lund 7
Sweden

May 19, 1983

"My PhD thesis in 1952[1] at the University
of Lund, Sweden, concerned fat absorption
in the rat. In 1954, I worked as a Rockefeller
Foundation Fellow at Johns Hopkins University in Baltimore in Al Lehninger's laboratory. At a lipid conference at the McCollum
Pratt Institute, I met Peter Ahrens from the
Rockefeller Institute for Medical Research
(now Rockefeller University), New York. We
had had a previous correspondence on the
methodology for separating 1- and 2 mono-
glycerides from intestinal content during
digestion and after some discussion decided
that I should come to work in his laboratory
later that year. Peter had developed a technique for sampling intestinal contents over
the length of the intestine in man with a thin
polyvinyl tubing. I returned to Sweden with
100 feet of the tubing, convinced that man
was the ideal experimental animal for digestion and absorption studies.

"Two years later I obtained a research
position with the Swedish Medical Research
Council for digestion studies in man. When
planning the experimental approach with

two graduate students (A. Dahlqvist and
G. Lundh), we realized what was desperately
needed was a nonabsorbable reference substance for the test meal. Figures for concentrations, *per se* rather noninformative, could
then be converted to meaningful figures for
dilution, concentration, and absorption. We
found such a substance in PEG-4000. Long-
chain polyethylene glycols became of interest to the pharmaceutical industry in the
1940s and were shown not to be absorbed
from the intestinal tract. They were used by
Sperber and Hyden in Uppsala, Sweden, in
1953 to study fluid movement in ruminant
digestion.[2]

"With these techniques at hand and the
analytical skill of Jan Sjövall, the only one
at that time who could determine bile salts
quantitatively, we proceeded to do studies
in man (the generous offer of relatives and
former friends to live transintestinally intu-
bated for weeks is in retrospect thankfully
acknowledged).

"This paper, in my opinion, has been cited
because it contained some 'firsts': 1) the first
use in man of a nonabsorbable reference
substance that allowed calculation of sites
of absorption and sizes of secretions, etc.; 2)
the first concentration profiles of bile salts
during digestion which made possible an estimate of number of cycling of the bile salt
pool and total amount of bile salt circulated
per day; 3) the first indication that dietary
fat undergoes considerable hydrolysis in the
stomach in man; and 4) less cited but not
less important were the results indicating
that the major locus of action of the enzymes (disaccharidases, peptidases) formerly ascribed to the *succus entericus* is intracellular.

"This work provided the basis for future
studies in man in health and disease by
Dahlqvist[3] and Lundh[4] (now professors in
nutrition in Lund and in surgery in Stockholm, respectively), myself,[5,6] and many
others. The importance of this work was recognized in 1979 when I was awarded the
William Beaumont Prize in Gastroenterology by the American Gastroenterological
Association."[7]

1. Borgström B. *Studies on intestinal fat absorption in the rat.* PhD thesis. Lund, Sweden: University of Lund, 1952.
2. Sperber I, Hydén S & Ekman N J. The use of polyethylene glycol as a reference substance in the study of ruminant
   digestion. *Ann. Agr. Coll. Sweden* **20**:337, 1953.
3. Dahlqvist A & Borgström B. Digestion and absorption of disaccharides in man. *Biochemical J.* **81**:411-18, 1961.
4. Lundh G. Pancreatic exocrine function in neoplastic and inflammatory disease; a simple and reliable new test.
   Gastroenterology **42**:275-80, 1962.
5. Hofmann A F & Borgström B. The intraluminal phase of fat digestion in man. The lipid content of the micellar and
   oil phases of intestinal content obtained during fat digestion and absorption. *J. Clin. Invest.* **43**:247-57, 1964.
6. Hildebrand H, Borgström B, Békássy A, Erlanson-Albertsson C & Helin I. Isolated co-lipase deficiency in two
   brothers. *Gut* **23**:243-6, 1982.
7. Presentation of the Beaumont prize to Bengt Borgström and Alan Hofmann. *Gastroenterology* **77**:948-66, 1979.

Walshe J M. Penicillamine, a new oral therapy for Wilson's disease.
Amer. J. Med. **21**:487-95, 1956.
[Thorndike Mem. Lab. and 2nd and 4th Med. Services, Boston City Hosp.;
Dept. Med., Harvard Med. Sch., Boston, MA; and Med. Unit, Univ. Coll. Hosp.
Med. Sch., London, England]

The ability of penicillamine to promote the excretion of a great excess of copper was first observed in a patient with Wilson's disease at the Boston City Hospital in May 1955. This observation led directly to the introduction of penicillamine for the treatment of Wilson's disease and other heavy metal intoxications. [The *SCI®* indicates that this paper has been cited in over 255 publications since 1961.]

J.M. Walshe
Department of Medicine
University of Cambridge Clinical School
Addenbrooke's Hospital
Cambridge CB2 2QQ
England

June 9, 1983

"The introduction of a new drug into clinical medicine is, today, the business of the multinational pharmaceutical companies. The cost of such a venture is now in excess of 50 million dollars—research and development on this scale is far beyond the reach of the individual physician. Yet this is the story of exactly how such a task was initiated.

"In 1955, I was working on the liver unit of the Thorndike Memorial Laboratory at Boston City Hospital. One spring morning we were asked by Denny-Brown to see a patient, in the neurological unit, who was suffering from Wilson's disease, an inherited metabolic disorder in which excess copper accumulates in the liver and brain with disastrous effects on both organs. While returning to Thorndike it occurred to me that penicillamine, a derivative of penicillin, had the correct structural formula to bind copper and might well promote its elimination from the body. This notion was based on an earlier observation that patients treated with large doses of penicillin excreted penicillamine (some at least in the -SH state) in their urine.[1] By good fortune, Charles Davidson, chief of the liver unit, was able to obtain a few grams of this rare compound, first from Merck Sharp & Dohme and later from John Sheehan, professor of chemistry at Massachusetts Institute of Technology. I took the first gram to prove this was safe and Denny-Brown's patient took the second. We both survived and I was rewarded by finding a tenfold increase in the urinary copper after these tests.

"Before returning to England, I bought all the available supplies of penicillamine in the Eastern US so that the project could be continued. I now had my next stroke of good fortune: further patients with such a rare condition as Wilson's disease might have been impossible to find had not my father been Britain's leading neurologist; he was able to persuade his colleagues to allow me to test this new treatment on their patients. In all cases it induced a spectacular increase in urinary copper excretion.

"This work was offered to and immediately accepted for publication by the *American Journal of Medicine*, but it only described the cupriuretic effects of penicillamine; it did no more than cautiously predict therapeutic benefit for the patients. Further, it suggested that the new drug might be of value in other heavy metal poisonings; its use in the management of cystinuria and rheumatoid arthritis was to be reported from other centres in the future.[2]

"If this work is frequently quoted in the literature (all too often it appears to be taken for granted), it is because it was the first description of a new therapy for a hitherto invariably fatal illness. It may be added that it was the first specific drug able to reverse any inherited metabolic disease and for the first time it became possible to reverse the course of a degenerative disease of the nervous system. A more recent account of the treatment of Wilson's disease with penicillamine can be found in *Metabolic and Deficiency Diseases of the Nervous System*."[3]

1. **Walshe J M.** Disturbances of aminoacid metabolism following liver injury. *Quart. J. Med.* **22**:483-505, 1953.
2. **Gordon D A,** ed. Proceedings: International Symposium on Penicillamine, 8-9 May 1980, Miami, FL. (Whole issue.) *J. Rheumatology* **8**(Suppl. 7), 1981. 181 p.
3. **Vinken P J & Bruyn G W,** eds. *Handbook of clinical neurology. Volume 27. Metabolic and deficiency diseases of the nervous system. Part I.* Amsterdam: North-Holland, 1976. 554 p.

CC/NUMBER 26
JUNE 27, 1983

Malkin R & Malmström B G. The state and function of copper in biological systems. *Advan. Enzymol. Relat. Areas Mol. Biol.* 33:177-244, 1970.
[Göteborg, Sweden]

The chemical properties of copper in proteins were correlated with biological function. Blue oxidases were shown to contain three types of copper. Nonblue Cu^{2+} (type 2) is spectroscopically normal, whereas blue Cu^{2+} (type 1) has unique properties owing to an asymmetric coordination forced upon it by the protein conformation. The third copper type is a metal pair which facilitates two-electron steps in dioxygen reduction. [The *SCI*® indicates that this paper has been cited in over 295 publications since 1970.]

Bo G. Malmström
Department of Biochemistry and Biophysics
University of Göteborg
and
Chalmers University of Technology
S-412 96 Göteborg
Sweden

April 28, 1983

"Richard Malkin came from the University of California, Berkeley, in 1967 to work with me and Tore Vänngård in Göteborg during a two-year postdoctoral period. He participated in experimental work which caused a reorientation of our concepts concerning copper in proteins, particularly in the blue oxidases. This was the reason that we accepted an invitation in July 1968 from the editor of *Advances in Enzymology* to write a review on copper proteins, despite the fact that the proceedings[1] of a symposium on the subject had been published only two years earlier.

"Vänngård and I had published a paper[2] in 1960, in which we showed, with the aid of electron paramagnetic resonance (EPR), that laccase and ceruloplasmin contain Cu^{2+} in a unique coordination environment. We suggested that the unusual coordination is the result of the protein conformation, and that it also is responsible for the anomalously strong blue color of these proteins.

Our report caused a storm of protest from inorganic chemists, who told us that only Cu^{1+} complexes could have such intense colors. Consequently, they argued, the same copper ion could not give both the EPR signal, which must stem from a Cu^{2+} ion, and the strong color. Our further experimental work,[3] however, showed that the original interpretation was correct, and Malkin and I discussed possible models which could explain the unique properties of blue or type 1 Cu^{2+} in our review.

"When Malkin started his work in our laboratory, it was generally believed that the blue oxidases do, in fact, also contain Cu^{1+}, as our group had shown[3] that only 50 percent of the total copper is detectable as Cu^{2+} by EPR. In 1968, I went to Rome to measure the kinetics of Cu^{2+} reduction in laccase, together with Eraldo Antonini, while Malkin stayed in Göteborg to study the thermodynamics of the same reaction, together with another postdoctoral fellow, James A. Fee. We exchanged experimental results by mail and initially expressed mutual skepticism about each other's findings. Eventually, we concluded that both sets of data showed that the EPR-undetectable ions are also present as Cu^{2+}, and we developed the model of an exchange-coupled Cu^{2+}-Cu^{2+} pair. The role of this pair in facilitating two-electron steps in the dioxygen reduction was discussed in the review, which was concluded with the suggestion that a similar mechanism may be operative in cytochrome c oxidase, the terminal respiratory enzyme in all aerobic cells. This hypothesis is now well proved by subsequent work.[4]

"There are, I think, several reasons why our review has been frequently cited. We provided the first comprehensive discussion of the classification of copper in proteins into three major types. More significantly is perhaps the explosive growth of bioinorganic chemistry during the 1970s. What first appalled the inorganic chemists became a great attraction, once they were convinced that blue proteins do indeed contain Cu^{2+} in a unique coordination environment. Blue copper has consequently become a favorite object of investigation for the bioinorganic chemist, as evidenced by a recent book."[5]

1. Peisach J, Aisen P & Blumberg W E, eds. *The biochemistry of copper.* New York: Academic Press, 1966. 588 p.
2. Malmström B G & Vänngård T. Electron spin resonance of copper proteins and some model complexes. *J. Mol. Biol.* 2:118-24, 1960.
   [The *SCI* indicates that this paper has been cited in over 175 publications since 1961.]
3. Broman L, Malmström B G, Aasa R & Vänngård T. Quantitative electron spin resonance studies on native and denatured ceruloplasmin and laccase. *J. Mol. Biol.* 5:301-10, 1962.
   [The *SCI* indicates that this paper has been cited in over 90 publications since 1962.]
4. Malmström B G. Enzymology of oxygen. *Annu. Rev. Biochem.* 51:21-59, 1982.
5. Spiro T G, ed. *Copper proteins.* New York: Wiley, 1981. 363 p.

This Week's Citation Classic

Mechoulam R. Marihuana chemistry. *Science* **168**:1159-66, 1970.
[Laboratory of Natural Products, Pharmacy School,
Hebrew University, Jerusalem, Israel]

This article critically reviewed the advances made in the isolation, structural elucidation, synthesis, chemical behavior, structural-activity relationships, and metabolism of the cannabinoids. Emphasis was placed on Δ^1-tetrahydrocannabinol (Δ^1-THC, or Δ^9-THC by the nomenclature mostly used in the US), the major active component which had been isolated in pure form only a few years before. It was stressed that the area was ripe for more sophisticated biological research. [The *SCI®* indicates that this paper has been cited in over 280 publications since 1970.]

R. Mechoulam
Department of Natural Products
Pharmacy School
Hebrew University
Jerusalem 91120
Israel

April 11, 1983

"In the early-1960s, I was at the Weizmann Institute in Rehovot back from a postdoctoral stay at the Rockefeller Institute in New York, and was looking for research topics of potential importance outside the heavily populated areas of current interest. I was surprised to find that the active component(s) of cannabis had never been isolated in a pure form and its structure was known only in a general way. A few cannabinoids had been reported but the structure of only one, the psychotropically inactive cannabinol, had been fully elucidated.[1-3]

"In 1962, I convinced the Israeli police to give me a few kilograms of confiscated hashish, and Yuval Shvo and I reisolated cannabidiol, a constituent which Roger Adams and Lord Todd had found in marihuana.[4,5] We elucidated its structure and stereochemistry.[6] A close friend and colleague, Yehiel Gaoni, became interested and we joined forces. An application for a National Institute of Mental Health (NIMH) grant was submitted, but was turned down. We were told that marihuana was not much of an American problem and that NIMH was not planning to support research in this area. The year was 1964! A year later, D. Efron, head of the pharmacology section of NIMH, flew over to see us and suggested that on second thought his institute should support our work. He told us that marihuana had now become an American problem and that even the son of a prominent US politician had been found in possession of the drug. Our research had become 'relevant.' Indeed, NIMH and later the National Institute on Drug Abuse generously supported our research for 16 years.

"In the meantime Gaoni and I continued the isolation, structural elucidation, and synthesis of natural cannabinoids. Δ^1-THC, the major active component, was isolated in 1964,[7] shortly to be followed by numerous others. Soon thereafter we published a synthesis of the *dl*-form of Δ^9-THC, and in 1967, after I had moved to Jerusalem, we found a simple synthetic route to the natural (-) form of Δ^1-THC.[8] A pharmacologist, Haviv Edery, took upon himself the testing of these compounds in monkeys. Other groups, in the US, Switzerland, England, and Germany, had also started publishing in this area.

"Most of the publications on cannabinoids until then were chemical in nature. However, it was evident that in the coming years much of the significant work would be on metabolism, pharmacology, and clinical aspects. Hence I decided to summarize the chemical background obtained till then for a biologically oriented general journal. I also introduced in this concise review speculations on the existence of active metabolites. This review apparently filled a gap. It also popularized the term 'cannabinoids' which we had introduced earlier.

"This paper has been highly cited for the following reason. Numerous groups in the early-1970s working on the various biological effects of THC, which had become available in part due to our contributions, preferred to cite this review rather than the original publications. As THC became a widely used term few of the thousands of papers describing its action continued to cite the chemical background. Cannabinoid research had come of age."[9,10]

1. **Cahn R S.** *Cannabis indica* resin. Part IV. The synthesis of some 2:2-dimethyldibenzopyrans, and confirmation of the structure of cannabinol. *J. Chem. Soc.* 1933:1400-5.
2. **Adams R, Baker B R & Wearn R B.** Structure of cannabinol. III. Synthesis of cannabinol, 1-hydroxy-3-*n*-amyl-6,6,9-trimethyl-6-dibenzopyran. *J. Amer. Chem. Soc.* 62:2204-7, 1940.
3. **Ghosh R, Todd A R & Wilkinson S.** *Cannabis indica.* Part IV. The synthesis of some tetrahydrodibenzopyran derivatives. *J. Chem. Soc.* 1940:1121-5.
4. **Adams R, Pease D C & Clark J H.** Isolation of cannabinol, cannabidiol and quebrachitol from red oil of Minnesota wild hemp. *J. Amer. Chem. Soc.* 62:2194-6, 1940.
5. **Jacob A & Todd A R.** Cannabidiol and cannabol, constituents of *Cannabis indica* resin. *Nature* 145:350, 1940.
6. **Mechoulam R & Shvo Y.** The structure of cannabidiol. *Tetrahedron* 19:2073-8, 1963.
7. **Gaoni Y & Mechoulam R.** Isolation, structure and partial synthesis of an active constituent of hashish. *J. Amer. Chem. Soc.* 86:1646-7, 1964.
8. **Mechoulam R, Braun P & Gaoni Y.** A stereospecific synthesis of (-)Δ^1- and (-)Δ^6-tetrahydrocannabinols. *J. Amer. Chem. Soc.* 89:4552-4, 1967.
9. **Mechoulam R.** Chemistry of cannabis. (Hoffmeister F & Stille G, eds.) *Psychotropic agents. Part III: alcohol and psychotomimetics, psychotropic effects of central acting drugs.* Berlin: Springer-Verlag, 1982. p. 119-34.
10. **Waller C W, Nair R S, McAllister A F, Urbanek B & Turner C T.** *Marihuana, an annotated bibliography.* New York: Macmillan Information, 1982. Vol. II.

Jerina D M & Daly J W. Arene oxides: a new aspect of drug metabolism.
Science **185**:573-82, 1974.
[Sect. on Oxidation Mechanisms and Sect. on Pharmacodynamics, Lab. Chemistry, Natl.
Inst. Arthritis, Metabolism, and Digestive Diseases, Natl. Insts. Health, Bethesda, MD]

The monooxygenase-catalyzed formation of phenols from aromatic substrates is a well-recognized biochemical pathway in animals, plants, and higher microorganisms. Arene oxides have been demonstrated as precursors of these phenols as well as numerous other secondary metabolites. The 'arene oxide pathway' constitutes an important and requisite part of normal cellular function. But for certain compounds, the metabolic formation of arene oxides represents the onset of toxic and carcinogenic processes within the cell. [The *SCI®* indicates that this paper has been cited in over 775 publications since 1974.]

Donald M. Jerina and John W. Daly
Laboratory of Bioorganic Chemistry
National Institute of Arthritis, Diabetes, and
Digestive and Kidney Diseases
National Institutes of Health
Bethesda, MD 20205

May 31, 1983

"Prior to 1968, the pathway by which monooxygenase enzymes convert aromatic substrates into phenols was unknown but was generally thought to consist of an insertion of oxygen into the carbon-hydrogen bond. Our entry into this area stemmed from attempts to develop rapid, simple radiometric assays for the enzymatic formation of phenols. The principle of these assays was the anticipated release of tritium during insertion of oxygen into a specific carbon-tritium bond. This approach had worked admirably for tyrosine hydroxylase.[1] *But for other 'aryl hydroxylases,' including the cytochromes P450, a far from stoichiometric release of tritium was observed compared to product formed.* In an attempt to develop an assay for phenylalanine hydroxylase with purported 4-tritiated phenylalanine, Gordon Guroff had become suspicious of the specificity of labeling of the substrate, since the 4-hydroxylated product tyrosine contained considerable tritium. The problem was resolved through the use of specifically 4-deuterated phenylalanine, which the enzyme converted to 3-deuterotyrosine.[2] This discovery of migration and retention of aryl ring substituents on hydroxylation by monooxygenases led to a stimulating and rewarding team effort involving members of both Bernhard Witkop's and Sidney Udenfriend's laboratories at the National Institutes of Health (NIH), and Udenfriend coined the imaginative term, the 'NIH shift,' to describe the phenomenon.[3]

"Research was directed toward establishing the mechanism of this novel reaction. Although discrete ionic species could be invoked, arene oxides, which are epoxides of formal aromatic double bonds, represented mechanistically attractive intermediates. K-region arene oxides of polycyclic aromatic hydrocarbons had been known for several years and were quite stable, while benzene oxide and other non-K-region arene oxides had just been synthesized and were quite unstable (reviewed in references 4 and 5). Despite the failure of our initial attempts to demonstrate that benzene oxide was the initial liver microsomal metabolite of benzene, a new microsomal enzyme, epoxide hydrolase, was identified and shown to catalyze the trans addition of water to arene oxides to form dihydrodiols. Cytosolic glutathione transferases were found to catalyze the addition of glutathione to benzene oxide. Studies with the more stable naphthalene 1,2-oxide allowed its characterization as the initial metabolite of naphthalene and provided the proof that it was the requisite intermediate in the formation of naphthol, the trans 1,2-dihydrodiol and a glutathione conjugate.[6,7] Numerous studies from these and many other laboratories followed which established the germinal role of arene oxides in the metabolism of aromatic hydrocarbons by monooxygenase enzymes.

"We believe that the extensive citation of our article stems from its impact on concepts in drug metabolism. The article not only drew attention to a new and major pathway by which drugs and other xenobiotic compounds are metabolized and excreted, but also emphasized the fact that highly reactive arene oxides could be responsible for the toxic and carcinogenic effects of certain aromatic hydrocarbons. Since writing the article one of us (J.W.D.) has concentrated his efforts in the area of neurochemistry, while the other (D.M.J.) has continued research in the field and has been the recipient of the 1982 B.B. Brodie Award for research in drug metabolism. A new review on the chemistry and biochemistry of arene oxides will shortly be available."[8]

1. **Nagatsu T, Levitt M & Udenfriend S.** A rapid and simple assay for tyrosine hydroxylase activity.
   Anal. Biochem. **9**:122-6, 1964.
2. **Guroff G, Reifsnyder C A & Daly J.** Retention of deuterium in p-tyrosine formed enzymatically from
   p-deuterophenylalanine. *Biochem. Biophys. Res. Commun.* **24**:720-4, 1966.
3. **Guroff G, Daly J W, Jerina D M, Renson J, Witkop B & Udenfriend S.** Hydroxylation-induced migration: the NIH
   shift. *Science* **157**:1524-30, 1967.
4. **Vogel E & Gunter H.** Benzene oxide-oxepin valence tautomerism. *Angew. Chem. Int. Ed.* **6**:385-401, 1967.
5. **Jerina D, Yagi H & Daly J W.** Arene oxides-oxepins. *Heterocycles* **1**:267-326, 1973.
6. **Jerina D, Daly J, Witkop B, Zaltzman-Nirenberg P & Udenfriend S.** Role of the arene oxide-oxepin system in the
   metabolism of aromatic substrates. I. In vitro conversion of benzene oxide to a premercapturic acid and a
   dihydrodiol. *Arch. Biochem. Biophys.* **128**:176-83, 1968.
7. --. The role of arene oxide-oxepin systems in
   the metabolism of aromatic substrates. III. Formation of 1,2-naphthalene oxide from naphthalene by liver
   microsomes. *J. Amer. Chem. Soc.* **90**:6525-7, 1968.
8. **Boyd D R & Jerina D M.** Arene oxides-oxepins. (Hassner A, ed.) *Small ring heterocycles.*
   New York: Wiley. Vol. 42. Part 3. In press, 1983.

This Week's Citation Classic

CC/NUMBER 47
NOVEMBER 22, 1982

Hansch C & Fujita T. ϱ-σ-π analysis. A method for the correlation of
biological activity and chemical structure.
J. Amer. Chem. Soc. 86:1616-26, 1964.
[Department of Chemistry, Pomona College, Claremont, CA]

This paper presents a mathematical model
which correlates the differences in physicochemical properties of a set of organic compounds produced in biochemical and biological systems. It shows that systems as
diverse as benzoic acids inhibiting mosquito
larvae, phenols inhibiting bacteria, phosphate esters killing houseflies, the analgesic
action of diethylaminoethyl benzoates on
guinea pigs, the action of thyroxine analogs
on rodents, and the carcinogenicity of polycyclic aromatic hydrocarbons could be correlated with two parameters—the Hammett
σ constant and a hydrophobicity parameter
derived from octanol/water partition coefficients. [The SCI® indicates that this paper
has been cited in over 535 publications since
1964.]

Corwin Hansch
Department of Chemistry
Pomona College
Claremont, CA 91711

June 30, 1982

"On coming to Pomona College in
1946, I soon became acquainted with
and joined forces with Robert Muir, a
plant physiologist in the botany department, in an attempt to understand how
changes in chemical structure affected
the potency of plant growth-regulators.
Shortly thereafter Muir moved to the
University of Iowa where he remains
today. At that time, Muir was among
the relative few who appreciated that
the phenoxyacetic acids probably
acted by the same mechanism as the
natural growth hormone indoleacetic
acid. The indole compounds were hard
to synthesize while the phenoxyacetic
acids were easy to make. Together, we
systematically studied the action of
phenoxyacetic acids on oat seedlings.

"We developed qualitative ideas
about the relationship of chemical
structure and biological activity, trying
to explain in terms of one variable the
electron distribution in the phenoxy
ring. Around 1960 it became clear that
this simplistic approach was worthless,
even as an approximation. Several variables were obviously involved and
their separate roles had to be delineated. We decided to use partition coefficients (the way a compound distributes
itself between an oily phase and water)
to model the way the growth-regulators
penetrate to their sites of action in
cells. Just as our model was developing, Toshio Fujita, a chemist from
Kyoto University, joined us to help, and
at about that same time an alumnus of
the college gave our department a
small computer. Being a synthetic
organic chemist, I had never in my life
expected to use a computer and probably would not have done so without
the help of Donald McIntyre, a Pomona geologist. We published our first
paper on quantitative structure-activity
relationships (this has now developed
into a field called QSAR) in 1962.[1]
Fujita and I then went on to apply our
mathematical model using numbers to
describe the partitioning, spatial, and
electronic properties of organic compounds, which resulted in this Citation
Classic.

"The reason for the paper being
cited so often is that this was the first
quasi-general mathematical approach
to structure-activity relationships. Today our approach has been shown to be
valuable in drug and pesticide design,
toxicology, reaction of organic compounds with enzymes[2] and other macromolecules, disposition of chemicals
in soil, and the bioaccumulation of environmental chemicals in fish, birds,
and other forms of life."

1. Hansch C, Maloney P P, Fujita T & Muir R M. Correlation of biological activity with Hammett substituent constants
and partition coefficients. Nature 194:178-80, 1962.
2. Smith R N, Hansch C, Kim K H, Omiya B, Fukumura G, Selassie C D, Jow P Y C, Blaney J B & Langridge R.
The use of crystallography, graphics, and quantitative structure-activity relationships in the analysis of the
papain hydrolysis of X-phenyl hippurates. Arch. Biochem. Biophys. 215:319-28, 1982.

Ussing H H & Zerahn K. Active transport of sodium as the source of electric current in the short-circuited isolated frog skin.
Acta Physiol. Scand. 23:110-27, 1951.
[Laboratory of Zoophysiology, University of Copenhagen, Denmark]

By aid of an adjustable electromotive force in series with the isolated frog skin, the skin potential can be totally short-circuited. It is thus possible to determine simultaneously the current which can be drawn from the skin and — using radioactive sodium — the influx and outflux of sodium. With Ringer solution on both sides, the short-circuit current is exactly equal to influx minus outflux, i.e., the net active transport of sodium. [The *SCI®* indicates that this paper has been cited over 1,250 times since 1961.]

Hans Ussing
Institute of Biological Chemistry A
University of Copenhagen
DK-2100 Copenhagen Ø
Denmark

July 21, 1981

"The origin of bioelectric potentials and bioelectric currents had been a matter of dispute since the discovery of the phenomena in the middle of the 19th century.[1] For individual cells, like those of muscle and nerve fibers, the potential had been related to the uneven distribution of potassium between cells and surroundings. When isotopes became available it was realized that both sodium and potassium exchanged readily across most cell membranes, and the idea of a 'sodium pump,' maintaining low sodium and high potassium concentration in cells, was advocated by several groups, including our own. Others maintained that the uneven ion distribution was due to colloid chemical properties of cytoplasm rather than to active sodium transport. Therefore, we turned to the study of certain epithelia where net transport of sodium chloride was known to take place.

"In 1949, I was able to demonstrate that chloride uptake through the frog skin could be a consequence of the electric potential difference between inside and outside, whereas sodium transport took place against both concentration and potential gradients, thus being due to active transport.[2] It then became desirable to demonstrate, not only that the sodium transport was active (which I considered proven beyond doubt), but also that active sodium transport was the sole source of the electric asymmetry of the frog skin.

"Partial short-circuit could be achieved by connecting inside and outside solutions with a copper wire, using reversible electrodes. I did a quick and dirty extrapolation which indicated that the influx of sodium as measured with 24-Na would be of the same order of magnitude as the estimated true short-circuit current, but the resistance of solutions and electrodes made it impossible to test the hypothesis. At the International Physiology Congress in Copenhagen, 1950, I was to give one of the main lectures on active sodium transport, and I badly needed some striking experiment to prove my point. Shortly before the deadline for summaries, it suddenly dawned on me that a battery and a variable resistance in series with the skin could be used to create the desired situation. I asked our mechanic to build a chamber from a piece of celluloid tubing. Zerahn wired up the circuit, while I did the glass blowing, and within less than a week we knew that the hypothesis was correct.

"Our paper had an immediate impact because it provided a firm basis for the concept of active sodium transport. Later the short-circuit method became a standard procedure in the study of ion transport phenomena in epithelia, and this is probably why the paper is still being cited frequently. More recent work is reported in 'Transport pathways in biological membranes.'"[3]

1. DuBois-Raymond E. *Untersuchungen über Tierische Elektrizität I-II.* Berlin: G. Reimer, 1848. 1154 p.
2. Ussing H H. The use of tracers in the study of active ion transport across animal membranes. *Cold Spring Harbor Symp.* 13:193-200, 1948.
3. Ussing H H, Erlij D & Lassen U. Transport pathways in biological membranes. *Annu. Rev. Physiol.* 36:17-49, 1974.

Avron M. Photophosphorylation by Swiss Chard chloroplasts.
Biochim. Biophys. Acta **40**: 257-72, 1960.

The article describes a detailed study for optimizing the conditions for preparation of chloroplasts from leaves of higher plants and of the reaction conditions for measuring photophosphorylation. Under the conditions specified, the world record rate (which still holds) of light-induced ATP formation, approaching 2500 μ moles $\times$ mg chl^{-1} $\times$ hr^{-1}, was attained. A simple assay system for following the incorporation of radioactive inorganic phosphate into ATP is described and is used to evaluate several mechanistic aspects of the process. [The *SCI*® indicates that this paper was cited 449 times in the period 1961-1976.]

Mordhay Avron
Department of Biochemistry
Weizmann Institute of Science
Rehovot, Israel

December 11, 1977

"This paper reports on my first effort as an independent scientist to evaluate for my own conviction the optimal capacity of chloroplasts, isolated from leaves of higher plants, to catalyse the then new and exciting process of photophosphorylation. I had just been appointed as an assistant professor at the Weizmann Institute of Science, after spending a post-doctoral period with Prof. A.T. Jagendorf at the Johns Hopkins University. During the latter period we succeeded in confirming and extending the report of Prof. D.I. Arnon that chloroplasts do indeed possess a new process by which light energy is used as the driving force for ATP formation. We also showed that the process was essentially irreversible with no significant ATP breakdown under any conditions. Commencing my work in Israel, I decided initially to devote some time to optimizing the methodologies used in such studies. Part of the impetus for this study came from the fact that the favorite material used in all studies of photophosphorylation until then was spinach leaves, which were unavailable in Israel. After checking several sources, I decided to do the study on Swiss Chard leaves.

"Several improvements in the preparative technique were developed, in particular the realization of the importance of maintaining reduced conditions during isolation. A simple radioactive assay method was devised, based on techniques previously used in other systems.

"The role of many parameters, such as light intensity, nucleotide and phosphate affinity and specificity were separate studies. All of these were summarized in a recommended 'standard assay conditions' under which consistent very high and stable rates of photophosphorylation were observed.

"This paper is not a classic in the common sense of the word. Its popularity, as indicated by its frequent citation, reflects, I believe, mostly its usefulness to many workers in the field in evaluating and defining in clear and concise terms the techniques, capacity, limitations, and problems in the then new and still highly important process of photophosphorylation."

Jensen R G & Bassham J A. Photosynthesis by isolated chloroplasts.
Proc. Nat. Acad. Sci. US 56:1095-101, 1966.
[Lawrence Radiation Laboratory, University of California, Berkeley, CA]

A method is described for rapid isolation of intact chloroplasts from spinach leaves capable of high rates of complete photosynthesis with CO_2. Not previously possible, the rates with isolated chloroplasts now resembled those of the intact leaf. [The _SCI®_ indicates that this paper has been cited in over 485 publications since 1966.]

Richard G. Jensen
Department of Biochemistry
University of Arizona
Tucson, AZ 85721

June 30, 1983

"After finishing my graduate research in 1965 at the C.F. Kettering Research Laboratory, Ohio, I went to the biodynamics laboratory on the University of California, Berkeley campus, under Melvin Calvin and his group to continue my studies on photosynthesis. Calvin had received the Nobel prize in 1961 for his elucidation of the path of carbon in photosynthesis. I elected to work with J.A. Bassham who was directing the research on photosynthetic carbon assimilation. Starting with the isolation procedures of D.A. Walker[1] with pea chloroplasts, we optimized the conditions for isolation of chloroplasts from spinach which retained their ability to do complete photosynthesis with CO_2. It was apparent that Tris buffer was inhibitory so we used one of the then new N.E. Good zwitterion buffers. Rapid isolation and the use of fresh spinach also increased the rates of light dependent CO_2 uptake. Although not sustained for more than ten to 15 minutes, the rates were greater than half of the rate of CO_2 uptake by the intact leaves and did not require added metabolites. The products labeled with ^{14}C were those expected for the Calvin cycle.

"At that time, the rate of CO_2 fixation with isolated chloroplasts was less than five percent compared to the intact leaf. Our vigorous rates of CO_2 fixation in the absence of any added metabolites gave us confidence that our observations were more like those expected of chloroplasts in the intact leaf. We later published observations on the effects of added cofactors and intermediates, the diffusion of labeled photosynthetic intermediates out of the chloroplast, and the apparent effect of light on the carboxylation reaction.[2-4] An overview of the biochemistry of the chloroplast has recently appeared.[5]

"At that exciting time our improvement was considered a breakthrough because it was the first demonstration that the entire process of photosynthesis from O_2 evolution to CO_2 fixation could operate at significant rates outside of the intact plant cell. Since then, each laboratory has added their particular improvements to the isolation with the elucidation of many important mechanisms pertaining to the regulation of photosynthesis. The excitement continues today as many in this field visualize the eminent potential for optimization of this key life process for increasing plant production."

1. **Walker D A.** Improved rates of carbon dioxide fixation by illuminated chloroplasts. _Biochemical J._ 92:22C-3C, 1964.
2. **Bassham J A, Kirk M & Jensen R G.** Photosynthesis by isolated chloroplasts. I. Diffusion of labeled photosynthetic intermediates between isolated chloroplasts and suspending medium. _Biochim. Biophys. Acta_ 153:211-18, 1968.
3. **Jensen R G & Bassham J A.** Photosynthesis by isolated chloroplasts. II. Effects of addition of cofactors and intermediate compounds. _Biochim. Biophys. Acta_ 153:219-26, 1968.
4. ------------------------------, Photosynthesis by isolated chloroplasts. III. Light activation of the carboxylation reaction. _Biochim. Biophys. Acta_ 153:227-34, 1968.
5. **Jensen R G.** Biochemistry of the chloroplast. (Tolbert N E, ed.) _The biochemistry of plants. Volume 1. The plant cell._ New York: Academic Press, 1980. p. 273-313.

Abeles F B & Rubinstein B. Regulation of ethylene evolution and leaf abscission by auxin. *Plant Physiol.* **39**:963-9, 1964.

The authors found that auxin stimulated ethylene production, and that it was the gas which promoted abscission after the ability of auxin to delay aging was lost. This paper provided experimental proof for the principle that ethylene could act as a second messenger in some of the effects of auxin on plant growth and development. [The *SCI*® indicates that this paper was cited 139 times in the period 1964-1977.]

Fred B. Abeles
U.S. Army Medical Research Institute
of Infectious Diseases
Fort Detrick, Frederick, MD 21701

December 29, 1977

"When I arrived at Fort Detrick in 1963, all I knew of abscission was that it occurred in the fall and that I wasn't exactly sure how to spell the word. I had just finished graduate school at the University of Minnesota and had an obligation to serve in the U.S. Army. My coworker on this paper was another lieutenant, Bernie Rubinstein, who had just finished a master's thesis on leaf abscission at Purdue University. After the work reported here was completed, Bernie went to graduate school at the University of California, Berkeley, and then to the University of Massachusetts where he is currently Professor of Botany. Bernie had an excellent grasp of the literature and, inspired by an excellent review by Stanley Burg,[1] had ordered a gas chromatograph

to measure ethylene production from plants. At that time only a few labs had this instrument available for ethylene studies, or in fact were even interested in ethylene research. It was, and still is, hard to think of a simple two-carbon gas as a plant hormone. In my opinion, a good deal of our success was due to the simple fact that we had a new tool to study an old problem.

"Our research was devoted to studying a paradox. Auxin (indoleacetic acid, a plant hormone) prevented abscission of debladed petioles when it was applied shortly after removal of the leaf blade. However, when it was applied four hours later, it stimulated abscission. How could one hormone have such divergent effects over a four-hour time span? With the aid of the gas chromatograph, we were able to show that auxin caused the plant to produce a great deal of ethylene. Since it was already known that ethylene was a powerful defoliant, it was a simple matter to show that the abscission-promotive effect of auxin was probably due to the enhanced ethylene production. The abscission-retarding effect of auxin, on the other hand, was due to its ability to delay aging or senescence, and the paradox of auxin action on abscission could be explained by its dual ability to delay aging on one hand and stimulate ethylene production on the other.

"A reexamination of the literature revealed the fact that Crocker et al.[2] in 1935 originally discovered the fact that auxin stimulated ethylene and at that time they suggested that some of the effects of auxin were simply due to its ability to enhance ethylene production.

"Our paper provided experimental proof that ethylene was capable of acting as a second messenger in plants and that auxin-enhanced ethylene production explained the ability of auxin to stimulate abscission. In my opinion, our paper has been cited so frequently because we provided clear experimental verification of a general concept in hormone physiology by studying the special case of leaf abscission."

REFERENCES

1. **Burg S P.** The physiology of ethylene formation.
   Annu. Rev. Plant Physiol. **13**:265-302, 1962.
2. **Crocker W, Hitchcock A E, Zimmerman P W.** Similarities in the effects of ethylene and the plant auxins. *Contr. Boyce Thompson Inst.* **7**:231-48, 1935.

Hager A, Menzel H & Krauss A. Versuche und Hypothese zur Primärwirkung des Auxins beim Streckungswachstum. (Experiments and hypothesis concerning the primary action of auxin in elongation growth.) *Planta* **100**:47-75, 1971.
[Inst. Botany, Univ. Münster, Federal Republic of Germany]

The paper describes a mechanism whereby the growth hormone auxin can change properties of the plant cell wall and thereby induce enhanced elongation. The hormone is assumed to cause an increase of H^+-concentration in the cell wall compartment via an activation of ATP-dependent proton pumps in the plasmalemma. The increased H^+-level, in turn, leads to an increase of wall plasticity and thus to increased cell elongation. [The *SCI®* indicates that this paper has been cited in over 215 publications since 1971.]

Institute of Biology I
University of Tübingen
D-7400 Tübingen 1
Federal Republic of Germany

December 28, 1983

"This paper was preceded by some controversy. I was working at the Botanical Institute of the University of Munich as a *Wissenschaftlicher Assistent* and, in 1962, I produced a so-called *Habilitationsschrift*, an extensive scientific paper, which had to be submitted to the Faculty of Natural Sciences of the University of Munich. A positive vote on this *Schrift* by the professors on the faculty is necessary—according to German academic customs—to obtain the *venia legendi*, i.e., the permission to lecture at the university.

"In this *Habilitationsschrift*, entitled 'Analyses of cell elongation mechanisms inducible by H^+-ions,'[1] I showed, among other things, that the growth of sunflower stem sections or oat coleoptiles can be stimulated when transferred into acidic buffers (pH 4-5), even without addition of the growth hormone auxin. I proposed the term 'acid growth' (*Säurewachstum*)[1] for this kind of elongation, which was thought to be caused by pH-dependent changes of cell wall properties and which could be induced also under anaerobic conditions.

"Within the faculty, the opinions of the referees concerning these results were rather controversial. Some considered the findings as artifacts or else as of no relevance for the explanation of biological growth processes, others showed a sort of benign indulgence, and finally some thought the results to be very interesting. In the end, however, the work was judged positive and I got my *Habilitation*.

"In 1970, after more experimental results, obtained in collaboration with Anne Krauss and Helga Menzel, doctoral students at this time, I dared to propose that a pH-decrease in the cell wall is an essential step in the auxin chain of action and that this pH-decrease is caused by an active export of H^+-ions from the cytoplasm into the wall. The driving force for this H^+-secretion could be ATP-dependent H^+-pumps in the plasmalemma; the activity of the pumps could be controlled by auxin. I presented these ideas for the first time in a plenary lecture at the meeting of the Deutsche Botanische Gesellschaft in Erlangen, in 1970, and published them in *Planta* in 1971.

"These ideas stimulated many research groups to test them critically and to do further experiments. In this connection, I would like to mention R. Cleland and D.L. Rayle; without their intensive research, the acid growth hypothesis would not have obtained acceptance so rapidly. The work of E. Marrè, P.M. Ray, M.L. Evans, Y. Masuda, L.N. Vanderhoeff, R. Hertel, and many others can also be referred to in this context.

"However, it was not until the early 1980s that we could clearly demonstrate the transmembrane transport of protons and the postulated ATP-dependent proton pumps.[2,3]

"I think the reason why this paper has been quoted so often may be the new approaches offered for an analysis of the molecular mechanism of action of the well-known growth hormone auxin; the proposed hypothesis could explain a 40-year-old finding,[4] namely, an auxin-induced cell wall softening.

"On the other hand, it was shown for the first time that the mechanism proposed by the hypothesis of Mitchell[5,6] (ATP $\rightleftharpoons$ ΔpH) does play a role not only in energy coupling at the membranes of 'procaryotic' mitochondria and chloroplasts but also at membranes of the eucaryotic ('host') cell where it can control growth processes."

1. **Hager A.** *Untersuchungen über einen durch H^+-Ionen induzierbaren Zellstreckungsmechanismus.* Habilitationsschrift. Munich: University of Munich, Faculty of Natural Sciences, 1962.
2. **Hager A, Frenzel R & Laible D.** ATP-dependent proton transport into vesicles of microsomal membranes of *Zea mays* coleoptiles. *Z. Naturforsch. Sect. C* **35**:783-93, 1980.
3. **Hager A & Helmle M.** Properties of an ATP-fueled, Cl^--dependent proton pump localized in membranes of microsomal vesicles from maize coleoptiles. *Z. Naturforsch. Sect. C* **36**:997-1008, 1981.
4. **Heyn A N J.** Der Mechanismus der Zellstreckung. *Rec. Trav. Bot. Néerl.* **28**:133-244, 1931.
5. **Mitchell P.** Chemiosmotic coupling in oxidative and photosynthetic phosphorylation. *Biol. Rev. Cambridge Phil. Soc.* **41**:445-502, 1966.
6. --------------. Citation Classic. Commentary on *Biol. Rev. Cambridge Phil. Soc.* **41**:445-502, 1966. *Current Contents* (16):14, 17 April 1978.

Cathey H M. Physiology of growth retarding chemicals.
Annu. Rev. Plant Physiol. **15**:272-302, 1964.

Chemical growth retardants (8 families) block cell elongation of stems without affecting leaf formation, resulting in compact, stress-resistant plants. With many woody plants the treated plants initiate flowers earlier than typical for the species. Chemical growth retardants are one of the reasons for the house plant boom in America. They have permitted the sizing of plants to fit any space. [The *SCI®* indicates that this paper was cited 258 times in the period 1965-1977.]

Henry M. Cathey
Florist & Nursery Crops Laboratory
Beltsville Agricultural Research Center
Beltsville, MD 20705

January 17, 1978

"A research project dealing with small (plants) in an age of super (plants) has become a standard cultural procedure for many growers of horticultural plants. Horticulturists in the later '50's worked on the growth-regulating effects of a chemical named gibberellin (GA), which accelerated stem elongation and early flowering, and overcame various kinds of dormancies in plants. The plants had pale-colored foliage, required elaborate staking, and aged rapidly. As a counter activity, I looked for chemicals which would make the plants into compact, dark green, and long-lived ones. The first two (nicotinium, quaternary ammonium compounds) had already been reported by J.W. Mitchell and P.C. Marth at Beltsville. The chemicals retarded the stem growth of snapbeans and chrysanthemums. From the horticultural viewpoint, this permitted us to use the very best cultivars, regardless of their natural growth habits, as compact, flowering container-grown plants.

"The house plant boom was just beginning in the US in the early 1960's and many new types of plants were needed to meet the consumer interests. Agricultural chemical companies at that time did not have screening programs to find these types of growth responses. They were looking primarily for chemicals which inhibited plant or shoot growth. I thus faced a situation where we had only two chemicals, a very limited plant response range, and no new chemicals in sight.

"During the next years, the concepts became a reality. We found eight families of chemicals (nicotiniums, quaternary ammonium carbonate derivatives synthesized from thymol or carvacrol, hydrazines, phosphoniums, substituted cholines, succinamic acids, pyrimidine-methanol, and piperidinium bromide) which produce compact growth characteristics in a wide range of plants. Only conifers were nonresponsive. Most container-grown and bedding plants are now treated to help them retain their deep green foliage color and compact growth.

"As a result of the regulation of cell division in the subapical meristems, Malus and Rhododendron plants form flower buds, and this is the basis for controlling their flowering and improved storage. Also, the intercellular spaces in the leaves were reduced in response to the chemicals; also water loss and sensitivity to air pollution decreased. This side effect has become extremely important in the '70's when urban areas are experiencing air pollution. How the chemical retardants work remains a mystery. They counteract the action, not only of gibberellin but also auxins, cytokinins, and a whole range of synthetic growth regulators.

"Many analogs have been made to test alternative metabolic pathways, giving evidence that many routes are involved in stem extension and leaf development. However they exert their effects, the plants treated with chemical growth retardants are better and last longer than those grown without them."

Ehrlich P R & Raven P H. Butterflies and plants: a study in coevolution.
Evolution **18**:586-608, 1964.
[Department of Biological Sciences, Stanford University, CA]

The relationships of butterflies and their larval food plants were described and the patterns were hypothesized to result from a reciprocal evolutionary process for which the term 'coevolution' was coined. The primary function of secondary plant chemicals was claimed to be defense against herbivores. [The *Science Citation Index®* (*SCI®*) and the *Social Sciences Citation Index®* (*SSCI®*) indicate that this paper has been cited in over 295 publications since 1964.]

Paul R. Ehrlich
Department of Biological Sciences
Stanford University
Stanford, CA 94305

April 16, 1984

"Our work began over the coffee table when I remarked to Peter Raven that it seemed strange that the *Euphydryas* butterflies that were the subject of my ecological research fed on plants of the families Plantaginaceae and Scrophulariaceae. Peter thought that combination not strange at all, and we began to have daily discussions in which I would describe patterns of food-plant use in butterflies, and he would say what sort of botanical 'sense' they made.

"We began ransacking the literature for data on which plants were eaten and for information on the common characteristics of those plants. The diets of butterflies turned out to be better documented than those of any other large group (12,000-15,000 species) of herbivores. Something was known of the food plants of roughly half of the genera, largely because of the interest of amateurs in raising butterflies in order to get perfect specimens for their collections. It was not long before we realized that the so-called 'secondary compounds' of the plants played a major role in the interactions.

"From that point on, it was a matter of brainstorming between two close colleagues, both evolutionists, one with much experience with butterflies and the other with plants. We did the work with a rising sense of excitement, as we suspected that coevolution was generally an underrated process. Zoologists tended to view plants almost as part of the physical environment; too many parasitologists did not consider the evolution of hosts; and so forth.

"I believe that our paper has been so widely cited because it provided for the first time a detailed discussion of the evolutionary relationships between two large, ecologically intimate groups of organisms. While various of the ideas can be found as far back as the writings of Darwin, and other people had suggested the defensive nature of plant chemicals, no one had put the picture together in this way before and discussed its manifold implications.

"The paper certainly helped spark the development of the now vast field of plant-herbivore coevolution and interest in the process of coevolution in general. Some idea of the ways in which this area of population biology has developed over the past two decades can be gained from a perusal of the excellent new volume edited by Futuyma and Slatkin.[1]

"Quite naturally, some of the ideas in our paper have been criticized, and some were probably quite wrong. Nonetheless, it seems to have stimulated the thinking of a great many people. It is probably the most-cited article either Peter or I have ever published, but that is not the thing that interests us most about it. Unlike our other work, it was done entirely around the coffee table and in the library—neither of us looked at an organism, living or dead, in the course of the work. Therefore our advice to young scientists, should they wish to publish a highly cited paper, apparently ought to be 'study books, and not nature!'"

1. Futuyma D J & Slatkin M, eds. *Coevolution*. Sunderland, MA: Sinauer Associates, 1983. 555 p.

Fraenkel G S. The raison d'être of secondary plant substances.
Science 129:1466-70, 1959.
[Department of Entomology, University of Illinois, Urbana, IL]

The secondary plant substances (allelochemicals) determine the acceptance of plant food by insects (and other organisms) by acting as repellents or attractants. This at once explains the raison d'être of these myriads of chemically unrelated compounds with no obvious nutritional function, and the specificity of host plants for their insects. [The *SCI®* indicates that this paper has been cited in over 200 publications since 1959.]

Gottfried Fraenkel
Department of Entomology
University of Illinois
Urbana, IL 61801

November 11, 1983

"This paper had a long and varied incubation period. A comprehensive study of the basic food requirements of many insects during the war years convinced me that they were essentially identical, and similar to those of 'higher' organisms. At the same time, I became involved in a study of human nutritional needs which also emphasized the importance of green vegetables in a national diet as seen under conditions of food shortages in wartime Britain. Subsequently, at the entomological congress in Amsterdam, 1951, I presented data to the effect that green leaves contained all the nutrients necessary for their insect predators, in excellent quantities and proportions, and there was no *a priori* reason why insects should not develop on any plant provided they ate them.[1]

"By what now seems a coincidence, during the war a then lieutenant of the Canadian Army turned up in my laboratory in England and became engaged in a PhD thesis on the role of the glucosinolates in cruciferous plants as feeding attractants for certain insects,[2] confirming similar earlier results.[3] Thus we had a situation in which all plants were potentially equally nutritious but were

only very selectively eaten, suggesting a role for obviously nonnutritious plant substances in the food selection of certain insects. What could be simpler than putting these two premises together: the enormous variety in the distribution and composition of the secondary plant substances, for which no comprehensive and plausible explanation then existed, accounted for the equally staggering variety of insect/food-plant relationships, by their acting as repellents and attractants for insects (and other organisms). This I first stated in a lecture given at the zoological congress in Copenhagen in 1953.[4]

"It took, however, another five years before these ideas found coherent expression in the paper under discussion. The reception of these views, judging by the annual citations (0 in 1959, six in 1960, one in 1961, two in 1962, one in 1963, one in 1964, three in 1965, seven in 1966, ten in 1967, eight in 1968, four in 1969, six in 1970, nine in 1971, six in 1972, seven in 1973, seven in 1974, nine in 1975, 15 in 1976, 24 in 1977, 16 in 1978, 13 in 1979, 14 in 1980, eight in 1981, 13 in 1982, and ten in 1983) now seems surprising—almost icy silence, and what comments there were were mostly negative during the first six years. Then the number of citations increased during the next five years, probably influenced by an important paper by Ehrlich and Raven,[5] but it was not until five years later that the field suddenly broke wide open. Since that time there has been an ever increasing avalanche of papers; almost annually occurring symposia; about 15 full-length books; and the creation of a virtually new discipline (chemical ecology) with its own journal and international society, now forming. Why this long delay in acceptance and prodigious explosion? The delay could not have been caused by a lack of exposure, with the paper in *Science* and a title which should have compelled equally the attention of organic chemists and plant scientists. Perhaps it seemed implausible that such a simple explanation could be virtually new, and at the same time correct."

1. **Fraenkel G.** The nutritional value of green plants for insects. *Transactions of the IXth International Congress of Entomology. Amsterdam, August 17-24, 1951.* The Hague: W. Junk, 1953. Vol. 2. p. 90-100.
2. **Thorsteinson A J.** The chemotactic responses that determine host specificity in an oligophagous insect (*Plutella maculipennis* (Curt.) Lepidoptera). *Can. J. Zool.* 31:52-72, 1953. (Cited 65 times since 1955.)
3. **Verschaeffelt E.** The cause determining the selection of food in some herbivorous insects. *Proc. Acad. Sci. Amsterdam* 13:536-42, 1910.
4. **Fraenkel G.** Insects and plant biochemistry. The specificity of food plants for insects. *Proceedings of the XIV International Congress of Zoology, Copenhagen. 5-12 August 1953.* Copenhagen: Danish Science Press, 1956. p. 383-7.
5. **Ehrlich P R & Raven P H.** Butterflies and plants: a study in coevolution. *Evolution* 18:586-608, 1964. (Cited 275 times.)

This Week's Citation Classic ™

Rothschild M. Secondary plant substances and warning colouration in insects. (van Emden H F, ed.) *Insect/plant relationships.* Oxford: Blackwell Scientific Publications, 1972. p. 59-83.
[Ashton, Peterborough, England]

The relationship between brightly coloured insects, toxic secondary plant substances and toxic self-secretions is described and discussed. The results of the chemical analysis of approx. 80 species are tabulated and it is shown that the model butterflies in two famous mimicry complexes sequestered and stored toxins from plants and were thus provided with powerful chemical defence mechanisms. [The *SCI®* indicates that this paper has been cited in over 80 publications since 1972.]

Miriam Rothschild
Ashton
Peterborough
England

March 26, 1984

"From field observations and experiments (during 1950-60) with captive birds, I became convinced that many brightly coloured (aposematic) insects contained toxic defensive substances, some of which were sequestered from their food plants and stored. At this period of my life I hero-worshipped Tadeus Reichstein, whom I regard as one of the most gifted chemists of all time and also one of the most creative and honourable men of his day. I was determined to persuade him to join me in a research project concerned with aposematic insects. Reichstein, however, insisted that he was a steroid chemist and pointed out that up to that time (1956) steroids were unknown in insects. 'In order to have the privilege and pleasure of working with you,' I told him, 'I would find steroids in old boots.'

"The first insect I discovered which measured up to his requirements was a desert grasshopper. This insect (*Poekilocerus bufonius*)[1] sequestered cardenolides from its food plant and also concentrated the material in the secretions of a defensive spray. I got a good mark for the fact we could collect this material on filter paper, thereby avoiding extraction of the whole insect. Reichstein identified calactin and calotropin and traces of other cardenolides in the fluid ejected. After this propitious start we identified sequestered cardiac glycosides in 23 aposematic insects from 6 different Orders. We also paid special attention to the supposed models of two of the most famous mimicry situations, the Monarch (*Danaus plexippus*)[2] and the *Aristolochia*-feeding Swallowtail *Pachlioptera aristolochiae*.[3] In both these models we found stored toxic secondary plant substances sequestered from their food plants, cardenolides in the former and aristolochic acids (not steroids!) in the latter, thereby resolving a hundred-year-old controversy.[4,5] The hitherto hypothetical defence of the models was thus placed on a sound chemical basis. All our results were tabulated in the paper in question, negative as well as positive findings.

"The popularity of this publication I believe rested on the following points:
1) The defensive role of the secondary plant substances had recently been pointed out by Fraenkel,[6,7] and had aroused considerable interest. The first proof of their sequestration and storage by insect herbivores proved topical and stimulating.
2) Reichstein's matchless chemistry (for instance he identified 10 cardenolides in the Monarch Butterfly) could be depended on and other workers besides myself could follow his lead with complete confidence.
3) The general discussion and review of the relationship of warning colouration in insects with toxic plants and toxic self-secretions provided a useful basis for further research.

"An up-to-date discussion of aposematic Lepidoptera will be published in the autumn of 1984 in a chapter by the author in 'The Moths and Butterflies of Great Britain and Ireland,' Vol. 2, Cossidae, Heliodinidae. (Heath, J. ed.) (Colchester, Harley Books, 1984 in press)."

1. von Euw J, Fishelson L, Parsons J A, Reichstein T & Rothschild M. Cardenolides (heart poisons) in a grasshopper feeding on milkweeds. *Nature* 214:35-9, 1967. (Cited 50 times.)
2. Reichstein T, von Euw J, Parsons J A & Rothschild M. Heart poisons in the monarch butterfly. *Science* 161:861-6, 1968. (Cited 75 times.)
3. von Euw J, Reichstein T & Rothschild M. Aristolochic acid-I in the swallowtail butterfly *Pachlioptera aristolochiae* (Fabr.) (Papilionidae). *Israel Journal of Chemistry* 6:659-70, 1968. (Cited 20 times.)
4. Bates H W. Contribution to an insect fauna of the Amazon valley. Lepidoptera: Heliconidae. *Transactions of the Linnean Society of London (Zoology)* 23:495-566, 1862. (Cited 35 times since 1955.)
5. Poulton E. Mimicry in N. American butterflies: a reply. *Proceedings of the Academy of Natural Sciences, Philadelphia* 1914:161-95, 1914.
6. Fraenkel G. The *raison d'être* of secondary plant substances. *Science* 129:1466-70, 1959.
7. ----------. Citation Classic. Commentary on *Science* 129:1466-70, 1959. *Current Contents/Life Sciences* 27(11):18, 12 March 1984.

Chapter

8

Neurobiology

The inclusion of a substantial number of Citation Classics in this chapter was intentional and probably inescapable. Inescapable because numerous Classics have been on topics such as stress, anesthesiology, neurotransmitters, memory, learning, sleep, etc. Intentional because these Classics should serve as an introduction of the uninitiated to what is predicted to become the most dynamic subject in the life sciences in the upcoming decades. Nerve growth factors, enkephalins, positron emission tomography, and novel therapy for psychiatric disease are already among us, and the future has yet no limiting horizon. It is also an area of neglect for many biologists because neurobiology welds anatomy, pharmacology, psychology, biochemistry, and other subjects into one, and relatively few life scientists are skilled in such diverse subjects. Fortunately, it is not necessary to recall exactly the description or definition of locus coeruleus, catecholaminergic, slow-wave sleep, or avoidance learning to enjoy these "This Week's Citation Classic" (TWCC) commentaries.

Among the forty-nine Classics listed here, only three exceeded citation 1,000 times. The article by Seeman is a review on the membrane affinity of many anesthetics and tranquilizers. The disordering of cell membrane constituents by these compounds is based on their great solubility in lipids, and disruption of neurotransmitter receptors is a primary site of action of these molecules. The TWCC by Hughes refers to his Classic with five coauthors on the structure of two enkephalins, a Classic quoted 1,460 times. These molecules function as natural painkillers or opioids, and this paper was one of the first important reports on these molecules. The most cited paper in this unit is that of Hodgkin and Huxley, with 1,970 citations to its credit. Their 44-page article correlated the movement of Na^+ and K^+ with changes in membrane electrical potential, a conclusion derived from tedious neurosurgical procedures on giant nerve fibers of the squid. The giant nerve will accept cannulation with objects of 80 to 100 μm in diameter, so this nerve is giant, but only in the sense of the dimensions of nerves.

Thirteen Classics are founded on the physiological role of the neurotransmitters acetylcholine, dopamine, and closely related compounds, a topic that spawned four reviews of Classic status. These reviews focused on the biochemistry of mental disease (p. 293), supersensitivity to catecholamines (p. 295), the role of calcium and the ever-involved cAMP in the release of the neurotransmitters (p. 299), and the chemical basis of excitatory and inhibitory transmitters (p. 303). Of these TWCCs, the last by Krnjević is a particularly straightforward and easily understood description of his Classic.

The neurobiologists' descriptions of their work tend to emphasize the scientific side of their work and not the sociotechnologic conditions surrounding it. Thoenen's TWCC is one exception, including as it does a statement about the paper's rejections by *Nature* and the fatherly editorial advice about processing nerves for electron microscopy.

Pain and anesthesia are the subjects of the next five Classics, which include the review by Seeman and the popular article described by Hughes, both of which are referred to in an earlier paragraph.

The next eleven Classics are behavioral in orientation and assay stress, anxiety, learning, mental illness, and IQ. None of these Classics was published more recently than 1973, and it must be suspected, despite their classic stature, that the original articles are suffering from antiquation. The examples of learning and memory are examples in point. These Classics examine the relationship of stress to learning (p. 314) with the outcome that weakly stressed persons performed better in intelligence tests. The influence of electroconvulsive shock on memory was found to produce amnesia (p. 317). Numerous attributes of memory were described in another Classic (p. 319). Contrast the approach of these reports to the contemporary studies of learning and memory published by Sashoua, Krnjević, Alkon, and others. These investigators, aside from showing that older individuals (goldfish) have only a short-term retention of a learned behavior, have shown that learning is associated with a transient accumulation of calcium in specific cells and that unique proteins termed ependymins are associated with learning and memory. Calcium is important in regulating the interconversion of ependymins from a polymerized to a dissociated state. Thus the earlier descriptive articles on memory–learning have been superseded by studies of its biochemical and biophysical basis.

Within this group, the TWCC by Selye should be read. There is probably no other TWCC in this volume that is as informative about the author.

The construction of stereotactic atlases of rabbit (p. 324) or human brain (p. 325), like the aforementioned, might be considered to a large degree antiquated by modern developments such as CAT or PET scanning. This is true to a large extent, but treatment of cranial cancer with radioactive microneedles relies heavily on topographic data of this sort.

The aforementioned paper by Hodgkin and Huxley demonstrated that Na^+ and K^+ permeability in nerves is correlated with the movement of membrane charges. Narahashi et al. utilized this information to prove that tetrodotoxin from the puffer fish is toxic because it blocks Na^+ channels in axons. Electrical activity of the brain is also the subject of the two following Classics that examine the effect of drugs (p. 331) and physical movement (p. 332) on electrical brain waves.

Of the several remaining Classics, the one on the hypothalamus has special impact on modern society. Separate eating and fasting centers control the hunger response. Perhaps the application of stereotactic mapping will someday permit control of these centers and help us all with our weight problem.

Coppen A. The biochemistry of the affective disorders.
Brit. J. Psychiatry 113:1237-64, 1967.

This paper was a review of investigations into the role of neurotransmitters, endocrine factors and electrolytes in the aetiology and treatment of depressive and manic illness. [The *Science Citation Index*® (*SCI*) and the *Social Sciences Citation Index* ™ (*SSCI* ™) indicate that this paper was cited a total of 178 times in the period 1967-1976.].

A. Coppen
MRC Neuropsychiatry Laboratory
West Park Hospital,
Epsom, Surrey, England

November 29, 1977

"The 1950s saw the development and use of various groups of drugs, including the monamine oxidase inhibitors, the tricyclic antidepressant drugs, and lithium salts. In addition to their powerful antidepressant action, these treatments were a stimulus to neurochemists, pharmacologists and psychiatrists to investigate their mode of action and consequently develop more powerful weapons to combat the prevalent conditions of severe depression and mania, which occur so frequently in middle-aged and elderly people.

"A very important difficulty in working out the modes of action of these compounds and investigating the causes of these illnesses was that they had no satisfactory animal models. In the 1960s I was one of the relatively few (and today there are still lamentably few) psychiatrists who attempted to parallel the enormous efforts being made in the non-clinical laboratories by performing investigations on patients. The practical difficulties of carrying out these investigations under sufficiently standard conditions to make scientific sense was, and still is, considerable. However, by 1967 it seemed to me there was enough data available to make an interesting review for psychiatrists, who were administering powerful and prolonged doses of these antidepressant drugs, but who often had very little knowledge about their mode of action and the rationale for their administration.

"Clinical and non-clinical investigators in the field of biogenic amines, endocrine and electrolyte research often knew little of each other's work. Fortunately, as I had worked in each of these areas, I was able to draw together the various strands and, as I hoped, encourage communication between the various disciplines that were being applied to the study of depressive and manic illnesses.

"I must admit that at times I was not very hopeful about the success of my venture and I was thus gratified by the good reception of the review and the interest it engendered in many branches of the life sciences. There is no doubt that today psychiatrists are much better informed about the biological basis for depression and mania and their treatment.

"I hope that my paper contributed to this and to the realisation that there must be effective collaboration between the non-clinical scientist, the clinician and the patient for those investigations to be fruitful. It is the very real difficulties inherent in this collaboration that must be overcome if we are to achieve the full potential of our considerable technological progress in the neurosciences."

This Week's Citation Classic

CC/NUMBER 36
SEPTEMBER 6, 1982

De Robertis E, Pellegrino de Iraldi A, Rodríguez de Lores Arnaiz G & Salganicoff L. Cholinergic and non-cholinergic nerve endings in rat brain—I. Isolation and subcellular distribution of acetylcholine and acetylcholinesterase. *J. Neurochemistry* 9:23-35, 1962. [Inst. Anat. General y Embriol., Fac. Ciencias Médicas, Univ. Buenos Aires, Argentina]

This paper gave the first detailed description of the isolated nerve endings, i.e., synaptosomes, from the brain. The methodology introduced allowed the separation of myelin, mitochondria, and two types of cholinergic and noncholinergic synaptosomes. These fractions were characterized by electron microscopy and by several biochemical markers. [The *SCI®* indicates that this paper has been cited in over 610 publications since 1962, making it one of the most cited ever published in this journal. Its continued relevance is reflected in the existence of numerous *ISI/BIOMED™* research fronts on synaptosomes.]

Eduardo De Robertis
Instituto de Biología Celular
Facultad de Medicina
Universidad de Buenos Aires
1121 Buenos Aires
Argentina

April 26, 1982

"At the end of the 1950s, and after a long exile, I returned to my country to become chairman of the Institute of Cell Biology with the task of starting a research group in the neurosciences. By that time, the electron microscope had revealed the extraordinary complexity in the ultrastructure of the central nervous system. Inside the nerve endings, the synaptic vesicles had been discovered,[1] and many structural details of the synaptic membranes, perikarya, dendrites, and glial cells were described. I thought that to simplify the structural and biochemical analysis of the brain, it was essential to develop new methods of cell fractionation adapted to its complex ultrastructure and fragile nature, and to obtain subcellular fractions as homogeneous as possible.

"This project involved an interdisciplinary approach in which investigators from the biochemical field (G. Rodríguez de Lores Arnaiz and L. Salganicoff) joined efforts with others (A. Pellegrino de Iraldi and myself) from the field of ultrastructure. We made the homogenization milder by increasing the clearance between the tube and the pestle, so as to protect the nerve endings from disruption. We separated the crude mitochondrial fraction by differential centrifugation, and from this we isolated five subfractions on a discontinuous sucrose gradient. A systematic electron microscope investigation revealed that the lighter fraction was myelin; the sediment free mitochondria and the three other fractions contained isolated nerve endings. For these structures the name 'synaptosomes,' suggested in 1964 by V.P. Whittaker et al.,[2] prevailed in the literature. We also used several biochemical markers and found that two of these fractions were rich in acetylcholine and acetylcholinesterase and the other poor in these biochemical markers (i.e., cholinergic and noncholinergic synaptosomes).

"The high citation of this paper appears to be justified because the isolation of the synaptosome started a whole new field of research, which was pursued in many laboratories. This work provided the foundation for the isolation of synaptic vesicles which, since our early work, were considered to be the structural units for the storage and release of the neurotransmitter.[3] It also led to the separation of the synaptosomal membranes and the localization of pre- and postsynaptic receptors.[4]

"The synaptosome is a self-contained particle having all the structural and many of the functional characteristics of the synaptic region. Within its exiguous limits it contains, in a miniature form, all the molecular constituents, and the complex structural and biochemical machinery, needed for the transmission of the nerve impulse. In addition, the synaptosome is probably able to execute other less known functions related to neurogenesis, plasticity, memory, and learning. It is good fortune that, after two decades, the work on synaptosomes is still wide open for research."[5]

1. **De Robertis E & Bennett H S.** Some features of submicroscopic morphology of synapses in frog and earthworm. *J. Biophys. Biochem. Cytol.* 9:229-35, 1955.
2. **Whittaker V P, Michaelson I A & Kirkland R J A.** The separation of synaptic vesicles from nerve-ending particles ('synaptosomes'). *Biochemical J.* 90:293-303, 1964.
3. **De Robertis E, Rodríguez de Lores Arnaiz G & Pellegrino de Iraldi A.** Isolation of synaptic vesicles from nerve endings of the rat brain. *Nature* 194:794-5, 1962.
4. **Criado M, Aguilar J S & De Robertis E.** Action of detergents and pre- and postsynaptic localization of ³H-naloxone binding in synaptosomal membranes. A structural approach. *J. Neurobiology* 12:259-67, 1981.
5. **De Robertis E.** *The synaptosome. Two decades of cell fractionation of the brain.* New York: Raven Press. In press, 1982.

Trendelenburg U. Supersensitivity and subsensitivity to sympathomimetic amines.
Pharmacol. Rev. 15:225-76, 1963.
[Department of Pharmacology, Harvard Medical School, Boston, MA]

Denervation supersensitivity turned out to involve two entirely different mechanisms on the one hand, a 'site of loss' (neuronal uptake) is lost; on the other hand, effector cells adapt to the loss of sympathetic tone [The SCI® indicates that this paper has been cited in over 580 publications since 1963.]

Ullrich Trendelenburg
Institut für Pharmakologie
und Toxikologie
Universität Würzburg
D-8700 Würzburg
Federal Republic of Germany

July 13, 1984

"Four weeks after removal of the superior cervical ganglion, the cat's nictitating membrane responds to 1/1,000 of the dose of noradrenaline that was needed to elicit a similar response in the innervated side.[1] Why? I fell in love with this fascinating problem during my training in Oxford (J.H. Burn), and a systematic study was carried out at the department of pharmacology, Harvard Medical School (O. Krayer). Help came from experienced colleagues, N. Weiner and J.R. Crout, who provided the sadly missing biochemical knowhow, and also from an international mix of young trainees, J.S. Gravenstein, W.W. Fleming, B. Gomez Alfonso de la Sierra, and A.J. Muskus.

"Virtually all earlier explanations of denervation supersensitivity attempted to find *one* explanation.[2] The realization that there are two entirely different types of supersensitivity did not come as a sudden flash of inspiration — it grew slowly.

"One type of supersensitivity (later termed 'prejunctional'[3] or 'deviation' supersensitivity[4]) involved the loss (denervation) or the inhibition (cocaine) of a site of loss (neuronal uptake). This leads to an increased concentration of the agonist at the receptors. The other type of supersensitivity[4] (later termed 'postjunctional'[3] or 'nondeviation' supersensitivity[4]) reflects the ability of the effector cells to (slowly) adapt to any interruption of the flow of tonic impulses; the responsiveness of the cells to a given agonist concentration increases. Once we realized that we were dealing with two entirely different types of supersensitivity, the experimental facts of several decades fell into a meaningful pattern — and this is what the review was about.

"It was Fleming who inherited the nondeviation supersensitivity which continues to pose the intriguing question whether charges in receptor populations provide the *full* explanation.[5] My own interest was captivated by a second 'deviation supersensitivity' to catecholamines, namely, that induced by inhibition of extraneuronal uptake or catechol-O-methyl transferase.[6] This type proves that we need *both* nomenclatures, since it turned out to be 'postjunctional deviation supersensitivity.'"

1. Langer S Z, Draskoczy P R & Trendelenburg U. Time course of the development of supersensitivity to various amines in the nictitating membrane of the pithed cat after denervation or decentralization.
   J. Pharmacol. Exp. Ther. 157:255-73, 1967. (Cited 105 times.)
2. Cannon W B & Rosenblueth A. *The supersensitivity of denervated structures.* New York: Macmillan. 1949. 245 p.
   (Cited 295 times.)
3. Trendelenburg U. Mechanisms of supersensitivity and subsensitivity to sympathomimetic amines.
   Pharmacol. Rev. 18:629-40, 1966. (Cited 390 times.)
4. Fleming W W. Supersensitivity in smooth muscle. Introduction and historical perspective.
   Fed. Proc. 34:1960-70, 1975.
5. Fleming W W, McPhillps J J & Westfall D P. Postjunctional supersensitivity and subsensitivity of excitable tissues to drugs. *Rev. Physiol. Biochem. Exp. Pharmacol.* 68:55-119, 1973.
6. Trendelenburg U. A kinetic analysis of the extraneuronal uptake and metabolism of catecholamines.
   Rev. Physiol. Biochem. Pharmacol. 87:33-115, 1980.

Poirier L J & Sourkes T L. Influence of the substantia nigra on the catecholamine content of the striatum. *Brain* 88:181-92, 1965.
[Dept. Sci. Neurologiques, Univ. Montreal and Allan Mem. Inst. Psychiatry, and McGill Univ., Montreal, Canada]

The evidence that a well-identified group of neurons (originating in the substantia nigra) exert through their efferent fibers a direct role in the elaboration of a specific substance (dopamine) in a distant structure (the striatum) contributes to a better understanding of brain circuitry. Moreover, this intracerebral 'dopaminergic' pathway is directly involved in Parkinson's and related diseases. [The *SCI*® indicates that this paper has been cited over 265 times since 1965.]

Louis J. Poirier
Université Laval and
Hôpital de l'Enfant-Jésus
Pavillon Notre-Dame
Québec, Canada G1J 5B3
and
T.L. Sourkes
Department of Psychiatry
McGill University
Montréal, Québec H3A 1A1
Canada

August 5, 1980

"This paper represents a typical example of novel conclusions derived from the pooling of biochemical and morphological expertise of two independent investigators. Without the conjunction of the two approaches the results could have been dealt with only speculatively. We began our collaboration in May 1963. Poirier, who was working as a neuroanatomist in the department of neurological sciences, University of Montreal, had a few years before suggested a close relationship between the dysfunction of the substantia nigra and the development of motor disorders associated with Parkinson's disease. He did so on the basis of the histopathological changes found in the brains of monkeys in which features of Parkinson's disease had been experimental-

ly reproduced.[1] Sourkes, head of the Neurochemical Laboratory at the Allan Memorial Institute of Psychiatry at McGill, had been studying the chemistry of extrapyramidal disorders.[2] The finding that dopamine concentration was low in the striatum of the brains from Parkinsonian patients,[3] together with Poirier's observation, led to the search for an intracerebral nervous pathway that plays a role in the synthesis of dopamine and which is at fault in Parkinson's disease.

"By making selective lesions of the brain and correlating the histopathological and biochemical data we disclosed a direct relationship between morphological changes in the substantia nigra and biochemical impairment in the striatum, thus leading us to deduce the existence of a nigrostriatal dopaminergic pathway functioning in the control of motor activity. By the fall of 1963, we were able to report some of our results at the Second Conference on Parkinson's Disease in Washington, DC.

"The data derived from this and related investigations provided the experimental basis for tying together the alterations in nigral histology and striatal dopamine with the pre-mortem existence of clinical Parkinsonism. Further, the work stimulated studies aimed at the identification of neurochemically defined pathways and their role in behavioral and psychoneuroendocrinological phenomena.

"This paper underlines the importance of collaborative work between investigators in different disciplines. By establishing a proper dialogue each researcher materially increases his contribution but, more so, the rate of progress of knowledge. In addition to the obvious advances contributed by specialized researchers and the multipolar experts, there is still a great need for investigations based on the combined efforts of scientists who agree to share their skills in order to achieve a synthesis that neither could attain working alone."

1. Poirier L J. Experimental and histological study of midbrain dyskinesias.
   J. Neurophysiol. 23:534-51, 1960.
2. Sourkes T L. Cerebral and other diseases with disturbances of amine metabolism.
   Progr. Brain Res. 8:186-200, 1964.
3. Ehringer H & Hornykiewicz O. The diffusion of nonadrenalin and dopamine (3-hydroxy-tyramin) in the human brain and their suppression in diseases of the extrapyramidal systems.
   Klin. Wochenschr. 38:1236-39, 1960.

Porte D, Jr., Graber A L, Kuzuya T & Williams R H. The effect of epinephrine on immunoreactive insulin levels in man. *J. Clin. Invest.* 45:228-36, 1966.
[Department of Medicine, University of Washington, Seattle, WA]

The paper describes the effects of an epinephrine infusion to inhibit plasma insulin responses to glucose, glucagon, and tolbutamide in man. [The *SCI®* indicates that this paper has been cited in over 615 publications since 1966.]

Daniel Porte, Jr.
Diabetes Center
University of Washington
and
Division of Endocrinology and Metabolism
Veterans Administration Medical Center
Seattle, WA 98108

June 6, 1984

"In 1963, I arrived at the laboratory of R.H. Williams as a postdoctoral fellow. Having studied with Richard Havel in San Francisco, I was interested in the role of the sympathetic nervous system in the regulation of free fatty acid (FFA) mobilization from adipose tissue.[1] To evaluate this system, Alan Graber and I gave prolonged infusions of epinephrine to a male subject. We found that FFA levels rose and then returned to basal despite continued administration of the amine. Since hyperglycemia and tachycardia persisted, we concluded that reesterification of FFA in adipose tissue or inhibition of lipolysis by insulin might be involved. Therefore, we asked Kuzuya (now at Jichi Medical School, Japan), another postdoctoral fellow, to apply the newly developed radioimmunoassay for insulin to samples from our catecholamine infusions to determine which mechanism was most likely. To our surprise, insulin levels did not change during the epinephrine infusions but rose dramatically upon their termination.

"Since the concept of the autonomic nervous system regulating the peripheral endocrine system was not considered likely at that time, we performed a number of control studies with glucose and tested other insulin secretogogues; all were inhibited. This finding had important implications

for metabolic regulation and suggested that many older studies in which insulin secretion had been assumed to parallel glucose levels would need to be reexamined. Before submission of the work, I had the opportunity to present it at a major national meeting. Afterward, I received a letter and reprint[2] from Colwell pointing out that he had predicted such a finding 30 years earlier when he observed inhibition of glucose metabolism during an epinephrine infusion. So much for the originality of my scientific finding!

"Nevertheless, the concept was new to most scientists and opened up enough new questions that I have spent the next 20 years studying its implications. What has developed is the idea that the peripheral nervous system regulates many hormones, not just those of the endocrine pancreas, and that the central nervous system in turn is regulated by peptide hormones secreted by peripheral endocrine cells.[3] A critical role for the neural control of islet function in the development of stress hyperglycemia has also been delineated.[4]

"The newness of the idea and its applicability to many other endocrine glands, plus the large number of diabetes-related investigators, are the most likely explanations for the large number of citations.

"In 1970, I was fortunate to be selected by the American Diabetes Association to receive the Eli Lilly Award for scientific achievement; and, in 1984, I received the David Rumbaugh Award of the Juvenile Diabetes Foundation. Both of these honors I relate in part to this early study and its findings.

"It is always pleasing to look back on research studies such as this one that had such a positive influence on a field as well as on one's own career. The finding added an entirely new area to my research and began my interest in neural-endocrine interactions and the regulation of carbohydrate metabolism. Since the finding did not represent conventional wisdom at the time, I am grateful that my mentor, Williams, and the scientific system allowed me considerable freedom to pursue these studies. I hope our system of scientific evaluation retains such flexibility in the future."

1. **Havel R J & Goldfien A.** The role of the sympathetic nervous system in the metabolism of free fatty acids. *J. Lipid Res.* 1:102-8, 1959. (Cited 345 times since 1959.)
2. **Colwell A R & Bright E M.** The use of constant glucose injections for the study of induced variations in carbohydrate metabolism. IV. Suppression of glucose combustion by continuous prolonged epinephrine administration. *Amer. J. Physiol.* 92:555-67, 1930. (Cited 10 times since 1955.)
3. **Woods S C, Smith P H & Porte D, Jr.** The role of the nervous system in metabolic regulation and its effects on diabetes and obesity. (Brownlee M, ed.) *Handbook of diabetes mellitus: intermediary metabolism and its regulation. volume III.* New York: Garland Publishing Co., 1981. p. 209-71.
4. **Porte D, Jr. & Woods S C.** Neural regulation of islet hormones and its role in energy balance and stress hyperglycemia. (Ellenberg M & Rifkin H, eds.) *Diabetes mellitus: theory and practice.* New York: Medical Examination Publishing Co., 1983. p. 267-94.

Thoenen H & Tranzer J P. Chemical sympathectomy by selective destruction of
adrenergic nerve endings with 6-hydroxydopamine.
Naunyn-Schmied. Arch. Pharmakol. Exp. Pathol. 261:271-88, 1968.
[Dept. Experimental Medicine, F. Hoffmann-La Roche & Co. Ltd., Basle, Switzerland]

This paper describes the use of 6-hydroxydopamine (6-OHDA) as a tool for the selective and extensive destruction of sympathetic nerve terminals in rats and cats. The specificity of the destructive action is based on the fact that 6-OHDA is selectively accumulated in sympathetic nerve terminals with high efficiency and that the highly reactive oxidation products of 6-OHDA undergo covalent binding with nucleophilic groups of macromolecules leading to the destruction of the nerve terminals. [The *SCI®* indicates that this paper has been cited in over 645 publications since 1968, probably the most-cited paper published in this journal.]

H. Thoenen
Abteilung Neurochemie
Max-Planck-Institut für Psychiatrie
8033 Planegg-Martinsried
Federal Republic of Germany

July 19, 1984

"In the context of a project designed to replace the physiological transmitter norepinephrine (NE) by 'false transmitters,' the late Jean-Pierre Tranzer and I investigated the ultramorphological manifestations of this transmitter exchange. By test-tube experiments, we could to some extent predict whether this replacement would result in apparent 'empty' or 'dense core' vesicles. Dense core vesicles were to be expected if the false transmitter substance, after reacting with glutaraldehyde, still reduced osmium tetroxide. Thus, the replacement of NE by 5-hydroxydopamine (5-OHDA) and 6-hydroxydopamine (6-OHDA) was predicted to result in dense core vesicles in adrenergic nerve terminals. This indeed was the case for 5-OHDA. This substance became a valuable tool for the unambiguous identification of catecholaminergic nerve terminals, i.e., after treatment with 5-OHDA all the vesicles contained an intense dense core.[1]

"The effect of 6-OHDA was quite different. Dense core vesicles could be demonstrated only at a very early stage of NE depletion, i.e., a few hours after administration of 6-OHDA.[2] However, this replacement of the physiological transmitter was very soon followed by signs of degeneration and finally destruction of the adrenergic nerve terminals. This explained the very long lasting NE depletion.[2] In adult animals, the destroyed nerve terminals regenerated, and a complete morphological, biochemical, and functional restoration occurred. In newborn animals, however, 6-OHDA resulted in a complete destruction of the peripheral sympathetic nervous system in a manner identical to administration of anti-nerve growth factor (NGF) antibodies.[3] After demonstrating that NGF acts as a retrograde messenger between effector organs and innervating sympathetic neurons, it became evident that the destruction of the sympathetic neurons in newborn animals by 6-OHDA resulted in an interruption of the supply of endogenous NGF from the periphery and that chemical sympathectomy with 6-OHDA and immunosympathectomy with antibodies to NGF both result from the abolition of the supply of NGF from the periphery to the sympathetic cell bodies.[4]

"6-OHDA became a standard experimental tool for general or local destruction of peripheral and central catecholaminergic nerve terminals.[2] Moreover, experiments with 6-OHDA led also to the detection of the transsynaptic induction of tyrosine hydroxylase, the first demonstration that nerve impulses regulate the synthesis of specific neuronal macromolecules.[4,5]

"When we submitted the first observation on the destruction of adrenergic nerve terminals by 6-OHDA to *Nature*, the paper was rejected because a reviewer came to the conclusion that our observation was an artifact. We also obtained the fatherly advice that we learn how to process tissue samples appropriately for electron microscopy in order to avoid future artifacts.

"After publication of the initial morphological observation,[6] we published this more extensive paper, in which the treatment schedule for as complete a destruction of sympathetic nerve terminals was given. I think that this was the major reason for the frequent citation. Moreover, we also offered a first explanation for the mechanism of action of 6-OHDA, namely, that the oxidation product(s) of 6-OHDA underwent covalent binding with nucleophilic groups of neuronal macromolecules resulting in their denaturation and, consequently, in degeneration of the nerve terminals. The reaction as such is nonspecific, but the specificity of the destruction results from the efficient accumulation of 6-OHDA by the transport system for biogenic amines localized in the plasma membrane of sympathetic nerve terminals."

1. Tranzer J P & Thoenen H. Electronmicroscopic localization of 5-hydroxydopamine(3,4,5-trihydroxy-phenylethylamine), a new "false" sympathetic transmitter. *Experientia* 23:743-5, 1967. (Cited 230 times.)
2. Thoenen H & Tranzer J P. The pharmacology of 6-hydroxydopamine. *Annu. Rev. Pharmacol.* 13:169-80, 1973. (Cited 120 times.)
3. Angeletti P U & Levi-Montalcini R. Sympathetic nerve cell destruction in newborn mammals by 6-hydroxydopamine. *Proc. Nat. Acad. Sci. US* 65:114-21, 1970. (Cited 145 times.)
4. Thoenen H, Otten U & Schwab M E. Orthograde and retrograde signals for the regulation of neuronal gene expressions: the peripheral sympathetic nervous system as a model. (Schmitt F O & Worden F G, eds.) *The neurosciences: fourth study program.* Cambridge, MA: MIT Press, 1979. p. 911-28.
5. Mueller R A, Thoenen H & Axelrod J. Adrenal tyrosine hydroxylase: compensatory increase in activity after chemical sympathectomy. *Science* 163:468-9, 1969. (Cited 125 times.)
6. Tranzer J P & Thoenen H. An electron microscopic study of selective, acute degeneration of sympathetic nerve terminals after administration of 6-hydroxydopamine. *Experientia* 24:155-256, 1968. (Cited 360 times.)

Rubin R P. The role of calcium in the release of neurotransmitter substances and hormones. *Pharmacol. Rev.* **22**:389-428, 1970.
[Department of Pharmacology, State University of New York, Downstate Medical Center, Brooklyn, NY]

This paper summarizes existing evidence regarding the actions of calcium and other cations on secretory systems. The nature of calcium's role in the secretory process is also considered, and parallelisms are drawn between stimulus-secretion coupling and excitation-contraction coupling in muscle. [The *SCI®* indicates that this paper has been cited in over 735 publications since 1970.]

Ronald P. Rubin
Department of Pharmacology
Medical College of Virginia
Virginia Commonwealth University
Richmond, VA 23298

June 7, 1983

"In 1960, I began working as a graduate student in the laboratory of William Douglas in the department of pharmacology at the Albert Einstein College of Medicine in New York City. The project on which I embarked concerned the role of calcium and other cations in the mechanism of catecholamine release from the isolated perfused cat adrenal gland.

"We found that calcium, but not sodium or potassium, was the crucial cation required for eliciting catecholamine secretion with the physiological neurotransmitter acetylcholine or with excess potassium. Calcium was not only a necessary but also a sufficient factor for supporting secretion. The close correlation between the concentration of extracellular calcium and evoked secretion, taken together with the demonstrated increase in radiocalcium uptake into chromaffin cells with acetylcholine,[1] supported the concept that 'acetylcholine stimulation enhances the entry of calcium into the medullary chromaffin cells.'

"The role of calcium in synaptic transmission had been defined in the 1950s and 1960s by Sir Bernard Katz and his colleagues during their analysis of acetylcholine release from the frog neuromuscular junction. So, the work emanating from the laboratories of Katz[2] and Douglas[3] provided the impetus for others to investigate the role of calcium in secretory phenomena. Confirmation of calcium-regulated secretion was obtained first in adrenergic synapses and then in a number of other secretory systems. The 1970 paper in *Pharmacological Reviews* reviews these findings. I believe that the reason this review has been cited so frequently reflects the prodigious interest generated by the original work of Douglas and myself[4,5] and the extension of these findings consummated in Douglas's laboratory. Today, the pivotal role of calcium in *stimulus-secretion coupling* is even more entrenched in scientific thought than it was in 1970. A book that presents a timely account of advances in this ever-expanding field has recently been published.[6]

"But the nature of calcium's role is much more complex than that envisioned in 1970. Activation of secretory cells was then viewed from a narrow perspective, involving the influx of extracellular calcium caused by an increase in membrane permeability. Our present view must encompass multiple cellular pools of calcium participating in the secretory response of many cells—a concept that was only briefly alluded to in the 1970 review.

"We are still permitted to draw parallels between stimulus-secretion coupling and excitation-contraction coupling in muscle, and to consider the molecular mechanism of calcium's action as being somehow linked to exocytosis, as we did in 1970. Calcium is still viewed as the progenitor of intracellular signals, but cyclic nucleotides, prostaglandins, and other arachidonic acid metabolites must now be portrayed as putative messengers that mediate the actions of calcium in secretory cells. Paradoxically perhaps, we have even progressed to the point where one may acknowledge the possible existence of noncalcium-dependent mechanisms, possibly acting synergistically with calcium, to activate the secretory process.[7] But as in 1970, a clear picture of the fundamental action of calcium on the arcane machinery that controls the secretory apparatus still remains elusive."

1. **Douglas W W & Poisner A M.** On the mode of acetylcholine in evoking adrenal medullary secretion: increased uptake of calcium during the secretory response. *J. Physiol.—London* 162:385-92, 1962. [Cited 130 times.]
2. **Katz B.** *The release of neural transmitter substances.* Springfield, IL: Charles C. Thomas, 1969. 60 p. [Cited 465 times.]
3. **Douglas W W.** Stimulus-secretion coupling: the concept and clues from chromaffin and other cells. *Brit. J. Pharmacol.* 34:451-74, 1968. [Cited 830 times.]
4. **Douglas W W & Rubin R P.** The role of calcium in the secretory response of the adrenal medulla to acetylcholine. *J. Physiol.—London* 159:40-57, 1961. [Cited 495 times.]
5. ——————————————. The mechanism of catecholamine release from the adrenal medulla and the role of calcium in stimulus-secretion coupling. *J. Physiol.—London* 167:288-310, 1963.
6. **Rubin R P.** *Calcium and cellular secretion.* New York: Plenum Press, 1982. 276 p.
7. **Kaibuchi K, Sano K, Hoshijima M, Takai Y & Nishizuka Y.** Phosphatidylinositol turnover in platelet activation; calcium mobilization and protein phosphorylation. *Cell Calcium* 3:323-35, 1982.

Clement-Cormier Y C, Kebabian J W, Petzold G L & Greengard P.
Dopamine-sensitive adenylate cyclase in mammalian brain: a possible site of action of antipsychotic drugs. *Proc. Nat. Acad. Sci. US* 71:1113-17, 1974.
[Dept. Pharmacology, Yale Univ. Sch. Med., New Haven, CT]

Dopamine produced a twofold increase in striatal adenylate cyclase activity (half-maximal increase in activity with 5 μM dopamine). Selected phenothiazines were potent and competitive inhibitors of enzyme activity with low inhibition constants (nanomolar range). The butyrophenones were competitive but weak antagonists; nonpsychoactive phenothiazines were without effect. [The *SCI®* indicates that this paper has been cited in over 475 publications since 1974.]

Yvonne C. Clement-Cormier
Department of Pharmacology
Medical School
University of Texas
Health Science Center
Houston, TX 77225

March 14, 1984

"By the early 1970s, substantial evidence had accumulated for a functional role for cyclic nucleotides in the nervous system. When I joined Paul Greengard's laboratory as a graduate student, the presence of a dopamine-sensitive adenylate cyclase in the caudate nucleus of rat brain had just been reported.[1,2] This observation was significant because it focused attention on dopamine-sensitive cyclase as a possible biochemical marker for the dopamine receptor.

"As part of the research for my PhD thesis, I chose a pharmacological approach to investigate the potential usefulness of the dopamine-sensitive adenylate cyclase assay as a marker for the dopamine receptor. First, several studies were performed which documented the presence of the enzyme exclusively in dopaminergic brain areas and in a variety of mammals including humans. In addition, the enzyme activity was found to be enriched in subcellular fractions of the caudate associated with postsynaptic structures. These studies also demonstrated that the action of dopamine was directly on a receptor and not via metabolites of the neurotransmitter. Thus, on the basis of the results of these experiments, we theorized that the 'dopamine receptor' of the caudate nucleus, as described physiologically, was closely associated with dopamine-sensitive adenylate cyclase activity.

"At the time I was performing these studies, it was known that drugs of the phenothiazine and butyrophenone classes could assume a conformation similar to that of dopamine.[3] Thus, the dopamine-sensitive adenylate cyclase assay provided a readily accessible tool to evaluate the hypothesis that part of the mechanism of action of the antipsychotic drugs involved a direct interaction with the dopamine receptor.

"Over 15 different antipsychotic drugs (representing the phenothiazines, butyrophenones, and dibenzodiazepines), as well as nonpsychoactive drugs, were tested to determine if they blocked the effect of dopamine on brain adenylate cyclase. The results demonstrated that the antipsychotic drugs were competitive inhibitors with extremely high affinities (in the nanomolar range) for the enzyme. More importantly, drugs which had little or no antipsychotic or extrapyramidal actions clinically had relatively high inhibitory constants. Similar studies which confirmed our observations were reported by Leslie Iversen's group.[4]

"Not surprisingly, there were some discrepancies between the results of various test substances observed in this enzyme system with the results of clinical trials and laboratory studies *in vivo*. Whereas the data fit well for the phenothiazine class of antipsychotic drugs, the butyrophenones, which were known to be potent antipsychotics, were weak antagonists of dopamine-sensitive adenylate cyclase activity. At about this time, studies using radioligand binding as a tool for tagging dopamine receptors also showed differences between dopaminergic sites which were identified by labeled butyrophenones and those coupled to adenylate cyclase.[5,6] These observations were at the forefront of describing what is now the widely accepted view of the heterogeneity of the dopamine receptor.[7,8]

"There are several reasons why this paper has been highly cited. First, it provided strong support for the conclusion that some physiological effects of dopamine may be initiated by increases in intracellular cyclic AMP. Recent studies have substantiated this point of view.[9] Second, for the first time a rapid, accurate, and inexpensive pharmacological tool for identifying a drug as a direct agonist or antagonist at dopamine receptors coupled to adenylate cyclase was established, as well as a useful procedure for calculating the affinity of the drugs for the receptor. Third, the identification of additional categories of dopamine receptors arose, in part, as a consequence of the comparison of the efficacy of antagonists upon the dopamine-sensitive adenylate cyclase and other biochemical models of dopamine receptor sites."

1. **Kebabian J W, Petzold G L & Greengard P.** Dopamine-sensitive adenylate cyclase in caudate nucleus of rat brain, and its similarity to the "dopamine receptor." *Proc. Nat. Acad. Sci. US* 69:2145-9, 1972.
2. **Kebabian J W.** Citation Classic. Commentary on *Proc. Nat. Acad. Sci. US* 69:2145-9, 1972. *Current Contents Life Sciences* 26(11):18, 14 March 1983.
3. **Snyder S H.** Catecholamines in the brain as mediators of amphetamine psychoses. *Arch. Gen. Psychiat.* 27:169-79, 1972. (Cited 215 times.)
4. **Iversen L K, Horn A S & Miller R J.** Actions of dopaminergic agonists on cyclic AMP in rat brain homogenates. *Advan. Neurol.* 9:197-212, 1975. (Cited 35 times.)
5. **Snyder S H, Creese I & Burt D R.** The brain's dopamine receptor: labeling with [³H] dopamine and [³H] haloperidol. *Psychopharmacol. Commun.* 1:663-73, 1975. (Cited 55 times.)
6. **Seeman P & Lee T.** Antipsychotic drugs: direct correlation between clinical potency and presynaptic action of dopaminergic neurons. *Science* 188:1217-19, 1975. (Cited 230 times.)
7. **Clement-Cormier Y C & George R J.** Multiple dopamine binding sites: subcellular localization and biochemical characterization. *J. Neurochemistry* 32:1061-9, 1979. (Cited 25 times.)
8. **Kebabian J W & Calne W B.** Multiple receptors for dopamine. *Nature* 277:93-6, 1979. (Cited 775 times.)
9. **Walaas S I, Aswad D W & Greengard P.** A dopamine- and cyclic AMP-regulated phosphoprotein in dopamine-innervated brain regions. *Nature* 301:69-71, 1983.

Crow T J. Catecholamine-containing neurones and electrical self-stimulation: 1.
A review of some data. *Psychol. Med.* 2:414-21, 1972.
[Dept. Mental Health, Univ. Aberdeen, Aberdeen, Scotland]

Intracranial self-stimulation was obtained with electrodes located close to dopamine-containing cell bodies in the ventral midbrain and to the noradrenaline-containing cells of the locus coeruleus; earlier findings could be explained as activation of one or the other of these two systems. [The *Science Citation Index®* (*SCI®*) and the *Social Sciences Citation Index®* (*SSCI®*) indicate that this paper has been cited over 115 times since 1972.]

T.J. Crow
Division of Psychiatry
Clinical Research Centre
Northwick Park Hospital
Harrow, Middlesex HA1 3UJ
England

October 14, 1981

"Intracranial self-stimulation described in 1954 by James Olds and P.M. Milner[1] with electrodes in septal and lateral hypothalamic regions demonstrated the existence of powerful central reward mechanisms but the neural identity of these was obscure. In 1962, Larry Stein suggested a noradrenergic system was involved.[2] I thought this was an exciting theory and in 1969 I visited the Karolinska Institute where K. Fuxe, U. Ungerstedt, and colleagues had recently completed the first comprehensive maps of central catecholamine neurones. Fibres of these systems in the medial forebrain bundle might well have been activated in previous self-stimulation experiments.

"I embarked on a study of self-stimulation sites in relation to cell body groups in the brainstem. In the midbrain such sites were close to dopamine cell bodies, a quite unexpected finding.[3] Systematic mapping revealed a second group of dorsally located sites just lateral to the central grey substance. This suggested two systems were involved: the second system might be the noradrenergic fibres ascending from the locus coeruleus in the mid-pons. With Gordon Arbuthnott and Jane Spear, I devised a technique for implanting electrodes in this region; with difficulty and after a number of attempts we were able to obtain the behaviour here also.[4]

"The *Psychological Medicine* paper reviewed these findings and argued that much of the previous literature (including the studies of Olds) could be explained by the two catecholamine system hypothesis. Curiously, subsequent studies have strongly supported what at the time seemed most controversial — that activation of dopamine neurones has rewarding effects. The involvement of the locus coeruleus system remains disputed although I think this is rather strongly suggested by evidence that with such electrodes the behaviour is associated with increased turnover of noradrenaline in the ipsilateral cerebral cortex.

"The paper may have been well cited because it put anatomical flesh on Stein's humoural hypothesis, though with a dopamine twist that he took time to assimilate. Also the theory has remained controversial, particularly with respect to the view that the functions of the two pathways can be related to the learning theory concepts of 'incentive' or 'drive induction' (the dopamine system) and 'reinforcement' or the 'results of action' signal (the locus coeruleus noradrenaline system), and may be phylogenetic derivatives of olfactory and gustatory pathways, respectively.[5,6] Most gratifying was the interest which Olds later took in this theory.[7] His tragic death in 1976 prevented him from pursuing it with the electrophysiological techniques of which he was master.

"I nearly forgot Olds's role in the original papers.[3,4] When I first met him three years after their publication he told me he had refereed them and recommended rejection; then he wrote to the editor to say he thought he had made a mistake. Catecholamine neurones might indeed play a critical role. The editor replied he had already decided to publish."

1. Olds J & Milner P M. Positive reinforcement produced by electrical stimulation of septal area and other regions of rat brain. *J. Comp. Physiol. Psychol.* 47:419-27, 1954.
2. Stein L. Effects and interactions of imipramine, chlorpromazine, reserpine and amphetamine on self-stimulation: possible neurophysiological basis of depression. (Wortis J, ed.) *Recent advances in biological psychiatry.* New York: Plenum, 1962. p. 288-308.
3. Crow T J, Spear P J & Arbuthnott G W. Intracranial self-stimulation with electrodes in the region of the locus coeruleus. *Brain Res.* 36:275-87, 1972.
4. Crow T J. A map of the rat mesencephalon for electrical self-stimulation. *Brain Res.* 36:265-73, 1972.
5. ———————, Catecholamine-containing neurones and electrical self-stimulation: 2. A theoretical interpretation and some psychiatric implications. *Psychol. Med.* 3:66-73, 1973.
6. ———————. A general catecholamine hypothesis. *Neurosci. Res. Program Bull.* 15:195-205, 1977.
7. Olds J. Reward and drive neurones: 1975. (Wauquier A & Rolls E T, eds.) *Brain stimulation reward.* New York: Elsevier-North Holland, 1976. p. 1-27.

Krnjević K & Schwartz S. The action of γ-aminobutyric acid on cortical
neurones. *Exp. Brain Res.* 3:320-36, 1967.
[Wellcome Dept. Res. Anaesthesia, McGill Univ., Montreal, Canada]

When γ-aminobutyric acid is applied to
single cortical neurones, it causes
changes in membrane potential and
conductance that are similar to the effects of synaptic inhibition. It is
therefore concluded that this normal
constituent of the brain could be the
physiological transmitter at inhibitory
synapses in the cerebral cortex. [The
SCI® indicates that this paper has been
cited over 240 times since 1967.]

Krešimir Krnjević
Departments of Physiology and
Research in Anaesthesia
McGill University
Montreal, Quebec H3G 1Y6
Canada

January 14, 1981

"For the last 25 years, there has been
almost unanimous agreement among
neurophysiologists that at points of
junction ('synapses') nerve cells communicate with each other mainly by
releasing specific excitatory or inhibitory substances. Until the 1960s little was
known about the identity of the
postulated central transmitters. In
1963, John Phillis and myself had
reported that, in the cerebral cortex,
L-glutamate and γ-aminobutyric acid
(GABA) have very powerful and rapid
excitatory and inhibitory actions
respectively; since both agents are normally present in the cortex in large
amounts, we suggested that they may
well be physiological transmitters.[1]
Without stronger evidence that these
effects are similar to the natural synaptic actions, this suggestion was not
taken very seriously, especially since
there was a wide consensus that these
agents could not be neurotransmitters.

"More critical tests required recording from *inside* a cortical neurone while

applying the postulated transmitter just
outside. In previous attempts, co-axial
microelectrodes had not been very
satisfactory, because huge 'coupling'
artifacts largely obscured any significant effects. A much better possible
solution was suggested to me by Robert
Werman during my visit to his laboratory in Indianapolis (in 1964): this was to
fix microelectrodes side-by-side, thus
greatly reducing electrical coupling.
He was moreover kind enough to give
me two of his Narishige micromanipu-
lators for this purpose.

"Shortly afterwards, I was joined in
Montreal by Susan Schwartz (who was
then a recent graduate of the Albert
Einstein School of Medicine) and we
started using these to prepare double
micropipettes. There were many technical difficulties, compounded by the
scarcity of equipment. I had only just
moved into the new, but empty,
anaesthesia research laboratories situated in the newly-opened McIntyre
Medical Building. The frequency of
useful intracellular penetrations was
much reduced by the presence of the
second external pipette; so it was
doubly frustrating when cells were successfully impaled but GABA could not
be released in sufficient quantity. On
the other hand, GABA's action proved
to be exceptionally favourable, since
its hyperpolarizing effect became even
more evident as cellular potentials
deteriorated, while its rapid time
course ensured that even during brief
intracellular recording there was some
chance of making significant observations. Moreover, the large increases in
membrane conductance evoked by
GABA could be detected even under
unpromising recording conditions.

"So we were very fortunate in being
able to show that GABA consistently
imitates the synaptic inhibitory action
and therefore fulfills what is generally
accepted as one of the most important
criteria by which one identifies a
transmitter."

---

1. Krnjević K & Phillis J W. Iontophoretic studies of neurones in the mammalian cerebral
cortex. *J. Physiology* (London) 165:274-304, 1963.

Krnjević K. Chemical nature of synaptic transmission in vertebrates.
Physiol. Rev. 54:418-540, 1974.
[Department of Research in Anaesthesia, McGill University, Montreal, Canada]

A comprehensive survey is presented of mechanisms of synaptic transmission in both the peripheral and central nervous system, with particular emphasis on the identity of various excitatory and inhibitory chemical transmitters, and their specific modes of action. [The *SCI®* indicates that this paper has been cited in over 910 publications since 1974.]

K. Krnjević
Departments of Physiology
and
Research in Anaesthesia
McGill University
Montreal, Quebec H3G 1Y6
Canada

April 11, 1984

"The idea that nerve cells might communicate with each other by releasing specific chemicals with excitatory or inhibitory properties can be traced back to the late nineteenth century. But it began to be taken really seriously only about 50 years ago, when Dale[1] and his colleagues, in England, first obtained strong evidence that acetylcholine was the transmitter of excitation at nerve-muscle and ganglionic junctions. Like all new concepts, it met with a great deal of resistance. Most of the opposition came from electrophysiologists who were quite happy — indeed much preferred — to think of the nervous system (especially the brain) in terms of purely electrical signals. They did not want to have to be bothered with chemistry and drugs — what was often referred to as 'soup physiology.'

"At the beginning of the 1950s, however, Eccles,[2] who had been one of the most prominent opponents, became an equally vocal advocate of chemical transmission at central synapses. His experiments on the spinal cord started a new search for possible transmitter substances. By more precise electrophysiological techniques, it became possible to release minute amounts of various substances from microelectrodes inserted into the brain and spinal cord and thus examine their effects on individual nerve cells. Numerous such 'iontophoretic' studies revealed widespread powerful and rapid (and therefore transmitter-like) actions of some well-known naturally occurring amino acids (glutamate, aspartate, gamma-aminobutyric, and glycine). These findings came as a great surprise. It had been expected that peripheral transmitters, such as acetylcholine and noradrenaline, would have a prominent role in the central nervous system (CNS). Amino acids, with well-identified roles in cellular metabolism, protein synthesis, etc. — and found abundantly throughout the brain — did not fit then current notions about transmitters. By the end of the 1960s, however, in the absence of any serious alternative, the amino acids were generally becoming accepted as the most widespread rapidly acting excitatory and inhibitory transmitters in the CNS.

"The invitation to write a review on this topic — in which I had been much involved for some years — thus came at an opportune time. What success the review has had I would ascribe to its convenient summarizing of an already vast literature (it achieved some notoriety by listing over 1,200 references), as well as to the fact that the field had reached a certain level of maturity — the status of a 'normal science.' This was before the neuropeptides burst upon the scene, initiating another revolution — but this is another story (cf. D.T. Krieger[3])."

1. **Dale H H.** Acetylcholine as a chemical transmitter of the effects of nerve impulses. *J. Mt. Sinai Hosp.* 4:401-29, 1938.
2. **Eccles J C.** *The neurophysiological basis of mind.* Oxford: Clarendon Press, 1953. 314 p.
3. **Krieger D T.** Brain peptides—what, where, and why. *Science* 222:975-85, 1983.

Jouvet M. Biogenic amines and the states of sleep. *Science* **163**:32-41, 1969.
[Department of Experimental Medicine, School of Medicine, Lyon, France]

The inhibition of the biosynthesis of 5-hydroxy-tryptamine (5 HT) by *p*-chlorophenylalanine is followed by a total insomnia which can be reversed into physiological sleep by a secondary injection of 5-hydroxytryptophan, the direct precursor of 5 HT. The destruction of 5 HT-containing neurons of the raphe system is also followed by the suppression of sleep. These results suggest a relationship between brain 5 HT and sleep. [The *SCI®* indicates that this paper has been cited in over 690 publications since 1969.]

M. Jouvet
Department of Experimental Medicine
Claude Bernard University
69373 Lyon
France

February 11, 1983

"When this paper was published in the first issue of *Science* for 1969, there was an explosive growth of sleep research which was due to the changes of the paradigm concerning the mechanisms of sleep. It has been admitted only recently that sleep is an active phenomenon, which contradicts the long held belief that sleep is a passive state of the waking system. Moreover, two different states alternating periodically during behavioural sleep had been discovered about ten years before—slow-wave sleep (SWS) and paradoxical sleep (PS) or REM sleep.[1] At first, the sleep mechanisms were studied with the so-called 'dry neurophysiology' (microelectrode recordings, stimulation, etc.). However, the time constant of the rebound of PS which may last for weeks could not be explained by classical synaptic physiology. It was then time for pharmacology.

"A first step in the direction of monoamines had been taken in 1964 by Jungi Matsumoto from Tokushima in my laboratory in the department of experimental medicine in Lyon. He demonstrated that after pretreatment with reserpine, a monoamine depletor drug, 5-hydroxytryptophan (5 HTP) or dopa could restore SWS and PS respectively.[2] In 1966, thanks to Weissman's courtesy, I obtained a small amount of

p-chlorophenylalanine. This drug was most interesting since it could suppress the biosynthesis of serotonin (5 HT) by inhibiting tryptophan hydroxylase[3] and induce a total insomnia. However, secondary injection of 5 HTP could still restore 5 HT and sleep.

"At about that time, the pharmacology of monoamines acquired some anatomical dimension thanks to histofluorescence and the mapping of monoaminergic neurones by Dahlström and Fuxe.[4] Thus, it became possible to directly attack 5 HT neurons by lesion of the entire raphe system. These lesions could suppress SWS or PS and biochemically decrease 5 HT in the brain. This was the ground for the monoaminergic theory of sleep which was first described in this paper and was further developed in a long paper in 1972.[5] At first, this theory could predict the effect of drugs acting upon monoamines in the sleep-waking cycle. However, since that time, some new experimental evidence seems to contradict the hypothesis that 5 HT is a sleep neurotransmitter. On the contrary, 5 HT neurons are more active during waking than during sleep and 5 HT release is increased during waking. These findings led to the hypothesis that during waking, 5 HT could act both as a transmitter and a neurohormone in inducing the synthesis and/or the liberation of some sleep factor(s). This hypothesis—and the discovery of numerous putative sleep factors—returns the sleep mechanism back to Pieron sleep hypnotoxin.

"I think that this paper has been cited so often for several reasons. First, it was one of the first reviews describing in detail PS and its ontogeny and phylogeny. Secondly, the states of sleep were described in quantitative terms which could be used as dependent variables for any pharmacological or surgical interventions. Thirdly, multidisciplinary approaches (polygraphic, pharmacological, biochemical, histochemical) were followed in altering 5 HT metabolism and quantitative data could correlate the amount of alteration of a biochemical system in the brain with a physiological state."

1. **Jouvet M.** Recherches sur les structures nerveuses et les mécanismes responsables des différentes phases du sommeil physiologique. *Arch. Ital. Biol.* **100**:125-206, 1962.
2. **Matsumoto J & Jouvet M.** Effets de réserpine, DOPA et 5 HTP sur les deux états de sommeil. *C.R. Soc. Biol.* **158**:2137-40, 1964.
   [The *SCI* indicates that this paper has been cited in over 80 publications since 1964.]
3. **Koe B K & Weissman A.** *p*-Chlorophenylalanine, a specific depletor of brain serotonin. *J. Pharmacol. Exp. Ther.* **154**:499-516, 1966.
   [The *SCI* indicates that this paper has been cited in over 1,270 publications since 1966.]
4. **Dahlström A & Fuxe K.** Evidence for the existence of monoamine-containing neurons in the central nervous system.
   1. Demonstration of monoamines in the cell bodies of brain stem neurons. *Acta Physiol. Scand.* **62**(Suppl. 232):5-55, 1964.
5. **Jouvet M.** The role of monoamine and acetylcholine-containing neurons in the regulation of the sleep-waking cycles. *Ergebnisse Physiol.* **64**:166-307, 1972.
   [The *SCI* indicates that this paper has been cited in over 490 publications since 1972.]

Koella W P, Feldstein A & Czicman J S. The effect of *para*-chlorophenylalanine
on the sleep of cats. *Electroencephalogr. Clin. Neuro.* 25:481-90, 1968.
[Worcester Foundation for Experimental Biology, Shrewsbury, MA]

In cats chronically prepared for EEG, EMG, and EOG recordings, p-chlorophenylalanine (PCPA) reduces in a dose-dependent manner sleep (SWS and PS about equally) and brain 5-HT. Sleep and 5-HT return toward control only slowly but not in a parallel fashion, suggesting the involvement of a negative feedback link. [The *SCI®* indicates that this paper has been cited over 125 times since 1968.]

Werner P. Koella
Division of Pharmaceuticals
Research Laboratories
CIBA-GEIGY AG
CH-4002 Basel
Switzerland

August 31, 1981

"Early in my career I worked with W.R. Hess, in Zürich. I learned from him that sleep is not a passive 'falling into Morpheus's arms' but rather an actively induced state. So a good part of my research activities was directed toward finding the central nervous structures that control sleep. After some early 'dry' attempts to pinpoint these structures, I became interested, in the late-1950s (at the Worcester Foundation), in serotonin which Brodie and Shore[1] assumed to be *the* central controlling instrument of trophotropic-endophylactic activities. In 1960, I demonstrated that intracarotid 5-HT exerts a modulatory influence on evoked potentials.[2] With Czicman,[3] I showed that cats react to 5-HT (intracarotid) with a transient arousal response (with mydriasis) followed by long-lasting

slowing of the EEG (with miosis) and that injection of small amounts of 5-HT into the fourth ventricle induces EEG slowing and miosis.

"What I aimed at then was the reversal experiment, namely, evidence that sleep would be reduced if 5-HT was taken away from the brain; reserpine was not specific enough to do this job. So it came as a godsend when Koe and Weissman[4] discovered p-chlorophenyl-alanine (PCPA) that allowed (almost) specific depletion of brain 5-HT. Late in 1966, we began a new, rather broad study on a dose-response basis, with long (up to four weeks) single experiments, on a large number of cats, including parallel biochemical probes (done by Aaron Feldstein). In that study we could show indeed that one shot of PCPA in a dose-dependent manner reduces sleep for several weeks *and* brain 5-HT levels. Because of the 'puristic' way in which we performed the experiments we were—unfortunately or fortunately— late with the publication of our results. But it was worthwhile waiting. We had results allowing good and reliable analysis of the quantitative and temporal relations between brain 5-HT levels and sleep (this paper). At that time we were convinced that 5-HT is important for sleep although it is also involved in other jobs such as pain and temperature regulation, drug habituation, and, peripherally, gut activity, blood clotting, and pupillary adjustment.

"Why would a paper like this one be cited so often? Perhaps because it is a manifestation of the 'endpoint' of a sequential and logical flow of ideas and experimental results; or, because it added by that time to our understanding of the organization and regulation of sleep; or, because by its very results it offered new possibilities for the development of new (and so badly needed) hypnotics. I have recently published in this field."[5]

1. Brodie B B & Shore P A. A concept for a role of serotonin and norepinephrine as chemical
   mediators in the brain. *Ann. NY Acad. Sci.* 66:631-42, 1957.
2. Koella W P, Smythies J R, Bull D M & Levy C K. Physiological fractionation of the effect of serotonin on
   evoked potentials. *Amer. J. Physiol.* 198:205-12, 1960.
3. Koella W P & Czicman J. Mechanism of the EEG-synchronizing action of serotonin.
   Amer. J. Physiol. 211:926-34, 1966.
4. Koe B K & Weissman A. p-Chlorophenylalanine: a specific depletor of brain serotonin.
   J. Pharmacol. Exp. Ther. 154:499-516, 1966.
5. Koella W P. Neurotransmitters and sleep. (Wheatley D, ed.) *Psychopharmacology of sleep.*
   New York: Raven Press, 1981. p. 19-52.

Domino E F, Chodoff P & Corssen G. Pharmacologic effects of CI-581, a new dissociative anesthetic, in man. *Clin. Pharmacol. Ther.* 6:279-91, 1965.
[Depts. Pharmacology and Anesthesiology, Univ. Michigan Medical Center, Ann Arbor, MI]

The first clinical pharmacological effects of ketamine in human volunteers are described in this report. This drug has a unique spectrum of actions including analgesia, anesthesia, cardiovascular stimulation, and only minimal respiratory depression. Recovery from anesthesia is moderately rapid. Its use as an intravenous anesthetic is recommended. [The *SCI®* indicates that this paper has been cited in over 290 publications since 1965.]

Edward F. Domino
Department of Pharmacology
Medical School
University of Michigan
Ann Arbor, MI 48109-0010

May 11, 1984

"It is interesting how during one's early professional career certain events shape subsequent research interests and endeavors. My involvement with the initial research on the clinical pharmacologic effects of CI-581 in man, now better known as ketamine, is certainly such an example for me. Twenty years ago a series of events culminated in a request by Alex Lane, then at Parke Davis, that I study the clinical pharmacology of a new chemical which had never been given to human beings. With my colleagues, Pete Chodoff and Gunther Corssen, and the collaboration of many colleagues at Parke Davis, including Graham Chen and Duncan McCarthy, we were able to study this unique new potential intravenous anesthetic. None of us shall ever forget the amazing spectrum of clinical pharmacological effects that this agent produced in the volunteers we studied. So unique were these effects that we had to invent a new set of words to describe its anesthetic properties. The drug produced 'zombies' who were totally disconnected from their environment, with their eyes open, and yet in a complete anesthetic and analgesic state. The observation of being disconnected from the environment gave rise to the term 'dissociative anesthesia.'

"It is of interest that August 3, 1984, will be the twentieth anniversary of the administration of ketamine to human beings. It remains a unique and safe anesthetic agent. However, its major problem in humans is an emergence delirium which this first study clearly described. The reason our paper has been cited frequently over the years is because this study with ketamine was the first of its kind. Those early events in my research career have caused me to return again and again to study ketamine and other arylcyclohexylamines.[1,2] Since ketamine has some actions clearly related to phencyclidine, we have tried to find ways to reduce the 'bad effects' of ketamine—or to 'tame the tiger' with diazepam premedication.[3] I remain convinced that some day new ketamine analogues will be synthesized which will be much more useful than ketamine itself.

"Perhaps the most exciting new development in the 1980s has been definitive evidence that there are mammalian brain receptors for phencyclidine, ketamine, and other so-called *sigma* receptor opioid agonists.[4,5] Even more exciting, there are now a number of endogenous peptides that have *sigma* receptor agonist actions.[6] What are these endogenous peptides doing in the brain? Why should mammalian organisms have specific brain receptors for arylcyclohexylamines? Ketamine and phencyclidine have had an indelible influence on my professional life for they have stimulated many collaborative studies with medical chemists, anesthesiologists, psychiatrists, and psychologists. I even have a few friends in the pharmaceutical industry who, although not convinced to invest their company's research dollars in new ketamine analogues, still tolerate that 'dissociative' pharmacologist who just might have a good idea in attempting to find an even better ketamine, perhaps an endogenous one."

1. Domino E F, ed. *(PCP) phencyclidine: historical and current perspectives.* Ann Arbor, MI: NPP Books, 1981. 537 p.
2. Kamenka J M, Domino E F & Geneste P, eds. *Phencyclidine and related arylcyclohexylamines.* Ann Arbor, MI: NPP Books, 1983. 690 p.
3. Zsigmond E & Domino E F. Ketamine—clinical pharmacology, pharmacokinetics and current clinical uses. *Anesthesiol. Rev.* 7:13-33, 1980.
4. Vincent J P, Kartalovsky B, Geneste P, Kamenka J M & Lazdunski M. Interaction of phencyclidine ('angel dust') with a specific receptor in rat brain membranes. *Proc. Nat. Acad. Sci. US* 76:4678-82, 1979.
5. Zukin S R & Zukin R S. Specific [³H] phencyclidine binding in rat central nervous system. *Proc. Nat. Acad. Sci. US* 76:5372-6, 1979. (Cited 90 times.)
6. Quirion R, O'Donohue T L, Everist H, Pert A & Pert C B. Phencyclidine receptors and possible existence of an endogenous ligand. (Kamenka J M, Domino E F & Geneste P, eds.) *Phencyclidine and related arylcyclohexylamines.* Ann Arbor, MI: NPP Books, 1983. p. 667-83.

This Week's Citation Classic

Seeman P. The membrane actions of anesthetics and tranquilizers.
Pharmacol. Rev. 24:583-655, 1972.
[Pharmacology Department, University of Toronto, Canada]

This review integrated data to create the concept that membranes are expanded and fluidized by lipid-soluble drugs, altering many membrane functions. These nonspecific anesthetic-like effects occur whenever the drug is approximately ten millimolar *in the membrane phase*, a value obtained directly from the drug's partition coefficient. This rule helps to distinguish between receptor and non-receptor mechanisms for membrane-active drugs. [The *SCI®* indicates that this paper has been cited in over 1,300 publications since 1972.]

Philip Seeman
Department of Pharmacology
Faculty of Medicine
University of Toronto
Toronto M5S 1A8
Canada

December 8, 1982

" 'How does one decide whether a membrane-active drug acts specifically on receptors or nonspecifically on the membrane? For instance, do tranquilizers (now called neuroleptics) or all anesthetics act on receptors or do they act nonspecifically to depress membrane excitability indirectly?' These questions preoccupied me while I was a graduate student at Rockefeller University in New York City between 1961 and 1966. My wife, Mary, was at that time a psychiatric resident at Manhattan State Hospital, working with many patients whose diagnosis was schizophrenia. This was my motivation to try to understand the molecular mechanism for the antipsychotic action of neuroleptic drugs. I felt that, if I could understand how the drugs exerted their effects, it would provide me with a research strategy for studying the presumably abnormal brain chemistry of schizophrenia. I first wanted to determine whether the site of neuroleptic action was specific or nonspecific. Knowing this, I could then go on to study it in the mentally disordered brain.

"I started by testing the effects of neuroleptics on erythrocyte membranes. As had been previously noted by others,[1] these drugs inhibited hemolysis. I soon found that virtually all lipid-soluble drugs, including anesthetics, alcohols, etc., protected erythrocytes from osmotic hemolysis by expanding the area of the membrane by only one to two percent.

"Membrane expansion of the kind I found should theoretically loosen, fluidize, or disorder the membrane constituents. Such an alcohol-induced fluidization of the membrane was discovered by Jim Metcalfe, Arnold Burgen, and myself[2] when I was a postdoctoral fellow at the University of Cambridge in 1966.

"Returning to Canada in 1967, I started measuring the binding of radioactive neuroleptics and anesthetics to membranes in order to establish a relation between membrane effect and membrane occupancy by the drug. Working with Wim Kwant and Sheldon Roth, we found that these and other lipid-soluble drugs all expanded and fluidized membranes whenever the drug attained a molarity *within the membrane phase* of about ten millimolar. This universal rule had been predicted by Meyer and Overton in 1901.[3,4] They had worked with olive oil as a model for the membrane.

"Since most biologists use lipid-soluble drugs on their systems, this review article has served as a convenient reference on the molecular mechanisms for the nonspecific membrane actions of anesthetics and other lipid-soluble drugs. The importance of the review is that it correlates the membrane effects with drug concentrations. This entailed a lengthy and detailed examination of the anesthesia literature since 1896, a voluminous task done on weekends amidst ice hockey and touch football with my three sons. The further-reaching importance is that the review prompted the subsequent discovery that alcohol-tolerant tissues have membranes which are more resistant to fluidization by ethanol.[5,6]

"The review shows how to calculate the membrane concentration for a particular drug from the partition coefficient. If the drug is membrane-active at much less than ten millimolar *within the membrane*, then a specific membrane receptor mechanism must be postulated. It was this result which subsequently led me to determine directly that dopamine receptors were the primary sites of neuroleptic action."[7]

1. **Halpern B N & Bessis M.** Action anti-sphérocytaire de certains corps synthétiques dérivés de la phénothiazine. *C.R. Soc. Biol.* 144:759-60, 1950.
2. **Metcalfe J, Seeman P & Burgen A S V.** The proton relaxation of benzyl alcohol in erythrocyte membranes. *Mol. Pharmacol.* 4:87-95, 1968.
3. **Meyer H.** Zur Theorie der Alkolnarkose. Der Einfluss wechselnder Temperatur auf Wirkungsstärke und Theilungscoefficient der Narcotica. *Naunyn-Schmied. Arch. Exp. Pathol. Pharmakol.* 46:338-46, 1901.
4. **Overton E.** *Studien über die Narkose zugleich ein Beitrag zur allgemeinen Pharmakologie.* Jena: Verlag von Gustav Fischer, 1901. 195 p.
5. **Curran M & Seeman P.** Alcohol tolerance in a cholinergic nerve terminal. Relation to the membrane expansion-fluidization theory of ethanol action. *Science* 197:910-11, 1977.
6. **Chin J H & Goldstein D B.** Drug tolerance in biomembranes: a spin label study of the effects of ethanol. *Science* 196:684-5, 1977.
7. **Seeman P, Wong M & Tedesco J.** Tranquilizer receptors in rat striatum. *Neurosci. Abstr.* 1:405, 1975.

Albe-Fessard D & Rougeul A. Activités d'origine somesthésique évoquées sur le cortex non-spécifique du chat anesthésié au chloralose: rôle du centre médian du thalamus. *Electroencephalogr. Clin. Neuro.* 10:131-52, 1958.
[Centre d'Études de Physiologie Nerveuse du Centre National de la Recherche Scientifique, Paris, France]

In chloralose-anaesthetized cats, stimulation of different body areas evokes bilateral positive convergent activities in cortical localised foci. Messages to these foci do not relay in the primary thalamic or cortical zones. The centre median (CM) of the thalamus exhibits similar convergent evoked responses and its stimulation produces responses in the cortical foci; the CM may be a relay to these foci. [The *Science Citation Index®* (*SCI®*) and the *Social Sciences Citation Index®* (*SSCI®*) indicate that this paper has been cited over 130 times since 1961.]

D. Albe-Fessard
Laboratoire de Physiologie des
Centres Nerveux
Université Pierre et Marie Curie
75230 Paris
France

December 15, 1981

"My first main interest, the organization of electric fish discharges, led me to use intracellular recordings just after this technique was developed by Eccles and his associates.[1] With the same technique, I later studied with P. Buser the activities evoked in the mammalian cortex cells. To pursue this work I needed to delimitate the primary areas in the cat anesthetized with chloralose, the anaesthetic agent used in Europe at that time. I performed this mapping out with A. Rougeul. One day, by mistake, we searched for the evoked activity not on the cortex contralateral to the stimulated limb, but on the ipsilateral. We obtained large responses and realized that in some cortical foci similar responses appear whatever the side and the site of peripheral stimulation. We called these foci convergent because each of them was activated from extensive body areas (see also

reference 2). Ablation techniques had shown us that the messages to these cortical foci do not need the integrity of primary thalamo-cortical projection. A thalamic convergent relay was thus to be found. We searched for it, using deep bipolar macroelectrodes mapping in medialis dorsalis, because this nucleus was thought in the 1950s to be the important associative thalamic nucleus. It was not there but underneath, in the CM, that we observed large convergent evoked responses resembling those in cortical foci. Convergences were also found at the unitary level using glass microelectrodes.

"I see two main reasons to explain why, in spite of this paper having been written in French, it was widely read and quoted. First, the evoked activities we described in the medial thalamus were not observed by workers using barbiturate anaesthesia.[3] In fact, under barbiturate anaesthesia, thalamic and cortical convergent responses are reduced or disappear. Thus, in spite of the fact that CM evoked responses were long before described by Magoun and McKinley,[4] our results were considered to be a chloralose artifact, and work was done by others[5] to verify our findings and by us to understand the difference in action of the anaesthetic agents.[6] In particular, from that time on, as a way of avoiding criticism, I used to frequently record in chronically awake animals. Second, the medial thalamic region we recorded from, and called CM in the cat, relying on the Jasper and Ajmone-Marsan atlas,[7] is made in fact of different nuclei, among them CM. However, the name CM taken from human anatomy was denied to this region by some anatomists and W. Mehler[8] in particular, who, on a cytoarchitectonic basis, called it CL. Our publication was the beginning of a long, friendly dispute with him and the anatomists who were searching for a relay of painful messages and had found a termination of spinothalamic pathways in the medial thalamus of man and monkeys. As a consequence, I became involved in research on pain pathways, work in which I am still active."

1. **Brock L G, Coombs J S & Eccles J C.** The recordings of potentials from motoneurones with an intracellular electrode. *J. Physiology* 117:431-60, 1952.
2. **Amassian V E.** Studies on organization of a somesthetic association area, including a single unit analysis. *J. Neurophysiology* 17:39-58, 1954.
3. **Mountcastle V & Henneman E.** Pattern of tactile representation in thalamus of cat. *J. Neurophysiology* 12:85-100, 1949.
4. **Magoun H W Y & McKinley W A.** The termination of ascending trigeminal and spinal tracts in the thalamus of the cat. *Amer. J. Physiol.* 137:409-16, 1942.
5. **Thompson R F, Johnson R H & Hoopes J J.** Organization of auditory, somatic sensory, and visual projection to association fields of cerebral cortex in the cat. *J. Neurophysiology* 26:343-64, 1963.
6. **Albe-Fessard D.** Organization of somatic central projections. (Neff W D, ed.) *Contributions to sensory physiology.* New York: Academic Press, 1967. Vol. 2. p. 101-67.
7. **Jasper H H & Ajmone-Marsan C.** *A stereotaxic atlas of the diencephalon of the cat.* Ottawa, Canada: National Research Council, 1954. 15 p.
8. **Mehler W R.** Further notes on the center median nucleus of Luys. (Purpura D P & Yahr M D, eds.) *The thalamus.* New York: Columbia University Press, 1966. p. 109-22.

CC/NUMBER 23
JUNE 7, 1982

Melzack R & Wall P D. Pain mechanisms: a new theory. *Science* 150:971-9, 1965.
[Dept. Psychol., McGill Univ., Montreal, Canada and Dept. Biol., Massachusetts Inst. Technol., Cambridge, MA]

The theory proposes that the dorsal horn of the spinal cord acts like a gate which modulates the flow of nerve impulses from the peripheral fibers to the central nervous system. The gate is influenced by peripheral fiber activity and by descending influences from the brain. [The *Science Citation Index®* (*SCI®*) and the *Social Sciences Citation Index®* (*SSCI®*) indicate that this paper has been cited over 975 times since 1965.]

Ronald Melzack
Department of Psychology
McGill University
Montreal, Quebec H3A 1B1
Canada
and
Patrick D. Wall
Department of Anatomy
University College London
London WC1E 6BT
England

February 15, 1982

"When we proposed the gate-control theory in 1965, we hardly expected the astonishing increase in research studies and new therapeutic approaches that was stimulated by it. Fortunately, the theory came at a time when the field was ripe for change. A small number of original thinkers had fought hard to replace the old concept of a specific pain pathway by a more dynamic conception in which pain is determined by many factors in addition to injury—by past experiences, culture, attention, and other activities in the nervous system at the time of injury. This small band of courageous people hammered away at the established, traditional theory. But despite occasional lip service to their ingenuity, the field continued unchanged, holding tenaciously on to Descartes' idea, proposed in 1664, that pain is like a bell-ringing alarm system whose sole purpose is to signal injury to the body.

"In the 1960s, a wave of new facts and ideas that had evolved gradually[1-3] was beginning to crest, and the gate-control theory rode in on the wave of the times. No one was more astounded at its success than we were. Naturally, acceptance was not immediate or total, but in spite of continuing controversy about details, the concept that injury-signals can be radically modified and even blocked at the earliest stages of transmission in the nervous system is now virtually universally accepted. A fortunate aspect of our publication in 1965 is the use of the phrase 'gate control.' It evokes an image that is readily understood even by those who do not grasp the complex physiological mechanisms on which the theory is based. The fact that the theory had relevance to a wide variety of fields in medicine, psychology, and biology also led to its frequent citation.

"The theory's emphasis on multiple determinants of pain has also provided a conceptual framework for the recent recognition of the complexity of clinical pain. Until the middle of this century, pain was considered primarily to be a symptom of disease or injury. We now know that chronic, severe pain is a problem in its own right that is often more debilitating and intolerable than the disease process which initiated it. The problem of pain has therefore been transformed from a mere symptom to be dealt with by the various medical specialties to a specialty in its own right which is now one of the most exciting, rapidly advancing fields of science and medicine. Happily, the main beneficiary has been the suffering person. The new concept provided the foundation and framework for a host of novel, exciting approaches to the treatment of pain. For a recent survey the reader can refer to J.J. Bonica et al."[4]

1. Melzack R, Stotler W A & Livingston W K. Effects of discrete brainstem lesions in cats on perception of noxious stimulation. *J. Neurophysiology* 21:353-67, 1958.
2. Melzack R & Wall P D. In the nature of cutaneous sensory mechanisms. *Brain* 85:331-56, 1962.
3. Wall P D. Presynaptic control of impulses at the first central synapse in the cutaneous pathway. *Progr. Brain Res.* 12:92-118, 1964.
4. Bonica J J, Liebeskind J C & Albe-Fessard D. *Advances in pain research and therapy.* New York: Raven Press, 1979. Vol. 3. 956 p.

Hughes J, Smith T W, Kosterlitz H W, Fothergill L A, Morgan B A & Morris H R.
Identification of two related pentapeptides from the brain with potent opiate
agonist activity. *Nature* 258:577-9, 1975. [Unit Res. Addictive Drugs and Dept.
Biochem., Univ. Aberdeen, Scotland; Pharmaceutical Div., Reckitt and
Colman Ltd., Hull; and Dept. Biochem., Imperial Coll., London, England]

This paper describes the structure, chemical synthesis, and actions of two endogenous opioid peptides, methionine-enkephalin and leucine-enkephalin, from pig brain. It also notes the sequence homology between methionine-enkephalin and the pituitary hormone β-lipotropin. [The *SCI®* indicates that this paper has been cited in over 1,460 publications since 1975.]

John Hughes
Department of Biochemistry
Imperial College of Science and
Technology
London SW7 2AZ
England

April 27, 1982

"As a lecturer in the newly formed pharmacology department at the University of Aberdeen, my research centred on adrenergic release mechanisms. My chairman, Hans Kosterlitz, worked on opiate modulation of acetylcholine release and quantitative aspects of opiate receptor interactions. We shared a common interest in neuromodulatory mechanisms and in 1972 our research interests converged with our discovery of opiate receptor mediated inhibition of adrenergic transmission in the mouse vas deferens.[1] The vas was to become, along with Hans's guinea-pig ileum preparation, a standard assay for opiate action; it also provided the means of testing an idea developed over many discussions about the function of opiate receptors. We reasoned that these receptors might form part of a neurochemical system subject to activation by a specific chemical signal. The effects of morphine could then be viewed as mimicking the endogenous opiate ligand in the same way as nicotine mimics some actions of acetylcholine. The opportunity to test this hypothesis came on Hans's retirement in 1973 when he invited me to join him as deputy director in establishing a drug research unit. I had barely moved when Eric Simon,[2] Sol Snyder,[3] and Lars Terenius[4] demonstrated the existence of specific opiate binding sites. These findings provided additional support for our hypothesis.

"Serendipity plus acquired Scottish parsimony gave an early lead in October 1973. Before throwing out some 'unsuccessful' frozen extracts I retested them and this time obtained a small but positive response. The initial negative result was due to interfering nucleotides which had degraded on storage allowing the detection of the more stable enkephalin.

"By spring 1974 the peptide nature and properties of our material had been established and a paper was submitted[5] although editorial processes delayed this for a year. Meanwhile Lars, who had obtained similar positive results with his receptor binding technique, and I disclosed our findings at a Neurosciences Research Programme meeting in Boston. The cat was out of the bag and we knew that we could expect strong competition to identify the 'endogenous ligand.' Lars declined to participate in such a race and decided to concentrate on the clinical aspects of the discovery.

"By the following spring, Linda Fothergill had obtained sequence data that proved ambiguous. We surmised but could not prove that this was due to the presence of a second similar peptide. However, at a seminar I had given in Cambridge, I had met and discussed the problem with Howard Morris. I prepared a further 100 nmoles of material for Howard, who then used his elegant mass spectrometric technique to unequivocally identify both methionine- and leucine-enkephalin. The resulting paper marked the beginning of a vast research effort in neurobiology involving many scientific disciplines. This probably explains the paper's high citation rate.

"Hans and I have received a number of honours including the Lasker Prize, and Howard the BDH Gold Medal for this and other work on biological structures. We owed much to the excellence of our collaborators and to laboratory camaraderie which ensured that many a heated scientific argument was settled over a good malt, the endogenous Scottish ligand."

1. Henderson G, Hughes J & Kosterlitz H W. A new example of a morphine sensitive neuroeffector junction: adrenergic transmission in the mouse vas deferens. *Brit. J. Pharmacol.* 46:764-6, 1972.
2. Simon E J, Hiller J M & Edelman I. Stereospecific binding of the potent narcotic analgesic [³H]etorphine to rat-brain homogenate. *Proc. Nat. Acad. Sci. US* 70:1947-9, 1973.
3. Pert C B & Snyder S H. Opiate receptor: demonstration in nervous tissue. *Science* 179:1011-14, 1973.
4. Terenius L. Stereospecific interaction between narcotic analgesics and a synaptic plasma membrane fraction of rat cerebral cortex. *Acta Pharmacol. Toxicol.* 32:317-20, 1973.
5. Hughes J. Isolation of an endogenous compound from the brain with pharmacological properties similar to morphine. *Brain Res.* 88:295-308, 1975.

Brown K T. The electroretinogram: its components and their origins.
Vision Res. **8**: 633-77, 1968.

This paper summarizes the author's microelectrode studies of the ERG (electroretinogram) in the vertebrate retina. The resulting identification of ERG components, and their cellular origins, has provided a basis for new applications of the ERG to both physiological and clinical problems. [The *SCI®* indicates that this paper has been cited 153 times since 1968.]

Kenneth T. Brown
Department of Physiology
University of California
San Francisco, California 94143

February 27, 1978

"The ERG (electroretinogram) has always offered the advantages of a readily recorded electrical activity of the retina that can tell much about retinal functions. But to fully exploit these advantages, it is necessary to analyze the ERG into its components and to identify the cells that generate each component. It appeared that microelectrode techniques would be required to clarify this subject, but early results in lower vertebrates had yielded conflicting interpretations. I began studying these problems in 1955, when they seemed especially crucial to further progress in retinal physiology. This work was initiated at The Johns Hopkins School of Medicine with the collaboration of Torsten Wiesel. After moving to San Francisco in 1958, my collaborators were Kyoji Tasaki, Kosuke Watanabe, Motohiko Murakami, Geoffrey Arden, and Peter Gage.

"This work proceeded by a series of steps. Techniques were first developed for using microelectrodes within the retina of an intact cat eye. A variety of methods was then used to analyze the ERG into 4 major components, and the amplitude of each component was plotted as a function of electrode depth in the retina. One component was thus shown to be generated by the pigment epithelium, while another was generated by the receptors, and the remaining two were from second-order cells of the inner nuclear layer.

"In San Francisco the microelectrode techniques were further developed and applied to the macaque monkey. Then in 1962 the receptor component of the ERG was successfully isolated by selectively clamping the retinal circulation and by using mild light adaptation to remove the pigment epithelial component. This provided the first clear isolation and identification of receptor potentials in the vertebrate retina. Beginning in 1962, study of these isolated receptor potentials also showed that cone and rod responses have characteristically different time courses, a fundamental aspect of the duplicity theory. In 1964, a very rapid light-evoked receptor response was then discovered. This was called the 'early receptor potential' and was proved to be generated by the visual photopigment. In 1965, similar rapid responses were shown to be generated when light was absorbed by melanin in the retinal pigment epithelium.

"The paper discussed here represented a final review and summary of the work just described. I believe that the frequent citing of this paper results from its summarization of work providing both techniques and results that have proved useful in a variety of subsequent studies. Some of this further work has concerned retinal physiology, especially that of the photoreceptors and pigment epithelium, while other work has applied the ERG to the diagnosis of retinal disorders."

Cigánek L. The EEG response (evoked potential) to light stimulus in man.
Electroencephalogr. Clin. Neuro. **13**:165-72, 1961.
[Dept. Clinical Electrophysiology, Inst. Experimental Medicine of the Slovak
Academy of Sciences, Bratislava, Czechoslovakia]

The visual evoked potential is a very small bioelectric potential complex which can be registered, by means of sophisticated methods, on the human scalp as a response of the brain to a light flash. It represents an objective, biophysical correlate of the visual afferentation and perception. [The *Science Citation Index®* (*SCI®*) and the *Social Sciences Citation Index®* (*SSCI®*) indicate that this paper has been cited over 280 times since 1961.]

L. Cigánek
1st Neurologic Clinic
Academician L. Dérer's Hospital
Comenius University Medical School
809 46 Bratislava
Czechoslovakia

November 25, 1981

"In human neurophysiology it has always been a most exciting problem to search for objective correlates of the subjectively experienced states. Visual sensations are such states and visual evoked potentials (VEPs) represent their objective, biophysical correlates. The work presented in the cited publication was inspired by my teacher, Professor Černáček. The task was not easy. The amplitudes of the VEPs ranged within microvolt order and their duration was only about 300 ms. Moreover, they had to be detected in the 'noise' of the continually ongoing background electroencephalographic activity with amplitudes of tens to hundreds of microvolts. Today this work is done in a most sophisticated way by means of computers (we are also going this way now), but from 1950 to 1960 this technique was not yet available. We used the photographic superimposition method introduced by Dawson in 1947[1] and thousands of measurements and computations realized by hand by Vera,

my wife and technician, who supported me in my work in any way. She was perhaps less efficient than the modern computers but surely much more reliable.

"The reason for the frequent citation seems to be quite clear. The publication presented a rounded knowledge about the basic parameters of the VEPs at a time when little was known in this field. In the carefully collected bibliography there were only six articles dealing directly with the VEPs studied accidentally in limited groups of subjects while the cited publication represented a systematic study of a rather large group of normal subjects. In the same year the same results were published *in extenso* in a German monograph,[2] the first monograph about the evoked potentials (EPs) in the literature at all. Later, throughout the world, a tremendous surge of work in the field of EPs brought extensive knowledge of the physiology and pathophysiology of the human brain and even knowledge about the mechanisms of the mind. The work is still going on and the discoveries of the last years also represent an important contribution to the clinical diagnosis of nervous diseases. In 1973, I was invited to read a didactic lecture about VEPs at the Eighth International Congress of Electroencephalography and Clinical Neurophysiology in Marseille, and later I was asked to write a chapter on the same topic for the first *Handbook of Electroencephalography and Clinical Neurophysiology*, which appeared in 1975.[3]

"It is a good feeling now to be a member of the extensive intellectual community of scientists working in this field. Many of them are my good friends. We meet time after time at congresses and symposia and visit one another in our laboratories, believing that future research into EPs will still teach us more about the way in which the human brain works. To the best of our knowledge there are no unknowable things, only those which are not yet known."

1. **Dawson G D.** Central responses to electrical stimulation of peripheral nerve in man.
   J. Neurol. Neurosurg. Psychiat. **10**:137-40, 1947.
2. **Cigánek L.** *Die elektroencephalographische Lichtreizantwort der menschlichen Hirnrinde.*
   Bratislava: Publishing House of the Slovak Academy of Sciences, 1961. 152 p.
3. ------------. Visual evoked responses. (Rémond A, ed.) *Handbook of electroencephalography and clinical neurophysiology.* Amsterdam: Elsevier, 1975. p. 8A-33-8A-59.

Citation Classics

Selye H. The general adaptation syndrome and the diseases of adaptation.
Journal of Clinical Endocrinology 6:117-231, 1946.

The general adaptation syndrome is defined as the sum of all non-specific, systemic reactions of the body which ensue upon long continued exposure to stress. The paper calls attention to the possible connection between the adaptation syndrome and various diseases. If this linkage can be proven, the author contends, then it follows that some of the most common fatal diseases of man are due to a breakdown of the hormonal adaptation mechanism. [The SCI® indicates that this paper was cited 167 times in the period 1961-1975.]

Hans Selye, C.C., M.D., Ph.D., D. Sc.
Universite de Montreal
Institut de Medecine
Case Postale 6128
Montreal 101, Canada
November 22, 1976

"It is gratifying to learn that my 1946 article has been so frequently cited, but it is even more encouraging to consider the progress in stress research since then. This article represented an attempt to review the status of the stress concept at the time. As such, it was both a continuation of my first tentative essays on this topic and a springboard for my later publications on the implications of stress in health and disease....

"I could not answer why it was this article, among the 1600 papers and 33 books I have published, that attracted the most attention. I suppose it might be because it gave the first holistic survey and description of the diseases of adaptation (or 'stress diseases'). Undoubtedly, it was also very helpful that an entire issue of *The Journal of Clinical Endocrinology* was devoted to my article, thanks to the editorial decision of Kenneth Thompson to print my paper as a whole. This was not my first nor, in my opinion, my most important publication; I would say that my book *The Stress of Life*[1] would be my most popular volume, having been written in 1956 and kept continuously in print unrevised--and translated into sixteen languages--until an updated edition was completed in 1976. But perhaps the success met by this 1946 article can be explained by the fact that it was the first attempt to survey the field

of stress in its entirety and, although much has been learned since then, every word in the paper still applies today.

"I wrote *The Stress of Life* in the belief that because the general public was becoming keenly aware of the role played by stress in their own lives, they would like to understand just what stress is and what it does to us. At the end of that volume I inserted a few philosophical musings on a code of behavior designed to meet and constructively deal with the stress of life.

"I went on to write another volume, *Stress Without Distress*,[2] in which I expanded what I had called a 'philosophy of gratitude' into a code of behavior named 'altruistic egoism' and based on the conviction that by *earning* our neighbor's love and becoming necessary to him, we can satisfy our own selfish needs while helping others. In this way we avoid creating interpersonal stress situations, and instead can make stress work *for* us. Looking back at the most cited 1946 article, I can see that the ideas that inspired this philosophy were already implied in it. Towards the end of that essay I wrote: 'Adaptation to our surroundings is one of the most important physiologic reactions in life; one might even go so far as to say that the capacity of adjustment to external stimuli is the most characteristic feature of live matter.'

"On a more personal note, it amuses me to see that even with all this progress since 1946 I am, in some ways, right back where I started. When I wrote that article I had just left McGill University to accept a position as Director of the University of Montreal's new Institute of Experimental Medicine and Surgery. I was putting together a documentation service and a laboratory, secretarial, and administrative staff that was to serve me well for the next thirty years in the same location. Today I am once again in the position of a beginner organizing a research, teaching, and library center, having just created the International Institute of Stress. But this time my scope is larger; I envision a network of institutions around the world to be affiliated with the Montreal headquarters. Still, my goals are the same: to help in the art of healing or, better, preventing disease, at the same time satisfying my insatiable curiosity about the mysteries of life and happiness."

1. Selye H. The stress of life. New York:McGraw-Hill, 1956.
2. ---------. Stress without distress. Philadelphia: Lippincott Co., 1974.

Mandler G & Sarason S B. A study of anxiety and learning.
J. Abnormal & Soc. Psychol. **47**:166-73, 1952.

A high anxiety (HA) and a low anxiety (LA) group were tested on two intelligence test tasks. LA subjects performed better than HA subjects; failure reports improved the performance of LA subjects, whereas no further reference to the test situation was optimal for the HA subjects. [The *Science Citation Index®* *(SCI®*) and the *Social Sciences Citation Index*™ *(SSCI*™*)* indicate that this paper was cited a total of 275 times in the period 1961-1977.]

George Mandler
Department of Psychology
University of California, San Diego
La Jolla, CA 92093

December 16, 1977

"This paper—the first of a series of four—was my initial serious effort in psychological theory and research. I was a second-year graduate student at Yale, working with Seymour Sarason, then a junior faculty member in clinical psychology. Sarason, whose work in mental retardation had led to an interest in the dynamics of the test situation, wanted to investigate the effect of anxiety on test performance. I was in the throes of fascination with Hullian theory (an infatuation soon abandoned and later vehemently rejected). Sarason evolved the idea of measuring individual differences in anxiety reactions to test situations and drafted a questionnaire (the Test Anxiety Questionnaire—TAQ). I wanted to apply drive theory to the expected results, and came upon the idea of distinguishing between task-relevant and task-irrelevant responses. Task-relevant responses arise out of the anxiety drive and reduce anxiety by leading to successful completion; the irrelevant responses are not specific to the task and interfere with the performance of complex tests.

"Sarason's highly supportive and encouraging attitude toward a fledgling psychologist was primarily responsible for two years of happy and intensive collaboration—and also my becoming first author on the initial paper.

"Our main interest was to demonstrate systematically in the laboratory an effect long known by teachers and students anecdotally, and to develop a useful theoretical framework. The result—as indicated by the frequency of citations— was that hundreds of studies used our questionnaire as well as the more influential Manifest Anxiety Scale developed by Janet Taylor Spence. In retrospect I am sure that neither Sarason, Spence, nor I was too happy with the indiscriminate use of our tests and the less than discriminate use of our theories. However, our studies did help to bring the investigation of complex human emotional and motivational phenomena into the laboratory.

"Years later—now a reformed cognitive psychologist and far from my behaviorist origins—I returned to the notion of cognitive interference stemming from preoccupations with irrelevant and anxiety-directed thought processes. Thus, in the long run the original purpose maintained its momentum. The anxiety scales are still being used, but hopefully in a more theoretical context. Sarason and I invented a shotgun—but it spread its effect wide and made some hits."

Rogers C R. The necessary and sufficient conditions of therapeutic personality change. *J. Consult. Clin. Psychol.* 21:95-103, 1957.

What conditions make possible constructive change in personality and behavior? Operational definitions are given of such conditions. These are: that the troubled client is in relationship with a therapist who is genuine, real; is experiencing an unconditional caring for, and a sensitive empathy for the client. When these conditions are experienced by the client, constructive change occurs. [The *Science Citation Index*® (*SCI*®) and the *Social Sciences Citation Index*™ (*SSCI*™) indicate that this paper was cited a total of 218 times in the period 1961-1977.]

Carl R. Rogers
Center for Studies of the Person
1125 Torrey Pines Road
La Jolla, California 92037

December 1, 1977

"I remember very well the comfortable new auditorium at the University of Michigan where I first presented this paper.

"I have a clear memory of the audience—academic psychologists, clinicians (both psychologists and psychiatrists), all psychoanalytically oriented. They were expectant and critical. I was at that time a controversial figure.

"I certainly remember my apprehensive feelings. I had been working for two years on a very concise and rigorous statement of client-centered theory.[1] This formulation was not an arm-chair theory. It was based on more than twenty-five years of experience and observation—my own and that of my students and colleagues. My aim had been to make every statement in that theory testable by research means. I had distributed this longer statement to the staff of the Counseling Center of the University of Chicago, and had profited from their perceptive criticisms and suggestions.

"Now, for the first time I was about to present a small but very important segment of that theory to an audience of critical peers. I included explanations and illustrations of each rigorous point, hoping to make them clear.

"The presentation aroused much open discussion. I believe I was regarded as presumptuous for having set forth what I regarded as *sufficient* conditions to account for effective psychotherapy. I know that I shocked many by stating that 'special intellectual professional knowledge' is not essential in psychotherapy. Another controversial statement was that a psychological diagnosis is only helpful in making the therapist feel secure. The major audience reaction was that traditional views had been sharply challenged, and that time was needed to assess the significance of my talk. It has been gratifying to see that the paper has gained increasing acceptance over the years.

"This article has sparked more research investigations than any other I have written. This is because every condition hypothesized as necessary for effective therapy is operationally defined, and therefore testable. I also made suggestions as to methodology by which each could be validated or invalidated. Consequently research workers have found here a basis for all kinds of research in psychotherapy, education, and personality development. These have taken place in this country and in many foreign countries. The most recent example is a summary by Prof. Reinhard Tausch of seventeen studies completed at the University of Hamburg, West Germany, during the 1970s.[2]

"All in all, I suspect this is the most significant paper I have ever written, and the most far-reaching in its effects."

1. Rogers C R. A theory of therapy, personality, and interpersonal relationships, as developed in the client-centered framework. *Psychology: A Study of a Science*, Vol. 3. (Koch S, ed.) New York: McGraw-Hill, 1959, p. 184-256.
2. Tausch R T. Facilitative dimensions in interpersonal relations: verifying the theoretical assumptions of Carl Rogers in school, family education, client-centered therapy, and encounter groups. *College Student Journal* 12(1), Spring 1978 (in press).

Szasz T S. The myth of mental illness. *Amer. Psychol.* **15**:113-18, 1960.
[State Univ. New York, Upstate Medical Center, Syracuse, NY]

Strictly speaking, the term 'disease' denotes a pathoanatomical or pathophysiological lesion. Because the mind is not a bodily organ or tissue, it can be diseased only in a metaphoric sense. This paper calls attention to the false, but widely accepted, belief that 'mental illness is like any other illness' and to the practical implications of that belief. [The *Science Citation Index®* (*SCI®*) and the *Social Sciences Citation Index®* (*SSCI®*) indicate that this paper has been cited in over 190 publications since 1961.]

Thomas S. Szasz
Department of Psychiatry
State University of New York
Upstate Medical Center
Syracuse, NY 13210

August 1, 1983

"As far back as I can remember, I was intrigued, puzzled, and disturbed by what is now conventionally called mentally ill behavior. Why do some people act in such strange, repellent, and frightening ways? And why does everyone say that these people are 'mentally sick' even though they do not claim or seem to be sick?

"I certainly did not know the answers to these questions when I was a child. But one thing was clear to me, even before I reached my teens: namely, that neither so-called mental patients nor psychiatrists behaved like ordinary patients and doctors. This insight has always struck me as similar to the classic Hans Christian Andersen story about the emperor's clothes. Everyone knows that mental illness is not 'really' an illness, that psychiatrists are not 'real' doctors, that mental hospitals are not 'real' hospitals. But when one is a grown-up person, one is not supposed to know this, and surely one is not supposed to say it.

"Sometime during my high school years I decided that I was going to say something about this subject. That is what I did in my 1960 paper in the *American Psychologist* (and in my book, *The Myth of Mental Illness*,[1] published a year later). Strictly speaking, then, this paper was not the product of any research. Rather, it was an attempt to tell a truth which, I was certain, everyone knew, but which they were too polite, too timid, too opportunistic, or too uncaring to articulate.

"I wrote 'The myth of mental illness' in 1957, soon after settling in Syracuse. For the next two years it made the rounds of psychiatric journals, from the *American Journal of Psychiatry* down. I must have submitted it to at least a half dozen psychiatric journals — and they all rejected it. That is how this paper ended up in the pages of a nonpsychiatric journal.

"After this paper was published, two things happened rather quickly. First, it was reprinted in a score of anthologies, and the phrase 'myth of mental illness' became both the stimulus for serious criticism of accepted psychiatric principles and practices and the banner under which many of its critics rallied. Second, academic and professional authorities — in my medical school and elsewhere — accused me of 'not believing in mental illness,' and did their best to harass me for this heresy in the ways available to such officials.

"During the more than two decades since the publication of 'The myth of mental illness,' I have written many papers and books elaborating on the implications of its thesis for mental health practices and for activities in related fields.[2-5] There has ensued a lively debate on the medicalization of life — some supporting and others opposing the 'medical model' of mental illness and the psychiatric coercions it justifies.[6] It is encouraging to note that, with the passing of years, the ideas expressed in 'The myth of mental illness' have become more widely accepted, especially outside of psychiatric circles."

1. **Szasz T S.** *The myth of mental illness: foundations of a theory of personal conduct.*
   New York: Hoeber-Harper, 1961. 337 p.
2. --------------. *Law, liberty, and psychiatry: an inquiry into the social uses of mental health practices.*
   New York: Macmillan, 1963. 281 p.
3. --------------. *The manufacture of madness: a comparative study of the Inquisition and the mental health movement.*
   New York: Harper & Row, 1970. 383 p.
4. --------------. *The myth of psychotherapy: mental healing as religion, rhetoric, and repression.*
   Garden City, NY: Anchor Press/Doubleday, 1978. 236 p.
5. --------------. *Sex by prescription.* Garden City, NY: Anchor Press/Doubleday, 1980. 198 p.
6. **Vatz R E & Weinberg L S,** eds. *Thomas Szasz: primary values and major contentions.*
   Buffalo, NY: Prometheus Books, 1983. 253 p.

McGaugh J L. Time-dependent processes in memory storage.
Science 153:1351-8, 1966.
[Department of Psychobiology, University of California, Irvine, CA]

This paper summarized the then recent evidence that processes underlying memory storage are time-dependent. It presented new evidence supporting the view that electrical stimulation of the brain produces retrograde amnesia, as well as evidence that memory is enhanced by posttraining administration of stimulant drugs. [The *Science Citation Index®* (*SCI®*) and the *Social Sciences Citation Index®* (*SSCI®*) indicate that this paper has been cited in over 490 publications since 1966.]

James L. McGaugh
Center for the Neurobiology of Learning
and Memory
University of California
Irvine, CA 92717

July 5, 1984

"I wrote this paper shortly after I arrived in the department of psychobiology at the newly (or almost newly) established Irvine campus of the University of California (UC). The work of my own that was included in the paper began a dozen years earlier when I was a graduate student at UC, Berkeley. The research was stimulated by findings, published in the 1940s and 1950s, indicating that electroconvulsive shock (ECS) impairs recently acquired memories.[1,2] These findings suggested that ECS interfered with time-dependent processes underlying memory consolidation. This interpretation was (and to some still is) controversial. The findings of my own studies of the effects of ECS on memory in rats and mice strongly supported this hypothesis. In addition, I had found, in a series of studies,[3] that memory could be enhanced by low doses of central nervous system stimulants if the drugs were administered shortly after training. The evidence suggested that the drugs improved memory through influences on memory consolidation processes. Much of the earlier work on experimentally induced retrograde amnesia had been reviewed within a few years prior to the time that I wrote this paper.[4] There were, however, many new findings concerning ECS effects on memory and the studies of drug enhancement of memory had not yet been reviewed in the context of controversial issues in consolidation theory.[3] Thus, when I received an invitation to contribute a paper to *Science*, I eagerly accepted.

"The major aim of this paper was to consider recent findings of my own (as well as other) studies concerning the basis of the memory-impairing effects of ECS and to summarize the work from my laboratory concerning drug enhancement of memory. The paper also examined the trade-offs (or interactions) between time and repetition in influence on the strength of memory. This attempt was, alas, largely ignored. Finally, the paper addressed the question of whether more than one memory system (e.g., short-term and long-term memory) is required to account for the findings of these experimental studies of memory.

"The paper was intended simply as an 'update.' Thus, it seems clear why it was frequently cited shortly after it was published. However, since it was neither a pure experimental paper nor a comprehensive review, it is less clear to me why it has continued to be frequently cited. Some of the following possibilities seem to be reasonable explanations. First, the paper appeared at a critical time in the history of memory consolidation research. Some attempt to address the controversial issues was needed. Second, it appeared in a prominent and readily accessible journal. Third, this area of research has remained active and has continued to spawn controversy.[5,6] It seems that the paper has continued to provide a convenient citation for all sides of all issues."

1. **Duncan C P.** The retroactive effect of electroshock on learning. *J. Comp. Physiol. Psych.* 42:32-42, 1949.
   (Cited 130 times since 1955.)
2. **Gerard R W.** Physiology and psychology. *Amer. J. Psych.* 106:161-73, 1949.
3. **McGaugh J L & Petrinovich L F.** Effects of drugs on learning and memory. *Int. Rev. Neurobiol.* 8:139-96, 1966.
   (Cited 140 times.)
4. **Glickman S E.** Perseverative neural processes and consolidation of the memory trace.
   Psychol. Bull. 58:218-33, 1961. (Cited 90 times.)
5. **McGaugh J L & Herz M J.** *Memory consolidation.* San Francisco: Albion, 1972. 204 p.
6. **McGaugh J L.** Hormonal influences on memory. *Annu. Rev. Psychol.* 34:297-323, 1983.

Kendler H H & Kendler T S. Vertical and horizontal processes in problem solving. *Psychol. Rev.* **69**:1-16, 1962.

The authors proposed a model that postulated two modes of behavior, associative and mediational, to explain the behavior of rats and college students in a discrimination-shift task involving either a reversal or extradimensional shift. The model also made assumptions about the operation of symbolic, perceptual, and motivational processes in specific problem solving tasks. A variety of data, including human developmental changes, were presented in support of the model. [The *Science Citation Index®* (*SCI®*) and the *Social Sciences Citation Index*™ (*SSCI*™) indicate that this paper was cited a total of 337 times in the period 1962-1976.]

Howard H. Kendler
Department of Psychology,
University of California,
Santa Barbara, California 93106
December 12, 1977

"This paper proposed a pretheoretical model to guide a research program, which began in 1954 and which was primarily concerned with understanding developmental changes in human problem-solving. The program was initiated by my inability to resist an offer of governmental support for research in human problem-solving at a time when I was deeply immersed in theoretical problems of animal learning. My initial attempts to investigate human problem-solving failed to yield any exciting leads until some data suggested that Kenneth W. Spence's famous discrimination-learning theory of animal organisms could serve as a good jumping-off place for excursions into human problem-solving. According to this theory, in a discrimination-shift problem an extradimensional shift should be easier to execute than a reversal shift; a prediction that was later confirmed with rats by one of my Ph.D. students. Before that, I and another investigator independently discovered that college students found a reversal shift easier. 'So what!' was a typical reaction to the discrepancy between the behavior of rats and humans. Is it surprising that rats and humans behave differently or that Spence's formulation is irrelevant to human behavior?

"At this point my wife became my research collaborator and it might be mentioned that there is no greater testimony to the sturdiness of our marriage than that it was able to withstand the stress and strain of our theoretical and methodological disputes. We perceived the discrepancy between the behavior of rats and humans more as a challenge to be explained than merely as evidence to be accepted. The findings posed two significant questions: (1) How can the difference between the behavior of rats and college students be conceptualized? (2) How would organisms 'in between' rats and college students behave?

"In answer to the first question it was postulated that rats behaved in a discrimination-shift study in a manner similar to the conditioning (associative) model suggested by Spence's theory; stimuli, defined in terms of their physical attributes, became directly linked, in a figurative sense, to the choice responses. College students, in contrast, behaved according to a mediational model; incoming stimulation was transformed (processed) into some symbolic representation that guided subsequent behavior. The second question was answered by an ontogenetic analysis of discrimination-shift behavior. The results supported the notion that in discrimination-shift studies the probability that a child's behavior will fit the conditioning or mediational model will depend on his age; the younger he is the more likely it is that he will behave according to the associative model, whereas the older he is the more likely it is that his performance will be consistent with the mediational model.

"This paper has been frequently cited because it (1) helped expand the empirical realm of theories of learning, particularly in the direction of developmental changes in problem solving, (2) offered a simple and effective experimental methodology to investigate such problems, (3) posed a variety of interesting theoretical questions about the associative and mediational modes of behavior as well as the manner in which the transition is made between them, and (4) justified a theoretical perspective that encompasses both human and infrahuman organisms.

Underwood B J. Attributes of memory. *Psychol. Rev.* 76:559-73, 1969.

The article proposes that a memory is a collection of attributes. These serve to discriminate one memory from another and to act as retrieval mechanisms. The attributes identified are temporal frequency, modality, orthographic, associative nonverbal, and associative verbal. [The *Science Citation Index®* (*SCI®*) and the *Social Sciences Citation Index ™* (*SSCI™*) indicate that this paper was cited a total of 141 times in the period 1969-1976.]

Benton J. Underwood
Department of Psychology
Northwestern University
Evanston, Illinois 60201
December 8, 1977

"What is the basic constituent of a memory? What is lost when we say that we have (alas) forgotten? Almost from the beginning of recorded thought, answers to such questions revolved around the concept of an association. Associations between words and between ideas were generally held to be the heart of a memory for an event. My paper, which described a memory as a collection of attributes (a collection of different types of information), represented a clear departure from the classical position. Associations were not abandoned; rather, they became only one of several types of information which were said to be constituents of memories. The theoretical problem was to describe the role played by each attribute in memory functioning.

"In the early 1960s I had come to realize that the human memory system was very sensitive to repeated events. Specifically, it appeared that quite unintentionally the system 'counted' repeated events, and if we interrogated the system the frequency information would be made manifest. Thus, when we asked college students to estimate the relative frequency with which various words appear in printed discourse, we found their judgments to be quite valid. Such information could not be classed as associative information except in a rather trivial sense. Each word in our vocabulary has frequency information associated with it, but this was not like the classical idea of associations between words or between ideas. Repetition may strengthen associations between words, but frequency per se has a representation in memory that is independent of the strength of the association.

"It did not seem to me that nature would provide us with such a remarkable counting mechanism without this mechanism having a role in memory functioning. We then offered a theory about the role played by frequency discrimination in recognition tests of memory. Having broken the associative rampart with this one concept, it seemed probable to me that other concepts could be found to join frequency in the assault. My article represents a summary of such concepts. These concepts or attributes were educed from a wide variety of studies, most of which had been published by other investigators. That this paper has been cited frequently probably stems from the fact that most investigators in the area were already thinking about memories in terms of multiple types of information. My paper simply identified a greater number of different types than had previously been identified."

Rescorla R A & Solomon R L. Two-process learning theory: relationships between Pavlovian conditioning and instrumental training. *Psychol. Rev.* **74:**151-83, 1967.

This paper examines the historical development of, and the evidence for, the separation of Pavlovian conditioning and instrumental training. It suggests a paradigm for examining their interaction in generating behavior and details some predictions from various instances of this paradigm. [The *Science Citation Index® (SCI®)* and the *Social Sciences Citation Index™ (SSCI™)* indicate that this paper was cited a total of 213 times in the period 1967-1977.]

R.A. Rescorla
Department of Psychology
Yale University
New Haven, Connecticut 06520

February 1, 1978

"This paper grew out of the joining of a long-standing interest of the second author with some empirical results obtained by the first author. It was written while R.A. Rescorla was a graduate student in R.L. Solomon's laboratory. In many ways it reflects the thinking of many people who were active in that laboratory over a period of about 10 years.

"Since the 1930's there had gradually been emerging a widely accepted distinction between two kinds of associative learning in animals: Pavlovian conditioning resulting from the arrangement of a contingency between a signal and a reinforcer, and instrumental learning resulting from a contingency between the response and the reinforcer. This paper presented a review of the emergence of this distinction, and evaluated some of the evidence supporting it. It further presented a theory of how these two kinds of learning might interact in producing learned behavior. On that theory Pavlovian conditioning plays an important role in the learning of motivations, which in turn govern the exhibition of instrumental, goal-directed behavior.

"Three of its principal contributions were the making explicit of a paradigm for the study of these interactions, the elaboration of various empirical consequences of those interactions, and an emphasis upon a modern view of the role of inhibition in Pavlovian conditioning. The paradigm involved the superimposition of stimuli given Pavlovian conditioning on ongoing instrumental behavior. That paradigm, although examined earlier, was largely unexploited and inexplicit. It turned out both to yield information about the interaction among learning processes and to provide a valuable tool for the study of Pavlovian conditioning itself. The view of Pavlovian conditioning espoused really represented in part a return to Pavlov's own views, from which American psychology had strayed.

"The principal reasons that this paper has been widely cited have to do with its codifying a modern version of a theory dependent upon two learning processes, its laying out of a wide variety of empirically testable predictions, and its foreshadowing a revolution about to come in our thinking about Pavlovian conditioning. In the first role it has seemed representative of a particular set of theories and is often cited in that context. In its second role it led to many empirical studies, not all of which yielded results favorable to the theory. Its third role has been somewhat preempted by subsequent papers.

"Our greatest personal satisfaction about the way in which this paper has been received stems from the comment of a teacher and friend, Francis Irwin. He found in this paper a reason for thinking that Pavlovian conditioning was not 'all spit and twitches' but actually governed important psychological processes."

Jensen A R. How much can we boost IQ and scholastic achievement?
Harvard Educ. Rev. **39**:1-123, 1969.

Individual differences in intelligence, as a scientific construct, can be reliably measured and indexed as IQ. IQ is the best single predictor of scholastic achievement, and has other occupationally, economically, and socially important correlates. Individual variation in IQ is largely genetic, as shown by heritability analysis of kinship data. Social class differences involve genetic factors, and many lines of evidence suggest it is a reasonable *hypothesis* that genetic factors may be strongly implicated in the one standard deviation average IQ difference between whites and blacks. [The *Science Citation Index*® *(SCI*® *)* and the *Social Sciences Citation Index*™ *(SSCI*™*)* indicate that this paper was cited a total of 638 times in the period 1969-1977.]

Arthur R. Jensen
Institute of Human Learning
University of California
Berkeley, CA 94720

December 15, 1977

"The question of genetic factors as a partial cause of the well established fact of a 15-point average difference between the IQs of blacks and whites in the US was only a small (less than 10%) part of this lengthy article. But it was that, more than anything else, that sparked the incredible academic and public uproar which so quickly followed publication.

"Although unfortunately the heat and fumes of the reaction to my position have not yet entirely abated, even after nine years, I believe that some scientific good has resulted from my having pulled this most taboo question out from under the carpet, whither it was swept decades earlier by the wake of Hitler's holocaust. Today, geneticists and psychologists and other behavioral scientists, in small but increasing numbers, are publicly acknowledging that the cause of the black IQ deficit, with all its socially significant correlates, is scientifically an *open question*.

And that is all I stated in my article, or have ever insisted on. We advance scientific knowledge by formulating and empirically testing *competing* hypotheses, and I have since presented more complete evidence and argument that a genetic hypothesis of the black-white IQ difference is a worthy contender in the scientific arena.[1]

"In the nine years following my controversial article, there has been a major revival of scientific interest, vigorous debate, and new research on the nature and measurement of intelligence and its genetic aspects. But the topic of genetic racial differences, *treated as an open question*, is still generally a taboo on most college campuses, both here and abroad, and those who dare to suggest that genetic factors are a plausible part of the explanation still must pay a price that is not only unappealing, but understandably intimidating to many academicians or their families. The risk of being labeled 'racist,' even though patently false, is dreaded by many.

"My unrelenting research in the so-called 'IQ controversy,' for example, has resulted, over the years, in the loss of the friendship of a number of my colleagues; in near-riotous demonstrations by student activists at many colleges where I have been invited to speak; and in last-minute cancellation of invited lectures, vilification in student newspapers and leaflets, and physical threats to me and my family, occasioning the need for police protection, even as recently as a month ago, and as far away as Australia. One may imagine subtler penalties, too, such as the loss of academic status and respectability, but this is more difficult to assess. It does not worry me perhaps as much as it should. The fact that I am not only alive and well, but reasonably happy and unstintingly carrying on my research on all aspects of human intelligence will no doubt be attributed to personal eccentricity. But I hope it will also be encouraging to others. From my experience I can say that, in the long haul, the consequences of sticking your neck out when you think you should, are not too bad. It is an exercise in conscience and self-respect, in which neither suffers, given the faith that the scientific pursuit of the currently most tabooed question will prove worthwhile to humanity."

REFERENCE

1. **Jensen A R.** *Educability and group differences.* New York: Harper & Row, 1973. 407 p.

Bolles R C. Species-specific defense reactions and avoidance learning. *Psychol. Rev.* **71**:32-48, 1970.

The paper starts with the assumption that animals have innate defensive behaviors, such as freezing, fleeing and fighting. It is proposed that if a particular avoidance response is rapidly learned, then that response must necessarily be one of the animal's species-specific defense reactions, or part of such a reaction. Some reinforcement-produced learning does occur with more slowly learned avoidance responses, but it is slow, uncertain, and not based on the conventional mechanism. [The *Science Citation Index® (SCI®)* and the *Social Sciences Citation Index™ (SSCI™)* indicate that this paper was cited a total of 260 times in the period 1971-1977.]

Robert C. Bolles
Department of Psychology
University of Washington
Seattle, Washington 98105

February 5, 1978

"The trick in writing a paper that will be frequently cited lies partly in saying the right thing, of course, but it also depends upon saying it at the right time. Say the right thing but say it a few years too soon, and (if the paper is published at all) it is likely to attract little attention. Say the same right thing just a few years too late, and it is likely to be suitable only for a textbook. I was fortunate to have this particular paper appear at just the right time.

"Psychology has always had a strong empiricist bias: a conviction that everything is explained by learning through experience. Once the basic, universal laws of learning were discovered, it was assumed, then everything else—all behavior, personality, social organization—would fall into place. That we could not find, or agree upon, any set of universal laws did not seem to matter; the bias persisted. Clear evidence of inhomogeneities in learning, implying a failure of learning laws to be universal, did not matter. The empiricist assumption swept everything before it like the incoming tide. Further evidence from ethology of genetic determinants of behavior was simply washed away. Nothing availed against the tide.

"My own research had indicated that animals are more able to learn avoidance behaviors in some situations than in others. This troublesome fact was already recognized, but it was not considered anything more than a peculiarity of avoidance behavior (long a problem for learning theorists) or of the specific situations that had been studied. In 1966 I conducted a seminar on ethology. Our discussion of this point of view led us to the idea that the inhomogeneities in avoidance learning, which were so troublesome for learning theory, might tell us something important about the animal and how it avoids the natural dangers in its particular environmental niche.

"This idea was promptly supplemented with further data and then polished up in a few colloquium presentations. When the paper appeared it was at that happy moment when Garcia, Seligmen, and Brelands, all acting quite independently, had each begun to question the universality of general laws of learning. And now as we splashed about we could see, incredibly, the tide was going out."

Archer J. Tests for emotionality in rats and mice: a review.
Anim. Behav. 21:205-35, 1973.
[Sch. Biological Sciences, Univ. Sussex, Falmer, Brighton, Sussex, England]

This paper reviews tests used to assess 'emotionality' or fearfulness in laboratory rats and mice. It is concluded that the various behavioural and physiological measures do not represent a single dimension and more detailed behavioural analysis is suggested. [The *Science Citation Index*® (*SCI*®) and the *Social Sciences Citation Index*® (*SSCI*®) indicate that this paper has been cited over 160 times since 1973.]

John Archer
School of Psychology
Preston Polytechnic
Preston, Lancashire PR1 2TQ
England

September 22, 1981

"Following the work of C.S. Hall[1] in the 1930s, psychologists interested in experimental animal research have sought to measure the relative levels of fear or 'emotionality' in laboratory rodents by simple, rapidly administered tests. The best-known of these involves placing the animal into a novel arena and recording its movement ('ambulation') and amount of defecation during a short time period. P.L. Broadhurst[2] standardised the test in the 1950s after which it was used extensively. It provides a quick, easy assessment of the animal's behavioural state, and serves the same function as some rapidly-administered human personality tests. In fact, both represent a psychological tradition which seeks to characterize human and animal behaviour in terms of a small number of dimensions amenable to measurement by simple tests.

"I was originally trained as a zoologist, and therefore come from a very different background, that of ethology, which recognizes the complexity and variety of animal (and human) behaviour. Nevertheless, I was interested in the same subject matter as the psychologists who were using the emotionality tests. When I was a postdoctoral worker at Sussex University (1969-1975), I began to have reservations about how accurately simple measures could provide a meaningful analysis of animal behaviour. I have always preferred writing review papers to carrying out research, and as there appeared to be no critical reviews of the subject, I set out to write one. This took longer and was more difficult than I anticipated. I sent the original version of the paper to a psychological journal and it was rejected. I then tried to make it more comprehensive and better organized. By the time I was ready to submit it again, the paper was over three times the previous length. I chose a more ethologically-slanted journal this time. One possible difficulty in getting the paper reviewed fairly was that researchers in this field were likely to have built careers and reputations using the very tests I was subjecting to critical scrutiny. The reviewers' comments were indeed lukewarm (though not necessarily for this reason!) but the editor was sufficiently favourable to accept it. He pointed out one aspect which must be a difficulty with all works of criticism, namely, that 'It is all very well to find fault with others, but what have you got to offer instead?' I did try to address the problem of alternative approaches in my final version, and I have since followed this up in a more recent article.[3]

"I think that the main reason the paper has been cited frequently is that it provided a useful appraisal and catalogue of studies involving tests which were widely used and yet gave rise to misgivings in many users. In this sense, my review was published at the right time, but opinions about my judgement remain divided to this day."[4]

1. Hall C S. Emotional behavior in the rat. I. Defaecation and urination as measures of individual differences in emotionality. *J. Comp. Psychol.* 18:385-403, 1934.
2. Broadhurst P L. Determinants of emotionality in the rat. I. Situational factors. *Brit. J. Psychol.* 48:1-12, 1957.
3. Archer J. Behavioural aspects of fear. (Sluckin W, ed.) *Fear in animals and man.* New York: Van Nostrand Reinhold, 1979. p. 56-85.
4. Jones R B, Duncan I J H & Hughes B O. The assessment of fear in domestic hens exposed to a looming human stimulus. *Behav. Process.* 6:121-33, 1981.

Sawyer C H, Everett J W & Green J D. The rabbit diencephalon in stereotaxic coordinates. *J. Comp. Neurol.* **101**:801-24, 1954.
[Dept. Anatomy, Univ. California, Los Angeles, CA; Dept. Anatomy, Duke Univ. Med. Sch., Durham, NC; and Investigative Med. Service, Veterans Admin. Hosp., Long Beach, CA]

A rabbit head-holder was designed to fit a standard Horsley-Clarke stereotaxic instrument carriage. Stereotaxic atlas drawings at 1 mm intervals of this histologically sectioned diencephalon of a 3.5 kg New Zealand rabbit were prepared and the coordinates corrected from findings on several experimental animals. [The *SCI®* indicates that this paper has been cited in over 650 publications since 1955.]

Charles H. Sawyer
Department of Anatomy
and
Brain Research Institute
Laboratory of Neuroendocrinology
UCLA School of Medicine
Los Angeles, CA 90024

February 23, 1984

"The female rabbit has long been recognized as a 'reflex ovulator' requiring the neural stimulation associated with coitus to induce an ovulatory surge of pituitary gonadotropin. At Duke University in the 1940s, we had shown that this neuroendocrine reflex involved an adrenergic mechanism probably localized in the brain.[1] As a result, we became interested in studying the hypothalamic sites and mechanisms of pituitary activation, an investigation involving localized electrical stimulation and the precise placement of cannulae and lesioning and recording electrodes. The program required a stereotaxic atlas, or map, of the brain, and since none was available for the rabbit, Everett and I set out in the summer of 1952, while he was a visiting professor at the University of California, Los Angeles, to produce our own brain atlas. The work was done at the Long Beach Veterans Administration Hospital, the temporary research space for the future UCLA Brain Research Institute.

"We built a primitive head-holder to fit into the cat-monkey stereotaxic frame, and with advice from H.W. Magoun implanted steel needle markers at precisely measured intervals into the brains of deeply anesthetized New Zealand rabbits. Microprojection of histological sections of the perfusion-fixed brains permitted the localization of nuclei and fiber tracts in transverse sections of the diencephalon and the graph-paper reconstruction of the midsagittal plane. John Green, who with Geoffrey Harris had done so much to establish the neurovascular concept of pituitary control, joined us at the end of the summer. He immediately designed a head-holder so much better than our original model that we adopted it and made him a coauthor of the atlas publication. (An improved version of the head-holder is available from TrentWells Inc., 8120 Otis Street, Southgate, CA 90280.)

"We have used the rabbit atlas in many brain lesion, stimulation, recording, implantation, and infusion studies such as those described or reviewed in the references below.[2-5] These include investigations of the feedback actions on the brain exerted by pituitary peptide hormones and gonadal and adrenal steroids, as well as the actions of brain amines and pharmacological agents which affect hypothalamo-pituitary function. Obviously, others have found the atlas useful in rabbit brain research and have cited it frequently.

"Green's brilliant career was cut short by his untimely death in 1964. He used the rabbit brain and its atlas in many electrophysiological experiments described in his posthumous review.[4] These include the first unit-recording studies on single neurons of the hypothalamus. His co-worker Barry Cross later used the rabbit brain in developing the technique of antidromic identification of hypothalamic neurons.[5]

"Everett was already a leading authority on the neuroendocrinology of reproduction before we published the atlas, and he has remained at the forefront of the field during more than 50 years of research. In 1977, the British Society of Endocrinology awarded him its prestigious Dale Medal for his research on the control of ovulation, corpus luteum function, and prolactin secretion.[6] In the 1970s, we each received the Hartman Award of the Society for the Study of Reproduction, and, in 1973, shared the Koch Award of the Endocrine Society."

1. **Markee J E, Everett J W & Sawyer C H.** The relationship of the nervous system to the release of gonadotrophin and the regulation of the sex cycle. *Recent Prog. Hormone Res.* 7:139-63, 1952. (Cited 95 times since 1955.)
2. **Sawyer C H, Kawakami M, Markee J E & Everett J W.** Physiological studies on some interactions between the brain and the pituitary-gonad axis in the rabbit. *Endocrinology* 65:614-68, 1959. (Cited 125 times since 1959.)
3. **Sawyer C H, Kawakami M & Kanematsu S.** Neuroendocrine aspects of reproduction. *Res. Publ. Assoc. Res. Nerv. Ment.* 43:59-85, 1966. (Cited 20 times.)
4. **Green J D.** Neural pathways to the hypophysis: anatomical and functional. (Haymaker W, Anderson E & Nauta W J H, eds.) *The hypothalamus.* Springfield, IL: Thomas, 1969. p. 276-310.
5. **Sawyer C H.** Some recent developments in brain-pituitary-ovarian physiology. *Neuroendocrinology* 17:97-124, 1975. (Cited 135 times.)
6. **Everett J W.** The timing of ovulation. *J. Endocrinology* 75:1P-13P, 1977.

This Week's Citation Classic

Spiegel E A, Wycis H T, Marks M & Lee A J. Stereotaxic apparatus for
operations on the human brain. *Science* 106:349-50, 1947.
[Dept. Experimental Neurology, Temple Univ. Sch. Med., Philadelphia, PA]

A stereotactic apparatus is described permitting
production of exactly placed subcortical lesions in
the human brain with minimal impairment of
other areas. The use of intracerebral reference
points and application in psychosurgery are reported. Further indications (intractable pain, involuntary movements, subcortical tumors) are envisaged. [The *SCI*® indicates that this paper has been
cited in over 100 publications since 1961.]

Ernest A. Spiegel
Department of Experimental Neurology
Temple University School of Medicine
Philadelphia, PA 19140

July 27, 1983

"In the winter of 1947, I watched a prefrontal
lobotomy being performed and I was appalled by
the resulting extensive brain damage and by the
severe personality changes. It occurred to me that
a reduction of the emotional and behavioral
disturbances attempted by lobotomy could be obtained also by small lesions of the thalamic dorso-
medial nucleus that forms a circuit with the frontal
lobe. Such a lesion would avoid severance of the
association fibers caused by lobotomy. H.T. Wycis,
a neurosurgeon participating in research in my
department, enthusiastically accepted my proposal of such thalamotomies.

"These procedures required not only the building of a stereotaxic guiding apparatus similar to
Horsley and Clarke's[1] employed in animals, but
also the preparation of a special atlas of the
human brain. Due to the great variability in the
shape of the human skull, reference points on the
skull as used by Horsley and Clarke[1] on animals
were unreliable. Intracerebral reference points
had to be chosen such as a calcified pineal gland
or parts of the circumference of the cerebral ventricles visualized roentgenologically, e.g., the anterior and posterior commissures. The procedures
using intracerebral reference points have been
called stereoencephalotomy,[2] stereotaxic neuro-

surgery, or stereotaxy. This special atlas had to
show brain sections in frontal, sagittal, and
horizontal planes. The illustrations were supplied
with millimeter scales, so that the coordinates of
subcortical structures in relation to the above
mentioned reference points could be measured.
Their variability also was determined. The tedious
work of preparing such an atlas and of measuring
the coordinates was greatly facilitated by the
faithful cooperation of Wycis. He also followed
meticulously my suggestions during these operations. The first stereotaxic thalamotomies were
performed in spring 1947.

"At an American Medical Association exhibit of
labyrinthine studies (June 1947), several visitors
surprisingly asked me about details of our unpublished thalamotomies. This alerted me to the importance of an early publication. This paper describing a stereotaxic instrument for man and its
application in psychosurgery was published in
Science in 1947. Here some further possible applications of stereoencephalotomy were mentioned
(intractable pain, involuntary movements, subcortical tumors). Later, convulsive disorders refractory to conservative treatment became an additional indication.[2-4] The preoperative, roentgeno-
logical visualization of the ventricles and/or of the
pineal gland also was mentioned in the original
publication.

"This technique has been widely used because it
permitted one to produce exactly localized subcortical lesions with minimal injury of other areas.
Thus, neurosurgeons in the US and Canada (151 according to Gildenberg[5]), in practically all European countries,[3,4] in Mexico, South America,
Japan, India, and Thailand became interested in
stereotaxic neurosurgery. Societies for this new
discipline were founded — an international, an
American, a European, and a Japanese one.

"I was awarded honorary MD degrees by the
Universities of Zürich and Vienna; honorary
presidency of the International Society for
Stereotactic and Functional · Neurosurgery;
honorary membership of the American EEG Soci-
ety, the German Neurosurgical Society, and the
Mexican National Society of Medicine; the
Foerster Medal (German Neurosurgical Society),
and the Erb Medal (German Neurological
Society)."

1. Horsley V & Clarke R H. The structure and function of the cerebellum examined by a new method.
      Brain 31:45-124, 1908.
2. Spiegel E A & Wycis H T. *Stereoencephalotomy: thalamotomy and related procedures.*
      New York: Grune & Stratton, 1952-62. Vols. I & II.
3. Riechert T. *Stereotactic brain operations.* Bern: Huber, 1980. 387 p.
4. Schaltenbrand G & Walker A E. *Stereotaxy of the human brain.* Stuttgart: Thieme, 1982. 700 p.
5. Gildenberg P L. Survey of stereotactic and functional neurosurgery in the United States and Canada.
      Appl. Neurophysiol. 38:31-7, 1975.

Cowan W M, Gottlieb D I, Hendrickson A E, Price J L & Woolsey T A. The
autoradiographic demonstration of axonal connections in the central nervous
system. *Brain Res.* 37:21-51, 1972. [Dept. Anat., Washington Univ. Sch. Med.,
St. Louis, MO and Dept. Ophthalmol., Univ. Washington, Seattle, WA]

In this paper we described a technique for tracing
connections in the brain that is based on the incorporation of isotopically labeled amino acids by
nerve cells and the subsequent transport of the
labeled proteins along their axons where they can
be visualized autoradiographically. [The *SCI*[*] indicates that this paper has been cited in over 745
publications since 1972.]

W. Maxwell Cowan
The Salk Institute
Post Office Box 85800
San Diego, CA 92138

February 26, 1982

"Until the early-1970s, the only effective way to
trace connections in the central nervous system
(CNS) was to map out the degenerative changes
that occur when the region of interest was injured.
For many purposes this approach is perfectly adequate and has revealed much of what we know
about the connectivity of the brain. However, in
many situations the results it gives are difficult to
interpret, and it is of only limited use in the developing nervous system. An alternative approach
is to label nerve cells with a radioactively labeled
amino acid. Since amino acids are incorporated into protein only in the perikarya of nerve cells, and
as many of the proteins so labeled are actively
transported down axons to their synaptic endings,
it is possible to map the distribution of the labeled
fibers autoradiographically. A. C. Taylor and
P. Weiss[1] had tried this approach in the amphibian
visual system. A few years later R. Lasek,
B.S. Joseph, and D.G. Whitlock[2] at the University
of Colorado used it effectively to determine the
central connections of spinal ganglia. But for
some unknown reason (perhaps because it was
thought that if labeled amino acids were to be injected directly into the CNS they would spread so
widely that it would be impossible to label small,
localized populations of nerve cells), apparently
no one had tried it in the brain.

"About this time, Anita Hendrickson[3] of the
University of Washington was exploring the usefulness of axonal transport for studying the central
connections of the retina at the electron
microscope level. She and I began to study the
changes that occur in axonal transport during
development. Her finding that some proportion of
the radioactively labeled proteins that are rapidly
transported down axons is delivered to their syn-

aptic terminals was reported in a seminar she gave
in St. Louis in the spring of 1970. This encouraged
my colleagues, David Gottlieb, Joel Price, and
Tom Woolsey, and me, at Washington University
in St. Louis, to see if the same approach could be
used in other parts of the CNS. We began by making injections of tritium-labeled amino acids into
several regions of the brain whose connections we
had previously analyzed with conventional
neuroanatomical methods to see if we could
reproducibly label the relevant connections and
identify the sites in which they terminate. Our first
experiments were fairly discouraging. The injections were far too large and resulted in so much
spread of isotope that the developed autoradiographs were almost black. Fortunately, one of
the systems of connections that was labeled was
so distinctively organized that even in those early
preparations the underlying pattern of the connections could be recognized. Thereafter, it was
simply a matter of scaling down the size of the injections, refining the isotope delivery system, determining the optimum post-injection survival
times, and appropriately modifying the autoradiographic procedure. Within a few weeks we had
clear evidence in four or five different neuronal
systems that the technique could be used to label
entire fiber pathways in the CNS and to rather
precisely define their modes of termination. We
presented some of the results of our work at the
annual meeting of the American Association of
Anatomists in the spring of 1971. In view of
the interest that was aroused, I suggested that we
write a paper documenting the use of the method.
Hendrickson spent a few days in St. Louis on her
way back to Seattle. As I recall, the outline of
the paper was written during a picnic in Forest
Park across the street from the medical school.

"In retrospect, it seems that the paper was well
received because at the time new techniques for
tracing pathways in the CNS were sorely needed
and the autoradiographic method clearly offered
several advantages. It was uncommonly sensitive,
it was not complicated by the labeling of fibers of
passage or by species and age differences, it lent
itself to certain types of developmental and quantitative analyses, and its use had been critically
evaluated in a number of neural systems. In addition, since we provided a straightforward and
detailed experimental protocol that others could
follow, within a relatively short time it was adopted by several laboratories around the world.
E. G. Jones and B. K. Hartman have published a recent review in this field."[4]

1. Taylor A C & Weiss P. Demonstration of axonal flow by the movement of tritium-labeled protein in mature optic
       nerve fibers. *Proc. Nat. Acad. Sci. US* 54:1521-7, 1965.
2. Lasek R, Joseph B S & Whitlock D G. Evaluation of a radioautographic neuroanatomical tracing method.
       Brain Res. 8:319-36, 1968.
3. Hendrickson A. Electron microscopic radioautography: identification of origin of synaptic terminals in normal
       nervous tissue. *Science* 165:194-6, 1969.
4. Jones E G & Hartman B K. Recent advances in neuroanatomical methodology. *Annu. Rev. Neurosci.* 1:215-96, 1978.

This Week's Citation Classic

Petsche H, Stumpf C & Gogolak G. The significance of the rabbit's septum as a relay station between the midbrain and the hippocampus. I. The control of hippocampus arousal activity by the septum cells.
Electroencephalogr. Clin. Neuro. 14:202-11, 1962.
[Insts. Neurology and Pharmacology, Univ. Vienna, Vienna, Austria]

Hippocampus theta waves of unanesthetized, curarized rabbits are triggered by cells in Broca's diagonal band: these bursting cells are phaselocked with the theta waves. Phase angles are different for different cells. The burst activity keeps going on when the theta rhythm is temporarily replaced by another activity. [The *SCI®* indicates that this paper has been cited over 165 times since 1962.]

H. Petsche, C. Stumpf, and
Gertrud Gogolak
Institute of Neurophysiology
and
Institute of Neuropharmacology
University of Vienna
A-1090 Vienna
Austria

August 3, 1981

"When we wrote this paper, hardly anything was known about the nature of the hippocampus EEG. For us, the hippocampus was no more than a useful model to study a regular electrical activity. One of us (Petsche) was interested in studying the phase angles between different recording electrodes by a toposcopic display system.[1] In these studies, theta activity turned out to consist of travelling waves diverging from a place in the midline and a few millimeter rostral the hippocampus. Since we hoped to find a sort of pacemaker in this region we stimulated this part electrically and indeed came across a zone in the diagonal band by the stimulation of which theta activity was either 'driven' (at low frequencies) or abolished (at higher frequencies).[2] This was the reason for studying this nucleus with microelectrodes; what we found were cells bursting at the theta frequency. The position of these cells coincided with the position of the lowest threshold for theta suppression by electrical stimulation. These findings suggested the idea that a pacemaker for the theta is in the septum.

"In recent years a sort of renaissance of theta research set in. This seems to us one main reason why our paper has been cited frequently.

"As far as the present level of theta research is concerned, there remains one essential unsolved question: in our original paper we assumed the existence of only one type of theta rhythm originating in CA_1 and being triggered by one pacemaker, whereas several authors postulate two different theta rhythms and also two generators (a second in the dentate gyrus).[3] Personally we cannot contribute to the solution of this question as we have since turned to other topics (Petsche to the neocortex and Petsche and Gogolak to the cerebellum and spinal cord). We only would like to make use of this opportunity to recall another old paper of our group[4] in which we found surprising similarities between the histograms of septum cell discharges and the shape of the CA_1 theta. Unfortunately, this paper seems to be forgotten. In our opinion, it may give essential hints to an understanding of the formation of theta waves."

1. Marko A & Petsche H. The multivibrator-toposcope, an electronic multiple recorder. *Electroencephalogr. Clin. Neuro.* 12:209-11, 1960.
2. Brücke F, Petsche H, Pillat B & Deisenhammer E. Ein Schrittmacher in der medialen Septumregion des Kaninchengehirnes. *Pflügers Arch.* 269:135-40, 1959.
3. Robinson T E. Hippocampal rhythmic slow activity (RSA; theta): a critical analysis of selected studies and discussion of possible species-differences. *Brain Res. Rev.* 2:69-101, 1980.
4. Gogolak G, Stumpf C, Petsche H & Sterc J. The firing pattern of septal neurons and the form of the hippocampal theta wave. *Brain Res.* 7:201-7, 1968.
[The *SCI* indicates that this paper has been cited over 55 times since 1968.]

CC/NUMBER 40
OCTOBER 5, 1981

Moruzzi G & Magoun H W. Brain stem reticular formation and activation of the
EEG. *Electroencephalogr. Clin. Neuro.* 1:455-73, 1949.
[Dept. Anatomy, Northwestern University Medical School, Evanston, IL]

Stimulation of the brain stem reticular formation evokes generalized desynchronization of the EEG, simulating the arousal reaction of sensory stimuli. The electrocortical arousal is mediated by an ascending system, which is still active after midbrain interruption of the classical sensory paths. [The *SCI®* indicates that this paper has been cited over 840 times since 1961.]

Giuseppe Moruzzi
Instituto di Fisiologia
Università di Pisa
56100 Pisa
Italy
and
Horace W. Magoun
Brain Research Institute
University of California
Center for the Health Sciences
Los Angeles, CA 90024

July 15, 1981

"In 1948, a visiting professorship supported by the Rockefeller Foundation brought one of us (G.M.) to Chicago and prompted our collaboration. The scientific background of the original project[1,3] may be found in works started during World War II: 1) demonstration of an inhibitory reticulospinal system,[4] involved in the paleocerebellar control of posture and movements;[1,2] 2) prolonged abolition by paleocerebellar stimulation of the clonic twitches elicited in the chloralosed cat by local strychninization of the motor cortex, an observation suggesting the existence of an ascending inhibitory influence.[5] We started from the working hypothesis that ascending reticular pathways might explain the paleocerebellar effects on the motor cortex. Our approach was a study on the effects of fastigial and bulboreticular stimulations on the electrical activity of the motor cortex of the chloralosed cat,

before and after local strychnine. Both conceptually and technically we were concerned with a simple problem, and our hope to reach a conclusion within a short time appeared justified. But the results were unexpected[1-3] and only at the end of the academic year was our work completed.

"The first experiment was made in December 1948. The EEG of the motor cortex became completely flat during stimulation of the inhibitory bulboreticular formation. By recording from other cortical areas we realized in the same day that the hasty statement written at the beginning of our protocol book ('activity of the motor cortex completely inhibited') gave a distorted picture of the reality. However, a few other experiments on unanesthetized *encéphale isolé* preparations were necessary in order to realize that the well-known phenomenon of the EEG arousal could be reproduced by reticular stimulation. In the next months the physiological significance of the *ascending reticular system* became clear. Our work was mainly concerned with parallel investigations on the phasic aspect of the arousal phenomenon, but it was realized that 'a steady background of less intense activity within this cephalically directed brain stem system' might contribute to the maintenance of the waking state. The first results of reticular lesions were reported by Lindsley, Bowden, and Magoun[6] in the same issue of the journal.

"The concept of structures responsible for the control of the general *level* of cerebral activities led to new approaches and views in several fields of the neurosciences: *neurophysiology* (mechanisms of EEG arousal and of the orienting reaction; levels of central activity during attentive and relaxed wakefulness, drowsiness, and sleep; and sleep-waking cycle); *neuropharmacology* (barbital narcosis); and *clinical neurology* (coma following midbrain lesions). The predicted ascending reticular pathways were found in neuroanatomical investigations. This convergence of scientific interests and several interdisciplinary symposia on the ascending reticular system may explain why the original paper was highly cited. A more recent review has been prepared by Hobson and Brazier."[7]

1. Magoun H W. The role of research institutes in the advancement of neuroscience: Ranson's Institute of Neurology, 1928-1942. (Worden F G, Swazey J P & Adelman G, eds.) *The neurosciences: paths of discovery.* Cambridge, MA: MIT Press, 1975. p. 515-27.

2. Dell P C. Creative dialogues: discovery, invention, and understanding in sleep-wakefulness research. (Worden F G, Swazey J P & Adelman G, eds.) *The neurosciences: paths of discovery.* Cambridge, MA: MIT Press, 1975. p. 549-68.

3. Moruzzi G. Le développement des connaissances sur l'organisation et les fonctions de la substance réticulaire. *J. Physiol. Paris* 70:681-93, 1975.

4. Snider R S, McCulloch W S & Magoun H W. A cerebello-bulbo-reticular pathway for suppression. *J. Neurophysiology* 12:325-34, 1949.

5. Moruzzi G. Sui rapporti fra cervelletto e corteccia cerebrale. III. Meccanismi e localizzazione delle azioni inibitrici e denamogene del cervelletto. *Arch. Fisiol.* 41:183-205, 1941.

6. Lindsley D B, Bowden J & Magoun H W. Effect upon the EEG of acute injury to the brain stem activating system. *Electroencephalogr. Clin. Neuro.* 1:475-86, 1949.

7. Hobson J A & Brazier M A B, eds. *The reticular formation revisited.* New York: Raven Press, 1980. 564 p.

CC/NUMBER 28
JULY 13, 1981

This Week's Citation Classic

Hodgkin A L & Huxley A F. A quantitative description of membrane current and its application to conduction and excitation in nerve.
J. Physiol. London 117:500-44, 1952.
[Physiological Laboratory, University of Cambridge, Cambridge, England]*

The changes in sodium and potassium permeability which underlie the nerve impulse can be described quantitatively by equations which are consistent with the idea that movement of membrane charges or dipoles, produced by changes in electric field, control gates to Na^+ and K^+. When these equations are solved, they account quantitatively for the shape and velocity of the propagated impulse and the associated conductance change and ionic movements, as well as several puzzling subthreshold phenomena. [The *SCI®* indicates that this paper has been cited over 1,970 times since 1961.]

Alan L. Hodgkin
Physiological Laboratory
University of Cambridge
Cambridge CB2 3EG
and
Andrew F. Huxley
Department of Physiology
University College
London WC1E 6BT
England

June 16, 1981

"This paper concludes a series describing experiments which we carried out with B. Katz in the summers of 1948 and 1949 on the giant nerve fibres of the squid. The method we used was introduced in a simplified form by K.S. Cole, who started experiments with it in 1947.[1,2] Our version required us to insert two metal wires down the axis of a nerve fibre—to avoid electrode polarization effects, separate wires were used for passing current and for sensing the resulting changes of potential. We did this by winding two 20 μm silver wires in a double helix round an 80 μm glass rod and then pushing the assembly down a track predrilled with a smooth capillary by the method that we introduced in 1939.[3] A feed-back amplifier sends through one wire whatever current is needed to make the other wire undergo a step-like change of potential, and the time course of this current through the membrane is recorded. This is much easier to interpret than the results of simpler procedures such as recording voltage at constant current or, as in most previous experiments, of applying a stimulus and letting the nerve take charge.

"Carrying out a good voltage-clamp experiment is rather like climbing a beautiful mountain by a difficult route. Of course there is no danger, but you have to go through a whole series of tricky operations over five or six hours and a single mistake can wreck all the hard work done by your colleagues and yourself. A cleaned squid nerve fibre with a double spiral of wire inserted correctly down its axis is really a most satisfying object and it is a pity that so few people have the opportunity of repeating these experiments. Perhaps some of the aesthetic pleasure and excitement associated with the experiments have crept into the dry pages of a scientific journal and have attracted a wider range of readers than usual.

"Another reason why our paper has been widely read may be that it shows how a wide range of well-known, complicated, and variable phenomena in many excitable tissues can be explained quantitatively by a few fairly simple relations between membrane potential and changes of ion permeability—processes that are several steps away from the phenomena that are usually observed, so that the connections between them are too complex to be appreciated intuitively. There now seems little doubt that the main outlines of our explanation are correct, but we have always felt that our equations should be regarded only as a first approximation that needs to be refined and extended in many ways in the search for the actual mechanism of the permeability changes on the molecular scale."

1. Cole K S. Dynamic electrical characteristics of the squid axon membranes. *Arch. Sci. Physiol.* 3:253-8, 1949.
2. ------------. *Membranes, ions and impulses.* Los Angeles, CA: University of California Press, 1968. p. 241-69.
3. Hodgkin A L & Huxley A F. Action potentials recorded from inside a nerve fibre. *Nature* 144:710-11, 1939.

* The experimental work on which this article is based was largely done at the Laboratory of the Marine Biological Association, Plymouth, during 1948 and 1949.

This Week's Citation Classic™

Narahashi T, Moore J W & Scott W R. Tetrodotoxin blockage of sodium conductance increase in lobster giant axons. *J. Gen. Physiol.* 47:965-74, 1964.
[Dept. Physiology, Duke Univ. Medical Center, Durham, NC]

The highly specific and potent action of tetrodotoxin, a puffer fish poison, in blocking nerve membrane sodium channels was demonstrated for the first time using the voltage clamp technique. This finding triggered a widespread use of this and other toxins as tools for the study of ionic channels. [The *SCI®* indicates that this paper has been cited in over 630 publications since 1964.]

Toshio Narahashi
Department of Pharmacology
Northwestern University Medical School
Chicago, IL 60611

June 29, 1984

"Whereas the experiments were performed in only two months, the above study was a result of my long-term dream and planning. In the early 1950s, I was an instructor at the University of Tokyo and was shocked, to say the least, by reading a series of publications by Hodgkin, Huxley, and Katz,[1] who used the voltage clamp technique to establish the ionic theory of nerve excitation. I clearly foresaw the applicability of this powerful technique to my study of drug action on nerve membranes, but the idea was too provocative to pursue at that time. The technique was difficult, and believe it or not, the use of chemicals as tools was almost unthinkable in neurophysiology.

"In 1959, I came across a truly fascinating action of tetrodotoxin (TTX), a toxic component contained in the puffer fish. Based on intracellular microelectrode experiments with frog skeletal muscle fibers, we proposed that TTX blocks the excitation through a selective inhibition of sodium channels.[2] On the day of my departure for the US, in January 1961, Norimoto Urakawa, a collaborator in the TTX study, slipped a small vial containing TTX into my pocket. We were hoping that some day we would be able to demonstrate our hypothesis by using the voltage clamp technique.

"It was not until December 1962 that I had a chance to do the long-awaited experiment. I was then an assistant professor at Duke University, and decided to stay in the US permanently. However, the situation forced me to go back to Japan temporarily to obtain an immigrant visa, and I had only two more months to work there before my departure. John W. Moore, an expert in the voltage clamp technique, and I thought that the TTX project could be carried out during that short period of time. Voltage clamp experiments with lobster giant axons were then conducted literally day and night throughout the Christmas holiday with the help of William Scott (then a medical student) using a double sucrose-gap technique.[3] The technique was far from satisfactory at that time, and countless experimental results had to be discarded because of poor membrane current records. Nevertheless, we were jubilant at finding that TTX blocked the sodium current without any effect on the potassium current. I took the films of oscilloscope records, which had barely dried, to Japan for analysis in January 1963. After submitting the above paper, I received the first request for a sample of TTX which was jotted down with the signature at the end of the referee's comments!

"Because of the highly specific and potent action, TTX has since become an extremely popular tool for the study of excitable membrane ionic channels. It has been used to estimate the sodium channel density, to identify and characterize sodium and other ionic channels, to study synaptic transmission, and to isolate and purify the sodium channels, to mention a few. I believe that this is the very reason for frequent citation of this paper which represents the first, clear-cut demonstration of the TTX action by voltage clamp. The paper has made another equally important contribution—it opened up the avenue to the use of specific toxins and chemicals as tools for the study of excitable membrane,[4-6] an enthusiasm generated only after that study. My ten-year-old dream had finally materialized. A series of studies initiated by this paper led Moore and me to receive, in 1981, the Cole Award in Membrane Biophysics, the most prestigious in the field."

1. **Hodgkin A L, Huxley A F & Katz B.** Measurement of current-voltage relations in the membrane of the giant axon of *Loligo. J. Physiol.—London* 116:424-48, 1952. (Cited 555 times.)
2. **Narahashi T, Deguchi T, Urakawa N & Ohkubo Y.** Stabilization and rectification of muscle fiber membrane by tetrodotoxin. *Amer. J. Physiol.* 198:934-8, 1960. (Cited 175 times since 1960.)
3. **Julian F J, Moore J W & Goldman D E.** Current-voltage relations in the lobster giant axon membrane under voltage clamp conditions. *J. Gen. Physiol.* 45:1217-38, 1962. (Cited 135 times.)
4. **Narahashi T.** Chemicals as tools in the study of excitable membranes. *Physiol. Rev.* 54:813-89, 1974. (Cited 425 times.)
5. **Catterall W A.** Neurotoxins that act on voltage-sensitive sodium channels in excitable membranes. *Annu. Rev. Pharmacol. Toxicol.* 20:15-43, 1980.
6. **Ritchie J M.** A pharmacological approach to the structure of sodium channels in myelinated axons. *Annu. Rev. Neurosci.* 2:341-62, 1979.

Bradley P B & Elkes J. The effects of some drugs on the electrical activity of the brain. *Brain* **80**: 77-117, 1957. [University of Birmingham, Department of Experimental Psychiatry, Birmingham, England]

Using animal preparations carrying chronically implanted electrodes and two acute preparations, the *encephale* and *cerveau isole,* it was possible to differentiate between drugs with similar effects on electrical activity and behaviour, and those which caused 'dissociation.' Certain actions could be related to the brain stem reticular formation. [The *SCI®* indicates that this paper has been cited over 220 times since 1961.]

Philip B. Bradley
University of Birmingham
Department of Pharmacology
Birmingham, B15 2TJ England

March 6, 1978

"It is very pleasing to learn that one's first major paper, published more than 20 years ago, has become a 'Citation Classic,' although it is difficult to think of reasons why this should be so. It is probably fair to say that at the time the research reported in this paper was being carried out we were pioneering, in the sense that there were very few others working in the same field. Thus, there were few centres working on the CNS, probably because of the lack of suitable techniques and almost none, at least in Europe, where centrally acting drugs were being studied. In England our work did not seem to meet with the approval of the scientific establishment and our presentations at the meetings of the Physiological Society provoked little discussion. At the time I found this very discouraging as I had doubts about whether or not I had done the right thing in leaving the discipline in which I had been trained (Zoology) for a completely new

area, which only later became known as 'Neuropharmacology,' and which at that time seemed to have a doubtful future.

"However, while I was busy trying to integrate different kinds of basic research, i.e., behavioural studies and electrophysiology, my colleague, chief, and co-author, Joel Elkes was occupied with developing the clinical aspects of our work. This was in the days before psychotropic drugs were even thought of, and we first heard about newly discovered drugs such as LSD 25 and chlorpromazine by word of mouth, usually from the representatives of the pharmaceutical industry, who also supplied us with our first samples of these new compounds. Suffice it to say our work caught the interest of many clinicians, psychiatrists, neurologists, neurosurgeons, etc. It was a result of a suggestion by Lord Brain, then Sir Russell Brain and editor of the journal *Brain,* that our first major publication was submitted to a clinical journal.

"In retrospect, considering the enormous increase in both interest and in the volume of active research, utilising highly sophisticated techniques, on the actions of drugs in the CNS, I find it surprising that our 21-year-old paper, which describes some relatively simple experiments, should still be quoted. Most of the findings were not unexpected, and only one, that of the 'pharmacological dissociation' between behaviour and the EEG when antagonists of acetylcholine were administered, has caused any controversy. Even this was not a new discovery, as similar effects had been reported in the dog. In my view the controversy still exists only because we have been *misquoted* and perhaps I might end with the plea, *not* to quote us without reading the original paper first."

This Week's Citation Classic

Goff W R, Rosner B S & Allison T. Distribution of cerebral somatosensory
evoked responses in normal man.
Electroencephalogr. Clin. Neuro. 14:697-713, 1962.
[West Haven Vet. Admin. Hosp. and Yale Univ. Sch. Med., New Haven, CT]

This paper describes the distribution over the scalp of electrical potentials generated in the brain of normal alert humans by median nerve or finger stimulation. The complex series of potentials was divided into components which had different scalp distributions, and by inference different neural origins. [The *SCI®* indicates that this paper has been cited over 145 times since 1962.]

William R. Goff
Neuropsychology Laboratory
West Haven VA Medical Center
West Haven, CT 06516
and
Departments of Neurology and Psychology
School of Medicine and Graduate School
of Arts and Sciences
Yale University
New Haven, CT 06510

August 14, 1981

"When I came to West Haven VA Medical Center in 1959, Burt Rosner (now at the University of Pennsylvania) and Truett Allison were developing a device for averaging the electroencephalogram (EEG) to derive stimulus evoked neuroelectrical potentials from the human scalp. They were pursuing Dawson's[1] demonstration that the signal averaging technique could be used to extract brain potentials occurring with a fixed temporal relationship to repeated stimuli from the higher-voltage EEG. These 'evoked potentials' (EPs) presumably reflected the brain's processing of sensory information. The exciting significance of Dawson's report was that such brain activity could be studied in man by a non-invasive technique.

"Dawson's averager was mechanical. Its memory was capacitors fed from a rotating switch. Rosner's idea was to use a loop of FM magnetic tape as a memory. The EP to the first stimulus was tape-recorded. When the loop came full circle, the first potential was played back into a summing amplifier synchronous with a newly evoked potential. The sum of these was rerecorded and the process repeated until reproducible potentials appeared on an oscilloscope. Simple in principle, but in practice, synchronizing the Nth and Nth + 1 potentials was difficult, and we spent nearly a year working out that and other technical problems.

"We christened the resulting Rube Goldberg apparatus 'ERA' (evoked response averager)[2] and showed that it properly summated square wave pulses. Nonetheless, we were uneasy about the validity of the complex waveforms we recorded to median nerve stimulation. Another EP computer, ARC-1,[3] had been developed at MIT by a group headed by Walter Rosenblith, a friend of Rosner's. We recorded somatosensory EPs (SEPs) from Allison on ERA in West Haven and on ARC-1 in Cambridge. The SEPs matched, and earned Allison the nickname 'calibrated brain.'

"Convinced by this curious validation that ERA worked—what to do with it? Preliminary recordings revealed that the total SEP was a complex series of potentials which lasted about 400 msec, and varied at different scalp locations. An obvious question was whether the topographic differences might tell us something about the location of the neural generators.

"Probably the resulting paper has been cited frequently because: 1) It first described SEP morphology and topography for the entire duration of the response. 2) With a companion paper,[4] it parceled the SEP into components and suggested that they arose from different neuroanatomical sources. 3) Scalp topographic studies continue to be one of the best methods of inferring the location of generators of EPs in humans,[5] in whom there are but few opportunities to record intracranially.[6]

"As co-workers for over 20 years, Allison and I have been afforded a privilege uncommon among scientists. Beginning with a home-built gadget early in the investigation of human EPs, we have witnessed an area of basic research develop into a clinically valuable procedure for evaluating a variety of neurological dysfunction."[7]

1. Dawson G D. A summation technique for the detection of small evoked potentials.
   Electroencephalogr. Clin. Neuro. 6:65-84, 1954.
2. Rosner B S, Allison T, Swanson E & Goff W R. A new instrument for the summation of evoked
   responses from the nervous system. *Electroencephalogr. Clin. Neuro.* 12:745-7, 1960.
3. Clark W A, Jr. *Average response computer (ARC-1).* Cambridge, MA: Research Laboratory of Electronics,
   Massachusetts Institute of Technology, 15 April 1958, Quarterly Progress Report. p. 114-17.
4. Allison T. Recovery functions of somatosensory evoked responses in man.
   Electroencephalogr. Clin. Neuro. 14:331-43, 1962.
5. Wood C C. Application of dipole localization models to the source of identification of human
   evoked potentials. *Ann. NY Acad. Sci.* In press, 1981.
6. Goff W R, Allison T & Vaughan H G, Jr. The functional neuroanatomy of event related potentials.
   (Callaway E, Tueting P & Koslow S H, eds.) *Event-related brain potentials in man.*
   New York: Academic Press, 1978. p. 1-79.
7. Desmedt J E, ed. *Progress in clinical neurophysiology. Vol. 7. Clinical uses of cerebral, brainstem and
   spinal somatosensory evoked potentials.* Basel: Karger, 1980. 352 p.

Garey L J, Jones E G & Powell T P S. Interrelationships of striate and extrastriate
cortex with the primary relay sites of the visual pathway.
J. Neurol. Neurosurg. Psychiat. **31**:135-57, 1968.
[Department of Human Anatomy, Oxford University, England]

Areas 17 and 18 of the cat visual cortex send independent retinotopic subcortical projections. Callosal fibres connect parts of the cortex related to the vertical visual meridian. Area 17 sends fibres to areas 18 and 19 ipsilaterally, and area 18 to areas 17 and 19. [The *SCI®* indicates that this paper has been cited in over 340 publications since 1968.]

Laurence J. Garey
Institute of Anatomy
University of Lausanne
CH-1011 Lausanne
Switzerland

June 25, 1984

"In 1963, after the preclinical part of a medical degree at Oxford, I undertook a research project in the department of human anatomy before continuing clinical studies. I was attracted to the visual system by undergraduate contacts with two leaders in the field, Sir Wilfrid Le Gros Clark, then professor of anatomy, and Tom Powell, my anatomy 'tutor.' Nineteen sixty-three was an exciting time for visual research: Hubel and Wiesel[1] had just published their paper on the cat visual cortex; James McGill had recently demonstrated the precise retinotectal projection in the pigeon;[2] and neuroanatomy was enjoying a resurgence thanks to the 'Nauta' technique for degenerating axons.[3] With Tom, I studied the connexions of the cat visual system from retina to cortex, and from cortex to subcortical centres, with particular attention to relationships with the superior colliculus. Max Cowan was also there, and I see from my records that Tom and Max introduced me to making discrete cortical lesions in October 1963.

"Using lesions in different parts of the visual cortex, we described the retinotopicity of descending cortical projections. However, virtually the whole of the geniculate and colliculus contained degeneration after lesions involving less than the total medio-

lateral extent of what was recognized as 'visual' cortex: its medial and lateral parts had independent subcortical projections. We therefore made small lesions restricted to the medial or lateral visual cortex, and even in the auditory, somatosensory, and motor cortex. The 1965 paper of Hubel and Wiesel[4] describing the organization of areas 17, 18, and 19 helped us define the separate subcortical projections from each area. The superficial laminae of the superior colliculus received information from the visual cortex, while other cortex projected to deeper layers. We also investigated ipsilateral and callosal cortico-cortical connexions, confirming that the cortex related to the vertical visual meridian projected callosally.

"It was not easy to cut frozen sections of the whole cat brain and then stain and mount the delicate sections. I well remember the 'dry ice,' used to freeze the brain, that would evaporate during the coffee break, and the dozens of little glass dishes into which the fragile sections were plunged one by one using tiny glass rods. Fortunately, we had the solid backing of Ron Brooke and his technical staff to help us. The Nauta technique sometimes worked—and sometimes did not! Was it the weather or the Oxford water? More likely it was our inexperience, for later its reliability improved and we were able to mass-produce consistent sections.

"In 1965, I left for St. Thomas' Hospital Medical School in London, leaving Tom with the unenviable task of making the relevant chapters of my thesis into a paper. At that time, Ted Jones arrived in Oxford from Otago and together they worked through the material, and added some; and so the paper was written. It gave anatomical support to contemporary work on the visual cortex, using an accurate and relatively reliable technique. The superior colliculus was emphasized as a cornerstone between the retino-thalamo-cortical visual system and oculomotricity. It also came at a time when attention was being paid to callosal and other cortico-cortical connexions. For Ted and me, it represented an important step in our introduction to experimental neuroanatomy."

1. **Hubel D H & Wiesel T N.** Receptive fields, binocular interaction and functional architecture in the cat's visual system. *J. Physiol.—London* **160**:106-54, 1962. (Cited 2,015 times.)
2. **McGill J I, Powell T P S & Cowan W M.** The retinal representation upon the optic tectum and isthmo-optic nucleus in the pigeon. *J. Anatomy* **100**:5-33, 1966. (Cited 110 times.)
3. **Nauta W J H & Gygax P A.** Silver impregnation of degenerating axons in the central nervous system. A modified technic. *Stain Technol.* **29**:91-3, 1954. (Cited 650 times since 1955.)
4. **Hubel D H & Wiesel T N.** Receptive fields and functional architecture in two nonstriate visual areas (18 and 19) of the cat. *J. Neurophysiology* **28**:229-89, 1965. (Cited 945 times.)

Atwood H L. Crustacean neuromuscular mechanisms. *Amer. Zool.* 7:527-52, 1967.
[Dept. Zoology, Univ. Toronto, Toronto, Ontario, Canada]

Speed and strength of muscular response in crustaceans is attributable to inherent electrical and mechanical properties of muscle fibers, and in part to quantitative aspects of transmitter release at nerve terminals. Peripheral inhibitory synapses modulate muscular contractions through pre- and post-synaptic mechanisms. [The *SCI®* indicates that this paper has been cited over 120 times since 1967.]

Harold L. Atwood
Department of Zoology
University of Toronto
Toronto, Ontario M5S 1A1
Canada

April 1, 1981

"Studies summarized in this paper were originally prompted by observations of G. Hoyle and C.A.G. Wiersma,[1] who measured contractions and electrical events evoked by stimulation of isolated motor neurons supplying crustacean limb muscles. Some muscles innervated by two excitatory motor neurons showed, with stimulation of a 'fast' motor neuron, large electrical events in individual muscle fibers, but a small contraction of the entire muscle. Conversely, identical stimulation of the companion 'slow' motor neuron elicited small electrical events in the same muscle fibers, accompanied by strong muscular contraction.

"They concluded that 'fast' and 'slow' axons released different transmitter chemicals that acted directly on the contractile machinery of the muscle fiber to produce characteristic contractions, independently of membrane electrical events.

"This bold idea, which was at variance with accepted dogma of excitation-contraction coupling in vertebrate muscles, attracted me into graduate work in comparative neuromuscular physiology. I went to work with Hoyle at the University of Glasgow in 1960, convinced that the hypothesis of Hoyle and Wiersma[1] was correct, and determined to strengthen it with further experimental evidence. However,

Hoyle soon left Glasgow for the University of Oregon. I stayed on; Peter Usherwood, another graduate student, became my chief mentor. I worked in relative isolation in a dark basement laboratory in Glasgow. There were very few distractions.

"While recording with microelectrodes from crustacean muscle fibers, I began to make findings at variance with my preconceptions. Although electrical events evoked by stimulation of the 'slow' axon were smaller than those of the 'fast' axon in accessible muscle fibers, a group of less accessible fibers showed the reverse pattern: much larger electrical events during stimulation of the 'slow' axon. These muscle fibers had distinctive membrane electrical properties[2] and were physiologically specialized for slow, powerful contractions. Other fibers in the same muscle were specialized for fast contractions. Thus, distinctive contractile responses evoked by the two motor axons were due to differential recruitment of specialized groups of muscle fibers, rather than to differences in transmitter chemicals. The mechanism of excitation-contraction coupling was found to be under electrical control, as in vertebrate muscle.[3]

"Few papers in comparative physiology become highly cited. Perhaps the reason is that comparative physiologists are more often concerned with divergent details than with generalities. The 1967 review counteracted this tendency by demonstrating features shared by neuromuscular systems of vertebrates and arthropods: elaboration of physiologically distinct muscle fibers for different functions, specialization of motor neurons into 'phasic' and 'tonic' types, matching of motor neurons and muscle fibers to form functionally diverse working units, and generally similar excitation-contraction coupling mechanisms. The reasons for the many citations of this review are several: it established some general features of crustacean neuromuscular systems, emphasized the utility of these systems for research on synaptic mechanisms akin to those in central nervous systems, and provided a timely summary of a rapidly developing field of research. A more recent review[4] has extended these themes."

1. Hoyle G & Wiersma C A G. Coupling of membrane potential to contraction in crustacean muscles. *J. Physiol. London* 143:441-53, 1958.
2. Atwood H L. Differences in muscle fibre properties as a factor in "fast" and "slow" contraction in *Carcinus. Comp. Biochem. Physiol.* 10:17-31, 1963.
3. Atwood H L, Hoyle G & Smyth T. Electrical and mechanical responses of single innervated crab-muscle fibres. *J. Physiol. London* 180:449-82, 1965.
4. Atwood H L. Organization and synaptic physiology of crustacean neuromuscular systems. *Progr. Neurobiol.* 7:291-391, 1976.

LaVail J H & LaVail M M. Retrograde axonal transport in the central nervous system. *Science* 176:1416-17, 1972.
[Dept. Neuropathology, Harvard Medical School, Boston, MA]

When the enzyme horseradish peroxidase is injected into the optic tectum of young chicks, the peroxidase is taken up and transported in a retrograde direction by the axons of retinal ganglion cells to the cell bodies in the retina. Likewise, when the enzyme is injected into the vitreal space of the eye, it is taken up and transported centripetally along efferent axons and is found histochemically in cell bodies within the isthmo-optic nucleus. This retrograde movement of protein from axon terminal to cell body suggests a possible mechanism by which neurons respond to their target areas. [The *SCI®* indicates that this paper has been cited in over 495 publications since 1972.]

Jennifer H. LaVail
Department of Anatomy
School of Medicine
University of California
San Francisco, CA 94143

May 11, 1983

"In 1971, Matthew LaVail and I were postdoctoral fellows in the Laboratory of Neuropathology at Harvard Medical School where we worked under the sponsorship of Richard Sidman. Sidman had assigned me a lecture on axonal transport during development, and in the course of preparing the lecture, I noticed a gap in the literature. Although there was growing evidence of the importance of transport of macromolecules from neuron cell body to axon, there was no information about what happened to macromolecules once they reached the axon terminal. Obviously, homeostasis in the cell required that material move back as well. In fact, that suggestion had been made in 1948 by Weiss and Hiscoe.[1] Other evidence, for example, that viruses might gain access to the central nervous system (CNS) by incorporation by the axon and transport to the cell body, had appeared in 1941 when Bodian and Howe described the dissemination of poliomyelitis virus in the nervous system.[2]

"One morning after I had given the lecture, Sidman left a copy of a brief report on my desk on the uptake and transport of horseradish peroxidase by peripheral nerve to motor neurons in the spinal cord by Kristensson, Olsson, and Sjöstrand.[3] Matt and I discussed the importance of the paper and decided it should be repeated in a site where no direct impalement of the axon was possible, and in the CNS where, we erroneously reasoned, the phenomenon of retrograde axonal transport might not exist, thus possibly explaining the absence of axonal regeneration in the CNS.

"From our graduate studies we were familiar with the structure of the retina and chick brain. Therefore, we chose to inject horseradish peroxidase intravitreally in the chick and look for evidence of its retrograde transport and accumulation in neuron cell bodies in the midbrain. In another series of experiments we injected the enzyme into the tectum and looked for its accumulation in retinal ganglion cell bodies. In both experiments, after retrograde transport the horseradish peroxidase was obvious in the tissue sections using a light microscope.

"The publication has been highly cited probably for two reasons. First, it demonstrated that the phenomenon of retrograde axonal transport occurred in the CNS in the absence of direct mechanical injury to the axons. Second, the fact that axon terminals in an injection site would pick up peroxidase and transport it to the cell somas meant that neurobiologists could determine whether or not a neuron cell body projected an axon to a particular site by simple light microscopic histochemistry. Such an experiment usually took only a day or two to complete as compared to the weeks or months necessary to obtain the same kind of information using other techniques. Previously we were handicapped by the fact that origins of long axonal projections could be determined only if the cell bodies died after axon transection, and often the neuron cell body would not respond in such a dramatic way. Moreover, visual detection of cell loss with the traditional methods often required that large populations of neurons die in response to axonal transection, whereas the retrograde transport method made use of a physiological property of the cell and individual neurons that transported a marker could be detected. Since then, advances in chromagens for horseradish peroxidase and additional probes have become available,[4,5] and we now have further information about the retrograde transport of endogenous macromolecules."[6]

1. Weiss P & Hiscoe H B. Experiments of the mechanism of nerve growth. *J. Exp. Zool.* 107:315-95, 1948.
2. Bodian D & Howe H A. The rate of progression of poliomyelitis virus in nerves.
   Bull. Johns Hopkins Hosp. 69:79-85, 1941.
3. Kristensson K, Olsson Y & Sjöstrand J. Axonal uptake and retrograde transport of exogenous proteins in the hypoglossal nerve. *Brain Res.* 32:399-406, 1971.
4. Mesulam M-M & Rosene D L. Sensitivity in horseradish peroxidase neurohistochemistry: a comparative and quantitative study of nine methods. *J. Histochem. Cytochem.* 27:763-73, 1979.
5. Kuypers H G J M, Bentivoglio M, Catsman-Berrevoets C E & Bharos A T. Double retrograde neuronal labeling through divergent axon collaterals, using two fluorescent tracers with the same excitation wavelength which label different features of the cell. *Exp. Brain Res.* 40:383-92, 1980.
6. Grafstein B & Forman D S. Intracellular transport in neurons. *Physiol. Rev.* 60:1167-283, 1980.

Taxi J. Contribution a l'étude des connexions des neurones moteurs du systéme nerveux autonome. (Contribution to the study of the connections of motor neurons in the autonomic nervous system.)
Ann. Sci. Natur. Zool. 7:413-674, 1965.
[Laboratoire de Biologie Animale, Faculté des Sciences, Université Pierre et Marie Curie, Paris, France]

This is a light and electron microscopic study of: 1) the ganglionic synapses in vertebrates, which emphasized their particular features in amphibians; 2) the significance of the 'interstitial cells' of Cajal in the neurovegetative effector pathway, rejecting their neuronal or glial nature; 3) the neuromuscular junctions in various mammalian muscles, leading to the existence of two main types of innervation. [The *SCI*® indicates that this paper has been cited in over 205 publications, making it the most-cited paper published in this journal.]

Jacques Taxi
Laboratoire de Cytologie
Faculté des Sciences
Université Pierre et Marie Curie
75230 Paris
France

May 5, 1984

"This work was initiated at Couteaux's suggestion when I first went to his laboratory in the 1950s. His original aim was to clarify the very confused situation prevailing at the time concerning the cellular relationships in the autonomic nervous system, since several leading morphologists like J. Boeke[1] and P. Stöhr[2] were defending reticularist conceptions of the organization of the autonomic circuits. The role of the primitive 'interstitial neurons,' which Cajal[3] confined to the peripheral autonomic nervous system, was extended by Boeke to the entire nervous system.

"Our paper, which appeared in 1965, as a DSc thesis, condensed about 15 years of histological and cytological observations. They started in light microscopy, but a veritable mutation occurred in the prospects for this work when electron microscopy became routinely feasible for us around 1956. The three main conclusions reached were: 1) the synapses of the autonomic ganglia studied in various vertebrates are very similar to those described in the central nervous system, with only a few particularities, especially in amphibians; 2) Cajal's interstitial cells (ex-neurons) should be clearly distinguished from both nerve and Schwann cells. We suggested that they were homologous to the connective cells of Henle's sheath enveloping the motor nerve endings; and 3) the relationships between autonomic nerve fibers and smooth muscle cells can be classified as 'fascicular' and 'individual' innervation, which fits well with the physiological distinction of Bozler[4] between 'unitary' and 'multiunit' muscles. This last part of our work was especially hard and time-consuming because, among other things, it required 400 grids of semiserial sections for electron microscopy of the intestinal muscle innervation.

"Conceptually, these conclusions were not really new, as they had already been formulated with varying degrees of clarity. However, objective evidence was lacking, due to the limitations of light microscopy methods. The new situation created by electron microscopy was indicated by a letter I received in 1966 from Stöhr, who had then just retired. He wrote, in reply to my article, that the continuity of nerve fibers and effectors could no longer be supported in view of the evidence provided by electron microscopy.

"If our contribution is still of some interest, it is perhaps because it was one of the first synthetic approaches to the ultrastructure of the autonomic nervous system, and was abundantly illustrated. Since 1965, many important new observations have been made in this field, but it nevertheless seems to us that our paper has been completed rather than challenged. Those interested will find that the topics discussed have been brought up to date in the recent reports by Gabella,[5] Thuneberg,[6] and Elfvin."[7]

1. **Boeke J.** The sympathetic end formation, its synaptology, the interstitial cells, periterminal network and its bearing on the neurone theory. Discussion and critique. *Acta Anat.* 8:18-61, 1949. (Cited 60 times since 1955.)
2. **Stöhr P, Jr.** *Mikroskopische Anatomie des vegetativen Nervensystems.* Berlin: Springer-Verlag, 1957. 678 p.
3. **Cajal S R.** *Histologie du systéme nerveux de l'homme et des vertébrés.* Paris: Maloine, 1911. Vol. II. p. 923.
4. **Bozler E.** Conduction, automaticity and tonus of visceral muscles. *Experientia* 4:213-18, 1948. (Cited 200 times since 1955.)
5. **Gabella G.** *Structure of the autonomic nervous system.* London: Chapman and Hall, 1976. 214 p.
6. **Thuneberg L.** Interstitial cells of Cajal: intestinal pacemaker cells? (Whole issue.) *Advan. Anat. Embryol. Cell Biol.* **71**, 1982. 130 p.
7. **Elfvin L G,** ed. *Autonomic ganglia.* New York: Wiley, 1983. 527 p.

This Week's Citation Classic

Kety S S & Schmidt C F. The nitrous oxide method for the quantitative
determination of cerebral blood flow in man: theory, procedure and normal
values. *J. Clin. Invest.* 27:476-83, 1948.
[Dept. Pharmacology, Univ. Pennsylvania, Philadelphia, PA]

Cerebral circulation and oxygen consumption of
the human brain were measured, using a new
method based upon the blood:tissue exchange of
an inert gas. In normal young men, mean values of
54 ml and 3.3 ml per minute per 100 g of brain
were obtained for blood flow and oxygen consumption, respectively. Assumptions on which the
theory was based were examined experimentally
and found to be tenable. [The *SCI®* indicates that
this paper has been cited over 565 times since
1961.]

Seymour S. Kety
Harvard Medical School
Mailman Research Center
McLean Hospital
Belmont, MA 02178

August 7, 1981

"This paper presented the theory and
validation of a new technique which permitted the first quantitative measurements of
human cerebral blood flow and energy metabolism in physiological states and in
disease. The method was based upon the uptake by the brain of a diffusible inert gas
supplied by way of the arterial blood. The
unique qualities of the human brain and the
lack of appropriate animal models of human neurological and mental disease made
the development of a method that could
safely be applied to unanesthetized man
particularly desirable.

"Earlier investigators had measured the
arteriovenous oxygen difference across the
brain as an index of blood flow or of oxygen
consumption under a variety of conditions.
What had limited the acceptance of that approach, however, was that the arteriovenous
oxygen difference, being determined by
both blood flow and oxygen consumption,
was not a valid measure of either. The oxygen consumption of the brain could not be
measured independently or even assumed
to be constant, since it would be expected to
vary with the states of activity or disease
which were the object of investigation.

"The brain, however, absorb by
physical solution an inert gas such as nitrous
oxide which reaches it by way of the arterial
blood. The accumulation of such a tracer in
the brain should be independent of mental
state and neuronal function, since it would
be determined instead by physical characteristics, such as its rate of diffusion or
solubility in brain, which would not be expected to vary with functional activity. It
seemed likely also that cerebral blood flow
could be calculated by monitoring the concentrations of the inert gas in arterial and
cerebral venous blood from the onset of its
inhalation until equilibrium was achieved,
at which time the partial pressure of the gas
in the brain would be equivalent to that in
its effluent blood.

"Then began a theoretical examination of
the dynamics of distribution of diffusible,
nonmetabolized substances,[1] reinforced by
a number of experimental studies with many
collaborators to test the assumptions involved, develop a practical technique validated by comparison with direct measurement in the monkey, and to employ the new
technique in gaining some understanding of
the circulation and energy metabolism of
the human brain in a wide variety of
physiological and pathological conditions.
Later, with Landau, Freygang, Rowland, and
Sokoloff,[2] I applied the theory to measurement of regional circulation in the brain of
the cat, using an internally calibrated
autoradiographic technique for measuring
the regional distribution of a radioactive
tracer.

"This early work has been cited probably
because it made possible contributions by a
large number of other investigators on the
blood flow, vascular resistance, and oxygen
and glucose utilization of the human brain
in health and disease,[3] and because the
theory on which it was based led directly or
indirectly to the development of current
methods for the measurement of regional
blood flow, metabolism, and the visualization of functional activity throughout the
human brain.[4,5] It has been the basis of
several awards, the first being the Theobald
Smith award in 1949 and the most recent,
the Passano award in 1980."

1. **Kety S S.** The theory and applications of the exchange of inert gas at the lungs and tissues.
   Pharmacol. Rev. 3:1-14, 1951.
2. **Landau W M, Freygang W H, Rowland L P, Sokoloff L & Kety S S.** The local circulation of the living
   brain; values in the unanesthetized and anesthetized cat. *Trans. Amer. Neurol. Assn.* 80:125-9, 1955.
3. **Lassen N A.** Cerebral blood flow and oxygen consumption in man. *Physiol. Rev.* 39:183-238, 1959.
   [Citation Classic. *Current Contents/Clinical Practice* 23(10):12, 10 March 1980.]
4. **Ingvar D H & Lassen N A.** Regional blood flow of the cerebral cortex determined by Krypton[85].
   Acta Physiol. Scand. 54:325-88, 1962.
5. **Sokoloff L.** Relation between physiological function and energy metabolism in the central nervous
   system. *J. Neurochemistry* 29:13-26, 1977.

This Week's Citation Classic™

CC/NUMBER 17
APRIL 23, 1984

Harper A M, Deshmukh V D, Rowan J O & Jennett W B. The influence of
sympathetic nervous activity on cerebral blood flow. *Arch. Neurol.* 27:1-6, 1972.
[MRC Cerebral Circulation Res. Group, Wellcome Surgical Res. Inst., and Inst. Neurological
Sciences, Univ. Glasgow, Scotland]

The effect on the cerebral blood flow (CBF) of
stimulation of the cervical sympathetic trunk was
explored in anaesthetised baboons at normocap-
nia and hypercapnia. Sympathetic stimulation
produced a significant reduction in CBF during
hypercapnia. It is argued that there is a dual control of CBF — the extraparenchymal vessels being
influenced by sympathetic nerves while the intraparenchymal vessels are under local intrinsic
metabolic regulation. The pial vessels are possibly
influenced by both systems. [The *SCI*® indicates
that this paper has been cited in over 175 publications since 1972.]

A. Murray Harper
Wellcome Surgical Institute
University of Glasgow
Glasgow G61 1QH
Scotland

February 3, 1984

"After qualifying in medicine, I was
awarded a research scholarship in the embryonic cardiovascular department at Glasgow Royal Infirmary in the late 1950s. At
that time, it seemed a good idea to study the
effects of cardiopulmonary bypass on the
cerebral circulation. Before doing this, I had
to set up techniques for measuring blood
flow through the brain (in experimental animals) which avoided the artefact of 'extra-
cranial contamination,' an artefact which
bedevilled so many of the methods in use at
that time. The effects of changing various
physiological parameters on cerebral blood
flow (CBF) had to be measured. Of particular importance were the blood gases, the
arterial blood pressure, and their interrela-
tionship. As far as I am aware, I was the first
to demonstrate quantitatively the presence
of cerebral 'autoregulation' — that is, the

maintenance of a relatively constant CBF in
the face of alterations in arterial blood pressure.[1] Interspersed with the development of
methods which involved the first use of
^{133}xenon for the measurement of CBF in
man[2] and other clinical studies, these 'control' studies (performed at the Wellcome
Surgical Institute, University of Glasgow)
took ten years.

"I was then convinced that the generally
accepted view of the cerebral circulation — namely, that it was controlled by the
metabolic needs of cerebral tissue — was
correct, when a bombshell arrived. A paper
published in 1969[3] suggested that extrinsic
nerves (sympathetic and cholinergic) supplying the neck arteries and the carotid sinus
played a major part in the control of cerebral circulation. My colleagues (a PhD student, V.D. Deshmukh; a physicist, J.O.
Rowan; and a professor of neurosurgery,
W.B. Jennett) and I decided that the observations were surprising on morphological
grounds and, over several years, devised a
series of experiments to test the effect of
neurogenic influence on CBF.

"The paper is only one of a series, but I believe it is cited so frequently because it contains an important hypothesis — namely, that
the sympathetic nerves in the neck can modulate the response of the cerebral blood vessels to changes in blood gases and blood
pressure, but are not the ultimate controlling factor in the amount of blood perfusing
any given area of the brain. It opened the
way to the rational interpretation of the effects of carotid artery stenosis and ligation
on CBF in animals and man which had considerable clinical benefits.[4] The paper also
made one suggestion which led to a proliferation of studies throughout the world —
namely, 'Sympathetic innervation may help
to protect the intraparenchymal vessels during acute hypertensive episodes....' This was
subsequently shown to be true.[5]

"Although I am still studying cerebral metabolism and blood flow, 26 years later I still
have not got around to starting my original
project!"

1. Harper A M. The inter-relationship between a PCO$_2$ and blood pressure in the regulation of blood flow through
   the cerebral cortex. *Acta Neurol. Scand.* 41(Suppl. 14):94-103, 1965.
2. Glass H I & Harper A M. Measurement of regional blood flow in cerebral cortex of man through intact skull.
   Brit. Med. J. 1:593, 1963.
3. James I M, Miller R A & Purves M J. Observations on the extrinsic neural control of cerebral blood flow in the
   baboon. *Circ. Res.* 25:77-93, 1969. (Cited 215 times.)
4. Jennett B, Miller J D & Harper A M. *Effect of carotid artery surgery on cerebral blood flow.*
   Amsterdam: Excerpta Medica, 1976. 170 p.
5. MacKenzie E T, McGeorge A P, Graham D I, Fitch W, Edvinsson L & Harper A M. Effects of increasing arterial
   pressure on cerebral blood flow in the baboon: influence of the sympathetic nervous system.
   Pflügers Arch.—Eur. J. Physiol. 378:189-95, 1979.

Anand B K & Brobeck J R. Hypothalamic control of food intake in rats and cats.
Yale J. Biol. Med. 24:123-40, 1951.
[Lab. Physiol., Yale Univ., New Haven, CT]

Bilateral destructions of a well localised area in lateral hypothalamus led to complete cessation of eating, while lesions involving ventromedial nuclei or the region between these and lateral area produced hyperphagia and obesity. The lateral region (hunger mechanism) was designated 'feeding center' and the medial (inhibitory control) 'satiety center.' [The *SCI®* indicates that this paper has been cited over 400 times since 1961.]

Bal K. Anand
Department of Physiology
All India Institute of Medical Sciences
New Delhi 110 016
India

June 23, 1980

"Development of obesity in animals following bilaterally placed destructive lesions in or ventrolateral to the ventromedial nuclei of the hypothalamus had been established through studies of Hetherington, Brobeck, and others,[1,2] primarily from a marked increase in food intake.

"In 1950, while working at Yale University and trying to stereotaxically place such hypothalamic lesions in the albino rats to make them hyperphagic, I was much disconcerted to find that my rats immediately after such lesions completely stopped eating and would die of starvation. Little did I realise at the moment that this may lead to the discovery of an important hypothalamic area controlling our hunger mechanism, which discovery would change our concepts about the nervous mechanisms concerned with our feeding behaviour and regulation of energy balance.

"To better understand the hypothalamic mechanisms we carried out investigations to discover the effects on food intake of small electrolytic lesions placed in the different areas of the hypothalamus which resulted in the discovery of a well localised area in the lateral hypothalamus, the bilateral destruction of which by very small lesions resulted in a complete loss of feeding behaviour to the point where the rat would die of starvation. If, however, this area was destroyed on one side only normal feeding persisted. This small hypothalamic area is so precisely located that even lesions as much as 0.5 mm away from it do not disturb feeding.

"We confirmed that bilateral lesions involving ventromedial nuclei produce hyperphagia and obesity. We also showed that destructive lesions between these nuclei and the lateral area also result in overeating, thus demonstrating laterally projecting inhibitory effects of the medial hypothalamic mechanism over the lateral area. As the hyperphagia resulting from medial lesions was converted into aphagia by subsequent lateral lesions, it was considered that the lateral area is the primary mechanism for hunger resulting in the urge to eat and hence constitutes the 'feeding center.' As the medial area provides an inhibitor mechanism for the lateral, this constitutes the 'satiety center.'

"Further studies carried out after my return to New Delhi not only confirmed the presence of these 'centers' in other animals; these also elucidated the nervous mechanisms for controlling and regulating our entire feeding behaviour which regulates the body's energy balance. They also established that the hypothalamic centers provide the basic urge of hunger and satiation, operating through the *level of utilisation of glucose within the nerve* cells of these regions. Sensory afferents from the stomach and other intestinal regions also activate the satiety center.

"The publication of this paper in 1951 resulted in the initiation of many similar experimental studies almost all over the world. The interest generated by such studies has also resulted in a number of special international symposia and seminars, etc., and the establishment of an international society for the study of feeding and drinking mechanisms.

"As the paper cited above was the first one to describe the presence of the hypothalamic control mechanisms for the feeding behaviour, it explains its high citation frequency."

1. Hetherington A W & Ranson S W. Hypothalamic lesions and adiposity in the rat.
   Anat. Rec. 78:149-72, 1940.
2. Brobeck J R, Tepperman J & Long C N H. Experimental hypothalamic hyperphagia in the albino rat.
   Yale J. Biol. Med. 15:831-53, 1943.

Lisk R D. Estrogen-sensitive centers in the hypothalamus of the rat.
J. Exp. Zool. 145:197-207, 1960.
[Dept. Biology, Harvard Univ., Cambridge, MA]

The findings demonstrate that estrogen implantation at specific brain sites (medial basal hypothalamus) blocks gamete maturation in the gonads and results in atrophy of the reproductive tract. Thus, regulation of gonadotrophin secretion from the pituitary is via hormone sensitive neural mechanisms located within the medial basal hypothalamus. [The *Science Citation Index®* (*SCI®*) and the *Social Sciences Citation Index®* (*SSCI®*) indicate that this paper has been cited over 175 times since 1961.]

Robert D. Lisk
Department of Biology
Princeton University
Princeton, NJ 08544

February 22, 1982

"This paper was part of my PhD dissertation based on studies carried out in the biological laboratories at Harvard University. During summers, while an undergraduate, I worked as a field assistant at the L. Opinicon Field Station of Queen's University, Kingston, Ontario. I was responsible for collecting data for a study of population dynamics in the white-footed mouse. This sparked an interest in reproductive biology which developed into my lifework. I quickly became convinced that one had to work in the laboratory to gain a deeper insight of how reproduction was regulated. First, I searched the literature to identify 'leaders' in reproductive biology. The name of Frederick Hisaw kept popping up and he appeared to be a 'father figure' in the field.[1] I was interested not just in the physiology of reproduction but also in the behavioral changes necessary for mating to occur. Hisaw was at Harvard as was Donald Griffin, who was doing exciting behavioral studies on bats' discrimination ability.[2]

"I was accepted at Harvard and prepared a literature review which demonstrated that destruction of a specific brain site could block the ovulatory cycle while lesioning of a separate site blocked the display of mating behavior without interfering with the ovulatory cycle. Therefore, I argued that the steroid hormones made in the ovaries must feed back onto specific sites in the brain to regulate the ovulatory cycle as well as facilitate mating behavior. My mentors, particularly Griffin, were convinced that I had something and asked what I needed to test this hypothesis. I replied that I needed a stereotaxic apparatus so that I could implant the steroid hormones at specific sites in the brain. In a few weeks a wooden box arrived; I opened it and there was a stereotaxic apparatus. All I had to do was learn how to use it and decide how I would secure the hormone at the brain sites of interest. This was solved by carefully melting the crystalline hormone and drawing it up by capillary action into stainless steel tubing. The tubing was lowered into the brain through a hole drilled in the skull and the tubing cemented in place with dental acrylic. The findings are summarized above.

"This study provided the first direct evidence that a gonadal hormone could act at the brain to regulate physiology. Such a radical conclusion was too giant a leap for the establishment. Thus, my paper was turned down by *Endocrinology* and came to be published in the *Journal of Experimental Zoology*. This paper has been highly cited for the following reason. Even though the findings appeared in a journal less likely to be scanned by physiologists and psychologists, the power of the technique for mapping hormone-sensitive brain areas in relation to their function was recognized.[3] Today many laboratories employ this technique to study hormone feedback control of physiology and behavior and the findings appear in the major journals both in biology and psychology."

1. Hisaw F L & Astwood E B. The physiology of reproduction. *Annu. Rev. Physiol.* 4:503-60, 1942.
2. Griffin D R. *Listening in the dark: the acoustic orientation of bats and men.*
   New Haven, CT: Yale University Press, 1958. 413 p.
3. Lisk R D & Barfield M A. Sites and mechanisms of steroid effects on behavior. (Stumpf W A & Grant L D, eds.)
   Anatomical neuroendocrinology. Basel: S. Karger, 1975. p. 232-44.

Singer M. **The influence of the nerve in regeneration of the amphibian extremity.**
Quart. Rev. Biol. **27**:169-200, 1952.
[Harvard Medical School, Boston, MA]

x

This paper is a review of works on the relation of the nerve supply to regenerative capacity. It includes the oft-cited stages in regrowth of an amputated limb. It also presents theories on the nature of the nerve action, the relative importance of different nerve components, the importance of quantity of axons for initiating the regeneration process, the problems of induction of regeneration in nonregenerating vertebrates, and the nature of the trophic neuronal quality. [The *SCI®* indicates that this paper has been cited in over 260 publications since 1961.]

Marcus Singer
Developmental Biology Center
Case Western Reserve University
Cleveland, OH 44106

August 4, 1983

"The ability to regenerate external body parts is widespread among animals, particularly in invertebrates and lower vertebrates, although higher ones including man can regenerate liver, bladder, and some other internal parts. Illingworth[1] has recently observed that children can regenerate amputated fingertips, leading one to support the prediction made 200 years ago by the eminent zoologist, Spallanzani, that innate within higher forms is the power to regrow external body parts requiring only useful circumstances (manipulations) to express the capacity.

"The most popular animals for the study of replacement of limbs, tails, and other parts are amphibians, especially the newt and salamander. These studies have taken various forms: cytological and histological analyses, and experimental manipulations of the various tissues at the wound surface. In recent years, an interesting and productive group in H.J. Anton's laboratory[2,3] at the University of Cologne, Federal Republic of Germany, have subjected the cytological analyses to quantitative computerized methods which detect the earliest cell changes at the wound surface preparatory to formation of the new part. The work on the newt also influenced the direction of the work by Jacqueline Geraudie of the University of Paris, who demonstrated a similar dependence of fish fin regeneration on the nerve.[4]

"In regeneration of the limb, what apparently occurs is first a wound-healing process with the liberation of cells which appear embryonic-like. These collect to form a bud of rapidly dividing cells which later differentiate into the various tissues of the limb. The bud is invaded by numerous new nerve fibers which eventually make connections to give a completely functional (sensitive and motile) limb.

"The study of two tissues (nerve, epithelium) has been especially intense for they play a leading part in the regrowth. The epithelium (epidermis) grows over the wound surface from the old skin bordering the amputation wound. The laboratory of the late C.S. Thornton,[5] University of Michigan, defined the epithelial role without which regeneration cannot occur. Indeed, regeneration of the fingertips in children does not ensue if the wound is surgically closed showing that bare epithelium in contact with the wound is a precondition of the regrowth. As for the nerve, it is known from many works that denervated amputation stumps in the newt will not regrow. Partially denervated ones will, but the speed of regeneration depends on the number of fibers available at the wound surface, also as shown well in the fish fin.[4] It is widely held now that the nerve fibers emit a trophic substance, the nature of which is being pursued in a number of laboratories, needed to stimulate and maintain the growth. At the moment, it appears to be a protein or peptide of low molecular weight.

"A major regeneration problem is the limited growth capacity of central nervous structures in higher vertebrates including man. Studies show that individual nerve fibers of the central, as in the peripheral, nervous systems are capable of growth but for some reason are blocked or suppressed in their growth. This is not the case in the lower animals. The newt and fish can regenerate the spinal cord and parts of the brain and recover complete function. They can do it because certain cells can recall their embryonic history and reproduce the part. It is a major challenge to find out how they manage to do this for the information may someday be applied to reconstructing the damaged nervous system of man. Many subsequent works, a sampling of which are cited here, have been built upon this classic review paper.[6-8]

"The reasons why this work has been highly cited are: 1) it is one of the first to give a historical perspective on the trophic nerve quality; 2) it provides a theory on the quantitative and qualitative action of the nervous action on regeneration; and 3) it opened the field to numerous subsequent studies testing these various hypotheses."

1. **Illingworth C M.** Trapped fingers and amputated finger tips in children. *Clin. Phys. Physiol. Meas.* **1**:87-9, 1974.
2. **Anton H J & Elsen W.** Cell activation and nuclear volume. *Microsc. Acta* (Suppl. 3):205-11, 1979.
3. **Anton H J & Bourauel M.** Volumen und Proteinsynthese der Kerne des Stratum Basale in der Stumpfepidermis während des Wundverschlusses nach Amputation der Vorderextremität bei *Triturus vulgaris*. *Develop. Growth Differ.* **24**:173-82, 1982.
4. **Geraudie J & Singer M.** Relation between nerve fiber number and pectoral fin regeneration in the teleost. *J. Exp. Zool.* **199**:1-8, 1977.
5. **Thornton C S.** Amphibian limb regeneration. *Advan. Morphogenesis* **7**:205-49, 1968.
6. **Singer M.** The trophic quality of the neuron: some theoretical considerations. *Prog. Brain Res.* **13**:228-32, 1964.
7. ------------. Neurotrophic control of limb regeneration in the newt. *Ann. NY Acad. Sci.* **228**:308-22, 1974. (Cited 95 times.)
8. **Kriegler J S, Krishman N & Singer M.** Trophic interactions of neurons and glia. (Waxman S G & Ritchie J M, eds.) *Demyelinating diseases: basic and clinical electrophysiology.* New York: Raven Press, 1981. p. 479-504.

Appendix A

A random selection of 100 papers from ISI®'s study of the 1,000 most-cited papers, 1961–1982, *SCI®*, for which a *Citation Classic®* commentary has not yet been published. The number to the left of the bibliographic data is the number of citations the paper received, 1961–1982. The reader will note that the selection has not been limited to papers in the life sciences.

812 **Anden N-E, Rubenson A, Fuxe K & Hokfelt T.** Letter to editor. (Evidence for dopamine receptor stimulation by apomorphine.) *J. Pharm. Pharmacol.* 19:627–9, 1967.

1,295 **Anson M L.** The estimation of pepsin, trypsin, papain, and cathepsin with hemoglobin. *J. Gen. Physiol.* 22:79–89, 1938.

1,060 **Arunlakshana O & Schild H O.** Some quantitative uses of drug antagonists. *Brit. J. Pharmacol.* 14:48–58, 1959.

2,161 **Bardeen J, Cooper L N & Schrieffer J R.** Theory of superconductivity. *Phys. Rev.* 108:1175–204, 1957.

860 **Berghuis J, Bertha I J, Haanappel M, Potters M, Loopstra B O, MacGillavry C H & Veenendaal A L.** New calculations of atomic scattering factors. *Acta Crystallogr.* 8:478–83, 1955.

767 **Bingham R C, Dewar M J S & Lo D H.** Ground states of molecules. XXV. MINDO/3. An improved version of the MINDO semi-empirical SCF-MO method. *J. Amer. Chem. Soc.* 97:1285–93, 1975.

719 **Bjorken J D.** Asymptotic sum rules at infinite momentum. *Phys. Rev.* 179:1547–53, 1969.

1,419 **Boas N F.** Method for the determination of hexosamines in tissues. *J. Biol. Chem.* 204:553–63, 1953.

1,000 **Bogdanski D F, Pletscher A, Brodie B B & Udenfriend S.** Identification and assay of serotonin in brain. *J. Pharmacol. Exp. Ther.* 117:82–8, 1956.

828 **Boyden S.** The chemotactic effect of mixtures of antibody and antigen on polymorphonuclear leucocytes. *J. Exp. Med.* 115:453–66, 1962.

866 **Brown H C & Okamoto Y.** Electrophilic substituent constants. *J. Amer. Chem. Soc.* 80:4979–87, 1958.

909 **Burgess R R.** A new method for the large scale purification of *Escherichia coli* deoxyribonucleic acid-dependent ribonucleic acid polymerase. *J. Biol. Chem.* 244:6160–7, 1969.

699 **Burton W K, Cabrera N & Frank F C.** The growth of crystals and the equilibrium structure of their surfaces. *Phil. Trans. Roy. Soc. London A* 243:299–358, 1951.

839 **Cahn R D, Kaplan N O, Levine L & Zwilling E.** Nature and development of lactic dehydrogenases. *Science* 136:962–9, 1962.

713 **Cantor H & Boyse E A.** Functional subclasses of T lymphocytes bearing different Ly antigens. II. Cooperation between subclasses of Ly cells in the generation of killer activity. *J. Exp. Med.* 141:1390–9, 1975.

1,043 **Cerottini J-C & Brunner K T.** Cell-mediated cytotoxicity, allograft rejection, and tumor immunity. *Advan. Immunol.* 18:67–132, 1974.

732 **Clementi E & Roetti C.** Roothaan-Hartree-Fock atomic wavefunctions. *At. Data Nucl. Data Tables* 14:177–478, 1974.

790 **Dement W & Kleitman N.** Cyclic variations in EEG during sleep and their relation to eye movements, body motility, and dreaming. *Electroencephalogr. Clin. Neuro.* 9:673–90, 1957.

1,291 **Dittmer J C & Lester R L.** A simple, specific spray for the detection of phospholipids on thin-layer chromatograms, *J. Lipid Res.* 5:126–7, 1964.

2,024 **Dixon M.** The determination of enzyme inhibitor constants. *Biochem. J.* 55:170–1, 1953.

2,580 **Dulbecco R & Vogt M.** Plaque formation and isolation of pure lines with poliomyelitis viruses. *J. Exp. Med.* 99:167–82, 1954.

1,350 **Edelhoch H.** Spectroscopic determination of tryptophan and tyrosine in proteins. *Biochemistry* 6:1948–54, 1967.

1,000 **Elson L A & Morgan W T J.** A colorimetric method for the determination of glucosamine and chondrosamine. *Biochem. J.* 27:1824–8, 1933.

795 **Engvall E & Perlmann P.** Enzyme-linked immunosorbent assay, ELISA. III. Quantitation of specific antibodies by enzyme-labeled anti-immunoglobulin in antigen-coated tubes. *J. Immunol.* 109:129–35, 1972.

1,913 **Falck B, Hillarp N-A, Thieme G & Torp A.** Fluorescence of catechol amines and related compounds condensed with formaldehyde. *J. Histochem. Cytochem.* 10:348–54, 1962.

757 **Fatt P & Katz B.** An analysis of the end-plate potential recorded with an intra-cellular electrode. *J. Physiol.—London* 115:320–70, 1951.

956   **Feshbach H.** A unified theory of nuclear reactions. II. *Ann. Phys.—NY* 19:287–313, 1962.

946   **Feynman R P.** Very high-energy collisions of hadrons. *Phys. Rev. Lett.* 23:1415–7, 1969.

1,252   **Fletcher R & Powell M J D.** A rapidly convergent descent method for minimization. *Comput. J.* 6:163–8, 1963.

13,974   **Folch J, Lees M & Sloane Stanley G H.** A simple method for the isolation and purification of total lipides from animal tissues. *J. Biol. Chem.* 226:497–509, 1957.

1,019   **Folin O & Ciocalteu V.** On tyrosine and tryptophane determinations in proteins. *J. Biol. Chem.* 73:627–50, 1927.

681   **Franks F & Ives D J G.** The structural properties of alcohol-water mixtures. *Quart. Rev.* 20:1–44, 1966.

685   **Fuoss R M.** Ionic association. III. The equilibrium between ion pairs and free ions. *J. Amer. Chem. Soc.* 80:5059–61, 1958.

979   **Garen A & Levinthal C.** A fine-structure genetic and chemical study of the enzyme alkaline phosphatase of *E. coli*. I. Purification and characterization of alkaline phosphatase. *Biochim. Biophys. Acta* 38:470–83, 1960.

1,182   **Giles K W & Myers A.** An improved diphenylamine method for the estimation of deoxyribonucleic acid. *Nature* 206:93, 1965.

3,669   **Gilman A G.** A protein binding assay for adenosine 3′:5′-cyclic monophosphate. *Proc. Nat. Acad. Sci. US* 677:305–12, 1970.

1,528   **Glowinski J & Iversen L L.** Regional studies of catecholamines in the rat brain. I. The disposition of [^{3}H] norepinephrine, [^{3}H] dopamine, and [^{3}H] dopa in various regions of the brain. *J. Neurochem.* 13:655–69, 1966.

742   **Gomori G.** A modification of the colorimetric phosphorus determination for use with the photoelectric colorimeter. *J. Lab. Clin. Med.* 27:955–60, 1942.

665   **Gorer P A & O'Gorman P.** The cytotoxic activity of isoantibodies in mice. *Transplantation Bull.* 3:142–3, 1956.

725   **Granato A & Lucke K.** Theory of mechanical damping due to dislocations. *J. Appl. Phys.* 27:583–93, 1956.

848   **Gray W R.** Dansyl chloride procedure. *Meth. Enzymology* 11:139–51, 1967.

747   **Guggenheim E A.** On the determination of the velocity constant of a unimolecular reaction. *Phil. Mag.* 2:538–43, 1926.

935   **Hogeboom G H.** Fractionation of cell components of animal tissues. *Meth. Enzymology* 1:16–9, 1955.

1,029   **Kane E O.** Band structure of indium antimonide. *J. Phys. Chem. Solids* 1:249–61, 1957.

1,721   **Kohler G & Milstein C.** Continuous cultures of fused cells secreting antibody of pre-defined specificity. *Nature* 256:495–7, 1975.

671   **Kory R C, Callahan R, Boren H G & Syner J C.** The Veterans Administration-army cooperative study of pulmonary function. *Amer. J. Med.* 30:243–58, 1961.

972   **Kubo R & Tomita K.** A general theory of magnetic resonance absorption. *J. Phys. Soc. Jpn.* 9:888–919, 1954.

1,329   **Lennox E S.** Transduction of linked genetic characters of the host by bacteriophage P1. *Virology* 1:190–206, 1955.

889   **Linsmaier E M & Skoog F.** Organic growth factor requirements of tobacco tissue cultures. *Physiol. Plant.* 18:100–27, 1965.

706   **Lowdin P-O.** On the non-orthogonality problem connected with the use of atomic wave functions in the theory of molecules and crystals. *J. Chem. Phys.* 18:365–75, 1950.

884   **MacPherson I & Stoker M.** Polyoma transformation of hamster cell clones—an investigation of genetic factors affecting cell competence. *Virology* 16:147–51, 1962.

1,564   **Mandell J D & Hershey A D.** A fractioning column for analysis of nucleic acids. *Anal. Biochem.* 1:66–77, 1960.

1,075   **Maniatis T, Jeffrey A & Kleid D G.** Nucleotide sequence of the rightward operator of phage λ. *Proc. Nat. Acad. Sci. US* 72:1184–8, 1975.

925   **March S C, Parikh I & Cuatrecasas P.** A simplified method for cyanogen bromide activation of agarose for affinity chromatography. *Anal. Biochem.* 60:149–52, 1974.

1,050   **Maroko P R, Kjekshus J K, Sobel B E, Watanabe T, Covell J W, Ross J & Braunwald E.** Factors influencing infarct size following experimental coronary artery occlusions. *Circulation* 43:67–82, 1971.

4,862   **Marmur J.** A procedure for the isolation of deoxyribonucleic acid from micro-organisms. *J. Mol. Biol.* 3:208–18, 1961.

1,766   **Mejbaum W.** Uber die Bestimmung kleiner Pentosemengen insbesondere in Derivaten der Adenylsaure. (Estimation of small amounts of pentose especially in derivatives of adenylic acid.) *Hoppe-Seylers Z. Physiol. Chem.* 258:117–20, 1939.

1,416   **Moore S & Stein W H.** Chromatographic determination of amino acids by the use of automatic recording equipment. *Meth. Enzymology* 6:819–31, 1963.

1,227   **Morrison W R & Smith L M.** Preparation of fatty acid methyl esters and dimethylacetals from lipids with boron fluoride-methanol. *J. Lipid Res.* 5:600–8, 1964.

1,085   **Murphy J & Riley J P.** A modified single solution method for the determination of phosphate in natural waters. *Anal. Chim. Acta* 27:31–6, 1962.

739   **Nakane P K & Kawaoi A.** Peroxidase-labeled antibody—a new method of conjugation. *J. Histochem. Cytochem.* 22:1084–91, 1974.

751   O'Malley B W & Means A R. Female steroid hormones and target cell nuclei. *Science* 183:610–20, 1974.

3,236   Omura T & Sato R. The carbon monoxide-binding pigment of liver microsomes. 1. Evidence for its hemoprotein nature. *J. Biol. Chem.* 239:2370–8, 1964.

916   Onsager L. Crystal statistics. I. A two-dimensional model with an order-disorder transition. *Phys. Rev.* 65:117–49, 1944.

1,363   Ouchterlony O. Antigen-antibody reactions in gels. *Acta Pathol. Microbiol. Scand.* 26:507–15, 1949.

1,127   Oyama V I & Eagle H. Measurement of cell growth in tissue culture with a phenol reagent (Folin-Ciocalteau). *Proc. Soc. Exp. Biol. Med.* 91:305–7, 1956.

833   Politzer H D. Reliable perturbative results for strong interactions? *Phys. Rev. Lett.* 30:1346–9, 1973.

693   Reed P W & Lardy H A. A23187: a divalent cation ionophore. *J. Biol. Chem.* 247:6970–7, 1972.

1,103   Reid R V. Local phenomenological nucleon-nucleon potentials. *Ann. Phys. NY* 50:411–48, 1968.

1,623   Richardson K C, Jarett L & Finke E H. Embedding in epoxy resins for ultrathin sectioning in electron microscopy. *Stain Technol.* 35:313–23, 1960.

763   Roseman S. The synthesis of complex carbohydrates by multiglycosyltransferase systems and their potential function in intercellular adhesion. *Chem. Phys. Lipids* 5:270–97, 1970.

3,941   Sabatini D D, Bensch K, Barrnett R J. Cytochemistry and electron microscopy. *J. Cell Biol.* 17:19–58, 1963.

696   Sawardeker J S, Sloneker J H & Jeanes A. Quantitative determination of monosaccharides as their alditol acetates by gas liquid chromatography. *Anal. Chem.* 37:1602–4, 1965.

873   Schultz S G & Curran P F. Coupled transport of sodium and organic solutes. *Physiol. Rev.* 50:637–718, 1970.

1,150   Schwert G W & Takenaka Y. A spectrophotometric determination of trypsin and chymotrypsin. *Biochim. Biophys. Acta* 16:570–5, 1955.

674   Scott T A & Melvin E H. Determination of dextran with anthrone. *Anal. Chem.* 25:1656–61, 1953.

1,478   Seldinger S I. Catheter replacement of the needle in percutaneous arteriography. *Acta Radiol.* 39:368–76, 1953.

777   Silber R H, Busch R D & Oslapas R. Practical procedure for estimation of corticosterone or hydrocortisone. *Clin. Chem.* 4:278–85, 1958.

877   Silber R H & Porter C C. The determination of 17,21-dihydroxy-20-ketosteroids in urine and plasma. *J. Biol. Chem.* 210:923–32, 1954.

730   **Slater J C & Koster G F.** Simplified LCAO method for the periodic potential problem. *Phys. Rev.* 94:1498–524, 1954.

823   **Smith J B & Willis A L.** Aspirin selectively inhibits prostaglandin production in human platelets. *Nature: New Biol.* 231:235–9, 1971.

1,162   **Somogyi M.** A new reagent for the determination of sugars. *J. Biol. Chem.* 160:61–8, 1945.

8,813   **Spackman D H, Stein W H & Moore S.** Automatic recording apparatus for use in the chromatography of amino acids. *Anal. Chem.* 30:1190–206, 1958.

679   **Staehelin L A.** Structure and function of intercellular junctions. *Intl. Rev. Cytol.* 39:191–283, 1974.

902   **Steers E, Foltz E L, Graves B S & Riden J.** An inocula replicating apparatus for routine testing of bacterial susceptibility to antibiotics. *Antibiot. Chemother.* 9:307–11, 1959.

988   **Swank R T & Munkres K D.** Molecular weight analysis of oligopeptides by electrophoresis in polyacrylamide gel with sodium dodecyl sulfate. *Anal. Biochem.* 39:462–77, 1971.

709   **Velicky B, Kirkpatrick S & Ehrenreich H.** Single-site approximations in the electronic theory of simple binary alloys. *Phys. Rev.* 175:747–66, 1968.

1,662   **Vogel H J & Bonner D M.** Acetylornithinase of *Escherichia coli:* partial purification and some properties. *J. Biol. Chem.* 218:97–106, 1956.

667   **Vogt V M.** Purification and further properties of single-strand-specific nuclease from *Aspergillus oryzae. Eur. J. Biochem.* 33:192–200, 1973.

2,302   **Warburg O & Christian W.** Isolierung und Kristallisation des Garungsferments Enolase. (Isolation and crystallization of the enzyme enolase.) *Biochem. Z.* 310:384–421, 1941.

689   **Wasserman E & Levine L.** Quantitative micro-complement fixation and its use in the study of antigenic structure by specific antigen-antibody inhibition. *J. Immunol.* 87:290–5, 1961.

773   **Weibel E R, Kistler G S & Scherle W F.** Practical stereological methods for morphometric cytology. *J. Cell Biol.* 30:23–38, 1966.

2,428   **Weinberg S.** A model of leptons. *Phys. Rev. Lett.* 19:1264–6, 1967.

729   **Weintraub H & Groudine M.** Chromosomal subunits in active genes have an altered conformation. *Science* 193:848–56, 1976.

809   **Wettstein F O, Staehelin T & Noll H.** Ribosomal aggregate engaged in protein synthesis: characterization of the ergosome. *Nature* 197:430–5, 1963.

1,880   **Wilkinson G N.** Statistical estimations in enzyme kinetics. *Biochem. J.* 80:324–36, 1961.

785   **Winzler R J.** Determination of serum glycoproteins. *Meth. Biochem. Anal.* 2:279–311, 1955.

804   **Witkin E M.** Ultraviolet mutagenesis and inducible DNA repair in *Escherichia coli. Bacteriol. Rev.* 40:869–907, 1976.

1,383   **Wroblewski F & LaDue J S.** Lactic dehydrogenase activity in blood. *Proc. Soc. Exp. Biol. Med.* 90:210–3, 1955.

964   **Zacharius R M, Zell T E, Morrison J H & Woodlock J J.** Glycoprotein staining following electrophoresis on acrylamide gels. *Anal. Biochem.* 30:148–52, 1969.

Index of Authors

All authors of Citation Classics are listed. An asterisk after a page number indicates that a commentary by the author appears on that page.

Index of Subjects

Index of Institutions

The institutions listed are those at which the work reported in the Citation Classic was done.

WALKING WITH HIM

WALKING WITH HIM

A Biblical guide through thirty days of Spiritual Exercises

Josef Neuner, S.J.

GUJARAT SAHITYA PRAKASH
ANAND, 388 001
GUJARAT
INDIA

Imprimi Potest: John F. D'Mello, S.J.
                   Provincial of the Patna Province
                   Patna, October 13, 1984

Imprimatur: † Benedict Osta, S.J.
                   Bishop of Patna
                   Patna, October 14, 1984

CONTENTS

A powerful current of fresh enthusiasm for the Spiritual
Exercises of Saint Ignatius of Loyola is one of the characteristics
of this age of renewal in which we live. As compared with a
generation ago, there is a notable increase in the number and
variety of people making the retreat, in their fidelity to the original
Ignatian design and, it would seem, in the profit that is derived.

This is due in large measure to the present understanding
of the Exercises, in the light of Ignatius' own spiritual experience
rather than simply according to the letter of the text. After all,
it is he himself who tells us in his autobiography, "that he had not
composed the Exercises all at once, but that when he noticed some
things in his soul and found them useful, he thought they might
also be useful to others, and so he put them in writing."

Looked at in this way—as a painstaking effort to communicate
an experience, and not just a piece of religious literature—the
Exercises reveal a vision of reality and a dynamic of personal
growth that are very much in tune with the mind and heart of the
Church in this post-conciliar period of her history. Hence it is
that committed Christians today find them admirably suited to
meet their actual spirital needs. But this supposes that there is
someone at hand to translate the authentic intuitions of Ignatius
into the current idiom and to place them in the context of contemporary life.

Father Josef Neuner is eminently qualified to render precisely
such a service, for he had both the vast fund of knowledge and the
delicate sensitivity that are required for the task, besides being
well practised in the art of teaching. Long before the Second
Vatican Council, he had already introduced his students to the
theological style that is now associated with that great ecclesial

event; during its sessions he had the opportunity to breathe its spirit and to collaborate in the shaping of its pronouncements; ever since, he has been a zealous promoter of the movement for renewal that it launched, often using the Ignatian retreat as an effective instrument.

In fact, Father Neuner could say of this book, in which he presents the Spiritual Exercises in a biblical perspective, what Ignatius said of the original. It is not a piece of literature but a painstaking effort to communicate an experience—an experience that has matured through his constant concern to share it with others, and particularly with that privileged group of his disciples: the many young men whom he has prepared for priestly ordination.

Those of us who have known and loved Father Neuner these many years will immediately recognize our revered master in these pages, and be thrilled at this fresh encounter. Everything is here, except the eloquent gestures: the solid scholarship enlivened by a contagious enthusiasm for God's marvellous plan of salvation, centred in the person of Christ; the struggle for adequate self-expression; the sense of mission that makes him put his whole heart into his utterances.

But what matters most—what Father Neuner would surely most desire, and indeed Saint Ignatius himself must want—is that in this book an earnest seeker can meet Christ, so as to know him more intimately and thus love him more ardently and follow him more faithfully.

Parmananda Divarkar, S.J.

Foreword to the Second Edition

Soon a second edition of "Walking with Him" was needed. The structure of the book remained unchanged, but some texts have been re-written and a good many minor changes and corrections have been made in the hope of improving the usefulness of the book.

J. Neuner S. J.
Epiphany 1987.

INTRODUCTION

The Spiritual Exercises of Saint Ignatius are based on his unique mystical experience at Manresa in which he saw the whole of creation in a totally new and comprehensive vision, as it originates from God, as it moves towards its fulfilment in God; Jesus Christ is the centre of this vision. This vision gave orientation to his whole life and work. He still had to grope towards the concrete form of his mission, and the various stages of this search often seemed to be conditioned by arbitrary circumstances which finally brought him to Rome and to founding the Society of Jesus. But in each of these stages the vision of Manresa remained his guiding light. It became also the uniting bond that gave to his first companions a common orientation, and a sense of intimate spiritual communion. Though they were soon scattered over many countries and engaged in very different works—which ranged from teaching children at the roadside to advising heads of kingdoms and working as theologians at the Council of Trent—they remained united by the same spirit which they had imbibed from the Spiritual Exercises.

This powerful influence has astonishingly not diminished over the course of four centuries. In countless documents the Society of Jesus has recognized them as the spiritual basis of Jesuit life and mission. The latest General Congregation, the 33rd, states in a matter-of-fact way that the movement of renewal in the Society of Jesus in the past years "manifests itself especially in the new impetus given to the Spiritual Exercises and to apostolic discernment" (G.C. 33, Decree 1, N. 10).

However, we cannot preserve the treasures of the past unless we discover them anew in our world, for our world. Bernanos wrote: "One cannot preserve the grace of the past unless one receives a new one." Thus it is imperative to rediscover the Spiritual Exercises again in the Church and world of today. How can we avail ourselves today of the treasures contained in the Spiritual Exercises?

Saint Ignatius had a unique capacity to *give* the Spiritual Exercises and so to convey the impact of his vision of Manresa to those who came close to him, many of whom became his companions. He then *wrote* down the guidelines of these Spiritual Exercises in stages. The booklet consists of (1) directions both for the one who gives and the one who makes them. These guidelines are scattered throughout the book; (2) meditations with definite themes, some fully developed, others merely sketched; (3) various rules for different needs and situations.

His work is intended as a practical guide, not as a textbook. Some passages, mostly the central meditations, offer a comprehensive vision of life and world, the vision which had ever guided Ignatius after Manresa. But even these great meditations are sketched in an austere style. Most parts of the book make dry reading and sometimes seem strangely pedantic. It is difficult to read this text, and the reasons are quite obvious.

Ignatius did not have the genius of self-expression as we find it, for instance, in his great contemporary, St. Teresa of Avila, who was able to spell out her mystical experiences within eight brief weeks (not counting some interruptions); she could write with an inspired flow of language and had no need to cancel or correct what she had written. For Ignatius, however, the *language* remained a barrier to convey his inner experiences.

Further, the *social and cultural milieu* which moulded his thinking and determined the symbolism of his writings is feudal. His ideals are sometimes expressed in terms of court life, and motivations are taken from the code of honour of medieval knights. The background of our twentieth century life and the spheres of our mission today have radically changed. We need different symbols, different motivations.

Ignatius naturally had to express his comprehensive vision within the *theological framework* of his time. He understands sacraments and liturgy in terms of scholastic theology. His ideas of Church and hierarchy were inherited from the Middle Ages. His vision of apostolate and mission reflects the era of the sixteenth century, of Renaissance and Reformation. The more we enter into his spiritual world the more we become aware of the inappropriateness of these conceptual tools to express his message effectively today. It is new wine which calls for new wineskins.

Thus the text of the Spiritual Exercises offers a challenging hermeneutical problem if we are to discover its riches and to translate it into contemporary language. Is it possible to distil from this text (obviously, with the help of other sources) the vision which guided Ignatius in his life and mission? Is this vision a meaningful orientation also for later eras, also for our time and world? Shall we be able to express this vision in a manner which is related to our radically changed social, economic, religious world? Will it be possible through a deeper understanding of the Bible, and with the categories of present theological language, authentically to express the Ignatian vision?

In the past years many efforts have been made to find new ways of giving and making the Spiritual Exercises. The long established tradition of "preached retreats", of conferences with an overstress on moral and spiritual teachings, has given way to "directed retreats", with emphasis on both personal awareness of God's design for our life and the guidance of the Holy Spirit. The results of solid biblical studies have been used in these new approaches.

This book, then, is an attempt to make a contribution to these new approaches. It is the result of a great number of thirty-days-retreats given mostly to theologians in preparation for their priestly ordination. These were directed retreats following the four weeks of the Spiritual Exercises. Every day is centred on one specific theme. This background explains both the book's *structure*, and its *orientation*.

The book is divided in four parts, corresponding to the four weeks of the Spiritual Exercises. In every chapter the

theme is unfolded by a general exposition, which is followed
by "short points for meditation", which are meant for the
personal use of the retreatant. Such seems to be the mind of
Ignatius. On the one hand he wishes to convey to the retreatant
a comprehensive vision of our life and world. At the same time
Ignatius does not wish the retreatant to be the mere passive
recipient of doctrines and directions; he expects his active
involvement. Hence the one who gives the retreat is told to
give only a factual exposition of the content of the meditations,
and to "add only a short, summary explanation" (n. 2). In
this way the active reflection and creativity of the retreatant
come into play. Ignatius foresees that this active involvement
"produces greater spiritual relish and fruit than if one in
giving the Exercises had explained and developed the meaning
at great length." Through personal reflection and prayer the
retreatant is meant to find the "intimate understanding and
relish of the truth that fills and satisfies the soul." (Ibid).

Also the *orientation* and the approach of this book are
greatly influenced by the specific situation of young men,
who after their theological studies stand on the threshold of
their ordination and pastoral and missionary ministry. It
seems that this focus on young men who are at a decisive
moment of their lives does not necessarily limit the broader
usefulness of the book. Four of its main features may be
identified here.

The presentation of the meditations is *biblical*. Not only
are the themes day by day presented in biblical perspective,
but the entire dynamism of the Spiritual Exercises is seen
in the perspectives of the Bible, beginning from Exodus
and Israel's journey through the desert up to the full
revelation of God in Jesus Christ. The need of a biblical
orientation in all spirituality is demanded by the Council
which places God's word at the centre of all Christian life:
"Such is the force and power of the Word of God that it
can serve the Church as her support and vigour, and the
children of the Church as strength for their faith, food for
the soul and a pure and lasting fount of Spiritual life" (n. 21).

Obviously no technical exegesis of texts should be expected in this book but it attempts to present and interpret the
biblical texts in their authentic sense as it is intended by the

xii

authors and to relate these texts to the actual situations of our life and mission. The Biblical orientation leads to the key concept of Jesus' Message, the Kingdom of God. A renewal of hearts must bear fruit in the transformation of society. The *social* dimension is an integral part of the biblical message and therefore vital in the Spiritual Exercises.

Though this book does not deal explicitly with *theological* problems, it does incorporate an understanding of revelation, of the person and mission of Jesus Christ, of Church and sacraments, which corresponds to the perspectives of Vatican II and to sound trends of modern theology. This is necessary for young theologians and, in fact, for all modern Christians. As believers we cannot live with a dichotomy between the spiritual values and practices on one side and apparently conflicting theological positions on the other. This appears obvious, but in reality such a dichotomy often exists. This book may in some way contribute to bring the spheres of theological reflection and spiritual praxis closer together.

The presentation of the Spiritual Exercises in this book, especially in the last section, is strongly *ecclesial*. This is a feature which perhaps is not often found in books on the Exercises, though Saint Ignatius had an extraordinary love for the Church and was totally committed to its mission. The renewed self-understanding of the Church in Vatican II as the people of God and as sacrament of Christ's saving presence in our world invites us to see the ecclesial dimension in every sphere of our Christian life.

This book has its *limitations* which readers should keep in mind lest they be disappointed. It is *not* a commentary of the text of Saint Ignatius. Naturally many texts are interpreted but without entering into any controversy. It is simply a positive presentation of the meditations through themes as they are proposed day by day.

Nor does this book deal with the many practical directions and rules of the Ignatian text, though some of these guidelines are referred to here and there. A serious explanation of these directions, however, would imply thorough historical studies and an assessment in the light of modern psychological insights.

How to use this book

This book is meant as a guide for the Spiritual Exercises over a period of thirty days. Each chapter contains (1) the theme of the day in a coherent exposition, (2) short points for meditation and (3) an indication for further readings; (4) some leading questions are added to help the retreatant relate the theme to his/her actual life situation.

From the outset of the retreat the retreatant should firmly decide to remain each day within the theme designated for the day. He will find in it sufficient material for thought and inspiration for prayer. It will be important to enter deeply and personally into each days' theme. It will also be valuable to repeat important meditations in order to deepen their impact. There is no need to cover all the points for meditation or the additional readings as they are indicated for every day. They are given mainly to offer a wider choice but this choice should not lead to superficiality. Saint Ignatius also insists that, while dealing with one topic or subject we should not think of other themes. (cf. Sp. Ex n. 11). To do so would take away concentration and therefore depth. Each day's theme should be experienced with its full impact. In a thirty days' retreat a "day of rest" ought to be kept after every Week or at the convenience of the retreatant.

Besides being used for the "long retreat" this book can be of help also for retreats of shorter duration, though a warning must be added. It would be futile to attempt to compress these themes of 30 days into a period of eight days by covering, say, several themes every day. By doing this we would deprive ourselves of that depth-experience to which we are meant to come during Spiritual Exercises. But it may be feasable in a retreat of eight days to select single themes from the four parts of the book. Another option, especially for those who make retreats regularly, is to concentrate on one Week only and to go through its corresponding themes thoroughly.

Acknowledgements

The Spiritual Exercises are the common inheritance of the Society of Jesus. Every attempt to interpret and articulate their meaning grows from the manifold interactions with those who live in the spirit of the Exercises. The author would be happy if many would recognise something of their own aspirations and ideas in the following pages.

The picture on the title page of the two disciples with Jesus on the way to Emmaus is the work of Shri Jyoti Sahi of Bangalore. The original is kept in the chapel of Navjyoti Niketan. It illustrates vividly the theme of the book, "Walking with him."

Special thanks have to go to members of my Navjyoti community for the support in this work and for going through the manuscript chapter by chapter to improve the text.

Finally, my thanks to the Gujarat Sahitya Prakash of Anand for their prompt and encouraging interest in printing this book.

As this book is the result of many retreats given to theologians, I must also gratefully remember them, since their sharing and questioning helped much in the search for ways to convey the treasures of the Spiritual Exercises to a young generation.

The book of the *Spiritual Exercises* is quoted by the numbers as they have been accepted in modern editions.

Documents of the *Vatican Council II* are quoted by the Initials of the Latin text:

AG = *Ad Gentes,* Decree on the Church's missionary activity.

CD = *Christus Dominus,* Decree on the pastoral office of bishops in the Church.

DH = *Dignitatis Humanae,* Declaration on religious liberty.

DV = *Dei Verbum,* Dogmatic Constitution on divine revelation.

GS = *Gaudium et Spes,* Pastoral Constitution on the Church in the modern world.

LG = *Lumen Gentium,* Dogmatic Constitution on the Church.

NA = *Nostra Aetate,* Declaration on the relation of the Church to non-Christian religions.

OT = *Optatam Totius,* Decree on the training of priests.

PC = *Perfectae Caritatis,* Decree on the up-to-date renewal of religious life.

PO = *Presbyterorum Ordinis,* Decree on the ministry and life of the priests.

SC = *Sacrosanctum Concilium,* Constitution on the sacred liturgy.

UR = *Unitatis Redintegratio.* Decree on Ecumenism.

PART I

MOVING OUT

Chapter One

IN THE DESERT

*"The Spirit immediately drove Jesus into the wilderness.
And he was in the wilderness forty days" (Mk 1,12f).*

The "Spiritual Exercises" of St. Ignatius are not a random
sequence of spiritual reflections. They have the definite goal
of helping us to find and realize the meaning of our life and
so to lead to practical decisions, decisions which affect the
personal life and work of the retreatant. Is it meaningful to
make such decisions at a distance from actual reality, in an
atmosphere of silence, in the desert? Should we not be in
touch with reality to be concrete and practical in our options?
Each one is placed by God into a concrete life situation to
fulfil his/her task; our personal life cannot be separated from
the world to which we belong and for which we bear responsibilities.

Indeed the retreat is not meant to be an escape from the
realities and problems of life. But experience teaches that at
times we must withdraw from actual tensions and pressures
in order to arrive at balanced decisions, not to be carried away
by "inordinate affections" (in Ignatius' terminology) i.e. by
spontaneous reactions of our temperament, by natural likings
or dislikes or by pressures from outside, subtle or open.
Decisions must come from the depth of the heart, the centre
of the personality, where we are touched by God. They must
mature in silence. Thus we go into the "desert". We have
the biblical examples for it—Jesus himself who went into the

wilderness for forty days before he embarked on his mission, and the wider biblical background of the desert-experience of Israel for forty years. *The decisive battles of God's kingdom are fought in solitude.*

The first day of the retreat should be spent in the awareness of the peculiar situation of the coming month. We call it a "desert" though we may find ourselves in pleasant surroundings and in a peaceful atmosphere with all our material needs fulfilled. Still, it is our desert.

To Withdraw

Desert means, first, to withdraw. This may be the first experience of the retreat: we are away from the usual surroundings in which we feel at home, from the people to whose company we are used, from the routine of work and engagements which fill our normal timetable and calendar. This distance from our ordinary life comes as a relief. We enjoy the freedom from pressures and tensions and from immediate responsibilities. We experience something of the "rest" of which the Bible speaks, God's own rest after the work of creation was completed (Gen 2,1-3). God is more than his work. This rest is meant also for us, it is our Sabbath, to share in God's silent presence as it is promised to his people: "There still remains for God's people a Sabbath rest, for whoever enters God's rest also ceases from his labours as God did from his" (Hebr 4,9f). This is the rest which Jesus offers his disciples: "I will give you rest" (Mt 11,28).

To be alone

But we are not used to this rest. We have our holidays, but they are different. On holidays we exchange the usual occupations with other engagements; we plan journeys, meet friends, read books, enjoy music. In the retreat we are alone. We are not used to being only with ourselves. Soon we discover how much we depend on the many things that keep us busy and the small talk that fills the gaps between our activities. Our daily occupations help us to get more easily over spells of disappointment and frustration. We are carried along, mingle with the activities of others and so are able to forget ourselves. In the desert we are alone, we are meant to find ourselves. Instinctively we reach for a book or turn on the transistor to escape the desert. It may be the first decision of these days of the retreat to accept the solitude. After a while you may begin to love it.

Among angels and demons

Our exterior occupations and contacts are useful also as selfprotection. We are afraid of loneliness and feel threatened by the darkness that lingers below the surface of our conscience. In the routine of our daily life it cannot come to the surface, but in days of solitude it rises as the mist rises from the valleys and clouds the clear vision of the hills. The deeper movements of our heart, which mostly remain hidden, come into the mind and fill it. These are different and often conflicting powers. There are dark and destructive powers, old hurts come up with new vigour, bitterness against people who have wronged us; there are feelings of emptiness, frustration, rejection. The passions and urges of the body assert themselves. Fantasies of ambition, vanity, self-glory flood the mind, at times under the pious disguise of doing great things for others and for God. But there are also the invitations of grace: God's joy and peace, the call to deep love and surrender, the desire for a truer and richer life. Many of these movements, which usually lie low underground come to life and we have to face them.

In the Bible the desert is the place of temptation and struggle. There are angels and demons, God's loving care and the allurements of Satan. St. Ignatius lays greatest stress on the recognition and discernment of these movements of the heart during the retreat (SP. Ex. nn. 313-336). It is the time to get in touch with the deeper layers of our mind, the disguised temptations, the silent invitations of God. We must learn to cope with these movements. Silence is needed to listen to the puzzling and disturbing voices that come from the depth of the heart, to discard masks and to see our true face. We renounce shortlived stimulants to spur our activity and search for the springs of true life, which flow from God.

With God

It is the very goal of the Spiritual Exercises to seek and find the will of God (nn. 1,21), over and beyond all the movements of our heart. We pray with confidence: "I have relied on you since I was born" (Ps 22,10). God's Promise is ours: "No longer will you be called forsaken" (Ps, 62,4), and "the Lord, our God, will lead you and protect you on every side" (Is 52,12).

Thus we enter the retreat with the firm determination to accept the desert situation, the aloneness, the confrontation with ourselves and all that is in us, the readiness to make a full commitment to God. Ignatius articulates it in two words: *Magnanimity*, i.e. a wide and open mind which is not confined to narrow horizons of small interests, and *generosity*, the readiness to give without conditions and limits (n. 5). We shall resist the temptation to fill the emptiness of these days with trifles or also serious occupations however useful they be in themselves. It will be our time in the desert, searching for God, with Jesus.

With Jesus in the wilderness (Mk 1,12f)

Jesus also needed the desert. He has a unique place in God's plan of salvation, but this eternal plan has to be realized and carried out in Jesus' earthly life. He had to find his mission and to make decisions. The gospels indicate situations that demanded decisive options on the part of Jesus. Most important among them is the account of the forty days in the desert and the temptation (Mt 4,1-13; Lk 4,1-11). It marks the transition from his hidden life to the public ministry.

Jesus' temptation was not a single, isolated event, it lasted through his whole life, till the prayer in the garden, till the cross. But in the account of the desert, the temptation becomes the theme of the biblical narration. It condenses the basic decision which had to be realized in the life situations as they unfolded. — This is also the situation of the retreatant. Our entire life is a continuous option, but there are situations which are decisive, when basic orientations have to be found and choices to be made. The Spiritual Exercises are such a situation.

So we turn to Jesus: we may start with the peaceful days in Nazareth in rural surroundings, where Jesus grew up and became increasingly aware of his surroundings: of the political humiliation of the nation under the Roman yoke and the growing restlessness of the people; of the social and economic inequalities and injustices; of the sterility of the religious legalism; of the apocalyptic expectations of God's kingdom which would come with power and destroy all earthly empires (cf. Dan 2,44f); of the anxiety of the people who looked out for something new to come. Jesus saw the large crowds that flocked to the Jordan to the Baptizer, to listen to his warning of the impending judgement and to be baptized by him as a sign of repentance.

He too went with the crowds to the Jordan and joined the movement. With the many who waited for a new turn in Israel's history Jesus accepted John's baptism. But John's message was only a preparation. Jesus' vision is different: God's love does not threaten and destroy; God wants a new creation, a world in which God's love conquers the powers of darkness and destruction. Jesus is consumed by his own mission. The Evangelists express it through the narration of the theophany at his baptism, the descent of the Spirit of Jesus and the voice from heaven: "You are my beloved Son, with you I am well pleased" (Mk 1,11). He has to proclaim God's reign, the coming of a new humanity. In this proclamation, and in his entire life, work and preaching he is guided by the Holy Spirit.

With his message of God's reign Jesus would have to face the people, the leaders of the nation, the politicians, the priests, the scholars. He was deeply aware that his message was the fulfilment of all that God had ever told them in the past, through the law and the prophets. All those who really understood the Torah and had listened to the prophets would rejoice in his proclamation. But he realized that it was different from what they practised, and it was a far cry from their actual expectations and dreams. Their dreams were different: either of a revolutionary leader who would fight against the Romans or of a holy worshipper of the law who would lead the people in their old observances. How will

Jesus face them? Will he make compromises and associate himself with their parties? With the supporters of government, or with the revolutionaries? With whom will Jesus side?

Jesus has to find that inner clarity and firmness which characterize his entire life and work. He has to find it in solitude, in a bitter struggle against the temptation to fall in line with any of these dreams. He goes into the desert: "The Spirit immediately drove him into the wilderness" (Mk 1,12). He has to make his free and personal option. Once the inner struggle is over he is ready to meet the people, "in the power of the Spirit Jesus returned to Galilee" (Lk 4,14), and proclaimed the good news of God's reign.

Thus Jesus was led into the desert by the Spirit to face the great issues of his life and mission before he entered into his ministry. (We shall return to the concrete content of Jesus' temptations at a later stage of the Spiritual Exercises). We too enter into the retreat to face the great issues of our life, God's invitations and the evil allurements. In the solitude of the desert we are meant to find the clarity and firmness which we shall need in the turmoil of actual life.

The desert in Israel's history

The forty days of Jesus in the desert are the climax of the ancient biblical tradition of Israel's desert experience. Yahweh led the people from the bondage in Egypt through the desert into their own land. A glance at the map shows that there would have been a shorter route from Egypt to Palestine. It obviously cannot be our purpose to examine the historical reasons why the Israelites took the long route through the wilderness and remained in the desert for forty years. In biblical perspective this is God's own design. The Bible says simply: "God led the people round by the way of the wilderness" (Ex. 13,18). He tested his people. They had to experience their dependence on him and live under his immediate protection and providence. He prepared them for the covenant, to become his own people.

This journey will forever be the pattern by which God liberates and leads his own; it is the pattern also of the journey of this retreat. We get used to our own life and world where we are held by many bonds of duty, of affection or at least of habit, where also our thinking and planning is confined to the accustomed patterns of the society in which we are at home. In the retreat we move out of this world into solitude. God beckons us towards new horizons: "I will bring them up out of that land into a good and broad land, a land flowing with milk and honey" (Ex 3.8). The Spiritual Exercises should make us attentive to God's invitation to the real life.

But this journey leads through the desert. Like the Israelites we have to learn to rely on the Lord and not on our resources. We may miss the comforts of the "flesh pots of Egypt" with which we used to console ourselves. At times we too feel tired and thirsty, but the Lord will refresh us in ways beyond expectation, with manna and with waters flowing from dry rocks. In a strange transformation we too may see the desert blossom and fertile rivers flow between barren hills. (cf. Is. 41,16).

The Meditations of this day

This day is devoted to the realization of the retreat situation, to find God, to find myself, to find the meaning of my life, to find it in the desert. Our reflections on the desert and the following points for meditation may be helpful. Their purpose, however, is not to make us contemplate Jesus in the wilderness or Israel on its journey through the desert, but to begin to live in our own desert, to be aware of, and accept, the situation and challenge of this retreat and to face its struggle. We pray that God be with us as he was with Jesus, as he accompanied the Israelites: "In a pillar of cloud by day to lead them along the way, and by night in a pillar of fire to give them light, that they may travel by day and by night" (Ex 13,21).

SHORT POINTS FOR MEDITATION

1. The beginning of the New Testament in the desert.

— The New Testament begins in the desert. John the Baptist is "a voice in the wilderness" (Mk 1,3) calling to repentance. As in the ancient times of Exodus

once more the desert becomes the scene of renewal for
the people who receive the baptism of repentance.

— The desert is for them not an escape from the realities
of life. John sends them back into their life situations,
but renewed: the people, the tax collectors, the
soldiers (Lk 3,10-14).

— Jesus, too, joins the movement of renewal and is
baptized (Mk 1,9).

— Then Jesus is guided by the Spirit into the desert.
Before proclaiming his message of God's kingdom,
meeting the people, facing hostile adversaries, he is
alone.

— He, too, experiences the loneliness after the sheltered
life in Nazareth, the insecurity, the homelessness,
the hunger.

— He is among wild animals, he, too, meets angels and
devils (Mk 1,13).

— Throughout his life Jesus renews the desert experience
and retires from the people to lonely places (Mk 1,35,
45; 6,31; Lk 6,12; Jn 6,15).

2. *The birth of God's people in the desert.*

— The departure from Egypt remains the great day
in Israel's history: "It was a night of watching by
the Lord to bring them out of the land of Egypt; so
this same night is a night of watching kept for the
Lord by all the people of Israel throughout the
generations" (Ex 12,42).

— When they were without shelter against their enemies
they complained; "Were there no graves in Egypt that
you have taken us away to die in the wilderness?"
(Ex 14,11). But Yahweh delivers them and drowns
the army of Pharaoh: "On that day the Lord saved
the people of Israel from the Egyptians" (Ex 14,30).

— They suffer thirst, but God makes bitter waters sweet
(Ex 15,22-25) and gives them water from the rock
(Ex 17, 3-7).

— They are hungry and God gives them food: "At twilight you shall eat flesh, and in the morning you shall be filled with bread" (Ex 16, 2-15).

— God makes the covenant with his people: "If you will obey my voice and keep my covenant, you shall be my own possession among all peoples; for all the earth is mine, and you shall be to me a kingdom of priests and a holy nation" (Ex 19,5f).

— They are in God's singular care: "He found Israel in a desert land and in the howling waste of the wilderness; he encircled him, he cared for him, he kept him as the apple of his eye. Like an eagle...that flutters over his young...bearing them on his pinions, the Lord did lead him" (Deut 32,10f).

— But they still have to cross the desert: "And we set out from Horeb and went through all that great and terrible wilderness" (Deut 1,19).

3. Israel's desert memories: The unique time of grace.

— Throughout the centuries Israel kept nostalgic memories of the desert. In the desert it was born as God's people, it will always find renewal by returning to the desert.

— "Once more God will send us his Spirit. The waste land will become fertile..." (Is. 32,15). "The desert will rejoice, and flowers will blossom in the wilderness..." (Is 35,1-10. cf. Is 41,17-20)

— Hosea sees Israel's corruption as the consequence of the comfort they found in Canaan. They began to worship Baal, the god of fertility and prosperity, therefore Yahweh will punish Israel (Hos 2,2-13). But then he will lead her into the desert again: "I will allure her and bring her into the wilderness and speak tenderly to her...And I will betroth you to me for ever" (Hos 2,14-23).

— Read the touching story of the Rechabites recorded by Jeremiah: They came to Jerusalem as refugees from the Assyrian invasion and there tell the story of their traditions. They had kept the life style of the desert literally. They refused to drink wine and to enjoy the comforts of a settled life (Jer 35, mostly vv 6-11).

QUESTIONS:

> *N.B.   These questions should be answered first personally; it will be useful to formulate the answers in writing. Where there is a group it may be helpful to share the answers in common.*

1. Why did I decide to make this retreat? What do I expect of it?
2. How do I intend to face the solitude of these days?

FOR REFLECTION

"The desert does not mean the absence of men, it means the presence of God."

"The richness of a novitiate in the Sahara is undoubtedly the solitude, and the joy of solitude, silence: the true silence that penetrates everything, fills every being, which speaks with wonderful, new power to the soul, a power of which distracted people know nothing. There you live in silence, you learn it in stages: Silence of the Church, the cell, the work, the interior; silence of the soul, silence of God".

CARLO CARRETTO

Chapter Two

ENCOUNTERING GOD

*"O God, you are my God, for you I long,
for you my soul is thirsting"* (Ps 63).

Waiting for God: this is the meaning of the Spiritual
Exercises, of the entire Christian life. From childhood we
have heard about God, at home, in school, and mostly in
the Church. We also speak about him in daily life; for many
it is part of their profession to teach about him. We also
pray to God with solemn words in the liturgy and also in
personal prayer. Priests and religious are called men and
women of God. There are others with no religious education,
or those who have been estranged from God, but the quest
for him, the ultimate goal of our life, arises from the depth
of the heart with irresistable power. All have to face a painful
question, a question which cannot be pushed aside: What do
we really know about God? or, more correctly: we may
know much about him, but do we really know him, find him,
realize his presence and power? Who could ever claim with
assurance that he has found him? There is a ring of deep
sadness in John's gospel when he sums up the search and
longing of endless generations, of men and women in their
quest for God: "No one has ever seen God" (Jn 1, 18).

The Search

The quest for God in our time has become ever more
difficult. We are affected by the climate of our secular culture.

People of our time have no real place for God in their world, they don't need him any more. They have found explanations for the universe, for nature, for human history and society. Natural sciences, sociology and psychology have replaced philosophy and theology. In former times we needed God because we knew little about the mechanisms of life, the working of the human mind and the laws that govern human society, God had to fill the gaps of our knowledge. Today more and more phenomena are explained by natural causes. The margin of unexplained phenomena shrinks daily, and the success of our scientific approach gives us the confidence that also the remaining puzzles will one day be resolved. No place seems to remain for God, he looks like a relic of the past.

Thus in our modern thinking no longer God but man has become central. The Vatican Council admits: "Believers and unbelievers agree almost unanimously that all things on earth should be ordained to man as to their centre and summit" (G.S. n. 12). Religion has lost its central place in modern society. Atheism has established itself with massive power not only in individuals but in large sections of human society. For others religion has become marginal, a social glamour added to important dates in their calendar without a major impact on actual life.

Believers too are affected by this climate of the modern world, also priests and religious. We are absorbed in secular affairs, drowned in an abundance of involvements and occupations, flooded by modern mass media. People who are deeply committed to their vocation often tend to orientate their life and work towards "practical" tasks like social problems. They know they are right in doing so — but where is God? Many accept the absence of God as something inevitable. It seems not so difficult because there are so many things to do and to care for. They may continue their prayers and they may even teach about God, but there is an emptiness in them, and their words have no longer the ring of truth and life.

We cannot resign ourselves to God's absence. In the depth of the heart remains the longing for the living God. This desire also is a sign of our times. The more we know about our world, control its hidden powers and fathom the depth of the human mind, the more we need God in whom all is contained and fulfilled. We need God more than the people of the past because our life has become richer and our world larger. But we need the *living* God: we are tired of words and formulas which have been used through centuries and have become like vessels, precious and venerable, but empty. What the mystics meant when they spoke about the ineffable mystery of God must become true again in our heart, in our world, for us.

Words and doctrines are not enough. Everywhere people begin to search for an *experience of God*, for the personal awareness of his presence: "Show us your face, O Lord". This is the prayer not only of the Psalms, but the quest of the entire spiritual tradition of India. Conceptual knowledge is of little use, it may even turn us away from God. We need a personal encounter. Only God can give it. But a warning must be added to the search for the awareness of God. What are you really looking for? For God? Or, for the experience of God? If you wait for God you will have to empty your whole being so that he takes possession of your life. This is a radical demand, the Saints knew it. John of the Cross taught us that we have to go through the night of all our senses, of memory, intellect, affection, and will, so that God alone takes possession of them all and fills them. If we speak of the experience of God we are in danger of seeking rather the fulfilment and enrichment of ourselves than God, we are bound to fail, because we have made ourselves the centre of our concern, and make use of God for our own growth. It is an insult against God. What would you tell a man who comes to you with the wish to experience a friendship? He would, in other words, make use of you for the fulfilment of his emotional needs. You would probably throw him out. It is entirely different if someone comes to you and asks you to be his friend; he reaches out to you and in doing so he surely will experience true friendship. So it is with our encounter with God: seek him with your heart, open yourself to him, allow him to take possession of your whole being and you will experience God.

We encounter God mostly in prayer. There are **many**

ways to deepen and enrich our prayer life and we may use them as they are helpful to each one. But one more warning may be added. There are people who read many books on prayer, attend seminars, get initiated into Indian ways of meditation or into Zen practices, and still they fail. All these approaches may be useful, but there is no absolute recipe. Each one has to find his and her own way in a humble search and in faithful perseverance. Butterflies taste from every blossom and rest on none; we pray not to find God's sweetness but to find *God* — and to give ourselves to him.

The meditations of this day are meant to help us in our search for God. We contemplate the great figures of the Old Testament, and Jesus himself in their encounter with God. However, in their figures we should find ourselves, our own lives with our desire to find God and to open our whole being to the mystery of the living God.

Moses' encounter with God and his mission

Exodus is a religious epic which narrates historical data in the religious context, in which they were experienced by Israel. It is the account of the birth of Israel as a nation. The lifespring of this beginning is God who encounters Moses and singles him out for his mission. This encounter, for which there are no words, is presented in powerful symbols that have kept their meaning through the ages and speak also to us in our search for God.

So we return to the desert, to Moses who encounters God in the burning bush (Ex 3,1-12). It is a unique personal experience, yet it concerns the whole people. It is the starting point of his mission. This is important also in these Spiritual Exercises. Our search for God is not merely the desire for a personal spiritual fulfilment. We can never expect to find God and our true life in isolation, apart from the world in which we live and for which we bear responsibilities. In this retreat the personal search for God is at the same time the search for our true place in the church, in society, in the world.

Moses encounters God at the lowest point of his life. The second chapter of Exodus tells us about the special providence that watched over his childhood. When as an infant he was doomed to die, he was found by Pharaoh's daughter and educated at the royal court. In spite of his privileged position, Moses remained loyal to his people and raised his hand against the Egyptians who exploited Israel. He tried to arouse them to resistance but they did not want it. Disowned by his own people he fled into the desert, a man of shattered dreams, resigned to his helplessness. So the account begins: "When he was feeding the sheep of his father-in-law...."

Who is God?

Moses sees the burning bush. From ancient times fire is the symbol of God with its consuming power, its elusive form, its force beyond control. This fire seems different: it does not destroy or consume, it blazes through the thorny branches but nothing is destroyed, or seems to change. This is the mystery of God that Moses—that all believers through the ages—have to learn: *God is power*, present, all pervading, but nothing is destroyed, nothing seems to change.

The unbeliever will ridicule you when you speak of God. "Show me his deeds; where can you see a change, his interference in our world? Have we not learned the laws according to which nature, life, also human history and society develop? Nowhere can you find God!" The believer has no simple answer to these questions, but he knows that God is here. Nothing seems to change, yet everything is different. Moses will be a new man, Israel's history begins. But you need eyes of faith to be aware of God's power and presence.

"Take off your shoes"

The natural response to a mysterious phenomenon is to come closer and to explore it: "I will see why the bush is not burnt." It is man's right and duty to explore earthly realities with his ingenuity. But God cannot be explored. Coming to God means a personal encounter. God calls Moses by name: "Moses, Moses"—so he calls also me. My name is my concrete existence, all that I am, all that God wants me to be. If God calls my name, the only answer can be, with Moses: "Here I am". In God's presence my real life begins, my mission. It is taken up by God, it is no longer mine.

With Moses I must accept my life and destiny in adoration: "Take off your shoes from your feet for the place on which you are standing is holy ground". Why do we wear shoes? To keep our feet clean, to protect ourselves against stones. Shoes make us stronger to climb mountains and swifter in sports. Shoes enhance the appearance, they can even become a status symbol. In God's presence we take off the shoes. There is no pretence or status, we are bare-foot with nothing to protect us. We encounter God as we are: "Here I am".

The living God

Yahweh calls himself "the God of your fathers, the God of Abraham. Isaac and Jacob", God as he was remembered in the ancient memories of the people. Is this God still alive? He now speaks to Moses. It is different to hear *about* God than to hear *him*, to encounter *him*. We too know the God of bible history which we learnt as children, the God of class-rooms and books, also of sermons. Has he ever touched me, spoken to me? Was I ever in his presence? If in faith I come into God's presence, I also may do like Moses: "And Moses hid his face and he was afraid to look at God."

God lives, and he is concerned with his people: This is no longer history of the past but actual presence, Moses' own situation, the situation of his people: "I have seen the affliction of my people...I have heard their cry...I know their sufferings...I have come down to deliver them, to bring them out of that land to a good, broad land flowing with milk and honey." God's people are not abandoned. God lives and cares. What seemed a situation of desolation and despair is now assured in God's promise. God is alive, he can make come true what proved beyond human power, the liberation of his people.

"I will be with you"

What follows is bewildering. If Yahweh is about to save Israel why does he send Moses? This is the way God brings about salvation, through human agents: "Now I will send you to Pharaoh that you may bring forth my people". Moses objects: "Who am I that I should go to Pharaoh and bring the sons of Israel out of Egypt?" He had tried and failed. God's only answer is the assurance of his presence: "I will be with you."

God works through us. His deeds of salvation are, and for ever remain, his own work, but they are, and forever will be, our own responsibility. There is no division between God's part and our contribution, it is all his, it is all ours. Exodus is God's great deed for his people and through all the following centuries Israel sings God's praise for it. Yet Moses has to carry it through in forty years of struggle, anxiety and often frustration. He receives no tangible help from Yahweh, no resources, no army, only the assurance of his presence: "I will be with you." Nothing has changed, yet everything is different. Such will be also our encounter with God.

SHORT POINTS FOR MEDITATION

1. *Moses encounters God.*

Go through the above reflections. In the meditation on Moses and the burning bush we recognise our own encounter with God in this retreat, in faith. It should not be a meditation on Moses but an initiation into our own coming into God's presence, barefoot, allowing our life to be filled with his presence, power and love. Also my world will apparently be the same after the retreat, nothing seemingly will be changed, but God will be with me.

2. *How young Samuel found God* (1 Sam 3,1-18).

This is the beautiful story of the boy Samuel's first encounter with the living God. He naturally knew about God and shared in the prayers in the temple, just as we know about God from childhood through parents and priests; he faithfully carried out his duties in the temple, but knew nothing of the living God.

— "The word of the Lord was rare in those days":

the decadence in Israel's religion under Heli is the background of the story (cf. 1 Sam 2,12-26).

— "Samuel was ministering the Lord": he is the faithful boy who, however, is still ignorant of the living God.

— "He was lying down in his own place": Be present on that peaceful evening in the temple, the last evening of the sheltered life of the boy. From now on God will take hold of his life.

— "The Lord called: Samuel, Samuel": God calls each one of us by name.

— "Here I am": The boy does not know to whom he really speaks, he runs to Heli, because "he did not yet know the Lord". Many of our prayers may be like this.

— "Speak Lord, your servant listens": This is Samuel's first real prayer which marks the beginning of his life as a prophet. It is the prayer of this whole retreat.

— "Then the Lord spoke to Samuel.." a message of punishment for Heli, for Israel. Their sin will not be expiated by sacrifices, on which they relied, only by a genuine return to God.

— "Samuel was afraid to tell the vision to Heli": how can he change his role of the dutiful servant boy to that of the prophet?

— "But Heli called Samuel": It is Heli's greatness that he desires to know God's message, and "Samuel told him everything." "It is the Lord, let him do what seems good to him". Heli surrenders himself to God who spoke through Samuel. The priest recognizes the prophet, and he bows before God.

This is the beginning of Samuel's mission as a prophet. Again Samuel's encounter with the living God becomes a turning point in Israel's history.

3. *Jesus enters his mission* (Mk 1,9-11).

Also the beginning of Jesus' mission is marked by

the evangelists with a theophany, a manifestation of the living God. Heaven opens, the Spirit descends, the Father acknowledges Jesus as the beloved Son. Jesus carries out his mission in the humble style of a wandering rabbi, yet it has dimensions which are expressed in the theophany of its beginning, and will be fully recognized only after the resurrection.

— "Jesus came from Nazareth...and was baptized": he begins his mission by joining the movement of John the Baptizer and identifying himself with the people who in repentance search for a new beginning.

— "He saw heaven opened": His mission is not limited to earthly horizons, to social changes. It has its origin in God, he proclaims God's reign on earth, the beginning of the new creation.

— "The Spirit descended upon him like a dove": in his preaching and works Jesus will be guided by God's Spirit only. He takes him into the desert, leads him back to Galilee, inspires his message to the poor, gives him power against demons, sin, death.

— "You are my beloved Son, in you I am well pleased": Jesus not only preaches, but his person, his life, work, struggle, his rejection and death and his resurrection are God's revelation: In him God reveals himself as the God of saving love. In him God tells us what we are meant to be. Jesus' life and mission are embraced by his Father's power and love. He is the beginning of a new humanity which is transformed into God's own people.

— "You too will see heaven opened" (Jn 1,51): this is Jesus' promise to the disciples. They are called to share in Jesus' life, mission, glory.

I pray that in my retreat, and every day of my life, heaven be opened; that I learn to live beyond the smallness of daily life, guided by Jesus' Spirit, included in Jesus' sonship, sharing in his obedience and in the assurance of the Father's love.

FURTHER READINGS: (They too may be used for meditation)

— Ps 8, psalm of God's glory and human dignity.
— Ps 63, longing for God.
— Ps 95, an invitation to come into God's presence.
— Wis 9, 1-14, a prayer for true wisdom.
— Eph 1, 1-14; Col 1, 9-14, we are chosen by God from eternity.

QUESTIONS:

1. How did I encounter God in my life? Was I ever personally aware of his power, guidance and love?

2. What can I do to enrich and deepen my prayer life?

FOR REFLECTION:

"Thou hast made me endless, such is thy pleasure. This frail vessel thou emptiest again and again, and fillest it ever with fresh life...Thy infinite gift comes to me only on these very small hands of mine. Ages pass, and still thou pourest, and still there is room to fill." (TAGORE, *Gitanjali*.)

Chapter Three

THE STARTING POINT

The desert is only the passage way to real life. We must encounter God and respond to his invitation in actual life situations. Thus it will be helpful right from the beginning of the retreat to take stock of the world in which we live, its demands and its opportunities. This world is different for each retreatant. However, in the following reflections an outline is given of the three main states in the life of the Church: The lay-person, the priest and the religious. It may serve as a frame work for the more personal and concrete reflections of the retreatant.

A. ENCOUNTERING GOD AS LAY-PERSONS

Lay-persons have often been described in negative terms, as Christians who are neither priests nor religious. We ought to understand the place of lay-persons in the church and world not in comparison with special ecclesiastical offices, nor with the religious state of life, but on its own merits: Lay-people are the believers who live their Christian life and mission in the concrete situations of secular society. Jesus' message is addressed not to special groups but quite simply to the world; lay-people have to realize the meaning of this message in the actual situations of daily life and to live it in their social and professional surroundings.

Jesus, a layman

In his Jewish community Jesus was simply a lay-person. He did in no way belong to the priestly class which was made up of certain families with particular functions in the temple. Nor did Jesus ever join any of the exclusive religious groups of his time like the monks of Qumran at the shores of the Dead Sea, secluded from profane life and living as a community in an ascetical life style. Jesus grew up as a simple village boy in an unknown hamlet of the backward countryside of Galilee. Later he was a wandering teacher who proclaimed his message without any institutional sanction or support, merely on the strength of his prophetic mission. His only support was his Father in heaven.

As a lay-person Jesus was close to the people. He joined their joyous feasts (cf Jn 2,1-12); he was close to the poor, but he moved also among the higher social rank and smiled about the social games that were played among them (cf. Lk 14,7-14). He was aware of the sufferings of the people, the oppression through the Romans, the arbitrary tax system, the helplessness of people in their family problems when a lonely widow loses her only son or a distressed father sees his younger boy walking away into the treacherous city. He also was keenly aware of the general lack of concern for the common people. In spite of good laws and established religious structures in temple and synagogue, he felt the people were like sheep without a shepherd (cf. Mt 9,36). Jesus spoke the language of the people; he did not give elaborate lectures but told them stories which everybody could understand.

Jesus died as a lay person, one may even say because he was a lay-person. He had no academic sanction for his teaching, no official recognition from the religious leaders. How could a simple lay-person claim to know and to interpret the Torah? How could he, with a whip in his hand, drive the merchants out of the temple without a permit of the temple police, who were responsible for the order in the holy place? And still Jesus had power, an irresistible power which came all from within, and so the authorities were afraid of him. He was a threat to their official position. Had not God established them in their authority? Lay people, they thought, should be silent. With the entire hierarchy against him, without any official support, Jesus stood no chance. He was condemned and executed.

Jesus, message for lay-people

In this closeness to the people Jesus effectively proclaimed his message of "the reign of God", of a new human society, where God and his love rule supreme. Jesus was not concerned with reforms in the temple nor with the elaborate regulations of the law. He called for a true conversion of hearts: from these renewed hearts a new society will be born, based not on power and exploitation but on mutual acceptance and love. The world will be renewed not by legislation and institutions, however necessary they are, but by people in the world who know that God is Father of all, and wants that justice and love be the bonds by which the people are united as brothers and sisters to build true community. It is for lay-people in their actual life situations to realize this message of Jesus in the world, also in our world.

The laity in the Church

For centuries the Church has been ruled too exclusively by clergy and hierarchy. The laity was reduced to passivity, and in particular women were kept in a subordinate position. The ideals of Christian holiness and perfection were derived from monasticism, and Christians who strived for holiness were expected to come as close as possible to the standard of monks, keeping regular hours of prayer and imitating their ascetical life-style. No genuine lay spirituality could develop under these conditions.

In the second Vatican Council the Church has seen herself in new perspectives. She embodies Jesus' mission and message for all people. The Church is the "sacrament of salvation" for the world, she is the visible sign and instrument through which Jesus' saving mission for us is to be continued. Thus the Christian community has to make Jesus' message, the vision of God's kingdom and a new society of men real and relevant for the modern world. In their life and work people of our time must realize that our life has meaning and hope, and our society can be renewed. This obviously is the task of the laity, of men and women who have their place in the midst of civic society and are involved in its needs, problems and aspirations. Thus the Constitution on the Church tells lay-people that their Christian vocation consists not in withdrawal from secular affairs, rather "by reason of their special vocation it belongs to the laity to seek the kingdom of God by engaging in temporal affairs and directing them according to God's will....Being led by the spirit of the Gospel, they may contribute to the sanctification of the world as from within like leaven" (L.G. n. 31).

Thus the kingdom of God as it was proclaimed and lived by Jesus must be lived by the laity in today's world. It is based on the unique dignity which we have from God, before whom we all are brothers and sisters in a new relationship of justice, equal basic rights, mutual acceptance, respect and love. Through Jesus the wounds of division and hatred are healed and a new community emerges.

SHORT POINTS FOR MEDITATION

1. *Living stones in God's temple*

Peter writes to Christians who are scattered, overawed by the superior culture of their pagan surroundings, often despised. How can they live their Christian life in confidence? How will they be able to share it with their surrounding world? He shows them their new identity in Christ:

— "Come to him, that living stone, rejected by men, but in God's sight chosen and precious": Peter has before his mind the Jewish temple of Jerusalem, powerfully built of stones. The Christian community also is a temple, firmly built, but not of material stones but a community of believers with Jesus, the Risen Lord, as their cornerstone, i.e. the centre which gives the community firmness and confidence, in whose life and love all share. Jesus too was rejected because his message of God's reign conflicted with the narrow, selfish, self-righteous world in which he lived. But God raised him and made him the corner-stone, the lifespring of the community of believers.

— "Like living stones be yourselves built into a spiritual house to be a holy priesthood": built on Jesus Christ, the corner-stone, united to him and sharing in his life, the community of Christians continues Jesus' presence in the world and fulfils his mission through the ages. Like him they have no sheltered security but have to live the gospel in a world of totally different values. Like him they also will realize conflict and experience rejection. But they are built into a house and become members of the community, which has the promise of fulfilment.

— "To offer spiritual sacrifices acceptable to God through Jesus Christ": sacrifice means a total self-gift to God. Jesus offered no ritual sacrifices as they were performed in the temple; they were symbols and substitutes for our true self-gift to God. But his entire life, his very being, was a self-gift to his Father. So also the faithful, united to Jesus Christ, belong to God with their whole being: "Present your bodies as a living sacrifice holy and acceptable to God" (Rom 12,1). In their life and work they have to live out their consecration to God in baptism.

— "That you may declare the wonderful deeds of him who called you out of darkness into his marvellous light": through a life that is consecrated to God they will also share in Jesus' mission and proclaim the good news, not so much in words as through their life. In the Christian lay-person the richness and joy of the Christian vocation must be visible, a life no longer shrouded in darkness, empty and without hope, but lived in God's love in genuine brotherhood, in the midst of our world,

2. I am baptized.

I have been born into my human life, into my family and social world. In baptism I have been reborn, not to add another life to my human existence but to live this life of mine in God, in its depth and richness, in its full responsibility. In this meditation I reflect on the ritual of baptism and its significance.

— The blessing of the water: "Lord, make holy this

water which you have created so that all those you have chosen may be born again by the power of the Holy Spirit and may take their place among your holy people." Be aware of the natural symbolism of water: it cleanses, refreshes, all life is born from it; it becomes the sign of your own life in God.

— The basic option, renouncing the powers of darkness and destruction, and embracing God's love and life: "Do you reject sin so as to live in the freedom of God's children?" Add your own concrete areas of conflict, which you reject—casteism, bribery, corruption, etc. "Do you believe in God the Father Almighty, in Jesus Christ his only Son and in the Holy Spirit?"

— The baptismal consecration: I am baptised in the name of the Father who holds me in his power and love; and of the Son through whom I share in God's sonship, who is Way, Truth and Life to me; and of the Holy Spirit who dwells in me and is the assurance of eternal life.

— The white garment is given as sign of Christian dignity: May we bring this dignity unstained into the everlasting life of heaven. "Live a life worthy of your calling to which you have been called" (Eph 4,1).

— The lighted candle; "Receive the light of Christ". Jesus, the light shining in darkness, tells his desciples: "You are the light of the world" (Mt 5,14).

— *Epheta:* The priest touches our ears and mouth: "The Lord Jesus made the deaf hear and the dumb speak. May he touch your ears to receive his word, and your mouth to proclaim his faith to the praise and glory of God the Father."

3. *My world.*

In your meditation return to the various spheres of your life: family, neighbourhood, friends, the professional world, and ask yourself these questions:

— What is my place in this sphere?

— What are its needs, problems, aspirations? Am I alive to them?

— How can this world of mine be oriented in the spirit of Jesus Christ?

FURTHER READINGS:

— Jn 17, 9-19; In the world, not of the world.
— Council, Constitution on the Church, n. 10; 30-38: The Laity in the Church.
— Mission Decree n. 21.

ENCOUNTERING GOD AS PRIESTS

More than other faithful the priests have been shaken by a crisis of identity in the wake of the Council. Their traditional image as the sole responsible leaders of the community was questioned: the entire people of God shares in Jesus' priesthood and is entrusted with his mission. Further, in the Constitution on the Church in the Modern World this mission is spelled out as responsibility for the world to help in the solution of the problems of modern society: safeguarding human rights, proper social and economic structures, peace between nations. Where then is the role of the priest? Should he shed his dignity and walk along with the pilgrim people in search of solutions in the pressing problems of our world? Where is the meaning of his preaching, the celebration of the Eucharist and the administration of sacraments?

It will be necessary for the priest at the beginning of his retreat to reflect on some of the perspectives of his priestly ministry and life.

Brother among brothers

Priests are first of all Christians, members of the community of the believers and sharing in the common vocation and mission of Jesus' followers: "In common with all who have been reborn in the font of baptism, priests are brothers among brothers and members of the same body of Christ which all are commanded to build up" (P.O. n. 9). Lay-people should look up to priests not as men in a privileged position but "have as brothers those in the sacred ministry who by teaching, sanctifying and ruling with the authority of Christ so nourish the family of God that the new commandment of love may be fulfilled by all". St. Augustine's words are then quoted: "I am frightened by what I am to you, but I am consoled by what I am with you, To you. I am the bishop, with you I am a Christian; the first is an office, the second is grace; the first is danger, the second salvation" (L.G. n. 32).

Sharing in Jesus' mission

However, the brotherly relation of priests and faithful should not obscure the specific task of the ministerial priest in the Church. From the beginning of the Church it was the responsibility of the Apostles to guide the Church in Jesus' spirit. This responsibility was exercised in various forms till the present structure of the triple ministry of bishop, priest and deacon became universally accepted in the 2nd and 3rd centuries. All Christian communities become more and more conscious of the significance of authority and ministry in the Church. The commission of "Faith and Order" states in its document on ministries addressed to all members of the World Council of Churches: "In order to fulfill its mission, the Church needs persons who are publicly and continually responsible for pointing to its fundamental dependence on Jesus Christ, and thereby provide, within a multiplicity of gifts, a focus of its unity. The ministry of such persons, who since very early times have been ordained, is constitutive for the life and witness of the Church". (n. 8). "The Church has never been without persons holding specific authority and responsibility" (n. 9 Faith and Order Paper 111, 1982).

Thus the representation of Jesus' mission belongs to the very essence of the Church. In her social structure she must contain both: 1. The word of God in its authenticity and authority, teaching, assembling and sanctifying the faithful, 2. the community of the believers which is gathered in response to this divine word, and in which God's reign begins to take shape in our world. This is how the Synod of Bishops in Rome (1971) formulated the place of priests in the Church:

"To the original and unalienable structure of the Church belong, according to the New Testament, (1) an Apostle and (2) a community of faithful united with one another by a multual link under Christ as Head and the influence of the Holy Spirit". This structure remains the same through the ages: "This essential structure of the Church consisting of the flock and of pastors constituted for this purpose...always remains the norm." In this structure "the priestly ministry alone...perpetuates the essential work of the Apostles: by effectively proclaiming the Gospel, by gathering together and leading the community, by remitting sins and especially by celebrating the Eucharist it makes Christ, the Head of the community, present in the exercise of his work of redeeming mankind and glorifying God perfectly" (Synod Document n. 4).

Jesus' priesthood

Thus the priest of the New Testament embodies Jesus' mission in and for the Church, in and for the world of today. He will understand his priesthood to the same extent as he understands Jesus' mission.

Jesus' priesthood is entirely different from that of the old covenant. No one would have called him a priest in his lifetime as he neither belonged to a priestly family nor exercised any priestly function in the temple. Only in retrospect, once his disciples had encountered him as the Risen Lord and began to grasp the impact of his mission, they realised that in him the true meaning of priesthood was fulfilled. Thus in the letter to the Hebrews Jesus is presented as the High Priest, in whom the levitical priesthood is transcended and abolished.

Priesthood means *mediation.* In the levitical priesthood mediation between God and man was an office that was performed in the temple by well regulated rituals. Jesus is mediator by his very being. In the depth of his person he is God-with-man and man-with-God. In him we have the redemption from sin (Hebr 9,12). He establishes the new and everlasting covenant (Hebr 8, 7-13; Jer 31, 31-34). Rituals of the Law are abolished. Jesus' body, his earthly life and existence with us, is the final and total sacrifice to the Father.

This is his morning prayer: "When Jesus came into the world he said: "Sacrifices and burnt offerings you did not desire... But a body you have given to me...Behold I come to do your will, O God' " (Hebr 10,5-7; cf. Ps. 40,6-8).

Thus Jesus' sacrifice, his self-gift to the Father, is spelled out in his entire life. His mission begins with the proclamation of the Good News: "The Spirit of the Lord is upon me...to proclaim the good news to the poor" (Lk 4,18). Jesus gathers disciples and forms them. He is concerned with the people who are like sheep without a shepherd (Mt 9,36). But his mission leads further to realms which no longer can be expressed in terms of human leadership: Jesus' earthly mission fails, he is rejected by those in authority and the people drift away. The seeds of his word have fallen on barren ground—so it seems. What he has gathered and bound together into a community—the first beginnings of the Church—is now scattered. Jesus underlines the ultimate mystery of his mission: the grain of wheat must die to bear fruit; only when he is exalted on the cross will he draw all people to himself. This is Jesus' total abandonment to God. This is also his final self-gift to us. He gives his life as a ransom for many (Mk 10,45). He is the shepherd who lays down his life for the sheep. (cf. Jn 10,11).

Priesthood in the New Testament means realizing Jesus' mission for the community of believers. The priest embodies this mission in his prophetic message of God's reign in the world and society of today, through the leadership in the community, and in the total self-gift of his life to God in the service of his brothers and sisters.

Priestly holiness

If the ministerial priesthood represents Jesus' mission in the Church, it follows that priestly holiness does not consist in the multiplication of spiritual practices—though it is obvious that a priest must plan his life of prayer in a meaningful manner—but quite simply in living out his mission. The Council says: "Priests will acquire holiness in their own distinct way by exercising their functions sincerely and tirelessly in the spirit of Christ" (n. 13). They will grow in the Spirit by proclaiming the Good News, by involvement in the life and struggle of the people, by celebrating the Eucharist—which is Jesus' sacrifice and our sacrifice — with them and for them.

How, then, does a priest encounter God? God has many ways of encountering his own. But a priest, for his part, will never be alone before God. For he can encounter the living God only with and for his faithful. Committed to his mission in Jesus' spirit, he is united to God by being united with God's people.

SHORT POINTS FOR MEDITATION

1. Jesus calls the Twelve (Mk 3,13-19).

— "Jesus went out into the hills to pray, and all night he continued in prayer to God" (Lk 6,12): this night in prayer indicates the decisiveness of the hour in which Jesus entrusts his mission to the Twelve.

— "He called those whom he desired": this vocation is not the choice of a career or a profession, it is Jesus' invitation. His mission must be shared.

— "They came to him": the free response of the Apostles is needed. Their vocation is a mutual commitment.

— "He appointed twelve to be with him": the first requirement for the apostolic mission is continued discipleship, the lifelong union with Jesus.

— "To send them out to preach": to continue Jesus' proclamation of the Good News of God's reign in word, deed and life.

— "To have authority to cast out demons": as Jesus came to over-throw the powers of darkness, the Church has to continue the struggle against all evil, all sin and all that cripples and enslaves human life and human society.

— The names of the twelve express the variety of people called by Jesus, not to constitute a homogeneous group but to be the new Israel, the new comprehensive community of believers in which personal and social differences are transcended.

— The inclusion of Judas in the list indicates the risk of entrusting the message of God's reign to human hands and hearts; God continues to trust us through the ages.

2. *Responsible shepherds* (Acts 20, 17-35).

The entire address of Paul to the elders of Ephesus is the portrait of the true apostle. We pick out the verses which point to apostolic responsibility (vv 28-31).

— "Be on guard for yourselves": the responsibility for the community begins with the self-examination of the elders themselves.

— "And for all the flock": the elders are shepherds, after the example of Jesus himself who was concerned for his own.

— "Of which the Holy Spirit has made you guardians": whatever the way in which ministers are chosen or selected, their authority is from above, they are responsible to God.

— "To feed the church of the Lord": in teaching, guiding, serving.

— "Which he obtained with his own blood": the community belongs to the Lord, never must the minister consider it as his own.

— "Fierce wolves will come": there are the dangers from outside, persecutions, destructive and dissolving influences, ideologies.

— "From among yourselves will arise men speaking perverse things": from the beginning till today the Church is threatened by forces of disintegration.

— "Therefore be alert": it is the watchword of priestly responsibility.

3. *The Spirit of the shepherd* (1 Pet 5, 1-4).

— "I exhort the elders among you as a fellow elder": Peter writes in a spirit of collegiality.

— "As a witness of the sufferings of Christ as well as a partaker in the glory that is to be revealed": his title to speak is not formal authority but his experience of Jesus' passion and the firm hope by which he is rooted in Christ.

— "Tend the flock of God that is in your charge": fulfil your task not with autocratic authority but as stewards of Jesus Christ who alone is Lord of the Church.

— "Not by constraint but willingly": it is not a burden but a charism, a service to which the minister has committed himself freely.

— "Not for shameful gain": it is the temptation of all authority, also in the church, that it be used for selfish interests, material or social.

— "Not domineering", with overbearing, oppressive, possessive authority.

— "Being an example to your flock", embodying the message of God's kingdom in their own life, as Jesus did, and so to be transparent to Jesus' presence in the community.

— "When the chief shepherd is manifested you will obtain the unfading crown of glory": the minister shares both in the mission of Jesus on earth and in the final glory.

FURTHER READINGS

— Lk 22,31-34: Jesus prays for Peter that he be the support of his brothers in spite of his weakness.

— 2 Cor 4,1-15: God's power in earthen vessels.

— 2 Tim 1,3-14: assurance in crisis.

— Constitution on the Church, n. 18a: The office in the Church as service to the community.

— id. n.28: The priestly ministry.

— Decree for priests; n.13: priestly holiness through the exercise of the prophetic, pastoral and priestly mission.

QUESTIONS:

1. Trace the development of your own understanding of the priesthood from the early days of the seminary till now. How has it changed?

2. Have I grown spiritually through my priestly ministry?

3. What are the priorities in my priestly ministry?

FOR REFLECTION:

"Since once again, Lord, I have neither bread nor wine nor altar, I will raise myself above these symbols, up to the pure majesty of the real self. I, your priest, will make the whole earth my altar and on it will offer you all the labours and sufferings of the world" *Hymn of the Universe:* "The mass of the world".) TEILHARD DE CHARDIN

HOW RELIGIOUS ENCOUNTER GOD

Religious seem to be a privileged group in encountering God. Their entire life is a consecration to God either in contemplative communities or in the active service of their brothers and sisters. Yet they too have their problems. Here we do not intend to speak about the difficulties they share with other Christians, their prayer life, the temptation to mediocrity and shallowness, their feelings of frustration and aimlessness; but we are concerned with thsoe problems which belong to the specific nature of religious life.

Key Problems of religious life

There is, first, the emphatic insistence of the Council that all christians are called to holiness: "The followers of Christ...have been made sons of God in the baptism of faith and partakers in the divine nature". Hence "all Christians in any state or walk of life are called to the fulness of Christian life and to the perfection of love" (L.G. n.40). The Council speaks no longer of religious life as a "state" of perfection. What then is the specific nature of religious life as compared to the ordinary condition of Christians in the world? What is the meaning of their special consecration through the vows of the evangelical counsels? What is their place in the Church, especially their relation to the community of believers?

Even the meaning of the consecration through the evangelical counsels is today questioned. Jesus came that we should have life, life in its fulness: "Because I live, you will live also" (Jn 14,19). The Council says that through "holiness a more human manner of life is fostered also in earthly society" (L.G. n. 40). But it would seem that religious life, and in particular the three evangelical counsels, diminish human life.

Human life grows in community, in the enriching and challenging communication from person to person. In the encounter with the "other" each one becomes himself/herself. We grow beyond ourselves through love, sharing and service. The fullest expression of human relation is the union of man and woman in the bond of marriage—this is God's design. In their complementarity they are God's image (cf. Gen 1,27). Is not the renouncement of marriage in religious life a radical diminution of human life, reducing it to isolation, depriving it of the creative power of love?

Food and shelter are the material basis of human existence. We need them to live and to grow. Personal and social life develop in the struggle for our existence. Such was God's design when he put man in command over the earth to cultivate it (cf. Gen 1,26ff). In taking possession of the earth and transforming it through his work man becomes fully man. John Paul's encyclical on Human Work says, that through work "man not only transforms nature...but he also achieves fulfilment as a human being, and indeed, in a sense, becomes more human" (n.9). Through the possession and control over the material world man extends, as it were, his personality into the material sphere. He is a farmer through the land he possesses and cultivates; he is an engineer through his tools and machines. Can we renounce through a vow of poverty the material basis of our existence without giving up our self-responsibility and self-assurance?

Finally, man is created free, different from all creatures which are but part of nature, woven into its cycles, without a life and destiny of their own. Man on the contrary, holds his life in his own hands. Though dependent in a thousand ways on the material world and his social surroundings man is responsible for himself to shape his life. This is his dignity. Never can he renounce his freedom and shift his responsibility on to other shoulders. Were he to do so he would give up the core of his personality. How then can man make a vow of obedience?

Such questions cannot be answered on a merely theological level. They concern very much the practice of religious life. We do find religious whose life-spring of love has dried up, whose attitude towards people is formal, professional, perhaps outwardly friendly but without the real, personal touch that gives warmth and colour to human relations. Other religious remain immature, without confidence to make personal decisions; their consecration has not become a source of life for them. Many reasons may account for such failures but often they imply a wrong and negative understanding of the evangelical counsels. What is their meaning?

Witness to the Christian vocation

The Mission Decree of the Vatican Council speaks only in a short paragraph on religious life, yet in a concise formula it epitomizes the lengthy debates of the Council on religious life. The decree says that religious life should be fostered right from the beginning in young Christian communities. The reason for this urgency is not only the important contribution of religious for many missionary activities, but "through the deeper consecration made to God in the Church religious life clearly shows and signifies the intimate nature of the Christian vocation" (A.G. n. 18).

A young church needs not only the Christian message and doctrine; it needs the witness of Christian life. What does it mean to be a Christian? In religious life, through the individual and through the community, the meaning of the Christian vocation becomes tangible. Religious are people who have understood what it means to be a disciple of Jesus and they commit themselves in earnestness to a life according to the gospel. Jesus' message of God's reign is not first of all a doctrine about God but the good news of a new creation, coming from God, to be realized through us in our world. How can people ever understand this message if they do not see it actually working, realized in people, in a life that is meaningful and convincing?

To give credibility to the Christian message it must be lived in a full commitment. True, every Christian is consecrated to God by baptism and called to holiness, but often this response remains something legal, and leads merely to a degree of conformity with laws and observances without an impact on actual life. If God's reign is to come, if our world is to be changed and renewed, conformism will not do. A wholehearted, radical response is needed. There must be people convinced and committed.

For two thousand years there have been Christians totally committed to Jesus, in the midst of the world. But from early times this total commitment has also been lived in an organized manner, in response to special needs of the times. The evangelical counsels are not meant so much to be additional obligations to which religious bind themselves beyond the commandments, but a radical form of living the commitment to Jesus Christ, a witness to his message.

The three evangelical counsels as witness to the Christian life

This, then, is the meaning of *consecrated chastity:* Every Christian is called to love God with his whole heart and soul (cf. Deut 6,6-8; Mk 12, 28f). Religious realize that the absorbing power of divine love can inspire their entire life and work, give them personal fulfilment and lead them to renounce marriage. For them this renouncement is not a diminution of life and love but an opening out beyond the limited concern of their own home and family to all those who are their brothers and sisters, especially to those to whom love and care are denied. This reaching out becomes for the religious a source of joy and ever greater love.

Concerning poverty: all followers of Jesus have to learn that the value and security of human life are not based on material wealth. We do need food and shelter; Jesus himself is deeply concerned with people who are hungry and destitute. Yet Jesus never tires of telling his disciples that the security of wealth is deceptive. The parable of the rich fool demonstrates the folly of losing our heart in possessions (Lk 12,13-21; cf Mt 6,21). In Jesus' encounter with the rich young man we see how wealth becomes a powerful snare that holds us back from a full commitment (Mk 10, 17-22). God alone is the ultimate support of our life; we must be free from all anxious concern (Mt 6,25-34).

It is for religious to show through the vow of poverty that it is possible freely to renounce possession and security and so to become rich in God. True, in our world of destitution it will not be easy to become witnesses to evangelical poverty. But surely religious can witness to the Christian attitude towards wealth where they cease to live a secluded life of their own apart from society, where they are truly available for the service of the people and, through their style of living, show the freedom and detachment from such goods which in a consumer society are held dear.

Obedience has become problematic in our world. Modern man has become deeply conscious of the dignity of the human person and the need of his genuine self-expression. At the same time we are aware that freedom must be exercised not for the advancement of each individual's status and power but in dependence on God and for the good of human society. Jesus proclaimed God's kingdom as a renewed society of justice, peace and love. Human life is fulfilled if we live in this new community: first seek God's reign and his righteousness and all the rest will be yours (cf Mt 6,33).

Through the vow of obedience religious life is totally oriented towards the proclamation and realization of God's kingdom. It is a witness to all believers, to all men, that a personal career is not the ultimate goal of human life. Real life is more. In God's reign, i.e. in the human society as Jesus visualized it, all relations, all bonds of communication, become channels of God's love realized in human brotherhood. Thus, religious obedience, rightly understood, is not an abdication of human freedom and responsibility but the radical integration into God's own vision of human society. It is possible to commit our whole life to Jesus' mission for the salvation and renewal of the human family. This is how Vatican II sees religious obedience: through it religious "are united more permanently and securely with God's saving will" (P.C. n. 14). Such obedience is the ultimate freedom and dignity of the human person.

How, then, are religious actually encountering God?

They have not a life of their own apart from the Christian community. Religious are consecrated to God not for their own sake only but in the Church, for the Church and for the world. Their presence and work among the people is meant to become an inspiration and assurance for others, to live their Christian vocation. *For religious, encountering God is not a private affair.* It is their charism to live the gospel in its integrity and in full earnestness so as to show in their lives its richness and challenge, to be convincing witnesses to God's kingdom and to give credibility to the Christian message.

SHORT POINTS FOR MEDITATION

1. *The radical following of Jesus* (Mk 10, 17-31).

Christian life is more than the legal fulfilment of the commandments. It leads to the personal response to Jesus' call.

— "What must I do to inherit eternal life?": eternal life consists in knowing the one true God and him whom he has sent, Jesus Christ (cf. Jn 17,3). 'Knowing' is more than fulfilling the commandments. It means sharing in God's own life.

— "All these I have fulfilled from my youth": the young man refers to the faithfulness in observing the Torah which regulated the life of the Jews.

— "You lack one thing": the legal observance must be completed through the personal surrender to God. God wants the whole person.

— "Sell...give...and follow me": these are three stages of surrendering to God—(1) ridding oneself from attachment and possession, (2) sharing with the poor, (3) following Jesus Christ in freedom.

— "At this saying his countenance fell and he went away sorrowful": many are ready to give their work, few are prepared to give themselves.

— "How hard it will be for those who have riches to enter the kingdom of God": riches become a block. They may consist in material wealth or in any form of power and social status.

— "With men it is impossible, not with God": the religious vocation is God's free gift.

2. *The life of the Spirit* (Jn 3, 1-8).

Religious life is charismatic, the response to the movement of the Holy Spirit who makes us see life and the world in a new light.

— Nicodemus, a pharisee, a ruler of the Jews. He was a man of social status, moral standards and deep honesty, the best disciple whom Jesus could find in his surroundings.

— "Unless one is born anew, he cannot see the kingdom of God": Nicodemus has to learn that God's kingdom is more than righteousness; it is a new life.

— "How can a man be born anew?" Nicodemus sees his life in the frame of his cultural and religious tradition. Jesus speaks of the life that comes from God. Man must be born anew from water and the Spirit.

— "The wind blows where it wills. You do not know whence it comes or whither it goes": we realize its power but we are unable to trace its origin and its goal. The spiritual man has his orientation not from within the framework of natural causes. His life cannot be understood by rational analysis nor assessed by merely practical standards. It is God's own life and gift.

3. *God's light reflected in his messengers* (2 Cor 4, 1-6)

Read the whole text reflectively. Paul defends himself before the Corinthians against accusations of dishonesty. He has to present the gospel through the witness of his life.

— "We have renounced disgraceful, underhanded ways...By the open statement of the truth we would commend ourselves to every man's conscience in the sight of God": honesty and earnestness in the commitment to God are the first demands of religious life.

— "The gospel is veiled...The god of this world has blinded the minds of the unbelievers": Jesus himself speaks of the gospel as a hidden treasure (Mt 13,44). Ever more the joy, richness and challenge of the gospel remain hidden.

— "What we preach is not ourselves, but Jesus Christ, as Lord, with ourselves as his servants": the gospel is revealed through the life of the Apostle. The only meaning of religious life is to reveal Jesus Christ and his message.

— "Let light shine out of darkness:" God shines in our hearts to reveal himself, to reveal his saving message to our world.

FURTHER READINGS

— Lk 14, 25-35: Counting the cost when building the house.
— 1 Cor 2,1-9: Wisdom of the world, wisdom of God.
— 1 Cor 7,32-35: Concerned with the things of the Lord.
— 2 Cor 4,7-12: Strength in weakness.

QUESTIONS:

1. Have I felt the challenge of the gospel? On what occasions? In what passages?
2. Do I feel the conflict between my life and the demands of Jesus? Between my religious life and the gospel?
3. How do I intend to put the gospel into the context of my daily life and work?

FOR REFLECTION:

Saint Teresa of Avila joined a modern convent with relaxed rules of enclosure, open to visitors, with reasonable comfort... "But once she was touched interiorly by the presence of Christ, and the inexorable reality of the gospel, free from all disguising phraseology, had come before her mind, she saw in all this the unbearable flight from the real task, the escape from the conversion demanded of her. She got up and was converted. This means she threw away the 'aggiornamento' and created a renewal that was not a compromise but a demanding claim, to commit oneself to the eschatological expropriation of Jesus Christ; to allow oneself to be fully expropriated with the crucified Lord, to belong totally to him, to his body".

J. Ratzinger, on the *Conversion of St. Teresa of Avila.*

Chapter Four

MY LIFE IN THE WORLD

*"Let us make man in our image, after our likeness; and
let them have dominion....over all the earth"* (*Gen 1,26*).

The first days of the retreat are devoted to the basic
orientation: the meaning of the desert experience for our
life, the encounter with God, the social frame in which we live
and work, as lay-persons, priests, religious. On this day we
reflect in a comprehensive way on the meaning of our
existence, of our life in the world.

This question has led men of all ages and cultures to
the deepest, never ending search for truth. At the beginning
of the Spiritual Exercises we feel the need for such an orientation, but not so much as part of a philosophical system of
thought but for an authentic understanding of what we are,
and are meant to be. A theoretical reflection would lead us
into the realm of concepts; *what we need to see is life as a whole.*
Can we fathom the meaning of our existence? Are we able to
articulate it?

Ignatius has done it, and has placed it at the outset
of the Spiritual Exercises under the title "the First Principle
and Foundation" (n. 23). It is a concise text, worded in
scholastic language and therefore not easy to fathom in its
existential depth. Yet for anyone who has tried to enter into
the spiritual world of the Saint, it vibrates with the power of
his mystical experience. For him it is anything but an abstract

formula; it is the condensed expression of his comprehensive vision of life and world. We try to follow the keywords of the text and to capture the personal experience expressed in them.

Man is created

These words are found in every catechism, and at first sight they look like a common-place statement. For Ignatius they are the expression of the deepest and richest mystical experience of his life which had come to him at Manresa. There he realized "with great spiritual joy how God created the world". From that time onward he had an entirely new understanding of life and world. The world is "from above" and all creatures are "transparent" to the mystery of God. Hugo Rahner sees in these terms two key concepts of Ignatian spirituality: "All earthly beauty, wisdom and righteousness were merely the reflected splendour of what he had already grasped in the immediacy of his mystical contemplation of God himself" (H. Rahner, *Ignatius the Theologian*, Herder, 1963, p. 4). No creature has ultimate meaning in itself and human life is empty unless it is filled and embraced by God. God is not merely the cause of the world, its hidden origin, but in him only all creatures, their harmony and history, have meaning. Not that we are able to "explain" all things— explaining means analysing the various components and influences of which the world of our experience is made up. But everything falls in place, is seen in God, just as all colours we see are contained in the brightness of the sun. Of his experience at Manresa Ignatius said: "I beheld, sensed within myself, and penetrated in spirit, all the mysteries of the Christian faith". He still had to study his theology in later years to spell out, as it were, his faith, but the intuition, the comprehensive vision comes before his studies; he had found it in God: the world is from above. It is transparent to him.

This, then, is the underlying view of man and world as it is embodied in the Spiritual Exercises. We naturally cannot claim the same experience for ourselves, but it would be wrong to dismiss it as a singular mystical gift of a Saint, which is beyond our grasp. What Ignatius perceived in overwhelming brightness, we know by faith in the dim awareness of God's love and presence. It may grow to greater lucidity in the following weeks of the Spiritual Exercises as their entire dynamism is based on God's word and the guidance of his Spirit. More and more we realize that world and life have meaning only in God, from above.

The meaning of human life is then spelled out in three words: praise, reverence and service.

To Praise God

Man is created to glorify God. Glory is not something added to the person, a halo of dignity and importance. In biblical language God's glory is Yahweh himself as he is with his people and reveals himself in power for their salvation. The Israelites perceived the presence and power of Yahweh tangibly during their journey through the desert, e.g. in the manna, Moses tells the people: "In the morning you shall see the glory of the Lord" (Ex 16,7) and they found the manna with which God nourished them. The visible sign of God's presence is the cloud which fills the holy tent (Ex 40,34) and goes before them leading them on their way. It is God himself who is with his people.

Through Jesus God's presence and glory are with us in a new way. In him the disciples encounter God's care and love: "We have seen his glory" (Jn 1,14). Even more, through Jesus God's glory becomes our own. God makes us share in his life and glory. "We rejoice in our hope of sharing in the glory of God" (Rom 5,2). At the end of time also our body "is raised in glory" (1 Cor 15,43). The transforming power of God is at work in us even during our earthly life: "We all, beholding the glory of the Lord, are being changed into his likeness from one degree of glory to another" (2 Cor 3, 18).

Thus glorifying, praising God means recognizing God's saving love, power and presence. We do it in prayers and hymns which are "the sacrifice of praise to God, the fruit of lips that acknowledge his name" (Hebr 13,15). But words have value only if they come from the heart and ring with the truth of life. The real glory of God is our life. It is not what we speak, not even what we do but what we are. This seems important in our secular society where we are liable to value people only by what they produce, not by what they are.

Thus praising God, the first meaning of our life, consists in living it in its richness and depth, transparent to God's presence and love. Life has no meaning outside itself. True, with a thousand threads it is woven into history and society; yet it has its value and dignity in itself. Never ask a flower why it blossoms; this is its very nature. Today it unfolds its matchless beauty and fills the air with its scent, though tomorrow it may be thrown into the oven. Jesus saw flowers, and he sees our lives. Are we not better than birds and lilies? (cf Mt 6,25-33). We praise God by living the life that he gave us, in which we unfold God's own life and love.

To Reverence God

The Bible speaks of the fear of God. The better word would be "*awe*", the sense of the infinite mystery of God who remains beyond our grasp, beyond our control. Awe gives us the awareness of total dependence on him and at the same time of total assurance, because God encompasses our life, past, present and future. This awe is totally different from "fear", which is the instinctive reaction to actual dangers and threats, real or imagined. It would be very wrong, indeed it is most harmful, to speak of the fear of God in this sense as if God threatened us. We ought never to be afraid of God.

But awe of God increases as we grow in understanding. Only superficial people think and act in their relation to God as they might deal with a good natured neighbour whose help we need at times, whom we get round to our ideas if needed, whom at times we also may fool. Such people know nothing of God. God is Lord. Ignatius likes to speak of him as the Divine Majesty, which, however, has nothing to do with oppressive domination but means the all encompassing presence and power, the all knowing wisdom and all embracing love of God that enfolds our life and gives us the ultimate assurance that no creature can give.

Thus reverence is not merely a form of behaviour as we express it in liturgy, but the awareness of the beyond, of the power from above that guides our life and pervades the universe. It is an attitude of silence and adoration before the wonder of God.

To Serve God

We may feel some gratification that at least this last word, by which St. Ignatius expresses the meaning of our life, seems to have found wide recognition in our time. Service is the key-word of the Council when it speaks about offices and responsibilities in the Church. It has even become a slogan in secular society when political parties wish to demonstrate their concern for the people. It is the necessary attitude for renewing the life of the Church and building human society.

Yet the word implies possible misunderstandings. It is a word of action; service is measured by functions which are offered to the community. So it may be understood as a multitude of activities in which God has no real place. We may even use it as a justification for such slogans as "my work is my prayer."

Surely service implies work, yet it is more. Jesus lived among his disciples as one who served, but his impact is measured not by the number of services rendered to them but by his attitude of giving and building up, of reconciling, healing and forming community. He fulfils in himself the ancient figure of the "Servant of Yahweh" whose mission consists in bringing about justice and righteousness in Israel, even to the ends of the earth (cf Is. 42, 1-4; 49,5f). His service consists in doing God's work, in bringing about reconciliation, the new creation (cf. 2 Cor 5, 17f).

Serving God, therefore, means integrating our life into the work of God. God is with us not for idle contemplation but in his work of creating and saving, of uniting and fulfilling in himself this vast world of ours. This small life of mine is assumed into this movement; I am called to take part in it with all my being within the limited sphere of my own world, to live not with and for myself only, but in God's kingdom.

And so to save his soul

How then, shall we find our fulfilment as human beings, the wholeness and fulness for which we are searching and striving? There is no fulfilment apart from the true and only meaning of our life: live our life transparent to God's presence ...*praise God*, in adoration accepting our situation and tasks as they are given to us from above...*reverence God* and see ourself with all that we are and have woven into God's own work of creation and salvation...*serve God*. This is the only way of saving our soul, i.e. of finding and living our true life.

The other things on the face of the earth are created for man

Such words invite us to deep philosophical and theological reflections about man, the crown of creation. However, the Spiritual Exercises are not a course in systematic theology but a spiritual initiation. Our relation to God and to the world is expressed in simple language in the parable of the prodigal son where the father speaks to the elder son: "Son, you are always with me, and all that is mine is yours" (Lk 15,31). These words of loving tenderness express our place in the world better than doctrinal explanations. Once we have realized the meaning of our life, the wonder of God's presence, and the whole world as coming from God, it is obvious that created things should never stand between God and us, they should never be a hindrance on the way towards our goal. They are God's majestic, beautiful, bountiful creation which he gives us to be contemplated, and to be perfected and completed. They are the medium through which God cares and provides for us and in which we are meant to find him. In this world that God gives us, God's reign is to be realized.

God wanted it to be our responsibility to realize the goal of creation, his kingdom, "to direct temporal affairs according to God's will" (L.G. n. 31). This demands deep insights and ever new personal decisions. Our life is a continuous movement towards our destiny, woven into world and society. This movement is a free response to God's invitation, given to us from above, by God. Whatever we do, every option should be in line with the meaning of our life, the praise, reverence and service of God. Thus the norm for our options is contained in the brief formula:

Man must make use of created things in as far as they help him in the attainment of this end; he must rid himself of them in as far as they prove a hindrance to him.

This formula may look too rational for actual life; it seems to stifle creativeness. Is it possible constantly to reflect on our actions? In doing so we seem to place ourselves outside real life, we are like managers who organise their business with cool detachment, or like gardeners who plan their garden and till it—should we not be rather like the flowers which blossom? In fact it has been an objection against Ignatian spirituality that it reduces life to rational planning and hinders creative spontaneity.

This surely is a danger. However, what Ignatius expresses here is only an abstract principle which has to be realized in the context of actual life. In fact, according to the Spiritual Exercises, decision-making does not take place at the brain level, though the rational element always remains included, but in the depth of the person, in that innermost sphere where God's Spirit is at work. In a mature person decision-making becomes a spontaneous expression of life that by its innate dynamism moves towards its realization with unwavering consistency, yet always it is adjusted to actual situations. We may compare it with a mountain path: it does not force its way through the rocks but passes them by, sometimes it seems to lose its direction, often the peak is out of sight, yet with consistent determinaton it leads towards the goal. This is the art of Christian living. It is genuinely human, woven into the world and responding to ever changing situations, still it is never drifting but always moving towards its fulfilment. Every step is an opening towards God.

These norms apply to all spheres of life: to habits of food, sleep, dress, to work and leisure. They affect our involvement in the very personal spheres of our life, from prayer and study, to the way we inter-relate with people. Such norms help us in finding the right attitude towards persons, and finally they help us make major choices, which may have a lasting impact on our life.

We must make ourselves free in relation to all created things

It will not be possible to realize this basic norm in our options and to live in continuous openness towards God unless we have found an inner freedom. The word "indifference" which is often used to express this attitude, is not very inspiring: it is too cold and could easily be associated with aloofness which would be a radical distortion of the Ignatian attitude towards created things. Also the word "detachment" may be misleading as it expresses only the negative side of the attitude, i.e. not to be bound by attachment. We need a positive and constructive attitude which is best expressed as *"freedom."* It is a gospel word which has the ring of victory over crippling and fettering forces, over the powers of darkness which threaten to engulf our life and lure us away from our goal into harmful compromises. It conveys at the same time the joy and assurance of a life committed to God: "Christ has set us free. Stand firm, therefore, and do not submit again to the yoke of slavery" (Gal 5,1).

This gospel freedom must be found and tested with regards to the various spheres of life. It implies the joyful acceptance of the ups and downs in life, of conditions of health, of life situations and surroundings, of work, success and failure. This freedom gives peace, it is the atmosphere in which we make the right decisions. It knows of the beauty and the values of created things, of all that is human, yet it has its centre in God, we may say it is in orbit with God, and in God's presence all creatures become weightless. Only one ultimate desire remains which is expressed in the concluding sentence of the text:

Our one desire and choice should be what is more conducive to the end for which we are created.

The accent lies on "more". True, the whole of creation has but one end. In the Spiritual Exercises this end becomes a personal concern and commitment of the retreatant; it is the core of his life and work. Thus the apparently detached, rational text of the "Principle and Foundation" ends with a dynamic and personal thrust, not to fall back into sluggish complacency, but to *remain ever open and alert to God.*

SHORT POINTS FOR MEDITATION

Ignatius gives no meditation on the Principle and Foundation. The retreatant is expected to enter into the dynamics of this text and to relate it to his life-situations. He ought to become aware of the coherence and comprehensiveness of this vision of life and world.

Every paragraph of the text could be substantiated from the Bible; we take up some aspects in these short points for meditation.

1. *Man, crown and Lord of creation* (Gen 1, 26-2, 3).

The power of the key words of this text should impress itself on our mind. See yourself at the dawn of creation; God gives all creatures to you: "Eat freely of every tree" (Gen 2, 16). Then meditate on these texts:

— "Let us make man in our image, after our likeness": images are terrestrial representations of the unseen. The Israelites were forbidden to make images; they detested the Canaanite images of divinities. The only image of God—the living God—his terrestrial representation is man. God makes him in his image, transparent to his glory. God gives him dignity above all creatures.

— "He let them have dominion...over all the earth": man has to impress his image on the earth. It must become a human world, reflecting God's creative love.

— "Male and Female he created them": in mutual complementarity, to live in family and community; only in mutual communion can we grow and build a human world.

— "Multiply and subdue the earth": it is man's unique task to continue God's creation, to humanize nature. The cradle of mankind is nature, the garden; its fulfilment is culture, the holy city (Rev 21,2), a world on which man has imprinted his stamp. If man ceases to be God's image, instead of humanizing the earth, he destroys it.

— "God saw everything that he had made, and behold it was very good": evil, real evil and destruction, comes only from man, when he ceases to be God's image.

— "God rested on the seventh day from all his work": true, God is present in his creation. We have to find him in his works, and do our work in obedience. But God is more than his works, and we too have to grow beyond creation, to enter into his rest, and in him to find our final goal (cf. Hebr 3,7-19).

2.   *The world as temptation*

God's creation is good. God saw that it was *"very good"* (cf. Gen 1,31). But its very goodness can make it a temptation, and so pervert its end.

— Creation in its majestic greatness and beauty has a fascination for man. From ancient times it absorbed man's mind and let him forget the Creator from whom it comes: "They did not recognize the craftsman while paying heed to his works" (Wis, 13,1-9). It is more so today.

— The possession of wealth gives wrong security and diverts man from his reliance on God: "Do not heap up treasures" (Mt 6,19); examples are the rich fool (Lk 12,13-21), the rich man who is prevented by his wealth from following Jesus (Mk 10,17-22).

— The entire gospel of John depicts Jesus' coming into the world as the conflict between light and darkness: "The light came into the world, but people loved darkness better than light" (Jn 3, 19).

— Jesus prays for the disciples: "They are in the world; Holy Father, keep them safe by the power of your name" (Jn 17, 11).

We pray that God be more real in our life than his creatures; that with pure eyes we find him in all creatures, and see all creatures as coming from him.

3. *Jesus' message of God's kingdom* (Mt 6, 24-33).

The message of God's kingdom implies the orientation of the entire life to God.

— "No one can serve two masters": Jesus takes up the ancient alternative between death and life, evil and good (cf. Deut 30,15f). We have to base our life either on earthly and material support or on God.

—· "Therefore do not be anxious about your life": Once we have recognized God as source of all life, we find freedom; work and labour remain ours, our worries are in God's hands.

— "Is not life more than food, and the body more than clothing?": Our attention is absorbed by daily needs, food, clothing, etc. Jesus asks us to turn our concern to the core of our life which comes from God.

— "Look at the birds of the air...consider the lilies of the field": the beauty and harmony we find in nature comes from the implicit assurance that each creature has its place in God's plan and fulfils its task in carefree obedience. Nature knows no lazy carelessness; there is dynamism and work, but all move in ultimate assurance.

— "Are you not of more value than they?": how much more must each person accept his/her place and task in the assurance that our real life cannot be lost. Our concerns often go beyond daily needs to a career, professional advance, social status. All this may turn us away from the core of our life, from its gravitational centre in God.

— "Seek first God's kingdom and his righteousness, and all these things will be yours as well": if our radical orientation towards God is realized in human society, in true brotherhood, also the material needs will be fulfilled.

It is obvious that this passage is not an easy, "supernaturalistic" evasion of the socio-economic problems of society. Jesus stands radically for human brotherhood against social injustice, in continuation of the entire

prophetic tradition. In genuine human communion all
material needs can be fulfilled. Thus this text is concerned
with the basic orientation of our life. If we are absorbed
by material cares in the selfish pursuit of social superiority,
economic power and domination, in competition with the
neighbour and at the expense of the weak, then we lose
the basic orientation of our life and ruin human society.

FURTHER READINGS:

— Ps 8 and 104: God's greatness in creation, man's dignity.
— Sir 16,24-17,32: The majesty of God's creation and his
        providence.
— Vatican II. The Church and the Modern World, nn.
        4-10: the situation of man in the world
        today. nn. 36 and 59: the recognition of
        human values.

QUESTIONS:

1.   Reflect on the basic orientation of your life: On what
     basis do you make your decisions? What are the
     criteria in arranging your timetable, calendar, etc?

2.   Reflect on your surroundings, your work, the people,
     etc. In what way do they help you to be closer to
     God, in what way do they hinder you?

FOR REFLECTION

"Everything you love for its own sake, outside God alone,
blinds your intellect and ruins your judgment of moral values
and vitiates your choices so that you cannot clearly distinguish
good from evil, and do not truly know God's will".

THOMAS MERTON

Chapter Five

THE BROKEN WORLD

"The time is fulfilled, the kingdom of God is at hand, repent and believe the good news" (Mk 1,15).

Once more we search for the starting point of our Spiritual Exercises. We have reflected on the world as God wanted it, and on our life and its meaning in God' design. But this is not the world in which we live and the life as we experience it. No lengthy arguments are needed to show that there is something wrong in the created world. We have witnessed the steady increase in our resources and the accelerating pace in our production, far exceeding all our needs. Still, there are the millions of people, hungry and homeless. Our libraries are overflowing with factual knowledge. We are offered every possible comfort; we are able to fill our homes with beauty and music. Our markets are flooded with remedies against all kinds of ills and evils. Still, we seem not to have become wiser, happier, healthier. We have woven a tight network of transport and communication over continents and oceans, and yet our human family is more sharply divided than ever, and we understand each other less and less.

Here, then, is the real starting point of our Spiritual Exercises. If we want to be realistic we must begin with our broken world. Also the Gospel does not begin with the beauty and order of creation but with Jesus "who will save his people from their sins" (Mt 1,21).

Why meditate on sin?

Some preliminary remarks may be helpful. First, *sin is not the centre of the Christian message*. At times it might have appeared to be so. Some preachers considered it the main theme of their sermons to warn Christians against sin and its dreadful consequences. Often churches had an atmosphere of somberness and fear. This surely is not the spirit of Jesus who devoted his entire life to the proclamation of the good news of God's reign, of God's love and concern for us, and of the new creation of peace and brotherhood. His preaching has a new ring, different from that of John the Baptizer who had accused the people of their complacent self-righteousness and the reliance on being children of Abraham, and had warned them of the impending judgement: The axe is laid to the root of the tree (cf Mt 3,7-10). Jesus speaks a new language; his key-words are *love* and *joy* so that John can summarize his message: "These things I have spoken to you that my joy may be in you and that your joy may be full" (Jn 15,11). This is the atmosphere of the entire Christian life, it is the spirit also of our meditation on sin.

As Christians we look at sin with the eyes of God himself who "so loved the world that he gave his only Son that whoever believes in him should not perish" (Jn 3,16). *God is greater than sin*. Our broken world is embraced by his love and power. At times we are overwhelmed with a sense of helplessness in the midst of a world of doom and division. This is not Jesus' spirit, who with undaunted courage struggled against the powers of darkness and finally conquered them as the Risen Lord. We need this joyful assurance in our meditation on sin. We cannot face the darkness of evil unless we are assured of God's creative power which will heal us, which will renew the world.

It would, however, be equally dangerous to play down the *seriousness of sin*. Some are inclined to reduce it to a failure in social adjustment or to psychological blocks which prevent us from integrating ourselves into society. We surely have to learn much from social and psychological studies which have broadened our understanding of human behaviour and enriched it enormously. We can no longer be satisfied with the simplistic classification of human actions as good or evil.

Still, in all our actions and attitudes we articulate something of the ultimate orientation of our life. They have a depth dimension which belongs to the core of our personality, through which we are related to our neighbour and to society, which makes our life constructive in our world, or destructive and divisive. With this depth dimension we are concerned. Have we allowed the powers, which ruin our life, to penetrate into our hearts and make us partners to the destructive evils which overwhelm our world?

There is one more danger. Many keep in their heart a *secret admiration for sin*. Read the papers and study the mass media, and compare the space given to crime and violence with that given to goodness and peace. Of course often enough crime is condemned, but its presentation in the media caters to the expectation of the people who associate crime with boldness, courage, strength, creative imagination; whereas virtue and good works are praised mostly in obituaries; and in the minds of many, virtues are associated with boredom, conformism and traditional attitudes. It has been stated maliciously that, while heaven must be a good place and hell rather unpleasant; still, the company in hell will be much more interesting than that in heaven! This contrasting, romantic image of sin over goodness is dangerous mostly for young people. It may be that through unearthly and unrealistic biographies we have presented an unattractive image of the Saints. It would be important to show how much of crime and violence actually is a cowardly escape from constructive work and commitment, and how, on the other hand, real goodness is creative and challenging. In fact, Jesus himself could be our supreme example. He was accused of many faults, of lacking observance of the law, of drinking and enjoying meals, but no one ever called him boring. Had he not been a disquieting challenge, he would not have been killed.

What is sin?

The real answer to these false perspectives must come from a deeper insight into what sin really is. Sin has been defined as breaking God's commandments. This is true; but it sounds too individualistic and legal, and it does not express the full reality of sin. A deeper spiritual awareness is expressed if sin is described as the refusal of God's love. God is not a law-giver who imposes commandments with sanctions, but he surrounds and leads us with his love to which we ought to respond. However, even this description remains within the sphere of the individual relationship to God, which does not include our whole life. For many people it remains elusive. This may be one of the reasons why the sense of sin is dwindling among so many people. We must see sin again with the eyes of the gospel: sin is the destructive and vitiating power that pervades our world, that penetrates into our heart, that embodies itself in our society, in social structures, in politics, in the economic system; it becomes the sin of the world.

It has its root in the human heart, in our freedom. Man makes himself autonomous—this is the core of the account of the first sin in the Book of Genesis—and it remains true today: "Out of the heart come evil thoughts, murder, adultery, fornication, theft.." (Mt 15, 19). Through sin man separates himself from God and so divorces himself from the source of life. This is the radical isolation of sin. Man has broken out of the order of God's love and creation; he makes himself the goal and centre of his life and actions.

Sin does not remain hidden in the heart, it moves on in its destructive course. It first affects the sinner himself. Centred on himself, his vision is blurred: "They became futile in their thinking and their senseless minds were darkened. Claiming to be wise they became fools" (Rom 1, 21f). In his blindness the sinner begins to justify himself and so makes his sinful option ever more his own. It is difficult to acknowledge our failures and to turn from them.

Centred on himself, the sinner asserts himself against his surroundings. The relation to his neighbour is no longer solidarity but competition. Sin becomes a social reality. Justice, truth, peace are changed into hatred, envy, revenge. In the Book of Genesis the sin in paradise is followed by the murder of Abel. (chap. 4).

Sin not only affects personal relationships but embodies itself in society at large. Sections of people group together in their common interests against others: social and economic groups, ethnical and cultural units, nations, often confronted in hardened antagonism with hostile frontiers, blind to their mutual obligations, to the other's needs and interests. Wars are waged with murderous weapons and with equally deadly mass media. Sin enwraps the human family and poisons the atmosphere in which we live and breathe.

This is how sin must be seen; not as individual failure only, but as the ruin of our world and society. Thus St. Ignatius wants us to begin the meditation on sin with (1) the ancient tradition of the revolt of the angels against God (cf. 2 Pet 2,4; Jude 6); (2) then to turn our eyes to the sin of our first parents and the corruption that came from it over the whole of mankind; (3) finally to the sin and condemnation of others who may have sinned less than myself. This meditation on the "triple sin", on its origin and its ever growing powers as it spreads over the world, is in no way meant as a diversion from myself, from the awareness of myself as a sinner. On the contrary, in these meditations I see myself before God as sinner, belonging to a sinful race. This flood of sin which comes from untraceable origins and has broken into our society and history and pervades our world, has entered also my own heart.

Never should the Spiritual Exercises be reduced to a merely subjective spirituality which is concerned only with the personal relation to God. We belong to a broken world; we are called to a new creation. But this solidarity with the world does not diminish my personal responsibility. Only if I recognize myself as a sinner can I respond to God's word of grace and offer myself to Jesus in his mission for the renewal of the world.

Before Jesus crucified

Yet, with all these reflections it still often remains difficult to come to a deep personal awareness of sin. We do see its devastating impact on the world; we may also have realized its blinding, paralysing and disintegrating influence on ourselves. But all this does not touch the core of our being. God himself has to speak to us, to make us realize what sin is, what it means to be a sinner. God has spoken in his Son, in Jesus crucified.

Ignatius concludes the meditation on sin with a "colloquy" with Jesus crucified (n. 53). This conclusion seems hardly connected with the preceding reflections on the sins of the angels, the first parents, and other sinners, and the consequence of these sins. Yet we feel that with this colloquy, where he leaves behind rational reflections and enters into Jesus' presence, he comes into his own. This colloquy contains the core of the meditation on sin.

What is sin? Once our lifeline with God is broken we are alone. We can fill our life with many things and activities, but they cannot sustain us. We drift into desolation and death. There is no support or hope.

Yet we are not alone. Jesus is with us to share our desolation. He too dies, not because he sinned but because he became our brother, the brother of our broken family. He shares our desolation and abandonment.

Why does he come? This is the question we ask him in the colloquy: "How is it that, though he is the creator, he has stooped down to become man and to pass from eternal life to death here in time that thus he might die for our sins?" Jesus does not answer, he does not speak—God speaks in Jesus his Son. This is God's love for us that he gave his Son so that we should not be alone in our exile, to give us the assurance that we have a home, that God's love is with us in all our darkness. So Ignatius makes the retreatant ask three questions.

1. *"What have I done for Christ?"* If God so loves me, how is it that I turn from him and close myself into the narrowness of my own life?
2. *"What am I doing for Christ?"* This is an hour of decision, of response to God's love and invitation.
3. *"What ought I to do for Christ?"* I have to look towards the future, to prepare myself for a commitment. In these spiritual exercises I should find the orientation for my life.

Thus the meditation on sin leads to the first encounter with Jesus Christ. He is with us in our broken world where we are abandoned, overwhelmed by the powers of darkness. He is not a mighty Saviour but our brother, sharing our

life and dealth to lead us into a new life. In him, in his
agony and death, I see where we really are, what sin is, where
the proud autonomy of sin has led us. In him we awaken
to the joyful awareness that we are not alone but that all
sin is embraced by God's love and healed in Jesus' wounds.

SHORT POINTS FOR MEDITATION

1. *What is sin? Election — refusal — destruction*
(Mt 23, 37-24, 2)

Chapter 23 of Matthew contains Jesus' reproach to
the leaders of the people. In the concluding paragraph
he sums up the bitter experience of his rejection:

— "Jerusalem, how often would I have gathered your
  children together as a hen gathers her brood under
  her wings": this is God's saving love extended to
  Israel from ancient times, offered in Jesus to the
  people as a final invitation. It is God's love reaching
  out also to me in a thousand ways.

— "But you would not!": Israel's history is a sequence
  of refusals, of murmuring and faithlessness. Also
  Jesus is rejected by the leaders of the people. Every
  sin is a refusal of God's invitation. This refusal is
  at times open, more frequently it is hidden and
  disguised. As Israel had "reasons" to reject Jesus,
  we have reasons, excuses and explanations. But
  in the depth of the heart we know that we refuse
  God's call.

— "Behold your house is forsaken and desolate": the
  holy city, the temple will be destroyed. We should
  not think of Jerusalem's destruction, or any other
  calamity, as acts of God's revenge. It is not God
  who condemns, but "he who does not believe is
  condemned already" (Jn 3, 18): he condemns himself
  by divorcing himself from the source of life. God gives
  freedom, respects our freedom, allows us to choose
  and to go our own way. This is my dignity, this is
  my responsibility, to choose between death and life.
  (cf. Deut 30, 15).

— "Not a stone will be left upon another that will not
be thrown down": in a transition to the following
chapter the disciples draw Jesus' attention to the
magnificent structures of the temple and the city.
But they are doomed. Exteriorly Jerusalem still
shines in its splendour, but it has lost its life-spring,
it will be destroyed. Whatever my status in the
Church, my achievements, my deeds, my efforts,
they will prove empty unless my heart is faithful.

2.   *How sin came into the world* (Gen 3)

A full understanding of this text requires deeper
exegetical studies. We reflect only on the key issues of
this account of the origin of sin — also of our sin.

— "Did God say: 'You shall not eat of any fruit in the
garden": this is the deceptive question. Does God
limit our freedom, envy our growth? Is God a rival
for man? Much of the modern revolt against God is
based on this misapprehension.

— "We may eat of the fruit of the trees of the garden":
the whole of creation is given to us to use it, to
develop it. (cf. Gen 1, 28f).

— "You shall not eat of the fruit of the tree which is
in the midst of the garden," (which is) "the tree of
the knowledge of good and evil, lest you die" (Gen.
2, 17): human freedom remains under God. Our
life is inserted into God's plan of creation and salvation.
He who destroys this order, destroys himself.

— "You will not die...your eyes will be opened and you
will be like God knowing good and evil": this is the
fallacious promise of human autonomy and emanci-
pation ever tempting, though proved destructive
in life and in history.

— "The woman saw that the tree was good for food...
a delight to the eyes...to be desired to make one
wise": the fascination of sin.

— "The eyes of both were opened and they knew that
they were naked": with their eyes opened   they

see the truth, not their likeness of God but the shame
of their nakedness.

— "The man and his wife hid themselves from the presence
of the Lord God": God's presence is no longer an
assurance but a threat.

— "The woman whom you gave me...gave me the
fruit...the serpent begiled me and I ate":
this is the beginning of the litany of excuses, of shifting
responsibility for failures.

— "God sent him forth from the garden of Eden": the
exile begins, man estranged from himself, from his
destiny and happiness. Yet he is not abandoned.
God's love follows him; Jesus joins him in his exile.
On his cross he promises all sinners: "today you
will be with me in paradise" (Lk 23, 43).

3. **Sin in human society** (Mt 25, 31-46).

Sin turns not only against God but becomes the
destructive power of human society. The attitude towards
the neighbour will be the ultimate criterion of good and
evil.

— "He will separate all the nations, one from another":
it is the final judgement in which the outcome of our
human life will be revealed. The gulf between good
and evil, which is hidden in our world, will be manifest.

— "Come, o blessed of my Father, inherit the kingdom
prepared for you from the foundation of the world":
this is God's kingdom, the new creation, the community of all men united in love.

— "I was hungry and you gave me food...": to this
kingdom belong all those who build this community
in actual sharing in all spheres of mutual support:
food, drink, receiving strangers, etc.

— "As you did it to one of the least of my brethren, you
did it to me": this is not merely social service; God
himself is present and active in our relations. His
love flows into our human society. It is embodied in
Jesus and becomes the bond that unites all. In

your neighbour you meet Jesus, in your service to
him you give yourself to God. This is God's kingdom.

— "Depart from me, you cursed, into the eternal fire...
For I was hungry and you gave me no food...":
those who refuse solidarity and responsibility for the
neighbour, do not belong to God's kingdom. Where
this refusal becomes a persistent option, it excludes
from the communion of love and salvation. To be
excluded means to be cursed, outside the realm of
grace and peace.

— "They will answer: Lord, when did we see you
hungry...?": they closed their eyes to God's kingdom,
they blinded themselves to the light that came into
the world.

FURTHER READINGS:

— Is 5, 1-7:   The vineyard planted by the Lord that
bore sour grapes (cf. Ps 80, 8-19).
— Ez 16: Israel's vocation, sin, salvation.
— Mt 22, 1-14:   The invitation to the marriage feast;
the refusal.
— Lk 13, 6-9:   The fig-tree that bears no fruit.
— Rev 2, 1-7:   The first love that was lost.

QUESTIONS:

1.   Did I ever encounter sin with its undisguised destructive power, in individuals and families? Recall the
details.
2.   How do I realize sin's impact in my own life?

FOR REFLECTION:

"Before you (O Lord) the abyss of human conscience lies open.
What would be hidden from you even if I would not confess
it before you? I would hide you from me, not me from you."
AUGUSTINE, *Confessions* 10,2.

Chapter Six

MY SIN

"You say: 1 am rich, I have prospered, and I need nothing; not knowing that you are wretched, pitiable, poor, blind and naked" (*Rev. 3, 17*).

In the meditation on sin we turn to ourselves. We are not concerned with reflections on our sinful world or on the evils of society — they are only the mirror in which we have to recognise ourselves, to find the real, the only true starting point for renewal. *I must look into myself — I must* have the honesty and courage to do it.

True and false humility

Again a warning is needed. These reflections on sin are not meant to darken or diminish our self-image. Many people, often very serious Christians, have a poor image of themselves. They may consider this as an attitude of humility. They are very conscious of their limitations and deficiencies; they may ignore their talents and gifts and discount or even deny the serious efforts of their life and work. Various reasons may account for this tendency to self-depreciation. But it should be clear that such denigration has nothing to do with true humility, which consists in the grateful and courageous acceptance both of God's gifts and of our limitations, and the faithful use of these gifts for God's service among the people with whom we live. Depressing depreciation and paralysing feelings of self rejection have no place in a Christian life.

Nor is this meditation on sin meant to arouse guilt feelings, which have nothing to do with true contrition for our sins. Guilt-feelings are the expression of the wounded ego. I feel insecure and threatened because I have failed in fulfilling the commandments. I have not been able to live up to the standards which I have set for myself. I have lowered myself in the eyes of others. All these feelings are centered on the ego; I am concerned about *myself*, *my* security, *my* image. Scruples also belong to the sphere of guilt-feelings. A scrupulous person is concerned not about God, the neighbour or the world around but about *his own security before God*. It is not easy to overcome such preoccupations and feelings. Clear insights, patient and persistent efforts and often prudent guidance will lead beyond these experiences of deep anxiety. But it ought to be clear that they have little to do with a true renewal of our Christian life. *God calls us beyond ourselves* and the narrow concerns of our ego—seek first God's reign and his righteousness, and he will take care of all your anxieties. Our justice does not consist in a faultless record of our life but in our faith and self-gift to God. This was Paul's personal experience and message (cf. Rom 3, 21-23; Phil 3, 8f). It is the aim of the Spiritual Exercises to liberate us from a self-centered ego, to lead us to a genuine commitment to God in Jesus Christ.

Know yourself

However, in these meditations on sin we ought to come to a true knowledge of ourselves and free ourselves of self-righteousness and illusions, which come from our ego, from the deep-seated need to protect and foster it. Who likes to recognise and accept his faults? All life has an inbuilt system of self-protection and self-defence. Where our body is concerned, we instinctively avoid dangers and defend ourselves against aggression. Our mind is doing the same. We find it difficult to admit that we are unable to control our tongue, that we are lazy and greedy, or envious and jealous of our neighbour and harbour feelings of revenge. All these and so many other movements of our heart, we consider improper and in conflict with what we want to be, and so we suppress them, we would not admit them as even possibly our faults. We disguise and excuse these feelings and link them with lofty principles. So we call ourselves "honest" people who always speak the truth while in reality we are rude and easily lose our composure. If we are fond of comforts, we condone it with the need of avoiding tension and stress. When we are in conflict with others, we convince ourselves of *their* vanity, ambition and politics! Jesus said: "You see the speck in your brother's eye but do not notice the log that is in your own eye" (Mt 7, 3). How much of our ingenuity goes into finding excuses, in creating illusions, to avoid seeing ourselves as we really are.

The temptation to protect ourselves and to build a world of illusions affects also groups of people: social, ethnic, religious and national groups. They all have their collective ego, are concerned with safe-guarding their common interests and projecting a favourable image of themselves. We are here more concerned with religious groups.

The gospels tell us how Jesus was confronted not so much with individual hostility and rejection as with the collective prejudices of religious groups in his surroundings. These are familiar to us by the names of Pharisees and Sadducees, and of the theologians of his time, the Scribes. It would be misleading to take Jesus' invectives against them as our only source of information; for in fact serious and committed people were among them. But many of the accusations as they are recorded in the gospels are a common feature of religious groups also today. Jesus' criticisms should be used as a mirror of regular self-examination today mostly by priests and religious. It is their profession too to teach and to preach. Jesus says of his enemies: "They teach but do not practice. They bind heavy burdens hard to bear and lay them on men's shoulders; but they themselves will not move them with their finger" (Mt. 23, 3f). They must keep their prestige in society and so "they do all their deeds to be seen by men...They love places of honour...and salutations in the market-places" (Mt 23, 5-7). They cultivate an image of faithful observance by fulfilling legal prescriptions while their heart harbours evil: "You cleanse the outside of the cup and the plate but the inside is full of extortion and rapacity...you are like whitewashed tombs which outwardly appear beautiful but within are full of dead bones and uncleanness" (Mt 23, 25-27). They glorify the prophets of past ages but resist God's message which comes to them in the present; so they really are "sons of those who murdered the prophets" (Mt. 23, 29-31). All this has been repeated in the history of the Church; it remains the temptation of all who hold official and honoured positions in the Church also today.

We may go a step further and reflect on the peculiar temptation which seems to be connected with the Christian message itself. The Christian community is based on faith in the Risen Lord, Jesus, who has conquered all evil, has taken away our sins and has poured out the Holy Spirit on his Church with the wealth of his gifts. We are the new creation and enjoy its peace and blessing. So it can happen—it has happened too often—that our praying and singing of the wonderful life and of the communion of love in the Church become unrealistic. We create a world of illusion as if all this had already been achieved. The Vatican Council tells us that "in the earthly liturgy we take part in a foretaste of that heavenly liturgy which is celebrated in the Holy City of Jerusalem towards which we journey as pilgrims" (S.C. 8). True, it is a foretaste, but we are still pilgrims. It is our part in this earthly life to remain with Jesus in his laborious mission, struggle and rejection. The new life which he gives us through the Holy Spirit must be lived in this earthly city in patience, often in contrast and conflict with our surroundiugs.

The temptation to escape from the earth-bound duties of a Christian life is as old as the Church. Read Paul's letter to the Corinthians who were so proud of the glorious gifts of the Spirit, of tongues and prophecy, which they loved to display in their assemblies. They had listened to modern theologians who had lectured in terms of contemporary philosophy on a much higher level than the simple proclamation of faith as they had heard it from Paul. So their sense of superiority had swept them off their feet. They had forgotten the elementary demands of a Christian community; they were split into parties (cf. 1 Cor 1, 10-13) and the way they celebrated their gatherings of love, the agape, with its climax in the Eucharist had become a grave scandal (1 Cor 11, 17-34). Paul knows only one remedy: return to Jesus as he lived his earthly life in humility and service, giving up his life for us on the cross. When his life was being drowned in hatred and betrayal, he gave himself to us in the ultimate sign of his love, in the Eucharist: "on the night when he was betrayed" (1 Cor 11, 23).

We have to do the same: All the great words of the Bible, *joy*, *peace* and *love*, have to be interpreted in the light of Jesus' earthly life. When we preach joy we cannot forget the millions who still carry Jesus' cross and break down under its weight. Peace in the language of Jesus does not mean compromise or the cowardly avoidance of conflict. Peace includes resistance to the powers of oppression; Jesus also brings the sword of judgment. Finally love cannot be without justice. Love is not a sheltered harbour for rest but the divine power that brings about a new creation in spite of conflict and defeat.

This, then, is the soberness of the Christian life. Conversion must take place in the depth of our heart; God's reign must come in our real world. God's promises are not cheap gifts for children who do not know their price. Jesus bought us with his precious blood: "You were bought with a price. So glorify God in your body" (1 Cor 6, 20). No one can share in the joy, peace and love of the Risen Lord unless he is united with Jesus in his life and death.

How to know ourselves?

There are many ways by which we can make progress in our self-knowledge, and we should faithfully and sincerely make use of them. We must listen to others and allow, even encourage, them to share their observations; we may gradually learn to see and hear ourselves through the eyes and ears of others. This will necessarily balance our self-image. We have to watch and assess the ups and downs of our emotional life and develop a sense of detachment and humour towards ourselves. There are also special occasions to look into our life: the daily examinations of conscience, regular celebration of the sacrament of reconciliation, days of recollection and our annual retreat.

However, none of these approaches promises true success unless it is based on, and accompanied by sincere, earnest prayer. We need the light of the Holy Spirit to penetrate into the depth of our heart; we need courage to face ourselves.

Thus St. Ignatius' wants us in these meditations on our sins to pray for a triple grace: "1. *A deep knowledge of my sins* and a feeling of abhorrence for them. 2. *An understanding of the disorder of my actions*, that, filled with horror of them, I may amend my life and put it in order. 3. *A knowledge of the world* that, filled with horror, I may put away from me all that is worldly and vain." These three petitions, he advises, should be addressed: (1) to Mary that she implore their fulfilment from her son; (2) then to Jesus that he may obtain them from the Father; (3) finally to the Father himself that he, the eternal Lord, may grant them to me (n. 63).

So, we first pray for the grace of a deep knowledge of our sins, the recognition and acceptance of failures and mistakes, of what in the light of the Gospel is wrong in us, without excuses and explanations.

But our words and deeds flow from our heart. We pray further for the understanding of the disorders in our heart. We cannot successfully fight against symptoms unless we find the root-cause of our ills. Much insight, and often the experience of others is needed to probe into the deeper layers of the human heart. Many of our attitudes towards others stem from deep-rooted feelings of insecurity and rejection, which may have their origin in long forgotten days of childhood. The Spiritual Exercises are not meant to become a study in psychology; deep-rooted problems of personality ought to be tackled outside the retreat. Still, the silent days of retreat lend themselves to deeper insights into our heart, to open it fully to the Lord.

Our life is woven into the context of the world. We are children of our time; the movements, aspirations and anxieties of the surrounding world affect us. Also our thinking and the values which determine our life are constantly influenced by the people and social structures around us. Hence, the third prayer to know the world. Ignatius speaks here about the world in the pejorative sense, as it is found in John's Gospel: the world which does not receive Jesus Christ, which loves darkness better than light. To renew ourselves we must have insight into the values, trends and movements which shape the world of today, the destructive powers, but also the promising aspirations of a young generation.

This triple grace of a deep insight into myself and the influences that shape my life should be implored through the intercession of Mary who gave herself to God free from all reserve, through Jesus Christ whose entire life was surrendered to the Father, from God the Father himself, the source of all life.

SHORT POINTS FOR MEDITATION.

1. *My sin*

Make your meditation in deep union with Jesus Christ. His message rings in your heart: The Kingdom of God is at hand. A new world is in the making; you are meant to belong to the new creation!

How will your life be?

Take Jesus to the various places of your life and work, to the people with and for whom you are living and working. See your life, work, relationships through his eyes.

— Go to church/chapel, in the privileged presence of Jesus reflect on: your prayer, the earnestness of your participation in the Eucharist, the depth of your consecration to God, your personal prayer.

— Go in imagination to the places of your work and reflect on: your duties and responsibilities, your professional involvement, your efforts for a deeper understanding of the meaning of your work.

— Meet the people of your professional world and review your relations with them. Are they too critical? compassionate and understanding? full of tension? How do you speak to them? about them?

— Meet the people of your personal world: family, community, friends. Review your correspondence. What is the human depth of these relations? Are they channels of God's presence and love, for others, for yourself?

In these reflections on the wide range of your life,
the silent presence of Jesus is essential: is every detail
of your life and work inspired by him?

2. *The soberness of the Christian life:* The letter of
James.

This letter is a down-to-earth reflection on our
Christian life in its various dimensions:

— The Gospel is the mirror of the Christian life: "Be
  quick to hear, slow to speak, slow to anger...
  Be doers of the word, not hearers only" (1, 19-25).

— Faith must bear fruit in works: "What does it profit
  if a man says he has faith but no works? Can his faith
  save him?" (2, 14-24).

— Let your teaching be a challenge to yourself: "You
  know that we who teach shall be judged with greater
  strictness" (3, 1).

— Guard your tongue: "If anyone makes no mistake
  in what he says he is a perfect man...The tongue
  is a fire...No human being can tame the tongue...
  With it we bless the Lord and with it we curse
  men" (3, 2-12).

— Avoid anger, bitterness, jealousy: "If you have bitter
  jealousy and selfish ambition in your hearts do not
  boast and be false to the truth...The wisdom from
  above is pure, peaceable, gentle, open to reason, full
  of mercy and good fruits" (3, 13-18).

— No big dreams; do good today. "All such boasting is
  evil. Whoever knows what is right to do and fails
  to do it, for him it is sin" (4, 13-17).

3. *The blindness of the heart* (Rev 3, 14-20).

You may contemplate one of the healings of blind
people by Jesus, e.g. Bartimaeus (Mk 10, 46-52), the
symbol of the blindness of the heart, the desire to see;
when he is healed, he follows Jesus on the way to Jerusalem.

The Book of Revelation contains the letter of warning
against the blindness of the heart:

— "I know your works: you are neither cold nor hot...
I will spew you out of my mouth": the resignation to
mediocrity has no place in God's kingdom.

— "You say, I am rich, I have prospered, I need nothing":
the more we drift away from God, the less we are
aware of our poverty; we are able to divert ourselves
with plans, activities, relations, but we lose the
spring of life.

— "Buy from me gold...and white garments to clothe
you...and salve to anoint your eyes that you may
see": God alone can restore our life and make us
see what we need.

— "Those whom I love I reprove and chasten": to those
who follow Jesus God gives peace. Also those who
turn from him he follows with his love, he makes
them restless. (cf. Rules for the Discernment of
Spirits, n. 314).

— "I stand at the door and knock:" it is for me to
be attentive, to open, to let him in.

FURTHER READINGS:

— Mt 7, 15-20: Be on your guard against false prophets.
— Mt 12, 33-37: Healthy trees are distinguished by
their fruit.
— Lk 13, 22-30: Enter through the narrow door.

QUESTIONS:

1. Where do I see the danger for my commitment to
the Lord?
2. What are the excuses and disguises of my faults? What
are my excuses to avoid duty and responsibility?

FOR REFLECTION:

"Take from me all that separates me from thee,
give me all that draws me to thee,
take me from me and give myself wholly to thee"
(Prayer of ST. NICOLAUS VON DER FLUE).

Chapter Seven

A NEW HEART

"Create in me a clean heart, O God" (Ps 51, 10).

The good news of the forgiveness of sin is already part of the prophetic message: "I have swept away your transgressions like a cloud and your sins like mist (Is 44, 22); "I will forgive their iniquity and I will remember their sins no more" (Jer 31, 34).

Jesus does more. He takes away the sin of the world (cf. Jn 1, 29). It is not enough that sins are forgiven, they must be destroyed. What would it help us if God covered our sins with his mercy but their poison would remain in us and continue its work of destruction? Our world must be freed of the crippling, paralysing and divisive powers of evil, God's creation must be renewed. First of all our heart must be renewed by God's creative power.

The promise of a new creation

This actually has been the age-old expectation of the nations all over the world: the hope of return to a golden age, a renewed world as it came from God at the dawn of creation. This expectation is enshrined in the Bible as the hope for the Messianic age. It broke forth most forcefully at the time of exile: "I will take you from the nations and gather you from all the countries and bring you into your own land. I will sprinkle clean water upon you...A new heart I will give you and a new spirit I will put within you and I will take out of your flesh the heart of stone and give you a heart of flesh." (Ez 36, 24-26).

The expectation of this renewal remained alive also in the early Church. The first Christians had recognized Jesus as the Messiah, but it was obvious that his work was not yet fulfilled and so they expected his final coming in glory for the fulfilment of his mission. "Then when all things are subjected to Jesus Christ, the Son himself will also be subjected to him who put all things under him that God may be everything to everyone" (1 Cor 15, 28). So it must be. If God created all things and "saw everything that he had made, and behold, it was very good" (Gen 1, 31), then this world cannot just glide out of his hand and end up in destruction and death. God's work cannot fail.

How will our world be renewed? The Bible tells us that God wants our world to be a paradise. But we lose it through sin; and more and more we turn our world into a desert. Greed and exploitation make our world less inhabitable and our society less hospitable. Blind selfishness, ambition and rivalry destroy our community and enslave millions. This world will always be our world. If there is to be a new and better world, *there must first be new human hearts*. This is our theme of today.

Surely human behaviour is greatly influenced by the social and economic structures in which people live, and there is a pressing need to transform and humanize these structures, to give all people the opportunity to live a truly human life. But it would be wrong, it would lead to illusion, if all our hopes for a better world were pinned on only political or social changes. The Christian message is rather that first our heart must be changed. Jesus calls people to repentance because God's kingdom is coming. In our time Pope Paul VI echoes this message in his call to evangelization: "There is no new humanity if there are not first new persons" (*Evangelli Nuntiandi* n. 18).

Born anew

This renewal must come from God, it is God's reign. Only the creator, God, can say: "Behold I make all things new" (Rev 21, 5). This we find difficult to understand. Our technical age has learned to take things firmly in hand and to solve problems with the resources at our disposal. We do not expect solutions to fall from heaven.

But to solve problems we need a new heart, we must be born anew. John writes in the beginning of his gospel: "To all who received him, who believed in his name, he gave power to become children of God, who were born not of blood, not of the will of the flesh nor the will of man, but of God" (Jn. 1, 12f). A new life is needed with new senses, with a new understanding. This is what Jesus told the startled Nicodemus: "Truly, I say to you, unless one is born anew he cannot see the kingdom of God." (Jn 3,3,).

In the Synoptics Jesus' language is more simple. Jesus sees in children the simplicity required to understand and accept the message of God's love: "Whoever does not receive the kingdom of God like a child shall not enter it" (Mk 10, 15). Children are still able to receive gifts. They have not yet adopted the commercial attitude that dominates most of our human relations: everything has its price...a favour for a favour. Children live on love and gifts. So is God's kingdom. We cannot earn it. We cannot bring it about with our human efforts. We can only open our hearts for its coming, for God's gift. We must be born anew as God's children.

This is not easy. It actually implies the most radical and difficult self-surrender. In Jesus' time the Pharisees were ready to make any sacrifice for the sake of their religious commitment— inconveniences with no end in the fulfilment of the law, prayers and frequent fasts. But they were not ready to surrender themselves to God, to be children before him and to accept his kingdom as the gift of his love. For Paul it was the break-through in his life, his "conversion", when he realized that his righteousness did not come from the fulfilment of the law and the traditions but from God alone. "I have no righteousness of my own based on law but that which is through faith in Christ, the righteousness from God that depends on faith" (Phil 3, 9). It is God's free gift of love received with an open heart. Thus the renewal of our life does not begin with practical resolutions (which may and must come in due time) but with the deep awareness that God's love and power are moving in my heart and drawing me beyond myself towards a new life. I allow him to take possession of my thinking and planning, of myself. This is my task during these Spiritual Exercises.

My task

However, reliance on God's love does not diminish my own responsibility. In his creation God is alone; He speaks and there is light, firmament, earth and waters and all creatures. But he entrusts his creation to us. For ever this world will be our world, our human world. *God's world has become our task.* This is our dignity and responsibility.

This task begins in my own heart. Too often we try to change our neighbour, we turn to some action plan, a scheme for the betterment of our society. All this may be very necessary, but surely it is not the beginning, and it may simply be an escape. I must turn to myself, the seriousness of my commitment to God, my many relations to people, the realm of my professional work.

The sacrament of the new heart

In the renewal of our heart the sacraments of the Church hold a special place, especially the sacrament of reconciliation, a source of healing. The renewed understanding and the more meaningful practice of this sacrament have been a matter of deep concern for the Church in these past decades. We are aware of its decline. Its administration and reception had often become too mechanical and routine-like, without a deeper impact on the life of the faithful and the Christian Community as a whole. The frequency of its use has dropped alarmingly; attempts to restore the traditional practice have largely failed.

This obviously is not the place for a theological and pastoral study of the problems involved. However, our reflections on the renewal of heart which is both God's gift and our human response towards a new life, lead us directly to the real meaning of this sacrament and thus may be helpful for its meaningful celebration during the retreat and in the course of our life. This sacrament is the liturgical celebration of the new heart, the celebration of on-going Christian conversion.

Jesus was sent not only to proclaim God's forgiving and saving love but to take away the sin of the world. His mission is to be continued, it is part and parcel of the life of the Church. So Jesus entrusted this mission to his disciples. All four gospels conclude with the final mission to proclaim salvation (Mk 16, 16), to make disciples, to baptize them for a new life (Mt. 28, 19) and to proclaim the message of repentance and forgiveness to all nations (Lk 24, 47). Most articulate is St. John: "Receive the Holy Spirit. If you forgive the sins of any, they are forgiven; if you retain the sins of any, they are retained" (Jn 20, 22f). This is Jesus' Easter gift to the Church. She is sent not only to preach forgiveness, but to make God's forgiving and renewing love real and effective in our life.

This, then, is the core of sacramental absolution: Once more the Holy Spirit with his lifegiving and healing love is poured out on us. We are restored to full communion with the faithful and again share in the dignity of God's children, just as the son in Jesus' parable who returned to the father. He was embraced, a ring was put on his finger and sandals on his feet so that it should be clear that he was son, not one of the servants, and a feast was celebrated with music and dancing (Lk 15, 11-32). This is what Jesus wants to be realized by all who turn to him in the sacrament of reconciliation.

God's welcoming and lifegiving love, however, must be received in our hearts with repentance. This has often been misunderstood as if God wanted us to be punished and to bear burdens of suffering. We cannot think of a God who likes to see his children in pain, but God respects our freedom and dignity. We have to make our decision and to break the fetters of our selfish isolation. This is painful. We have to act against attitudes and habits which may be deeply engrained in us. In this struggle we are moved by the Holy Spirit and supported by the entire community and the prayer of the Church.

In the early centuries this effort of the penitent was actually the main concern of the ecclesial reconciliation of sinners. Through prayer and penance penitents had to express the seriousness of their conversion. Today hardly anything of the ancient austerity is left in the practice of the Church. But this lack of legal injunction cannot dispense us from the seriousness of a personal commitment. Sacraments are not a substitute for personal decisions and efforts. The actual fruit of a sacrament is measured by our disposition, the capacity and readiness to receive God's grace. The brightness of the sun does not give us light if our eyes are blind; streams of living water will not quench our thirst unless we fill our cup.

In the Middle Ages the main stress in sacramental reconciliation was shifted to the actual verbal confession of sins, as a sign of true repentance. Why have we to confess our sins? Already at the time of St. Augustine people argued that as they had sinned before God, so they would confess their sins before him, but not before a priest. This argument is not conclusive for we always are members of our human community. Whatever good we do and whatever sin we commit, even our most secret failures, have their repercussions on the surroudings. What a difference it would make in our life if we were conscious of the deep, often imperceptible bonds which link us with these surroundings. Thus it is right that in the Eucharist we confess our sins before God and *our* brothers and sisters. In confession, then, we confess before the priest as representative of the Church. The Council has asserted anew the community character of sacramental reconciliation in which sinners "obtain pardon from God's mercy for the offence committed against him and are, at the same time, reconciled with the Church which they have wounded by their sins" (Lumen Gentium n. 11). Besides, experience tells us that we must exteriorly profess what is hidden in our heart if our renewal is to be genuine. If we do so the words of absolution and reconciliation have their full meaning. The isolation of sin is thus broken, we are again living members of the body of Christ.

We see then that the sacrament-of-the-new-heart is the celebration of the encounter of God with the sinner. God's love, his leading grace, his life giving power come to us. We on our part open our hearts to God in confession and whole-hearted repentance, ready to accept the pain and struggle implied in this new beginning, the renewed life in Jesus Christ.

A meaningful renewal of sacramental reconciliation, therefore, will lay all emphasis on the *personal encounter* with God in Jesus Christ. New ways of doing this have proved helpful; they can easily be used in the context of the Spiritual Exercises. The confession will consist not merely in the enumeration of sins, but in the grateful praise of God for the guidance and love we have experienced. It will include the actual situations of our life, the difficulties we face, our failures and sins. It will lead to the prayer for forgiveness, new light and guidance, for the healing and renewing power of God, and conclude with an absolution, the peace of the Church, the peace of Easter.

Thus the first week of the Spiritual Exercises leads to the renewal of our life in Jesus Christ. In him we shall discover the richness of our new life. This will be the content of the second week.

SHORT POINTS FOR MEDITATION:

1. **The promise of the New Covenant** (Is 49, 14-16; Jer 31, 34; Ez 36, 22-36).

Read these texts meditatively. They are addressed to Israel in the distress of exile and contain the promise of salvation. (Is *49, 14-16*) God's concern for his own.

— "Zion said: the Lord has forsaken me": the experience of desolation in our world—in the Church—in my personal life.

— "Can a woman forget her sucking child?": all human love and concern are only the limited participation in God's concern for his own.

— "Your builders outstrip your destroyers": God's power to build and bring about a new creation is stronger than the forces of destruction.

— "Lift up your eyes and see: They all gather, they come to you": Your own renewal is part of the coming of God's reign in our world.

Jer 31, 31-34: The New Covenant

— "I will make a new covenant with the house of Israel":

God's covenant with his people will never be abandoned; and my own vocation, to which he called me, will not be lost in spite of my failures.

— "I will put my law within them and I will write it upon their hearts": the old law was written on tablets of stone; God's new law will be written in our hearts. It will not be imposed from outside or from above but the indwelling Spirit will guide us from within.

— "I will be their God and they shall be my people": God made Israel his own people (Ex 19 5f); this union is deepened and extended to all, to be fulfilled in the final glory (Rev 21, 3).

— "They shall all know me from the least of them to the greatest": also my knowledge of God will not be second hand; it is God's gift to all, to me.

Ez 36, 22-36: Israel's renewal among the nations.

— "The nations will know that I am the Lord when through you I vindicate my holiness before their eyes": Israel's renewal will be a witness to the nations; so will all life be that comes from God. "Let your light shine before people" (Mt 5, 16).

— "I shall clean you from all uncleanness": God will not only forgive, but take away sin and give new life.

— "A new heart I will give you and a new spirit I will put within you": the people will be renewed not by new laws but sharing in God's own life.

— "And you shall dwell in the land that I gave to your fathers": the inner renewal will bear fruit in the renewal of society of the world.

2. *Jesus' invitation*

Jesus' message of God's reign, of the new heart and the new community of brothers and sisters, is being unfolded in the second week of the Spiritual Exercises. In this meditation we turn to some of the warnings attached to his message. Read the following texts attentively:

— "If your right eye causes you to sin, pluck it out..."
(Mt 5, 29f): God's kingdom demands clear and
often hard decisions. Jesus rejects half-hearted compromises.

— "Someone said to him: Lord, will those who are
saved be few? He said to them: strive to enter the
narrow door..." (Lk 13, 23): Jesus does not
answer questions of curiosity. His message is a call
to commitment, to basic decisions which may run
contrary to the views and expectations of the majority.

— The unclean spirit who had been driven out "brings
with him seven other spirits more evil than himself,
and they enter and dwell there..." (Mt 12,43-45):
Jesus rebukes the Pharisees who glory in their
righteousness, but in their self-righteousness they
are worse than before. Any form of self-righteousness
before God is the sin most radically opposed to God's
kingdom.

— The parable of the talents: "To everyone who has,
more will be given and he will have abundance;
but from him who has not, even what he has will be
taken away" (Mt 25,14-30): the parable speaks of
our responsibility and accountability for the gifts
we receive from God. The new creation is God's
gift entrusted to us. Those who respond go on
growing. Those who do not respond go on losing.

3. The new life in Jesus Christ   (Eph 4,23-5,2).

This is the vision of the new life which comes from
God in Jesus Christ:

— "Be renewed in the spirit...Put on the new nature":
Conversion is a radical renewal from within.

— "Putting away falsehood, let everyone speak the
truth, for we are members one of another": God's
reign is community in truthfulness and love.

— "Be angry but do not sin; do not let the sun go down
on your anger, and give no opportunity to the devil":
anger is a necessary, often constructive reaction to

evil. It is at the same time a source of frequent sin and of serious disturbance. Have I learned to channel anger into a constructive power in my life and work?

— "Let the thief do honest work with his hands so that he may be able to give to those in need": this is the radical reversal from taking to giving, from sinful greed to loving sharing.

— "Let no evil talk come out of your mouths but only such as is good for edifying": our language betrays our heart. A new heart filled with Jesus' spirit will have words that make for community. (cf Jas 3,1-13).

— "Let all bitterness, wrath and anger...be put away from you": compare these poisoned attitudes to the gifts of the Spirit in Gal 5, 22f.

— "Walk in love as Christ loved and gave himself for us, a fragrant offering and sacrifice to God": God's kingdom is the community of self-giving love which is rooted in Jesus' total self-gift to God, to us.

FURTHER READINGS:

— Ez 37, 1-14:   The vision of the valley of the dry bones which are brought to life.

— Lk 7,36-47:   The sinful woman is received by Jesus.

— Lk 15:   The three parables of the lost sheep, the lost coin, the lost son.

— Rom 6,5-11:   Dead to sin, alive to God.

— Rev 22, 1-5:   The stream of living water that heals the nations. (cf Ez 47,1-12).

QUESTIONS

1. What are the areas in which I need the healing power of Jesus?

2. In what areas do I need a breakthrough, a new outlook and commitment?

3. How do I plan the ongoing renewal of my life in the future?

PART II

ON THE WAY

Chapter One

JESUS, MY LORD

"All things are delivered to me by my Father (Mt 11,27).

The meditation on Jesus Christ stands by itself in the Spiritual Exercises, between the first and the second week. It leads to the central theme not only of the retreat but of the entire Christian life. God wants that this broken world of ours be renewed through Jesus his Son and so become the new creation, his own kingdom. Even more: Jesus is not only a remedy for our sick world, but in him God tells us what man is really meant to be in his plan, our human life with its struggle and death and with its ultimate destiny, in God: "In these last days God has spoken to us through a Son".
In the destiny of man the meaning of all creation is summed up. Through Jesus Christ it will be included in God's love: "God made Jesus heir of all things; through him he created the world" (Hebr. 1,2).

The Personal Encounter with Jesus

Thus the Christian message does not consist in a doctrinal system but is contained in the person of Jesus. Seeing him means encountering God (cf. Jn 14,9). In him, according to words of the Council, "revelation is completed, perfected and confirmed with divine guarantee. He did this by the total fact of his presence and self-manifestation, by words and works, signs and miracles, but above all by his death and glorious resurrection from the dead, and finally by sending the Spirit of truth" (Dei Verbum n. 4).

The following weeks of the Spiritual Exercises will be spent on the meditations on Jesus Christ, his life and message. To do this we need, to be sure, a safe basis of historical knowledge and critical assessment of the biblical texts. However, this historical knowledge should be acquired independently of the Spiritual Exercises. In these weeks of the retreat we are concerned with the person of Jesus Christ who is "the Way, the Truth and the Life" (Jn 14,6). This personal encounter with Jesus must take place in our own world and situation. It is meant to become the transforming inspiration and power of our life.

The greatest obstacle in our encounter with Jesus is our familiarity with him. We think we know him. As children already we have learnt from our parents to believe in him. We have read the gospels, and heard them read, a hundred times. We know the ancient formulas by which the Councils define him: true God and true man. Daily we pray to him in the liturgy and in meditation. Surely Jesus is the most familiar figure in our spiritual household. But so it may have happened that we never really met him personally, never were challenged by him.

It was different with his first disciples: Sometimes we may envy them for their unique privilege of having been with him at the lake of Galilee and having been allowed to listen to his words, and we might imagine that it must have been easy for them to find personal faith in him. This is surely not true. Try to live with them through their experience.

Jesus exercises a powerful attraction on the people. They were drawn to him by the way he spoke about his Father with an absolute assurance that our life is sheltered in his care; by his love for the people, for all, even those who had lost all hope; by his call to a new communion among all men and women as brothers and sisters, in the love of God. He had the power to heal body, mind and spirit. In his presence the entire world seemed changed, the daily worries appeared small, people flocked to him from the villages, they forgot their meals and followed him into deserted places. He was truly "anointed with the Holy Spirit and with power, and went about doing good and healing...God was with him" (Acts 10,37f). Whatever he spoke was simple, real; it all came from the depth of his being and had the unmistakable ring of truth, that unique truth which comes only from God and touches every heart, which does not need arguments but must be accepted in trusting faith. They began to see what true life is, what we all are meant to be.

But the leading circles of the Jews turned against him. They felt threatened. The politicians were disturbed on account of the mass-movement among the people whose expectations always had a political bent. They were apprehensive of the Roman authorities who might tighten their military control.

Also religious minded people were disturbed: What was the authority of this man to speak about matters of tradition and law? He never quoted any of the great rabbis who were renowned for their knowledge of the Torah. He spoke only on his own authority without any academic preparation or official sanction. The trends of his teaching seemed dangerous: all considered the strict observation of the Sabbath a mark of faithfulness towards the law. Jesus had very liberal ideas about it and provoked open conflicts when he healed people during the Sabbath service in the synagogue as if he were the Lord of the Sabbath. He spoke lightly about ritual purity concerning food and other injunctions as if only the purity of the heart were important. He welcomed sinners, shared meals with tax-collectors and declared that there was greater joy in heaven for a converted sinner than over all the faithful observers of the law. His conception of God's absolute love and of the common brotherhood of all was a threat to their feelings of superiority and to their power position in society.

Could one man stand against the entire sacred tradition of the past, one against all others, politicians, scholars, religious leaders? Soon it became evident that he was isolated. A handful of disciples still went with him fascinated by his personality. But when his defeat became evident they too abandoned him, and he was alone. His enemies still gave him a last chance when he was hanging on the cross: if he was the Messiah...God would surely save him: "Let the Christ, the King of Israel, come down now from the cross that we may see and believe" (Mk 15,32). Nothing happened, and he himself confessed, according to the gospels: "My God, why have you forsaken me?" (Mk 15,34).

Do you still feel that it would have been easier to follow Jesus in his earthly life, easier than today? Would you have been the only disciple who, along with Mary his mother, remained standing under the cross, the disciple whom Jesus loved? Truly, it seems almost impossible to stand alone against the whole world: political power, public opinion, religious authority, and just to believe in Jesus, the Son of God, with no other argument except his word, his person.

Today many find it easy to believe in Jesus because they do not encounter him in person; without much thinking they repeat the formula of the Creed. It would not be so easy if also today we were ready to feel his challenge.

Jesus came to proclaim and inaugurate the kingdom of God, a new world and a new human society, different from this broken world of ours which is full of division and oppression, as we have made it with our selfish and sinful hearts. Jesus offers an alternative, the only real alternative to our society: how God has seen our world, how we have to live our life personally and in mutual communion. Are we ready to accept Jesus and to listen to his call?

Thus, to find Jesus it is not enough to study, to remain in the realm of concepts. Concepts insulate truth and protect us from its impact; with our concepts we can safely handle the most dangerous realities: we can speak about life and death, world and eternity without danger, just as we can handle the deadly electric currents that come from the power-house without danger of a electric shock, if only we take care of effective insulation. We have been used to keep a safe distance when we read the gospels and study theology. So we could speak about Jesus as if we knew him, without danger, because we met him only in the well protected sphere of liturgy and doctrine. We must again find him, his person. This is the meaning of the three following weeks of the Spiritual Exercises to encounter the person of Jesus, to be open to his impact, without protection, ready to allow his power and light to flow into our life and to transform it.

The Meditations of This Day

In the meditations of this day, and through the following three weeks, the goal of the Spiritual Exercises is firmly kept in mind. Ignatius wishes that the preparatory prayer before each meditation remain the same: "that all my intentions, actions, works be directed purely to the praise and service of his Divine Majesty" (n. 46). This prayer, however, will be more concrete in the meditations of Jesus' life because the praise and service of God is realized through our union with Jesus Christ. Thus in these meditations we ask "for an intimate knowledge of our Lord who has become man for me, that I may love him more and follow him more closely" (n. 104). Knowledge, love and service are the triad of the personal commitment to Jesus. Knowledge in biblical language is more than the intellectual grasp of an object; it means the personal encounter, leading to love, which, in its turn, does not consist in words or emotions but expresses itself in deeds, in service. (cf. n. 230).

This day is meant to lead to a deep awareness of Jesus' unique place in God's plan of salvation and to inspire the retreatant to a wholehearted commitment to him and to his mission. To prepare him for this commitment, Ignatius uses the parable of an earthly king chosen by God who calls his people to a crusade against the infidels. He applies this parable to Jesus Christ who is sent by his Father to conquer the powers of darkness and sin in this world and calls his followers to join him in this struggle (nn. 91-100).

As this parable reflects the distant era of a feudal age, it may be easier for us to remain on biblical grounds and to contemplate Jesus' person and mission in a global and comprehensive manner as it is presented in the gospels to inspire us to a radical commitment to Jesus.

The first encounter with Jesus (Jn 1,29-31).

"The next day John saw Jesus coming toward him". Live through this hour with John who had been waiting for the Messiah, preparing the people and his disciples. It is the hour of fulfilment.

It is also the hour of surrender: on the following day two of his disciples go over to Jesus; soon he realizes his own decline: "He must increase, I must decrease" (Jn 3,30).

So is also my encounter with Jesus. Ancient longings for the Saviour, for 'Emmanuel' are waking up in my heart. Will Jesus fulfil them, what will be his demands on me? I first greet him with John's words: "Behold the Lamb of God who takes away the sin of the world". My sinful life will be renewed, our broken world will be healed.

Jesus, the prophet

The Synoptics spell out Jesus' mission in more concrete terms. His prophetic proclamation is summed up, according to Mark, in four brief lines:

> "The time is fulfilled,
> the reign of God is at hand,
> repent,
> and believe the Good News."

Jesus' coming is indeed the decisive hour, the time of searching and waiting is fulfilled, the new creation dawns. Jesus calls the new creation the reign of God. The later prophets had spoken about the era of God's reign as a realm of righteousness, justice and peace. They had seen it in the great vision of the holy city of peace, where God will "gather those who have been driven away", and we, his people "will walk in the name of the Lord our God for ever and ever", because "the Lord will reign over them in Mount Sion" (Mic 4,5-7).

Of this reign of God Jesus says that it is "*at hand*", in the coming, not yet established but already present. It is not a dream of future glory but takes place in our world, in families, villages, among the people in actual life. It will not fall from heaven as a miracle but has to be realized in every human heart, in our human community, and from there it will flow from person to person to renew and transform human society in its many spheres of culture, economics, politics.

Thus the prophets never spoke about God's reign without a call to *repentance*, to a change of heart. The expectation of

God's reign was not allowed to become a soothing dream. The prophets denounced injustice and exploitation (e.g. Amos 2,6-8), and destroyed the illusion that through sacred rituals in the temple they could find God's favour (e.g. Is, 1,12-17). John the Baptizer had taken up the prophetic message in his call to repentance. He challenges the people not to blame outside enemies for their misfortune. Israel, God's own people, has to be renewed. As a sign of earnestness they are called upon to accept the baptism of repentance because the very essence of their vocation is at stake: "The axe is laid to the root of the tree" (Mt 3,10).

Thus also Jesus calls men to repentance. The new creation, God's reign, will not come as a sensational manifestation of power from heaven but must blossom in human hearts. Still, there is a difference in Jesus' preaching: he has reversed the sequence in his demands: God's reign stands first before repentance. Jesus message is Good News, the coming of the new creation. It is coming in love, it will not be forced on anyone, but it must be received and realized in our life, personal and social. It must become the beginning of a new creation, which is at once God's gift and our task. Jesus' proclamation ends with the call to *believe the Good News*, to open our hearts to God's coming.

Jesus as teacher

This prophetic message of Jesus must be explained as it has its implications in all spheres of life and society. Thus Jesus is not only the prophet but also the teacher. In Mark's narration of Jesus' public life the first scene takes place in Capernaum: "And immediately on the Sabbath he entered the synagogue and taught" (Mk 1,21). Nothing is said about the content of his teaching, but Mark points out its characteristic note: "They were astonished at his teaching, for he taught them as one who has authority, not as the scribes" (Mk 1,22). Also the scribes had authority among the Jews, the authority of scholars who knew their sources and were able to quote other rabbis to confirm their opinions. Jesus was different. He never relies on any authority. Even the Scriptures he quotes with sovereign freedom: "The men

of old have been told...But I say to you" (Mt. 5,21-48.)
He speaks from the depth of his consciousness. "We speak
what we know" (Jn 3,11). All words of Jesus have the ring of
truth, his simple teaching, his conversations, the controversies,
the parables. It is our lifelong task to listen to Jesus' authentic
teaching.

Jesus the master

Jesus is more than teacher, he is master. He has not
'students'—they learn from their teachers—but 'disciples', who
allow their lives to be formed and transformed by their master.
This was the experience of all of Jesus' disciples: Their
background and education were poor, and their humble
social status as fishermen held no promises. Yet from the
outset the gospels see them in a new perspective. Jesus calls
them with a promise: "I will make you become fishers of
men" (Mk 1,17). Whoever comes into contact with Jesus
receives a new identity, a new horizon for his life and work.
Jesus invites them to share in his own mission.

Jesus the friend

Even more, Jesus calls them friends: "You are my
friends if you do what I command you". This friendship
implies total trust. A servant carries out errands but does
not share in the master's plans and intentions. Jesus keeps
no secret from his disciples, he shares with them his message,
the very purpose of his life and work: "No longer do I call
you servants, for the servant does not know what his master
is doing; but I have called you friends, for all that I have
heard from my Father I have made known to you" (Jn 15,14f).
Jesus is conscious of the risk involved in this trust: one of his
disciples became a traitor. He also realized the slowness of their
understanding, the lack of depth in grasping the real meaning
of his words, the unreliability of their attachment, the worldly
outlook and the personal ambitions prevailing in their minds
and hearts. Still, he trusts them. It is a trust which lasts
through the centuries of the Church's mission, as long as
Jesus' messengers proclaim his word. It is the incredible risk
which he takes in entrusting his message to human beings.

Perhaps some feel it would have been safer to entrust it to a book (as Mohammed did). But it is the very essence of God's reign that it is to be realized by men and women, and proclaimed by Jesus' disciples, by those he calls his friends.

Jesus Prays

Jesus can take his risk because his work and mission are sheltered in God's own power. This is Jesus' prayer for his own: "I have given them the words which you gave me... they have believed that you didst send me...I am praying for them, for those whom you have given me: Holy Father, keep them in your name...that they may be one...keep them from the evil one...sanctify them in the truth...I consecrate myself that they also may be consecrated in truth" (Jn 17,6-19). Jesus prays to the Father for the ultimate guarantee of truth and unity, the Holy Spirit: "I will pray the Father and he will give another counsellor, to be with you for ever, the Spirit of truth" (Jn 14,16f).

Jesus Our Life

John's gospel reflects the intimate, lasting union with Jesus Christ as it was experienced in the community of believers. During his earthly life Jesus had invited the disciples to come, to see, to follow, to believe. The Christian community realizes a deeper union: "Abide in me and I in you. As the branch cannot bear fruit by itself unless it remains in the vine, neither can you unless you abide in me...Abide in my love." (Jn 15,4-11). It is the union which we celebrate day by day in the Eucharist. We have the assurance "I am with you always, to the end of the age" (Mt 28,20).

This, then, is the meaning of the meditations of this day: to encounter Jesus Christ and to realize his world-wide mission. This new creation has become visible, real in him. It must be realized now, in our world, through his disciples.

Thus we are called to a commitment, a new life and a new vision: a world renewed and united in God's love, in Jesus Christ.

SHORT POINTS FOR MEDITATION

1. The first disciples according to John (Jn 1,35-51).

Live once more through this hour in John's life, as he points at Jesus, the Lamb of God, and how he sees the two disciples leaving him and following Jesus.

— "What do you seek?": it seems a trivial question, just to start a conversation. Yet it is the basic question of their life—of my life, of every life.

— "Where are you staying?": again a common place question of the disciples, yet it is the question of their discipleship: where will they find him? where shall I find him? in books? sacraments? neighbour? I need the personal encounter.

— "Come and see": this simple answer contains key-words of the Gospel. There is no way of finding Jesus but to follow him in actual life. Neither study nor activities lead to Jesus unless the commitment to him is born out in true discipleship; the more we follow Jesus the more we shall see.

— "Andrew...found his brother Simon. He brought him to Jesus." The new community is being formed, the growing cell, through human contacts, to become Christ's body.

— "You shall be called Cephas": Each one who follows Jesus receives a new destiny, a new name (cf. Rev. 2,17).

2. Knowing Jesus. (Mt 11, 25-30; Lk 10, 21)

What is the disposition required to come to know Jesus?

— Matthew gives as background to the text Jesus' lament over the cities who have not received him. (Mt 11,20-24); Luke gives as immediate occasion the joyful return of the seventy after their first proclamation of the Good News, with the victory over the powers of evil. Why is the reception of the gospel so different?

— "Father, you have hidden these things from the wise": from those who know already, do not want to learn, have their own fixed expectations of God's reign (political, eschatological, etc), measure Jesus by their own ideas and wish to reduce God's word to the framework of their plans. People at all times are tempted to do so and thus become deaf to his message.

— "You have revealed it to little ones": to children in the sense of the bible: whose mind is not yet programmed; people who are open to truth and wonder; who do not rely on their own resources, do not calculate, are still able to receive gifts of love, God's own love.

— "All things are delivered to me": Jesus has no secular power, but God's reign of love, embodied in him, will outlast earthly empires.

— "No one knows the Son except the Father": People could see Jesus and listen to him, but no one can fathom the depth of his person without God's light. To Peter, who confesses him as Messiah, Jesus replies: "Not flesh and blood have revealed this to you but my Father who is in heaven" (Mt 16,17).

— "No one knows the Father except the Son and those to whom the Son reveals him": Through Jesus we are called to enter the inner life of God, what God is, what he is to us.

— "Come to me all who labour...I will give you rest". Jesus offers the Sabbath rest, the ultimate assurance and peace beyond travail and anxieties, work and struggle, the dawn of the new creation, of God's reign.

— "My burden is light": Jesus does not add burdens of the law. His only commandment is love.

OTHER READINGS:

— Phil 3, 1-14:   Jesus Christ is the only support of the life and mission of his disciples.

— Eph 1, 15-2:10   The grace to know Jesus Christ.
— 1 Jn 1,1-4:   God reveals himself in Jesus Christ.
— Col 1,12-23:   Jesus Christ the centre of creation.

QUESTIONS:

1. Reflect on Jesus' many titles: Which of them is most meaningful for you?

2. How shall I be able to deepen my knowledge of, and my commitment to Jesus Christ?

A PRAYER OF SELF OFFERING TO JESUS CHRIST

"Eternal Lord and king of all creation, humbly I come before you. Aware of the support of Mary, your mother, and all your saints, I am moved by your grace to offer myself to you and to your work. I deeply desire to be with you in accepting all wrongs and all abuse and all poverty, both actual and spiritual, and I deliberately choose this, if it is for your greater service and praise. If you, my Lord and king, would so call and choose me, then take and receive me into such a way of life".

(Paraphrase of ST. IGNATIUS' prayer for this day).

Chapter Two

JESUS OUR BROTHER

"When the time was fulfilled, God sent his Son, born of a woman, born under the law, to redeem those who were under the law, so that we might receive adoption as sons" (Gal 4, 4).

Jesus is the beginning of the new creation. How will this new world come about? How will God establish his reign?

For the Jews of Jesus' time the coming of God's kingdom was part of their faith. It was never shaken. In spite of decay and the disasters of their history they were sure of God's faithfulness. Jerusalem will be the city of peace to which the nations will flock. (cf. Is 2,1-4).

How the Jews expected God's Kingdom

The coming of the kingdom was conceived in two opposite ways: many pinned their hope on a political upheaval, an armed revolt, through which the foreign rule of the Romans would be overthrown and the national sovereignty of God's own people over and against the pagan nations restored. Through all his life Jesus is surrounded and challenged by these expectations.

Different was the expectation expressed in the apocalyptic writings which abound at the time of Jesus: God's reign was seen as the end of all history. The book of Daniel had already indicated the pattern of this expectation in the interpretation of Nebuchadnezzar's dream of the statue of gold, silver, bronze, iron and clay, which was smashed by a stone from above. As the stone smashed the statue, so "the God of heaven will set up a kingdom which will never be destroyed.. it shall break in pieces all these kingdoms and bring them to an end, and it shall stand for ever" (Dan 2,44). This is the apocalyptic expectation: human history will not be redeemed. It goes its way of decay and will be destroyed, but over the ruins of a destroyed world God's eternal kingdom will be established.

Both expectations had in common the theme of faithfulness in fulfilling the law; if Israel is loyal in the observance of the law, God will renew his people.

Jesus belongs to his people and is affected by their hopes and aspirations. His message, the understanding of God's reign in his preaching, contains elements of both trends, but is distinctly his own. The political model was based on the firm conviction that God's reign must be realized in our world. The prophetic vision of the new Jerusalem (Is 2,1-4; Micah 4,1-6; Is 60,1-7) is not an utopian dream but it means a better world of harmony with nature and of peace among the nations. At all times, also today, we are tempted to write off our world and history as evil and pin all hopes on a world to come. This is not Jesus' mind: he proclaims a world redeemed, a new community of brothers and sisters. But the danger, the actual distortion of the political conception of God's reign lies in its identification with a concrete national and political structure. The Jews at Jesus' time identified God's kingdom with the revival of national sovereignty and the triumpth over the nations. In the Middle Ages the Church attempted to realize it through the dominion over all secular powers. The same temptation and distortion still lingers in her reliance on institutional support, financial security and political influence; none of which figures in Jesus' own proclamation of God's reign.

The apocalyptic expectation, on the other hand, is based on the deep awareness that God alone can bring about the new creation. The bible knows Yahweh as the God who gives life, who tells the down-trodden people: "I will open

your graves...I will bring you home into the land of Israel"
(Ez 37,12). These apocalyptic visions saw the restoration of
Israel no longer as part of our history. God's reign comes
only on the ruins of the present world because, so they felt,
this world of ours cannot be redeemed. It will be destroyed
and out of its ashes God's kingdom will rise.

This conception of God's reign implies a failure on the
part of God. If he were to create a new world, which would
replace this world of ours, he would write off his first creation
of which it had been said in the beginning that "it was good."
Our world is God's work which cannot be abandoned.

And further: could God save man and our world without
our own involvement? He created us in freedom and placed
our life and future into our hands; this is our unique dignity
above all other creatures. If God were to bring about a new
creation alone, bypassing our freedom, it would not truly
be salvation, restoring human wholeness; it rather would
be a declaration of human bankruptcy. It would be a work
of power, a triumphant victory over darkness, but it would
not be a work of love. Man would stand humiliated, deprived
of his dignity to shape his own destiny.

The Incarnation

What, then, is God to do? Only God can heal, renew
and fulfil our human life and this world of ours. But this
saving work of God is the work of his love—he *loved* the world—
and love can never destroy. It cannot humiliate man. "He
will not break the bruised reed or quench the dimly burning
wick" (Is 42,3). Thus salvation will be God's own work,
more so even than the work of creation. And still it must be
also the achievement of man. How will this be possible?

The *meditations* of this day are not meant to lead to
theological speculations; rather should they be the silent
contemplation in wonder and love of God's unique gift to
us: "He so loved the world that he gave his only Son"
(Jn 3,16). Let them be centred on the words of Paul where
he speaks about the incarnation:

"When the time was fulfilled
God sent forth his son,
born of a woman,
born under the law,
to redeem those who were under the law,
so that we might receive adoption as sons" (Gal 4,4).

When the time was fulfilled

Of what time does Paul speak? Is there a time on earth which offers better opportunities for God's work of salvation?

In Palestine, where Jesus is born, it is a time of distress: of political impotence under foreign rule, of social injustice, of religious decadence, under high-priests who used their power for family interests and political convenience. The voice of the prophets had long been silent, it was replaced by that of the scribes who administered legal norms to the people. Jesus will have compassion on these people "because they were harassed and helpless, like sheep without a shepherd" (Mt 9,36).

The wider world at Jesus' time was the Roman empire under Augustus. Peace had come to the world, but it was the peace of oppressive power, guaranteed by the iron legions of Rome; peace at the expense of freedom and dignity.

It was a world of prosperity which, however, was based on the labour of millions of slaves who were helpless though their number was more than half the population.

The timings of God's coming do not depend on events of history. No time is worthy to be blessed by God, no time is excluded from salvation. It comes in God's own time "at the date fixed by the Father" (Gal 4,2). This remains true throughout history, in every life. To find God we need not wait for special constellations of favourable circumstances, for suitable psychological or sociological conditions. Wherever we are and whatever our situation, God comes, he touches our heart and enters our life when his time has come, when his love reaches out to us.

God sent his Son

It is God's own initiative for our salvation, to heal our

wounds and the wounds of our society, to establish his reign on earth. Yet it is more than an initiative of creative power: God himself wants to be with us. His eternal life and love begin to flow through the channels of our world and society. He lives this life of ours, a human life with its joy and beauty, with its insecurity, anxiety, pain, with conflicts and rejection, with death. He is with us not as our ruler but as our brother in unbreakable solidarity.

In Jesus he shows us what it means to be man, not in the distortion of sin, in pride or fear, as we see it all around, as each one is inclined to live it, but as sons and daughters of God: Jesus lives his eternal sonship as a human being, in an earthly life. In him we see what in our life could be new and how our world could be different. We are not meant to be slaves of oppressive powers neither from outside nor from within. In spite of poverty and weakness we can be free and live a life of love. In Jesus we also see that this freedom is not an illusion but God's promise. Even when Jesus seems to be drowned in the darkness and hatred of the world, when he dies, he remains God's Son. He has the assurance of eternal life, he will rise.

Born of a woman

Jesus' earthly life begins in the poverty of Bethlehem as a helpless child. Old theology often attributed to Jesus special prerogatives: as he was Son of God he must have had extraordinary knowledge even as a baby. Such conceptions are not borne out by the biblical accounts. It was the very meaning of his coming that he should be our brother and live our human life with all its limitations. He truly "emptied himself" (Phil 2,7), and "though he was rich, for your sake he became poor (2 Cor 8,9). He was the child of poor parents and grew up as a village boy. His first life experience was that of rural Palestine, and his religious world was that of pious Jews. What stands out in his consciousness, however, as far as we can trace it through the gospel accounts, is his unique closeness to God his Father. It accompanies him throughout his life. He called God *"Abba"*. This is a word that children speak to their father, an address devoid of all liturgical solemnity, never used by Jews in praying to God. It expresses an intimacy which was entirely new, and so it remained untranslated in the Gospels (Mk 14, 36). It is the core of his being: Son of God, but now living his sonship in the limitations of his human life: born of a woman.

Born under the law

In the actual context of the epistle to the Galatians Paul speaks of the Jewish law, the religio-social world of Israel based on the covenant. This is the framework of Jesus' life, which he accepts, in which he lives, which he brings to perfection. Every human life is shaped by its social surroundings and must develop within a concrete cultural milieu, which means both an enrichment and a limitation: it offers the opportunity of growth and the challenge of a critical encounter with the surrounding world; but it also narrows the horizon of human possibilities to one culture and to a short period of human history and development. Thus Jesus lives as a Jew, in the cultural and social setting of the distant past of Israel. This is part and parcel of the incarnation. He must live his life and fulfil his mission within these limitations.

But how then is Jesus' life and mission relevant to people of other times and cultures? How can Jesus speak to modern people?

To redeem those who were under the law

This is the meaning of Jesus' life and message: to redeem his own people, to redeem all men. Redeeming means breaking fetters, setting free, restoring the fulness of life. Every person is in need of redemption, of breaking out of the prison and bondage imposed on him through society in a thousand bonds of dependence of body, mind, heart, to live his true life in its three dimensions: before God, with fellow men in society, in the surrounding world.

Jesus lived his life and fulfilled his mission in the narrow world of Palestine, yet in *freedom*. This is the most characteristic feature of his personality: He is totally himself, assured, fearless, not dependent or relying on material resources, on people or on their attitude towards him. Whether Jesus is loved and acclaimed, or rejected and condemned, he is always himself, authentic, with an indestructable assurance in his decisions. He is sure of his Father in heaven and of his mission for the people on earth — love of God, brother of all.

In this freedom Jesus is more than a Jew: he belongs to all people. He does not adopt other cultures, nor does he wipe out social differences, but he shows how we all can be truly human in every culture and social setting. He has this freedom because his life is rooted in his Father. Man can be truly free only if he has his roots in God's absolute power and love.

So that we might receive adoption as sons

Full redemption implies many dimensions of our life: economic security, social acceptance, removal of psychological blocks which prevent true growth and free relations to others. All this has to be taken care of, yet all this is secondary; it actually would be worthless if the core of our whole being were not redeemed. I must be accepted by God, renewed in his love. My life must be sheltered in him, the ultimate truth, power and love, because I need the assurance that my life is precious and rich and holds a promise of fulfilment. Only so can I be free in spite of the many bonds by which I am tied in this earthly life, in spite of the narrow borders that surround me on all sides and limit the horizons of my vision and my actions.

This is Jesus' gift: I belong to God, we are sons and daughters of God because he became our brother to make us share in his sonship. So Jesus makes us also members of a new community of brothers and sisters which is based not on natural ties of family or cultural closeness, but on the common sonship of God. This is God's kingdom, comprising all cultures and ages.

SHORT POINTS FOR MEDITATION

1. "God so loved the world". (Jn 3, 16-21).

This is John's vision of Jesus coming:

— "God loved the world": Do we really believe it?

Many feel that God, if he exists at all, is infinitely distant from our world while our life goes its own way. Others, spiritual people, have in their mind the image of an angry God who must be pacified and Jesus reconciles him. But Paul tells us that it was not God who was reconciled, rather we are reconciled to God through Jesus (cf. 2 Cor 5, 18). God's love alone is the lifespring of our salvation.

— "He gave his only Son": The eternal bond of love, that unites the Father to his Son, is extended to us, that we be included in this love, that with Jesus as our brother we too become sons and daughters; he gives Jesus to us that he should share in our life, and we in his.

— "Not to condemn the world": Light dispells darkness, love redeems hatred, life awakens death. Jesus' coming is the crisis of the world, we feel threatened, "men loved darkness better than light". Yet love never condemns, it redeems; those who do not open their heart to the light condemn themselves: "He who does not believe is already condemned".

— "That whoever believes in him...has eternal life": In believing we open mind and heart to his light; we become disciples and allow him to transform us. His divine life, his life in the Father, begins to penetrate our being, becomes our own life with no limit or end.

— "He who does what is true, comes to the light": Thus Jesus is Saviour. Yet he does not dispense us from our free commitment, from doing what is true. This doing is not proud self-righteousness but the response to God's love, the covenant of God's love with human freedom.

2. *Jesus our brother*

Four texts from the Letter to the Hebrews.

— Jesus is God's final *revelation*: "In these last days God has spoken to us in a Son" (1,2): In Jesus Christ has spoken. He has given himself to us in irrevocable love. In him he also unfolds for us the meaning of our human life. Our world is included in God's love.

— Jesus is our *brother:* "He is not concerned with angels but with the descendents of Abraham. Therefore he had to be made like his brethren...Because he himself has suffered and been tempted, he is able to help those who are tempted" (2,14-18).

— *The compassionate* high priest: "We have a high priest...who in every respect has been tempted as we are, yet without sinning. Let us then with confidence draw near to the throne of grace" (4,14-16). The priest is mediator between God and man. Jesus is priest not through rituals but in his own person: He shares our struggles, yet in obedience is united to his Father: "Jesus offered prayers and supplications with loud cries and tears to him who was able to save him from death...He learned obedience through what he suffered" (5,7-10).

— *Following Jesus:* "Let us run the race...looking at Jesus the pioneer and perfecter of our faith" (12,1-2). As Jesus is with us, so we are invited to be with him. In his strength we are able to live our life in the assurance of God's power and love.

3. **The Church in Our World** (Vatican II, A.G. n. 11).

Read this text with a meditative mind. As Jesus is our brother and shares in our life, so the Church as sacrament of salvation, i.e. as visible sign of Christ's continued presence in our midst, must be intimately related to our world:

— through the example of Christians;

— manifesting the newness of life, the power of the Spirit;

— in respect and love of all people;

— sharing in their social and cultural life;

— familiar with the national and religious traditions;

— aware of the cultural transformations of our time, the loss of religious values;

— seeing in other religions the seeds of the Word;

— recognizing the riches God has bestowed on the nations;

— illumining these riches with the light of the gospel.

OTHER READINGS

— 1 Jn 1,1-4:   God's Word made visible in Jesus Christ.

— Vatican II: Constitution on the Divine Revelation, n. 4: God's revelation in Jesus Christ.

— Vatican II: Constitution on the Church in the modern world, nn. 1-3: The Church's insertion in the world.

QUESTIONS:

1. Reflect on Jesus: in what way was he similar to us, in what way different? What does this mean in terms of inculturation for the Church, for me?

2. The Church must be incarnate in the world: what are wrong ways of "incarnation" (worldliness, mere institutional presence, etc), what is the right presence?

FOR REFLECTION

The Christian faithful who have been gathered into the Church from every nation and are not marked off from the rest of men either by country, by language or political institutions, should live for God and Christ according to the honoured customs of their race. As good citizens they should sincerely and actively foster love of country and, while utterly rejecting racial hatred and exaggerated nationalism, work for universal love among men.

(*Mission Decree n. 15*).

Chapter Three

THE MOTHER OF JESUS

"Blessed is she who believed" (*Lk 1,45*).

This day is devoted to Mary, Jesus' Mother. In Mary we contemplate our own place in God's plan of salvation.

Mary's Place In God's Plan of Salvation

God speaks in Jesus Christ. This speaking is different from the word of creation where God is alone. Now he speaks to us. The new creation is to come about not by God alone but by God and man. It is covenant, bridging the abyss between heaven and earth, God and man.

Also in the work of salvation God acts with sovereign power — but now it is the power of love. Love includes the partner, it never overrules, never bypasses the beloved. Love cannot help becoming dependent on him, on his freedom and response. Love, therefore, is a risk, not just one of the many risks we have to take every day when we put our foot on the highways of life and the world. Love is a total risk, because in love the very core of the person is at stake. Can God ever take a risk? Can he hand over his work of creation and salvation to human hearts and hands and entrust himself to his creatures?

God's risk begins when he creates man in freedom and allows him to take his life and destiny into his own hands, to shape the world and to build human society according to his own designs. This is the drama of creation, culture, history.

God's risk, however, becomes radical in the plan of salvation when he gives up his own Son to the world to save it not through a display of divine power but through an invitation of love.

The history of Israel, the inconclusive account of their responses and refusals, foreshadows this drama of salvation. It comes to its climax in Jesus. He is the new covenant, God's irrevocable gift to man.

But how will Jesus be received? Will he remain alone, or is he the beginning of a new creation? If there were no one to receive him, the revelation would be futile; it would remain, as it were, a divine monologue of God with himself. Will Jesus find a place in this world of ours where he is welcome and finds faith?

Luke tells us of the response God's word of salvation receives in our world through Mary, the first of believers. In faith and love she surrenders her life to God's saving plan. Through her the Son of God becomes a member of our human family. She allows her own life to be drawn into God's designs and to be shaped by Jesus, her Son, and by his mission. So she is also the first to have her share in his work of salvation. Vatican II sums up her vocation. "Committing herself wholeheartedly and impeded by no sin to God's saving will, she devoted herself totally, as a handmaid of the Lord, to the person and work of her Son, under and with him serving the mystery of redemption, by the grace of Almighty God" (L.G. n. 56).

Thus Mary has her place in the Church, subordinate to Jesus her Son, yet irreplaceable. She is not a mere ornament of Christian life, nor is her veneration an accidental form of popular devotion which is dispensable. She has her place already in the Gospels. True, the earliest kerygma does not speak of her as it is totally centred on the person and mission of Jesus, who is "Lord and Messiah" (Acts 2,36). But the gospels of Luke and John were written at a time when Christian faith had matured. The life of the Christian is not only God's gift but includes also the human response. Thus Luke tells us of Mary's surrender to God's call in the Annunciation and her gradual growth towards an ever deeper understanding of her vocation until she is in the midst of the Apostles awaiting the outpouring of the Holy Spirit on the young Church (Acts 1,14). John sees her standing under the cross with the disciple whom Jesus loved, united with Jesus in the fulfilment of his mission (Jn 19,25-27).

These are not marginal additions to the Gospel message but its genuine unfolding. In her response to God's word Mary is the model of the whole Church which in continuation of her obedience, receives God's word and "keeps the faith in its entirety and purity...imitating the mother of the Lord, and so, by the power of the Holy Spirit it keeps intact faith, firm hope and sincere charity" (Lumen Gentium n. 64). Thus Mary's role as handmaid of the Lord, who commits herself freely to God and allows her life to be absorbed in the life and mission of her Son as a model of the Church, belongs to the substance of the Christian message.

There is more. The early Church saw Mary in the context of humanity's salvation: as Eve stands at the beginning of our history of sin, so Mary is the beginning of the history of salvation. The Council quotes Epiphanius: "The knot of Eve's disobedience was untied by Mary's obedience; what the virgin Eve bound through her disbelief, Mary loosened by her faith" (Lumen Gentium n. 56).

Thus Mary has also her special place in the *Spiritual Exercises*. It is their goal to give guidance in the search for God's will, to lead us to a total commitment to God free from delusions, and to realize God's plan in our actual life. Thus time and again Ignatius invites us to turn to Mary in prayer for her intercession because it is she who was unconditionally committed to God's will. In her presence we are meant to become aware of the disorientations of our heart and the deceptive allurements of the surrounding world (cf. n. 63). On this day, however, all our meditations should be centred on Mary to find in her the model and pattern of our own life in response to God's word.

Our *meditations* are based on Luke's gospel. His account of Jesus' conception and birth are written for the community of believers: Salvation comes from God and must be received with faith and love. Mary is the model of the full surrender to God. Every word in these accounts reflects the unspeakable mystery of God's encounter with his creature in the language and the symbolism of Jewish tradition. It has inspired religious art through the centuries.

In our meditation on the Annunciation we search for the spiritual message of the narration.

The Annunciation. (Lk 1, 26-38).

"God sent the angel".

Salvation is God's own work, his own gift. Yet, God sends an angel: angels are messengers. God is no longer alone as in the dawn of creation. God offers us a covenant of love, and love needs a partner. God makes himself dependent on the free response of his creature. So it will be till the end of time, so it is also in my own life. God needs our response. He invites, but he respects our freedom.

"To a city of Galilee named Nazareth"

To whom is God's message sent? What does he expect on the part of his creatures? Though God wants to establish his reign against the powers of darkness he does not seek an alliance with human power; the angel is not sent to Rome, the concentration of political and military might. Though God begins to speak his word of divine wisdom and truth, the angel is not sent to the proud centres of human learning in Greece and Egypt. Though God wants to sanctify the world and lead it to true worship in spirit and in truth, the angel is not sent to the holy city of Jerusalem where psalms are chanted and sacrifices are offered every day; salvation has nothing to do with ritual performances which have been emptied of spirit.

The angel is sent to Galilee, a backward *country* politically, culturally, spiritually, "people who live in darkness" (Is 9, 1; Mt. 4, 16), to Nazareth, a village which cannot be found on any map of the ancient world. "What good can come from Nazareth?" (Jn 1, 46). On what do we rely in our mission? What does God really expect on the part of his creature when he comes to save us?

"To a virgin whose name was Mary"

Mary is a virgin; it is the only title attached to her name, the only explanation for her being chosen to be mother of the Saviour. In biblical language virginity means first of all faithfulness to God. Israel is called the virgin daughter of Sion as long as the people are faithful to the covenant, listen to Yahweh, and turn away from idols. The worship of the Baals of the Cananites is called adultery. This meaning of the word is kept up in the New Testament. The 144,000 of the Apocalypse, who are saved from final destruction are virgins "who follow the Lamb wherever it goes", men and women who have not stained themselves with idolatry (Apc 14.4). Paul calls on the Corinthians to remain faithful to the unadulterated gospel because he has betrothed them to Christ as "a pure virgin". Let them not be led astray "from a sincere and pure devotion to Christ, as the serpent deceived Eve by his cunning" (2 Cor 11, 2f). Virginity concerns mind and heart. Physical integrity and purity are its symbol.

This, then, is Mary's prerogative: She is a virgin; she will not listen to other voices and allurements, only to God's angel. Her entire life and being will depend on God, she will be guided by his word. This in fact is the basic requirement of all who wish to find God's will. The Spiritual Exercises are meant to lead precisely to this singleminded attentiveness to God, not to be diverted by the multitude of voices which claim our attention.

"Hail, Favoured by God"

Thus Mary is prepared to receive the angel's message, she will respond to it wholeheartedly. This readiness, however, is not her own achievement, it is itself God's gift, the most intimate gift which God can bestow on us: the power to respond to his love. When we translate this line as "full of grace", easily a misunderstanding is created that grace is something that can be measured. Grace is a living bond of love. It is love, but not as it is hidden in the depth of God but as it becomes creative in our heart and brings about our faithful response. It is totally God's gift, but realised in our freedom. God awakens his creature from its earthbound blindness, he moves the heart, he draws us to himself.

It is Mary's privilege that from the beginning her life is embraced by God's love, ever responsive to him. In God's design she belongs totally to Jesus who in every way is similar to us except sin. With him she belongs totally to God, totally against the powers of evil. In her life God's love is always stronger than any selfish desire. Her readiness to commit herself to God's will is the fruit of God's own choice to link her life in a unique way with Jesus. Mary will be his mother.

It will be the experience of the retreatant that the openness of heart and the generosity of self-gift, to which he/she is invited are not his/her own achievement but God's gift.

"The Lord is with you"

So she may now enter without fear into her real life as God has destined it for her, to be the mother of the Saviour. Even before the message is pronounced, she is assured of the guarantee God ever offered to his chosen ones that he will be with her. So he was with Moses (Ex 3, 12), with Gideon (Judg 6, 12), with Jeremiah (Jer 1, 8); so he will finally be with the apostles in their impossible mission to the whole world (Mt 28, 20). It is the only assurance given to all who follow Jesus.

"You will conceive in your womb and bear a Son".

This is Mary's unique privilege, to be Jesus' mother, to give the Son of God a human mind and body, to make him a citizen of this world and a member of our human family. It is the pride of every mother when she recognizes her own features in the face of her child, her own character in the child's reactions, and her voice in his first stammerings. Jesus will be her child.

Yet even in this unique privilege Mary is not alone. The conception of Jesus in her womb is only the *beginning* of his lasting presence in our world and society in and through his believers. The body of Christ, his presence in our world, will expand. All believers will become his body, his continued presence in human society and bear the responsibility to make his coming real and effective in our world. Thus the Church herself "by receiving the word of God in faith becomes mother" and through the fulfilment of her mission gives life to others "who are conceived by the Holy Spirit and born of God to a new and immortal life" (Lumen Gentium n. 64).

"You will call his name Jesus"

Every mother must learn, often with pain, to allow her child to grow and to live his own life. Mary too has to learn it in a unique way. Her Son has a mission which is expressed already in his name 'Jesus', which means *Saviour*. She herself has (according to Luke) to give him the name and so, as it were, ratify his destiny not to belong to her but to the people. She will experience the implications of this name when he leaves his home in Nazareth. She will set out to see him in Capernaum, the centre of his activities, but has to learn that Jesus' true family, his "mother and brethren are those who hear the word of God and do it" (Lk 8, 21). It is based not on blood relationship but on faith in him. So in fact she is the first to belong to his family, as she was the first who believed. But it is hard for her to realize that he does not belong to her. He does not belong to her, but she belongs to him. She has to give him to the world, and she will have to widen her own heart till it is as open as Jesus' heart with a love that embraces all peoples.

"He will be called Son of the Most High"

This is the title of Jesus' unique dignity, but it implies also Mary's total surrender: Jesus belongs to his Father, his loyalty is with him, not with his mother. She experiences it when at the age of twelve he remains in the temple and she asks the painful question: "Why have you done this?" Jesus has the simple answer of his only loyalty: "Did you not know that I must be in my Father's house?" But she did not understand. (Lk 2, 48-51). The full impact of his obedience to God falls on her under the cross.

It takes Mary her whole life to realize the full implications of being the mother of Jesus, from her encounter with

Simeon in the temple, who speaks of Jesus' destiny to be "for the rise and fall of many" to her presence on Calvary and in the cenacle at Pentecost. Every one of her experiences draws her closer to the mystery of his person and mission. God's designs are disclosed to her in stages, in the events of Jesus' life, of her own life, in the life of the apostolic community.

"Behold, I am the handmaid of the Lord"

This is Mary's surrender giving up her life to God, the security of her small world in which she is at home, her humble but assured place in her village. She surrenders herself to the unknown, to a future which is beyond her control and comprehension, with no one to guide her and to answer her questions, to God's inscrutable mystery, with no other assurance but God himself, his word.

"Let it be to me according to your word"

God's word came to her through the message of the angel; God continues to speak to her in all stages of her life. Her word of obedience is total, but as the events of her life take place it will be deepened and enriched; her marriage with Joseph — the birth of the child — the years in Nazareth — Jesus' farewell — the agony under the cross — the bliss with her Risen Son. In all these events her 'fiat' is not only repeated but tested and fulfilled till her life and love are blended with the glory of her Son. So are our commitments an ongoing surrender to God as our life unfolds. No one knows what is implied in his or her baptismal vows, in one's religious profession or priestly commitment. Each self-gift to God must be ratified in the ongoing unfolding of our life.

"The angel departed from her"

This now will be her life: She has received the angel's message, now she is left alone, to herself. The Lord is with her silent faithfulness.

SHORT POINTS FOR MEDITATION

1. Sharing God's gift.   The Visitation: (Lk 1, 39-45).

In the visit to Elizabeth the significance of the Annunciation is spelled out.

— "Mary arose and went with haste": She is drawn into Jesus' own mission; God's love is present in our world, reaching out, serving.

— "She greeted Elizabeth": Where people are touched by God their meeting has new significance. Human contacts become channels of God's working; a simple greeting has new depth. (Look at the greetings in Paul's letters).

— "Elizabeth was filled with the Holy Spirit": In the new community of believers life is shared, transformed.

— "Blessed are you among women": Mary's role in the plan of salvation is that of the woman. She is virgin, spouse, mother in her relation to God. She realizes the ultimate richness and depth of womanhood.

— "Blessed is she who believed": Mary's greatness is not based on her physical motherhood, but on her *faith*.

— "It will be fulfilled what was spoken to her": her life, her future lies in God's hands; he alone is the assurance of the fulfilment.

2. Mary's praise, The Magnificat (Lk 1, 46-55).

God's gift becomes a hymn of praise in Mary's heart. It contains the great themes of Luke's Gospel.

— "My soul glorifies the Lord": God's work of salvation is echoed in Mary's heart. Her contemplation consists in encountering the God of salvation in his deeds and praising his love.

— "All generations will call me blessed". This is God's greatness that the creature who encounters him is not humiliated by his majesty but is rised to share in his glory.

— "Holy is his name": In drawing creatures to himself, God loses nothing of his infinite holiness; in her most intimate relation to God Mary bows before his holiness.

— "He has shown the strength of his arm": God's reign begins; it consists in raising the lowly, in filling the

hungry with good things. The Good News comes to
the poor; those who glory in their earthly power
and wealth will be scattered and go hungry.

— "As he spoke to our Fathers": In the work of
salvation the ancient promises are fulfilled: God is
faithful. God is ever present in our world, "from
generation to generation." My own vocation and
my trust in God are woven into the history of God's
faithfulness which starts with creation, which will
last to the end.

OTHER READINGS
— Lk 1,5-25:   The annunciation of John's birth to Zachary,
                 his doubt and punishment.
— Mt 1, 18-24:   Jesus' origin acc. to Matthew. The story
                 of Joseph.
— Heb 11:   Faith through the ages.
— Eph 1, 3-14:   The mystery of God's election.

QUESTIONS:

1. What is Mary's place in my life? Can I realize the
    significance of the Marian tradition of the Church
    beyond devotional practices? Can I learn true faith
    from her?

2. Has my faith been tested and challenged by my personal
    experiences or by my surroundings? Has it been
    deepened and enriched through the years or has my
    faith become dim and shallow?

FOR MARIAN REFLECTION:

"You came down from your throne
    and stayed at my cottage door.
I was singing all alone in a corner,
    and the melody caught your ear.
You came down and stood at my cottage door.
Masters are many in your hall
    and songs are sung there at all hours.
But the simple carol of this novice struck at your love.
One plaintive little strain
    mingled with the great music of the world,
and with a flower for a prize
    you came down and stopped at my cottage door."

                              (TAGORE, *Gitanjali n. 49*).

Chapter Four

COMING IN POVERTY

"This will be a sign for you: You will find a child, wrapped in swaddling cloths and lying in a manger" (Lk 2, 12).

How will the new creation, for which the world is waiting, come? The Jews of Jesus' time believed in God's power and expected the miraculous revelation of his kingdom. When the Pharisees asked about its coming, their question was not concerned with its nature—they took its coming in power for granted—they merely asked about the time (Lk 17, 20). Jesus' disciples also expected a glorious revelation in power and asked for privileged positions in Jesus' kingdom, at his right and left (Mk 10, 37). When after the collapse of their hopes because of Jesus' death they encountered him anew as the Risen Lord they asked once more: "Will you at this time restore the kingdom of Israel?" (Acts 1, 6). Jesus was constantly confronted by a deep-rooted misconception of God's kingdom; it persisted also in the early Church and lingers on through the centuries into our own days.

This is Jesus' answer: "The kingdom of God is not coming with signs to be observed, nor will you say 'here it is ' or 'there,' for behold the kingdom of God is in the midst of you" (Lk 17, 21). There will be, it is true, a final manifestation of God's reign in glory, but for this ultimate coming there is no date in our calendars. It will be God's own day, "no one knows it, not even the angels in heaven, nor the Son, but only the Father" (Mk 13, 32). But it is coming now: There will be

no harvest unless the seeds are sown in human hearts and
ripen in our world. God's reign must come now, but not in
a mighty display of splendour but with the silence of life;
all true life comes in silence.

How then is the seed for the final harvest sown in the
furrows of our world? Of this wonder Luke tells us in the
Christmas story, which many claim to be the most popular
story of world literature. But it is more than poetry. Every
word has significance. It is our meditation of this day.

Christmas

Precisely in the beauty and popularity of this story lies
its danger. Too many take it as a story belonging to a dream
world. All know that this world of ours should be different:
there ought to be peace, good will and joy. So for a day we
behave as if we lived in a better world. We do something like
a role play. We greet each other with joy and exchange Christmas cards; we share gifts and avoid clashes. If it happens
that there is a war we sometimes agree on an armistice for one
day — because it is not nice to kill each other on Christmas
and so to spoil the feast; but tomorrow we shall continue the
battle. Good business is always alert to the mood of the
people and realizes the chances of profit on Christmas under
the title of selfless love; it uses Bethlehem as an advertisement
and accompanies profitable transactions with the tunes of
carols. How many have been disillusioned with Christmas
because they feel it is just not true—no one really wants a
different world; at heart we remain the same, also on Christmas
day.

But Christmas is a new beginning. We reflect on the
three sections of Luke's account: the hard reality, its meaning, the response.

The hard reality (Lk 2, 1-7).

Luke speaks of facts. Though some of his data may be
questioned as to their historical accuracy, surely he depicts
the crude reality of poverty and helplessness. Mary and Joseph
on their way to Bethlehem; they belong to the masses of

people whose life depends on those in power. Augustus, the emperor, decrees a census of the empire which affects millions who have to set out to be registered in their ancestral towns. They are dragged into this mass migration with which they have nothing to do, irrespective of their own needs and conditions, mere objects of a political regulation. Is there no divine providence to protect Mary in her advanced pregnancy against such arbitrary interference? No place is found for them in the inn; there is never place for poor people. She delivers her child in helpless poverty, lays him in a manger where the fodder for animals is kept and wraps him in pieces of linen. This is the way poor people have to struggle for survival.

This is the way the reign of God is coming: nothing seems to change, God seems infinitely far away, no one cares. Life, politics, business go on as ever, and there is no sign of hope for a change. Life goes on and follows its cruel course: Mary, Joseph and the child are part of this harsh and empty world.

What does all this mean? (Lk 2, 8-14).

The second part of Luke's account contains the interpretation of these facts, the wonder of God's saving love as it is celebrated by the community of believers. It is God's own interpretation. When angels speak in the gospels it means that more is needed than human understanding; we have to listen to God himself and open our hearts in faith.

The angel appears to the shepherds; they are poor, living outside the comfortable cities, illiterate, far away also from the temple and the synagogues, not very holy people, and so they are despised not only by the rich but also looked down upon by the religious elite. And yet the gospel comes *first to the poor:* this is the core of Luke's gospel. He will make it the theme of Jesus' key address in Nazareth: "The Spirit of the Lord is upon me, he has anointed me to preach the good news to the poor" (Lk 4, 18).

The angel who speaks to the shepherds is surrounded by God's glory: in Jesus' birth God himself reveals his power and love. In ancient times the glory of God surrounded Moses the powerful leader of the Exodus; it guided the people on their journey through the desert; it filled the wonderful temple of Solomon. But who would ever expect the glory of God to shine in this poor world of shepherds!

The Christmas message of God's saving presence among us remains true through the ages. It must be celebrated by the community of today: "Today a Saviour has been born to you".

They need a sign. All believing Jews expected signs for the coming of God's reign; they looked out for wonders or frightening events. The sign God offers in Bethlehem is different: "This will be a sign for you: You will find a child, wrapped in swaddling cloths, lying in a manger". This is not power but weakness; it is not frightening but inviting! Is it a sign at all? Children and poverty are part of our everyday world. How can they signify the dawn of salvation?

This is God's new way of speaking to us.   Prophets had spoken powerful words, threats of disaster, glorious visions of a new Jerusalem. We need such words to awaken us; even Jesus himself will challenge us with his hard denunciations of evil and with his inspiring words of assurance and promise. But at Bethlehem God does not speak with words; he speaks in his Son, in this child.   His Son Jesus begins to live a human life, our human life.   He lives it not in the disfigured form which he finds in his surroundings, which we see all around, far from God, relying on money, social status and political power, where neighbour lives in competition with and envy of his neighbour. Jesus is different: He is the Son of his Father, and brother to all; his reliance is not on money or power but on God alone; he will not dominate but invite all people to reconciliation, brotherhood and community.   This child of Bethlehem is the sign, the beginning of a new humanity, not yet programmed for a particular role in society, not yet entangled in the network of a competitive society.   He is simply a child, unprotected, helpless, yet the future belongs to him, "appointed heir of all things" (Hebr 1, 2).   He is the dawn of the new creation.

A multitude of the heavenly host appears praising God: "Glory to God in the highest, and on earth peace among men with whom he is pleased". Heaven opens over this child, his coming has world-wide dimensions. Where is God's glory in all creation if it is not summed up and turned into praise through man? The heavens and stars could not proclaim God's glory if it were not echoed in human hearts, in God's children, who in Jesus are born to a new life. And if God is the centre and source of human life, they will be brothers and sisters again, and there will be peace on earth.

The response (Lk 2, 15-20).

"Let us go over to Bethlehem". The gospel account widens. The shepherds represent the millions of all times and nations who are drawn to Jesus. What do they expect? Popular imagination, in a deep insight into the mystery of Christmas, sees the shepherds bringing gifts to the child. Salvation does not mean receiving gifts. It is the awakening to God's love, to receive his love and to share it.

"They made known the saying which had been told them concerning this child". They become messengers. Those who respond to God's word and follow the invitation must share it.

"Mary kept all these things, pondering them in her heart". This is Mary's contemplation: it does not take her away from the realities of her life and from her tasks. She lives her life and responds to its demands, both joyful and painful and often surprising. In them she encounters God's love and will. So she grows ever deeper into the unfathomable mystery of God.

"The shepherds returned glorifying and praising God". The Church is a community of praise. In the Old Covenant God was praised for his mighty deeds in Israel; in the New Testament we praise him for the coming and the work of salvation of Jesus his Son. Day by day we also praise him for the events in our own life, for each one's personal history of salvation. This is the rhythm of Christian contemplation and praise: to encounter God in his deeds, as they are recorded in the Bible, as they happen in our own life, to recognize his presence and love in them, to praise him.

SHORT POINTS FOR MEDITATION

1. *Jesus' birth according to Matthew* (Mt 2, 1-12).

Also Matthew's account is a meditation. In Jesus' birth he sees the breakthrough from national narrowness towards the universality of God's Kingdom. Political and national prerogatives are abolished. Barriers of selfishness and self-righteousness must fall in our own day as well.

— "Wise men from the East came to Jerusalem": They represent the nations. In Luke's gospel the message of salvation is addressed to the socially depressed; in Mathew to those outside the Jewish world.

— "Where is he who is born king of the Jews?": The ancient expectation of a new era of peace, of a Saviour-king, rings in this question. It is the question of all who are waiting for a renewed world, for the new creation.

— "We have seen his star": They were called by the sign God gave them, they were guided by the inner light of which the star is the sign. It leads them to Jerusalem, the holy city, where they expect to find the Saviour-king. Also today people come to the Church, the sacrament of salvation, to find the saviour. Do they always find him?

— "Herod was troubled, and all Jerusalem with him": Is Jerusalem no longer the city where God will establish his kingdom? Herod thinks only of his political power and is afraid of a rival. But the fear is deeper: All are afraid when God's reign dawns. It is the end of human autonomy; power is replaced by love. Where are our thrones, where is our self-righ teousness? Also today we are afraid of losing positions of power and social status in the church, in society. We are afraid of our salvation.

— "They went their way, and lo, the star which they had seen in the East went before them". They follow the direction given to them by the scholars of Jerusalem, but their inner assurance comes not from scholars but from the star God had given them: "they rejoiced exceedingly with great joy when they saw the star."

— "They saw the child with Mary his mother!": it is the fulfilment of their search. Many are the insights of Eastern wisdom which they represent; Jesus does not compete with them. What these wise men seek and need is the person of Jesus born of Mary. This is the new way of God's speaking, in human life, through his Son.

— "They offered him gifts: gold, frankincense and myrrh": This is our salvation, not to receive but to give, to share what we are and what we have, freed from ourselves.

2. *Poverty*

Bethlehem is the scene to reflect on poverty, in particular the poverty of the Church, of priests and religious. The many problems and aspects of this poverty must ultimately be related to the poverty of Jesus; from him we may also find orientation in our concrete questions.

Jesus encountered poverty in its many forms: the destitution of beggars, the personal helplessness and poverty of the sick, the social poverty of the rejected and marginalized, the spiritual poverty of sinners. All this poverty Jesus heals by becoming himself poor, sharing our condition. By accepting our poverty freely in love and solidarity Jesus saves us: "Though he was rich, yet for your sake he became poor so that by his poverty you might become rich" (2 Cor 8, 9). He was rich, endowed with divine power. But becoming man he divested himself of power; his mission is based on love. This is the model of Christian poverty to be realized in the Church, and most of all in priests and religious. It must be spelled out in its various aspects:

— *Detachment:*— free from clinging to material goods or other supports on which we tend to rely for our security; the life and mission of the Church is guaranteed by God alone.

— *Acceptance of actual situations:*— Paul knew how to be grateful for abundance and how to suffer want— food, house, surroundings, social situation, companions (cf. Phil 4,12).

— *Simplicity of life style:* — free from excess and luxury, in personal life and in institutions.

— *Hard work:* — the lot of people at large, tiredness.

— *Availability:*— time, skills, talents, whatever is at my disposal is for the people.

— *Service:*—In my work, in my relationships, the main view point is not my individual interests but the needs of my neighbours.

— *Solidarity with the poor:* — awareness of their situation and needs, assistance in their struggle.

— *Accountability:* —responsible use of money and all things of material value; clear accounts: the middle between stinginess and lavishness.

OTHER TEXTS:

Tit 3, 1-7:   God's loving kindness appeared.
—John 1, 1-18: The Word was made flesh.

QUESTIONS:

1. Have I celebrated Christmas as my own birthday, the beginning of my life in Jesus Christ?
2. What is my attitude towards material resources?
3. Have I learnt to outgrow social, linguistic, national barriers?

FOR REFLECTION:

"Our poverty should be true Gospel poverty. Gentle, tender, glad and open-hearted, always ready to give an expression of love. Poverty is love before it is renunciation. To love it is necessary to give. To give it is necessary to be free from selfishness."

(from MOTHER TERESA's *Constitutions for the Missionaries of Charity.*)

Chapter Five

THE PERSONAL VOCATION

"I will give him a white stone with a name written on the stone, which no one knows except him who receives it"
(Rev 2, 17).

The Personal Vocation

With the following meditations we enter into the decisive stage of the Spiritual Exercises, the actual search for God's will. Speaking about God's will may arouse negative feelings as if God had prepared a set of commands and decisions to impose them on us. This is not the gospel message. Already the Prophets describe the New Covenant as a time of freedom where laws will be written no longer on tables of stone but implanted into our hearts: "I will put my law within them and I will write it upon their hearts" (Jer 31,33). Jesus came to bring life, not laws: "Because I live, you will live also" (Jn 14,19). Life is not imposed, it is given, it becomes my own. It is God's will that I find true life, my life, and live it.

But life has many forms. Even in nature every flower is produced anew and has its own time to blossom and to ripen its seeds. And so it is, and much more so, with each human life. It is rich and has ever new features. We are different today from the people of the past; we are different also within our present world, using many languages and living in different cultures. Each human life is shaped by the inheritance of the past and by the social surroundings of its family and the society in which it develops. But the difference is still deeper. Each human person is unique. The uniqueness and distinctiveness of each person cannot be reduced to the arbitrary combination of blind forces by which it is formed and by which it will be dissolved again, like the waves on the surface of the ocean, which are shaped by hidden currents, changing winds or boats passing by, without lasting identity, dissolved again into the vastness of the ocean. Waves have no names.

But before God *we have names*. Israel had a name: "Fear not, I have redeemed you, I have called you by name, you are mine" (Is 43,1). Jeremiah, the prophet, who had to go through agonies of frustration, is assured from the beginning of his mission: "Before I formed you in the womb I knew you, and before you were born I consecrated you" (Jer 1,5). In the New Testament we know that the personal love of God includes not only people of a special destiny, as the prophets, but extends to all: "God chose us in Christ before the foundation of the world that we should be holy and blameless before him" (Eph 1,4). To each one is given a new name "which no one knows except the one who receives it" (Rev 2,17).

It is the task of every one to discover his/her name, and personal identity, and to realise it in actual life, because the very meaning of one's life depends on it. The painting of an artist is more than the mixture of colours which anybody can buy in the bazaar. The arrangement of the colours, their tuning and blending together is nothing arbitrary; it is the work of love and skill of the artist. So is my life. It is more than the chance combination of genetic, psychological and social factors. What does God want of me? What is the meaning of my life, not only of human life in general but of the personal life that is mine, that is entrusted to me?

It is *entrusted to me*. It is not yet finished, it is given to me like a sketch to be completed. Shall I be able to complete it? I could spoil it and make it into a caricature. I can also misunderstand the intention of God and try to make something different out of it, according to my own ideas and ambitions. But my life will be a failure unless I grasp God's idea of it. How did he see me? I must become aware of his intentions which are enshrined in me and be ready to set aside my own instinctive likings. I must live in deep and lasting communion with God's own creative power and love, thus only can I become truly myself.

How to Find My Personal Vocation?

How can we discover this personal vocation? We are familiar with the natural development of a person, the growth from childhood through adolescence to adulthood. Childhood is the age of dependence, of learning, of conformity with the life of the family and the wider world. The child learns to be part of the society into which it is born, it must first become part of it. However, the child is meant to grow into a person and to live and unfold its own life. Already in early periods of childhood the first stirrings of self-assertion are felt, sometimes in periods of stubbornness and self will. The real crisis comes in adolescence, in the teens, when traditions are considered a burden which oppress personal freedom and prevent growth and authentic self-expression. Young men and women break away from existing structures and begin to shape their life and world independently. It is an age of transition, passing through insecurity with the desire for an authentic life, towards maturity. An adult is a person who has achieved his/her own identity, with the appropriate relation to world and society.

Also the life of the spirit has such stages. It begins in the Christian community, in the family and parish, with its worship, devotions and catechism classes, with its inspirations and limitations. For long years the religious praxis of many young people may be confined to such a framework even when they get married, or enter a seminary or join a noviciate. Faith and religious behaviour remain faithful to the routine of prayer, work and study in the rhythm of the institution to which they belong, according to the norms of approved and tested structures.

This is the framework in which the personal life of each one must be awakened and develop; it must come into its own and find its specific expression as a personal *response* to God, who calls each one by name and gives his life its own

features. Life thus becomes a truly personal response which
cannot be imposed from outside, not even from above, nor
can it be learned or copied from others. God calls each one
by his/her own name. Finding our personal identity is clearly
more important than choosing one's profession or career. Our
identity concerns not what we do but what we are; not what
people think or expect of us but what we are in God's presence.
A person's identity is the life-centre from which all relations
to others and all activities find meaning.

Teachers of the spiritual life call this breakthrough a
second conversion, or the vocation within the vocation. The
first conversion consists in the acceptance of the common
patterns of Christian life in one of its existing forms: as member
of a family, in a single life, in marriage, in priesthood, in
religious life. But the time comes when we become aware
of the inadequacy of a life which is based on conformism with
traditional patterns. The world is more complex, life is deeper
and richer, its challenges can no longer be ignored, and
problems cannot be covered over by accepted formulas of
the past. We have to read the signs of the time in the events
and situations of our surroundings. We have to listen to the
inner stirrings of the spirit. A personal response is demanded:
What does God want of me?

We do not speak here of those who are disappointed with
their life, Christians who gradually drift away from the
Church, or religious and priests who lose the joy of their
vocation. Here we think of those to whom the accustomed
forms of prayer have become shallow and no longer express
or fulfil the deep desire for God which stirs in the depth of
their heart; of those who wish to reach out to people beyond
the accepted channels in preaching, worship and activities;
who face a crisis of authority, situations of unjust structures;
who sense that there is a new urge of the Spirit all around us.
The gospels speak of a new wine which needs new wine-skins.
God's Spirit is leading us not merely by norms of the past or by
orders which come from institutional authority, but also from
within: where is my place in this newly emerging world?
Where is God leading me? What does he want me to be
and to do?

This is the question which the retreatant has to answer. More correctly, this is the question for which he has to ask for an answer from God. The answer must be based on God's word, on his revelation in Jesus Christ; it must be related to the context of his actual life; it must be realized in the depth of his heart.

The Personal Vocation in the Spiritual Exercises

Two situations are foreseen in the Spiritual Exercises: the retreatant may have to make a basic decision for the state of life which he wished to adopt, marriage, priesthood, religious life. This actually is the situation which originally is foreseen by St. Ignatius (nn. 169-189).

More frequently, however, the Spiritual Exercises are made by people who are already settled in their state of life, or at least have decided on it long ago and are engaged in their formation. They have come to a decisive stage in their life where they feel the need of a breakthrough to a new vision, to a deeper meaning of their vocation. For such persons two steps will now be important.

First, it will be very helpful for them to review and assess the decisions they have made earlier. Thus a religious who goes through the Spiritual Exercises in preparation for final vows will normally not question the earlier decision to become a religious; or seminarians who prepare themselves for ordination in most cases will make the retreat with the assumption that their vocation to the priesthood has been sufficiently tested (though at times a radical revision may be needed). Still, a serious assessment will be very important in order to deepen the understanding and the motivation of the earlier choice and to clarify doubts and problems of which they have subsequently became aware. Such an assessment should include the following questions:

1. What were the motives of my earlier decision? Have they been deepened and confirmed in the subsequent years? (The understanding of priesthood, for instance, may undergo great changes from the year of joining the seminary to the time of ordination).

2.　What are the concrete experiences concerning my vocation e.g. the feedback I get from the exercise of pastoral duties, teaching, preaching, etc; meeting priests who were happy, or unhappy in their priestly life and ministry; the attitude of people towards priests; appreciation, criticism, rejection; the personal experiences of pastoral contacts with people?

3.　How do I look forward to priesthood? Is there a fear in me, articulate or hidden, of personal frustration, of tensions, of pastoral failure, of aloneness in a celibate life? What are the hopes and aspirations to which I am looking forward?

Such an assessment is important: it clarifies the outlook on the future life; it also gives some assurance. When one day actual problems do come up, they are not simply unexpected if the life-situation has been, as far as possible, realistically foreseen and accepted earlier. Analogous assessments should be made in other life situations.

In a second step the retreatant is asked to identify and, as far as possible, to spell out clearly his 'vocation within the vocation', i.e. the concrete way in which his life ought to be lived. The answer to this question obviously includes the acceptance of the general norms and demands of his/her state of life—they are the frame of reference within which his/her personal vocation must be lived. But it must be lived in a personal manner corresponding to his/her individual qualities, to the needs and the situation of the surrounding world, and to inner inspirations.

The Biblical Pattern

The meditations of this day may follow again a biblical pattern as it actually is suggested by St. Ignatius himself (n. 138): he sees Jesus' hidden life as a life of "observing the commandments", i.e. as conformity to existing structures, but there is also Jesus who, at the age of twelve, unexpectedly remains in the temple—his personal response to God's call.

1.　*Jesus' Life in Nazareth*　(Lk 2,39.51f).

The texts of Luke contain nothing of a personal biography of Jesus. In their stereotype paleness they express best what these years really meant for Jesus: his insertion into our human conditions and limitations as a simple village boy. We call these years Jesus' hidden life in contrast to the public ministry. But this very word "hidden" could be misleading as if Jesus had something to hide and lived this life as a kind of role play—'as if' he were an ordinary boy. He *was* an ordinary boy, growing up within his small world in rural surroundings, in the religious traditions of his people.

Many of the parables of his later preaching stem from the rural world of his youth and reflect the early impressions of his boyhood: fields and vineyards, the seeds sown in the furrows and growing up along with the tares, rain, sunshine, harvest feasts, all become illustrations of his message. He is familiar also with the close-by lake, with the fishermen and their trade. He knows the people in the village, the children, the poor people who cannot get justice in the court, parents who see their children run away, burglars who break into houses. He is part of this world, obedient to his parents.

In this world "he grew up and became strong, filled with wisdom." He awakens to the social and religious situation which forms the background of his later teaching. He felt the lot of the poor people who were marginalized, the contempt in which they were held by the religious elite. He becomes aware of the divisions among the people, their judgmental attitudes, the lack of forgiveness and brotherhood, the emptiness of much of their religious ritual. He also witnesses the political unrest and the resistance against the Roman rule which occasionally flared up in armed revolt.

In this world of striking contrasts—rural beauty and peace, human poverty, oppression, helplessness and messianic hope—Jesus' vision of God's reign develops. He becomes himself. We have no records of this time of his growth, his prayer, his waiting and patience. The only thing we know is his anonymous identity with the world into which he is born and his obedient acceptance of it.

2. *Jesus' Personal vocation* (Lk 2, 41-50).

There is, however, one gospel account in which the personal and unique vocation of Jesus breaks through. This is an anticipation of what he is to be, of what grows in him during the silent years of anonymity. It is the account of Jesus at twelve years of age in the temple.

The age of twelve is significant for the Jews. It marks the beginning of personal responsibility in the observation of the law. It is also the beginning of the teens, the age of the awakening personality and self-assertion. How will Jesus assert his personal life?

Jesus is the Son of God, sent by the Father on his unique mission to reveal God's love and proclaim his reign. This call awakens in Jesus during the silent years of Nazareth. Luke's description of this awakening is full of deep insights into an adolescent mind. It is a teenage story with all the beauty and pain of this age.

"They went up (to Jerusalem) according to the custom": It all takes place in the setting of a customary pilgrimage, which, however, expresses something of the unique vocation of Israel. The centre of Jewish life is not the village but the temple, which is the visible symbol of God's presence among his people. Jesus sings the same hymns of pilgrimage as the other boys (PPss 120-134), but in his voice they have a new ring. He takes part in the same ritual of the paschal meal as all the people, but his heart is full of God's mystery. What does all this mean, these ancient rituals and traditions?

"He stayed behind": he must remain in this city. It is his home, the house of his Father. Here he must realize more and more the mystery of God and of his people. This is the world to which he really belongs.

"The parents looked for him among their kinsfolk and acquaintances": Mary and Joseph remain in the world of traditions, and expect their boy also to be there. They must realize that Jesus no longer belongs to them.

"They found him in the temple, sitting among the teachers, listening to them and asking them questions": these are the keywords of adolescence: he listened to the traditions which struck him with such power on this great feast of the passover. He listened to their explanations and comments. But these comments do not fathom the depth of the mystery.

What does God really want of his people? This boy has so many searching questions. What is going on in him? "And all who heard him were amazed at his understanding and his answers".

The parents find their son. Two worlds are meeting: Mary and Joseph who knew Jesus only as the boy of the small world of Nazareth. "They were astonished": they feel the pain which the child has inflicted on them. They still address him as *their* son: "Son, why have you done this?"

Jesus' answer seems to come from an infinite distance, from another world which had been growing in him: "Did you not know that I must be in my Father's house?" This is his real life, it is his future. His loyalty belongs not to his earthly family, it belongs only to his Father.

"They did not understand the saying": how could they? It is the pain of parents who see their child growing towards his own future. Jesus awakens to his mission. It sets him apart from all people, even his own family. They returned to Nazareth and "Jesus was obedient to them." He remains in the traditional setting of his life till his hour comes. But his life is sealed with his own destiny.

In Jesus the adolescent I may recognise the awakening of my own personal vocation, the name by which God called me, my true self. It is the aim of today's prayer that I may find it. No one can give it to me; it is hidden in the depth of my heart. Many may not understand it. But only if I respond to God's personal call shall I be able to fulfil my mission also for the people.

Short Points For Meditation

1. *"Behold, I come"* (Hebr. 10,5-7).

The letter to the Hebrews shows Jesus as the fulfilment of the religious traditions of the Jews. In the old covenant Yahweh had called Israel into a personal communion with himself: "I have called you by name, you are mine" (Is 43,1) and "you shall love the Lord your God with all your heart" (Deut 6,5). The covenant had been institutionalized, embodied in laws and rituals. At Jesus' time the code of rules had for many become more important than the love of God; it had become a substitute for the personal surrender to God in a new life of righteousness and justice.

In Jesus, in the New Covenant, the original meaning of Israel's election is renewed and fulfilled. It is not based on rituals but on the personal self-gift to God realized in his life and death. So Jesus becomes the pattern of our life. The text from the letter to the Hebrews is a paraphrase of Ps 40. The Psalm becomes Jesus' morning prayer in which the entire meaning of his life and mission is summed up.

— "When Christ came into the world, he said: Sacrifices and offerings you have not desired": rituals are symbols, which have meaning only as long as they represent and express the personal self-gift to God, otherwise they are empty. They can become a mockery if they are turned into a substitute for the personal surrender.

— "A body you have prepared for me": The biblical meaning of 'body' is the entire person, his life, his relation to the world, his contacts with the people, his task and mission, his struggles, conflicts, rejections. God has given Jesus his body, his concrete human life. In this body, in his human existence he belongs to God, comes to God.

— "I have come to do your will, O God": this is Jesus' life, all that he is from birth to his end on the cross, his deeds, his speaking, his invitations, rejections, failures, all this is his body, his coming to God: "I have come". Jesus is the model of the Christian vocation.

— "I appeal to you...to present your bodies as a living sacrifice holy and acceptable to God, which is spiritual worship" (Rom 12,1). This is the Christian vocation in Jesus Christ. It is never to be diverted into exterior actions or observances. The one life, the body, which God has given to me, is the only gift that God wants of me. It is the searching question of the Spiritual Exercises to find and to accept my life and to offer it to God.

2. *How to find my personal vocation*

This is a process of growing inner awareness. It may be helped:

— by retracing the divine guidance in my life, the growing understanding of the lights and experiences of past retreats;

— by inspirations coming from the world in which I live and move: the encounter with people, inspiring books, challenging happenings in my surroundings and in the wider world, the actual situations and responsibilities I have to face;

— by the inner inspirations during these Spiritual Exercises; compare them with past experiences and share them with the director of your retreat. It is in sharing that vague perceptions become more articulate;

— by prayer: only in God's presence, in the silent attentiveness to his inner guidance can I find my true self, the vocation within my vocation.

OTHER READINGS

— Col 3, 12-25: God's will in ordinary life.

— Jer 1, 1-10; 1 Sam 3: (or other vocation accounts) The personal call.

QUESTIONS

1. As Jesus was obedient and lived under the law, do I recognize God's will in my ordinary life: surroundings, companions, occupations, responsibilities, etc.?

2. Have I become aware of my own unique, personal call, the vocation within my vocation?

FOR REFLECTION:

A Prayer of Abandonment:

"Father, I abandon myself into your hands, do with me what you will. Whatever you may do, I thank you, I am ready for all, I accept all.

Let only your will be done in me and in all creatures. I wish no more than this, O Lord.

Into your hands, O Lord, I commend my soul. I offer it to you with all the love of my heart, for I love you, Lord, and so must give myself, surrender myself into your hands without reserve and with boundless confidence, for you are my Father."

CHARLES DE FOUCAULD

Chapter Six

THE CONFLICT

*"Do you think that I have come to give peace on earth?
No, I tell you, but rather division"* (*Lk 12-51*).

Such words sound hard and seem to belie the Christmas
message of peace. What is peace, what is conflict in the bible?
The peace of Christmas, and once more the peace of Easter,
is not capitulation before the powers of darkness. Christian
peace does not avoid conflict. It is not pacifism at any price.
It is the divine assurance of final victory over all evil, not a
victory by power but a victory by love.

Jesus' entire life is a conflict, a relentless struggle against
the powers of darkness, of evils which are entrenched in human
society and have their ally in every human heart. It is a
conflict in which Jesus seems to go down and to suffer defeat,
and still it ends in the glory of Easter. All authentic disciples
of Jesus have to share in this conflict and will be with him in
his final victory: "He who conquers, I will grant him to sit
with me on my throne, as I myself conquered and sat down
with my Father on his throne" (Rev. 3,21).

John's gospel sees the entire mission of Jesus as a struggle
against the powers of evil. The world has lost its original
beauty in which it was created; John sees it closed against the
light. Jesus is God's light entering into the darkness of the
world, but the world resists: "The light came into the world
and men loved darkness rather than light because their deeds

were evil" (Jn 3.19). Still, light remains victorious: "The light shines in the darkness, and the darkness has not overcome it" (Jn 1,5).

The synoptic gospels describe this struggle between light and darkness as it took place in Jesus' own life and unfolded in his mission. This conflict of Jesus is the pattern of every Christian life and must be faced by all disciples of Jesus.

This then is the theme of the meditations of this day: the conflict in Jesus' life, the temptation at the outset of his mission, which continues in his own life and in the life of the Church, in every human life. His conflict with the powers of darkness must be faced by all who wish to be his disciples.

Jesus' Temptation (Mt 4,1-11; Lk 4,1-13).

"Jesus was led by the spirit into the wilderness to be tempted by the devil."

The account of Jesus temptation, which in the synoptic gospels is put at the beginning of his public ministry, is not the report of a chronicler but the thematic anticipation of the conflict that pervades his entire life. The people expect a victorious Messiah who will restore Israel's glory. Will he fall in line with their dreams? This is his temptation that their hopes become his own dreams, alluring him to a life of success, glory, power.

Jesus was truly tempted. What the gospels tell us is not merely a pedagogical demonstration of how to face and resist temptations. Any such demonstration would be useless had not Jesus himself really been tempted. This is what the gospels tell us, and it has been repeated with special emphasis in the letter to the Hebrews: "Because he himself has suffered and been tempted he is able to help those who are tempted" (Hebr 2,18), and "he was tempted as we are yet without sinning" and so He is able "to sympathize with our weaknesses" (Hebr 4,15).

Jesus was tempted "by the devil". This seems puzzling. His temptations are very human. In all expectations of the

people the Messiah is seen as a ictorious leader, who would
be concerned with the actual needs of the people; he would
be acclaimed and he would wield power over the enemies. Is
this wrong? Why should Jesus' mission be different and
lead him to ruin?

The human heart is the battlefield between light and
darkness, love and selfishness, God and Satan. Sin is selfishness. It finds its expression in every sphere of human life.
Each one is engaged in building up his own world and tries
to expand it: comfort, reputation, influence, power over others.
It all seems so human, yet it is in this human sphere that light
comes into conflict with darkness, God's reign with the powers
of evil. It is Jesus' mission to proclaim God's reign, the law
of love against selfishness, of self-gift against possessive power.
He has to carry his message into every sphere of human life;
wherever his message is preached, there will be conflict.

With this message Jesus stands alone. All will love his
vision of a new world but they will not be ready to change
their hearts. Will Jesus be able to hold out against the common
trend, against public opinion, accepted standards and values?
It would be so much easier for him to be "realistic", to conform with the thinking of the people and speak their language.
They would welcome him and make him their leader. But
Jesus is on a mission from his Father, he must be different
and offer a real alternative, an altogether new outlook on
world and life. Will he be able to do this, all his life, even when
they reject him and put him to death?

Thus Jesus' entire life is the persistent struggle against the
temptation to conform himself to his surroundings and to
make a compromise with the expectations of the people, to
surrender to the attractions and powers of selfishness which
have a hold on every heart and so to give recognition to
Satan's reign.

"If you are the Son of God"

This is tempting language: can there be anything greater,
more assuring than to be Son of God? The Father's voice
from heaven had recognized Jesus as "my beloved Son".

The world seems to lie open before him: God will make him great. "If you are God's Son..." It is a word that arouses dreams of a future of glory and power; Jesus will be the messianic king as so many Jews expected him.

For Jesus "Son of God" has an entirely different meaning. For him God the Father is love, intimate communion, but also total authority to whom all life belongs: "I am the Lord, your God...you shall have no others gods before me" (Ex 20.2f) and "the Lord our God is one Lord and you shall love the Lord your God with all your heart" (Deut 6,4). At last the people have to understand it: this is what Jesus has to proclaim, the total, all embracing reign of God. He will not speak of a God in the high heavens but of his love which pervades and transforms our world. All other powers—possessive, oppressive, divisive, destructive—will be overcome. There will be a new community of brotherhood, of love and sharing; this is God's reign. It will be Jesus' mission to proclaim and live this new life and to bear the conflict and the rejection on the part of the people, because they will feel threatened. Many speak about God and dream of being God's children, but what they expect from God is an increase of their own comfort, glory and power. Will they ever understand what God's reign really means?

"Command these stones to become loaves of bread"

Bread is our basic need. Jesus knows it after the forty days of prayer and fasting. If God is Father, he must give us *bread* as he did to the fathers in the desert.

Now the dream begins, the temptation: Jesus is Son of God—does this not mean unlimited power? Cannot he himself do what God himself did in the past to the people who were in distress, tortured by hunger? Jesus sees himself among the people who are hungry, far away from their villages. They trusted him, followed him into deserted places, and now they are tired and need food. He sees himself in their midst, he will make them sit down in groups and then take the few loaves which are there and have them distributed by the disciples. They will all eat and be satisfied; and they will wonder, their enthusiasm will rise. They will acclaim him king, thousands are there, and many more will follow and swell their ranks, and he will be the Messiah King, in power!

The dream goes on: they will believe in him, they will accept him as sent by God. But they will come to know that he is from that village—can anything good come from Nazareth? No, the Messiah must come in *glory*. They actually expect him to come from heaven. He sees himself on the pinnacle of the temple; below him are the masses of the people coming from villages, coming from distant countries, from all nations. Again the voice speaks: "Throw yourself down! Are you not God's Son? Angels will come and catch you, and the people will acclaim you." This is what they need — the Roman Emperors, all great kings knew it—bread to eat and spectacles, something for the eye to see, and for the imagination to feed upon. He will be carried by the enthusiasm of the masses, and then his reign will begin. This is the ancient promise that David's kingdom will be restored, *he will rule*. Not only this small country will bow before him, but all the nations, the kingdoms on the earth!

There is once more the tempting voice: "All this I will give you". From powerful countries and distant islands they will come and bring their gifts to the Holy City; Jerusalem will be the centre of the world—is it not all written in the sacred books of the prophets? Dreams are fascinating, you cannot get rid of them and you do not want them to disappear; king over the nations! There is only one condition. . . .

"If you fall down and worship me"

This is the real issue: *Who will rule?* Will Jesus proclaim and establish his own kingdom, a human kingdom, on the same basic pattern of the great empires as they come and go? All of them are built on the same foundation: bread, glory and power. Or, will he inaugurate God's reign, the everlasting kingdom, the beginning of a new creation? Once more the same issue is at stake as in the beginning of our race: is God the centre and goal of all things, or will men usurp his place? "You will be like God". This is what all people through the ages try to become by building their own kingdom in their own way. Most of them have to do it on a small scale within their little world, others succeed in extending it to ever wider dimensions. But the driving force and the root is the same: the human need for bread, the longing for glory and social status, the urge to dominate, the desire for power. Is this merely human? Or is it once again the serpent of the paradise, evoking the temptation of human autonomy, the revolt against God?

"You shall worship the Lord your God, and him alone shall you serve"

This is Jesus' answer. He is sent to proclaim, and bring into the world, the reign of God; he remains faithful to this mission to the end of his life. It will happen, as in the vision of his temptation, that he will indeed give bread to the hungry. And they will inevitably be aroused by his love and power and will want to make him king by force. The disciples will be thrilled by this movement. What will then happen? Jesus will send the masses away, order his disciples into a boat, and he himself will withdraw to a lonely place to pray (Jn 6,14f; Mk 6,45f). On the following day, when he meets the people once more in the synagogue of Capernaum, he will speak to them not of bread that fills the stomach but of the bread of life, and they will be disappointed and drift away. What finally can Jesus bring them?

His message is of God's reign. Not like human empires will it be based on bread, on economic, social and political foundations, but on God's word, on his love and power: "Man shall not live on bread alone but on every word that proceeds from the mouth of God." God's word can change our hearts, renew our society, make us brothers and sisters. This will be possible if we renounce our selfish sovereignty by which we try to influence and manipulate our neighbour and even our God. This is what we do when we tempt the Lord and try to make him subservient to our selfish schemes: "You shall not tempt the Lord your God." We must understand the meaning of God's reign, its contrast to our thinking and planning. Jesus will be the "Son of God", not as people expect and Satan suggests as though "Son of God" were a title of glory and power for himself. Jesus will be the Son of his Father who proclaims God's reign and reveals his love to save and renew the world: "You shall worship the Lord your God, and him alone shall you serve."

SHORT POINTS FOR MEDITATION

1. *Jesus' conflict with the disciples* (Mk 8,27-38).

The scene is the Galilean crisis, the period in Jesus' life when it becomes clear that Israel's leaders refused to accept him and his message. He withdraws outside the country near Caesarea Philippi. It is a decisive hour for the disciples. In your meditation live this hour, which is decisive also for you, as a personal encounter with Jesus.

— "Who do men say that I am?": the disciples are invited to assess the reaction of the people to Jesus. What do people say today about him?

— "Who do *you* say I am?": this is the personal question addressed to every disciple—addressed to me.

— "You are the Christ!": Peter recognizes Jesus as Messiah through whom God's reign is revealed and inaugurated. Speaking in the name of the other disciples, Peter stands for Jesus in this hour of crisis. But for Peter "Messiah" still means earthly king. Thus Jesus has to make it clear what his mission implies:

— "The Son of Man must suffer": Faith in Jesus implies acceptance of the conflict and rejection which are part of his mission. His final victory will not be won by power but through his death and resurrection.

— "Peter began to rebuke him": the idea of a suffering Messiah is totally inconceivable for Jewish thinking; it contradicts all their expectations and hopes.

— "Get behind me, Satan. You are thinking not God's thoughts but human thoughts": For Jesus Peter's words ring once more with the voice of the tempter in the desert.

— "He called the multitude": Jesus wants no ambiguity:

followers must know that his own way of the cross will be the pattern of all Christian life.

— "If any man would come after me, let him deny himself and take up his cross and follow me": Jesus' disciples will share in his conflict, rejection and cross, to have part in his saving mission.

— "Whoever would save his life will lose it; whoever will lose his life for my sake will save it": all must find life. Jesus came to bring it in its fulness. But those who try to find it in the narrowness of their own world and interests will lose it; those who learn to surrender it in love to God, and to spend it in service to their neighbour, will find it. This is the paradox which faces all followers of Jesus.

2. *Two Banners*   (Spiritual Exercises nn. 136-148)

Ignatius presents the conflict between God's reign and the powers of darkness in the meditation on the two banners. As other texts in the Spiritual Exercises, so also this meditation is framed in the feudal setting of his time. However, the conflict described in it concerns every person. Read it in the original text.

The entire world is divided into two camps: Jesus in Jerusalem, the city-symbol of God's reign, and Lucifer in Babylon, the city-symbol of the power of darkness. These are the two cities which Augustine described as the rallying centres of good and evil, "the city of God built on God's love, the city of Satan built on self-love". The dividing line between the two cities runs not through the visible world — there is no black and white demarcation running through our history or society—it runs invisibly through the whole world, also through the Church, through every human heart, wherever God's invitation and the allurements of Satan conflict with one another.

Satan sends his messengers into every corner of the world. His threefold program is to lead people "to covet riches...to attainment of honours...to come to pride:" his allurement consists of wealth, glory, power and pride.

Jesus too sends his messengers into the entire world to invite all people to spiritual and actual poverty, to readiness to suffer insults and humiliation, and thus to come to humility, which "leads to all other virtues". These are the three steps in contrast to Satan's triple allurement: poverty, insults and humility as against riches, honours and pride.

It is the special prayer of this meditation to obtain "knowledge of the deceits of the rebel chief and help to guard myself against them", and a "knowledge of the true life" exemplified in Jesus.

The fruit of these meditations should be a deeper and clearer insight into the movements of my own heart, the orientations and values which influence my life, my relations to others and my work. All spheres of my life are scenes of the conflict between love and selfishness, ultimately between God and Satan; they ought to be transformed by the Spirit of Jesus.

OTHER READINGS:
— Mt 7, 15-20:   False prophets.
— Mt 13, 24-30:   Tares and wheat.
— Eph 6, 10-18:   The spiritual armour.

QUESTIONS:
1. What are the areas of conflict, open or hidden, between the spirit of Jesus and the allurement of evil in your surroundings? Do you find this conflict in your own heart?
2. Have you experienced the isolation that results from your determined stand in conflict sitations?
3. Reflect on your understanding of peace and conflict: there is *false* peace based on ignorance and blindness in face of evil; of laziness which does not want to be disturbed; of cowardice which is afraid of conflict and commitment. *True* peace is the assurance of God's guidance and power in the struggle against evil.

There is a *false* conflict — the world is full of it. It is based on selfishness, oppression in its many forms. There is the *true* conflict between light and darkness, true love and selfishness.

Chapter Seven

JESUS ALONE

Three Degrees of Humility Sp. Ex. nn. 165-168.

"For his sake I have suffered the loss of all things and count them as refuse in order that I may gain Christ and be found in him... that I may know him and the power of his resurrection and may share in his sufferings becoming like him in his death that, if possible, I may attain the resurrection from the dead" (*Phil 3, 8-11*).

The second week of the Spiritual Exercises is meant to lead the retreatant to the threshold of an option for his life. Where is the norm for this option? Is there an ultimate criterion by which human decisions can be guided, which help us to assess the happenings around us and to discern also the movements and trends in the life of the Church? We have the divine commandments; we find norms and guidelines in the gospels and in Jesus' own words; for more immediate guidance the Church has formulated her laws and religious communities have worked out their Constitutions. But laws and norms are general, are for everyone, and still leave the most personal decisions to each one's free option, e.g. to get married or to remain single, to become a priest, to join a religious community, etc. Besides, laws are elusive and can easily be manipulated according to each one's likings. Is there a workable norm?

God's ultimate word to us does not consist in laws; it is not even fully contained in Jesus' message and teaching. God has spoken in his Son: Jesus himself, his person, his life, his proclamation and struggle, his rejection and death and finally his resurrection are God's final word to us. In Jesus, his Son, God shows us the meaning of our human existence, the destiny of man and the world. It is for each generation, for every culture, in ever changing situations that we have to grasp the meaning of this revelation and to translate it into the language of our time. Each person has to find for himself/herself what Jesus invites him/her to be, to become. The Spiritual Exercises are a guide in this search. The meditations of this day lead us to the ultimate norm in our options, to the person of Jesus.

Ignatius puts this theme into a wider context: he presents the norms for our options in three stages, which he calls three degrees of humility. Humility for Ignatius has nothing to do with self-abasement, with feelings of inferiority or self-destructive submissiveness. Humility here means the ready and joyful acceptance of my unique place in the world and in God's designs. The whole of the Spiritual Exercises is ordained to this goal, to find and accept this God-ordained place.

The first stage of humility consists in the radical acceptance of God's design as expressed in creation and its order, and as contained in his commandments. In the traditional language used by Ignatius this implies the firm commitment under no circumstances to commit mortal sin, i.e. never to break away from God and the divinely established order, even when this commitment would mean the risk of one's physical life. God is recognised as centre and goal of creation.

The integration into God's designs is deepened in the **second stage of humility.** Our life is woven into the context of world and history. We cannot choose place and circumstances of our life, the family, the nation into which we are born; the gifts and supports, or the blocks and burdens that affect our life and work. It is for us obediently and joyfully, with inner freedom, to accept situations and influences which are beyond our control, whether they favour our growth or seem to hinder it. We must keep this inner freedom also in planning our future: the options which lie before us should not be decided on the basis of merely personal enrichment, e.g. by the desire for wealth and comfort, for social prestige or any other consideration which would reduce them to a level of pure pragmatism and self-interest. Our only desire should be to serve God, trusting in his wisdom, power and love to give us what is best for us.

This second degree of humility is described by Ignatius also as a firm determination under no condition "to consent to commit venial sin". Venial sin implies an ambiguity, a compromise, as we experience it day by day. On the one hand we wish to remain within the frame of God's will and plan; God is not rejected. On the other hand there is the human weakness to follow our natural attractions and impulses. Thus the second degree of humility means the readiness and determination for the total integration of our life and work into God's designs. Could there still be a higher degree of humility, a still more radical commitment to God's will and love?

The third degree of humility.　In fact, this second level with its total surrender to God in detachment of all personal desires would be the perfect response to God's will had God not already spoken to us in Jesus Christ. But in his Son, God has given us the concrete form in which his reign on earth is to be established. Jesus comes as Saviour into a world which is broken and God's reign must be realized in the midst of human autonomy, which includes self-assertion, competitiveness, exploitation, and self-righteousness. Jesus' message was rejected in his time and so it will be rejected also today. He had no weapons to assert or defend himself; neither had he material resources nor social or political power. He had only the good news of God's reign. We too have to learn not to rely on institutional power in the proclamation of Jesus' message. Jesus was, outwardly at least, defeated and ridiculed. Under the cross everyone laughed about this king who has no power; everybody could see that his message did not work in actual life. In the same way his disciples are also ridiculed: "If the world hates you, know that it has hated me before it hated you" (Jn 15, 18).

A glance at history tells us that the Church did not always follow the path of Jesus. Often in her work and mission she has relied on social, political, economic power. Today, however, she has recognized that in proclaiming the Good News we have to follow the path which has been shown to us by Jesus who "carried out the work of redemption in poverty and oppression so the Church is called to follow the same path...Although the Church needs human resources to carry out her mission, she is not set up to seek earthly glory, but to proclaim, and this by her own example, humility and self-denial" (Lumen Gentium n. 8). "Urged by the Spirit of Christ, the Church must walk the road Christ himself walked, a way of poverty and obedience, of service and self-sacrifice even to death, a death from which he emerged victorious by his resurrection" (Mission Decree n. 5).

It will be hard to follow this path. Jesus, too, was afraid of suffering and death and "offered up prayers and supplications with loud cries and tears to him who was able to save him from death" (Hebr 5, 7). As to ourselves, no one expects us to love pain and death. Still, we know that in suffering we are close to Jesus, similar to him, united in his saving passion as his true disciples. So it may be possible, as Ignatius suggests to "desire and choose poverty with Christ poor rather than riches; insults with Christ loaded with them rather than honours; to be accounted worthless and a fool for Christ rather than to be esteemed as wise and prudent in this world, in order to imitate and be in reality more like Christ our Lord" (n. 167).

Thus there is an ultimate norm for our decisions: God has spoken in Jesus, his Son. His life of love, his struggle against injustice and sin, his rejection and his passion are the model in which his disciples see their own life and mission.

SHORT POINTS FOR MEDITATION

1. *The Transfiguration* (Mk 9, 2-8; Mt 17, 1-8; Lk 9,28-36).

All Synoptics link the Transfiguration with the preceding scene in which Jesus speaks about the passion and Peter rebukes him. Once more Jesus is tempted and rejects Peter: "Get behind me, Satan." Jesus cannot withdraw

his words about the passion and the need for all his disciples to follow him in carrying the cross; but Peter does
not come any closer to an understanding of the mystery
of suffering. There are things which cannot be explained
in words. How will Jesus resolve the impasse?

This is the context of the scene in the gospels. The
account itself breaks through the frame of Jesus' earthly
life and reflects the post-Easter experience of the disciples.
It is placed into this context because the readers of the
gospels, the Christian community, must see Jesus' passion
in the light of Easter, of fulfilment. They must listen to
the whole Christ, who suffered, who is risen.

— "After six days": Jesus' passion and glory are
inseparably connected, yet distinct like question and
answer. Often we seem to be left alone with our
questions; Peter had to wait six days.

— "Jesus took Peter and James and John with him and
led them up a high mountain apart by themselves":
the answer to this problem must be very personal.
Each one is called by name to come apart from the
crowd, from the pressure of daily surroundings. In
aloneness, in God's presence, on a high mountain,
God's Spirit teaches and transforms human hearts.

— "He was transfigured before them": God speaks
in Jesus. In their daily experience the disciples—and
all of us—encounter only the earthly reality, the
inscrutable world, the powers of darkness and death.
Faith sees more: Jesus is transfigured before their
eyes. The light of eternity breaks into our night.

— "Elijah and Moses appeared to them": they are the
representatives of the old covenant, the law and the
prophets. Jesus' coming and message are not an
isolated event but the climax of God's revelation
in the past; he is the new and final covenant. He
sums up the voice of the prophets and brings the
law to fulfilment in his death and resurrection.

— "Master, it is well that we are here": This is once
more Peter, who had not understood Jesus' words

of the passion. He sees the glorious Lord who comes to life through death, who is the promise of our glory. God can give us hours when the dimness of our faith is filled with the light of God.

— "And a cloud overshadowed them": The cloud is the sign of God's presence, before him we become silent.

— "This is my beloved Son, listen to him": God speaks, as at the baptism in the Jordan, in words of the prophet (Is 42,1). In Jesus, his Son, God speaks to us. It is for us to listen.

— "They saw no longer anyone but Jesus alone": they return to their normal surroundings. There is Jesus again in his everyday appearance. Life with him goes on along the same path leading towards Calvary. Nothing has really changed, not even their minds have been enlightened. Still, the paralysing spell is broken. They know that Jesus is more than what the eye can see. They continue to follow him.

The fruit of this meditation shall be the longing for a deeper union with Jesus. Growth in the spirit does not consist in intellectual insights but in the growing closeness to Jesus' person; in him I shall find the mystery of his life and of my own life.

2. **The Experience of an Apostle** (Phil 1, 12-2, 5).

St. Paul writes to the Philippians in a situation of utter insecurity. He is in prison, under trial. Will he be condemned to death? Will he be freed? What will happen to the young communities in Corinth, in Philippi? In reading these texts in a meditative manner, try to enter into the mind and heart of a man who knows human experiences of joy, of disappointment, of anxiety, but is deeply anchored in the peace and assurance of his faith in Jesus Christ.

— "What has happened to me has served to advance the gospel" (1, 12-14): his personal condition as a prisoner, the limitation of his freedom and the embarassment from his situation do not matter to him, if only Jesus Christ is preached.

— "Only that in every way, whether in pretence or in truth, Christ is proclaimed; in this I rejoice" (1,15-18): he is faced with the jealousy of other "apostles" who take advantage of his predicament to advance their own status and influence. Paul is beyond such rivalries; his only concern is the gospel of Jesus Christ.

— "Always Christ will be honoured in my body, whether by life or by death" (1, 19-26): he cannot foresee the outcome of his trial. On the one hand he desires "to depart and be with Christ, this is far better". At the same time he knows that "to remain in the flesh is more necessary on your account". He feels that he will remain, and his return to the community will give "ample cause to glory in Christ Jesus".

— "Let your manner of life be worthy of the gospel of Christ" (1, 27-30): the unity, firmness, confidence, readiness to suffer will be a witness to Christ.

— "Have this mind among yourselves which was in Christ Jesus" (2, 1-11): The community problems of Philippi, discord, ambition, rivalry are solved in the Spirit of Jesus, for which he quotes an ancient Christological hymn.

In reading such texts we may learn to see the concrete problems and challenges of our own life, big and small, in the new light of Jesus. We will find in him, in his person, the norm for our attitudes and decisions. The Christian life is based not on a code of rules or a set of principles but on the person of Jesus Christ. Paul sums up this supreme norm of Christian living: "Be imitators of me as I am of Christ" (1 Cor 11, 1).

Similarly, Peter teaches the faithful to look at Jesus, particularly in times of trial because "Christ also suffered for you giving you an example that you should follow in his steps" (1 Pet 2, 21). Jesus himself points to his example when he finds his disciples divided by jealousy and ambition: "The Son of man came not to be served but to serve and to give his life as a ransom for many" (Mk 10,45).

OTHER READINGS:

— Eph 3, 14-21: That Christ may live through faith in your hearts.

— Phil 3, 1-14: Jesus Christ is the only support of the Apostle's life.

— Hebr 12, 1-11: Look at Jesus as the model of faith.

QUESTIONS:

1. What are the criteria by which we judge the importance of events and enterprises in our surroundings, civic and religious? What are the criteria for making our own decisions?

2. Reflect on ways by which you learn to make Jesus the norm of your life and work. Exterior imitation is meaningless, because he lived in altogether different conditions; how can your life be guided and formed by his Spirit?

FOR REFLECTION:

When Paul VI addressed the Vatican Council for the first time, after having been elected Pope, he gave this basic orientation: "Christ is our beginning, Christ our leader and our life, Christ our end...The living and holy Church, i.e. we ourselves, are linked to Christ from whom we begin, through whom we live, towards whom we move...Our minds should not look for any other truth but the words of the Lord who is our daily teacher. At nothing else should we aim but in faithful obedience to follow his commands. No other confidence should support us but the one that strengthens our weakness when we trust his words: I shall be with you to the end of time."

Chapter Eight

LIFE ACCORDING TO THE GOSPEL

"Blessed are your eyes, for they see, and your ears for they hear. Truly I say to you, many prophets and righteous men longed to see what you see and did not see it, and to hear what you hear and did not hear it" (Mt 13, 16f).

The second week of the Spiritual Exercises has the twofold aim: through the meditations on Jesus' life and mission, (1) to lead to a deeper knowledge of Jesus and to a firmer commitment to him: and (2) to apply the Christic pattern to the life and work of the retreatant who has to find his personal vocation and must be prepared for the conflicts that are implied in the discipleship of the Master.

Some of the mysteries of Jesus' life are given special emphasis in the Spiritual Exercises, for example, the Incarnation and the Nativity. For the remaining meditations of this week various topics are proposed (n. 161); these are spelled out briefly at the end of the book (nn. 211-288). Much freedom is left in their use (n. 262). The purpose of such meditations is not to cover the entire life of Jesus but to deepen the personal contact of the retreatant with Jesus the Lord. In most of these meditations we use the Bible to come to a fuller understanding of the person of Jesus and of our own life. Thus this day is devoted to a more general reflection on the bible and its place in Christian life.

The Bible in our Spiritual life

Already before the Council, and even more in its wake,

the Church has experienced a renewal in the use of the Bible. The biblical texts in the liturgical celebration of the Eucharist and in the Divine Office are now read in the major vernaculars. Thus the word of God is understood and becomes food for the Christian Life. At last it has come true what Vatican II proclaims that in the liturgy the Church "offers to the faithful to partake from the one table of the word of God and the body of Christ" cn. 21). Besides, the liturgy both of the Eucharist and the hours has been enriched through many more texts of the Old and New Testament. To make the faithful more familiar with Scripture, courses on various levels are being organised. Prayer groups have based their spirituality on the word of God.

However, this renewed interest in the Bible has made us more keenly aware of the *problems* of its proper use. On this day of the Spiritual Exercises we reflect on these problems, on the meaning of Holy Scripture in our life, prayer and work.

The most tangible problem in the study of the Bible is the vast distance of the biblical world from our contemporary surroundings. We are led into foreign lands to people of ancient times. This is the case not only with regards the Old Testament but also with the Gospels and the letters of St. Paul which refer to concrete conditions of Christian communities which are not familiar to us today. The way of speaking and arguing, the entire symbolism of the language is often strange to us. Thus we need the continuous study of Scripture.

But these studies often lead into further problems. We can become engrossed in historical and literary problems, the historical facts which are narrated, the exact meaning of the texts. Obviously these investgations are important, because God spoke to us in the history of Israel and finally in the person of Jesus Christ. The study of these events, and the literary form in which they are transmitted to us, is fascinating and absorbing. Those who engage themselves in such a study with the intention of discovering in them the inexhaustable mysteries of divine wisdom and love will be amply rewarded.

There is, however, the danger of limiting or even diverting our interest from the core of the biblical message. The method of biblical research is adopted from secular historical research which aims mainly at the establishment of perceptible facts. Scripture also deals with facts but has a wider, more profound purpose: to find in and through the events of history God's designs for his people, for us. The biblical texts are given to us not simply to supply information about happenings in the past but to convey their significance. The Bible contains what God wants to tell us in Jesus Christ about our life in union with him.

How does the bible lead us to a personal knowledge of Jesus Christ? The importance and uniqueness of a person come to us mostly through his impact on the surroundings, on other people, on history. So it is with Jesus. Surely it was not in God's providence to give us mere factual information, which could easily lead us to superficial and passing impressions. The richness and depth of Jesus' message comes to us in a different way: through his disciples as they met him, through the impact he had on their lives and the way they shared his message with the nascent communities of believers. This message is contained in the books of the New Testament.

What Paul's letters teach us

The first writings come to us from Paul who had never met Jesus in his earthly life and so would seem a very inadequate witness. Yet he was called to be an Apostle. In fact he was the most powerful and influencial witness to Jesus Christ in the early Church. He encountered Jesus not in his earthly ministry in Galilee but as the Risen Lord. In him Paul found the answer to the deepest, most personal and disturbing problem of his life — of every human life: how can we find true peace and assurance in God? Paul calls it in his Jewish language "righteousness'. He had sought righteousness as a Pharisee through the observance of the Torah and the ancient traditions, but what he actually was striving for was *self-* righteousness, that sense of satisfaction and fulfilment that comes to us when we have done our duty. Paul had fulfilled his duties before God by observing the law and the ancient traditions. But can we human beings confidently stand before God with our own achievements? In the depth of our hearts is a law of sin and selfishness which is opposed to God. About this law of sin Paul would write later: "I know that nothing good dwells within me, that is in my flesh. I can will what is right, but I cannot do it. For I do not do the good I want, but the evil I do not want is what I do" (Rom 7, 18f). The law says that we must love God with our whole heart (Deut 6, 4) but we are always caught in the narrow circles of our selfishness.

Before his conversion Saul saw that the Christians were different. He probably knew only those Christians who were critical of the law, like Stephan, not the other group who under James' leadership clung faithfully to the traditions. These Christians formed a community of love and were free. They believed that God had revealed his unconditional love to us in Jesus Christ. If they respond to this love in true faith God's spirit renews and transforms their lives. The faith and the life of these Christians made Saul's own efforts to keep the law and the traditions meaningless and empty, almost ridiculous. He hated and persecuted them — until he himself encountered Jesus. Damascus was the turning point of his life. Now he knows that truly "God's justice has been manifested apart from the law:...the rightousness of God through faith in Jesus Christ for all who believe" (Rom 3, 21. f). Thus "if anyone is in Christ he is a new creation, the old has passed away, behold the new has come. All this is from God who through Christ reconciled us to himself" (2 Cor 5, 17f). Thus Paul does not find Jesus in the words and deeds of his earthly life; he finds him in the Risen Lord. God loves us and makes us a new creation in him.

Paul's letters are the earliest books of the New Testament. They convey his own experience of the crucified and risen Lord as the lifespring of the Christian community, of a renewed world. In this experience he writes to the various communities, he guides, exhorts and corrects them. He has no Canon Law at his disposal to guide him in settling problems about worship and discipline or to deal with scandals in the ranks of the believers. His only norm is Jesus Christ. This is what we learn from Paul: Jesus is the norm of our life, his Spirit must guide individuals and the churches. The life of the communities is not based on a set order. An order must gradually develop; but it must always be adapted to the changing situations. The norm of all legislation and organisation is Jesus Christ himself. Concrete instances are his reactions to the rivalries in the church of Philippi where Paul places Jesus' example of self-effacement before their eyes in form of the great Christological hymn. Christ's self-emptying is the pattern for mutual relationships (Phil 2, 1-11). Again there is Paul's angry disapproval of the disorderly way in which the Church of Corinth held its meetings. Here Paul puts before his Christians the example of Jesus' total self-gift during the last supper, on the night he was betrayed. (cf. 1 Cor 11, 17-34).

This, then, is the basic lesson we learn from Paul's letters: Christian life consists not in exterior observances but is based on a deep, personal awareness of Jesus' Spirit. "Be strengthened with might through his Spirit in the inner man, that Christ may dwell in your heart through faith and that you, rooted and grounded in love, may have the power to comprehend...the love of Christ which surpasses knowledge" (Eph 3, 17-19). None of the situations to which Paul addresses himself is simply repeated in our lives. We cannot copy him. Scripture is not meant as a collection of case studies for all times but presents to us a way of acting, reacting, deciding in ever new situations in the Spirit of Jesus, as Paul did. Similar reflections could be made concerning the other letters of the New Testament.

The significance of the gospels

Still, something more is needed. Paul himself felt the gap in his life of not having been a witness to Jesus' earthly ministry. The death and resurrection of Jesus must never be isolated from Jesus' whole life. Theology has often made the mistake of limiting its reflections to the Paschal mystery. Surely, Jesus, cross was the climax of his life, the supreme manifestation of his obedience to the Father and his love for us, and in the resurrection God put his seal on Jesus mission. The Paschal Mystery is the sum total of what God tells us in his Son.

But we need not only the climax, the sum total of God's revelation in Jesus, it must be spelled out. Thus from earliest times the Christian community preserved and narrated the details of Jesus' earthly life, his sayings, small episodes, the controversies, the miracle stories, the parables in which he explained God's reign. This actually corresponds to Jesus' own intention in the formation of his disciples. He gathered them into a group and lived with them in the way the rabbis used to share their life with their disciples, not merely teaching them but allowing them to witness their way of life, their dealings with people and their problems in various situations. So also Jesus' followers lived with him in daily life and allowed themselves to be influenced by him, his attitude towards the people, his reaction to the various groups in Israel's political and social life: the ruling group of the Sadducces, the government party of the Herodians, the religious renewal party of the Pharisees. So the mind of his apostolic community was tuned in, as it were, to Jesus' way of thinking and judging. This was a slow process of learning made up of many incidents and impressions which converged on a comprehensive, overall picture of Jesus' person in the actual context of his world. This precisely they had to learn: what God's reign means in actual life-situations.

The many incidents and words of Jesus, as they were preserved in the minds of the disciples, formed the content of the early catechesis, in the Christian communities. They were handed on and finally put together in the gospels as we have them today. But this was not done in a mechanical, factual way as we today might do it, keeping photos and tapes of events, in our archives. It was in a living tradition, always relating Jesus' words and deeds to the actual situation of the communities. Thus each of the gospels has its own perspective. Each is on the one hand an account of Jesus' life and message; each is at the same time an answer to the situation and needs of the people for whom it is composed, Jews, or Greeks, or a cosmopolitan hellenistic group. The purpose is always one, to build up the life of the faithful and of the communities. The Spirit of Jesus must be realized in the actual context of our world.

In this light we read the gospels today. In our liturgy the gospels are considered the most important section of Scripture, pointing to the person of Jesus in his relevance and challenge for us. We are meant to find in the gospels Jesus' message as it concerns the world in which we live and where we are meant to realize it as a beginning of God's reign.

Thus Bible texts are not simply God's word; they are the echo of his message brought about by the Holy Spirit in human hearts, in the Church. We have not only to hear or read them, but we must with discernment *translate* them into our life situations, which are very different from Palestinean or Hellenistic surroundings. Just as the same sun shines over all continents but brings forth a different and distinct climate and vegetation in each one, ranging from tropical exuberance to the stern aridity of the desert, so God is one, his message and reign are universal, but the gospel's actual realization differs from person to person, culture to culture, and from country to country. Through the Bible we learn to listen to God's word, to take it into our life and to be alert to its meaning and challenge today.

Reading the Bible with Humility

Thus the reading of the Bible opens our heart to God speaking to us in Jesus Christ. But we listen to him not as individuals but as members of the Church, of the community of believers. This is the way the Bible originated, not as the private initiative of individual writers but as the authentic expression of the faith of the apostolic communities, recognized as such by the universal Church. In this spirit it must be read and interpreted through the ages. There is the ever recurring danger that enthusiastic followers of Christ take the Bible as if it were exclusively their own and make themselves judges of its authentic meaning. This was the principle also of some forms of early Protestantism, which with cruel logic led to division within the Church and to the emergence of ever new groups and sects. God's word that should unite us, becomes the source of separation. Vatican II has gone deeply into the relation of the written word of God and the living tradition of the Church (De Verbum Ch. II) and sums up briefly: "The

Church has always regarded, and continues to regard the Scriptures, taken together with sacred tradition, as the supreme rule of her faith" (n. 21). No single book of the Bible can be isolated, no text can be taken out from the total context of scripture and by itself made the basis of faith. When we do this we make ourselves judges of God's word and message, we select what seems important to us and neglect what does not fit into our conceptions. Thus, God does not speak, but we speak using his word to support our human thinking.

The use of the Sacred Books in charismatic prayer groups is often beautiful in its spontaneous creativity and its alert readiness to listen to God's word. But this method too has its dangers. We have, to be sure, striking examples of the guidance God can give to individuals through the Bible: Francis of Assisi opened the Bible at a crossroad of his life and found the texts on poverty which became the inspiration of his life. St. Antony heard, according to ancient accounts, the gospel of the rich young man and felt shaken so deeply that he left the Church even before the Eucharist was ended and set out for a new life. Much inspiration and light come to us from the use of the Bible as source of help in our actual life situations. But we cannot consider this the ordinary way of God's guidance. It would be dangerous to consider the book of the Bible like a magic oracle. Too easily we begin to use texts to support our own ideas and desires, and finally we use biblical passages to defend and fortify our own opinions and decisions. It is the opposite of what God's word to us in Jesus Christ is meant to be: the beginning of God's own reign, a reign which cannot be manipulated.

We should reflect on this day of the Spiritual Exercises on our own use of the Bible, in daily life and in these days of retreat. It should help us to come to a deeper personal knowledge of Jesus Christ, to find our own life and personal vocation, and so conform ourselves and our work to Jesus, in the power of his Spirit who teaches us to discern the Scriptures' true meaning for our own lives.

SHORT POINTS FOR MEDITATION

The meditations of this day should help to put into

practice the proper way of using the Bible. It may be helpful to take texts with which we are familiar and to realize their threefold significance: to know Jesus, to find myself, to ordain my life in Jesus' spirit.

As example we propose here points on the Sermon of the Mount.

1.　*The Beatitudes* (Mt. 5, 1-10; Lk 6, 20-23).

Jesus proclaims and promises the unique gift: God's kingdom, the new life, the new creation of peace and fulfilment. It is not a theory, not a dream. He teaches what he lives. Luke gives the original, shorter form.

— "Yours is the kingdom of God": we allow the deep joy of this assurance to sink in. Then we ask: what on our part is required to receive this gift? Luke gives four conditions: You must feel the brokenness of our world, not in despair but in order to be ready for God's promise.

— "Blessed are you poor": the social and economic degradation of so many is not meant to last; they will see God's kingdom.

— "Blessed are you that hunger..." God does not want misery; all should have bread.

— "Blessed are you that weep.." God made us for joy; salvation means true life.

— "Blessed are you when men hate you.." this seems to be an addition of the early community which experiences God's rewarding love in the midst of hostile surroundings.

These four beatitudes are confirmed by the fourfold woe against the rich and self-satisfied.

Matthew broadens the beatitudes beyond the social frame, so that they include the general situation of Christians in the world: Blessed are the poor in spirit, etc.—He adds four beatitudes which express characteristic attitudes in God's kingdom:

— "Blessed are the meek": God's reign comes not as domination, with violence and self-assertion, but in love.

— "Blessed are the merciful": the compassionate, who are open to the need and pain of others, of the surrounding world.

— "Blessed are the pure in heart": in simplicity and sincerity, not double-minded and deceitful.

— "Blessed are the peacemakers": sharing God's peace with their neighbour, bringing reconciliation to conflicts, creating an atmosphere of acceptance and understanding.

2. Salt, light, city (Mt. 5, 13-16).

Jesus' proclamation of God's reign concerns the world. It consists in the renewal and transformation of human relations, human society.

— "You are the salt of the earth": God's reign is embodied in the disciples and must be realized in society and culture; it cannot remain in isolated seclusion, in a ghetto.

— "If salt has lost its taste...": God's reign is an alternative, a challenge to the standards and values of society. If the Church, the disciples, yield to conformism they lose the very meaning of their existence, their inspiration, their impact.

— "You are the light of the world": only God is light; Jesus is God's light shining in our darkness; we can only "walk in the light" (1 Jn 1,7). But here the disciples themselves are called the light of the world. They are meant to be transformed by Jesus, permeated by his Spirit.

— "A city set on a hill cannot be hid": the Church is to be shelter and assurance which only God can give, with open doors for all.

These words point far beyond the proclamation of Jesus. They express the experience and challenge of the young church in the fulfilment of her mission.

3. *Jesus and the Law* (Mt 5, 17-20)

This is a burning question in Jesus' own mission, as he is consistently attacked on account of his liberal interpretation of the Law. It is the problem also of the young church, the great controversy of the apostolic community in Jerusalem. Christians are free of the Judaic Law, yet face the danger of the misuse of freedom.

— "Think not that I came to abolish the law": the law gave orientation to the life of the people, united them in community, embodied a scale of values which cannot be given up.

— "but to fulfil it": it can deteriorate into exterior observance and become empty. Fulfilment consists in realizing its meaning, its spirit.

— "unless your righteousness exceeds that of the Scribes and Pharisees, you will never enter the kingdom of heaven": thus Jesus' disciples must surpass the most accurate observance of the law, because it is written in their hearts (cf Jer 31,33). They live it not as an imposition on them, but as their own life. They are reborn in the Spirit.

Read these and the other texts from the Sermon on the Mount not as general norms, but as addressed to *you*.

QUESTIONS:

1. How do I read the Bible? Has it become for me a personal link with the person of Jesus? Is it the norm of my actual life?

2. Recall passages which have special appeal to you. Why do you cherish them, what do they mean to you?

FOR REFLECTION:

J. RATZINGER describes the conversion of St. Teresa of Avila;

She joined a "modern convent" with relaxed rules of enclosure, open to visitors, with "reasonable" comfort... "But once she was touched interiorly by the presence of Christ, and the inexorable really of the gospel, free from all disguising phraseology, revealed itself to her mind, she saw in all this the unbearable flight from the real task, an escape from the conversion demanded of her. She got up and was converted. This means she threw away the 'aggiornamento' and created a renewal that was not a compromise but a demanding claim, to commit oneself to the eschatological expropriation of Jesus Christ, to allow oneself to be fully expropriated with the crucified Lord, in order to belong totally to him, to his body".

Chapter Nine

DISCIPLES OF JESUS

"You did not choose me, but I chose you and appointed you that you should go and bear fruit, and that your fruit should abide" (*Jn 15,16*).

This last day of the second week of the Spiritual Exercises is meant to sump up the reflections, prayers and decisions of the whole week in a personal offering to Jesus the Lord. This takes the form of commitment to full Christian discipleship.

In India the relationship *Guru-chela* is part and parcel of a spiritual tradition going back to Upanishadic times. It comprises more than teaching and learning. The disciple lives with his master and is formed more by the impact of his person than by his words. Thus the disciple is not simply a student who has to absorb a certain syllabus but he is a person who has to *grow*. The guru initiates him stage by stage into deeper insights as the disciple is ready and able to perceive them. The disciple receives not so much knowledge but *truth*, which is not found in books and cannot be tested by examinations. In the fulfilment of this task the master is God's own instrument; at times in some traditions he is actually divinised, since through him the Ultimate Truth, God Himself, is communicated to the disciple.

All this is consummately realized in Jesus' relationship to his disciples; but it receives its own distinct character, which arises not merely from the different cultural setting but from the person and mission of Jesus.

Discipleship in Jewish Tradition

In the Old Testament the idea of discipleship is almost absent. We do find followers of prophets as e.g. Elisha attaches himself to Elijah (1 Kings 19,19). However, a Jewish prophet is only the herald of God's message to the people, he does not become the centre of a group of disciples. God alone assembles his people and is the bond of their community. Only in the post-exilic period we find the idea of discipleship. It is centred not around prophets—there are no prophets any longer, they have been replaced by teachers of the law—but around rabbis, who embody various traditions, as they had developed in the Jewish community.

These rabbinical communities, then, were the pattern adopted by Jesus in gathering and forming his disciples. He initiated them into their new life not primarily by instructions but through the contacts of daily life. His followers were considered by the people as belonging to Jesus, their master. Thus people turn to his disciples in their misgivings about Jesus when for example, they see him having meals with tax-collectors (Mk 2,16). Similarly, Jesus is held responsible for the behaviour of his disciples, when they do not fast as John's disciples had done, or when they pick grains on the Sabbath (Lk 5,33; 6,2).

"I have chosen you"

Still, there are deep differences between the disciples of Jesus and the disciples of the rabbis. The first distinctive feature in Jesus' discipleship is the manner of selection of followers. Among the Jews it was the disciple who chose the master who seemed best to correspond to the disciples' expectations. In the gospels, however, it is Jesus who chooses. In every one of the apostolic vocation accounts, the initiative of Jesus is emphasized. He calls the fisherboys from the shore of the lake. He tells Levi to follow him. When the twelve are elected from the larger group of followers to be "the twelve" we are told explicitly: "He called to him those whom he desired" (Mk 3,13). In John's gospel we read Jesus' words: "You did not choose me but I chose you and appointed you that you should go and bear fruit and that your fruit should abide (Jn 15,16).

The reason for this contrast lies in the very nature of Jesus' discipleship. For a young Jew the study of the law in the school of one of the rabbis was equivalent to choosing a career which gave him a place in society and a corresponding social status. For Jesus the call to discipleship is a strictly personal call, to be with him and to share his life and mission (cf. Mk. 3, 13f). It is not the opening to a special profession or office (all this is to develop at a later stage and unfortunately often blurs the basic conception of a christian vocation). Jesus simply says: "Follow me!" (Mk 1,17; 2,14). The personal commitment to Jesus is the core of the Christian vocation.

The total surrender

From this follows the second difference in Jesus' discipleship: the absoluteness of the commitment to him. Rabbis did not separate their disciples from their families nor disturb their place in society. The groups of their disciples were accepted in the social and professional structure of the people. Rabbinical discipleship never implies a conflict with social traditions or accepted values.

It is different with Jesus and his followers. His outlook on life, on religious traditions and on society, sharply differs from that of his surroundings. His followers must be ready to stand by him against public opinion. Thus, when a large crowd follows him he is not happy with their enthusiasm. What they want is a hero according to their own mind and expectation but they will fall away when conflicts with family and traditions arise. Thus the hard words: "If anyone comes to me and does not hate his own father and mother and wife and children and brothers and sisters, yes even his own life, he cannot be my disciple" (Lk 14,26). "Hating" here means simply: Whatever comes between Jesus and his disciple to divert him from his commitment must be discarded. Nothing can take preference to Jesus, no natural ties of family or society. This does not imply a negative attitude towards life; Jesus loves life and came to bring it to fulness. He is not antisocial but builds community, heals rifts, reconciles, and praises peacemakers. But his communion with people is based not on the natural ties of family or of social groups. Such ties bind together only limited sections of society, and, unless they are transcended, they lead to conflicts with other groups and become divisive forces. Jesus builds a new community of brothers and sisters for God's kingdom based on God's love: "My mother and brothers are those who hear the word of God and do it" (Lk 8,21).

This road leads Jesus and his disciples into an aloneness which is both rich and painful. It gives them the freedom to be themselves, free from fettering conventions and social prejudices. Jesus communicates with all, rich and poor, religious people and the rejected group of the tax-collectors. He is little concerned about the comments of the people. He also keeps his freedom in the observation of the law, because human needs and concern for the neighbour are more important than traditional customs. But Jesus has also to accept the consequences. He loses the social acceptance that seems so indispensable in our life. He is no longer really at home with tradition-bound people. He warns a would-be follower that, as his disciple, he has to forgo the basic need of all creatures to have their place of shelter: "Foxes have holes and birds of the air have nests, but the Son of man has nowhere to lay his head" (Lk 9,58). Those who follow him must be ready resolutely to walk a lonely path: "No one who puts his hand to the plough and looks back is fit for the kingdom of God" (Lk 9,62).

Sharing in Jesus' passion

The radicality of these demands follows from the nature of Jesus' mission: He invites his disciple not only to an action programme to collaborate in a team, but to share in his life, in the realization of God's reign. It implies a radically new orientation of the entire person as it is embodied in Jesus himself. He never described this orientation systematically but simply invited the disciples to follow him, to "come and see' (cf Jn 1,39), and in patient, often painful learning to enter into the meaning of his mission, his person. More and more his words to the disciples are concerned with the paschal mystery. Take the scene where two of the disciples ask for the first place in his kingdom at his right and at his left(Mk 10,35-45). He asks them only: "Are you ready to drink the cup that I drink?", which means to share in his life, rejection and death. They think they are ready—but how much did they really understand? But when the ten showed their annoyance with the two and felt jealous of them Jesus had to lay open the full meaning of discipleship: to share in his total self-gift to the Father. Jesus' disciples are not meant to strive for superiority or aim at places of domination. This is what all people do. It is the common feature of our society; "Those who are supposed to rule over the Gentiles lord it over them." Jesus demands the opposite: "It shall not be so among you. Whoever would be great among you must be your servant." This does not make sense in our competitive society, in the common struggle for survival and influence. People who refuse to take part in the struggle for a place in the sun will be pushed aside and trodden down. Jesus does not argue; it is impossible to argue about God's kingdom on merely logical grounds. He points to himself. He himself is the norm. His life, death and resurrection are the only guarantee that it is possible to live in this way: "The Son of man came not to be served but to serve and to lay down his life as a ransom for many". This is Jesus' only answer and explanation. It is not demonstrated in discussion but lived in his person.

Gradually the disciples must become familiar with the impending passion which concerns not only Jesus but also themselves: "I have chosen you out of this world, therefore the world hates you. The servant is not greater than his master. If they have persecuted me they will persecute you" (Jn 15,19). All this is summed up in the short-formula of discipleship which is recorded in various contexts and wordings by all evangelists: "Whoever would save his life will lose it; and whoever loses his life for my sake and the gospel's will save it" (Mk 8,35). This is the ultimate meaning of human life in Jesus' vision: as long as we shelter ourselves, close ourselves into our own interests, the richness and beauty of our life are lost. True life means to be open, to be shared, to be centred not on the narrow ego and the horizon of individual needs but on the neighbour, the world, on God. If you close your life, it is lost. If you find the courage to open your life, to share it, to give yourself to your neighbour, and ultimately to God, not only what you have but what you are, then certainly you shall find true life. This is Christ's uncompromising challenge.

Remain in me

Discipleship with Jesus never ends. This is the last difference between rabbinic discipleship and the following of Jesus. The time in the house of a rabbi was limited to the period of training. The goal of it was an independent life, when the young man would embark on his own work. This is natural, as each one has to live his own life in personal responsibility.

The discipleship of Jesus is different. In his words God speaks, and in his actions God is at work. He is the channel through which God's own life and love flow into our hearts, into our world. In Jesus we encounter God; we can never cease being open to God. Already in the days of Galilee the disciples felt close to God in Jesus' presence: so they asked him to teach them to pray (Lk 11,1). They experienced what Jesus said: "No one knows the Father except the Son and to whom the Son pleases to reveal him" (Mt 21,27). It is the core of his mission to reveal the Father. In John's gospel Philip asks Jesus to show them the Father. The question does not seem to make sense: "Have been with you so long and yet you did not know me, Philip? He who has seen me has seen the Father...Do you not believe that I am in the Father and the Father in me?" (Jn 14, 8-10). Whoever comes into the presence of Jesus cannot help realizing the unique bond, the absorbing union between Jesus and his Father. For Jesus all creation is merely a shadow compared with the overwhelming reality of his Father. Jesus lives in his presence in a unique awareness of his love and care and of the mission that is entrusted to him. Whoever comes to Jesus with an open heart, sensitive to his inner life, must be aware also of the source of Jesus' freedom and strength, of the peace and assurance with which Jesus lives and moves—the powerful bond of love between Jesus and the Father.

Also in his glory Jesus remains our way to the Father.
His disciples must remain united to him for ever; it is a bond
of love that can never be broken. "As the Father has loved
me, so I have loved you; remain in my love!" (Jn 15,9). The
parable of the vine and the branches is the last and fullest
description of our union with Jesus. In the beginning Jesus
invited the disciples to come and follow, to see and believe.
Now he says: "*Abide in me and I in you*" (Jn 15,4). This union
will have its fulfilment in the participation in Jesus' glory:
"When I go and prepare a place for you, I will come again
and will take you to myself, that where I am you may be
also" (Jn 14,3).

Jesus is the beginning of the new creation, a new communion of people as brothers and sisters. This new creation
does not grow through mere organisation. It is a life process;
it starts with the small cell of Jesus' disciples in Galilee. It is
meant to grow and embrace all men and women, the kingdom
of God. Its beginnings are in our world; its fulfilment is the
communion of saints in glory, with the Risen Lord.

SHORT POINTS FOR MEDITATION

1. *Jesus calls disciples*   (Mk 1,16-20).

This is the very first deed of Jesus after returning from
the wilderness. Mark places it immediately after the
proclamation of the coming of God's kingdom. Jesus'
message of a new world must be embodied in a community.
Jesus gathers his disciples as the first cell of the worldwide
communion of believers which will be his Body.

— "Passing along the sea of Galilee he saw Simon and
Andrew": this beginning seems very casual, somewhere at the lake side, like meeting any ordinary
fisherboys. This is God's Kingdom: not a display
of power or glamour, it grows from the grassroots and
takes place in our everyday world where Jesus passes
and sees us — "Follow" — "Follow me": Mark tells
us nothing about previous contacts (cf. Jn 1,35f).
The shocking brevity and abruptness of the call
expressed best the newness of their life with Jesus.

It will be centred on him only; henceforth their life will be formed and transformed by him.

— "I will make you fishers of men": Jesus not only teaches them but gives them a destiny of their own, a mission that will be totally beyond their own horizon. Discipleship of Jesus makes them part of God's reign which changes their heart and is meant to renew human society.

— "They left their nets and followed him": Jesus' call must be answered in freedom, in a commitment to him. For some it means a break with the past, a totally new beginning, others will be sent back into their own milieu to be witnesses of Jesus (cf Mk 5,19).

2. *The Cost of Discipleship* (Lk 14,25-35).

— "Great multitudes accompanied him": Jesus loved the people, had compassion on them, gave them bread. Thus, many people crowded round him. But discipleship is something different, it demands a basic decision.

— "Anyone who does not hate father and mother... cannot be my disciple": "Hating" means whatever would come in between Jesus and his disciple must be discarded because following Jesus means more than all human relationships. It is God's own claim on our life before whom all human claims have to yield.

— "Whoever does not bear his cross and come after me cannot be my disciple": discipleship means not only to be with Jesus, but to live our own life with its burdens and conflicts through the power of Jesus.

— "If you build a tower...sit down and count the cost": this advice sounds unlike Jesus, who never counts costs, who also of his disciples asks a full self-gift beyond rational calculations. Still, Jesus does not want his followers to come to him with illusions; they must know that the commitment to him is radical. Each one must ask himself whether he is ready for it.

— "When going to war measure first your own strength": in this second parable Jesus points to the conflict with the surrounding world, with its own selfishness for which his followers must be prepared.

— "Whoever does not renounce all that he has cannot be my disciple": this should be the prayer of this day of retreat to receive the clear insight into the joy, but also into the demands of discipleship, and to receive the strength to make and to live this commitment to the end.

— "If salt loses its taste, how shall its saltness be restored?": will Jesus' disciples live up to his call? Many of those committed to his service resign themselves to mediocrity. The Book of Revelation says: "Because you are lukewarm, neither cold nor hot, I will spew you out of my mouth" (Rev 3,16).

3. The Decisiveness in following Jesus
(Sp. Ex. nn. 149-157).

Orientations for our life are of little use unless they are followed by decisions. Thus to the meditations of the 2nd week, which give guiding norms (the two banners, p. 150; the three degrees of humility p. 153ff), a reflection on decision making is added. It consists in a case story taken from daily experience. A man has accumulated much money by not very fair ways and he is confronted with the problem what to do about it. The meditation tells us to reflect on the right way of facing decisive situations in our life.

The story seems quite ordinary, still, Ignatius gives it the most solemn setting: "to see myself standing in the presence of God Our Lord and all his Saints that I may desire and know what is more pleasing to his divine Goodness." In the concreteness of our daily decisions the very meaning of our life is at stake.

Three possible attitudes are proposed:

1. *Drifting and procrastination:* The owner of the money wants to do something about it but post-

pones the decision for "tomorrow". Nothing will ever happen. You may think of actual problems in your own life. You feel the need of coming to grips with them, but you drift along. Problems will not be solved by drifting and putting off a basic decision.

2. *Reservations and laying down conditions:* The owner wants to do something about the money but has decided already to keep it. This is not readiness to fulfil one's duty but the futile attempt to calm one's conscience and to manipulate God's will according to one's own intentions and desires. We do this in a thousand ways in daily life: we cling to our own plans, ambitions, aspirations, and then try to justify them and to camouflage them as God's service. Only a serious discernment in full openness to the guidance of the Holy Spirit can lead us out this impasse of self-deception.

3. *Surrender unconditionally to God's will:* "They seek only to will and not will as God our Lord inspires them, and as seems better for the service and praise of the Divine Majesty". The choice must be made, not avoided; it must be made in total openness to God's will.

This meditation is not an exercise in abstract principles but an ultimate, fundamental effort to come to grips with the concreteness of our life and to cut through the manifold entanglements that hold us back from the simple, total self-gift to God.

We are recommended in prayer to turn: (1) first to Mary who committed herself without reservation to God; (2) next to Jesus, who knew only the will of his Father even when it meant the way of the cross (3) finally, to God the Father whose love is the only norm and source of our life. We pray that we may be able to offer up our life to his love and service.

OTHER READINGS:

— Mt 8, 18-22:　　hesitant disciples.

— 2 **Cor** 2, 14-17:   the fragrance of Jesus spread by his disciples.

— Phil 2, 12-18:   the light shining in darkness.

QUESTIONS:

1.   Jesus found in his disciples ambition, a lack of faith, etc. what does he find in me?

2.   Examine yourself about the attitude of procrastination; of making conditions in your response to Jesus' call.

REFLECTION:

"If you would extol Christianity, oh, do not wish for yourself the tongues of angels nor the art of all poets, nor the eloquence of all orators. In the same degree in which your life shows how much you have forsaken for the sake of it, in the same degree you extol Christianity".

KIERKEGAARD

PART III

THE NARROW PATH

Before narrating Jesus' farewell from his disciples on the eve of the passion, John indicates the unique significance of the last section of his gospel. "When Jesus knew that his hour had come to depart out of this world to the Father..." (Jn 13, 1). John speaks of Jesus' hour. Jesus was conscious of this supreme hour (cf. also Lk 22,14). In biblical language the hour is not simply a period in the even flow of time but indicates the decisiveness of the situation. Jesus' life and mission reach their climax. In his passion, death and resurrection the meaning of his coming, his work and preaching is fulfilled and revealed. In the last events of his life we see who Jesus really is, what God's reign, proclaimed by him, means and in what way this reign of God is to be realized in our world.

In each gospel Jesus' passion and resurrection form a distinct part clearly marked off from the preceding narration. Yet at the same time the passover events are the continuation and conclusion of Jesus' life and mission, the outcome of the growing conflict with the leaders of the people. In fact, Jesus' condemnation and crucifixion would be unintelligible without viewing the growing estrangement and the ever widening separation of Jesus from the ruling class. For the meditations of the third week of the Spiritual Exercises both the continuity between Jesus' life and passion and the clear distinctiveness of the Paschal Mystery in the gospel narrative are important.

If Jesus' passion and death are considered in isolation from his earthly life they are deprived of their significance in Jesus' own life and of their relevance for Jesus' disciples. The trend towards an independent and isolated understanding of Jesus' death is found already in the New Testament, when it speaks about his blood in terms of Jewish rituals, of sacrifice and expiation.   But for the apostolic communities these terms were immediately understood as interpretations of the historical events in Jerusalem of which they were fully conscious. Later theology, however, tended to reflect on Jesus' death more and more independently of the actual events, of Jesus' conflict with the political powers of his time and of the religious legalism of his adversaries. Thus seen, his death was no longer an integral part, the final conclusion of his earthly mission but rather became a general theology of sacrifice and satisfaction.

For the retreatant the context of Jesus' passion with his earthly life is significant because the third week is not a general reflection on the meaning of suffering. It is a school in following Jesus on the way of the cross. For Jesus the death on the cross is the consequence of his commitment to God's reign against the powers of evil, the price he has to pay for his obedience to the Father and his love for us till the end. The retreatant too has made his option for God's kingdom in the second week. So he finds in the meditations on Jesus' passion the invitation to follow the master also into the darkness of suffering and rejection.

Yet, in spite of this continuity, there is something entirely new in the passion accounts, which again is significant for the retreatant. During his earthly life Jesus himself controlled the development of his mission.   There was a time of preaching and addressing the crowds, of indictments and of challenges to those in power. There was also a time of withdrawal because his hour had not yet come. Now a new situation arises. Jesus seems no longer master of his destiny. He gives up the control of events and abandons himself to powers which are outside him. The disciples still want to take up the fight against the enemies in the garden. Jesus refuses. He tells the soldiers: "This is your hour and power of darkness" (Lk 22,53). Yet even the powers of evil are subject to God's design: "Shall I not drink the cup which my Father has given me?" (Jn 18,11). God is also at work in the hour of darkness.

Thus Jesus shares in our life, even in the night of death. With us he enters into the inscrutability of God. God's reign prevails where human efforts are exhausted. Jesus surrenders to the ultimate mystery of God.

The retreatant has to do the same. The third week of the Spiritual Exercises has a different atmosphere. The retreatant finds himself in the presence of the ultimate mystery. In the second week he has made his option with Jesus. Now he discovers himself with Jesus in a realm where his own planning and reasoning have come to an end. He shares in the total aloneness of Jesus who is abandoned by all while he fulfills the ultimate self-gift of love—his surrender to God on the cross.

Chapter One

THE LAST SUPPER

When Jesus entered his passion the disciples were scattered. It is part of his passion that in this last stage of his earthly life he is alone without any human support. Still, precisely in this paschal mystery Jesus is more *intimately united with his own* than in any other situation during his whole earthly life. His passion concerns all. His followers and all people have to go through their own passion, to carry their own cross, and to learn from Jesus' passion the ultimate surrender to God, to find the door to true freedom and salvation.

Thus where and when Jesus is most alone he is closest to his own. This closeness will not be realized by the disciples in the dark hours of Good Friday, but they must experience it at the entrance into his passion during the last meal he had with them. Jesus wanted that in this last hour his friends realize once more in a unique way what he was to them. They should be aware of the depth and richness of their union with Jesus, and so, in this hour, they should have already a foretaste of that final union which they will find only after his death, when he is risen. This hour of the last supper will remain for ever in the memory of his disciples not only as a dear remembrance but as the ever new celebration of his self-gift for them and their union with him.

The Paschal feast

It is the national feast of the Jews on which they commemorated the great events in their history, the liberation from the slavery in Egypt, the day on which *Yahweh made them his own people* like "an eagle spreading out his wings over his young, bearing them on his pinions" (Deut 32,11). On this day they felt one nation, coming together from all towns and villages and from distant countries. They were conscious of their unique place among the nations because Yahweh had made them "His own possession among all peoples" (Ex 19,5). They needed this celebration, the day on which they forgot the drudgery of their daily life and the humiliation of living under the rule of the Romans. On this day they became aware with deep gratitude of God's gift. Jerusalem was vibrating with the joy and pride of the nation because they were conscious— this was their faith—that it was not only in olden days that God liberated them from slavery, but on this day God was setting them free.

What is Jesus' place in these celebrations? How will he take part in the rituals of the feast? Jesus lived in the Jewish traditions and cherished them. It pained him, and he spoke harsh words, when he saw how the sacred traditions were often reduced to empty ceremonies. For him Jerusalem was the holy city and the temple was his Father's house. The Passover was the celebration of God's love and power for his people. It was a feast with the promise of the kingdom, a victory over the powers of darkness, a new and lasting covenant, a new communion among all people as brothers and sisters.

It had been his mission to proclaim the coming of God's reign, God's new community. It had begun to take shape in the small group of Jesus' disciples. In this last celebration of the passover with them they should experience the nature of this new community.

The Community of the disciples

The passover celebration is particulary apt to convey to the disciples the spirit of *the new community* as they have to live it. It is a meal which for Jesus is the most telling symbol of community. When he describes the final fulfilment of the kingdom of God he visualises it as a great banquet, with Abraham, Isaac and Jacob presiding, and the nations coming together from East and West. Even after the resurrection his disciples recognized him in a meal, by the breaking of the bread. In this gesture of sharing they realized that he was truly with them again. Thus all four gospel accounts present the last supper as a celebration of community.

John centres his description on the *washing of the feet.* The meal is an obvious occasion for rivalries, and Luke refers to the renewed dispute among the disciples jostling for the better places (Lk 22,24). In this context he reports Jesus' sayings about authority which, in God's kingdom, does not consist in domination, as among Gentiles, but in service (Lk 22,24-27: Mk10,42-45). For John this beccmes the central message of the last supper. Jesus washes the feet of the disciples and wants his action to be understood as a norm for the disciples. He is "teacher and Lord"; and they recognise him as such. He has authority but exercises it only in service: teaching the people, healing, struggling against the evils in society, forming and inspiring his followers. Jesus sees his life as service, and the disciples must learn that mutual service, giving and receiving it, is the very essence of their relationship. This is Jesus' answer to Peter when he refuses that his feet be washed by him: "You have no part with me". All must learn it: "You also must wash each other's feet". Jesus gave the example that the disciples have to follow, because the servant is not greater than his master. It is the substance of the Christian community: "If you know these things, blessed are you if you do them."

The Eucharist

What John tells us in this scene of the washing of the feet, the Synoptics convey in their narration of the last supper. Jesus followed the traditional procedure. After preparatory rituals the meal is solemnly opened by the prayer of the father of the family, praising God for the gifts he has bestowed on his people in the great days of their history. God had taken from them the yoke of servitude in an alien land, had guided them through the desert and had provided for them and had given them the land he had promised to the fathers. This land was bread and prosperity for them. For Israel bread became the silent symbol of God's faithful love, unconspicuous, taken for granted in daily meals, but God's care is embodied in it. Bread must be *broken and shared*. In this way it becomes a symbol also of community. On this evening, when the Jews celebrated God's love and power, every father of the family held bread in his hands, said grace and shared it with his own.

Jesus is now performing this ritual at the beginning of the passover meal. He does it in a new community. New is this community because it is bound together in a new way, more powerful than by the ancient traditions, by the person of Jesus. In his company they had learned what God's kingdom really is, not an astounding miracle falling from heaven but the new togetherness in brotherhood with Jesus as the centre, a community with no barriers, comprising all people. Jesus had been with them with his daily concern, teaching and guiding them. In this hour of farewell they realized this communion in a unique way. His self-gift to them finds its ultimate expression.

Jesus, like all other Jews, holds the bread in his hands, and, according to tradition, says the blessing. But he ends the prayer in a new way with the astonishing conclusion: "Take and eat, *this is my body*". He gives himself. His "body" means his whole being, his life, as God had given it to him, as he gave it to his Father. "A body you have given to me... Behold, I come" (Hebr 10,5). This body, this life, which totally belongs to God, is now given to us. Jesus gave it to the people every day of his life, it will be given to them more fully in his death and finally in his glory. The bread, broken and shared, will for ever be the sign through which he gives himself to his disciples.

He tells them to take this bread, which is his body. "Taking" implies an active acceptance. Never can communion with Jesus consist in passively receiving him. It always is personal communion in faith and discipleship, in order to be transformed by him. Jesus' mision must be realized in his communities, in new life and new brotherhood, in cells of God's kingdom that are to pervade and transform the world. To take this bread is a commitment to Jesus' mission.

He tells them to eat it. Bread is to be eaten. In the bible there lives still something of the mystical meaning of eating (cf Rev 10,9-11). Ancient Indian wisdom knows this: *what you eat, you become.* In eating we allow the food to become part of ourselves. Those who eat of this bread make Jesus' life and mission their own: "You cannot partake of the table of the Lord and the table of demons" (1 Cor 10,21). Eating the Eucharist becomes the uniting bond of the community: "Because there is one bread, we who are many are one body, for we all partake of the same bread" (1 Cor 10,17).

The Paschal meal went on. Next Jesus and his disciples ate the roasted lamb. The feast moves towards its climax, the solemn conclusion which consisted of the great Hallel, Psalms 113 and 114, the joyful praise of God for the covenant. The third cup of wine was filled. This was called the "cup of blessing" to be shared by the participants in the meal. It was the cup of the covenant. With the coming of God's kingdom in Christ this covenant becomes new. God's people is no longer segregated from other nations. It is the communion of all in God's love without dividing walls. Jesus had lived this covenant in the union with his Father and the self-gift to the disciples, to all. Tomorrow he will seal it with his blood. Blood is life — this was the understanding of the Jews. Blood flows in my veins, the blood which is my life. Jesus' blood will be poured out, it is meant to become my life, the life of the world. This is the new covenant in his blood for the eternal life of all.

Jesus holds the cup in his hands: "*This cup*", he says, "*is my blood, the blood of the new covenant,* poured out for the multi- tude of people". And once more he adds: "Take it and drink it". It is not enough to receive Jesus' love. Love must become the pervading and unifying presence of God among the people. In drinking it we make Jesus' life and love our own.

This hour of the last supper sums up the entire mission and message of Jesus. It is the celebration of God's reign embodied and revealed in Jesus. It has to unfold down through the ages in many forms, but it will, it must, always keep the same features. God's love is revealed in Jesus. It is accepted by his disciples and realized in their lives. It binds them into a community with the assurance that our world and our history are moving towards the fulfilment of God's love. Thus Jesus adds: *"Do this as a memorial of me."* This hour is to remain alive in all communities of his disciples, not as a nostalgic remembrance of his farewell but in the consciousness of his continued presence in their midst to be the source and inspiration of their life, the bond of their community, the joy of their mission and the assurance of fulfilment.

This, then, is the meaning of the eucharistic celebration. It is the celebration of our union, of the union of the whole Church, with Jesus. We call it a *sacrifice* because Jesus' own life, his very being, is to go to his Father, to be with his Father. Whoever comes to Jesus and is united to him in the Eucharist is drawn by the power of Jesus into this movement to God. The surrender to Jesus in the Eucharist is at its very core a sacrifice, a self-gift to God. It can be conceived in two ways: it is the believer's self-gift to God in and through Jesus Christ, or it is the self-gift of Jesus to his Father realized anew in the community of the faithful. Just as a brooklet merges its water into the mighty current of a stream that irresistably moves towards the ocean, so believers merge their sacrifice with that of Jesus.

This union is also *sacrament:* God's gift to us of Jesus his Son comprises all the gifts of creation and salvation. It is even more: he who in true faith receives this sacrament and is united to Jesus becomes himself a bearer of God's saving love. Thus in the Eucharist the whole Church, the community of all believers, becomes "sacrament, sign and instrument of communion with God and of unity among all men" (L.G. n. 1).

SHORT POINTS FOR MEDITATION

1. The Eucharist in the community of Corinth
   (1 Cor 11, 17-34).

   The Eucharist is the centre of the Christian community. It is open to misuse. It can become a monotonous ritual. The greatest danger, however, is the loss of its innermost meaning as the source of unity among the believers.

   — "In this I do not praise you": Paul uses hard words. The Corinthians, who glory in their gifts of the Spirit,

do not know what community is. There are parties
(cf. 1 Cor 1,12); the poor are publicly humiliated.

— "I have received from the Lord..": Paul does not
give his private opinion but proposes Jesus' own
message.

— "On the night he was betrayed....": the attitude of
the Christian towards his neighbour has its pattern
in Jesus' total self-gift to his disciples even in the
face of their betrayal of him.

— "Everyone should examine himself": Participation
in the Eucharist is a commitment to Jesus' spirit,
to community, a call to mutual esteem, reconciliation,
brotherliness. Failures in charity, which are
unavoidable, must not become the rule, as in Corinth.

— 'Wait for one another...": the harmony of a Christian community is shown in mutual concern and
consideration.

2. *Food for the Pilgrim People*   (Jn 6).

God's care for his people is fulfilled in Jesus: Yahweh
had given them bread in the desert (Ex 16,6-27); so Jesus
feeds those who follow him (6, 1-14).

New is the way in which Jesus gives bread: he uses
the bread of the boy and has it distributed through the
disciples. God's gift comes through human hands. This
is God's reign.

— "I am the bread of life": (6.35) most of all, the bread
is new; it is Jesus himself, given to us not for the
transient pilgrimage only, but for eternal life.

— "This is the will of my Father, that everyone who sees
the Son and believes in him should have eternal
life," (6.40): to give us life, Jesus must be received
in faith.

— "He who eats my flesh and drinks my blood abides
in me and I in him": (6.56) the union in faith is
crowned in the sacramental union with Jesus. Every
eucharistic celebration begins with the word of God
in order to awaken the faith of the believers. This
leads to the eucharistic communion with Christ.

— "No one can come to me unless it is granted him by my Father" (6,65): God alone is our Saviour, but he speaks to us, and gives himself to us in Jesus, his Son.

OTHER READINGS:

— 1 Cor 10,14-17: To become one body in Christ.

— 1 Cor 13: Love is the highest gift.

— Phil 1,3-11: The Apostle's love for his community.

QUESTIONS:

1. What place has the Eucharist in my life? Have I understood it as the sacrament of union with Jesus and my brothers and sisters?

2. Has the frequent participation in the Eucharist made its celebration monotonous? How can I make it more personal?

FOR REFLECTION:

"You ask what you should offer. Offer yourself. What else does the Lord seek of you but yourself? In the whole of creation he has made nothing better than you. He seeks you from you, because you had lost yourself."

AUGUSTINE, *scrm. 48, cp. 2*

Chapter Two

JESUS ENTERS THE PASSION

Jesus' passion is the most frequent topic of meditation among his disciples. In the third week of the Spiritual Exercises these meditations should be made so that they help the retreatant in realizing the goal of the retreat, *finding God's will through a deeper union with Jesus*.

Saint Ignatius tells us in these meditations "to ask for sorrow, compassion and shame because the Lord is going to his suffering for my sins" (n. 193). Each of these attitudes, however, must be understood correctly. It would be entirely against the purpose of the Spiritual Exercises to reduce "sorrow" to feelings of sadness, or "compassion" to a sense of pity for Jesus in his sufferings. Jesus does not want to be pitied; he freely enters his passion. Ignatius tells us to contemplate Jesus who "desires to suffer" (n. 195). Compassion means partaking in Jesus' sufferings, entering into his darkness and pain which he shares with us because he becomes a member of our human family. For Jesus the passion is part and parcel of his struggle against the powers of darkness.

This, then, is the meditation of the day: to contemplate Jesus as his life moves towards the shadows of suffering and death, how he faces the powers of destruction.

Jesus' human experience

Jesus' passion does not begin at the end of his life; it grows from the human experience which he shares with all of us. In each stage of our life we leave something behind, its joy and beauty; so it was also in Jesus' life. At times Jesus is depicted as if he had lived from the beginning in the consciousness of this final sacrifice which awaited him. It is, of course, true that every human life moves irrevocably towards its own end. Augustine says: "When we are born, we begin to die." As to Jesus, we also know that he was born to be our Saviour through the death on the cross. Still, it would be wrong to think of Jesus living his life like the rehearsal of a ready-made programme, acting out, as it were, a role which he had learned beforehand. He lived a genuine human life, and there is no biblical evidence, nor any convincing theological reason for assuming that he knew his future from the beginning.

So we should think of Jesus not just as sharing only or mainly our sorrows but he fully shares our joys and the richness of our human life. He had a happy childhood sheltered in his home, in peaceful rural surroundings as they are reflected still in the stories and parables of his later teaching. Nature was the picture book which Jesus used to illustrate his teaching about the Father and his care for us. He could enjoy the peace of this world even more deeply with his unique awareness of God's presence and love when he observed the birds of the sky who never worry about tomorrow and the matchless beauty of the flowers though they blossom only for a day. How much more will the Father care for his own children?

Farewell to Nazareth

Jesus' passion begins with the farewell to Nazareth. He outgrows childhood, and becomes increasingly aware of his wider surroundings, of the hardships and sufferings of the people. He sees injustice and cruelty and the helplessness of the poor. Is this the world as God wanted it? He knows his Father, his love and concern. God wants a different world: God's kingdom must come, his love and power will overcome the narrowness and selfishness which dominate our world. The vision of a new humanity, of brotherhood and love takes shape in him. He has to set out from his hometown on his mission.

His mission separates him from his family. The impact of this farewell on Jesus is conveyed to us only in an indirect manner, in the reaction of Mary. She comes to Capernaum, the new centre of his activities, to see him. They gave the message to Jesus that his mother wants to see him. But Jesus lives in a new world. Mary has to learn that Jesus has a new family, the community of his followers: "Who is my mother and my brothers?" and looking round on those who sat around him, he said, "Here are my mother and my brothers. Whoever does the will of God is my brother, sister, mother" (Mk 3,31-35). This new family of Jesus is bound together not by human bonds but by God's word. Jesus' relation to Mary is not broken, it is reversed: no longer is Jesus the obedient child of Mary, but she becomes a disciple. In fact, Luke introduces Mary as the first who listens to God's message and allows her life totally to be guided by his word. In the first chapter of the Acts of the Apostles she is among Jesus' disciples in the prayer for the outpouring of the Holy Spirit (Acts 1,14) and John sees her under the cross together with only one disciple, the one whom Jesus loved (Jn 19,25). But in his earthly life Jesus walks a solitary path with no home for himself: "Foxes have holes, birds have nests. The Son of Man has nowhere to lay his head" (Mt 8,20). The departure from Nazareth was his first farewell.

Still, a new world opens before Jesus. It is hard. There is the ever recurring temptation to an easier life, a shortcut to success. But he has committed himself to his Father's will and mission. He never flattered people, but opens their eyes to a new vision of life and to a world where God's love flows through all the channels of human relations in a communion of brotherhood and freedom. There is a fascinating power in his proclamation of a new society where the poor will be blessed and the hungry will have to eat, and where the simple people, who have no power and had been bent low, will possess the land. What inner freedom and strength vibrates in his challenges to those who look down on others in hypocritical self-righteousness, against the ruling class of those who hold power and social status. It is difficult to imagine a happier, richer life, not in the sense in which most people think of happiness in terms of comfort and social prestige, but

of a man who lives his life fully, his own life, from the depth
of his being, sure of his mission and fearless of any power on
earth, because he knows God is with him.

The Crisis of his mission

In these first stages of Jesus' prophetic life the pain of
the farewell to Nazareth seems more than compensated by
the new horizons that open before him. Soon, however, he
realizes that resistance against him is stiffening. The leading
circles of the people are hostile and the masses waver and
drift away. Jesus finds his impact limited to a small group of
followers. The Synoptics describe this crisis in the scene
of Caesarea Philippi (cf. p. 149), John depicts it in the scene
in the synagogue at Capernaum. Jesus had fed the multitudes
with bread, and enthusiastic masses gather round him once
more. But now he begins to speak of his real mission, not
only to give bread for the body but bread from heaven. He
himself is this bread: "I am the bread of life". They do not
understand. How can he say, "I am the bread which came
down from heaven? They said: "Is not this Jesus the Son
of Joseph, whose father and mother we know?" How can
he say "I have come down from heaven"? (Jn 6, 41f). And
finally he speaks about himself as the living bread: "If anyone
eats of this bread he will live for ever." Now they begin to
leave him: Many of the disciples said: "This is a hard
saying, who can listen to it?" They go. "Many of the disciples
drew back and no longer went about with him." In this
distressing situation Jesus asks the twelve the courageous
and anguished question: "Will you also go away?" Peter
answers in the name of all: "Lord, to whom shall we go?
You have the words of eternal life" (Jn 6,60-68).

It is the hour of decline. In Mark's narration Jesus now
begins to speak about his passion. He realizes that by human
estimation his mission has failed, the mission entrusted to him
by his Father. It is the painful experience of being rejected
and deserted. His path becomes more narrow, steeper, leading
to death. But the twelve are still with him.

The loss of his disciples

The gospels tell us of a further stage in his farewell, even more painful: the breaking up of his intimate circle of the twelve which dramatically takes place in the passion itself, yet which has its forebodings already before. There are, for example, the repeated references to the "one who would betray him" (Mk 3, 18; Jn 6, 70f). In their passion stories all the Synoptics tell us about that most painful situation during the last supper, in this unique celebration of communion, when the traitor is still with them. "One of you will betray me, one who is eating with me" (Mk 14,18). John gives the scene in more detail, with a last assurance by Jesus of his love for Judas; but Judas opts out, to move into the darkness (Jn 13, 21-30).

The eleven who remain with Jesus are, according to the gospels, still under the spell and support of his personality, unaware of their own weakness. But Jesus sees it. A gulf begins to separate the disciples from Jesus. With full conviction Peter can dare to say: "Lord I am ready to go with you to prison and death". Jesus knows better: "The cock will not crow this day until you three times deny that you know me" (Lk 22,31-34). The others also will desert him: "You will all fall away, for it is written: I will strike the shepherd and the sheep will be scattered" (Mk 14,26f). The inner cohesion among the disciples and Jesus is broken. The assuring words spoken by Peter and the other disciples seem to come from an infinite distance. The more Jesus moves towards his end, the deeper is his aloneness. Only his Father remains. "The hour is coming, indeed it has come, when you will be scattered, every man to his home, and will leave me alone; yet I am not alone, for the Father is with me" (Jn 16,32).

This is Jesus' last journey with his apostles. He goes out with the eleven through the festive city which still rings with the joy of the feast, crosses the dark Kedron valley on the way to the garden of Gethsemane. He leaves the disciples behind: "Stay here while I pray". Even the inner three are left behind: "My soul is very sorrowful, even unto death, remain here and watch". When Jesus comes back he finds them sleeping. In utter loneliness he can now speak only to God: "*Abba*, my Father" (Mk 14, 32-41).

"For our sins"

This entry of Jesus into his passion is inseparably connected with his mission, the proclamation of God's reign, and the uncompromising indictment of all powers which marshal themselves against it. These powers have different names; each of Jesus' adversaries has his own reason to reject him. In fact, it is surprising that in the attack against Jesus the Sadducees and Pharisees combine together, whereas otherwise they are enemies. The common ground on which they meet against him is the rejection of God's reign. They would, of course, have raised a solemn protest against such an insinuation. They all pretended to stand on the side of God. The crime of which they accuse Jesus and the title under which they condemn him is blasphemy: Jesus insults God, when he claims divine authority for his teaching (Mk 14, 63f). They reject Jesus in the name of their God.

This precisely is the issue between Jesus and them all: which God? For the Sadducees God guarantees the stability of their power position. They exercise their authority in his name. For the Pharisees God is the support of their righteousness and superiority. This God gave the law and gave dignity to those who keep it. Each enemy of Jesus has made God the support of his own position: social, political, economic, religious. Each adversary has constructed his own little kingdom for which he claims God's divine authority. This is, however, their own kingdom, not God's reign. Jesus has no kingdom of his own. He knows only God's reign. So all of his enemies are troubled; they feel threatened by him. It is their sin that they have made use of God for their own purpose, to strengthen their ego, to assure their glory before men, to support their power in order that they may continue to rule over their neighbour. Sin has a thousand forms and disguises. In this festival of sin, in which everyone celebrates himself in the name of God, Jesus stands alone, becomes ever more alone with his Father.

My part in Jesus' passion

In the contemplation of Jesus' passion the retreatant is not an outside observer. He himself is involved. Ignatius

wants us to realise that the Lord is going to his suffering "for my sins" (n. 193).

There is no need to imagine (as it is done at times) that we took part in Jesus' crucifixion. It simply means that I am part of this sinful world in which Jesus was born and for which he has now to suffer. I too have allowed the powers of sin and selfishness to penetrate my life. I see myself with shame in the ranks of those who under whatever disguises reject God's reign and resist Jesus.

These meditations are meant to clarify and strengthen the decision against sin in the first week and to confirm and deepen the option for Jesus against the powers of darkness in the second week. In the light of Jesus' life, of his relentless struggle which leads him into the solitude of the passion, I begin to see more clearly the conflict in my own life and mission. Once I have made the option for Jesus I also see the implications for myself, to follow Jesus' lonely path and to have no ultimate support but God.

SHORT POINTS FOR MEDITATION

1. *Jesus' agony in the garden*
(Mk 14, 32-42; Lk 22, 40-46).

— "He began to be greatly distressed and troubled": this is the moment of Jesus' passion. He has ended his work and struggle. It is the time to allow the powers beyond his control, to take over and have their way. It is the threshold between life and death.

— "Sit here while I pray": Jesus leaves human support and companionship behind: his only support is his Father.

— "*Abba*, Father": Jesus' agony is the context in which Mark gives us this special word with which Jesus used to address God. In its intimacy "Abba" contains both his total obedience-unto-death and his total confidence that even death has its place and meaning in God's mystery.

— "Remove this cup from me": Jesus makes his own the prayer of millions who face pain and death. The letter to the Hebrews is even more explicit: "Jesus offered up prayers and supplications with loud cries and tears to him who was able to save him from death" (Hebr 5, 7).

— "Not what I will but what you will": this is his surrender to God's will and Jesus' response to that mystery which is beyond human understanding, the mystery of redemptive suffering.

— "And there appeared to him an angel from heaven strengthening him": the prayer of Jesus changes nothing of the coming events, but Jesus in prayer receives the strength to bear it.

— "In an agony he prayed more earnestly and his sweat became like great drops of blood falling down upon the ground": Jesus is broken but his union with the Father is unshakeably strong.

— "Simon, are you asleep? Could you not watch one hour?" Peter is addressed intimately by his own name, not with the name Jesus had given him. In this scene Simon Peter appears in his human weakness.

— "Watch and pray that you may not enter into temptation": Jesus is more concerned about the disciples than himself. In the impending hour of crisis watchfulness and prayer are needed.

— "It is enough, the hour has come. The Son of man is betrayed into the hands of sinners. Rise, let us be going, my betrayer is at hand": once more Jesus has renewed his option, left utterly alone by the sleeping disciples. Jesus has consistently faced his life and mission with firmness and clarity; so now he faces up to his death with this same single-minded fortitude.

2. **The Hour of darkness** (Mk 14, 43-51; Lk 22, 47-54).

— "A crowd arrived with swords and clubs": Jesus' inner struggle is over: the brutal reality of the passion begins. This is their answer to Jesus' proclamation of justice and love, of God's reign.

— "Judas, would you betray the Son of man with a kiss?": The sign which Judas had given the enemies was that of friendship and devoted discipleship. A kiss now becomes the sign of betrayal.

— "Lord, shall we strike with the sword?": Jesus knows no violence. God's reign has no other weapons but truth and love. Violence divides, creates new hatred, reduces the struggle between light and darkness to a duel of physical power. Luke adds Jesus' gesture of healing the ear of the wounded enemy.

— "And they all forsook him and fled": His disciples would have been ready to fight: this is their natural instinct. But they are not ready to suffer with Jesus. They have not yet learned that loyalty of faith which lasts also where exterior supports fail.

— "This is your hour and the power of darkness": Jesus surrenders to the powers against which he had been struggling. It seems like his defeat. "They seized him and led him away". In and through these powers of darkness, however, God's own design will be realised. In the surrender to his enemies, Jesus surrenders himself to God.

These reflections can easily be continued in the meditations:— on Jesus' trial before the High Priest and the Sanhedrin; on Peter's denial of knowing him; on the mockery of Jesus: on his condemnation.

OTHER READINGS:

— Any of the passion accounts will be suitable for this day.

— Is 52,13-53,12: The suffering Servant.

— Ps 22: Abandonment to God in suffering.

QUESTIONS:

1. Have I experienced in my life the 'going out' into darkness, leaving behind precious values?

2. Do I come across the powers of darkness and destruction in my personal life, in my surroundings: hatred, betrayal, cruelty, calumny? How do I face and encounter these?

FOR REFLECTION:

"Since the mission of the Church continues and unfolds in the course of history the mission of Christ himself who was sent to evangelise the poor, the Church, guided by the Spirit of Christ, must follow the same path that Christ followed, namely the way of poverty, obedience, service and self-sacrifice even unto death, from which he through the resurrection came forth as victor. This is the way the Apostles followed in hope, and by much trouble and suffering they filled up what was lacking in the sufferings of Christ for his body, which is the Church. Often enough blood was the seed of Christians" (Vat. II, *Mission Decree*, n. 5).

Chapter Three

THE CROSS

"I decided to know nothing among you except Jesus Chris, and him crucified" (1 Cor 2, 2).

It is impossible to philosophize on the cross, to comprehend it, or to "make sense" of it. On Good Friday we read the passion, we pour out our needs, the needs of the whole world before God, and then we adore, we kneel before the cross, we adore God. Nowhere is God closer to us than in the cross of Jesus, nowhere he seems so far away, so totally beyond our understanding as at Golgotha. This is Jesus' hour where all he is, all that God has to tell us in him, is summed up. Jesus no longer teaches. He forgives and he prays, till his voice dies. He too becomes silent before the cross.

Jesus reveals the Father

Who is God? No one has ever seen him. Still, God has revealed himself to us. He did so first in creation which is spread out before us as a gigantic exhibition of greatness, wisdom, power, of rich diversity woven together into harmony where all conflicts, all struggle for survival are blended together and redeemed in a mysterious order of growth. In the created world God has offered us not only "a constant evidence of himself" so that we could be aware of his divine mystery, but also his concern for us "opening up to us a way of heavenly salvation" (Vatican II D.V. n. 3). No one has realized this more deeply than Jesus, for whom the world and human life were like mirrors in which he recognized his Father with his love and concern. Jesus never ceases to propose it to us in his parables.

But Jesus knows more about God than what is reflected in nature. He prayed to him with unique intimacy, called him "*Abba*, dear Father," which was unheard of in Jewish religious tradition. God had come infinitely close to us in Jesus who could speak about him in ways no one had ever spoken to the poor and exhausted people. No hair would fall from their head without their Father's knowing. When they felt lost and abandoned, he would set out in search for them and bring them back. Truly, no one knew the Father as Jesus did. No one could convey the message of his love as Jesus did and in a language which touched the depth of their hearts. Jesus' coming was truly like a wedding feast. Jesus himself used this parable. At a wedding feast people are allowed to forget the drudgery of their daily toil, when they laugh and dance in an abundance of joy and music. This feast is not reserved for a privileged class but all are welcome to it, the poor and crippled, the blind and lame (cf Lk 14,21). In Jesus' presence all social distinctions and political labels are brushed aside, all find their God-given dignity. It is the only true human dignity we can ever find, with him, as God's beloved children.

This is Jesus' message of God. It is as new and redeeming today as in his time, the message of God's love and concern, deeper, richer, more faithful than any human love. We are somehow able to grasp this message and to express it in human language. But it is not yet the last word about God. *God speaks his last word through Jesus in his death on the cross*. We must follow Jesus into this ultimate mystery which is beyond human understanding, where there is only faith and love.

The mystery of God

From ancient times people have sensed this unspeakable mystery. They left the first fruits, which they needed most urgently, on the trees; these belonged to God. They burnt animals on altars, allowing them to go up in fire and smoke to the unknown. Does this make sense? Is it mere superstition, magic? Or is this mankind's dim awareness of the ultimate mystery of God which eludes human understanding and practical usefulness? God is greater than our mind, greater than our world, greater than our heart.

The cross is Jesus ultimate encounter with the mystery of God. His entire life is going to God, a pilgrimage towards the unknown, not in fear and anxiety, but in love. "A body you have given to me, behold I come." Jesus sees his body, his actual human existence, moving towards that ultimate destiny, stage by stage, until he reaches that last barrier where God is hidden and all understanding comes to an end. He enters into the darkness of the passion, the "hour of darkness", which becomes ever denser as the end approaches so that the evangelists have no better way of expressing it than that "darkness was over the whole land" when the sun stopped shining (cf Mk 15,33). Whatever Jesus knew about his Father seems lost; his last prayer is Psalm 22: "God, my God, why have you forsaken me?" (Mk 15,34).

Who is God? It seems, the more Jesus advances in his life and work, which is meant to be God's revelation to us, the more God is hidden. Jesus himself seems abandoned by his Father. Yet, Jesus is *with us*. He suffers *our* agony. He dies *our* death *for us*. This is God's final revelation: "Greater love has no man than this that a man lays down his life for his friends" (Jn 15,13). God is love! It remains true that God cannot be seen, and in Jesus' death it is awefully true. Who can discover God in this confluence of hatred, cruelty, blasphemy, falsehood and violence? Do not try to see him; but sense the closeness of Jesus, who became our brother to be with us in this world of darkness and death. This is love; and where there is love, there is God. God is no longer the elusive object of contemplation. He no longer speaks in words and deeds which can be analysed by human reason. He is here at Calvary on the cross with his very being. And I am in him, if I have love. Teresa of the Child Jesus at the end of her short life lived through agonies not only of body and mind, but of heart. She felt totally abandoned and doubts came to her so that she knew no longer whether she still believed in God. Only love remained: "I know nothing," she said, "I only know that I love." This is the end of her little way, to go on and on in God's love, until everything disappears, until all assurances and securities break down. Only love remains. Love is beyond all human wisdom (cf. 1 Cor 13).

We are saved through Jesus' Cross

The cross is the end of Jesus' life and mission. On Golgotha God speaks his final word to us, the word of salvation through Jesus crucified. For ever the cross will remain the centre of the Christian community. It is our hope and salvation. How are we saved through Jesus' cross?

Scripture speaks in many ways about our salvation, which comes to us through Jesus' death. In the key text of his letter to the Romans, Paul tells us that we are justified and live our true life before God not from the observation of the law but from God's grace, "as a gift through redemption, which is in Christ Jesus, whom God put forward as an expiation by his blood to be received in faith" (Rom 3,24f). It is God's own work who "delivered us from the dominion of darkness and transferred us to the kingdom of his beloved Son, in whom we have redemption, the forgiveness of sins" (Col 1,13). This redemption is through Jesus' blood (cf. Eph 1,7).

What is salvation? It is the renewal of human wholeness, the fulfilment of our destiny as God sees us from the beginning, as he wants us to become. God himself is love, eternal love. He made man according to his image and likeness. Man is truly himself only when he loves, in the true and full sense of love, as God himself loves, as God has revealed his love in Jesus his Son. As *Jesus loves:* He loves his Father with unbroken loyalty, in face of temptation, in intimate union, in confidence and obedience to the end, on days of joy and strength and in the darkness of the passion, where his Father seems infinitely distant. Jesus loves us, his brothers and sisters, with a love that cannot be extinguished by betrayal and hatred, by insult and cruelty. Never can the light of his love be devoured by the powers of darkness. Jesus loves to the end and gives himself to us, that we should live in and through him. This is Jesus, as God saw him: "This is my beloved Son in whom I am well pleased". This, at last, is man, as he is meant to be. Man can be truly himself only in love, when he loves God, when he loves his neighbour, unconditionally.

But in Jesus God tells us also how hard it is truly to love. How easily we misuse this word "love" to stand for superficial feelings, and often this word is a cover for selfish passion. *Love of God* leads beyond all limits of creation, in which God's wisdom and power have been revealed, into the mystery of God himself where reasoning ceases, where there is only silent adoration and surrender. *Love of neighbour*—all neighbours, including strangers and enemies, is entirely different from the spontaneous flow of affection by which human lives are woven together. It has its source in God and flows like a spring from his unfathomable depth into our world. It can never be embittered, because its origin is God himself. It can never be disappointed by the weakness and cowardice it encounters in our world. It can never be dried up in the desert of our loveless society. It can never be poisoned by human meanness and cruelty.

This love is our salvation. God gives us Jesus that we should have life. But how can Jesus' love save *me*, change *my* life, liberate me from the prison of selfish narrowness? His love puts me to shame; but how can it save me? I am incapable of such love. It is beyond my powers, beyond my boldest dreams ever to reach to such heights.

True, such love, such total freedom is beyond my reach. Still it is my only salvation, not as my achievement but "as God's grace, as a gift". God gives it to me in Jesus his Son, and I must "receive it in faith". God sees me, sees all people, in Jesus his Son and through him draws us to himself into the mystery of his love. If I cling to Jesus in true faith and invite him to enter into my life, I shall grow and be transformed by 'him. I shall receive the impact of his power by looking at him in meditation: "They shall look at him whom they have pierced" (Jn 19,37). I shall also receive him in the sacrament of his love, eat the bread, his broken body, and drink the cup, the blood that was poured out for me. If I do it in faith and love he will transform me. I shall find his life and love in the community of his believers. *If we are together in a fellowship of faith and love, the mystery of God's love will become real in our world and we shall experience his saving power.*

SHORT POINTS FOR MEDITATION

1. *Jesus' saving death* (Lk 23, 26-49).

Each of the passion narratives has its own perspective. Each should be read and meditated upon within its own context. Here we choose Luke's narration, in which the passion is narrated as the last manifestation of Jesus' saving love.

— "They seized Simon of Cyrene...and laid on him the cross": Mark identifies Simon by mentioning that he was the father of Alexander and Rufus, who obviously were Christians well known to his readers. Thus it appears that Simon's family became Christian. Though unwillingly drawn into Jesus' passion Simon must have felt the impact of Jesus' person.

— "Do not weep for me but for yourselves and for your children": Jesus speaks to the women who lament over him. He is compassionately concerned more about Jerusalem, its impending disaster, than for himself.

— "Father, forgive them, for they do not know what they do": this is Jesus' prayer for forgiveness in the midst of hatred, mockery and brutal cruelty. This prayer will be repeated throughout the centuries by his Spirit-filled disciples (cf. Stephen's prayer, Acts 7,60).

— "He saved others, let him save himself, if he is the Christ of God": this mockery expresses the deep truth that Jesus is Saviour for us. Jesus' love is concerned with others, not with himself.

— "This is the king of the Jews": this sacrilegious parody too contains the deep truth that Jesus' kingly power is based only on love, that in being defeated he begins to reign.

— "Today you will be with me in paradise": this is Jesus' last consolation, not to receive relief but to give the assurance of salvation to a lost man.

— "There was darkness over the world...and the curtain of the temple was torn into two": the hour of darkness opens out our final access to God. Through Jesus' total surrender to God in death the separation of man from God is overcome.

— "Father, into your hands I commit my spirit": this is Jesus' last prayer according to Luke. This is salvation: that our life finds its shelter in God beyond darkness and death.

— "The Centurion praised God and said, Certainly this man was innocent": the powers of darkness have had their way. Now the tide turns; there is an atmosphere of peace on Golgotha after Jesus' death. The Centurion is the first to be touched by it.

2.   *People in Jesus' passion*

Powers of darkness and powers of grace are at work in Jesus' passion. They are embodied in the people whom we encounter in the passion accounts. To understand more explicitly the struggle between light and darkness in our own life and in the world it is helpful to look at the persons in Jesus' passion.

The struggle between light and darkness goes on in history with ever new faces and names. We too are involved in this struggle. In the characters of the passion-account we see some of these powers at work.

Jesus' enemies:

— *Judas*, the frustrated disciple, whose hopes were not fulfilled. He was called. Has he ever opened himself to this call without reserve? Why did he die in despair?

— The *judges* and the *witnesses* who speak against Jesus. They have convinced themselves that they are acting in God's name, to protect the sacred traditions of the nation.

— The *soldiers* who abandon themselves to cruel instincts.

— *Pilate*, the politician, who is caught in the nets of blackmail.

— *Herod,* who does not care for justice but wants to see a wonderworker.

— the blind mass of the people with no convictions, swayed by hatred.

The wavering

— *Peter* has to learn how little he can rely on his strength.

— the tired, discouraged *diciples* who still have to find a faith which cannot be broken by power.

— The silent majority of the people who love Jesus but are not prepared to take risks.

The faithful believers

— *Mary of Bethany,* from whom Jesus accepts a last sign of devotion when she anoints him.

— *Simon,* who carries Jesus' cross.

— *Mary of Magdala* who follows Jesus even to Golgotha.

— The *thief* who dies at Jesus' side.

— The *Roman captain* who finds faith on Golgotha (cf. Mk 15,39).

— *Mary,* Jesus' mother.

QUESTIONS:

1. There are many ways of meditating on Jesus' passion, mainly two: 1) To learn from Jesus to go into darkness and to face trials and pain, 2) the sharing in Jesus' passion today, in the continued struggle between light and darkness in our world and Church. Have I learned these ways of meditating?

2. Jesus has suffered because he struggled against the unjust suffering of the oppressed and rejected. Am I ready to stand against injustice when it implies risks?

FOR REFLECTION

"O Lord, my eyes look into your dying countenance and my soul kisses the bleeding wounds. O Lord, there are people who put their trust in the innocence of their life, others in the practice of austerity, one in this, one in that. But all my confidence rests in your suffering, Lord, in your satisfaction and merit".

HENRY SUSO

Chapter Four

THE SILENT DAY

"On the Sabbath they rested as the Law commanded"
(*Lk 23,56*)

This is the day of silence. In Jerusalem it is the great Sabbath when all are in their homes. The faithful few had hurried Jesus' funeral so as not to interfere with the beginning of the Sabbath when the sun set. The battle is over, Jesus' enemies have achieved their goal. Jesus' disciples are in hiding. All is quiet. This is the end of Jesus' life and the beginning of his real life. But there is a day of rest in between.

It is a day of silence also in the Liturgy with no celebration; only the office is said. It is the suspension like a bridge, between Good Friday and Easter.

It should be a day of silence also in the retreat. No new theme should be introduced to distract the attention from what is central. Tnis great mystery must sink in, the mystery of Jesus' passion and love. Jesus' blood has been shed, the tomb is closed and sealed. What does it all mean in Jesus' own life? in my life? in the life of our world?

We need this rest. God gave the commandment: "Remember the Sabbath day, to keep it holy. Six days you shall labour and do all your work, but the seventh day is a sabbath to the Lord your God; in it you shall do no work... The Lord blessed the sabbath day and hallowed it" (Ex 20, 8-11). The entire retreat is a time of silence, but this day should be kept free even from all exterior exercises. The day is meant to be spent in prayer, contemplation, adoration.

Each one may find his/her own way of entering more deeply into Jesus' passion. On the following pages a few possible approaches are given.

1. The mystery of the passion as it was realzed by Paul

For Paul the paschal mystery became the source and centre of his life and mission. His experiences, spread all through his letters, are expressed with great force in some of the great texts. We may take the bible and allow these texts to sink into our mind and heart.

Jesus' self-gift and exaltation (Phil 2,6-11).

In the words of this ancient hymn, which is quoted by Paul, I see the descent of God's love into our world: *Jesus empties himself*, discards glory and power and fulfils his mission not in the form of domination but of service, *becoming obedient unto death on the cross.* This self-emptying is his greatness; it is the only title of his exaltation. Because he allowed his name to be trodden down, he was given a name above all other names. Because he discarded power, he is Lord, and all knees bend before him.

The Paschal Mystery must be realized in the Christian life (Rom 6,1-14).

Every Christian life is patterned after Jesus Christ. The contemplation of Jesus' passion is not the remembrance of past events but the realization of what is taking place today in our own life. We cannot gather the fruit of Jesus' death and enjoy sinful self-indulgence, "continue in sin so that grace may abound" (v. 1). Through baptism we share in Jesus' life, death and resurrection. His death must be realized in our life, so that we be allowed to share in his glory: "We know that our old self was crucified with him so that the sinful body might be destroyed and we might no longer be enslaved to sin" (v. 6). This is our hope that "if we have died with Christ, we believe that we shall also live with him" (v. 8).

"Let not sin reign in your mortal bodies to make you obey their passions...but yield yourselves to God as men who have been brought from death to life" (V. 12f).

Jesus' death and resurrection in the apostolate
(2 Cor 4,7-14).

This is Paul's personal experience in his work. The fruit of his apostolate does not come from his own power but it is God's gift. He carries the treasure of Christ's mission "in earthen vessels to show that the transcendent power belongs to God and not to us" (v. 7). In his persecutions and trials he is "always carrying in the body the death of Jesus" but also the power of the resurrection so that also "the life of Jesus may be manifested in our mortal flesh" (v. 10). For Paul the apostolate is the continuation of the paschal mystery in the Church, the self-gift of the apostle that bears fruit in the community, "so death is at work in me but life in you" (v. 12).

The cross is folly for man, but wisdom of God
(1 Cor 1,17-2,5).

Once more Paul shares his experience. The Corinthians had been impressed by eloquent language and philosophical wisdom as it was taught in Greece. God, however, has chosen to speak not in human wisdom but in the human life and death of Jesus' person. The message of Jesus is shared "not in plausible words of wisdom but in demonstration of the spirit of power" (2,4). But those, who open themselves to God's word in faith will find true wisdom, the "secret hidden wisdom of God, which God decreed before the ages for our glorification" (2,7).

Jesus' redemptive passion is continued in the Church
(Col 1,24).

At times we look at Jesus' saving death like a treasure, a deposit, from which the Church through the ages can draw grace and strength. True, it remains for ever the lifespring of the Christian community, but it must be renewed in the believers. Jesus' parable of the grain of wheat that must die in order to bring forth fruit remains true also in the Church. Jesus has fulfilled his own suffering; what is missing still is the participation in his death and resurrection by his followers. "I rejoice in my sufferings for your sake, and in my flesh I complete what is lacking in Christ's afflictions for the sake of his body, that is, the Church."

2. Meeting and interviewing those who were involved in Jesus' passion.

This is a play of fantasy which obviously does not claim historical accuracy but may help us in a living way to enter into the complexity of Jesus' passion and its ultimate meaning. Such faith fantasies have a truth of their own. Also in works of literature, in dramas or novels, the dialogues never took place in reality, still sometimes they give more insights into characters and motivations than mere factual accounts. In our meditations such fantasies can be an effective means to keep our mind and imagination centred on the topic of reflection.

Meeting Caiaphas: You have achieved your goal and are rid of the disturbing movement round Jesus. Are you so sure? Why did you ask Pilate to send a guard of soldiers to watch the tomb? (Mt 27, 62-66). Do you really believe that you are now in control of the situation? that Jesus' message and mission have come to an end? You surely have learned to handle critical situations firmly, without inhibitions of your conscience, and to carry public opinion with you. But was this merely another "critical situation"? You are the High Priest and have acted as God's representative for his people. Are *you* Lord of the people—or Yahweh! You should have seen more in Jesus than a "disturbance" of your political games. As High Priest you should have been alert to God's speaking. Priests are meant to represent God—it is terrible if they try to replace him.

Meeting Pilate: Your name has been blackened! You have come even into our creed as the one who was legally responsible for Jesus' death. It is true, in a way, but the really responsible people are among the Jews. Who would not see your dilemma? You had your obvious problem with Rome, the malignant blackmail of the Jews could do you much harm. Whatever the historical details in your trial of Jesus, through those accounts I still feel that you realized something of his greatness. Some even thought you were a Christian at heart (Tertullian). We read of the warning sent to you by your wife during the trial. She had suffered much in her dreams on account of Jesus (Mt 27,19), and in the end you washed your hands: "I am innocent of this righteous man's blood: see to it yourselves" (Mt 27,24). Why should we single you out to condemn you? Your struggle, your failure have been repeated a thousand times through the history of the Church. We are all sinners. Jesus had to die.

Meeting Judas: You have been called the most enigmatic figure in the New Testament, and you have become more and more a symbol of all malice. In John's gospel you are even called a devil (6,70). But I know that you have been loved and trusted by Jesus. You were a noble character. I shall never believe that you betrayed Jesus merely for the sake of some money. You yourself were ashamed of it and wanted to get rid of it. You were gifted and had high connections. You had great hopes. All your expectations for a better world, for a true national renewal, you invested in Jesus. Some people even think that you gave him away to his enemies only to force him at last to come out with firm and powerful leadership to change the distasteful situation of the nation. And when your calculations proved wrong, you fell into despair. But whatever really was in your mind? Why did you not surrender your hopes and dreams to Jesus? Could you not trust *him?* You felt you were more clever and saw things more clearly than the others who were hardly aware of that impossible situation. You saw it and took a decisive action; and then you realized you were terribly wrong. But could you not even after your deed turn to Jesus? You knew that he loved you and never would have rejected you. One last question: *Judas, have you ever really loved?* You died in despair. But I still believe that God kept you in his love. God is greater than your sins, my sins, all peoples' sins!

Meeting Peter: Your story has been told extensively in the gospels, almost too much for me. How many sermons have been preached on your denial of Jesus! I find it so easy to understand. You were fully part of the small group round Jesus and felt safe and sheltered in it. You even were their spokesman. You realized, of course, something of the difficulties and of the growing hostility against Jesus, but this did not affect that inner security which you felt in Jesus' presence. This was his unique power, to awaken confidence and to give assurance. You simply could not imagine what would happen once you were away from his assuring presence, left to yourself. So you made those bold statements of being ready to go to jail and even to death with Jesus. Who could blame you? But it was all so different in the court of the High Priest. All were there, the detachment of police that had brought Jesus from Gethsemane, the servants, the servant girls, all talked about Jesus and laughed about him. They had set out in strength to capture him, but there was no real resistence. They had brought him in easily. It was ridiculous! Now Jesus was questioned by the authorities; what incredible charges were brought against him! And those followers of him, stupid people from the North. How they had been fooled! They all ran away—but, wait! Here is one of them. All of a sudden one of the girls made a scene: "Of course you were among them." There was no way out. A whole crowd all at once came round: "You can't deny the way you talk. You are from Galilee..." And so it happened. How you felt it when Jesus came out and looked at you; and then you broke down. You never tried to explain or to excuse yourself. You simply owned up to it. You were terribly honest. Not many are. But I must tell you that I understand you so well. We are so dependent on the people around us, we just cannot stand alone. We need support and company. We need the Church so much; today more than ever. How many of us can stand alone in the midst of all our ideologies and mass media? We need people with whom we can share our faith. Jesus prayed for you, Peter, in particular, that you should *strengthen your brethren.*

Such fantasies can easily be multiplied, but they should never become a mere play of imagination. They are a means to get into inner touch with the wealth and depth of the gospel, a way of digging for the hidden treasure.

3. Holy Saturday with Mary, his Mother

This is another fantasy, a visit to Mary in the silence of that great sabbath day, when all the experiences of her life came back to her. In her life, as in a mirror, Jesus' life, mission, passion and death are reflected.

— Would you tell me, how it all began?

— The angel told me that I would be his mother. I began to feel his presence. My whole life, my body, my mind, my whole being began to be centred on him. I can think of nothing greater than to be his mother, to feel him grow in my womb. All God's promises would be fulfilled in him, and he would be my child.

— The angel told me to call his name *Jesus*, Saviour. He was my child but he had a mission, his own destiny. He would not belong to me. From the very beginning I had to accept that he belonged to the people, to the world.

— He would be called Son of the Most High. God would be his Father. It filled me with awe. But I had to learn that all his loyalty would belong to God alone.

— I did consent to be his mother, as the Lord's handmaid. God's love and power had taken hold of my life, it was no longer mine but it belonged to God. Yes, I gave my word, but it took me a whole life to realize its full meaning. Yesterday, under the cross, I spoke the same words, but I felt their full weight: *Be it done to me according to your Word.*

— In Bethlehem the angels sang of God's glory, and of peace among men. It was beautiful, and yet so different from what people expected. We were so terribly poor and helpless. More and more I began to ponder about all this in my heart; nobody guided me, but as life went on I began to see it better.

— I took him to the temple. Old Simeon held him in his arms: A light to the Gentiles and the glory to the people of Israel—but also a sign of contradiction, and a sword through my own heart. Slowly I began to understand it.

— Nazareth was peaceful, till he was twelve. But this was too much! How anxious Joseph and I were till we found him in the temple, in the house of his Father, as he called it. This was his real home, but I could not understand it. It was meant, I guess, as a preparation for later years.

— He left Nazareth and joined that movement of John at the Jordan. It was hard; I was alone. I felt he would not come back. He had his mission, no one could stop him.

— Some of our people felt that he was out of his mind and wanted to bring him back; it was useless. I also wanted at least to see him, but he was fully in his mission. Those who hear God's word were his new family.

— I watched him from a distance. He had some truly glorious days. But I felt the growing tension, the cold isolation. How lonely he must have been!

— I heard how finally they got hold of him and condemned him; all people talked about it. I had to be with him in this hour.

I stood under his cross, alone with the disciple whom he loved, in the midst of an ocean of hatred and cruelty. The crowd pointed at me; they were curious about his mother. It was all pain and shame.

— He entrusted me to his disciple to give me a home. It was not only to have a place to stay, it was my final place, with his disciples.

— I held him when they took him down, and cleansed the wounds of his broken body. We wrapped him in linen. It was all in a hurry on account of the feast. And I was very tired.

— Now he lies in the tomb. No one can understand it: My Son, God's Son. He gave his life into his Father's hands.

QUESTIONS:

1. Our life has hours of action, hours of celebration. Our life needs also hours of silence and reflection on the divine mysteries and on my own life and experience. Do I love silence? Can I fill it with God's presence?

2. Have I learnt to see my life in the mirror of Jesus' life, of the paschal mystery?

FOR REFLECTION:

"Nothing is more paiful than suffering, nothing more joyful than having suffered. Suffering is short pain, and long love".

HENRY SUSO

PART IV

WITH HIM

Chapter One

EASTER

"He is risen, he is not here" (*Mk 16,6*).

The Final Revelation

Easter is not a continuation of Jesus' earthly life, a new chapter, which would lead his mission and work to a "happy end." His earthly life ended on Golgotha. Much less is it something entirely new, different from his life and mission in the villages of Galilee and his conflict with the leaders of the people. There is not nor should there seem to be, some sort of gulf between Jesus preaching and working in his earthly life and the proclamation of the Christian message in the early communities, or between the Jesus of history and the Christ of faith and worship in the Church. Any such separation would destroy the oneness and comprehensiveness of God's revelation in Jesus his Son.

Nor should we understand Easter merely as a subsequent approval of Jesus' claims, of his mission and message. In his entire life Jesus stands on his own authority, with no legitimation from outside, neither academic qualifications nor any sanction from the religious authorities of the Jews. His authority lies in the authenticity of his life and preaching. From the beginning the people realized the difference between Jesus and the teachers of the law, who relied on traditions and the authority of others. Jesus taught with his own authority, with an inner clarity and genuineness which can never be proved. Nor do they need a proof. Jesus' ministry is an ongoing self-revelation in which his mission, God's revelation in him, becomes more and more articulate, becomes more and more perceptible and meaningful for the people who open their minds and hearts to it. But it always remains a dialogue, a self-communication of God to us, which demands a human response and acceptance. Every person is, indeed, able to withdraw, to close himself into his own world, to become an "objective observer" of Jesus' doing and speaking. Anyone may become Christ's enemy if he feels himself threatened by Jesus' message.

Thus also Jesus' miracles are not really meant as "proofs" for his teaching. They are not attached to his teaching as an exterior support, as if a miraculous action could substantiate the intrinsic truth of his message. His miracles are *signs;* they are a different way of speaking. Through them it has to become clear and tangibly perceptible that God's reign had truly come. Jesus' message was not an empty promise, a deceptive dream, which would soon vanish in the stark daylight of actual reality. Jesus heals and casts out demons; he utterly silences the powers of destruction. The new creation is in the making. However, these signs, just like Jesus' words, must be understood and accepted with open hearts. Even these powerful signs can be rejected, argued against and ridiculed. So they have been by Jesus' adversaries.

Also Easter is not simply an argument that was subsequently added to Jesus' life and mission to substantiate the truth of his teaching and the divine origin of his mission. Rather, the Easter Event is the full and final revelation of Jesus, of what he really is, of what his message of God's reign means. Once more the disciples encounter Jesus, but his life is no longer subject to the plots and the violence of his enemies. His mission has broken through the narrow limits of the Palestinian world. God has raised him and begins to show him to us in a new and definite way, as he himself has seen his Son from the beginning, the beloved Son in whom he is well pleased. The Apostles, and all believers, learn to see Jesus in the light to Easter.

The Easter Faith

Easter is not a continuation of Jesus' earthly life, but

everything the disciples had experienced when they were with him now appears in a new light. They "recognize" him, he is the same Jesus who had died. In their own way the Evangelists express this recognition when they narrate how Jesus showed them his wounds or ate with them. But now he is taken into the life and light of God.

Thus new light falls on their experiences of the past: they had been drawn to his person and had become his disciples; they had listened and learned; often they had failed to understand and yet they had perceived something of the depth of his words. Now, on Easter, they know that Jesus has words of eternal life. All that they had seen and heard and had stored in the recesses of their memory is now awakened. It is not merely remembered but seen in a new way, in its full meaning, as we often see the significance of early events of childhood only in a mature age. All seems to fall into place. Truly, Jesus has been sent by his Father to proclaim God's kingdom; it had been embodied in his person and lived out in his life; it had already begun to take shape in their small community. All this now comes to life in the Easter experience which Luke sums up: "He presented himself alive after the passion by many proofs, appearing to them during forty days and speaking of the kingdom of God" (Acts 1,3).

Our Easter

What Jesus gave to his disciples, the true and final understanding of his life and mission, the awareness of their own vocation, their community, their task, he gives to all his followers: Jesus came for all. We too must understand his life and message, must make his mission our own and proclaim God's reign in our world. How can we do it unless he gives us the light of Easter? We too — more than ever today — need the assurance of God's care and love and the promise of final fulfilment. The light of Easter guides us on our way through our world towards the fulfilment of our mission.

This, then, is the fourth week of the Spiritual Exercises. It is the school of living in the light of the Risen Lord, in the light of faith. It is the life every Christian has to live, in the midst of situations often painful and disturbing, which defy explanations. Faith is more than the intellectual assent to revealed truth. It is a new life with a new awareness: God is present in our midst in a way that cannot be observed like the various happenings in our surroundings. Faith gives a new dimension to all our experiences. It cannot be demonstrated by means of scientific methods and yet it is more real than what the eye sees and reason explains. What did Moses experience in that great hour of his life which the bible describes in the scene of the burning bush? He would find it difficult to prove that it was more than imagination, that he really had encountered God. Still, he knows that God has spoken, his life has been changed, Israel's history has begun. God is with him in his mission, and he has the courage to face the impossible task. His entire life is proof that God was with him.

So it is with the disciples after Easter. They live in faith. God had been with Jesus during his earthly life. They had sensed his hidden presence, now it has broken out with a power, that has conquered death. Through Jesus who lives in their midst God is with them and so they will be able to continue his mission. The Church begins.

So it is in our world. The believer is aware of God's silent presence which does not meet the eye and yet is very real. Exterior conditions do not change and apparently no new resources are given to him for his work. Yet there is the inner assurance, the peace and strength which filled the disciples from the beginning. God has pledged himself to our world in Jesus Christ and can never desert us. This is Paul's confidence: "I am sure that he who began the good work in you will bring it to completion" (Phil 1,6). This awareness of God's presence in power and love, hidden and yet real, is the wellspring of the Christian life. It changes our world and gives an assurance which can never come from outside, from any human support. In this fourth week of the Spiritual Exercises we are meant to deepen and enrich this life of faith.

The directions of St. Ignatius

Thus Ignatius tells us to begin the meditations of this week with the prayer "for the grace to be glad and rejoice intensely because of the great joy and glory of Christ the Lord" (n. 221). In the Risen Lord God tells us that our world is called out of darkness into his wonderful light and love. In Jesus' earthly life love had appeared in weakness. It is infinitely vulnerable. It has no weapons, it can never use violence. Love can die, it can never kill. In the Risen Lord, God's love is irreversibly victorious. It can no longer be touched by the powers of destruction. It outlasts hatred. It is the seed of the new creation.

We are further told to meditate how "the divinity, which seemed to hide itself during the passion, now appears and manifests itself so miraculously in its true and most sacred effects" (n. 223). We should not think of divinity and humanity as two agencies which work side by side. They are two levels of existence in Jesus' own life, and in all creation. Jesus' earthly life develops apparently like that of any other man: he is strongly influenced by his surroundings immersed in his work, in conflict with the various powers of Jewish society, and finally he is destroyed. All seems to take place within the realm of our own experience. There is already, however, the divine dimension shining through; but it is perceptible only to the eye of faith. It does not impose itself. On Easter, in the Risen Lord, this hidden reality shines forth in its full brightness. The Apostles experience profoundly this divine power that raises Jesus.

It is the sovereign power of God's love. The retreatant is invited now to "consider the office of consoler that Christ Our Lord exercises and compare it with the way in which friends are wont to console each other" (n. 224). Jesus' entire being is 'for us', for the world. In the Risen Lord God's love for us is no longer hemmed in by exterior conditions; the entire Easter event is God's unique gift to us. In fact we know nothing about Jesus' resurrection itself. We know only of Jesus encountering *his own*.

The time of Jesus' apparitions—Luke puts it as forty days— is not just a last stage of Jesus' communication with his disciples. This is, rather, the beginning of Christ's final and lasting presence in his church. The apparitions are, as it were, the initiation into our life of faith, into the continued awareness of his presence in our life. The Lord is with us as the light which dispels darkness, as the life that conquers death. We cannot see the light but through the light we see all things; we cannot touch life but it vibrates in every touch. This is our life of faith which unfolds in the fourth week of the Spiritual Exercises. In this precious time we learn to live our life in the presence and power of the Risen Lord.

SHORT POINTS FOR MEDITATION

1.   The Easter liturgy

The purpose of this day's meditations is to enter into the atmosphere of Easter, as the final revelation of God in Jesus Christ, which is to be lived and realized in the Church and world of today. Thus, a meditation on the liturgy of the Easter vigil in which we celebrate this mystery may be useful. It will be helpful to use the text as we find it in the missal.

The symbols used in the liturgy:

— Fire is the age old symbol of God: powerful, elusive, transforming. Fire breaks the darkness of the night, it shines forth from the tomb. In the Risen Christ God reveals his power: "Father, we share in the light of your glory through your Son, the light of the world".

— *The Easter candle* symbolizes Jesus in the midst of the Church. It is marked with the titles of the Risen Lord: "Alpha and Omega." Jesus is beginning and goal of all creation (Rev. 1,8). "All time belongs to him": thus, the number of the current year is engraved. As the Risen Lord transcends all time, so he is present in every hour. "By his holy and glorious wounds may Christ our Lord guard us and keep us": Christ is present in world and history *in his continued passion*. The candle is lit and carried into the Church: "Christ is our light!" "Thanks be to God!" All candles are lit from this Easter candle. Jesus becomes the light of the Church and of the world here and now.

— The "Exsultet" is sung. This is a hymn of praise and thanks for the new light of Easter.

— *The liturgy of the word:* in the readings from the Old Testament the divine deeds of creation and salvation are recalled. Central among these texts is the account of Israel's liberation from slavery and the crossing of the Red Sea (Ex 14,15-15,1). This exodus remains the type for God's saving power also in the New Testament, realized in baptism.

— *The baptismal font:* the cleansing, life-giving power of water receives its full meaning in the sacrament of baptism. "By the power of the Spirit give to the water of this font the grace of your Son."

The renewal of baptismal vows: once more we are faced with the alternative of life and death. We renew and deepen the basic option which was implied in our baptism, rejecting Satan and all that belongs to him, and confessing our faith in the Father, Son and Holy Spirit. "May God keep us faithful to Our Lord Jesus Christ for ever and ever."

The Easter vigil concludes with the celebration of the Eucharist, the banquet in which we celebrate the communion of the believers in Jesus Christ and anticipate its final fulfilment in God's glory.

2. **The Women on the way to the tomb** (Mk 16, 1-8).

The biblical account of this morning expresses in a unique manner the newness of Easter. For the women, who are on their way to the tomb, this morning is the continuation of Good Friday and the Sabbath; they wish to care for Jesus' dead body. *But Easter is not the continuation of the past.* It is the beginning of the new creation, the encounter with the living God. God had spoken in Jesus in our human language; he now speaks, in the same Jesus, his own language, in supreme power and love.

— "Very early on the first day of the week...", life returns to its normal course after the distressing events of Good Friday. All seems to be as before, nothing appears changed. Could anything ever really change? Jesus' death has apparently proved once more that this world cannot change, that Jesus' attempt to offer an alternative to our society was doomed to failure.

— "Who will roll away the stone?": they set out for an impossible task. And then, what would really be the meaning of enbalming Jesus, to preserve his human form, to perpetuate him as dead? But these women go out with faith and love; God will answer them in his own way.

— "They saw that the stone was rolled back": the scene has changed. The soldiers are no longer there, a new power is at work.

— "They saw a young man...in a white robe": angels appear where human reasoning has come to an end; here God begins to speak and act.

— "There was a great earthquake" (Mt 28,2): Matthew adds the Messianic sign of an earthquake, the manifestation of God's power, the birthpangs of a new creation.

— "You seek Jesus of Nazareth...He has risen, he is not here": they had known Jesus in his earthly life, which has come to an end. They will now find him in his true life. Where shall they find him?

— "Go and tell the disciples and Peter that he is going before you to Galilee": you will not find him where you have laid him; you find him where you are meant to be, where you need him.

— "You will see him as he told you": you will see Jesus and so be able to be his witnesses. But in him you will see more—in the light of Easter you will see the world, and your own life, in a new way, in its true meaning as God sees it, as the beginning of eternal life. Today is the first day of forever!

OTHER READINGS:

— 1 Cor 15,1-19:   The Risen Lord, the assurance of our own resurrection.

— Rom 6,1-11:   Dying and rising with Jesus in baptism.

— Col 3,1-4:   The new life of the Christian.

QUESTIONS:

1. Are you prepared to share in the life of the Risen Lord by renewing once more, with personal commitment, your baptismal vows?

2. Have you known by your life of faith, in your surroundings and in the events of your daily life, the hidden presence of God, his saving love and power?

FOR REFLECTION

"Without the resurrection love would not be authentic power, without the cross this power would not be love"

J. SOBRINO

Chapter Two

THE RISEN LORD WITH HIS OWN

"For forty days after his death he appeared to them many times" (Acts 1, 3)

In ancient visions the era of the Messiah was seen as the time of God's saving presence among his own. Isaiah had called him "Emmanuel", God-with-us (Is 7, 14). This name contained for Israel the assurance of victory over the enemies who threatened the nation. The abiding presence of God with his own is the fulfilment of the covenant. This will be perfected at the end of time: "Behold the dwelling of God is with men. He will dwell with them and they shall be his people" (Rev 21, 3).

Jesus with his own

This ancient expectation comes true already in Jesus' earthly life. He proclaims God's reign as coming, God's presence in our world to heal it, to renew, transform and unite it. He not merely proclaims this new creation; it is being realized in his own person. His miracles are signs of the kingdom, the tangible manifestation of the victory over Satan, sin and death. If Jesus casts out demons in God's Spirit, it is the sign that God's reign has come (cf. Mt 12, 28).

Further, God's reign means peace: rest from labour and rest from enemies. Israel celebrated this rest on the Sabbath, which was for the Jews like an anticipation of the final rest in God's power and love. Its observation had become a central part of the Law, a test of fidelity to the sacred traditions. At Jesus' time, however, Sabbath rest had been reduced to legal prescriptions full of casuistry. It is significant for Jesus' message that he restores the Sabbath to its true meaning. For him it is the day of God's presence among his people. The assembly in the Synagogue on the Sabbath were for him occasions for healing and saving life (Mk 3, 4) and for liberating a woman kept bound many years by Satan (Lk 13, 16). These deeds of Jesus were major scandals for the Pharisees. Jesus invited all who "labour and are heavy laden" to come to him: "I will give you rest" (Mt 11, 28). This then is the meaning of Jesus' claim to be "Lord of the Sabbath" (Mt 12, 8): God's final peace is embodied in his person. Through Jesus peace is given to us.

The transparency of Jesus' person — revealing God's own presence and love for us — is found also in the parables. Jesus speaks with compassion to the people who are "harassed and helpless like sheep without a shepherd" (Mt 9, 36) and he describes the concern with which a shepherd searches after the lost sheep (Lk 15, 4-7). Who is the shepherd? The Old Testament points to Yahweh himself. Through his prophets Yahweh accuses useless shepherds for allowing the sheep to be scattered. Yahweh through Ezekiel concludes his accusation with a promise: "I myself will search for my sheep, I will seek them out as a shepherd seeks out his flock when some of his sheep have been scattered abroad" (Ez 34, 11f). In his parable Jesus does not seem to speak about Yahweh but about himself. Jesus justifies his own attitude towards the tax collectors and sinners with whom he consorted and shared meals in terms of a shepherd searching for his sheep. In the same chapter Luke narrates the parable of the prodigal son. Who in this magnificent short story is this father who installs his son again in all his rights and celebrates an extravagant feast for his return? Is it Yahweh who receives sinners? Or is it Jesus who makes these sinners and all rejected, marginalized people feel at home and accepted? Is it not God, whose inscrutable love is revealed in Jesus his Son so that in his presence we should perceive the wonder of his Father's infinitely merciful love?

This, then, is the experience of his disciples: in and with

Jesus they are close to God. The entire gospel of John is written under the heading of God's presence among us. "The Word became flesh and dwelt among us, full of grace and truth" (Jn 1, 14). In literal translation the text reads: "He pitched his tent among us" — and became the companion on our pilgrimage. As God had been with his pilgrim people during their journey through the desert giving them manna to eat, so Jesus is with us himself as the true manna, the "bread of life" (Jn 6, 35).

Thus, for the disciples to live in his presence in day-to-day life was the core of their 'formation' during Jesus' earthly sojourn. They lived under the impact of his personality. Their understanding was often slow and their petty jealousies would not die out so soon. Jesus did not mind these faults so much, as long as they were ready to continue to follow him. When he picked his intimate circle of the twelve, the first thing he wanted of them was "to be with him" (Mk 3, 14).

However, his earthly presence had limitations. As all human relations it was dependent on local closeness and expressed through the inadequate means of human communication, through words and gestures. As with all human contact it was confined to a limited circle. Even among close friends human sharing remains imperfect. No one is able to speak out his innermost being, and even the most intimate communication leaves us with a feeling of distance that can never be bridged. Besides, Jesus' life with the disciples was very short and remained open-ended. Through the ages we are faced with ever new questions which were never asked nor answered by Jesus. The early Christian community realized it (cf. Jn 14,25). But the short time of being together was sufficient to give them the taste of his presence and to make them realize their deep dependence on him. In fact this profound dependence was the reason for the apostles' total collapse at his death. It seemed that the bond of communion was broken: "You will be scattered, everyone to his own home" (Jn 16, 32).

The presence of the Risen Lord

Easter means the renewed presence of Jesus with his own. This is the meaning of the apparitions. However, it is a presence of an entirely new nature.

Even a superficial glance at the apparition accounts tells us that they are not meant to give us a continuation of Jesus' life story. It is impossible to reconstruct the chronological sequence of the events. E.g. in his concluding chapter Luke places Easter and Ascension on the same day, whereas in the Acts the two events are spaced according to the Jewish calendar so that Jesus appears to his own for forty days, and the outpouring of the Holy Spirit falls on Pentecost, when the Jews celebrated the covenant of Yahweh with his people. All the apparition accounts have the one purpose to tell us of the renewed presence of Jesus with his disciples.

The first feature of these encounters is the *initiative on the part of Jesus:* He seeks them, comes to them. Just as their first vocation had been Jesus' own call to follow him — "You did not choose me but I chose you" (Jn 15, 16) — so is also their new encounter with Jesus in which their vocation is renewed. It is the resurrected Jesus' new gift and new call. The Easter experience is not due to the surviving faith of the disciples or the still lurking hope to see him again. Any such reconstruction of the apparitions ignores the basic experience of the disciples as described in these passages. The Easter event is due to God's power once more speaking to them in Jesus the Risen Lord, making them realize that Jesus is with them forever.

Further, after Easter Jesus is with them *in a new way.* During the earthly life his presence was dependent on places and hours. He was in Peter's house, or in a village, or at the seaside. People could search and find him in the temple or synagogue. It is different now. Nobody can trace his going and coming. The two disciples did not expect Jesus when they were on the way to Emmaus returning home after their tragic disappointment. Again Jesus watches his apostles from the shore as they drag their empty net out of the water, after a whole night of fruitless labour. Jesus suddenly stands in their midst in the upper room even though they had locked all doors out of fear. And as imperceptibly as Jesus comes, so he goes. His presence has its own logic: *He is where they are in need, He is for them.*

Another characteristic is that there is nothing conspicuous in his presence. There is no trace of the sensational apocalyptic wonders, which filled the imagination of the Jews at Jesus' time and against which he had warned them even in his earthly life (cf. Lk 17, 20f). His coming blends into their daily life and has only one purpose — to make them realize that he lives and is *with them*. His apparitions are, as it were, the training of the community of believers for a deeper awareness of God's presence. They shall have to live their life and carry out their work in normal surroundings; yet they must be conscious that God, who raised Jesus from death, is with them.

Finally, all apparitions end with a *mission*. Mission is the very meaning of Jesus' coming: to bring the good news of God's kingdom and to give us the richness and joy of our life and world, a world which often looks like a desert of drudgery and pain. We are sheltered in God's hands and we are intended to grow towards the light—just like a seed, hidden in the dark furrows, gropes towards air and light. In Jesus we know that there is true life and light: in him there is joy and a final promise. This is Jesus' message; it must be shared. No one may close it into his own heart; if we did so, this would be a proof that we have not understood him at all. Where there is Jesus, there is also the new creation which must renew our world. Thus, Jesus tells the women, who come to the tomb, to go and tell the disciples. To the twelve he says: "As the Father has sent me, even so I send you" (Jn 20,21). And finally Jesus sends the apostles to all nations. In Jesus' death and resurrection, therefore, all barriers have been broken. During his earthly life his mission had been limited to the "lost children of the house of Israel" (Mt 15,24). John tells us the story how shortly before his crucifixion the disciples were embarrassed when two Greeks approached them to see Jesus. This was the occasion for Jesus to raise his eyes to a vision of the future, when he will belong to all people: "When I am lifted up from the earth, I will draw all men to myself" (Jn 12,32).

SHORT POINTS FOR MEDITATION

1. *Jesus encounters Mary of Magdala* (Jn 20, 11-18).

This is the most personal encounter of the apparition narratives. Mary was the ardent follower of Jesus who had healed her (Lk 8, 2). In him she had found her true life, her very name, which means *beloved, exalted*. She had witnessed to his death and burial.

— "Mary stood weeping outside the tomb": in this meditation identify yourself with the utter desolation of her loss.

— "She saw two angels in white...": angels are messengers. In her situation messengers are of no use. No one can replace the Beloved.

— "She turned round and saw Jesus standing, but she did not know that it was Jesus": *she* can not find him. *No one* can find him, *unless he reveals himself*. But her whole being is in search for him: "tell me... I will take him".

— "Jesus said to her, Mary": through her own name, spoken by Jesus, he reveals himself to her. *He* only knew her name, he only had given her identity. In this personal encounter she recognises him.

— "She turned and said to him: *Rabboni*": He is her teacher, her master. In him she has found herself. This encounter is Jesus' Easter gift to her.

— "Do not hold me": this encounter still belongs to her earthly life. She still has her task in this world.

— "Go to my brethren and say to them, 'I am ascending to my Father and your Father, to my God and your God' ": the joy and assurance of the Risen Lord now begin to radiate.

2. *Jesus meets the disciples* (Jn 20, 19-23).

Easter is the feast not only of the personal encounter with the living Lord but of the community of Jesus' disciples. Once again, in a new way, he is in their midst, the source of their faith and mission. It is the experience of the early church.

— "The doors being shut where the disciples were for fear of the Jews": this is the situation in which Jesus encounters them, closed in, paralysed by fear.

— "Jesus came and stood among them": his presence among them depends no longer on doors. He is where they are in need of his presence.

— "Peace be with you": this was the greeting of the first Christmas. Its meaning is now unfolded in the minds of the disciples through the experience of conflict and death. *Jesus' peace* is not an escape from trouble but the *victory of God's power and love.*

— "He showed them his hands and his side": he is the Jesus who suffered but now is in his Father's glory. Easter is the assurance that love has outlived hatred: life is stronger than death.

— "As the Father has sent me, so I send you": Jesus' own coming was only the beginning of God's reign. His mission must be continued to renew our world.

— "He breathed on them and said: 'Receive the Holy Spirit' ": the new creation is God's work; God's Spirit will work through the disciples.

— "Whose sins you forgive, they are forgiven": this is the mission: to reconcile, to overcome destruction and division.

3. *Jesus, the Good Shepherd* (Jn 10, 11-18).

This parable reflects the life of the early Christian community sheltered in God's care. It speaks of Jesus as the fulfilment of the ancient promise of Yahweh: "I myself shall search for my sheep": (Ez 34, 11). Read meditatively the whole parable (Jn 10, 1-18).

— "I am the good shepherd": Yahweh's own care for his people is blended with Jesus' person. Jesus' care for his own stands in contrast to those who misuse their authority to misguide and exploit the people.

— "The good shepherd lays down his life for his sheep": in his ultimate self-gift unto death for our sake Jesus reveals his whole being, which is love.

— "A hireling...leaves the sheep and flees": this is the mark of the false shepherd that he is concerned more for himself than for the flock.

— "I know my own and my own know me": this is Jesus' intimate union with his own. Just as he knows his Father in intimate union (cf Mt. 11, 27), so he is united with us with unbreakable bonds as our Brother.

— "I have other sheep that are not of this fold... There will be one flock, one shepherd". Jesus comes to reconcile and unite, He belongs to all. God's reign proclaimed by him and embodied in him encompasses all people.

OTHER READINGS:

— Ps 23:   The Lord is my shepherd (cf. Is 40, 11).

— Ez 34:   Israel's shepherds: their negligence, the judgement on them. Yahweh himself will now be Israel's shepherd.

QUESTIONS:

1. Reflect on the presence of Jesus in your life. Am I aware of his concern and his silent company in all situations and times of my life?

2. Which features in the parable of the Good Shepherd strike me as most significant for me: He lays down his life? He knows his own?

FOR REFLECTION:

"Put on joy which always pleases God and is acceptable to him, and rejoice in it. The joyful man does good, thinks what is good and despises sadness. The sad man always acts evil... because he does not pray and does not praise God...Free yourself of the evil of sadness and you will live for God. All those who throw away sadness and array themselves with joy will live for God." *The Shepherd of Hermas*, 2nd. cent.

Chapter Three

GATHERING HIS OWN: THE CHURCH

"In Jesus Christ you who once were far off have been brought near in the blood of Christ. He is our peace who has made us both one and has broken down the dividing wall of hostility" (*Eph 2, 13*).

Yahweh had made Israel into one nation out of many tribes, his own people. When they were scattered again in exile he gave them the promise that he would "bring back the strayed" (Ez 34, 16). *God unites!* In the prophetic vision the unity is not limited to Israel but comprises all nations: "Many nations shall come and say, 'Come let us go up to the mountain of the Lord'...and they shall beat their swords into ploughshares and their spears into pruning hooks...They shall not learn war any more. They shall sit every man under his vine and under his fig tree and none shall make them afraid" (Mic 4, 1-4; cf Is 2,2-4). These are not apocalyptic dreams of a heavenly world, once the sordid history of our world is over. It is the prophetic vision of a renewed society which has to be realized in our world.

The community of Jesus' disciples

Jesus takes up this prophetic message in the proclamation of God's reign. A new society must be born, a new community of people. He does not start with a worldwide organization but at the grassroot level, in the small group which he gathers round himself. This will become the pattern for the Christian community. Can we say that Jesus founded the Church? When we speak of him as the origin of the Church, we cannot conceive his work in terms of organization or legislation, or mapping out the institutional structures for a worldwide organisation. Jesus creates a new community in which the true Israel is being realized in brotherhood. This group has its undisputed centre in his own person. It also points towards structures, a decision-making framework, which would have to develop, because no community can be without structures. Jesus actually sees in the twelve the leaders for a much wider group of his followers. But it is not his concern now to specify these institutional structures; this will be a matter of history under the guidance of the Holy Spirit. His concern is the community itself.

Thus the time which the disciples spent in Jesus' company was of unique significance. They had to realize what God's reign really means in actual life. In fact it would have been entirely useless to issue a proclamation or to make rules and programmes if this new way of life were not really lived. Jesus initiates his disciples into *a new way of living*, of being together and relating to each other, and so to have an impact on the surroundings and on society at large.

To begin with, they had to learn to live together in mutual acceptance and adjustment. The gospels tell us enough of the tensions and rivalries which plagued the small group. They have first to be a real *community*. But then they must overcome the temptation to become cliquish and to consider themselves as a closed group. *Jesus is for the people.* So must his disciples be. When the people crowd round Jesus and get hungry, the disciples suggest to send them to the bazaar before it gets too late. Jesus tells them: "You give them to eat" (Mk 6, 37). The disciples have further to learn to overcome social prejudices which were so marked in Jewish society. It was shocking for "self-respecting" Jews when Jesus shared his meals with tax-collectors. At such violations of their sensibilities the upper classes had shown their disapproval to the disciples (Mk 2,16). In Jesus' company his followers learn to be free with people of all strata of society while giving preference to the poor and the rejected. All this does not make Jesus popular.

As a result the disciples have soon to learn to live in an atmosphere of *conflict* with the surrounding world, to be misunderstood, rejected, ridiculed, to feel hostility and the loss of human support. All this they experience in Jesus' company. They learn what it means to be his Church. All these life-experiences of the disciples with Jesus are woven into the gospels and are meant to be read by future generations of Christians.

Easter means for the disciples the renewal of their community. Jesus appears to them once more "speaking about the kingdom of God" and renewing their common mission for the world (Acts 1,3-5). The community which they had experienced on their journeys through Galilee and on their way to Jerusalem must remain for ever and it must be deepened. Their group is meant to be the living cell from which the "body of Christ," the ever growing community of faithful, will develop. After them the Church will always be called "apostolic". She is so not only through the succession of the bishops in the apostolic ministry. The community itself must be apostolic and realize faithfully and in ever new ways the characteristics of Jesus' message and of the community on which Jesus imprinted his Spirit.

The universal Church

Thus the Church must first be universal. God's reign proclaimed by Jesus knew no barriers. The struggle for the full understanding of this universality and its implications began already in the very first generation and goes on through the centuries.

The early church was tempted to fall back into the particularism of the Jewish community. Could non-Jewish believers in Jesus Christ become full fledged members of the community without first accepting Jewish laws and traditions? Already Peter had been guided by the Spirit to admit a Roman officer to baptism (Acts 10). This was a single case. But Paul broke all barriers when he extended his mission beyond the Jewish synagogues and founded hellenistic communities. Some people in Jersualem protested: "Unless you are circumcised according to the custom of Moses you cannot be saved" (Acts 15,1). The Council of Jerusalem decided the issue in Paul's favour (Acts 15,23-29), but such problems are not settled by decrees. A conversion of heart is needed, the genuine acceptance of Jesus' Spirit and the readiness to face the social consequences. In Paul's letters still rings the passion with which he fought against any narrowness.

However, through the centuries the church somehow retreated, became at times enclosed and was far from universal. She became Western in culture, in structure, in liturgy and in theology. This was especially true in her missionary thrust when the gospel was carried to other nations. In Vatican II, however, she has again articulated herself as the Church for the world and of the world, ready to embrace all nations and cultures. But decrees alone cannot bring about a breakthrough. Our horizon must be widened. All christians have to ask themselves what Jesus' message means for their continent: for Africa, for Asia, for India. How can the community be structured, the liturgy celebrated, the message proclaimed, so that it has its impact today on our world?

The Church as community

The early church was community. The apostles did exercise their authority, and gradually the structures of the local organization developed. But the Church-experience of the early centuries was that of brotherhood. Granted that the picture presented in the Acts of the Apostles is surely over-optimistic as it leaves out the existing tensions; still it brings out just this sense of unity: "The company of those who believed were of one heart and soul and no one said that any of the things he possessed was his own, but they had everything in common" (Acts 4,32).

The Church had to grow and develop structures of authority, teaching, and service. No one should regret this development which for a community is as essential as the skeleton is for the body, to give it coherence and firmness. But all structures and laws in the Church are for the "good of the whole body" and authority is to be exercised only "in the interest of the brethren."

Not only priests and bishops, all the faithful have to renew their understanding of the Church and to see it first as community, as togetherness of brothers and sisters in mutual support, help, sharing both in material things and personal life. The Church can never effectively proclaim God's reign unless "she is on earth the seed and beginning of that kingdom." (L.G. n. 5) It will be a matter of serious reflection for the retreatant in these Spiritual Exercises how he looks at the Church and how he sees his place and role in the Christian community.

Jesus in his Church

All renewal of the Church begins with the recognition that Jesus is its lifespring and centre. In daily life we perceive its human reality, often helpful and supporting, at times rigid, impersonal, even hurting, bureaucratic, authoritarian, infested with politics and rivalries. As it is a well established institution it seems often too distant from reality, not sensitive to the struggles and agonies of the people, not responsive to the needs of our time in a fast changing world. Such experiences may come up and trouble us also in days of solitude during the retreat and it is important to reflect on these. A sober assessment of the problems may lead us to a greater awareness of the complexity of most of these problems and make us more tolerant in our judgements, it may also urge us to decisions to take a clear stand.

But we should not expect—during our retreat, especially— to find an answer to all problems! We must learn to live, and also to suffer with problems in the Church in deep faith. For ever Jesus will be the only centre of the Church. This was his final promise: "I am with you always to the close of the age" (Mt 28,20). This holds good also for the local community. "Where two or three are gathered in my name, there I am in the midst of them" (Mt 18,20)—a mystery eluding our comprehension.

Thus there are two ways of looking at the Church. We can see her as a human community with its beauty and vigour, but also with its limitations, failures, even sinfulness. Or we can look at her as the body of Jesus Christ through which he is present in our midst. This twofold aspect of the one Church is meant when the Council speaks with emphasis on the Church as "mysterium" or also with the equivalent term "sacrament of salvation". "Mysterium" should not be misunderstood as something elusive which defies clear description. It means: (1) something which is very real and fully belongs to our world. (2) This reality, however, has a meaning beyond what the eye can see — in and through this visible community of the Church Jesus Christ is present in our midst. Through her very being the Church points at the saving love of God which cannot be seen. (3) Thus the visible church becomes the vehicle, the medium through which God carries on the mission of Jesus his Son, the mission of truth, salvation and love. The Church functions in many ways like a secular community with its laws, offices and manifold activities. But through these manifold channels of human relations and functions God is at work.

This holds good for those who exercise authority in the Church. Through them "the Lord Jesus Christ, supreme high priest, is present in the midst of the faithful" so that through their word "*he* preaches the word of God" and "through their wisdom and prudence *he* directs and guides the people of the New Testament" (L.G. n. 21). So it is with the community of the faithful and with their manifold gifts which are meant "to bear witness to the wonderful unity in the body of Christ. This very diversity of graces, ministries and works gathers the sons of God into one" (ibid. n. 32).

In the solitude of the retreat we may become aware of the urgent need to deepen our sense of belonging and loyalty to the Church. This does not simply mean uncritical approval of all that takes place in our communities and in the Church at large. Loyalty does not mean conformity, blind assent. In fact the one-sided insistence on obedience and the acceptance of the guidance of the Church authorities has contributed much to the lethargy of large sections of Catholic Christians who are used to shift all responsibilities onto the clergy. No one can dispense mature men and women from thinking for themselves and making their own decisions. They have also the right, and at times the duty, to speak out in the proper way.

But it always must be done with truth and love. Criticism, however right, is destructive unless it springs from genuine love.

This love of the Church, the sense of belonging in spite of misgivings and disappointments, is the criterion of the true loyalty to Jesus Christ. Jesus came to build community; whatever destroys community is alien to Jesus.

SHORT POINTS FOR MEDITATION

1. *The disciples on the way to Emmaus* (Lk 24,13-25).

With Jesus' death the community of the disciples began to break up. The two are on their way home; Jesus brings them back to the community.

— "...talking with each other about all these things that had happened": this is not a dialogue. It is a combined monologue of two frustrated men. It never builds community.

— "Jesus drew near and went with them": the first gift of his presence consists in silent listening, allowing them to open their hearts.

— "What is this conversation...": this is the presence of concern with their experiences and situation.

— "He was a prophet...we had hoped...it is now the third day...": this is their disillusionment. Their expectations are shattered. They are unable to come to terms with their experiences. No one can if he remains alone.

— "O foolish men, and slow of heart to believe all that the prophets had spoken": the two had got stuck in their problems, unable to read the wider context which they would have found in the prophets, blind and deaf even to the happenings in the immediate surroundings. They do not listen, for example, to the accounts of the women and of the other disciples. So too do we behave whenever we isolate ourselves.

— "He interpreted to them in all the Scriptures the things concerning himself": they had read these texts often enough, just as we do; but they and we need Jesus himself to explain them.

— "Was it not necessary that the Christ should suffer these things and enter into his glory?": this cannot be explained. *It can only be accepted—in Jesus Christ.*

— "Stay with us, for it is towards evening....": this is the christian pilgrim prayer, the prayer that builds community with Jesus, with the Church.

— "He took the bread and blessed...": the meal is the sign of sharing oneness.

— "Their eyes were opened": Jesus is recognized by this gesture of community.

— "They rose that same hour and returned to Jerusalem": they join the eleven again, and in their communion they once more encounter the Risen Lord and receive their mission.

2. **Rebuilding the community:** At the lake of Galilee (Jn 21,1-17).

More than other stories this account bears the mark of everyday life with Jesus present inconspicuously, directing them in their work, cooking a meal for them, and no one would ask, "who are you?" In this contemplation be personally present at this scene, immerse yourself in it with all your senses, and in faith.

— "That night they caught nothing": work without Jesus!

— "Jesus stood on the beach": there is nothing conspicuous about him. He just blends into the scenery.

— "Cast the net on the right side of the boat": such advice could come from any experienced observer. Still, by the extraordinary results they recognize Jesus.

— "They hauled the net ashore, full of large fish, one hundred-and-fifty-three": whatever the mystical significance of the number, it means *universality*, all kinds of fish (Jerome). This was Jesus' promise: "I will make you fishers of men" (Mk 1,17).

— "Come and have breakfast": in your meditation share the informal togetherness with Jesus, where no questions are needed. He simply is there.

— "Simon, son of John, do you love me more than these?": this is the *only* qualification required for his task, for the service of God's people.

— "Feed my sheep": Peter has to continue Jesus' own mission, to teach, to guide, to give his life for the community.

— "Follow me!": Peter's life will no longer be his own. More and more Peter will realize the implications of the very first invitation of Jesus, to come and to follow him. (Mt 4:19).

3. **The bond of the community** (Phil 1,3-11).

Paul was linked to the church in Philippi with bonds of special affection. It was the first community on European soil founded by him (Acts 16,11-40). Of these community bonds he speaks in the first chapter of his letter.

— "I thank my God in my remembrance of you... making my prayer with joy": in spite of the problems of Philippi Paul thinks of the community with gratitude and joy.

— "Thankful for your partnership in the gospel": Paul is always conscious of his authority and responsibility, yet he sees his relationshp to the community as *partnership*.

— "He who began the good work in you will bring it to completion:" Paul's hope never fails. It is based not on favourable conditions (which hardly existed) but on the Risen Lord.

— "I hold you in my heart...I yearn for you with the affection of Christ Jesus": his love is human as Jesus' own love for us is human. True human love is the sacrament—the visible, effective sign—of God's own love and concern for us.

— "It is my prayer that your love may abound": mutual love is the bond and basis of the true community.

— "With knowledge and discernment so that you may approve what is excellent": love must be enlightened and lead to the practical judgements which regulate our daily life.

— "That you may be pure and blameless for the day of Christ": in and through the community we grow towards Christ "to the glory and praise of God".

OTHER READINGS:

— Jn 17, 20-23:   Jesus' prayer for unity
— Eph 2,11-22:   Walls of separation are broken.
— Eph 4,1-16:   One body in Christ.

QUESTIONS:

1. What is my attitude towards the Church? To bishops, priests, religious, laity, institutions? Critical, negative or positive? Do I think of them in terms of "*they*" or "*we*"?

2. How do I relate to companions, co-workers? Do I see myself as a community builder?

FOR REFLECTION:

"Christ invited men into relationship with him: 'I have called you my friends,' he said, in a clear emphasis on the fact that the Christian life reveals itself in the way we walk with one another. 'Believe in me' he said, strongly emphasizing the personal nature of Christian faith. In the same way he did not say: 'Follow these directions', but follow me'. Over and over he emphasizes the fact that those who drink from the wells of living water do not merely commit themselves to a set of theological propositions; they commit themselves to living in the way Christ did". E. KENNEDY, *The People are the Church p. 166*

Chapter Four

JESUS SENDS THE DISCIPLES

"I will send you the Holy Spirit from the Father, and he will be my witness, and you too are my witnesses because you have been with me from the beginning" (*Jn 15,26f*).

Good News

From the beginning, the Christian message was *Good News*. The angel came to Mary with the greeting of joy: "Rejoice (*chaire*), favoured by God" (Lk 1,28). For Elizabeth the coming of Mary was a unique joy: "When the voice of your greeting came to my ears the baby in my womb leaped for joy" (Lk 1,44). Mary sings her song of praise and joy over God's wonders (Lk 1,46). When Jesus is born the shepherds receive the message: "I bring you good news of great joy which will come to all the people" (Lk 2,10).

So Jesus began his mission with a message of *joy:* "The time is fulfilled, the kingdom of God is at hand" (Mk 1,14f). When he spoke to the multitudes his words aroused people to joy: "Blessed are you poor, yours is the kingdom of God." (Lk 6,20). Even when the shadows of death fell on his life and the apprehension of his impending departure gripped the disciples, he once more assured them: "These things I have spoken to you that my joy be in you and that your joy may be full" (Jn 15,11).

Jesus' mission must be continued. He called disciples from the very outset of his ministry with the purpose to make them "fishers of men" (Mk 1,17) and Luke illustrates this intention by connecting the call of the disciples with the account of the rich catch of fish (Lk 5, 1-10). Thus it is natural that all gospels conclude with the final mission of the disciples into the world with the message of salvation.

This is the message entrusted to the Church also today. It is a message of joy and salvation, to awaken the people from fatalistic resignation and despair. But the gospel has sometimes been represented as an imposition on the people, a burden, even as a threat for those who refuse to accept it. It is important to clarify such misunderstandings, because even in the gospels we find expressions which may sound bewildering. In his last chapter Mark tells us how Jesus sent the disciples into the whole of creation with his message of salvation. But then the words are added: "He who does not believe will be condemned". Such words sound harsh. However, they express the experience of the believers who know that they are saved through faith, that refusal of faith would leave them in darkness. These words contain no judgement on people who have no access to faith. God wants to draw all people to himself in love. Love will never use force; it only invites. If I do not respond to it, my life remains empty; it is not the fault of God who loves me but my own failure. John makes it very clear: "God so loved the world that he gave his only Son that whoever believes in him should not perish but have eternal life". God's intention is *only to save*, "not to condemn the world". "He who does not believe is condemned already, because he has not believed in the name of the only Son of God. This is the judgment that the light has come into the world and men loved darkness better than light" (Jn 3,16-19). God has placed life and death in our own hands; by his gift we are free and therefore responsible for our salvation.

Today, more than ever, the Christian message must have a ring of joy and hope, but also of urgency. We know that God can speak to people and nations in a thousand ways and by his own designs draw them to himself. But we are also aware of the terrible powers of destruction. They all rise ultimately out of the human heart (Mt 15,19). If we look into our world with open eyes we see that Jesus' message of God's love and human brotherhood is not only a matter of individual salvation but of survival for our world. Jesus must find his way into our society.

We may regret that often in the past mistakes have been made in the presentation of the Christian message, when the Church felt that she had to defend her faith with the sword, or when the preaching of the Gospel was too closely connected with political or colonial powers which ruled over the nations. There is, however, little use in blaming past generations for mistakes committed wittingly or unwittingly. We need today the new awareness of the urgency of Jesus' message for our world and the joyful conviction that we, the Church of today, are entrusted with the task to share it with the people.

Our task today

The place of the Church in our contemporary world is different from her role in the past. In many places the Church has lost positions of power; even where she still enjoys institutional strength catholics may feel threatened and weakened. This should not be a matter of regret. Power intimidates. Power reduces people to subservience. Power creates distance and brings about resentment. How much resistance and even hatred against the Church has been built up during past centuries when in certain countries she dominated politics and social life and controlled education and science!

The Church has recognized that the era of ecclesiastical domination belongs forever to the past. In an irreversable process the world and the nations have come into their own. Vatican II has found an entirely new language in the Church's relation to the world. It is no longer one of ruling or simply teaching, but of solidarity, concern and dialogue. The Constitution on the Church in the Modern World begins with words of love: "The joy and hope, the grief and anguish of the men of our time, especially of those who are poor and afflicted in any way, are the joy and hope, the grief and anguish of the followers of Christ" (G.S. n. 1). Thus the Council knows "no more eloquent expression of its solidarity and respectful affection for the whole human family, to which it belongs, than to enter into dialogue with it about all these different problems" (ibid, n. 3). So we are coming closer to Jesus' own approach. He never held any power-position; he never imposed anything on people. He had his message of love and the burning desire to convey it to the people. When they accepted it he rejoiced and praised the Father. When at the end of his life he realised his failure, "he wept over the city saying: 'Would that even today you knew the things that make for peace" (Lk 19,41). Jesus' message was an offer, an appeal, love can never demand.

In this new situation, however, the Church has to fulfil her inalienable mission for the world, in a way different from political parties or ideological systems. She actually has no clearcut political or socioeconomic programme to offer—just as Jesus had none and so disappointed many who looked for immediate shortcut solutions for the problems of his time. But from Jesus the Church has her basic orientations, and it is her duty to "clarify these problems in the light of the Gospel and furnish mankind with the saving resources which the Church has received from her founder under the promptings of the Holy Spirit" (ibid. n. 3). The Gospel must be read in relation to the world, and the world must be seen in the light of the Gospel.

More than ever it will be important in our world, and in particular in India, to be aware that social and political resources alone are inadequate to renew our society. Our world is God's own creation; only in his power and love can it live and grow. This, however, can be achieved only through people who have learned to listen to God's word and who are guided by his Spirit. Men and women of spiritual insight and commitment are needed in the Church at every level—in places of leadership and responsibility and on the local level, for immediate contacts with all people. Are we not too much a managing church, busy with day-to-day programmes and immediate pressing needs? India needs the praying, listening Church, sensitive to the signs of the time and the promptings of the Spirit.

New dimensions in the Church's mission

Once we have renounced positions of power, we shall understand more readily how much the realization of God's reign in our world depends on broad *collaboration* within the Church, with all non-Catholic Christians, with all people of good will. The era of isolation is past. Our thinking and planning were too much centred on the institutional aspects: to grow in numbers and influence. There has not been enough service to the world for which Jesus was born and into which the Church is sent. We must learn to see in our neighbours not rivals but allies.

With its concern for all people Vatican II has broken through the defensive walls of isolation. In particular the *Mission Decree* is emphatic on the need of opening out to all others: "Christians ought to insert themselves and collaborate with others in the right ordering of social and economic affairs". They should "share in the efforts of those people who, in fighting against famine, ignorance and disease are striving to bring about better living conditions and work for peace in the world. In this work the faithful, after due consideration, should be eager to collaborate in projects initiated by private, public, state or international bodies, or by other Christian or even non-Christian communities" (A.G. nn. 12).

In fulfilling the Church's responsibility for the world, the role of the *laity* is indispensable. The time has passed when the laity was considered an auxiliary force meant mainly to support the work of the clergy. Modern society can be influenced and renewed only by men and women who "belong to the nation into which they were born, share in its cultural riches by their education, are linked to its life by many social ties, contribute to its progress by personal efforts in their profession, feel the problems to be their own and try to solve them" (A.G. n. 21). There remains, to be sure, many a specialized task of pioneering, for which the hierarchy and particular religious institutes are responsible. But their work can never become a substitute for the mission of lay christians living in ordinary circumstances to make Jesus' message and love tangible and effective in their daily surroundings. The entire Church—laity, religious, and clergy—is entrusted with Jesus' mission, and in our secularized society it is only through the laity that Jesus' message can be communicated.

In the great document on the Church's mission today, the Apostolic exhortation *"Evangelii Nuntiandi"* (1975) Paul **VI** uses a *new language:* The word 'mission' is replaced by *"evangelization"*. Mission in the sense of going out to people who do not know Christ remains an essential task of the Church. However, it is only part of a more *universal responsibility* of all believers to realize God's reign in the whole of human society. All Christians need to open themselves to God's word and to respond to its invitation and challenge. We have to begin with ourselves: how could we ever offer ourselves to the Lord to go out to the people and to proclaim his message as long as our own heart is narrow, prejudiced, loveless, despondent, timid! The Church herself must be constantly purified of rivalries, jealousies, ambitions, lust for power and autocracy. Evangelization means the renewal of everyone: hierarchy, clergy, religious and laity. The awakening of all Christian communities is called for, a renewal in mutual acceptance, love, and support. We need the inspiration of the youth to pledge themselves for a new world; we need the revival of the Christian family as the cradle of Christian life.

In the Spiritual Exercises the retreatant will reflect on his/her own mission. Each one has to realize it in his/her own world, within the frame of his/her own social and professional surroundings. Each one, however, must be aware that no spiritual renewal is possible in isolation, apart from the world to which he/she belongs. All share in the one mission of the Church to build up the body of Christ.

SHORT POINTS FOR MEDITATION

1. *Jesus' own mission* (Mk 1,21-39)

If you ask yourself how your mission may look in daily life, read the first chapter of Mark which follows Jesus in his earthly mission:

— *Jesus teaches.* Mark does not give the content of the teaching but stresses the way of teaching: "With authority". This points up the absolute authenticity of Jesus' teaching.

— *He casts out a demon.* Jesus liberates from the powers of evil and destruction.

— In Simon's house Jesus *heals* the mother-in-law. Jesus has profound personal concern for people.

— In the evening "he healed many who were sick..." this is the touching scene of Jesus' universal concern.

— In the morning "he went out to a lonely place and there *he prayed*". His involvement with people is always balanced with the intimate closeness with his Father.

— "Everyone is searching for you"..."Let us go to the next towns." Jesus' mission is not confined to one place; he is alert and ready for ever new horizons.

2. The final mission of the disciples. (Mt 28, 18-20).

Jesus' final words to the eleven disciples remain the inspiration for the Church's mission through the ages.

— "All authority in heaven and on earth has been given to me": what was Jesus' authority in his earthly life? He had no power, no political backing; he never used coercion. He leaves his disciples free: "Would you also go?" (Jn 6,67). His only authority is truth and love. What is the authority of the Church? Do I give true freedom to all?

— "Go": it is the urge of love not to close in, to wait, but to go out, to meet, to serve.

— "Make disciples": the goal of christian mission is not primarily doctrinal teaching, nor organization. It is *building community*, bringing people together in Jesus, to embody in this community his Spirit, God's kingdom.

— "Of all nations": all people are brothers and sisters. Their communion is based not on social or racial affinities, but on God's all embracing love.

— "Baptize them in the name of the Father, and of the Son and of the Holy Spirit": baptism is not a ritual but a consecration. Those who commit themselves to God's kingdom have God as Father. They are brothers and sisters of Jesus; they are renewed and united in his Holy Spirit. The community of the Church is not man-made; it is born of God.

— "Teaching them to observe all that I have commanded you": share Jesus' message! All teaching and preaching of the Church must ring with the joy of Jesus' own message. This is the norm by which we ought to examine all our teaching.

— "I am with you always to the close of the ages": the ancient assurance given to the poor of Yahweh— Moses, Gideon, Mary—is now extended to the whole Church, for all times.

3. *"Evangelii Nuntiandi"* (1975) gives us the modern perspectives of evangelisation.

This document embodies the contributions made by the bishops in the Synod on Evangelisation, 1974. Pope John Paul II called it the testament of Paul VI. Read the main numbers of the document of which the key words are given here:

— Evangelization is not limited to preaching and baptizing, n. 17.

— It consists in "bringing the good news to all strata of humanity and through its influence transforming humanity from within and making it new". Its aim is "to convert the personal and collective consciences of people", n. 18.

— Its goal consists not in mere geographic expansion of the Church but in "affecting and as it were upsetting through the power of the Gospel mankind's criteria of judgement...", n. 19.

— The Gospel is independent of any culture but "capable of penetrating them all without becoming subject to any of them" n. 20.

— The primary content of evangelization is the "witness that in his Son God has loved the world", n. 26.

— The Gospel concerns "man's concrete life both personal and social", n. 29.

— In particular, it concerns liberation from all enslaving and destructive forces: "The Church has the duty to proclaim the liberation of millions of human beings... the duty of assisting the birth of this liberation, of giving witness to it, of ensuring that it is complete. This is not foreign to evangelization.", n. 30.

— However, evangelization must not be reduced "to a man-centred goal...to material well-being... forgetful of all spiritual and religious preoccupations...If this were so the Church would lose her fundamental meaning." n. 32.

FURTHER READINGS:

— Mt 9,35-10,9:   Jesus' mission must be extended and continued through the Twelve.

— Lk 4, 16-21:   Jesus' mission of liberation for the poor.

— 1 Cor 3,5-27:   Servants of the Lord.

— 1 Cor 9,12-27:   All things to all men.

QUESTIONS:

1. How do I understand the mission of the Church today? In India?

2. What is my specific task within this mission?

FOR REFLECTION

"We are fortunate to be living in an age that is truly great for the Church; for never in the past has mankind had such a chance to realize to what extent Christian love in all dimensions is the indisputable law for our very survival, the survival of humanity, for progress and for a peaceful organisation of the earthly city."

R. VOILLAUME

Chapter Five

OUR HOPE

*"All creation waits with eagerness for God
to reveal his sons" (Rom 8,19).*

The entire Bible is a history of promise and hope. God created an *unfinished* world and entrusted it to us— "Multiply, fill the earth and subdue it". All God's gifts are tasks and promises. When we failed, and the powers of evil engulfed us, his assuring promise gave us new strength. God will always remain with us in the struggle against evil; the seed of the woman "will bruise the head of the serpent" (Gen 3, 15). When Israel's history began Abraham was told to leave his own country with the promise of the land that God will show him (Gen 12,1). This land of promise remained the magnet that attracted Israel through the millennia. It was the hope of Israel when under Moses' guidance they moved out of Egypt. "I will bring them to a good and broad land, a land flowing with milk and honey" (Ex 3,8). Later, in the distress of exile, Yahweh speaks once more: "I will take you from the nations...and bring you into your own land" (Ez 36, 24). So it goes on until the Bible ends with the vision of fulfilment: "And I saw the holy city, the new Jerusalem coming down from heaven" (Rev 21, 2).

Man cannot live without hope; every day has its little expectations. Children wait for the next holiday; we look forward to promotions, to meeting friends. Young people plan their future and look out for their partners.

Hopes are dangerous. Many hopes remain unfulfilled, or the fulfillment does not come up to our expectations. We are disappointed, disillusioned. Many people are frustrated if obstacles against their hopes pile up too high, and life seems to deny the promised fulfilment. And so some become bitter, cynical; they close themselves in.

At the end of the retreat I must ask myself earnestly: What really are the hopes that sustain my life?

Jesus' Hope

The hopes of people are different, but have a common origin. All hopes grow from the inadequacy of the actual conditions in which we are living and reach out towards the unknown, into a world of desires and expectations. These hopes shape the entire life of people, of whole nations. Jesus knew it when he said: "Where your treasure is, there is your heart also" (Mt 6, 21).

Jesus lived in a restless world and was surrounded by different hopes, individual and collective. All Jews at his time felt the need of something new to come in the political, social and religious decadence of their nation, but they looked in different directions for an answer. There were those who still hoped for the political restoration of Israel under a new Son of David, a revolutionary hero who would lead them to armed victory. History tells us how these hopes finally led to the inevitable military confrontation with the Romans and to the destruction of Jerusalem. — There were, on the other side, the apocalyptic hopes, the expectation of God's powerful triumph over this evil world and the establishment of his own kingdom on the ruins of our world and history. This hope embodied the unshakable faith of the nation in Yahweh's power, but it did not change the actual world and led many to narrow self-righteousness. They gloried in the observation of the law and the traditions but did not touch the actual urgent tasks that lay at their doors.

Jesus lived in the midst of these hopes, dreams, illusions. He was not carried away by any of them. He had his own message of hope: God's reign is at hand, it is in the making, but it is not yet fully revealed. It still is a promise. It is like the seeds that still sleep in the furrows of the field with the promise of the harvest.

Jesus does not offer any spectacular hopes, nothing that can be seen and admired. Such would only be a diversion from his real message, from God's reign, the new creation, the new communion of brotherhood among all people, which does not come through human power, but from the Spirit. It must be born in our hearts and it has an assurance of fulfilment, which is based not on human resources but on God alone.

Jesus has to live out this message of hope in his own life. In his own person he had to realize what hope is. He has experienced both its joys and its agonies. No one could speak with greater assurance about the coming of God's reign than he. God's reign rings through every one of his parables: the seeds sown in the field will bear fruit, though many seem to be wasted; the good harvest will be brought into the barns in spite of the tares which were sown in the fields; what joy fills the heart of the lucky merchant who has discovered the precious pearl! And to those who earnestly seek God's kingdom, Jesus promises that all the rest will be given to them (Mt 6, 33). Yet, no one's faith has been tested as that of Jesus: He realizes the growing resistance against his message and person; even his disciples in their pettiness seem to be blind to the vision of true brotherhood. Where is God's power when Jesus faces the final breakdown of his efforts and he offers to God "prayers and supplications with loud cries and tears" (Hebr 5,7)! In going through his agonies Jesus becomes "the pioneer and perfecter of our faith", because "he endured the cross for the joy that was set before him, despising the shame" (Hebr 12, 2).

This is the hope Jesus communicates to the disciples during his earthly life. They too, like all of us, were tempted to look for shortcut solutions in the search for a better world. They felt like joining the political movement of the zealots, but in their heart they sensed that this was not the answer to the real problems of their world. God's reign is more than the change of political structures. Neither would Jesus allow his followers to yearn for spectacular events such as were expected by the apocalyptic writers. "The kingdom of God is not coming with signs to be observed, nor will they say 'here it is' or 'there'. For behold, the kingdom of God is in the midst of you" (Lk 17, 21). This is Jesus' sobering answer to all such dreams: never expect the world to be changed by spectacular events; God's reign has nothing to do with sensational happenings. Unless his reign comes here, in our midst, in the world of our daily life and work, it will not come at all.

Jesus then tells us that when God's reign in fact does arrive, we can hardly observe it. He tries to make it clear in some of his parables in which he uses striking contrasts: the reign of God is like a mustard seed, which is the smallest of all seeds. But it grows into a tree and gives shelter to the birds of the sky. Again it is like the leaven which a woman takes and mixes into three measures of meal till all is leavened. (cf. Mt 13, 33-35). There seems no proportion between these insignificant beginnings and the final outcome. This precisely is the coming of God's reign. Jesus therefore trains the disciples in this hope. They see this hope embodied in him: He gives the vision of a new world which comes from God and must be realized by us. In him we see the courageous struggle against evil. We see also his rejection, the agony, the patience, the unfailing faith. In his resurrection the final victory is revealed.

The hope of the Church

What hope really means could be understood by the disciples only after Easter. Whatever earthly expectations they had pinned on Jesus had died and had been buried with his dead body. In the encounter with the Risen Lord they learned that real hope which carried the pilgrim Church through the centuries of her history.

The paschal hope first gives them the assurance which only God, who is the source of all life, can give. The disciples encounter Jesus alive, he who had died and was buried. The Easter revelation comes while they are still under the shock of Golgotha's disaster. Now Jesus lives and stands in their midst, the same Jesus who had been nailed to the cross. Here God is at work, God Creator and Saviour, "who gives life to the dead and calls into existence the things that do not exist" (Rom 4, 17). From now on they know that both their life and God's kingdom are not based on their own resources but have their roots in the unfathomable mystery and power of God, in his love which they cannot explain, for which they have no claim. Yet, this love is with them; they can rely on it. This is the experience which Paul has in his apostolate: "We are afflicted in every way but not crushed; perplexed but not driven to despair; persecuted but not forsaken; struck down but not destroyed, always carrying in the body the death of Jesus so that the life of Jesus may also be manifested in our bodies." (2 Cor 4, 8-10).

As every great word of the Bible can be misinterpreted so also the biblical message of hope. Already among the first Christians at Thessalonica there were discipes who so banked on the early coming of the Lord in his Parousia that they thought there was no need to continue in their daily work and toil. Paul has hard words for such people: "We command you, brethren, in the name of the Lord Jesus Christ, that you keep away from any brother who is living in idleness and not in accord with the tradition" (2 Thes 3, 6). Never is the hope for the coming glory an excuse to cease from labour in this world or to avoid struggles and conflicts in our life. Jesus is for ever the sole model of our hope: his life was a struggle against evil, a struggle which was crowned in his resurrection.

Christian hope is not idle. It is the hope for a harvest which must be sown with confidence. It is the assurance of *God's* victory in us over evil, sin and death, but only after *we* ourselves have fought against the dark powers. Hope in the Risen Lord gives vigour to the work and struggle of every Christian and inspires the Church in her mission. It also gives strength to persevere when human efforts fail. God's reign is entrusted to us. It is for the Church to proclaim and realize it on this earth; it will be for God alone to reveal it in its fulfilment.

True, hope looks towards the future; it is waiting for fulfilment. But hope does not turn us away from the present, from our actual world. The future is already with us as the tree is already in the mustard seed. God is with us. The indwelling Holy Spirit is the "guarantee of our inheritance" (Eph 1, 14). Our world is pregnant with the Christ-life of salvation and our agonies are the travails of this new creation. In Jesus' own language: "When a woman is in travail, she has sorrow because her hour has come; but when she is delivered of the child she no longer remembers the anguish for joy that a child is born into the world" (Jn 16, 21). Our waiting for the new creation has nothing of the dull emptiness of a waiting room; it is filled with the active joys of Advent. The thoughts of a mother are with the child in her womb, while her body gives it shelter and builds its limbs. So it is with the Christian's hope for God's kingdom. We are constantly busy, building up Christ's body in our world, sharing also in his struggle and passion, so that we "may know him and the power of his resurrection" (Phil 3, 10).

From the very beginning the Gospel was a message of hope. Early Christians had nothing to boast about or to rely upon. They were a poor community, made up to a great extent of slaves — suspected, harassed, persecuted. But they could never be broken. They had an inner strength from the Risen Lord that made them face life and death without fear. In the midst of a society characterized by aimless drifting, cheap pleasures, and fatalistic resignation, Christians were united in their faith in Jesus Christ, who had conquered death.

SHORT POINTS FOR MEDITATION

1. The Ascension (Acts 1, 3 -12).

The Ascension is the concluding account of Jesus' life and the first scene in the history of the Church. It comprises both sides of the Christian hope: the earthbound mission of the disciples into the world, and the divine assurance of fulfilment.

— "He presented himself alive after his passion": this is the meaning of his apparitions — Jesus lives and is with his own.

— "Speaking about the kingdom of God": this was Jesus' message which is now entrusted to his Church.

— "You shall be baptized with the Holy Spirit": just as Jesus went about his mission guided by the Holy Spirit (Lk 4, 14:18), so the Church can fulfil her mission if she is guided, and allows herself constantly to be led, by the Holy Spirit.

— "Lord, will you at this time restore the kingdom of Israel?": even after the experience of Jesus' death and resurrection the disciples are still unable to forget their dreams of power and earthly dominations. Will the Church ever outgrow such dreams!

— "You shall be my witnesses in Jerusalem...Judea and Samaria and to the end of the earth": these are the ever widening circles of the Church's mission.

— "As they were looking on, he was lifted up": they experience the end of the first phase of their discipleship, physical companionship with Jesus.

— "A cloud took him out of their sight": for Jesus this separation means the fulfilment of his life in union with his Father. The cloud is the symbol of God's presence.

— "They were gazing into heaven as he went": This is the situation of the Church in the world. She comes from God, she goes to God. But her place is on earth.

— "Why do you stand looking into heaven?": their task is now *on earth*. Jesus will bring history to fulfilment.

— "Then they returned to Jerusalem": in your meditation be with them on the way back into the holy city, to wait for the Holy Spirit, to be ready for their great task. This way back to Jerusalem is the beginning of the Church through the centuries of her history and mission.

2. *The expectation of all creation* (Rom 8, 18-26).

Read this text with a deep sense of solidarity with the entire creation, with all the pain and agony of our world, with all who struggle for a better world and often lose hope.

— "The sufferings of this present time are not worth comparing with the glory which is to be revealed to us": this is the vision of faith which comprises the struggles and agonies of our world in the light of the final glory which shines forth in Jesus Christ.

— "Creation was subject to futility": it loses its very meaning if man, the crown of creation, betrays his destiny. World and history have meaning only in man. They lose their meaning if man is no longer God's image. The world is waiting, God has given us hope.

— "Also we ourselves, who have received the first fruit of the Spirit...wait for the adoption as sons, the redemption of our bodies": we are redeemed and sealed by the Holy Spirit, but still, along with the whole world, we wait for the full revelation of our sonship.

— "In hope we are saved": we *are* saved! Work and struggle are not lost but sustained by God's promise, to which we cling in hope.

"We hope for what we do not see": this is the very essence of Christian hope. All human hopes are based on empirical data, on our calculations; they are spelled out by computers. The hope of the Christian life, of God's kingdom, has no verifiable basis. It rests on God alone.

— "The Spirit intercedes for us with sighs too deep for words": it is not possible to rationalize and spell out the ultimate assurance of our life. The Spirit in the depth of our heart leads us on.

3. *The final splendour* (Rev 21, 1-5).

This is the vision of fulfilment in which human history is completed. It is expressed in apocalyptic language.

— "I saw a new heaven and a new earth" New means closeness to the origin. In the biblical sense heaven and earth are new because in their origin they come from God, they are sustained by God, and are fulfilled in God.

— "The first heaven and the first earth have passed away": the limitations, struggles, anxieties of this life have come to an end.

— "The sea was no more": the sea was considered the realm of chaos, danger and destruction. All these have been vanquished forever.

— "I saw the holy city, the new Jerusalem, coming down out of heaven from God": it bears its old name, it is the city which had been built by human hands, with the glories, anxieties, catastrophies of its history. Now it comes from God, it is the holy city as God meant, it, as it is fulfilled in his glory.

— "Prepared as a bride adorned to meet her husband": this is the ultimate destiny of all creation, to be fulfilled in God.

— "God will dwell with them, and they shall be his people": this recalls the old covenant with Israel on Sinai. His people now comprises all nations; they are fulfilled in glory, beyond human reckoning.

— "He will wipe away every tear from their eyes and death shall be no more": The pain of our broken world is healed.

— "Behold, I make all things new": I hear these words as spoken to me. I belong to the new creation.

OTHER READINGS:

— Jer 30: God restores Israel.

— Jer 31: 15-22: I have created something new.

— Rev 5: The mystery of history, the book with the seven seals.

— Rev 21 and 22: The vision of fulfilment.

QUESTIONS:

   1. Do I realise the difference between the many little hopes that fill my daily life, and my one big HOPE?

2. Does this HOPE give me vigour in my daily life and in my apostolic work? Does it inspire me with patience, perseverance, assurance, courage?

FOR REFLECTION

"The secret of prayer is a hunger for God and the vision of God, a hunger that lies far deeper than the level of language and affection."

THOMAS MERTON

"We could not seek God unless he had already found us. We cannot learn to desire God because he must find us first."

PASCAL

Chapter Six

THE SPIRIT OF JESUS

"If you love me, you will keep my commandments. And I will pray the Father, and he will give you another Counsellor, to be with you for ever, the Spirit of truth (Jn 14, 16).

What will happen after Jesus? It is the question asked in every new enterprise or movement which was conceived and built through the vision and personality of a great man: who will succeed him? Whom has he groomed to carry his work into a new era?

Jesus has no successor. He is and remains the only Lord of his Church. Offices in his Church are not meant to replace him but to make him, his own authority and saving grace, present and effective to all later generations. The community of disciples will for ever remain his own Church he has assured us. *"I shall be with you to the end of the ages"* (Mt 28,20).

But how will the authenticity of his teaching be guaranteed? Who will interpret it to future generations? Who will hold the disciples together and save them from the divisive forces which are bound to buffet them? Who will preserve among them the unity which will be threatened and attacked from the beginning and which will be assaulted more and more as numbers increase and as the Church will include ever more heterogeneous groups and cultures? Most important, who will uphold and foster the impact of his personality, the assurance he was able to give to his disciples even in critical situations? As long as he was with them they always felt sure, too sure, of their future and their mission, but as soon as he was taken from them on Mount Olive they were scattered. Who can take his place?

The other Advocate, the Counsellor

Jesus' discourses during the last meeting with the disciples (Jn 13-17) deal precisely with this problem. He assures them that he will remain with his own through his Spirit, the Advocate, the Counsellor, whom he will send from the Father. So powerful is this assurance that Jesus makes the startling statement: "It is to your advantage that I go away, for if I do not go away the Counsellor will not come to you; but if I go I will send him to you" (Jn 16,7). Jesus' presence among the disciples was needed to give them the concrete experience of his personality and of God's kingdom embodied in him. His disciples had to realize it in actual life, by the way Jesus related to them and to the people: they had to experience him in the face of adversaries, in suffering and death. There is no substitute for this existential encounter with his person. In Jesus God has revealed himself to us. In him he has revealed also the full meaning of our human existence and of true community.

But the encounter with Jesus in his human life should not become a limitation. The disciples are not meant just to repeat his words and to copy his deeds. God surely did not want that all followers of Jesus should belong to that small, narrow Palestinian world and culture in which Jesus lived. Thus Jesus had to go, but his message belongs to the whole world. He sends the Holy Spirit, his Spirit and the Spirit of the Father.

Who is this Spirit of Jesus? Perhaps we should ask first more simply: what is Jesus' spirit? What is the spirit of a person? It comprises the richness and depth of his life. It includes the vision from which his message and teaching spring, the sustaining power of his work and struggle, his attitudes towards nature, towards the world, towards people. But it is much more; it simply cannot be defined. It is the atmosphere people breathe when they come into Jesus' presence, the peace and rest he radiates in the midst of tensions, the joy he inspires and the attraction that draws people spontaneously to himself and to the Father. It is the lifespring of his actions and words which animated all encounters with people. It is the unspeakable wonder of God with which his life was filled, beyond understanding—because the spirit is like the wind; you do not know from where it comes and whither it goes. And yet, this spirit is so much more real than the single words or deeds which people perceived. It is the divine wonder which does not take him away from us, which makes him truly human, so that in his presence we actually begin to understand what it means to be human and what our life ought to be. At the same time it makes us feel closer to God than any sacred celebration, so that in him we begin to realise who God is.

This Spirit of Jesus is *God* himself—not God in the distant heaven whom no one has ever seen but God present in our world. God's spirit it is who moves over the primeval waters so that chaos becomes cosmos and order. God's Spirit enters into dead matter, so that it begins to breathe. It is the Spirit who comes and enlivens Israel's history, so that broken people suddenly become valiant and free. It is the Spirit who in times of despondency and confusion gives guidance and new confidence through the prophets. It is the Spirit through whose power Jesus was conceived. He descended on Jesus at the beginning of his mission and guided him in every stage of his life. It is the Spirit whom the disciples need. They will not be allowed by this Spirit merely to copy Jesus; nor will He let them just repeat what Jesus has done and taught. The world is wide open with its immense potential, and history must unfold creatively through the centuries. Our entire world must become God's kingdom; this can come about only through the Spirit of Jesus Christ. Jesus' earthly life must end; his Spirit must come.

Jesus' Spirit in the Church and in the World

Jesus' promise of the Holy Spirit is expressed in three key words: (1) The Father "will give you another Counsellor (Advocate), *to be with you for ever*, the Spirit of truth" (Jn 14,16). Jesus' mission is not confined to a particular situation. God's reign is co-extensive with the world and its history; it is meant to become the new creation. While Jesus goes, his Spirit must remain, the same divine presence and power that guided his earthly life in the villages of Galilee and in Jerusalem. He is the Spirit of truth, source and lifespring of all that exists and moves. He leads all creation to its destiny, through Him it is to be transformed into God's kingdom, to become the holy city of God. He will be with us *for ever*.

(2) Further, "*He will teach you all things* and bring to your remembrance all that I have said to you" (Jn 14,26). He will not add anything to Jesus' message but he will explain and interpret it. "He will not speak on his own authority, but whatever he hears he will speak" (Jn 16,13). This may sound bewildering. Has not God spoken through the ages with increasing clarity and depth, in Israel, and in all nations, through Moses, through prophets, through wise and holy men and women? Would God ever cease to speak to us? It is indeed difficult to conceive of Jesus as the consummate end of revelation as if in the future we would have only to hand on, to defend and clarify what has been told already in the past. Is not the whole of human history an ongoing revelation of God in and through man with ever wider horizons and ever deeper insights into the mystery of man, world, and of God himself?

So it is, and in fact each human being is called upon to have his and her share in this ongoing revelation. Every human life has its own depth and in every human person there is something new, something unrepeatable of God's wonder. Still, there is a finality in God's revelation through Jesus, something unique and unrepeatable. God tells us in Jesus that this world of ours is, and for ever will be, in his hand. This will be questioned a thousand times. People will be tempted to consider their life as lost and the whole world a gamble. Jesus does not discuss our problems, but in his entire life, death and resurrection he spells out the unspeakable truth: "God so loved the world that he gave his only Son, that whoever believes in him should not perish but have eternal life" (Jn 3,16). God gave us freedom and we have wandered not only through valleys of tears but also through the mysterious, dreadful abysses of wickedness and hopelessness. But no one can take away or change what God has told us in Jesus, that he loves this world of ours and that it will find its fulfilment in him.

All happenings, therefore, in this world are securely embraced in God's sovereign providence and incorporated into the final destiny of all creation. There is then no limit to growth in this world or to ever new insights. New cultures may be born and new philosophies develop. The explosion of knowledge will continue and human control over the resources of this world will be ever stronger. New problems will arise, evil will increase, and it will make use of the ever greater powers at our command. All this, however, cannot change our ultimate destiny: this world belongs to God. This is the revelation in Jesus Christ, it is the work of the Holy Spirit.

(3) Finally, the Counsellor, the Spirit of truth *"will bear witness to me; and you also are my witnesses* because you have been with me from the beginning" (Jn 15,26f). A witness is more than a source of information. He who gives witness in court not only contributes his knowledge but stands for its credibility. This is what Jesus' disciples need in their mission, what we today need more than ever. Jesus' message will be questioned, challenged, attacked. Right from the beginning it had been rejected by the leaders of the people and the majority of the Jews. Many of these seemed to be well intentioned people. Were the few disciples who believed in Jesus right? And, suppose Jesus was right in his time, is his message anything more than a reaction against the social and religious decadence of his time? Has it value in other cultures, in later ages, for *all* people, for *me?*

Such questions cannot be answered by mere rational arguments. They have been discussed through centuries. Those who reject Jesus' divine mission will continue to interpret the events of Jesus life and person by means of social and psychological categories. The believers in Jesus Christ will offer counter arguments. Such discussions are useful because the complexity of the problems comes more and more to the surface. However, we should not expect them to lead to cogent conclusions. What then makes us really see the truth? Where shall we find a final assurance?

Take an example from daily life: you have a friend. You know his mind, his aspirations, his 'spirit', and you trust him.

Someone tells you a malicious story about him, that he has committed some fraud. You simply laugh. "I know him, this is just not possible!" What are your arguments? You have studied the evidence which may, or may not be fully conclusive. But your conviction is ultimately based not on any exterior evidence but on the knowledge you have of your friend. You know his 'spirit'. This is exactly what Jesus tells us about the ultimate assurance of our faith. Jesus sends the Holy Spirit, the same Spirit that descended on him and guided him. He is the Spirit of God, who in the beginning hovered over the chaos. He is the Spirit of truth. The Spirit alone comprehends all that exists. He opens your eyes to the wider context of events, the fuller sense of words, to the hidden truth that can only be grasped intuitively by the heart. He does not dispense you from thinking, but He gives you the clarity and assurance, you may call it a *spiritual instinct*, that cannot fail you. So you know that Jesus is right and that his message is valid also today, for our world. You also know that the true meaning of your life depends on the faithful-lived-acceptance of his message. You will not be afraid of present or future ideologies. You will always remain ready to learn and joyfully recognize whatever is true and beautiful in the world of ancient religious and in modern movements. Never will you be a narrow fanatic, for God's Spirit will keep your heart open and at peace.

Thus will you also be able to be a witness to Jesus — as he told his disciples—because you too have been with him and are guided by his Spirit. You are not merely a teacher who is able to explain doctrine but a person of living faith and conviction. You will be able to stand by your words, for you carry out in your life what you speak and teach. The joy and strength of Jesus' message, of the Christian vocation, will be progressively embodied in you so that also your words will carry more and more conviction. You are his witness. This is what Peter told Christians who lived in totally non-Christian, hostile surroundings. Through your life you will "declare the wonderful deeds of him who called you out of darkness into his wonderful light" (1 Pet 2,9).

This is also the final gift of your retreat. You followed Jesus in your meditations through the days of his earthly life and have patterned your mind and heart after him. You need his Spirit and the assurance of his continued presence and guidance in your personal life and in your mission. His Spirit gives you the assurance that through Him your own life and work are woven into God's design, the coming of God's reign!

SHORT POINTS FOR MEDITATION

1. *God's Spirit in Israel's history*

The prophets hailed the messianic time as the era of God's Spirit.

— "I will pour out my spirit over all flesh" (Joel 2,28f; cf Acts 2,17f): no longer will the Spirit of God be given only to prophets and leaders of the people. God's entire people will be filled with him, nen and women, young and old.

— "I will put my spirit in you and I will see that you follow my law" (Ez 36,27): the laws of the old covenant were written on tablets of stone, and imposed on the people. The law of the new covenant will be imprinted into human hearts: God will lead his people from within, in freedom, through the Spirit.

— "I baptize with water, he will baptize you with the Holy Spirit" (Mk 1,8): through his baptism John the Baptizer led the people to repentance. Jesus not only prepares their hearts for God's reign but shares his own Spirit.

— "If anyone thirst, let him come to me and drink, who believes in me. As Scripture says, 'out of his (Jesus') heart shall flow rivers of living water'. This he said about the Spirit which those who believed in him were to receive" (Jn, 7, 37f). This is Jesus' own promise of the Holy Spirit. It is connected with the feast of the tabernacles on which the Jews commemorated God's mighty deeds during the journey through the desert. On the last day they carried water jars in procession to celebrate the water from the rock given to them by Yahweh.

This gift is fulfilled in Jesus' own person. The symbol is fulfilled in Jesus himself when through his death he gives us the Holy Spirit. It is symbolized on the cross when from the opened side of the expired Jesus blood and water flow (Jn 19,34).

— The most explicit promise is contained in the texts on the "Counsellor": — Jn 14,16f; 14,26; 15,26f; 16,7-15. cf. above.

2. *Pentecost* (Acts 2,1-21).

It is the harvest feast, which in later Judaism was associated with the Covenant on Sinai. Thus Luke sees in the Holy Spirit's outpouring on the Church the inauguration of the New Covenant.

— "They were all together in one place": Mary with them (Acts 1,14). It is the beginning of the Church and its mission.

— "A sound came from heaven like the rush of a mighty wind": the invisible power of the wind symbolizes the presence and working of the Holy Spirit in the Church (cf. Jn 3 8).

— "Tongues of fire which parted and came to rest on each of them'": fire is the ancient symbol of God (cf. the burning bush). It descends on each one individually. The church is the communion of the believers; each one is called personally. Each one has his gift of the Spirit, all are united in this same Spirit.

— "They began to speak in other tongues": all understood them. This is the sign of the new unity, based on the one message of salvation, understood and received by each one within his own life-situation.

— "All were amazed and perplexed saying to one another: 'What does this mean? This is something new'.": For those who believe, it is the wonder of the new creation for others, it is an occasion for mockery.

— "Peter, standing with the eleven, lifted up his voice": this is the first proclamation of the Christian message through the Apostles, leading up to the confession of Jesus as Lord and Messiah, and the gathering of the first community (2,36-42).

3. The gifts of the Holy Spirit

The entire Christian life is the unfolding of the life of the Spirit given to all believers. It is mostly in Paul's letters that the richness of this life has been spelled out. Read the following texts slowly, with a meditative mind.

— The *understanding* of God's word comes to us through the Holy Spirit: 1 Cor 2. Paul preached to the Corinthians the message of Jesus crucified, "that your faith might not rest in the wisdom of men but in the power of God" (2,5).

— The Holy Spirit reveals to us *our sonship of God:* Gal 4,4-6. God sent his Son "that we might receive adoption as sons". Sonship of God, however, is not merely a legal status; it is meant to penetrate the depth of our consciousness. Only the Holy Spirit, sent into our hearts, gives us the intimate awareness of this sonship and makes us cry out: "Abba, our Father." (cf. Rom 8,12-17). In the Spirit we also can pray (Rom 8,26).

— The Spirit of *unity,* Eph 4,3-6. "There is one body and one Spirit." The unity of the Church is rooted in the one Spirit of Jesus Christ, who unites all believers in one faith, hope and love. He brings us together also in the visible unity of the Church, into one body.

— The Holy Spirit is the inspiration for the Church for her *mission* (Acts 1,4-8). The first missionary enterprise started from the young community at Antioch: "While they were worshipping the Lord and fasting, the Holy Spirit said: Set apart for me Barnabas and Paul for the work for which I have called them" (Acts 13,2). All new endeavours in the Church are marked by the promptings of the Holy Spirit. Reflect on these promptings in your own heart, your own life.

— The *many gifts* of the one Spirit, 1 Cor 12, 4-13.   The unity of the Church is made up of the richness of the spiritual gifts which are freely given to all members of the community: "To each one is given the manifestation of the Spirit for the common good".

— The *guarantee of fulfilment*, Eph, 1,11-14.   On the pilgrimage to our final destiny and in the continuous struggle between grace and sin in the world and in our own heart, the Holy Spirit is "the guarantee of our inheritance until we acquire possession of it to the praise of his glory".

OTHER READINGS:

— Jn 3, 1-8:   Born anew of the Spirit.   The scene with Nicodemus.

— Gal 5,16-26: Life in the Spirit and its fruit.

QUESTIONS:

1. Try to remember, and re-live vividly, occasions when you realised the enlightening, life-giving power of the Holy Spirit in yourself. Think of others in whom you could observe similar experiences.

2. What is your special gift from the Holy Spirit, your main "charism"? Have you in gratitude recognized it, accepted it, fostered it? Do you use it for the good of others: your neighbour, the Church, people at large?

FOR REFLECTION

"Human beings can see Jesus as the Son, and draw closer to the Father only if they live a life in accordance with the Spirit of Jesus. We can come to know Jesus as the Christ only insofar as we start a new life, break with the past and undergo conversion, engage in Christian practice and fight for the justice of God's kingdom."

J. SOBRINO.

Chapter Seven

DIVINE LOVE

"The first commandment is this: Hear, O Israel, the Lord our God, the Lord is one and you shall love the Lord your God with all your heart and with all your soul and with all your mind and with all your strength. The second is this, you shall love your neighbour as yourself" (Mk 12,29-31).

The concluding day of the Spiritual Exercises is devoted to the "contemplation to attain love of God" (nn. 230-237). This day is no longer part of the fourth week but stands by itself. Just as the contemplation on the kingdom of Christ stands by itself, since it is meant to focus the mind of the retreatant on the person and mission of Jesus which is to be unfolded in the three following weeks, so the contemplation on divine love sums up the entire movement of the Spiritual Exercises and its dynamism: God's love for all men –for me, and my response to him. It is a movement which originates in God when he creates the world, when he sends his Son for our salvation and when he pours out his Spirit on all creation. This movement necessarily returns to God from the world—from me, through Jesus Christ in the Holy Spirit.

What is love?

The great words of our language have been used in many ways. Thus 'love' has been praised as the divine power that raises us to God's own realm. But the same word is used for mere emotional attachments and even for enslaving passion.

Love can be possessive, binding, smothering, clinging; it can become a prison. But authentic love liberates and sets us free for our true life.

When Jesus speaks about love he adds a significant qualification: "Love one another *even as I have loved you*" (Jn 13,34). In Jesus God tells us what love really means, God's love for us and our response to him, and the expansion of this love towards our neighbour, towards all people.

God the *Father's love* is creative and liberating. In the beginning he created light, stars, sun and moon and our earth with plants and animals. But all these have no real meaning as yet because they have no freedom of their own; they cannot respond to God. God saw that they were good, but he cannot love them. Love asks for a response. So he creates man after his image and likeness. He gives us all the treasures of the earth and all the hidden potentialities of the oceans, the universe, the atoms. He wants us to be rich and powerful, to enjoy and to celebrate the glories of his creation. Because he loves us he wants us to be free, and so he allows us also to go our own way. He even allows us to sin and to turn his paradise into a desert. Will we ever come back? He still loves—so much he loves the world that he gives his Son to us! Shall we accept him as our brother and learn from him to live as God's sons and daughters? He will never force us. His love is infinitely respectful. He does not display power, he only invites. He has placed our life into our own hands. This is his love. All those who share his final glory shall bear the mark of nobility and freedom.

So is also *Jesus' love*. It begins with the total abdication of all power: he empties himself. So he embarks on his mission of salvation. When he proclaims God's kingdom he has only truth and love on his side. He has no security of human resources; he has no earthly power.

So he deals with the disciples: he calls them, but they follow him in freedom. He would allow them to leave him. He does not prevent Judas from betraying him nor Peter from disowning him. He entrusts his entire mission to them—it is an incredible risk. It will be their task and responsibility. He loves and trusts them.

The final mark of his love for the disciples is the outpouring of the Holy Spirit; he will be their guidance and assurance in doubt and danger. Church structures will be there, authority and laws too. They are not meant to coerce or fetter the faithful but to channel and coordinate the life of the Spirit. Jesus himself, through the indwelling Spirit, will for ever be the strength, guidance and inspiration of his own, the guarantee that God's kingdom is coming.

This then is *God*'s *love* revealed in Jesus Christ: God shares and communicates himself completely. He recognizes us, his creatures, in our freedom and he wishes us to grow and to become ever more ourselves. He trusts us in spite of our failures and he is waiting for us to be with him to share his eternal life in glory, face to face, in ineffable joy.

In Jesus also *our response* to God's love is revealed. He sums up the love to his Father in the word "Abba". This way of speaking to God is born of deep love, simple, unhindered. Love knows no 'problems', the kind of pre-occupations that isolate us from people, from God. Jesus has many questions, anxious questions; his life is full of obstacles and temptations, yet it remains always open before his Father, in thanks and praise for his life and mission. "I thank you, Father, Lord of heaven and earth, that you have hidden these things from the wise and understanding and have revealed them to children" (Mt 11,25). He trusts his Father: "Not one sparrow falls to the ground without your Father's will' (Mt 10,29), and he obeys God to the very end: "Not what I will but what you will" (Mt 14,36).

This, then, is authentic love: finding others, finding God by going out of ourselves, giving selflessly to the end, and thus finding ourselves, our true self in God.

The contemplation to attain divine love

All this is implied in the preliminary remarks which St. Ignatius puts in the beginning of the contemplation: that *love must show itself more in deeds than in words*, that *love means sharing*.

We are asked to make this meditation not in the seclusion of a merely individual encounter with God but in the presence

of the angels and saints. This contemplation has a cosmic
horizon, it concerns the whole of creation as it comes from
God and is moving towards its divine destiny. This movement
must be realized in human hearts, in *my heart*. So I pray for
an intimate knowledge of all that I have received, a knowledge
not at the brain level but the deep personal awareness of
God's love. For once I should be gripped by the consciousness
that my entire life has been embraced by love, from before I
was conceived in the womb, yes, from the foundation of the
world (Eph 1,4).

This contemplation is made in four stages. (1) I recall
the blessings of *creation and redemption* and the special favours I
have received.   Let these graces pass through your mind as in
a flash-back, the surroundings of your childhood, the people
among whom your life developed.   From your early age God
has touched your heart in countless ways.   Each person has
his/her own history to remember, how his/her life took shape
and matured through the years up to the present hour.

(2)   The second stage of this contemplation leads us deeper:
"*How God dwells in creatures*".   In our human relations gifts
are often a substitute for personal concern.   We give something
to a beggar to get rid of him.   Even on a social level we fulfill
obligations through a present, but we have neither time nor
interest to meet on a personal level.   Real presents are valued
not by their price in the bazaar but by the degree of personal
concern expressed in them.   The first drawing of a child
presented to its mother would not fetch a high price, but the
mother knows its preciousness; it comes from love.

In each one of God's gifts he himself is present to us: with
the rays of the sun God touches my eyes; through the vibra-
tions of the air he fills my life with sound and music. Most
of all God is present in the movements of my spirit, my heart,
in my growing understanding of life, world and people, in the
awakening to my true value and destiny. With deep respect
God allows me to build my life with my own mind and
efforts, and in my growing vision, my courage, my joy of
moving on and reaching out, God is present. He does not
advertise himself. His apparent absence, his anonymity, is
part of the delicacy of his divine love. So he finally makes

myself his temple. He wants to be present in me, and through me present to the world in which I live.

(3) In a third stage of this contemplation I see *God at work in all creatures*. The work of creation does not consist only in a single act of God in the inscrutable past; creation lasts as long as creatures exist. God sustains them, he is at work in them, with his creativity so that they grow and unfold in immense variety. God intends them to grow beyond themselves into ever new forms and perfections. He is creative also in my own growth. He is present and at work as my consciousness deepens and my horizons widen, as I find my identity and the meaning and mission of my life. In this contemplation we may become aware of the process of our growth in the past, in these Spiritual Exercises, in this very hour: of the moments of discovery, of "opened eyes", of a new readiness to go out in an ever new response to his call. God is at work around me, in me, through me.

(4) Finally, in the fourth stage of this contemplation on divine love, I pray to become aware of the richness and beauty which I encounter in my world, in my life. Each such experience is a *glimpse of the divine mystery*. You need not think of spectacular events—Jesus used to brush them aside—God's presence and reign are realized *in daily life*. Have you never met a man who faithfully stands for honesty and right in the midst of corruption and betrayal, and he suffers for it? What would we know of God's own faithfulness if we never encountered such people? A woman is left alone by her husband with no resources, and yet she cares for the children, silent, cheerful, patient, unaccusing, uncomplaining. This is an image of the self-gift with which Jesus gives himself for us in patience and concern. Have you never been touched by true love, when you felt fully accepted and supported, when you began to realize your own worth and what you could and should be? God has touched you through such love. He wants his love and beauty to flow through his creatures, so that in the millions of channels of human life he can meet and touch you, and give you a glimpse of what he is, and of what you are, and what you shall be in his glory: "What no eye has seen, nor ear heard, nor the heart of man conceived, what God has prepared for those who love him" (1 Cor 2,9).

The prayer of surrender

Each of these four stages of the contemplation on divine love should be concluded with the prayer of surrender in which the movement of the Spiritual Exercises is summed up. Reflect on it word by word.

"Take and receive"

The answer to God's love consists in our own self-gift to him, in union with Jesus, the surrender to God our Father of whatever we are and have. This surrender is expressed through two verbs which must be understood in their distinct meaning: God *"takes"* what is his. Taking is the expression of sovereign power. So I pray now that God take what is *his*. I belong to him and have no claim to autonomy.

But God will never take possession of me *unless I freely* offer myself. Thus I add: *"receive"*. We receive from others only what they freely give. My coming to God is the response of my freedom, and I ask him to receive what I give to him. In these two complementary words we express the act of our surrender to God: *Take what is yours, receive what is mine*. God has placed my life into my own hands so that I can give it to him.

"All my liberty"

We have a new understanding of the richness of this word. Man is the only creature who holds his life in his own hands. It is for us to give it its orientation. This is hard. It demands options which may be painful. Sartre once said that man is "condemned to freedom." We cannot live carefree and unconcerned, like animals; we must answer the question of our life's meaning and we must therefore make decisions. In Jesus Christ we have found an inner awareness of true life, to live it moving towards God. We have made this most basic of options. It is our human dignity that we are able to make it and so we pray: "Take all my liberty".

"My memory, my understanding and my entire will"

This is the ancient triad which makes up human existence. *Memory* is more than my remembrances. It is the totality of influences which are stored in me and which make up my concrete existence. The atoms, which have a history of millions of years until now they form my body. The dawn of life on earth and the beginnings of the human family of which I am a member, I am one with all this, part of one living continuum linked inseparably with my ancestors. The combination of genes that finally gave shape to the uniqueness of my physical and psychic life is part of my memory, it is meant to be actuated in my life. Then comes my own life as it has developed under the complex influence of parents and family, of my social and economic surroundings, my education and the manifold experiences and challenges of my life. All these, and my reactions to them, constitute my true being. Most of it is submerged below the level of consciousness yet it actively influences me. This is my memory, what in this growing and evolving world I have become in unrepeatable uniqueness.

Animals too have "memory", but they have no *understanding*. In me all the treasures and legacies of the past, that are stored up in me, can become conscious. They become my life with its burdens, opportunities, with its many challenging puzzles. Only in man can creation come to itself in lucidity because it becomes conscious and transparent to the divine mystery. In every person this consciousness is realized in a unique, unrepeatable way. Everyone has his/her own most personal memory, of which he/she becomes conscious. No one is a copy of another person. Every one is God's image in a unique way and is meant to unfold it in his/her life. This is the task of our freedom, of our will.

To this life of mine I alone can give orientation. This is my *will*. I can choose to close in on myself and to manipulate things around me, and other people as well, all for my interests, or I can opt to open out to others, to God. It is for me to decide, my whole being is ambivalent. I contemplate my hand: it can be an instrument for taking, grabbing, keeping, beating; it is meant to be open, to give, share and bless. Every organ of my body can be used in opposite ways. My whole being can be directed either for self or for others, to God. The meaning of my existence is decided by my will.

"All that I have and possess, thou hast given to me, to thee, O Lord, I return it."

My self-gift to God is the full response to God's love. Whatever I am is his gift to me, whatever I am is my gift to him.

"All is thine, dispose of it according to thy will."

It is the purpose of these Spiritual Exercises to grow beyond individual desires, to find God's will and to realize it in my life, and in my mission.

"Give me thy love and thy grace, for this is sufficient for me."

We surely need many things in daily life for body and mind, and we trust that God will provide them. This is not my concern in this retreat. Daily necessities do not constitute my life. I need one thing: *God's love.* The assurance that I am in his love, that, with Jesus, I can call him "Abba", my Father. I need also *his grace:* God's love must become real and active in my own heart and evoke a response. Only he can enable me to respond wholeheartedly to his love. This is grace, which enables me freely and joyfully to make my full self-gift to him. Give me thy love and thy grace.

SHORT POINTS FOR MEDITATION

1. *Love in Christian life* (1 Cor 13).

Read this canticle of love as addressed personally to you. It speaks in three paragraphs of the meaning and richness of Christian love.

— "If I have not loved, I am nothing": love is the greatest of all charisms. Whatever our achievements, talents, graces, they all are empty unless they are inspired by love, to build up the body of Christ.

— "Love is patient and kind...": this is how love looks in daily life. It is clothed in working-day-garments, with no glamour and little hope for social recognition. It is the hidden presence of God. It is the way in which God's kingdom grows in our world.

— "Love never ends...": Other gifts are bestowed on us for our earthly life. Love is the beginning of eternity. Here on earth we begin to share in God's life that will fill our eternity.

2. *God is love* (1 Jn 4,7-21).

This is the ultimate message of the bible, summing up the mystery of salvation as the revelation of God's love and our response to it. Read this whole text with a meditative mind as a last message to you coming from the disciple whom Jesus loved.

— "He who loves is born of God and knows God": this is the ultimate mystery of God, that he is love. He shares his life in the eternal communication of Father, Son and Holy Spirit. He communicates his life to us in creation and salvation. If I begin to share in this mystery of love, I know God, I share in his life.

— "He who does not love does not know God, for God is love": he may know *about* God, but he does not know him, he is not in communion with God.

— "In this is love, not that we loved God but that he loved us and sent his Son": all comes from God, all life and salvation is the revelation of his love. Our love is response to God's love, its realisation in our world.

— "If God so loved us we also ought to love one another": love of neighbour is the expansion of God's love in our world and society. God's love penetrates our world through the channels of human love. It is meant to permeate our society and to transform it into God's kingdom.

FURTHER READINGS:

— Mt 5, 43-48:   Loving enemies.
— Mt 7, 1-5:   Do not judge.
— Mt 18, 21-35:   The parable of the unforgiving servant.
— James 2,14-26: Practical charity.

QUESTIONS:

1.   Am I able to see my life as a *dialogue of love:* God's love for me, my response to him?

2.   Have I realised the practical implications of love of my neighbour in attitudes and thoughts, in speech, action, reaction?

FOR REFLECTION:

In this context Teilhard de Chardin speaks of the creative powers which God has bestowed on us to penetrate and build our world. But also the destructive powers of life and world lead to God; they empty our hearts for God's love:
"My God, I deliver myself up with utter abandon to those fearful forces of dissolution which, I blindly believe, will this day cause my narrow ego to be displaced by your divine presence. The man who is filled with an impassionate love for Jesus... he will awaken in the bosom of God."
(TEILHARD DE CHARDIN, *Hymn of the Universe*).